INVITATION TO BIOLOGY

SECOND EDITION

INVITATION TO BIOLOGY

HELENA CURTIS

SECOND EDITION

WORTH PUBLISHERS, INC.

INVITATION TO BIOLOGY, SECOND EDITION

ILLUSTRATOR: SHIRLEY BATY

PICTURE EDITOR: ANNE FELDMAN

DESIGN: MALCOLM GREAR DESIGNERS

PRINTED IN THE UNITED STATES OF AMERICA

LIBRARY OF CONGRESS CATALOG CARD NO. 76–50677

ISBN: 0-87901-072-X

FIRST PRINTING, MARCH 1977

WORTH PUBLISHERS, INC.

444 PARK AVENUE SOUTH

NEW YORK, NEW YORK 10016

This book is dedicated to Caroline Rogers

Preface to the Teacher

This second edition of *Invitation to Biology* is in part a shorter version of *Biology,* second edition, and in part a wholly new book. Those familiar with *Biology* can see from the Table of Contents that *Invitation* follows the same general plan. However, as we discovered in working on the first edition of this shorter book, it is not possible just to cut and paste, inserting a graceful transition or two. The entire book must be reconsidered, sentence by sentence, rewritten to a large extent, and once again submitted to the icy gaze of the reviewers (a helpful but often merciless crew).

Writing a short book is harder, in some ways, than writing a longer one. One has to face, even more relentlessly, the question of what to put in and, much more difficult, what to leave out. What is modern biology? What do students *have* to know? (This question was particularly pertinent at the writing of this edition because of the wide range of material now being presented in first-year biology courses.) What will students remember 10 years from now? What will still be true?

It was in the course of pondering these questions that we came up with the new Introduction to the book, in which we venture to state the principles of biology. I found it helpful to think through these ideas, and I enjoyed writing it. I would like to hear about the Introduction from readers, especially from people who disagree. It occurs to me that the student should probably read it both before and after reading the book. In any case, to return to my earlier question as to what students should come away with from this course, I believe that the principles of biology are what students have to know and what they will re-member if we—you and I—have served them well.

After the Introduction, we are on firmer and more familiar ground. Once again, we have chosen to take the levels of organization approach, beginning with the atoms and molecules that make up living systems. Part I deals with cells, Part II with individual organisms, and Part III with groups of organisms. Each part has three sections. This time around we did not even consider elimi-nating the material that students woefully refer to as "the chemistry." One of the things the student *has* to know is that living systems are made up of atoms and molecules organized in particular ways, according to certain general laws. It is also for this reason that we have continued to present structural formulas in other places throughout the text. The students certainly should not be expected to remember the structure of testosterone compared to estrogen, but they should be reminded of what forces are at work. At the end of Section 1, we attempt to show the cell in action, replacing the conventional catalog of cell parts with something that we believe is much livelier and, with the help of *Paramecium* and *Didinium,* even amusing.

Before embarking on this second edition, we wrote to and telephoned many teachers, asking, among other questions, whether they would like more or less material on photosynthesis and respiration. About half said more and half said less. So we compromised. For those who want less, Chapter 5 provides an overall, general look at this subject of cellular energetics. For those who want more, Chapters 5, 6, and 7 offer a quite thorough discussion, comparable to that found in the larger book. These three chapters now form a separate section.

The third section, Genetics, was widely approved of by most of those who used the first edition and has remained much the same except for the last two chapters, in which some current and (to us) interesting work is presented, such as the latest ideas on the organization of the eukaryotic chromosome and the controversy concerning recombinant DNA. These are, as in the previous edition, optional chapters.

Part II, Organisms, has been the most completely revised. It begins with a long chapter on the diversity of living organisms. I apologize to those who are fascinated by insects for trying to deal with all arthropods in two to three pages, to those who care about mosses for the cursory treatment of mosses, and so on, taxon by taxon. I offer as my only defense, the facts that (1) it does give the student at least a glimpse of the broad panorama of living organisms, and (2) the photographs and the drawings are pretty spectacular.

The plant section has been extensively rewritten and reorganized. One of the several pleasures in working with experts such as Peter Raven and Ray Evert on a botany text* is that it keeps me informed about this important and sometimes neglected branch of biology. It is because of this fortuitous exposure to their knowledge and enthusiasm that I have resisted all pressure, intra- and extra-mural, to "integrate" plant and animal physiology. I have never seen this done in such a way that the beginning student could have any notion whatsoever of a plant as a complete, functioning, and interesting organism.

In the animal physiology section, the focus continues to be on the human, although some illuminating comparisons are made with other animals. This section has also been reorganized and revised with increased emphasis on physiological control systems and homeostasis. It is enriched by several new four-color drawings of human organ systems.

In Part III, Evolution and Ecology, we confronted again the question of which of these subjects to present first and this time chose Evolution. This section has been strengthened and expanded. It begins with an introduction that places evolutionary theory in an historical and cultural perspective. (Certainly no other scientific concept—even the Copernican Revolution—has had such social consequences.) Between editions, we have had cause to be reminded of the fact that there are literate, intelligent people, young and old, who are not convinced that evolution actually took place, a point of view that has even attracted the support of some state governmental agencies. With such persons in mind, we have added more of the evidence—such as the testimony of the fossil record— that Darwinian evolution is a fact.

All of the book has been greatly enhanced by the use of four-color photographs, but nowhere are they as important as in the next to last section, Ecology, where they truly bring to life the great landscapes of the world and their inhabitants. We have sometimes thought of reorganizing the book so that it would begin with this familiar and inviting material and proceed down rather

---

* *Biology of Plants,* 2d ed., Worth Publishers, Inc., New York, 1976.

than up the levels of organization. I would like to hear from anyone who thinks this might be a good idea.

Finally the book closes with the evolution of man. As the story of human evolution unfolds, it becomes more and more evident that the evolution of man is perhaps the best of all examples of the interactions of ecology and evolution—what G. E. Hutchinson so aptly termed the ecological theater and the evolutionary play.

At the end of each chapter, there are several questions designed to help students evaluate their understanding of the chapter. We also would like to call your attention to the excellent study guide prepared by Vivian Null to accompany the text.

At the end of each section is a short list of books for further reading. We have not included lists of *Scientific American* reprints, simply on the assumption that these excellent supplementary materials are widely known, widely available, and constantly increasing in number. We would like to urge that, if it is at all possible, students be encouraged to explore on their own, journals such as *Scientific American, Science,* and *American Scientist.* In this way, they will have the pleasure of discovering something for themselves and of seeing the process of the acquisition of scientific knowledge as it takes place.

As with the previous texts, we have been deeply dependent on reviewers and consultants. Dorothy Luciano reviewed and helped substantially with the rewriting of the Human Physiology section and we are especially grateful to her. We also have a continuing—and probably unrepayable debt—to David Shappirio of the University of Michigan. Despite our previous comments about merciless reviewers, Dr. Shappirio has been kind and sympathetic, though sometimes gently reproachful. I envy all students who have him as a teacher.

In addition, among others who have made major contributions to the book are: Paul G. Arnison, Northeastern University; Andrew Beattie, Northwestern University; Robert M. Beck, Sarah Lawrence College; Charles Brown, Santa Rosa Junior College; Roy Caldwell, University of California, Berkeley; Alice Chew, Santa Monica Community College; Norman Christensen, Duke University; Martin S. Cohen, University of Hartford; Richard Colby, Stockton State College, Pomona, New Jersey; Shirley A. Crawford, State University of New York, Morrisville; Mel Cundiff, University of Colorado, Boulder; Daniel Franck, University of Wisconsin, Madison; Douglas Futuyma, State University of New York, Stony Brook; Henry Galiano, American Museum of Natural History; Laurel Hansen, Spokane Falls Community College; Holt Harner, Broward County Community College; James D. Jamieson, Yale University; John Kirsch, Yale University; Richard C. Lewontin, Harvard University; Earl Manning, American Museum of Natural History; Jane Menge, Harvard University; Gene Meyer, New York University; Robert Sheffield; Frank Smith, Jr., Southern Oregon State College; Timothy Teyler, Harvard University; Richard F. Thompson, University of California, Irvine; A. Randolph Thornhill, University of New Mexico; Norton D. Zinder, Rockefeller University.

Finally, we must thank all of the many teachers who wrote to us about *Biology* with criticisms and suggestions. They helped us greatly in preparing the present text and we earnestly hope they will continue. There is no one whose opinion we value more.

East Hampton, New York
January, 1977

Helen Curtis

CONTENTS IN BRIEF

CONTENTS

INVITATION TO BIOLOGY

SECOND EDITION

INTRODUCTION

The Science of Life

Biology is defined as the "science of life." Its scope extends from the organization of the atoms and molecules that make up living matter through the interactions of whole organisms and groups of organisms with one another and with the environment in which they live. This is also the scope of this text, which begins with the submicroscopic world within the cell and ends with a view of the biosphere, the entire world of living things.

Science, biological and other, is a way of seeking principles of order in the natural world. (Art is another, as are religion and philosophy.) Science has two separate components: (1) objective evidence based on observation or experiment or a combination of the two—the data—and (2) the structuring of data by making meaningful connections among them. The great discoveries in science are not merely the addition of new facts but the perception of new relationships among the available facts, in other words, the development of new concepts and new ideas.

All ideas are not equal. It is customary to categorize the ideas of science in ascending order of credibility as hypotheses, theories, and principles or laws. Lower on the scale than the hypothesis is the hunch, or educated guess, which is how most hypotheses begin. A hunch becomes a hypothesis when it is stated in such a way that it can be tested. (As we shall see, a hypothesis may be stated verbally, mathematically, or in the form of a model.) Until it can be so stated, it is useless, scientifically speaking. When a hypothesis has been subjected to a number of tests, it becomes, strictly speaking, a theory. A theory that has withstood repeated testing over a period of time becomes elevated to the status of a law or principle, although not always identified as such. The "theory" of evolution is an example. It is not theoretical, in the common sense of the term, nor has it been for almost a hundred years. As far as scientists are concerned, it is a fact, just as the cell "theory" is. However, many of the details of cellular structure and of the evolutionary process are still in the stage of theory, or even hunches.

In the following few pages, we shall discuss four such principles of modern biology. They are so well established that biologists seldom discuss them among themselves. You could read voluminously in the literature of modern biology without seeing any of them explicitly mentioned. Yet it is impossible to understand biology without being aware of its foundations. These four principles will recur as major themes throughout this book and we shall call your attention to them from time to time.

ALL LIVING ORGANISMS ARE RELATED

The theory of evolution states that "all species descended from other species"; in other words, all living things share a common ancestor. Charles Darwin was not the first to propose this idea, but the credit is rightly his, for two reasons. First, his "long argument"—as *The Origin of Species* has been characterized—left little doubt that evolution had actually occurred and so marked a turning point in the history of biology. The second reason, which is closely related to the first, is that Darwin correctly perceived the mechanism by which evolution comes about. Evolution is a two-stage process. The first stage is the occurrence, absolutely at random, of genetic variations among individuals—that is, variations that can be inherited. These variations have adaptive values; that is, they may be more or less useful to the organism as measured by its survival and reproduction. The second stage is natural selection, which is a process of interaction between an organism and its environment. As a result of this interaction, some organisms, because of their inherited characteristics, leave more offspring than other organisms with other inherited characteristics. Given enormous amounts of time, evolution leads to the accumulation of slow changes that differentiate one group of organisms from another and results in the great diversity of living things that now inhabit this planet. An important corollary to the theory that all living things share a common ancestor is that life, as it now exists on this planet, arose only once. The fact that the genetic code is universal—from the simplest bacterial cell to the most complex organism—lends impressive support to these concepts.

Evolution is viewed as the greatest unifying principle in biology.

ALL LIVING ORGANISMS ARE MADE UP OF CELLS

A second important principle of biology is the cell theory. It states simply that the cell is the basic unit of living material and that all living organisms are composed of one or more of these fundamentally similar units.

The word "cell" was first used in a biological sense some 300 years ago. In the seventeenth century, Robert Hooke, using a microscope of his own construction, noticed that cork and other plant tissues are made up of small cavities separated by walls. He called these cavities "cells," meaning "little rooms." The word did not take on its present meaning, however, for more than 150 years.

In 1838, Matthias Schleiden, a German botanist, came to the conclusion that all plant tissues were organized in the form of cells. In the following year, zoologist Theodor Schwann extended Schleiden's observation to animal tissues and proposed a cellular basis for all life.

The cell theory took on an even broader significance when the great pathologist Rudolf Virchow generalized that cells can arise only from existing cells: "Where a cell exists, there must have been a preexisting cell, just as the animal arises only from an animal and the plant only from a plant. . . . Throughout the whole series of living forms, whether entire animal or plant organisms or their component parts, there rules an eternal law of continuous development." There is an unbroken continuity between modern cells—and the organisms which they compose—and the primitive cells that first appeared on earth more than 3 billion years ago. Thus the cell theory, though its origins were very different, becomes part of the more general theory of evolution.

Until comparatively recently, many prominent biologists believed that living systems are qualitatively different from nonliving ones, containing within them a "vital spirit" that enables them to perform activities that cannot be carried out outside the living organism. This concept is known as vitalism and its proponents as vitalists. (Do not dismiss this "foolish" idea too rapidly; you may discover you are a vitalist.)

In the seventeenth century, the vitalists were opposed by a group known as the mechanists. The French philosopher René Descartes (1596–1650) was a leading proponent of this group. The mechanists set about proving that the body worked essentially like a machine; the arms and legs move like levers, the heart like a pump, the lungs like a bellows, and the stomach like a mortar and pestle. By the nineteenth century, such simple mechanical models of living organisms had been abandoned, and the argument now centered on whether or not the chemistry of living organisms was governed by the same principles as the chemistry performed in the laboratory by man. The vitalists claimed that the chemical operations performed by living tissues could not be carried out experimentally in the laboratory, categorizing reactions as either "chemical" or "vital." The reductionists, as their opponents were now called (since they believed that the complex operations of living systems could be reduced to simpler and more readily understandable ones), achieved a partial victory when the German chemist Friedrich Wöhler (1800–1882) converted an "inorganic" substance (ammonium cyanate) into a familiar organic substance (urea). On the other hand, the claims of the vitalists were supported by the fact that, as chemical knowledge improved, many new compounds were found in living tissues that were never seen in the nonliving, or inorganic, world.

In the late 1800s, the leading vitalist was Louis Pasteur, who claimed that the marvelous changes that took place when fruit juice was transformed to wine were "vital" and could be carried out only by living cells—the cells of yeast. In spite of many advances in chemistry, this phase of the controversy lasted until almost the turn of the century.

However, in 1898, the German chemist Eduard Büchner (1860–1917) showed that a substance extracted from the yeast cells could produce fermentation outside the living cell. (This substance was given the name enzyme, from *zyme,* the Greek word meaning "yeast" or "ferment.") A "vital" reaction was proved to be a chemical one, and the subject was laid to rest. Today it is generally accepted that living systems "obey the rules" of chemistry and physics, and modern biologists no longer believe in a "vital principle."

Perhaps the greatest test for this principle of modern biology came about 25 years ago. As you know, one of the outstanding characteristics of living things is their capacity to reproduce, to generate faithful copies of themselves. By about 1950, this capacity had been shown to reside in a single type of chemical molecule, deoxyribonucleic acid (DNA). The race to discover the structure of this molecule began, and the question in everyone's mind was whether or not the structure of this one "simple" molecule could possibly explain the centuries-old question of the mysteries of heredity. As it turned out, and as we shall discuss in the course of this text, it could.

The next test for this concept seems to lie in the mechanisms of the brain. Even those of us who now willingly admit that the nature of heredity—so long a

Art and Science

The act of discovery in science engages the imagination (first of the man who makes it, and then of the man who appreciates it) as truly as does the act of creation in the arts.

J. Bronowski, Identity of Man, rev. ed., Natural History Press, Garden City, N.Y., 1971.

Science and Theory

On principle, it is quite wrong to try founding a theory on observable magnitudes alone. It is the theory which decides what we can observe.

A. Einstein, from J. Bernstein, "The Secrets of the Old Ones, II," The New Yorker, March 17, 1973.

The Scientific Method

Indeed, scientists are in the position of a primitive tribe which has undertaken to duplicate the Empire State Building, room for room, without ever seeing the original building or even a photograph. Their own working plans, of necessity, are only a crude approximation of the real thing, conceived on the basis of miscellaneous reports volunteered by interested travelers and often in apparent conflict on points of detail. In order to start the building at all, some information must be ignored as erroneous or impossible, and the first constructions are little more than large grass shacks. Increasing sophistication, combined with methodical accumulation of data, make it necessary to tear down the earlier replicas (each time after violent arguments), replacing them successively with more up-to-date versions. We may easily doubt that the version current after only 300 years of effort is a very adequate restoration of the Empire State Building; yet, in the absence of clear knowledge to the contrary, the tribe must regard it as such (and ignore odd travelers' tales that cannot be made to fit).

E. J. DuPraw, Cell and Molecular Biology, Academic Press, Inc., New York, 1968.

Scientists at Work

Scientists at work have the look of creatures following genetic instructions; they seem to be under the influence of a deeply placed human instinct. They are, despite their efforts at dignity, rather like young animals engaged in savage play. When they are near to an answer their hair stands on end, they sweat, they are awash in their own adrenalin. To grab the answer, and grab it first, is for them a more powerful drive than feeding or breeding or protecting themselves against the elements.

It sometimes looks like a solitary activity, but it is as much the opposite of solitary as human behavior can be. There is nothing so social, so communal, so interdependent. An active field of science is like an immense intellectual anthill; the individual almost vanishes into the mass of minds tumbling over each other, carrying information from place to place, passing it around at the speed of light.

There are special kinds of information that seem to be chemotactic. As soon as a trace is released, receptors at the back of the neck are caused to tremble, there is a massive convergence of motile minds flying upward on a gradient of surprise, crowding around the source. It is an infiltration of intellects, an inflammation.

There is nothing to touch the spectacle. In the midst of what seems a collective derangement of minds in total disorder, with bits of information being scattered about, torn to shreds, distintegrated, reconstituted, engulfed, in a kind of activity that seems as random and agitated as that of bees in a disturbed part of the hive, there suddenly emerges, with the purity of a slow phase of music, a single new piece of truth about nature. . . .

There is something like aggression in the activity, but it differs from other forms of aggressive behavior in having no sort of destruction as the objective. While it is going on, it looks and feels like aggression: get at it, uncover it, bring it out, grab it, halloo! It is like a primitive running hunt, but there is nothing at the end of it to be injured. More probably, the end is a sigh. But then, if the air is right and the science is going well, the sigh is immediately interrupted, there is a yawping new question, and the wild, tumbling activity begins once more, out of control all over again.

Lewis Thomas, "Notes of a Biology-Watcher," New England Journal of Medicine, 288:307–308, February 8, 1973.

mystery—is explained by a single molecule, an evolutionary accumulation of atoms organized in a particular way, are perhaps less willing to admit that our dreams and our ideals and our deepest emotions are based on "nothing but" the chemistry and organization of our brain cells.

LIFE RUNS DOWNHILL

Among the laws of physics that are pertinent to biology are the laws of thermodynamics. They state simply that (1) energy can be changed from one form to another but it cannot be gained or lost, that is, the total energy of the universe remains constant; and (2) all natural events proceed in such a way that concentrations of energy tend to dissipate or become random. A heated object, which is one example of concentrated energy, loses its heat to its surroundings. A living system, which is a concentration of energy of another kind, can maintain itself only by a constant intake of energy. The energy by which living systems maintain themselves enters the biosphere in the form of radiant energy—light—from the sun, and it is changed by photosynthesis to chemical energy in the form of sugar and other molecules. Chemical energy is then used by each individual cell to do the work of the cell, and in the course of this work, the chemical energy may be transformed to the energy of motion (kinetic energy), to heat energy, or even back to light energy again.

This flow of energy is the essence of life. Evolution may be viewed as a competition among organisms for the most efficient use of energy resources. Cells can be best understood as a complex of systems for transforming energy. At the other end of the biological scale, the structure of the ecosystems and of the biosphere itself is determined by the relationships in terms of energy exchanges of the groups of organisms within it.

Most organisms budget their energy day to day, or even hour to hour, with no reserves beyond that stored in their own fat, liver, or other tissues. Some few, such as the honeybee and the squirrel, set aside some minimal provisions. Only man has been able to live on deficit financing; as you will see in Chapter 40, it is now costing civilized man 10 calories in energy for each calorie of food that reaches the table. This fact is not without relevance to world politics and economics.

SCIENCE AND HUMAN VALUES

Before we close this introduction, we should mention one more characteristic of biology and other sciences, one that distinguishes it from art, religion, or philosophy, with which we compared science in our opening paragraphs. The raw materials of science are the phenomena of the natural universe. Science is limited to what is observable and measurable and, in this sense, is rightly categorized as materialistic. Hunches are abandoned, hypotheses superseded, theories shattered, but the facts endure and, moreover, they are used over and over again, sometimes in wholly new ways. It is for this reason that scientists stress and seek objectivity. (In the arts, by contrast, the emphasis is on subjectivity—experience as filtered through the individual consciousness.)

Because of this emphasis on objectivity, value judgments cannot be made in science in the way that such judgments are made in philosophy, religion, and the arts, and indeed in our daily lives. Whether or not something is good or beautiful, or right, in a moral sense, for example, cannot be determined by the scientific method. Such judgments are always subjective, even though they may be based on a consensus of opinion. They are not subject to proof by the scientific method.

At one time, sciences, like the arts, were pursued for their own sake, for pleasure and excitement and satisfaction of the insatiable curiosity with which mankind is both cursed and blessed. In the twentieth century, however, the sciences have spawned a host of giant technological achievements—the H-bomb, the Salk vaccine, DDT, indestructible plastics, nuclear energy plants, perhaps even ways to manipulate our genetic heritage—but have not given us any clues about how to use them wisely. Moreover, science, as a result of these very achievements, appears enormously powerful. It is thus little wonder that among members of this generation, as perhaps not in any previous one, there are many individuals who are angry at science, as one would be angry at an omnipotent authority who apparently has the power to grant one's wishes but who refuses to do so.

The reason that science cannot and does not solve the problems we want it to is implicit in its nature. Most of the problems we now confront can be solved only by value judgments. For example, science gave us the A-bomb and also some predictions as to the extent of the biological damage that its use might cause. Yet, it could not help us, as citizens, in weighing the risk of damage from fallout against the desire for a strong national defense. In fact, in the 1950s, the most outspoken advocate of atomic testing was one of the country's most prominent physicists, as was the most ardent opponent.

Science has produced the information for the construction of an artificial kidney and can explain how it works. Yet the application of scientific methods cannot help us to decide who, given a shortage of such machines, should be treated by them and who should be abandoned. Similarly, science can predict the possible extent of damage to a particular area from the use of DDT, but it cannot weigh this damage against the reduction of food crops or the increase in malaria that would occur were DDT prohibited. It is, indeed, one of the ironies of this so-called "age of science and materialism" that probably never before have ordinary individual men and women been confronted with so many moral and ethical problems.

At one time, it was traditional for scientists to abstain from such political and ethical controversies, maintaining, quite rightly, that scientists have no special qualifications in such matters. "Living in an ivory tower" became a well-worn phrase. Now there is an increasing tendency for scientists to take public stands, particularly on matters involving the applications of science. For example, a large number of distinguished scientists, including several Nobel laureates, recently issued a public statement on the safety of nuclear reactors. Also, in recent years, a small group of molecular geneticists took the unprecedented move of calling for a general moratorium on the research in which they themselves were involved. (We shall describe these studies in Chapter 17.) The extent to which this work shall continue and the conditions under which it will be carried out are still a matter of public debate.

THE STUDY OF BIOLOGY

You may have been persuaded to study biology because of the environmental problems now confronting us or because of a desire to know more about the mechanisms of your own body or an interest in the "green revolution"—in short, because it is "relevant." The study of biology is, indeed, pertinent to many aspects of our day-to-day existence, but do not make this your main reason for the study of biology. Above all other considerations, study biology because it is "irrelevant"—that is, study it for its own sake, because, like art and music and literature, it is an adventure for the mind and nourishment for the spirit.

One final word: In our enthusiasm for telling you all that biology has discovered, do not let us convince you that all is known. Many questions are still unanswered. More important, many good questions have not yet been asked. Perhaps you may be the one to ask them.

SUMMARY

Science is concerned with the organization of phenomena observed in the natural world. Among the important organizing concepts of biological science are (1) the theory of evolution, (2) the cell theory, (3) the assumption that biology operates in accordance with the principles of chemistry and physics, and (4) the laws of thermodynamics.

SUGGESTIONS FOR FURTHER READING

BAKER, JEFFREY J. W., and GARLAND E. ALLEN: *Hypothesis, Prediction, and Implication in Biology,* Addison-Wesley Publishing Company, Inc., Reading, Mass., 1971.*

An introduction, with interesting examples, of the nature of scientific inquiry in the biological sciences.

BRONOWSKI, J.: *The Ascent of Man,* Little, Brown and Company, Boston, 1973.*

An informal and illuminating history of the sciences, originally prepared as a television series. The emphasis is on science's relation to human culture, and the design and illustrations are beautiful.

DARWIN, CHARLES: *On the Origin of Species by Means of Natural Selection, or the Preservation of Favored Races in the Struggle for Life,* Doubleday & Company, Inc., Garden City, N.Y., 1960.*

Darwin's "long argument." Every student of biology should, at the very least, browse through this book to catch its special flavor and to begin to understand its extraordinary force.

GROBSTEIN, CLIFFORD: *The Strategy of Life,* W. H. Freeman and Company, San Francisco, 1965.*

A brief, provocative introduction to the fundamental ideas of modern biology.

KUHN, THOMAS: *The Structure of Scientific Revolutions,* 2d ed., University of Chicago Press, Chicago, 1970.

An analysis of the process by which scientific progress takes place.

* Available in paperback.

PART I   Cells

SECTION 1   # The Unity of Life

CHAPTER 1

The Origin of Life

Sagittarius, the archer, is a constellation that is visible in our summer skies. If you look toward Sagittarius on a clear, moonless night, you are looking toward the center of the Milky Way, the galaxy to which our planet belongs. There are 250 billion stars in our own galaxy, and within the range of our most powerful telescopes there are about 10 billion more such galaxies.

Astronomers calculate that the Milky Way was perhaps 10 billion years old when the star that is our sun came into being. According to current hypotheses, it formed, like other stars, from a condensation of particles of dust and hydrogen and helium gases whirling in space among the older stars. Now there are fewer such clouds of particles and gases in the Milky Way, but you can still sometimes see them as dark patches in the starlit night.

The immense cloud that was to become the sun condensed gradually as the hydrogen and helium atoms were pulled toward one another by the force of gravity, falling into the center of the cloud and gathering speed as they fell. As the cluster grew denser, the atoms moved faster and faster. More and more atoms collided with each other, and the gas in the cloud became hotter and hotter. As the temperature rose, the collisions became increasingly violent until the hydrogen atoms collided with such force that their nuclei fused, forming helium and releasing nuclear energy (the same series of reactions that produce the explosive energy of an H-bomb). Energy from this thermonuclear reaction, still going on at the heart of the sun, is radiated from its glowing surface. It is this energy, captured in the cells of green plants, on which all life on earth depends.

The planets, according to current theory, formed from the remaining gas and dust moving around the newly formed star. At first, particles were collected merely at random, but as each mass grew larger, other particles began to be attracted by the gravity of the largest masses. The whirling dust and forming spheres continued to revolve around the sun until finally each planet had swept its own path clean, picking up loose matter like a giant snowball. The orbit nearest the sun was swept clean by Mercury, the next by Venus, the third by earth, the fourth by Mars, and so on out to Neptune and Pluto, the most distant of the planets. The solar system, including earth, is calculated to have come into being about 4½ billion years ago.

1-1
Our sun is near the outer edge of a disk-shaped galaxy, the Milky Way, containing 250 billion stars. These stars, like the sun, formed from nebulae—clouds of gas and dust—such as this nebula in the constellation Sagittarius. The atoms of which our bodies are composed had their origins in interstellar events such as this.

THE EARTH AND THE BIOSPHERE

During the time earth and the other planets were being formed, the release of energy from radioactive materials kept their interiors very hot. When earth was still so hot that it was mostly liquid, the heavier materials collected in a dense core whose diameter is about half that of the planet. As soon as the supply of stellar dust, stones, and larger rocks was exhausted, the planet ceased to grow and began to cool. As earth's surface cooled, an outer crust, a skin as thin by comparison as the skin of an apple, was formed. The oldest known rocks in this outer layer are about 3.98 billion years old.*

Only 50 kilometers (30 miles)† below its surface, the earth is still hot—a small fraction of it is even still molten. We see evidence of this in the occasional volcanic eruption that forces lava, which is molten rock, through weak points in the earth's skin or in the geyser, which spews up boiling water that has trickled down to the earth's interior.

The biosphere is the part of the planet within which life exists. It forms a thin film on the outermost layer, extending only about 8 or 10 kilometers up into the atmosphere and about as far down into the depths of the sea.

WHY ON EARTH?

In our solar system, earth among all the planets is most favored for the production of life. A major factor is that earth is neither too close nor too distant from the sun. At very low temperatures, the chemical reactions on which life—any form of life—depends must virtually cease. At high temperatures, compounds are too unstable for life to form or survive.

Earth's size and density are also important factors. Planets much smaller than earth do not have enough gravitational pull to hold a protective atmosphere, and any planet much larger than earth may hold so dense an atmosphere that light from the sun cannot reach its surface.

* The process by which these dates are estimated is described on page 30.
† A metric table with English equivalents is in Appendix A.

1–2

Photograph of Mars taken on July 21, 1976, the day following Viking I's successful landing on the planet. The dusty, windblown soil is littered with rocks and boulders. The spacecraft landed with one foot on such a rock, leaving its spidery body tilted at an 8° angle. The red color of the surface of Mars may be due to iron oxides.

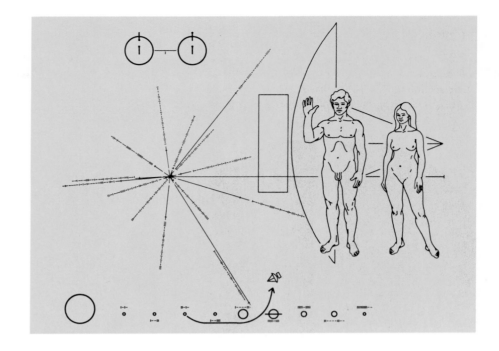

Many cosmologists believe that intelligent life exists in outer space. This small metal plaque was sent off in Pioneer 10, the first spacecraft to leave our solar system. It contains a message to the inhabitants of some far-distant planet: two members of the species Homo sapiens, *with a hand raised in peaceful greeting, and their home address. Pioneer 10 will reach the planetary system of another star in about 10 billion years.*

As recently as the late 1960s, serious students of exobiology (the science of life outside our own planet) were pessimistic about the discovery of life within our solar system. However, the wealth of data obtained from the Mariner 9 spacecraft, which made nearly 700 orbits around Mars in the early 1970s, reawakened interest in the possibility of finding some primitive, perhaps dormant, form of life on that planet. It seems possible that some of the canals and other surface irregularities that have long intrigued observers of the red planet were made by liquid water. Wherever liquid water and light are found on earth, life also exists. It was hoped that the Viking spacecrafts, which landed on Mars in 1976, would gather information to settle the question of whether there is now or ever has been life on that planet. The results so far are ambiguous.

Some biologists contend that given certain conditions—such as an energy source, water, a temperature range in which the water can exist in liquid form, and a long enough time—the evolution of some forms of life is inevitable. The discovery on another planet of a living organism, no matter how primitive, whose origin was independent of earth, would strongly support this hypothesis. There are approximately 10^{20} (the number 1 followed by 20 zeros) stars in the universe like our own sun that can provide energy for living things. At least 10 percent of these, according to astronomers, are likely to be surrounded by planetary systems such as our own. If only 1 percent were to have planets with environments roughly similar to those of earth, that would offer some 10^{18} possibilities for the existence of life in other solar systems.

Bolts of lightning in the steam boiling up from a volcanic crater. Such sources of energy were probably present on the primitive earth and might have contributed to the formation of organic molecules. This photograph, taken in 1963, shows the birth of the island of Surtsey off the coast of Iceland.

THE BEGINNING OF LIFE

Sometime between the time earth formed and the date of the earliest fossils discovered so far—an interval of about a billion years—life began. The chief raw materials for life were to be found in the atmosphere of the young earth. The principal component of the sun and hence of the solar system is hydrogen, and earth, when it formed, was probably surrounded by a cloud of hydrogen gas.

What do we mean when we speak of "the evolution of life," or "life on other planets," or "when life begins"? Actually, there is no simple definition. Life does not exist in the abstract; there is no "life," only living things. We recognize a system as being *alive when it has certain properties that are more common among animate objects than inanimate ones. Each of these photographs illustrates a property found in living systems.*

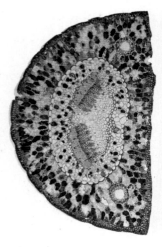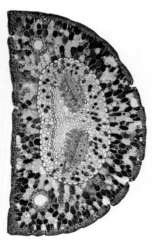

Living organisms are highly organized. Look, for example, at this cross section of pine needles. They represent the complicated organization of many different kinds of atoms into molecules and of molecules into complex structures. Such complexity of form is never found in inanimate objects.

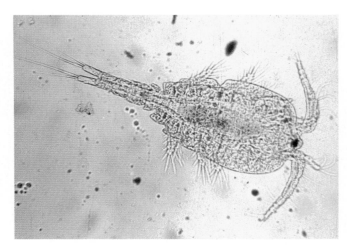

Living things are homeostatic, which means simply "staying the same." Even though they constantly exchange materials with the outside world, they maintain a relatively stable internal environment quite unlike that of their surroundings. Even this tiny, apparently fragile animal, a crustacean, has a constant chemical composition that does not change with changes in the environment.

Living organisms take energy from the environment and change it from one form to another. They are highly specialized at energy conversion. Here a bobcat is converting chemical energy to kinetic energy, thereby procuring a new source of chemical energy, in this case, a snowshoe hare.

Living things respond to stimuli. Here scallops, sensing an approaching starfish, leap to safety.

Living things reproduce themselves. They make more of themselves, copy after copy after copy, with astonishing fidelity (and yet, as we shall see, with just enough variation to provide the raw material for evolution).

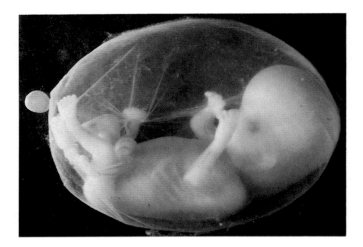

Living things grow and develop. Growth and development are the processes by which, for example, a single living cell, the fertilized egg, becomes a tree, or an elephant, or as shown here, a human infant.

Living things are adapted. Giraffes have long necks that enable them to reach the tops of trees. The length from heart to head in a giraffe is more than 2 meters, and, not coincidentally, they have very high blood pressures—higher than that recorded for any other animal. Adaptations are the inevitable consequence of the long, inexorable process of evolution.

Conditions believed to exist on the primitive earth are simulated in this apparatus. Methane and ammonia are continuously circulated between a lower "ocean," which is heated, and an upper "atmosphere," through which an electric discharge is transmitted. At the end of 24 hours, 45 percent of the carbon originally present in the methane gas is converted to amino acids, the building blocks of proteins, and other organic molecules.

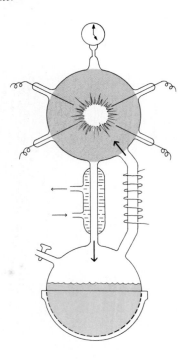

Oxygen was present mainly in the form of water (H_2O). Under conditions of an abundance of hydrogen, the available nitrogen, carbon, and oxygen atoms also present would tend to combine with hydrogen to form ammonia (NH_3), methane (CH_4) and other carbon-hydrogen gases, and water. These were the raw materials for living systems. Combinations of these four elements available in the primitive atmosphere—hydrogen, oxygen, carbon, and nitrogen—make up more than 95 percent of the tissues of all living things.

In order to break apart the simple gases of the atmosphere and re-form them into organic* molecules, energy was required. And energy abounded on the young earth. There was heat energy, both boiling (moist) heat and baking (dry) heat. Water vapor spewed out of the primitive seas, cooled in the upper atmosphere, collected into clouds, fell back on the crust of the earth, and steamed up again. Violent rainstorms were accompanied by lightning, which provided electrical energy. The sun bombarded the earth's surface with high-energy particles and ultraviolet light, another form of energy. Radioactive elements within the earth released their energy into the atmosphere. These conditions can be simulated in the laboratory, and scientists have now shown that under such conditions, organic molecules are produced from the simple gases ammonia, methane, and water vapor. Among these molecules are some of the amino acids, the important building blocks of proteins, which are basic components of living matter.

---

* The complex molecules of living things are generally referred to as organic molecules. The term "organic chemistry" was first applied to substances made by plants and animals, and by extension it came to include all compounds that contain both carbon and hydrogen, since these are the chief constituents of such compounds. Now many such carbon-hydrogen compounds are made routinely in the laboratory, and some of them, like synthetic fibers and plastics, are never found in organisms. The term "organic molecules" will be used in this book to mean the relatively complex carbon-containing molecules found in living systems.

In the next step in the sequence of events which led to life, as biochemists reconstruct the process, these compounds were washed out of the atmosphere by the driving rains and began to collect and become concentrated. Some organic molecules have a tendency to aggregate in groups and form droplets, similar to the droplets formed by oil in water. Such droplets may have been the forerunners of the first primitive cells.

The exact moment at which living systems first came into being is not known. The earliest fossils found so far, which are about 3.2 billion years old, are sufficiently complex so that it is clear that some little aggregation of chemicals moved through that twilight zone separating the living and nonliving some millions of years before.

THE FIRST CELLS

Living organisms, as we noted on page 2, are organized in small units, or compartments, called cells. One essential feature common to all cells is an outer membrane, the _cell membrane_ (sometimes called the plasma membrane), which separates the cell from its external environment. Another is the genetic material—the hereditary information—that directs a cell's activities and enables it to reproduce, passing on its characteristics to daughter cells. All cells also have a _cytoplasm_, which contains a large variety of molecules and also some _organelles._ Organelles are specialized structures, visible within the cytoplasm, that carry out particular functions within the cell, such as providing energy for the cell's activities and making enzymes and other proteins.

The earliest living organisms were probably single cells. According to the fossil record, these early cells were comparatively simple, resembling present-day bacteria. The genetic material of bacterial cells, unlike that present in more complex cells—such as the cells that make up the tissues of our bodies—is contained in one large molecule of a chemical known as DNA. (When the cells are growing and dividing rapidly, they may contain more than one molecule, but each contains all the hereditary information.) Bacteria are also surrounded by an outer cell wall that is manufactured by the cell itself. Their only organelles are ribosomes, small cytoplasmic particles on which protein molecules—the building blocks of the cell—are assembled. Such cells are known as _prokaryotes_ (from _pro_ meaning "before" and _karyon_ meaning "nucleus"). An example of a prokaryotic cell is shown in Figure 1–6.

1–6

Cells of Escherichia coli, _a modern prokaryote that is a common inhabitant of the human digestive tract. The DNA is in the less dense (lighter-appearing) material in the center of each cell. The most dense small bodies in the cytoplasm are organelles called ribosomes. The two cells in the center have just finished dividing and have not yet separated completely. These bacteria have been magnified 24,500 times._

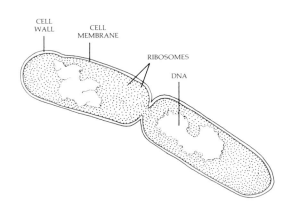

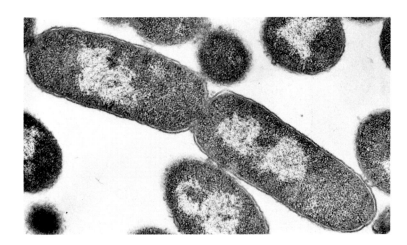

Heterotrophs and Autotrophs

As we have mentioned previously, life depends on a constant input of energy. The energy for the very first cells probably came from energy-rich compounds that formed in the primitive atmosphere and collected in the warm, shallow seas that covered the earth's surface. Such cells are known as *heterotrophs*, a category of organisms that today includes all living things classified as animals or fungi, many bacteria, and all the protozoans. *Hetero* comes from the Greek word meaning "other," and *troph* comes from *trophos*, "one that feeds." A heterotrophic organism is one that is dependent upon others—that is, upon an outside source of organic molecules—for its energy.

As the primitive prokaryotic heterotrophs increased in number, they began to use up the complex molecules on which their existence depended and which had taken millions of years to accumulate. Competition began. Under the pressure of this competition, cells that could make efficient use of the limited energy sources now available were more likely to survive than cells that could not. In the course of time, cells evolved that were able to make their own energy-rich molecules out of simple nonorganic materials. Such organisms are called *autotrophs*, "self-feeders." Without the evolution of autotrophs, life on earth would soon have come to an end.

The most successful of the autotrophs were those that evolved a system for making direct use of the sun's energy—the process of photosynthesis. The first photosynthetic organisms, also prokaryotes, resembled modern blue-green algae (Figure 1–7). The cells contain membranes in which chlorophyll and other pigments are embedded and in which photosynthesis takes place. With the advent of photosynthesis, the flow of energy in the biosphere came to assume its modern form: radiant energy channeled through photosynthetic autotrophs to all other forms of life.

1–7

Electron micrograph and diagram of a photosynthetic prokaryotic cell, a blue-green alga (Anabaena azollae). *In addition to the nuclear material and ribosomes, this cell contains a series of membranes in which photosynthetic pigments are embedded. This cell can synthesize its own energy-rich organic compounds by chemical reactions powered by the radiant energy of the sun.*

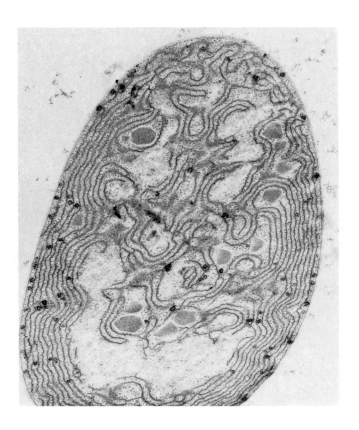

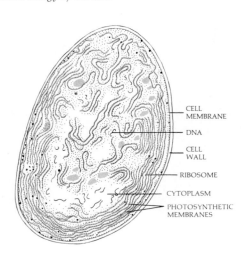

CELL MEMBRANE

DNA

CELL WALL

RIBOSOME

CYTOPLASM

PHOTOSYNTHETIC MEMBRANES

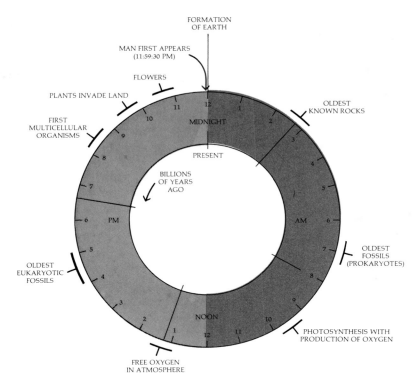

1-8

The clockface of biological time. Life first appears relatively early in the earth's history, before 7:00 A.M. on a 24-hour time scale. The first multicellular organisms do not appear until the twilight of that 24-hour day, and man himself is a late arrival—at about 30 seconds to midnight.

FORMATION
OF EARTH

MAN FIRST APPEARS
(11:59:30 PM)

FLOWERS

PLANTS INVADE LAND

FIRST
MULTICELLULAR
ORGANISMS

MIDNIGHT

PRESENT

BILLIONS
OF YEARS
AGO

PM

AM

NOON

OLDEST
KNOWN ROCKS

OLDEST
FOSSILS
(PROKARYOTES)

PHOTOSYNTHESIS WITH
PRODUCTION OF OXYGEN

OLDEST
EUKARYOTIC
FOSSILS

FREE OXYGEN
IN ATMOSPHERE

SCALE: 1 SECOND = 52,000 YEARS
1 MINUTE = 3,125,000 YEARS
1 HOUR   = 187,500,000 YEARS

EUKARYOTIC CELLS

The prokaryotes, the fossil record indicates, were the only forms of life on this planet for more than 1½ billion years until *eukaryotes* ("well nucleated" cells) evolved. Eukaryotes are usually larger than prokaryotes and have a number of features not found in the earlier cells. One of these—from which they get their name—is a *nucleus*, which is separated from the rest of the cell by a *nuclear envelope*, composed of membranes. The nucleus encloses the genetic material, which is organized into *chromosomes*, which contain both DNA and protein. (In prokaryotes, the DNA is not combined with protein.) Some eukaryotic cells, including plant cells and fungi, have a cell wall; others, including the cells of our own body and other animal cells, do not.

Organelles for Energy Exchanges

Eukaryotic cells are also characterized by complex organelles bound by membranes that are not present in prokaryotic cells. All photosynthetic eukaryotes have *chloroplasts*. Chloroplasts contain numerous membranes in which chlorophyll and other photosynthetic pigments are embedded. (Blue-green algae, as we saw, also have membranes that contain chlorophyll but they are not separated from the rest of the cytoplasm by an outer membrane.) Virtually all eukaryotic cells (including photosynthetic ones) have *mitochondria* (singular, mitochondrion), which are also membrane-bound organelles. In these organelles, energy-rich molecules are broken down, using oxygen, to provide the energy needed for cellular activities. (When we breathe, we are working for our mitochondria, in order to supply them with their required oxygen.)

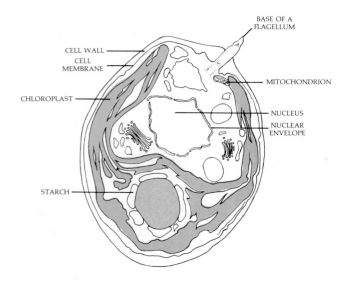

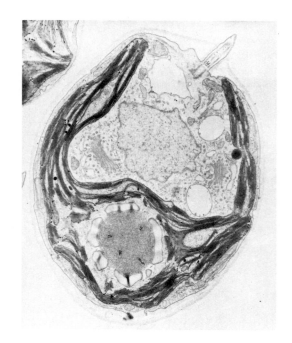

BASE OF A
FLAGELLUM

CELL WALL

CELL
MEMBRANE

CHLOROPLAST

MITOCHONDRION

NUCLEUS

NUCLEAR
ENVELOPE

STARCH

1–9

Electron micrograph and diagram of Chlamydomonas, *a photosynthetic eukaryotic cell,
which contains a membrane-bound ("true")
nucleus and numerous organelles. A single,
irregularly shaped chloroplast, surrounded
by a double membrane, fills most of the cell.
Numerous mitochondria are visible. The cytoplasm is surrounded by a cell membrane,
outside of which is a cell wall.*

Figure 1–9 gives an example of a single-celled photosynthetic eukaryote
(*Chlamydomonas*). *Chlamydomonas* is a common inhabitant of freshwater ponds
and aquariums. These organisms are small, bright green (because of the chlorophyll), and move very quickly with a characteristic darting motion. Being
photosynthetic, they are usually found near the water's surface. They have a
single, large chloroplast, which fills most of the cell. You will notice that the
mitochondria are found near the base of the *flagella* (singular, flagellum), the
whiplike extensions at the anterior of the cell. The mitochondria provide the
energy for the flicking movements of the flagella that propel the cell through the
water. The cell is surrounded by an outer wall composed mostly of cellulose.
Chlamydomonas is believed to resemble the sort of cell from which the plants
evolved.

THE ORIGINS OF MULTICELLULARITY

The first multicellular organisms, as far as can be told by the fossil record, made
their appearance a mere 750 million years ago. The various major groups of
multicellular organisms—such as the plants, the animals, and the fungi—presumably had their origins in different types of single-celled eukaryotes. All
truly multicellular organisms are made up of eukaryotic cells.

The cells of the modern multicellular organisms closely resemble those of
single-celled organisms (see Figure 1–10). They are bound by a cell membrane
identical in appearance to the cell membrane of single-celled organisms. Their
mitochondria and chloroplasts are constructed according to the same design.
Their cells differ from single-celled organisms in that they are characteristically
specialized to perform a relatively limited role in the life of the organism.
However, each remains a remarkably self-sustaining unit. The human body is
made up of trillions of different individual cells, composed of at least 100
different types of cells, each specialized for its particular function but each
working as a cooperative whole (Figure 1–11).

1-10

*Electron micrograph and diagram of cells
from the leaf of a corn plant. The nucleus
can be seen on the right side of the
central cell. The granular material within
the nucleus is a combination of DNA and
proteins. Note the many mitochondria and
chloroplasts, also enclosed by membranes.
The vacuole and cell wall are characteristic
of plant cells but are generally not found in
animal cells. Note the many similarities between this cell and that of the single-celled
alga* Chlamydomonas.

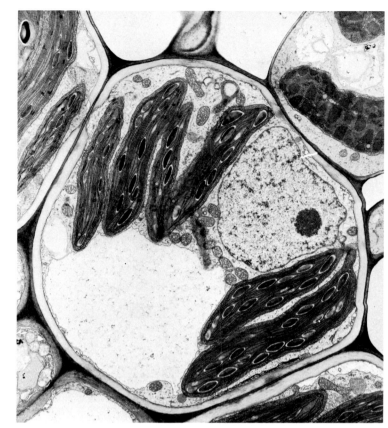

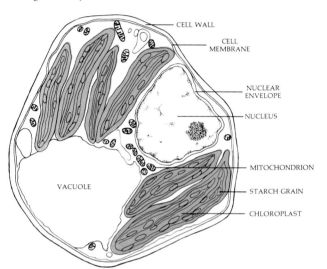

CELL WALL

CELL
MEMBRANE

NUCLEAR
ENVELOPE

NUCLEUS

MITOCHONDRION

STARCH GRAIN

CHLOROPLAST

VACUOLE

1-11

*An example of cellular specialization. In the
center of the electron micrograph are portions
of two cells that fit together to form a capillary, a tiny blood vessel. Blood flowing
through the lumens (passageways) of the capillaries carries oxygen and food molecules to
all the cells of the body, including the capillary cells themselves. Surrounding the capillary, portions of heart muscle cells—also
very highly specialized cells—can be seen.*

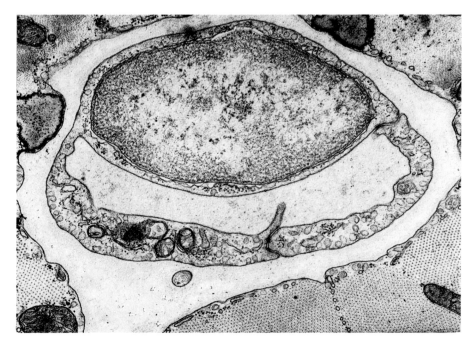

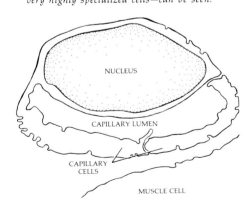

NUCLEUS

CAPILLARY LUMEN

CAPILLARY
CELLS

MUSCLE CELL

SUMMARY

The sun and its planets were formed 4½ billion years ago—the sun probably from the condensation and contraction of a hydrogen gas cloud, and the planets as accumulations of interstellar debris. Of the nine planets in this solar system, only earth is known to support life, but there are likely to be other planets in the galaxy with some form of life.

The primitive atmosphere of earth held the chief raw materials of living matter—hydrogen, oxygen, carbon, and nitrogen—combined in water vapor and gases. The energy required to break apart the simple gases in the atmosphere and re-form them into more complex molecules was present in heat, lightning, radioactive elements, and high-energy radiation from the sun. Laboratory experiments have shown that the types of molecules characteristic of living organisms—that is, organic molecules—can be formed under the conditions that prevailed on the young earth. These molecules gradually accumulated to form cells, the first living things.

The earliest fossil cells resemble modern bacteria and blue-green algae. Such cells are known as prokaryotes. They are distinguished from the larger, more complex cells of plants and animals—known as eukaryotic cells—chiefly by the fact that eukaryotes have a membrane-bound nucleus and many specialized structures (organelles) not found in prokaryotes.

The earliest cells were probably heterotrophs, that is, organisms that depend on outside sources of organic molecules for their energy. Organisms that can use the sun's energy directly to make organic molecules are known as autotrophs. The most familiar of the modern autotrophs are the green plants.

Multicellular organisms evolved comparatively recently—only about 750 million years ago. They are composed of eukaryotic cells specialized to perform particular functions.

QUESTIONS

1. Distinguish between the following: heterotroph/autotroph, eukaryote/prokaryote, chloroplast/mitochondrion.

2. It is now becoming possible to land electronic equipment on other planets. Using such equipment, what kinds of experiments and observations could you devise to test for the presence of living organisms?

3. The experiment pictured on page 16 was attempted at Stockton, California, in 1973, but instead of making amino acids, the electric sparks caused the apparatus to explode. (No one was hurt, fortunately.) What might be present in today's atmosphere that was not present in the primitive atmosphere and that would account for the explosion?

4. What kinds of nonliving objects—manmade and otherwise—exhibit one or more of the signs of life discussed on pages 14 and 15? Can you think of an eighth "sign" that would distinguish these objects from living organisms?

CHAPTER 2

The Composition of Living Matter: Water

In this chapter and the one that follows, we are going to be examining the atoms and molecules that living cells are made of. By far the most important of these molecules, at least in quantitative terms, is water. Water makes up 50 to 95 percent of the weight of any functioning living system and an equal or even greater proportion of the weight of any living cell. It is a common liquid, by far the most common. Three-fourths of the surface of the earth is covered by water. In fact, if the earth's surface were absolutely smooth, all of it would be $2\frac{1}{2}$ kilometers under water. But do not mistake "common" for "ordinary"; water is not in the least an ordinary liquid. It is, in fact, extraordinary in comparison with other liquids. If it were not, it is most probable that life—at least, our form of life—could never have evolved.

THE EXTRAORDINARY PROPERTIES OF WATER

Surface Tension

Watch water dripping from a faucet (Figure 2–1). Each drop clings to the rim and dangles for a moment by a thread of water; then just as it breaks loose, tugged by gravity, its outer surface is drawn taut, enclosing the entire sphere as it falls free.

2–1
The formation of water droplets.

2–2
Note how the water forms a continuous sheet, blanketing the head of this frog. This effect is the result of the surface tension of water, caused by the water molecules' clinging to one another.

Look at a pond in spring or summer; you will see water striders and other insects skating on its surface as if it were almost solid. Take a needle or a razor blade and place it gently on the surface of a glass of water. Although metal is denser than water, the object floats.

The phenomena just described are all the result of *surface tension*. Surface tension is caused by the cohesiveness of water molecules. The only liquid with greater surface tension than that of water is mercury. Atoms of mercury have so great an attraction for one another that they will not adhere to most other surfaces. (Adhesion is the holding together of unlike substances, and cohesion is the holding together of like substances. Water both coheres and adheres strongly.)

Capillary Action

If you hold two dry glass slides together and dip one corner in water, the interaction of cohesion and adhesion will cause water to spread between the two slides. This is *capillary action*. Capillary action similarly causes water to rise in glass tubes of very fine bore, to creep up a piece of blotting paper, or to move slowly through the tiny spaces between soil particles and so become available to the roots of plants.

Imbibition

2–3
A water strider's legs dimple the surface of a pond. The water strider, which lives on this surface, has specialized hairs on its first and third pairs of legs that enable it to rest upon the surface film, depressing it but not penetrating. The second pair of legs, which penetrate the film, serve as propelling oars.

Imbibition ("drinking up") is the movement of water molecules into substances such as wood or gelatin, which swell or increase in volume as a result of adhesion between them and the water molecules. The pressures developed by imbibition can be astonishingly large. It is said that stone for the ancient Egyptian pyramids was quarried by driving wooden pegs into holes drilled in the rock face and then soaking the pegs with water. As the wood swelled, a force

The germination of seeds begins with changes in the seed coat that permit a massive uptake of water. In these acorns, photographed on the forest floor, the embryonic roots have penetrated the seed coat and emerged through the tough outer layer of the fruit.

sufficient to break the slab of stone was created. Seeds imbibe water as they begin to germinate (Figure 2–4), often swelling to many times their original size as a result.

Water and Heat

The amount of heat a given amount of a substance requires for a given increase in temperature is its *specific heat.* One calorie of heat, by definition, is the amount of heat required to raise the temperature of 1 gram (1 cubic centimeter) of water 1°C. The specific heat of water is about twice the specific heat of oil or alcohol; that is, approximately 1 calorie will raise 1 gram of oil or alcohol 2°C. It is four times the specific heat of air or aluminum and nine times that of iron. Only liquid ammonia has a higher specific heat. In other words, it requires a high input of energy in the form of heat to raise the temperature of water.

What does this high specific heat mean in biological terms? It means that for a given rate of heat input, the temperature of water will rise more slowly than the temperature of almost any other material. Conversely, the temperature will drop more slowly as heat is released. Because so much heat input or heat loss is required to raise or lower the temperature of water, organisms that live in the oceans live in an environment in which the temperature is relatively constant. This constancy of temperature is important because, as we shall see, biological reactions take place only within a narrow temperature range. Similarly, land plants and animals, because of their high water content, are more easily able to maintain a relatively constant internal temperature.*

Water also has a high *heat of vaporization.* Vaporization—or evaporation, as it is more commonly called—is the change from a liquid to a gas. It takes more than 500 calories to change a gram of liquid water into water vapor, five times as much as for ether and almost twice as much as for ammonia.

Thus water, in comparison with other liquids, has both great stability and great fluidity. Both of these properties are essential to the role of water as the medium in which each cell, from *Chlamydomonas* to a human egg, lives its entire life.

* Another important property of water—that it expands as it freezes—is discussed on page 499.

THE STRUCTURE OF WATER

Why does water have these unusual properties? To answer this question, we must look at how a water molecule is put together. As you know, each water molecule is made up of two atoms of hydrogen and one of oxygen—H_2O. The hydrogen atom is the simplest of all atoms. Its nucleus contains a single positively charged particle, a *proton*, and outside the nucleus is a single negatively charged particle, an *electron*:

The electron moves in orbit around the proton. The path of movement of the electron is called the *electron shell*. The proton and the electron have equal charges; as a consequence, they balance one another out. If, however, the hydrogen atom loses its single orbiting electron, it will have a charge of $+1$, becoming H^+. Such charged atoms or groups of atoms are referred to as *ions*. A gain of one or more electrons results in a negatively charged ion, just as a loss of an electron results in the positively charged hydrogen (H^+) ion. (A hydrogen ion consists of a single proton and is often referred to simply as a proton.)

The oxygen atom is a little more complicated. In its nucleus are eight protons, and outside the nucleus are eight electrons:

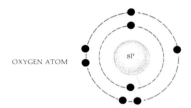

The inner shell has two electrons. (The innermost shell can never have more than two electrons, according to the "rules" of atomic structure.) There are six electrons in the outer shell when oxygen is electrically neutral, but according to another "rule," this outer shell is not complete unless it has eight electrons. For reasons which are not clearly understood, atoms "seek" to complete their shells, which they often do by sharing electrons with other atoms.

In order to complete its outer shell, an oxygen atom requires two electrons. A hydrogen atom requires one. Thus, when one oxygen atom shares electrons with two hydrogen atoms, each completes its outer shell:

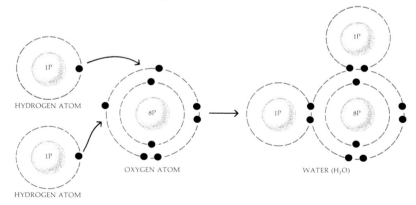

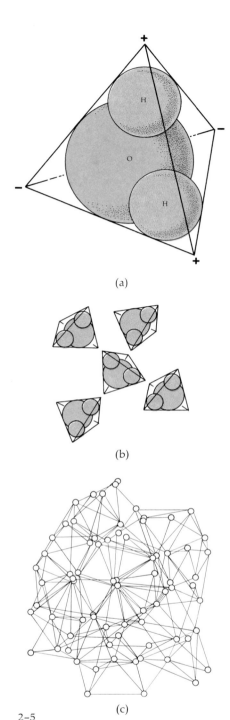

(a)

(b)

(c)

2-5

Water. (a) A single molecule. Note the four-cornered distribution of positive and negative charges. (b) As a result of these positive and negative charges, each water molecule can form hydrogen bonds with four other water molecules. (c) All the water molecules in a sample of liquid water are linked together by shifting hydrogen bonds. This is a diagram of water as "seen" by a computer.

This electron sharing creates forces of attraction—*bonds*—among the atoms. Bonds that involve the sharing of electrons are known as *covalent bonds*. Covalent bonds are relatively strong bonds, and most molecules important in biological systems are held together by covalent bonds.

The Water Molecule

The water molecule is electrically neutral, having an equal number of electrons ($-$) and protons ($+$). However, the single electrons of the two hydrogen atoms, which are shared with the oxygen atom, are attracted more strongly by the oxygen nucleus and are therefore located nearer to it than to the hydrogen nuclei. As a consequence, the molecule has two local positive charges, carried by the hydrogen atoms, and two local negative charges, carried by the oxygen atom. The water molecule, in terms of its electrical charges, is thus four-cornered, forming a tetrahedron with two "positive" corners and two "negative" ones.

A molecule with zones of negative and positive charge is known as *polar*, by analogy with a magnet, which has a negative pole and a positive pole.

When a positively charged hydrogen atom of a water molecule comes into juxtaposition with an atom carrying a sufficiently strong negative charge—such as the oxygen atom of another water molecule—the force of the attraction forms a weak bond between them, which is known as a *hydrogen bond*. A hydrogen bond forms between the negative "corner" of one water molecule and the positive "corner" of another. Every water molecule can therefore establish hydrogen bonds with four other molecules. Liquid water is made up of water molecues bound together in this way, as shown in Figure 2–5.

As we noted previously, hydrogen bonds are weak bonds. Furthermore, any single hydrogen bond has an exceedingly short lifetime; on an average, each such bond lasts only about 10^{-11} second, which is 1/100,000,000,000 second. All together, however, they have considerable strength. It is because of these multiple, shifting hydrogen bonds that liquid water is both fluid and stable.

These attractions between water molecules are responsible for its surface tension and also for its capillary action, which takes place as a result of the molecules clinging to one another (adhesion) as well as to the surface (cohesion) along which they are creeping. The movement of water through the plant body takes place in large part because of this attraction of the water molecules for one another, as we shall see in Chapter 20.

Hydrogen Bonds and Heat

The high specific heat of water is also a consequence of hydrogen bonding. Heat is a form of energy, the energy of movement (kinetic energy) of molecules. In order for the kinetic energy of water molecules to increase sufficiently for the temperature to rise 1°C, it is necessary to rupture a number of the hydrogen bonds holding the molecules together. When you heat a pot of water, much of the heat energy is used to break the hydrogen bonds holding the water molecules together and so a relatively small amount is available to increase molecular movement. (Note that heat and temperature are not the same. A lake may have a lower temperature than that of the bird flying over it, but the lake contains more heat because it has more molecules in motion. Temperature is measured in degrees and reflects average energy of the molecules. Heat is measured in calories, which reflect both molecular movement and volume.)

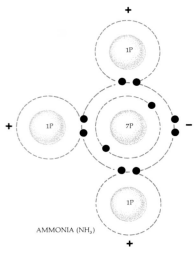

2-6

Ammonia is very similar to water in its chemical structure, and biologists have speculated about whether it might substitute for water in life processes. The ammonia molecule, like the water molecule, is made up of hydrogen atoms covalently bonded to nitrogen, which, like the oxygen in the water molecule, retains a slight negative charge. But because there are three hydrogens with slight positive charges to one nitrogen, ammonia does not have the cohesive power of water and evaporates much more quickly. Perhaps this is why no form of life based on ammonia has been found, although NH_3 was very common in the primitive atmosphere.

AMMONIA (NH_3)

"Ammonia! Ammonia!"

[*Drawing by R. Grossman;* © *1962 The New Yorker Magazine, Inc.*]

Hydrogen bonding is also responsible for water's high heat of vaporization. Vaporization comes about because the rapidly moving molecules of a liquid break loose from the surface and enter the air. The hotter the liquid, the more rapid the movement of its molecules and, hence, the more rapid the rate of evaporation. But, no matter what the temperature is, so long as a liquid is exposed to air that is below 100 percent humidity, evaporation will take place and will continue to take place right down to the last drop.

In order for a water molecule to break loose from its fellow molecules—that is, to vaporize—the hydrogen bonds have to be broken. This requires heat energy. As we noted previously, more than 500 calories are needed to change 1 gram of liquid water into vapor (the exact amount depends on the temperature of the water). The same amount of heat would raise the temperature of 500 grams of water 1°C. As a consequence, when water evaporates, as from the surface of your skin or a leaf, it absorbs a great deal of heat from the immediate environment. Thus evaporation has a cooling effect. Evaporation from the surface of a land-dwelling plant or animal is one of the principal ways in which these organisms "unload" excess heat.

WATER AS A SOLVENT

Many substances within living systems are found in solution. (A solution is a uniform mixture of the molecules of two or more substances in which one—the solvent—is a liquid.) The polarity of the water molecules is responsible for water's usefulness as a solvent. The charged water molecules tend to pull apart compounds such as sodium chloride (table salt) into their constituent sodium (Na^+) and chloride (Cl^-) ions. Then, as shown in Figure 2–8, the water molecules cluster around and so segregate the charged ions.

Many of the small molecules important in living systems are also polar, that is, they have areas of positive and negative charge, and so attract water molecules in the same way that water molecules are attracted to one another. These

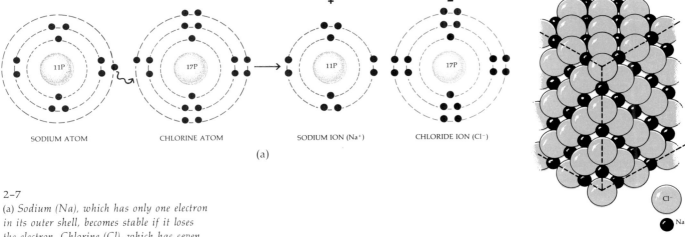

SODIUM ATOM   CHLORINE ATOM   SODIUM ION (Na⁺)   CHLORIDE ION (Cl⁻)

(a)

(b)

Cl⁻

Na⁺

2-7

(a) *Sodium (Na), which has only one electron in its outer shell, becomes stable if it loses the electron. Chlorine (Cl), which has seven electrons in its outer shell, becomes stable if it gains one. When sodium and chlorine interact, sodium loses its single electron and chlorine gains it; following this transaction, sodium has a positive charge and chlorine a negative one. Such charged atoms are called ions. (b) Oppositely charged ions attract one another. Table salt is crystalline NaCl, a latticework of alternating Na⁺ and Cl⁻ ions held together by their opposite charges. Such bonds between oppositely charged ions are known as ionic bonds.*

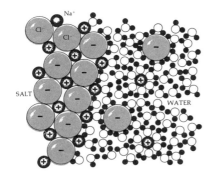

2-8

Because of the polarity of water molecules, water can serve as a solvent for ions and other polar molecules. This diagram shows sodium chloride (NaCl) dissolving in water as the water molecules cluster around the individual ions, separating them from each other.

polar compounds, such as sugars, therefore form hydrogen bonds with water molecules.

Because of water's tendency to dissociate certain compounds and to cluster around other atoms and molecules—that is, to dissolve them—virtually all of the chemical reactions of the cell take place in water. Moreover, as we shall see in the next chapter, water molecules participate directly in many of these reactions.

Ionization: Acids and Bases

Water molecules also have a slight tendency to ionize, separating into H^+ and OH^- (hydroxide) ions. In any given volume of pure water, a very small but constant number of water molecules will be dissociated in this form. The number is constant because the tendency of water to dissociate is exactly offset by the tendency of the ions to reunite; thus even as some are ionizing, an equal number of others are forming bonds. This can be expressed as:

$$HOH \; \rightleftharpoons \; H^+ + OH^-$$

The arrows indicate that the reaction can go in either direction. The fact that the arrow pointing toward HOH is longer indicates that the reaction is more likely to go in that direction. (In any sample of water, only a small fraction exists in ionized form.)

In pure water, the number of H^+ ions exactly equals the number of OH^- ions. (This is necessarily the case since neither ion can be formed without the other when only H_2O molecules are present.) A solution acquires the properties we recognize as acid when the number of H^+ ions exceeds the number of OH^- ions; conversely, a solution is alkaline (basic) when the OH^- ions exceed the H^+ ions.

Hydrochloric acid (HCl) is an example of a common acid. It is a strong acid, meaning that it tends to be almost completely ionized into H^+ and Cl^- ions, releasing additional H^+ ions into the solution. Sodium hydroxide (NaOH) is a common strong base; in water it exists entirely as Na^+ and OH^- ions. Weak acids and weak bases are those that ionize only slightly.

The most common form of hydrogen, as we noted, contains a single proton in its nucleus. Most atoms, in addition to protons, contain particles in their nuclei called neutrons. As their name implies, neutrons have no charge, although they do have mass. The mass of a neutron is approximately equal to that of a proton, and the atomic mass of an atom is about equal to the combined numbers of protons and neutrons. Atoms that contain the same number of protons but different numbers of neutrons differ, therefore, in their atomic mass. However, they are very similar chemically; that is, each can substitute for the other in chemical compounds or chemical reactions. Such chemically interchangeable atoms are called isotopes.

The most common isotope of hydrogen (H), then, contains a single proton in its nucleus. Another fairly common isotope of hydrogen, heavy hydrogen, or deuterium, contains one proton and one neutron and so has an atomic mass of about 2 (^2H). Another isotope of hydrogen, tritium, has one proton and two neutrons and so has an atomic mass of about 3 (^3H). As indicated previously, chemically each of the isotopes is similar. Physically, however, they behave differently. Deuterium and tritium are heavier than hydrogen. Moreover, tritium, like a number of other isotopes, is unstable, or radioactive. Tritium atoms become stable by emitting high-energy particles. As long as the nucleus contains only one proton, the atom is still a hydrogen atom—that is, it behaves chemically like hydrogen—even though the number of neutrons increases. If, however, the nucleus gains a proton, a different element is formed—helium. The common isotope of helium (He) contains two protons and two neutrons, and so has an atomic mass of 4.

The sun consists largely of hydrogen nuclei. At the extremely high temperature at the core of the sun, hydrogen nuclei strike each other with enough velocity to fuse. In a series of steps, four hydrogen nuclei fuse to form one helium nucleus. In the course of these reactions, energy is released, enough to keep the fusion reaction going and to emit tremendous amounts of radiant energy into space.

Biological Uses of Radioactive Isotopes

Many naturally occurring isotopes are radioactive. All the heavier elements—atoms that have 84 or more protons in their nuclei—are unstable and, therefore, radioactive. All radioactive elements emit radiation (as particles or rays) at a fixed rate; this process is known as radioactive "decay." The rate of decay is measured in terms of half-life: The half-life of a radioactive element is defined as the time in which half the atoms in a sample of the element lose their radioactivity and become stable. Since the half-life of an element is constant, it is possible to calculate the fraction of decay that will take place for a given isotope in a given period of time.

Atomic models of hydrogen, deuterium, tritium, and helium. Note that hydrogen, deuterium, and tritium each have only a single proton and a single electron, and so they behave similarly chemically although they differ in the number of neutrons. Helium, which has one more proton and one more electron, is chemically very different from hydrogen and its isotopes.

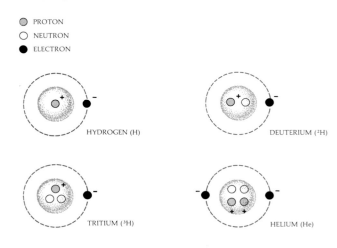

- ⊙ PROTON
- ○ NEUTRON
- ● ELECTRON

HYDROGEN (H)　　　　DEUTERIUM (^2H)

TRITIUM (^3H)　　　　HELIUM (He)

Half-lives vary widely, depending on the isotope. The radioactive nitrogen isotope ^{13}N has a half-life of 10 minutes, and tritium has a half-life of $12\frac{1}{4}$ years. The most common isotope of uranium (^{238}U) has a half-life of $4\frac{1}{2}$ billion years. The uranium atom undergoes a series of decays, eventually being transformed to an isotope of lead (^{206}Pb).

Isotopes have a number of important uses in biological research. One of these is in dating the age of fossils and of the rocks in which fossils are found. The proportion of ^{238}U to ^{206}Pb in a given rock sample, for example, is a good indication of how long ago that rock was formed. (The lead formed as a result of the decay of uranium is not the same as the lead commonly present in the original rock, ^{204}Pb.) A second use of isotopes is as radioactive tracers. A plant, for example, that is exposed to radioactive carbon dioxide ($^{14}CO_2$) will use the radioactive carbon to make sugar just as it would the more common isotope. The use of radioactive carbon dioxide has played an important role in enabling plant physiologists to trace the path of carbon in photosynthesis, as described in Chapter 6. A third use is autoradiography, a technique in which a sample of material containing a radioactive isotope is placed on a photographic film. Energy emitted from the isotope leaves traces on the film and so reveals the exact location of the isotope within the specimen.

Table 2-1 *The pH Scale*

	CONCENTRATION OF H⁺ IONS (MOLES PER LITER)		pH	CONCENTRATION OF OH⁻ IONS (MOLES PER LITER)	
	1.0	$= 10^0$	0	10^{-14}	
	0.1	$= 10^{-1}$	1	10^{-13}	
	0.01	$= 10^{-2}$	2	10^{-12}	
Acidic	0.001	$= 10^{-3}$	3	10^{-11}	
	0.0001	$= 10^{-4}$	4	10^{-10}	
	0.00001	$= 10^{-5}$	5	10^{-9}	
	0.000001	$= 10^{-6}$	6	10^{-8}	
Neutral	0.0000001	10^{-7}	7	10^{-7}	Neutral
		10^{-8}	8	$10^{-6} = 0.000001$	
		10^{-9}	9	$10^{-5} = 0.00001$	
		10^{-10}	10	$10^{-4} = 0.0001$	
Basic		10^{-11}	11	$10^{-3} = 0.001$	
		10^{-12}	12	$10^{-2} = 0.01$	
		10^{-13}	13	$10^{-1} = 0.1$	
		10^{-14}	14	$10^{-0} = 1.0$	

Chemists define degrees of acidity by means of the pH scale. For example, 1/10,000,000 mole* of hydrogen ions, which is 10^{-7} mole, is referred to simply as pH 7 (see Table 2–1). At pH 7, the concentrations of free H⁺ and OH⁻ ions are exactly the same; this is a "neutral" state. Any pH below 7 is acidic, and any pH above 7 is basic.

Table 2–2 lists pH values for some familiar solutions.

WATER MOVEMENT

Many biological processes involve water movement. Water moves in living systems in three ways: by bulk flow, by diffusion, and by osmosis. Of these three, bulk flow is the most familiar.

Bulk Flow

Bulk flow is the overall movement of water (or some other gas or liquid). It occurs in response to differences in the potential energy of water, usually referred to as *water potential.*

A simple example of water that has potential energy is water at the top of a hill (Figure 2–9). As this water runs downhill, its potential energy can be converted to mechanical energy by a water mill or to electrical energy by a hydroelectric turbine.

Pressure is another source of water potential. If we put water into a rubber bulb and squeeze the bulb, this water, like the water at the top of a hill, also has

Table 2-2 *pH Values for Various Solutions*

SOLUTION	pH
Gastric secretions	0.9
Cola drink	2.5–3.0
Vinegar	3.0
Grapefruit juice	3.2
Orange juice	2.6–4.4
Tomato juice	4.3
Urine	4.8–7.5
Saliva	6.6
Milk	6.6–6.9
Pure (distilled) water	7.0
Human blood	7.4
Tears	7.4
Intestinal secretions	7.0–8.0
Seawater	8.0
Egg white	8.0

* The mole is the principal unit of measure used for defining quantities of substances involved in chemical reactions. The number of particles in 1 mole of any substance (whether atoms, ions, or molecules) is always exactly the same: 6.023×10^{23}. For example, 1 mole of hydrogen ions contains 6.023×10^{23} ions, 1 mole of hydrogen contains 6.023×10^{23} hydrogen atoms, and 1 mole of water contains 6.023×10^{23} water molecules.

2-9

Bulk flow describes the overall movement of water from a position of higher water potential to one of lower water potential, as over this dam.

water potential, and it will move to an area of less water potential. Can we make the water that is running downhill run uphill by means of pressure? Obviously, we can. But only so long as the water potential produced by the pressure exceeds the water potential produced by gravity. Water moves from an area where water potential is greater to an area where water potential is less, regardless of the reason for the water potential. Your heart pumps blood to your brain against gravity by creating a greater water (blood) potential.

The concept of water potential is a useful one because it enables physiologists to predict the way in which water will move under various combinations of circumstances. Measurements of water potential are usually made in terms of the pressure required to stop the movement of water—that is, the hydrostatic (water-stopping) pressure—under the particular circumstances. This pressure is usually expressed in units of bars. One bar is the average pressure of the air at sea level.

Diffusion

Diffusion is another familiar phenomenon involving movement of molecules from one place to another (Figure 2-10). If you sprinkle a few drops of perfume in the corner of a room, the scent will eventually permeate the entire room even if the air is still. If you put a few drops of dye in one end of a rectangular glass tank full of water, the dye molecules will slowly distribute themselves evenly throughout the tank. (The process may take a day or more.) Why do the dye molecules move apart?

The answer to this question is found in the fact that all molecules are constantly in motion. If you could observe the individual dye molecules in the tank, you would see that each one of them moves at random and moves individually; looking at any single molecule—at either its rate of motion or its direction of motion—gives you no clue at all about where the molecule is located with respect to the others. So how do the molecules get from one side of the tank to the other? In your imagination, take a thin cross section of the tank, running from top to bottom. Dye molecules will move in and out of the section, some moving in one direction, some moving in the other. But you will see more dye molecules moving from the side of greater concentration. Why? Simply because there are more dye molecules at that end of the tank. Consequently, if there are more dye molecules on the left, the overall movement will be from left to right, even though there is an equal probability that any one molecule of dye will move from right to left. Similarly, if you could see the movement of the individual water molecules in the tank, you would see that their overall movement is from right to left.

What happens when all the molecules are distributed evenly throughout the tank? The even distribution does not affect the behavior of the molecules as individuals; they still move at random. And, since the movements are random, just as many molecules go to the left as to the right. But because there are now as many molecules of dye and as many molecules of water on one side of the tank as on the other, there is no overall direction of motion, although, provided the temperature has not changed, there is as much overall motion as before.

Substances that are moving from a region of greater concentration of their own molecules to a region of lesser concentration are said to be moving *along the gradient.* Diffusion occurs only along the gradient. (A substance moving in the opposite direction, toward a greater concentration of its own molecules, moves *against the gradient,* which is analogous to pushing something uphill.) In our imaginary tank, there are two gradients; the dye molecules are moving along one of them, and the water molecules are moving along the other. When the

THE WATER CYCLE

The earth's supply of water is constant and is used over and over again. Most of the water (98 percent) is in liquid form—in the oceans, lakes, and streams. Of the remaining 2 percent, some is frozen in polar ice and glaciers, some is in the soil, some is in the atmosphere in the form of vapor, and some is in the bodies of living organisms.

The water cycle is driven by solar energy. Energy from the sun evaporates water from the oceans, lakes, and streams, from the moist soil surfaces, from the leaves of plants (transpiration), and from the bodies of other living organisms. The water molecules are carried up into the atmosphere by air currents and eventually fall again as rain.

Some of the water that falls on the land percolates down through the soil until it reaches a zone of saturation. In the zone of saturation, all pores and cracks in the rock are filled with water. The upper surface of this zone of saturation is known as the water table. Below the zone of saturation is solid rock, through which the water cannot penetrate. The deep groundwater, moving extremely slowly, eventually emerges in lakes and streams and even in the ocean floor as seepages and springs, thereby completing the water cycle.

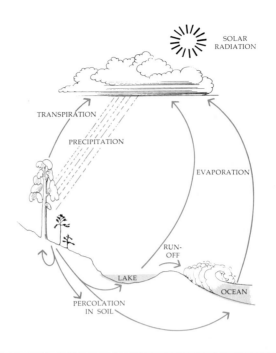

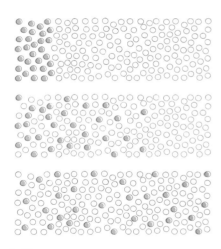

2-10

Diagram of the diffusion process. Diffusion is the random movement of molecules from a more concentrated to a less concentrated area. Notice that as one type of molecule (indicated by color) diffuses to the right, the other diffuses in the opposite direction. The result will be an even distribution of both types of molecules. Can you see why the net movement of molecules will stop when equilibrium is reached?

molecules have reached a state of equal distribution, that is, when there is no more gradient, they are said to be in *dynamic equilibrium.*

An aqueous system in which diffusion is occurring is a system that possesses water potential. Because water, like other substances, will move from a region of greater concentration to a region of lesser concentration, the area of the tank in which there is pure water has a greater water potential than the area containing water plus dye or some other solute.

The essential characteristics of diffusion are (1) that each molecule moves as an individual and (2) that these movements are random. The net result of diffusion is that the diffusing substance becomes evenly distributed.

Cells and Diffusion

Diffusion is a principal way in which substances move within cells. (This process may be hastened by cell streaming, the movement of cytoplasm within cells, which is readily visible in *Amoeba* and in growing plant cells.) This dependence upon diffusion is one of the factors limiting cell size. As you can probably deduce by studying the simple diagram (Figure 2–10), the process becomes increasingly slower and less efficient as the distance "covered" by the diffusing molecules increases.

Water, oxygen, carbon dioxide, a few other simple molecules, and some ions diffuse freely across cell membranes. Larger molecules and other ions are stopped by the outer cell membrane and also by membranes within the cell, such as the endoplasmic reticulum, the nuclear envelope, and the membranes surrounding the chloroplasts and the mitochondria. These substances may move across cell membranes by a process known as *active transport.* How the cell membrane performs this important function is not known, but according to most theories, the membrane must have specific molecules associated with it that, under certain conditions, "recognize" other molecules and escort them into the cell. By active transport, substances can be moved against the diffusion gradient. Movement against the gradient requires the expenditure of energy.

Osmosis

In order to move into and out of living cells, water must move through cell membranes. Cell membranes, like many other kinds of membranes, permit the free passage of water but inhibit or retard the passage of most solutes, that is, materials dissolved in the water. Such membranes are called selectively permeable, and the movement of water through them is known as _osmosis_. In the absence of other forces (such as pressure), the movement of the water in osmosis will be from a region of lesser solute concentration (and therefore of greater water concentration) to a region of greater solute concentration (lesser water concentration). In other words, the presence of solute decreases the water potential and so the water moves from the region of greater water potential to the region of lesser water potential (Figure 2-11).

The movement of water is not affected by what particular solute is present in the water, but only by how much solute it contains, that is, by how many molecules or ions of the solute are present in solution. The word _isotonic_ was coined to describe solutions that have equal numbers of particles dissolved in them and therefore have equal osmotic pressures. There is no net movement of water across a membrane separating two solutions that are isotonic to one another, unless, of course, pressure is exerted on one side. In comparing solutions of different concentration, the solution that has less solute is known as _hypotonic_, and the one that has more solute is known as _hypertonic_. (Note that _iso_ means "the same"; _hyper_ means "more"—in this case, more molecules of solute; and _hypo_ means "less"—in this case, fewer molecules of solute.)

In sum, solutes decrease water potential. Therefore, a hypotonic solution has a higher water potential than a hypertonic one. Thus, in osmosis, water molecules move through a selectively permeable membrane from a hypotonic solution into a hypertonic one until the water potential is equal on both sides of the membrane.

Osmosis and Living Organisms

The movement of water across the cell membrane from a hypotonic to a hypertonic solution causes some crucial problems for living systems. These problems vary according to whether the cell or organism is hypotonic, isotonic, or hypertonic in relation to its environment. One-celled organisms that live in salt water, for example, are usually isotonic with the medium they inhabit, which is one way of solving the problem. Similarly, the cells of higher animals are isotonic with the blood and lymph that constitute the watery medium in which they live.

Many types of cells, however, live in a hypotonic environment. In all single-celled organisms that live in fresh water, such as _Paramecium_, the interior of the cells is hypertonic to the surrounding water; consequently, water tends to move into the cell by osmosis. If too much water were to move into the cell, it could dilute the cell contents to the point of interfering with function and could even eventually rupture the cell membrane. This is prevented in _Paramecium_ and some other single-celled organisms by a specialized organelle known as a contractile vacuole, which collects water from various parts of the cell body and pumps it out with rhythmic contractions (Figure 2-12).

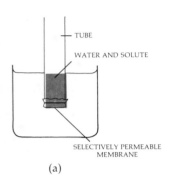

TUBE

WATER AND SOLUTE

SELECTIVELY PERMEABLE MEMBRANE

(a)

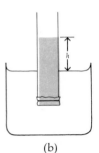

(b)

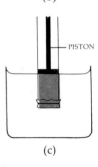

PISTON

(c)

2-11

Osmosis and osmotic pressure. (a) The tube contains a solution and the beaker contains distilled water. (b) The selectively permeable membrane permits the passage of water but not of solute. The movement of water into the solution causes the solution to rise in the tube until the osmotic pressure resulting from the tendency of water to move into a region of lower water concentration is counterbalanced by the height, h, of the column of solution. (c) The hydrostatic ("water-stopping") force that must be applied to the piston to oppose the rise in the tube of solution is a measurement of the osmotic pressure. It is proportional to h.

2–12

Paramecium *magnified 300 times. In the photomicrograph, two contractile vacuoles (indicated by arrows) are visible. These rosettelike structures beat rhythmically, pumping excess water out of the cell body.*

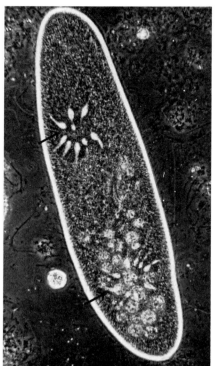

Eric V. Gravé

Turgor

Plant cells become larger by taking up water. Because plant cells are usually hypertonic to their surrounding solution, water tends to move into them. This movement of water into the cell creates pressure within the cell against the cell wall. The pressure causes the cell wall to expand and the cell to enlarge. Some 90 percent of the growth of a plant cell is a direct result of water uptake.

If the water potential on both sides of the plant cell membrane becomes equal, the net movement of water ceases. (We say "net movement" here because water molecules continue to diffuse back and forth across the membrane. However, these movements are in equilibrium; that is, as many water molecules are going in as are coming out.)

As the plant cell matures, however, the cell wall stops growing. Mature plant cells typically have large central vacuoles. These vacuoles often contain solutions of salts and other materials. (In citrus fruits, for example, they contain the acids which give the fruits their characteristic sour taste.) Because of these concentrated solutions, water continues to "try" to move into the cells. In the mature cell, however, the cell wall does not expand further, and equilibrium of solute concentration is not reached. As a consequence, the cell wall remains under constant pressure (Figure 2–13). This water pressure exerted on the cell wall from inside is known as *turgor*. Turgor keeps the cell wall stiff and the plant body crisp. When turgor pressure is reduced, as a consequence of water loss, a plant wilts.

Life and Water

Water has been called the cradle of life. The first living systems came into being, as far as we know, in the warm primitive seas, and today, almost everywhere water is found on this planet, life is present, too. There are one-celled organisms that eke out their entire existence in no more water than that clinging to a grain of sand. Some species of algae are found only on the melting under-

2–13

(a) *A turgid plant cell. The central vacuole is hypertonic in relation to the fluid surrounding it and so the cell gains water. The expansion of the cell is held in check by the* cell wall. (b) *A plant cell begins to wilt if it is placed in an isotonic solution, so that water pressure no longer builds up within the vacuole.* (c) *A plant cell in a hypertonic* solution *loses water to the surrounding fluid and so collapses, with its membrane pulling away from the cell wall.*

CELL WALL   CYTOPLASM

VACUOLE

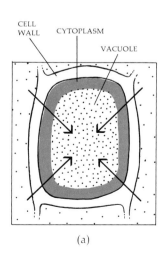

(a)

↘ NET MOVEMENT OF WATER

∴ SOLUTES

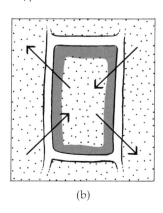

(b)

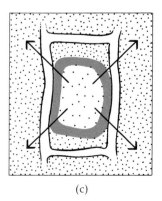

(c)

surfaces of polar ice floes. Certain species of bacteria and blue-green algae can tolerate the near-boiling water of hot springs. In the desert, plants race through an entire life cycle—seed to flower to seed—following a single rainfall. In the jungle, the water cupped in the leaves of a tropical plant forms a microcosm in which numerous small organisms are born, spawn, and die. We are interested in whether the soil of Mars and the dense atmosphere surrounding Venus contain or ever contained water, principally because we want to know if life is there.

There are, then, two essential ingredients in the biosphere. One of these is energy, supplied by the rays of the sun, and the other is liquid water. Wherever these are found, life is there, too.

SUMMARY

One of the chief components of cells is water. Water, which is the most common liquid in the biosphere, has a number of remarkable properties. These properties, which are responsible for its "fitness" for its role in living systems, include a high surface tension, unusual capacities for cohesion and adhesion, and a high specific heat. These properties of water are a consequence of the chemical structure of the molecule. Water is made up of two hydrogen atoms and one oxygen atom held together by covalent bonds. (Covalent bonds, which involve the sharing of electrons between atoms, are the most common bonds in organic chemistry.)

As a result of the arrangement of the electrons, the water molecule is polar, carrying two weak negative charges and two weak positive charges. As a consequence, weak bonds form between water molecules. Such bonds, which link a positively charged hydrogen atom of one molecule to a negatively charged oxygen atom of another molecule, are known as hydrogen bonds. Each water molecule can form hydrogen bonds with four other molecules, and although each bond is weak and constantly shifting, the total strength of the bonds holding the molecules together is very great. These strong intermolecular attractions account for the high surface tension of water, its capillary action, its high specific heat, and the fact that it evaporates relatively slowly.

The polarity of the molecule is responsible for water's capacity for adhesion and also for its capacity as a solvent for ions or other polar molecules.

Water moves by bulk flow, diffusion, and osmosis. The direction in which it moves is determined by water potental; that is, water movement takes place from where the water potential is greater to where it is less. Bulk flow is the overall movement of water, as when water flows downhill. Diffusion involves the random movement of molecules. Osmosis is the movement of water through a membrane that permits the passage of water but inhibits the movement of solutes (a selectively permeable membrane). In the absence of other forces, the movement of water in osmosis is from a region of lesser solute concentration (a hypotonic medium) to one of greater solute concentration (a hypertonic medium).

QUESTIONS

1. Sketch the water molecule and label the areas of positive charge and negative charge. What are the major consequences of the polarity of the water molecule? How are these effects important to living systems?

2. Define water potential, diffusion, osmosis, and turgor.

3. In diffusion, random movement of molecules continues (as long as the temperature remains the same). However, net movement stops. How do you reconcile these two facts?

4. Imagine a pouch with a selectively permeable membrane containing a salt-water solution. It is immersed in a dish of fresh water. Which way will the water move? If you add salt to the water in the dish, how will this affect the water movement? What living systems exist under analogous conditions? How do you think they maintain water balance?

5. Three funnels have been placed in a beaker containing a solution (see the figure below). What is the concentration of the solution? Explain your answer.

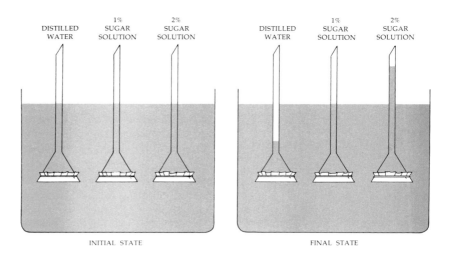

6. Surfaces such as glass or raincoat cloth can be made "nonwettable" by application of silicone oils or other substances that cause water to bead up instead of spread flat. What do you suppose is happening, in molecular terms, when a surface becomes nonwettable?

7. What is the relationship between heat and the breaking of hydrogen bonds; that is, why should heat cause these bonds to break? Why doesn't heat cause the breaking of covalent bonds?

8. Whereas pure (distilled) water has a pH of 7, water exposed for any time to the air dissolves some carbon dioxide gas (CO_2), and the following reactions occur:

$$\underset{\substack{\text{CARBON} \\ \text{DIOXIDE}}}{CO_2} + \underset{\text{WATER}}{H_2O} \rightleftharpoons \underset{\substack{\text{CARBONIC} \\ \text{ACID}}}{H_2CO_3}$$

$$\underset{\substack{\text{CARBONIC} \\ \text{ACID}}}{H_2CO_3} \rightleftharpoons \underset{\substack{\text{HYDROGEN} \\ \text{ION}}}{H^+} + \underset{\substack{\text{BICARBONATE} \\ \text{ION}}}{HCO_3^-}$$

What effect do you suppose this would have on the pH of the water? (The effect would certainly be seen in tap water.)

CHAPTER 3

The Composition of Living Matter: Some Organic Compounds

Almost a hundred different elements are present in the earth's crust, in the waters that cover much of the earth's surface, and in the thin veil of atmosphere surrounding it. Of these elements, only six make up some 99 percent of all living tissue. These six are carbon, hydrogen, nitrogen, oxygen, phosphorus, and sulfur—conveniently remembered as CHNOPS.

CHNOPS combine in different ways to make up the molecules found in living systems. Here again, although many combinations are possible, only a few different kinds of molecules are found in large quantities in living systems—five, to be exact. One of these is water. The other four are sugars, lipids (including fats and waxes), amino acids (which combine to form proteins), and nucleotides. Only the first three of these will be discussed in this chapter; we shall take up the nucleotides in later chapters when we discuss energy exchanges and genetic systems, in both of which nucleotides play key roles.

Each of the four kinds of molecules—sugars, lipids, amino acids, and nucleotides—contains carbon, hydrogen, and oxygen. In addition, amino acids contain nitrogen and sulfur, and nucleotides contain nitrogen and phosphorus.

Interactions of the six atoms CHNOPS and the five types of molecules they combine to form are responsible, in large part, for the special properties that we associate with living things.

THE CARBON ATOM

Just as knowledge of the structure of the water molecule helps us understand water's unusual characteristics, so we can better understand the central role of the carbon atom in the chemistry of living matter if we look at the way in which it is put together. A carbon atom has six protons in its nucleus and six orbiting electrons, two in the inner electron shell and four in the outer shell:

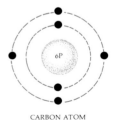

CARBON ATOM

Table 3-1 *Atomic Composition of Three Representative Organisms*

ELEMENT	MAN	ALFALFA	BACTERIUM
Carbon	19.37	11.34	12.14
Hydrogen	9.31	8.72	9.94
Nitrogen	5.14	0.83	3.04
Oxygen	62.81	77.90	73.68
Phosphorus	0.63	0.71	0.60
Sulfur	0.64	0.10	0.32
CHNOPS total:	97.90	99.60	99.72

This arrangement means that carbon, rather than losing or gaining electrons and becoming ionized like Na^+ and Cl^- ions (page 29), tends to share electrons, forming covalent bonds. With four electrons in its outer shell, carbon needs four additional electrons to complete the requirement of its outer shell for eight electrons. Therefore, each carbon atom can form four covalent bonds simultaneously. For example, one carbon atom can combine with four hydrogen atoms to form methane gas (CH_4):

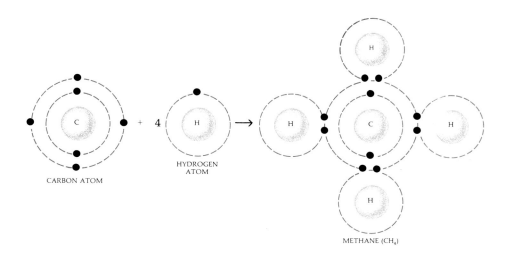

CARBON ATOM

HYDROGEN
ATOM

METHANE (CH_4)

Even more important, in terms of its biological role, carbon atoms can form bonds with each other. Butane gas, for instance, consists of a chain of four carbon atoms with ten hydrogen atoms attached:

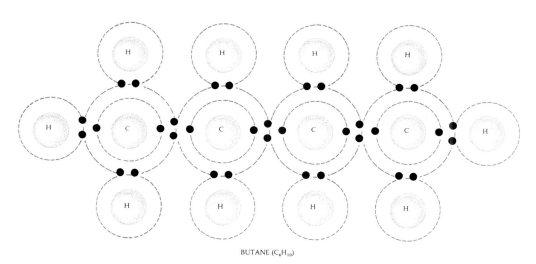

BUTANE (C_4H_{10})

Molecules that consist only of hydrogen and carbon, such as methane and butane, are called _hydrocarbons_. In the primitive atmosphere, it is believed, carbon atoms were present in the form of such hydrocarbon gases. In the modern atmosphere, carbon is present mostly as carbon dioxide. Carbon atoms are incorporated into living systems by the process of photosynthesis, in which the sun's energy is used to combine carbon, hydrogen, and oxygen into organic compounds.

MOLECULAR MODELS

There are many different ways to describe a molecule. Glucose, for example, has 6 carbon atoms, 12 hydrogen atoms, and 6 oxygen atoms, so $C_6H_{12}O_6$ is one way to describe it. If we want to emphasize the fact that it is a carbohydrate, we can make this point by describing glucose as $6CH_2O$, which, of course, says just the same thing in a slightly different way. However, fructose, another simple sugar, also contains 6 carbons, 12 hydrogens, and 6 oxygens and has a similar structure—a chain of carbon atoms to which hydrogen and oxygen atoms are attached. The symbol = in the drawings below indicates a double bond. Double bonds involve the sharing of four rather than two electrons between two atoms. The differences between glucose and fructose are determined by which carbons the other atoms are attached to. The molecules can therefore be described more precisely by drawing the carbon chains:

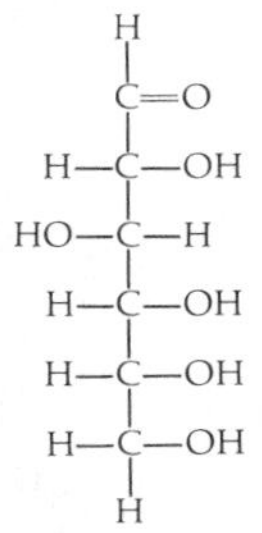

GLUCOSE

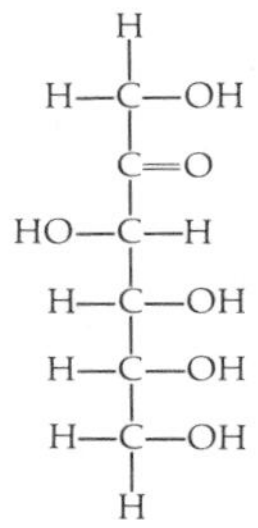

FRUCTOSE

When glucose and fructose are in solution, however, the chains tend to form rings:

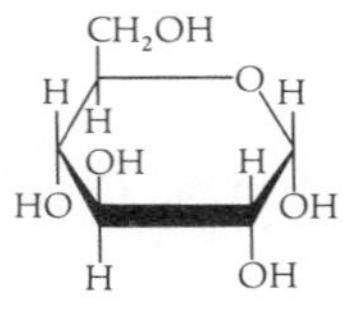

GLUCOSE
$C_6H_{12}O_6$

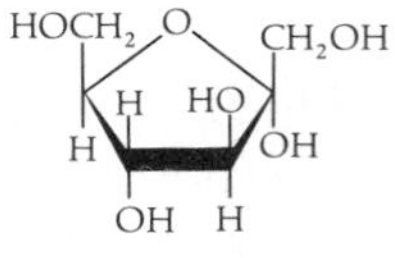

FRUCTOSE
$C_6H_{12}O_6$

The lower edge of the ring is made darker and thicker to emphasize the three-dimensional nature of the structure. By convention, the carbon atoms in the ring are "understood" and not labeled in this type of diagram. The position of each carbon atom in the ring is numbered:

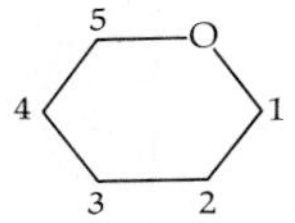

Another way of depicting molecules is by the "ball-and-stick" method, in which the balls represent atoms and the sticks represent the bonds between them. In these models, for instance, carbon atoms are large black spheres, oxygen atoms large white spheres, and hydrogen atoms small black spheres:

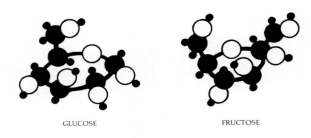

GLUCOSE

FRUCTOSE

These models emphasize three-dimensional shape more than the simpler ring models do. They also make it possible to show very clearly the lengths of the bonds (that is, the relative distances between the different atoms) and the angles of the bonds. For instance, the water molecule looks like this in the ball-and-stick model:

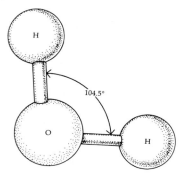

Finally, molecules can be diagramed by space-filling models:

GLUCOSE

FRUCTOSE

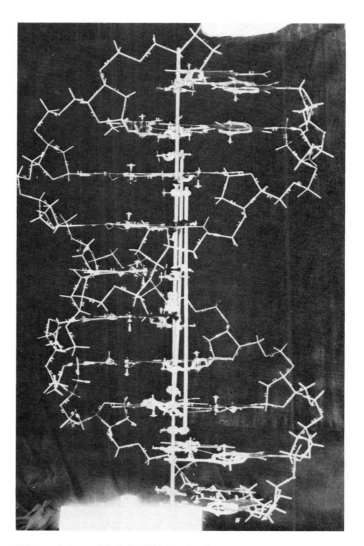

Wire and tin model of the DNA molecule.

Actually, molecules do not fill space in the same way that we think of a table or a rock as filling space. The atoms that make up molecules consist mostly of empty space. If the outer shell of electrons in an atom were the size of the perimeter of the Astrodome in Houston, the nucleus would be a ping-pong ball in the center of the stadium. What "fills" the space in these models are the paths of the circulating electrons, which set up areas of charge around the atoms. One molecule "sees" another molecule in terms of these areas of charge. As a consequence, for instance,

an enzyme that digests a glucose molecule will not have any effect on a fructose molecule because of the differences in shape. All the intricate biochemistry that goes on in the cell is based on this ability of molecules to "recognize" one another.

It is important to remember that none of these ways of showing the molecules attempts to tell what the molecule actually might look like if we could see it. Rather, they are models. A model, like a theory, is a way of organizing a particular set of scientific data.

SUGARS

Sugars, which are formed by photosynthesis, serve two extremely important functions in the cell: (1) they are the principal energy source for most living systems, and (2) they are the basic materials from which many other kinds of molecules are built.

Sugars are composed of carbon, hydrogen, and oxygen atoms, and are therefore known collectively as *carbohydrates*. In sugar, the proportion is one carbon atom to two hydrogen atoms to one oxygen atom, as indicated by the shorthand symbol CH_2O. There are three kinds of sugars: (1) monosaccharides ("single sugars"), such as glucose, ribose, and fructose; (2) disaccharides ("two sugars"), such as sucrose (table sugar), maltose, and lactose; and (3) polysaccharides ("many sugars"), such as cellulose and starch. A compound that is composed of many similar or identical subunits, as is a polysaccharide, is known as a *polymer* (from "many" and "part"). Each subunit is called a *monomer*.

Sugar as Energy

In photosynthesis, carbon dioxide and water are combined to produce sugar:

$$\text{Carbon dioxide} + \text{water} + \text{energy} \longrightarrow \text{sugar} + \text{oxygen}$$

When the cell breaks down (oxidizes) the sugar molecule, this process is reversed:

$$\text{Sugar} + \text{oxygen} \longrightarrow \text{carbon dioxide} + \text{water} + \text{energy}$$

Organic substances, such as wood, gas, and oil, also release their stored energy when they are oxidized, as occurs when they are burned for fuel. Sugar, too, can be burned to release energy. In these cases, however, the energy stored in the molecule by photosynthesis is converted to heat energy. In the cell, much of the stored energy is converted to other forms of chemical energy and there is comparatively little release of heat, which would not be useful to the cell and, in any large amount, would certainly be harmful. The way in which the cell breaks down the sugar molecule and repackages its energy in a different chemical form is described in Chapter 7.

The amount of energy a compound contains is measured in terms of the number of calories released when the compound is burned or oxidized. In such a reaction, chemical bonds are broken and new bonds form. Energy is released because the new bonds have less energy than the old bonds — that is, they are not as strong as the old bonds. There is more energy in the bond holding a hydrogen atom to a carbon atom than in the bond holding an oxygen atom to a carbon atom; so when H—C bonds are broken and O—C bonds form, there is some energy "left over," which is released as heat.

Energy changes taking place during chemical reactions can be measured very precisely by means of a calorimeter (Figure 3–1). A known amount of glucose or some other substance is placed in the crucible and burned. The heat released causes the temperature of the surrounding water to rise. From the known weight of the water and the rise in its temperature, the heat of combustion of the glucose can be calculated in terms of calories. When a mole of glucose is broken down to carbon dioxide and water, 686,000 calories (686 kilocalories, or "Calories," as commonly referred to in diets*) are released. The yield is the same whether this reaction takes place inside or outside the cell.

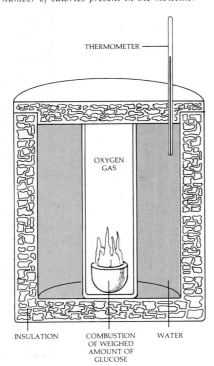

3–1

A calorimeter. A known quantity of glucose or some other material is burned; the rise in the temperature of the water indicates the number of calories present in the molecule.

THERMOMETER

OXYGEN GAS

INSULATION   COMBUSTION OF WEIGHED AMOUNT OF GLUCOSE   WATER

Disaccharides

Disaccharides consist of two sugar molecules coupled together. Many common sugars are found in nature in this form. Maltose (malt sugar) consists of two glucose units chemically linked to one another. Lactose, a sugar which occurs only in milk, is made up of glucose combined with another monosaccharide, galactose. Sugar is transported through the blood of insects as another type of disaccharide, trehalose. Sucrose is the sugar most common in plants and is our usual table sugar. Sucrose is made up of the monosaccharides glucose and fructose.

Disaccharides are formed by the removal of a molecule of water from the two monosaccharide molecules:

$$GLUCOSE \qquad FRUCTOSE$$

$$H_2O$$

$$SUCROSE$$

The addition of a molecule of water splits the sucrose molecule back into glucose and fructose. This splitting is known as *hydrolysis*, from *hydro*, meaning "water," and *lysis*, "breaking apart."

Hydrolysis is an energy-releasing reaction. The hydrolysis of sucrose yields 5,500 calories. Conversely, the formation of sucrose from glucose and fructose requires an energy input of 5,500 calories.

3–2

In addition to the sugar naturally present in fruits, vegetables, and milk, we consume large amounts of sucrose, glucose, and fructose extracted from plant products. Sucrose, which comes from sugar beets and sugarcane, is table sugar. Glucose and fructose are in the forms of sugar syrups. On the average, every man, woman, and child in the United States eats about 56 kilograms (125 pounds) of sugars per year as table sugar or in baked products, soft drinks, candy, breakfast cereals, or some other form.

Sugarcane, one of the grasses, regrows from underground roots and runners after it is cut down. It is usually harvested by hand-cutting, as shown here; machines are likely to tear up the plants by the roots. Hence it is a crop well suited to areas where labor costs are low and heavy-duty machinery not generally available.

3-3

In plants, sugars are stored in the form of starch. Starch is composed of two different types of polysaccharides, amylose (a) and amylopectin (b). A single molecule of amylose may contain 1,000 or more monomers of glucose in a long, unbranched chain, which winds to form a helix. A molecule of amylopectin is made up of about 48 to 60 glucose units arranged in shorter, branched chains. Starch molecules tend to cluster into granules. (c) Starch, unlike sugar, is insoluble and so provides a useful storage form. In this electron micrograph of a chloroplast in a tomato leaf, the large, pale objects within the chloroplast are starch grains.

AMYLOSE

(a)

AMYLOPECTIN

(b)

(c)

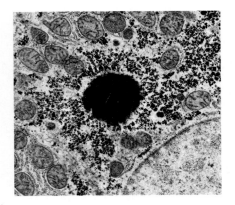

3-4

Glycogen, which is the common storage form for sugar in vertebrates, resembles amylopectin in its general structure except that each molecule contains only 16 to 24 glucose units. The dark granules surrounding the central fat droplet in this liver cell are glycogen. When needed, the glycogen can be converted to glucose. The large body in the lower right is the nucleus.

Polysaccharides

Polysaccharides are made up of monosaccharides linked together in long chains. They constitute a storage form for sugar. The principal storage polysaccharide in higher plants is starch (Figure 3-3), and in higher animals and fungi it is glycogen (Figure 3-4). Both starch and glycogen are built up of many glucose units. The differences between them are in the length of the polysaccharide chains (starch is made up of longer chains) and in the ways in which the glucose molecules are linked. Polysaccharides must be hydrolyzed before they can be used as energy sources or transported through living systems.

Polysaccharides also play structural roles. In plants, the principal structural polysaccharide is cellulose (Figure 3-5a). Although cellulose, like starch, is a polymer of glucose, the arrangement of its long-chain molecules makes it rigid, and so its biological role is very different. Also, the bonds linking the glucose units in cellulose are slightly different from those in starch or glycogen, and so the three-dimensional structure of the molecules differs. Because of this difference in the orientation of the glucose monomers, cellulose cannot be broken apart by the same enzymes that hydrolyze starch. As a consequence, once monosaccharides are incorporated into the plant cell wall in the form of cellulose, they are no longer generally available as an energy source. Only some fungi, bacteria, protists, and a very few animals (silverfish, for example) possess enzyme systems capable of breaking down cellulose. Certain organisms, such as cows and other ruminants, termites, and cockroaches, are able to utilize cellulose as a source of energy, but only because of the microorganisms that inhabit their digestive tracts.

Chitin, which is a major component of the exoskeletons of insects and other arthropods and also of the cell walls of many fungi, is another tough, resistant, common polysaccharide (Figure 3-6).

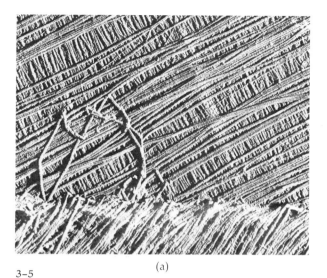

3–5

Eric V. Gravé

(a)

(b)

(a) *The plant cell wall is composed largely of cellulose. Each of the microfibrils you can see here is a bundle of hundreds of cellulose strands, and each strand is a chain of glucose monomers. The microfibrils, as strong as an equivalent amount of steel, are embed-* *ded in other polysaccharides. (b) Because the orientation of the glucose units in cellulose are different from those in starch or glycogen, very few organisms can digest cellulose and so utilize the sugar molecules for energy.* *One such organism is the protist* Trichonympha, *of which two are shown here.* Trichonympha *lives in the guts of termites and is entirely responsible for their well-known proficiency at digesting wood.*

3–6

Green darner molting. The shells, or exoskeletons, of insects contain chitin, a nitrogen-containing polysaccharide. Some types of insects, after molting, recycle their sugars and nitrogen by thriftily eating their discarded exoskeletons.

Sugar as a Starting Material

In addition to serving as the building blocks of polysaccharides, glucose and other monosaccharides also serve as starter materials for the construction of more complex molecules. Some of the sugar taken in in the diet is converted to glycogen for storage in the liver (about 24 hours' worth). Any excess is converted to fat for storage. The change from sugar to fat involves a partial breakdown and then a rebuilding of the molecule. These processes take place within the cell. The building blocks of protein and of nucleic acids are also synthesized within the cell, frequently using glucose or another monosaccharide as a starting material.

LIPIDS

Lipids are fatty or oily substances. They have two principal distinguishing characteristics: (1) they are nonpolar and so are generally insoluble in water, and (2) they contain a larger proportion of carbon-hydrogen bonds than do other organic compounds and, as a consequence, store a larger amount of energy than do other organic compounds. Fats yield more than twice as many calories as an equivalent amount of carbohydrate or protein. On the average, fats yield about 9.3 kilocalories per gram* as compared to 3.79 per gram of carbohydrate or 3.12 per gram of protein. The fact that lipids are insoluble and energy-rich determines their roles in living systems.

* 1,000 grams = 1 kilogram = 2.2 pounds, so oxidation of a pound of fat would yield about 4,700 kilocalories, more than the 24-hour requirement for moderately active adults.

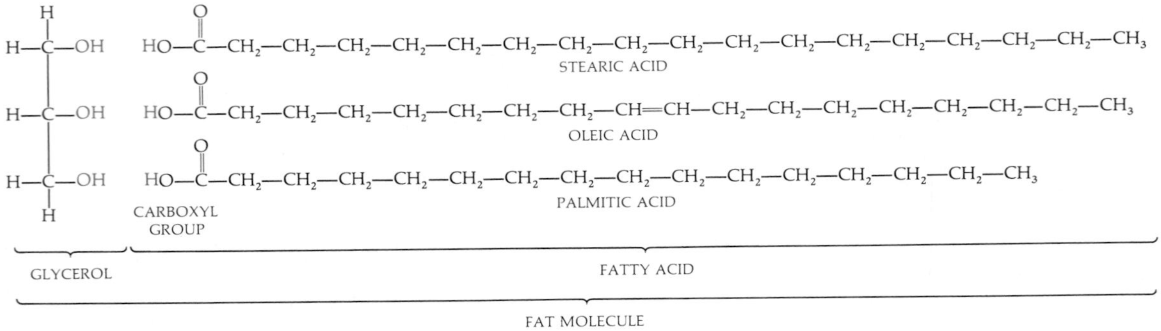

H—C—OH   HO—C—CH₂—CH₂—CH₂—CH₂—CH₂—CH₂—CH₂—CH₂—CH₂—CH₂—CH₂—CH₂—CH₂—CH₂—CH₂—CH₃

STEARIC ACID

H—C—OH   HO—C—CH₂—CH₂—CH₂—CH₂—CH₂—CH₂—CH₂—CH=CH—CH₂—CH₂—CH₂—CH₂—CH₂—CH₂—CH₂—CH₃

OLEIC ACID

H—C—OH   HO—C—CH₂—CH₂—CH₂—CH₂—CH₂—CH₂—CH₂—CH₂—CH₂—CH₂—CH₂—CH₂—CH₂—CH₂—CH₃

CARBOXYL GROUP   PALMITIC ACID

GLYCEROL   FATTY ACID

FAT MOLECULE

3–7

A fat molecule consists of three fatty acids joined to a glycerol molecule. The long hydrocarbon chains of which the fatty acids are composed terminate in carboxyl groups, which become covalently bonded to the glycerol molecule. Each bond is formed by the removal of a molecule of water. The physical properties of the fat—such as its melting point—are determined by the lengths of the chains and by whether its component fatty acids are saturated or unsaturated. Three different fatty acids are shown here. Stearic acid and palmitic acid are saturated, and oleic acid is unsaturated, as you can see by the double bond in its structure.

CH₃
HC—CH₃
CH₂
CH₂
CH₂
HC—CH₃
CH₃
CH₃

HO

CHOLESTEROL

(a)

OH
CH₃
CH₃

O

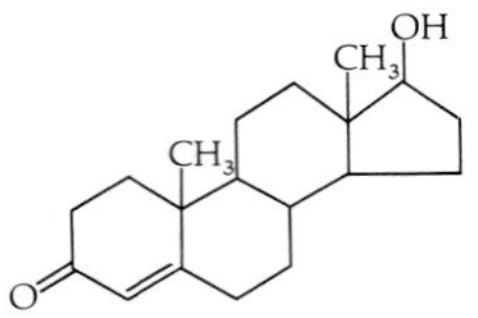

TESTOSTERONE

(b)

3–8

(a) The cholesterol molecule consists of four carbon rings and a hydrocarbon chain. (b) Testosterone, which is synthesized from cholesterol by cells in the testes, also has the characteristic four-ring structure but lacks the hydrocarbon tail.

Fats

Lipids are stored in the cell chiefly in the form of fat, which cells synthesize from sugars. A fat molecule consists of three molecules of a fatty acid joined to one glycerol molecule (Figure 3–7). As with the disaccharides and polysaccharides, each bond is formed by the removal of a molecule of water.

The physical nature of the fat is determined by the chain length of the fatty acid and by whether the acid is saturated or unsaturated. In saturated fatty acids, such as stearic acid, every carbon atom in the chain (except the last one) holds two hydrogen atoms, which completes the bonding possibilities of the carbon atom. Unsaturated fatty acids, such as oleic acid, contain carbon atoms joined by double bonds. Such carbon atoms are able to form additional bonds with other atoms (hence the term "unsaturated").

Unsaturated fats, which tend to be oily liquids, are more common in plants than in animals; examples are olive oil, peanut oil, and corn oil. Animal fats, such as butter and lard, contain saturated fatty acids and usually have higher melting temperatures.

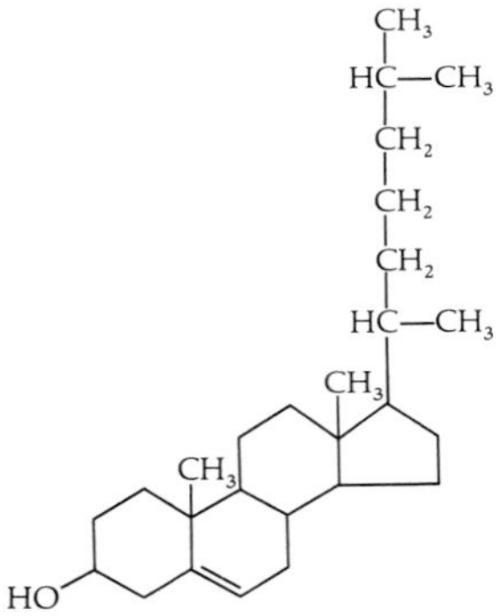

Cholesterol and Other Steroids

The steroids do not resemble the other lipids structurally, but they are grouped with them because of their properties—particularly their insolubility in water. Structurally, they are characterized by four carbon rings (Figure 3–8).

The steroids are an important group of compounds. Cholesterol, along with the phospholipids, is found in cell membranes of animal cells. The sex hormones and the hormones of the adrenal cortex, such as cortisone, are steroids; in the body they are synthesized from cholesterol. Cholesterol is synthesized in the liver from saturated fatty acids. It is also taken in in the diet, principally in meat, cheese, and egg yolks. High concentrations of cholesterol in the blood are associated with arteriosclerosis ("hardening of the arteries") and are found in fatty deposits on the interior lining of diseased blood vessels (see page 375).

Waxes

Waxes are also a form of lipid. They are found as protective coatings on skin, fur, feathers, on the leaves and fruits of higher plants (Figure 3–9), and on the exoskeletons of many insects.

3-9

Waxes are also lipids. This electron micrograph shows waxy deposits on the upper surface of a eucalyptus leaf. Biosynthesis of waxes, which protect exposed plant surfaces from water loss, is a property of all groups of higher plants.

Phospholipids

Like fats, the phospholipids are composed of fatty acid chains attached to glycerol. In the case of the phospholipids, the third carbon of the glycerol molecule is occupied not by a fatty acid but by a phosphate (phosphorus plus oxygen) group, PO_4^{3-} (Figure 3–10). The phosphate end of the phospholipid molecule is polar and therefore soluble in water; the fatty acid end is nonpolar and therefore insoluble in water. When phospholipids are added to water, they tend to form a film along its surface, with their soluble polar "heads" under the water and their insoluble fatty acid "tails" protruding above the surface (Figure 3–11a). In the watery interior of the cell, phospholipids tend to align themselves in rows, with their insoluble fatty acid tails oriented toward one another and their soluble phosphate heads directed outward (Figure 3–11b). Such configurations, which are formed spontaneously by these molecules, are probably involved in cellular structures, particularly membranes.

PROTEINS

Proteins, the substances that play such important and complicated roles in living systems, are made up of one or more chains of nitrogen-containing molecules known as amino acids. The sequence in which the amino acids are arranged in these chains determines the biological character of the protein molecule; even one small variation in this sequence may alter or destroy its function.

Twenty different kinds of amino acids are found in proteins, and since protein molecules are large, often containing several hundred amino acids, the number of different amino acid sequences, and therefore the possible variety of protein molecules, is enormous—about as enormous as the number of different sentences that can be written with a 26-letter alphabet. The single-celled bacterium *Escherichia coli*, for example, contains 600 to 800 different kinds of pro-

3-10

A phospholipid molecule consists of two fatty acids linked to a glycerol molecule, as in a fat, and a phosphate group (indicated by colored screen) linked to the glycerol's third carbon. It also usually contains an additional chemical group, indicated by the letter R. The fatty acid "tails" are nonpolar and therefore insoluble in water; the polar "heads" containing the phosphate and R groups are soluble.

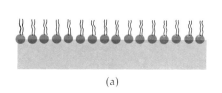

3-11

(a) Because phospholipids have water-soluble heads and water-insoluble tails, they tend to form a thin film on a water surface with their tails extending above the water. (b) Surrounded by water, they spontaneously arrange themselves in two layers with their heads extending outward and their hydrophobic (water-hating) tails inward. This arrangement is believed to be important in the structure of cell membrane.

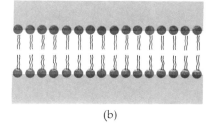

(a)

(b)

teins at any one time, and the cell of a higher plant or animal probably has several times that number. In a single complex organism such as man, there are thousands, possibly hundreds of thousands, of different proteins, each with a special function and each, by its unique chemical nature, specifically fitted for that function.

Amino Acids

Amino acids are made up of carbon, hydrogen, and oxygen, as are sugars. All also contain nitrogen. Every amino acid contains an amino group ($-NH_2$) and a carboxyl group ($-COOH$) bonded to a carbon atom:

$$H_2N-\underset{\underset{H}{|}}{\overset{\overset{R}{|}}{C}}-\underset{\underset{O}{\|}}{C}-OH$$

This is the basic structure of the molecule, and it is the same in all amino acids. The "R" stands for the rest of the molecule, which is different in the chemical structure of each different kind of amino acid (Figure 3–12). These differences are very important because they determine the different biological properties of the individual amino acids and, as a consequence, of the various kinds of proteins.

The amino "head" of one amino acid can be linked to the carboxyl "tail" of another by the removal of a molecule of water (Figure 3–13a). The linkage that is formed is known as a peptide bond, and the molecule that is formed by the linking of many amino acids is called a *polypeptide* (Figure 3–13b).

AMINO ACIDS AND NITROGEN

Like fats, amino acids are formed within living cells using sugars as starter materials. But while fats are made up only of carbon, hydrogen, and oxygen atoms, all available in the sugar and water of the cell, amino acids also contain nitrogen. Most of the nitrogen in the biosphere exists in the form of gas in the atmosphere. Only a few organisms, all microscopic, are able to incorporate nitrogen from the air into compounds—nitrites and nitrates—that can be used by living systems. Hence the amount of nitrogen available to the living world is very small.

Plants incorporate the nitrogen in the nitrites and nitrates into carbon-hydrogen compounds to form amino acids. Animals are able to synthesize some of their amino acids, using ammonia as a nitrogen source. The amino acids they cannot synthesize, the so-called essential amino acids, must be obtained either directly or indirectly from plants. For adult human beings, the essential amino acids are lysine, tryptophan, threonine, methionine, phenylalanine, leucine, valine, and isoleucine.

Until recently, agricultural scientists concerned with the world's hungry people concentrated on developing plants with a high caloric yield. Now, increasing recognition of the role of plants in supplying amino acids to the animal world has led to emphasis on the development of high-protein strains of food plants and of plants with essential amino acids, such as "high-lysine" corn.

Collecting pollen from a corn plant, one of the steps in a program to breed nutritionally superior strains.

3-12

Eight of the 20 different kinds of amino acids found in proteins. As you can see, their basic structures are the same, but they differ in their R groups. The other 12 amino acids—not shown here—are glutamine (gln), arginine (arg), cysteine (cys), histidine (his), leucine (leu), asparagine (asn), methionine (met), isoleucine (ile), aspartic acid (asp), lysine (lys), threonine (thr), and proline (pro).

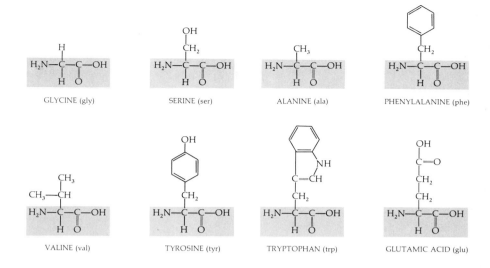

GLYCINE (gly) SERINE (ser) ALANINE (ala) PHENYLALANINE (phe)

VALINE (val) TYROSINE (tyr) TRYPTOPHAN (trp) GLUTAMIC ACID (glu)

3-13

(a) Formation of a peptide bond (shown in color) involves removal of a molecule of water. (b) Polypeptides are polymers of amino acids linked together by peptide bonds, with the amino group of one acid joining the carboxyl group of its neighbor. The polypeptide chain shown here contains six different amino acids, but some chains may contain as many as 300 linked amino acid monomers.

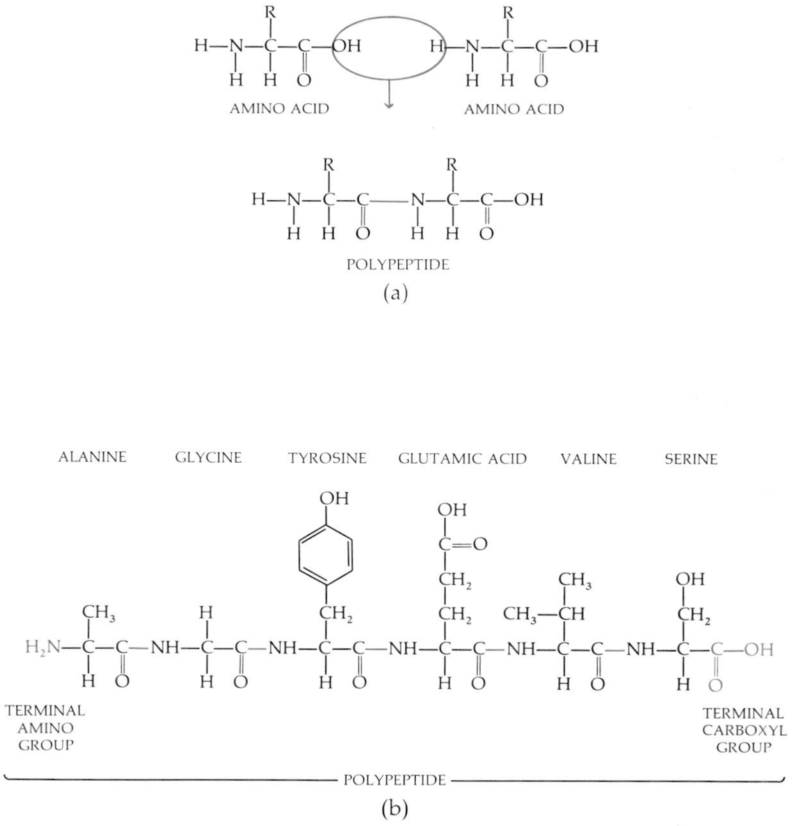

AMINO ACID   AMINO ACID

POLYPEPTIDE

(a)

ALANINE   GLYCINE   TYROSINE   GLUTAMIC ACID   VALINE   SERINE

TERMINAL AMINO GROUP   TERMINAL CARBOXYL GROUP

POLYPEPTIDE

(b)

3-14

Primary structure of the enzyme lysozyme. This enzyme, which contains 129 amino acids, is found in many different types of animal cells. The sequence of the amino acids constitutes its primary structure. One consequential feature of the primary structure is the presence of the sulfur-containing amino acid cysteine. Covalent bonds form between the sulfur atoms of different cysteine monomers, linking parts of the molecule together.

Lysozyme breaks down the cell walls of many types of bacteria. It was first detected by Alexander Fleming, better known for his discovery of penicillin, as a consequence of his having a bad cold. A droplet fell from his nose onto a glass slide containing bacteria, and Sir Alexander, as he was to become, noticed that the bacteria in the vicinity of the droplet were destroyed. Lysozyme was subsequently discovered in tears, saliva, milk, and various other body fluids of many different animals. The molecule shown here was extracted from egg white. Lysozymes from other organisms might differ in some of their amino acids.

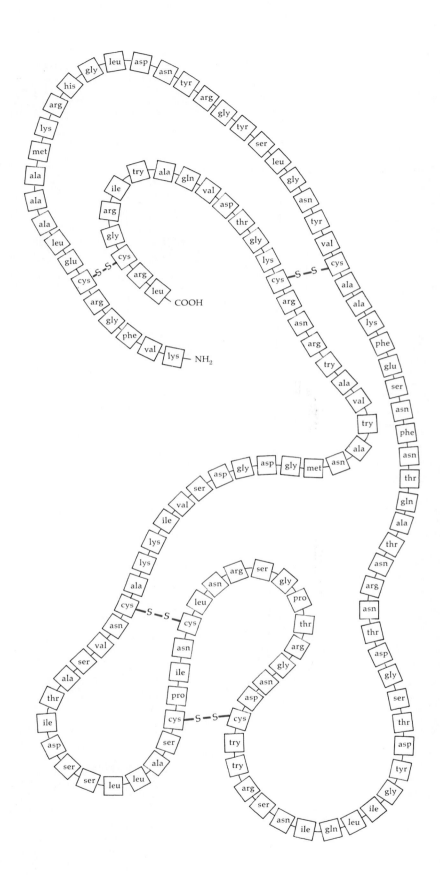

Secondary structure of a protein. The helix is held in shape by hydrogen bonds, indicated by the dashed lines. In this particular helix (the alpha helix), bonds form between every fourth amino acid. Asterisks indicate carbons bonded to R groups.

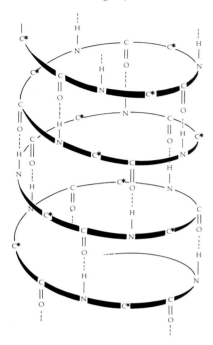

As you can see, in order to assemble amino acids into proteins, a cell must have not only a large enough quantity of amino acids but also enough of every kind—just as a typesetter, even though he may have a large supply of letters, cannot set a sentence if he is lacking e's. (This fact is of great importance in solving problems of human nutrition.)

The Levels of Protein Organization

Polypeptides are assembled in the cell on ribosomes. In this process, the head of one amino acid is linked to the tail of another, like a line of boxcars, always in a particular sequence. This linear assembly is known as the *primary structure* of the protein (Figure 3–14).

As the chain is assembled, it tends to fold in a simple pattern, known as its *secondary structure.* One common secondary structure is a coil or, as it is more commonly called by biochemists, a *helix.* The helix is held in shape by hydrogen bonds (Figure 3–15). The protein molecules of hair and of myosin (a muscle protein) tend to form helices and, as a consequence, are stretchable (elastic) because the hydrogen bonds break and re-form easily. (Straightening curly hair or curling straight hair involves breaking of hydrogen bonds or the making of new ones.)

Other proteins, such as silk, are made up of extended polypeptide chains lined up in parallel and linked to one another by hydrogen bonds. Polypeptides assembled in this way are smooth and supple but nonelastic.

Collagen has another type of secondary structure. Collagen molecules are made by specialized cells called fibroblasts; as the long polypeptide chains are synthesized, three of them wrap around each other to form a cablelike fiber. Collagen, which is a principal component of cartilage, bones, and tendons, consists of a collection of such fibers (Figure 3–16). Wool, fingernails, and horns are also structural proteins whose texture and other characteristics vary because of differences in the arrangements of the polypeptide chains.

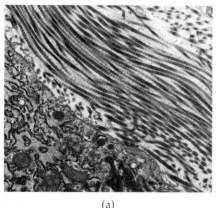

(a)

(b)

(c)

3–16

The structural protein collagen makes up about one-third of the protein in the human body and is probably the most abundant protein in the animal kingdom. (a) Collagen is composed of long polypeptide chains woven together to form fibrils, such as those shown

here. These fibers are a major constituent of skin, tendon, ligament, cartilage, and bone. (b) The capacity to produce the structural protein keratin is characteristic of all vertebrates. It is found in scales, horns, wool,

nails, and feathers. These are cross sections of human hairs as seen by the scanning electron microscope. (c) Keratin also is the chief component of the formidable armor of this member of the class Reptilia.

(a) *The primary structure of a protein is the linear sequence of its amino acids.* (b) *Because of interactions among these amino acids, the molecule spontaneously coils into a* secondary structure, such as the alpha helix shown here, and (c) into a tertiary structure, such as a globule. (d) *Many globular proteins, including hemoglobin and some en-* zymes, are made up of more than one amino acid chain. This structure is known as a quaternary structure.

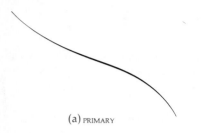

(a) PRIMARY

(b) SECONDARY

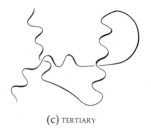

(c) TERTIARY

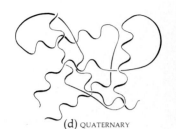

(d) QUATERNARY

In other proteins, the long helix folds back on itself to make a complex *tertiary structure.* Such proteins are called globular proteins; enzymes are an example. These globular structures also form spontaneously; they form largely as a result of attractions and repulsions between amino acids with different charges in their R groups.

Often two or more globular proteins will fit together in a *quaternary structure.* Hemoglobin, for example, the oxygen-carrying molecule of the blood, is made up of four intertwined protein chains, each about 150 amino acids long. (A diagram of hemoglobin is shown on page 386.)

In every protein, the primary structure determines the secondary, tertiary, and quaternary structures.

Enzymes

Enzymes are globular proteins that are specialized to serve as catalysts. Catalysts are substances that accelerate the rate of a chemical reaction but are not themselves used up in the process. Because they are not permanently altered, catalysts may be used over and over again, and so they are typically effective in very small amounts.

In the laboratory, the rates of chemical reactions are usually accelerated by the application of heat, which increases the force and frequency of collision among molecules. But in a cell, hundreds of different reactions are going on at the same time, and heat would speed up all these reactions indiscriminately. Moreover, heat would break hydrogen bonds and would have other generally destructive effects on the cell. Because of enzymes, cells are able to carry out chemical reactions at great speed and at comparatively low temperatures. A single enzyme molecule may catalyze as many as several hundred thousand reactions in a second.

The chemical, or chemicals, upon which an enzyme acts is known as its *substrate.* In the reaction diagramed in Figure 3–18, sucrose is the substrate, and invertase is the enzyme (note that enzyme names typically end in "-ase"). The site on the surface of the globular enzyme molecule into which the substrate fits—like a key in a lock—is the *active site.* Only a few amino acids of the enzyme are involved in any particular active site. Some of these may be adjacent to one another in the primary structure, but often the amino acids of the active site are brought into proximity to one another by the intricate foldings of the chains

3–18

A model of enzyme action. Sucrose is hydrolyzed to yield a molecule of glucose and a molecule of fructose. The enzyme involved in

this reaction is specific for this process; as you can see, the active site of the enzyme fits the opposing surface of the sucrose molecule.

For example, a molecule composed of two subunits of glucose would not be affected by this enzyme.

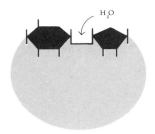

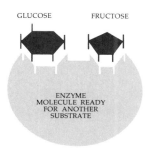

involved in the tertiary structure. Over a thousand enzymes are now known, each of them capable of catalyzing a specific chemical reaction.

Enzymes characteristically work in a series, like workers in an assembly line, each performing one small step in, for example, the synthesis of a molecule. The activities of enzymes are controlled in various ways. One of the simplest and most effective is what is known as an *allosteric* ("other shape") *interaction*. In this interaction, the end product of the enzyme assembly line combines with an enzyme in the series—usually the first enzyme—and changes its shape. So the active site no longer fits the substrate. As a consequence, the assembly line shuts down and no more of the product is produced.

SUMMARY

Living matter is composed of only a few of the naturally occurring elements, chief among which are oxygen, carbon, hydrogen, nitrogen, phosphorus, and sulfur (CHNOPS). The bulk of living matter is water. Most of the rest is made up of organic compounds. Sugars, lipids, and proteins are three of the principal organic compounds in cells. (Nucleotides, to be discussed later, constitute the fourth.)

Sugars serve as a primary source of chemical energy for living systems and also as important structural elements in cells. The simplest sugars are the monosaccharides ("single sugars"), such as glucose and fructose. Monosaccharides can be combined to form disaccharides ("two sugars"), such as sucrose, and polysaccharides (chains of many sugar monomers), such as starch, glycogen, and cellulose. These molecules can be broken apart again by hydrolysis, the addition of a water molecule.

Lipids are another source of energy and also of structural material for cells. Compounds in this group, which includes fats and waxes, are generally insoluble in water. Phospholipids are major components of cellular membranes.

Proteins are very large molecules composed of long chains of amino acids; these are known as polypeptide chains. There are 20 different amino acids in proteins, and from these, an enormous number of different protein molecules are built. Fibrous proteins serve as structural elements, and globular proteins play important dynamic and metabolic roles. The principal levels of protein organization are (1) primary structure, the amino acid sequence; (2) secondary structure, the coiling of the polypeptide chain; (3) tertiary structure, the folding of the coiled chain into various shapes; and (4) quaternary structure, the folding of two or more protein molecules around each other.

Because of enzymes, which are globular proteins, cells are able to catalyze reactions at high speed and at body temperature. The specificity of enzymes is due to the "lock-and-key" fit of the active site on the enzyme with the substrate molecule.

QUESTIONS

1. Distinguish between the following: carbohydrate/hydrocarbon, polysaccharide/polymer/polypeptide, glycogen/starch/cellulose.

2. What is the basis of the specificity of enzyme action? What is the advantage to the cell of such specificity? The disadvantage?

3. Sketch the arrangement of phospholipids when surrounded by water.

4. Knowing that chemical reactions have to be balanced right and left, fill the appropriate numbers (*hint:* from 1 to 6 in all cases) into the underlined spaces:
(a) __ H_2O + __ O_3 ⇌ __ H_2O_2
(b) __ $Ca(OH)_2$ + __ H_3PO_4 ⇌ __ H_2O + __ $Ca_3(PO_4)_2$
(c) __ O_2 + __ $C_6H_{12}O_6$ (sugar) ⇌ __ H_2O + __ CO_2

5. Health authorities agree that sugar, at least in the amounts presently consumed in the United States, is nutritionally harmful. Can you explain, then, in terms of evolutionary theory, why sugar is so pleasurable to the human species and to many other animals as well? (*A clue:* Many carnivores, including the domestic cat, cannot taste sugar at all.)

6. In all allosteric interactions, why is it economical for the product of the enzyme series to combine with the first enzyme? What would happen if, in a series of eight enzymes, the product combined with the third? What would happen if one enzyme of a series were missing totally? [This is the cause of a disease called phenylketonuria (PKU).]

CHAPTER 4

The Cell in Action

In the preceding chapters, we have looked at how atomic particles, such as electrons and protons, combine to form atoms; at how atoms combine to form molecules; and at how molecules make macromolecules. In Chapter 3, we examined how these macromolecules make up cellular structures and carry out cellular functions. In this chapter, we are going to try to glimpse how all these things together make up that unique entity—the living cell. We use the word "glimpse" because, despite the fact that a single work of cell physiology may fill several volumes, there is much that is still not known about the life of a cell, so our narrative is by necessity incomplete.

Although we can look at only one process at a time, remember that most activities of a cell go on simultaneously and influence one another. *Chlamydomonas*, for instance, is swimming, photosynthesizing, absorbing nutrient ele-

4-1

At the top, Paramecium; *at the bottom,* Didinium. Didinium *is stalking* Paramecium. Paramecium, *in defense, has discharged a barrage of barbs (visible as a cloud at the right of the micrograph). Didinium is about to eject a bundle of slender, poisonous strands (not visible), which will paralyze* Paramecium *in a matter of seconds. Each organism is a single cell.*

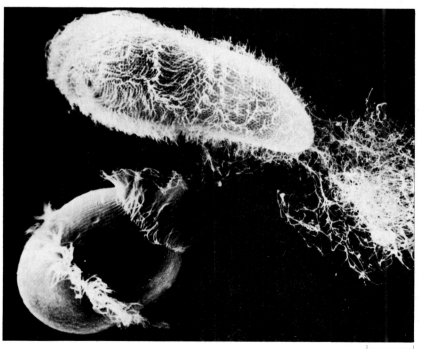

30 μm

Table 4-1 *Measurements Used in Microscopy*

1 centimeter (cm) = 1/100 meter = 0.4 inch*

1 millimeter (mm) = 1/1,000 meter = 1/10 cm

1 micrometer (μm)† = 1/1,000,000 meter = 1/10,000 cm

1 nanometer (nm) = 1/1,000,000,000 meter = 1/10,000,000 cm

1 angstrom (Å) = 1/10,000,000,000 meter = 1/100,000,000 cm

or

1 meter = 10^2 cm = 10^3 mm = 10^6 μm = 10^9 nm = 10^{10} Å

* A metric-to-English conversion table is found in Appendix A.
† Micrometers were formerly known as microns (μ).

ments from the water, building its cell wall, making enzymes, converting sugars to starch (or vice versa), and oxidizing food molecules for energy, all at the same time. It is also likely to be orienting itself in the sunlight, it is probably preparing to divide, it is possibly looking for a mate, and it is undoubtedly carrying out at least a dozen or more similarly important activities. In this chapter, we are going to look at just a few of these cellular functions.

Before we begin to examine individual cells, let us take a look first at the measurements used in cell biology (Table 4–1). The one most commonly used is the micrometer, one-millionth of a meter. The abbreviation for micrometer is μm, using the Greek letter μ (which corresponds to our *m* and is pronounced "mew").

Most of the cells that make up a plant or animal body are within a size range of between 10 and 30 micrometers in diameter. At the bottom of each micrograph in this chapter and the ones that follow you will see a short, straight line that provides a reference for size. The same system is used to indicate distances on a road map.

A principal restriction on cell size seems to be the relationship between volume and surface area. As you can see in Figure 4–2, as volume increases, surface area decreases rapidly in proportion to volume. Materials entering and leaving the cell have to move through the surface, and the more active a cell is, the more rapidly these materials must pass through. Also, oxygen, carbon dioxide, and other atoms and molecules important to cell metabolism* move in and out of the cell by diffusion, which, as we saw in Chapter 2, is effective only over short distances. Materials can move faster into, out of, and through small cells.

It is not surprising, therefore, that the most metabolically active cells are usually small. The relationship between cell size and metabolic activity is nicely

* Metabolism is simply the sum total of all of the chemical activities of a living system.

4-2

The single 4-centimeter cube, the eight 2-centimeter cubes, and the sixty-four 1-centimeter cubes all have the same total volume; however, as the volume is divided up into smaller units, the amount of surface area increases (for example, the single 4-centimeter cube has only one-fourth the total surface area of the sixty-four 1-centimeter cubes). Similarly, smaller cells have more surface area (more membrane surface through which materials can diffuse into or out of the cell) and less volume (shorter diffusion distances within the cell).

ONE 4-CENTIMETER CUBE

SURFACE AREA
(SQUARE
CENTIMETERS)   96

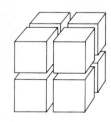

EIGHT 2-CENTIMETER CUBES

192

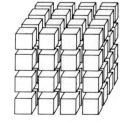

SIXTY-FOUR 1-CENTIMETER CUBES

384

Paramecium *and* Didinium *(continued).*
What appears to be a hole on the surface of
the Paramecium *(now paralyzed) is actually*
a groove, the oral groove. It ends in a stoma.
Bacteria are swept into the opening, where
they collect at the stoma before they enter the
Paramecium.

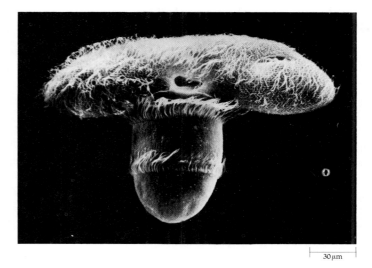

|— 30 μm —|

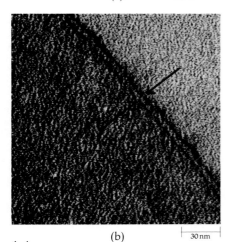

(a) |— 0.2 μm —|

(b) |— 30 nm —|

4–4

Electron micrographs of (a) a section of a
thin bacterial cell, showing the cell mem-
brane, outside of which is the cell wall, and
(b) a portion of the cell membrane of a
human red blood cell (the darker material on
the left is hemoglobin). The cell membranes
in each micrograph are indicated by arrows.
All cellular membranes have this charac-
teristic two-ply appearance.

illustrated by egg cells. Many egg cells are very large. A frog's egg, for instance, is 1,500 micrometers in diameter. Some egg cells are several centimeters across. (Most of this is stored nutrients for the developing embryo.) When the egg cell is fertilized and begins to be active metabolically, it first divides many times before there is any increase in volume or mass, thus cutting each of its cellular units down to an efficient metabolic size.

A second limitation on cell size appears to involve the capacity of the nucleus, the cell's control center, to regulate the cellular activities of a large, metabolically active cell. Again, the exceptions seem to "prove" the rule. In certain large, complex one-celled animals—the ciliates, of which *Paramecium* is an example— each cell has two or more nuclei, the additional ones apparently copies of the original.

HOW THINGS GET INTO AND OUT OF CELLS

As we saw in Chapter 1, a cell can exist as a separate entity because of the existence of the cell membrane, which keeps the inside in and the outside out. Cell membranes are basically the same for all living systems, from *Escherichia coli* to *Homo sapiens. Didinium* and *Paramecium,* like all other cells, are surrounded by cell membranes, and if a *Didinium* is going to "swallow" a *Paramecium*—and in Figure 4–3 one is clearly trying to—the *Paramecium* will have to cross the *Didinium's* cell membrane.

Because cell membranes are only about 90 Å in thickness, they are not visible in the light microscope. Before electron microscopy became available, cytologists presumed that the cell was surrounded by an invisible film, because if this hypothetical film were ruptured, the contents of the cell could be seen to spill out. Now, with the electron microscope, the membrane can be visualized as a continuous, thin double line (see Figure 4–4). (Note that the double line in each micrograph represents one membrane. A double membrane, which is found around chloroplasts, mitochondria, and the nucleus, would be visualized at high-enough magnification as four fine lines.)

According to current hypotheses, the membrane consists essentially of phospholipid molecules arranged so that their hydrophobic fatty acid "tails" are pointed inward (see Figure 3–10, page 47). The membrane is thus sometimes referred to as a "butter sandwich." (It is because the "butter" is electron-transparent that the membrane has its characteristic two-ply appearance.)

Unaided, the human eye has a resolving power of about 1/10 millimeter, or 100 micrometers. This means that if you look at two lines that are less than 100 micrometers apart, they merge into a single line. Similarly, two dots less than 100 micrometers apart look like a single blurry dot. To separate structures closer than this, optical instruments such as microscopes are used. The best light microscope has a resolving power of 0.2 micrometer, or 200 nanometers, or 2,000 Å, and so improves on the naked eye about 500 times. It is theoretically impossible to build a light microscope that will do better than this.

Notice that resolving power and magnification are two different things; if you take a picture through the best light microscope of two lines that are less than 0.2 micrometer, or 200 nanometers, apart, you can enlarge that photograph indefinitely, but the two lines will continue to blur together. By using more powerful lenses, you can increase magnification, but this will not improve resolution. The limiting factor is the wavelength of light. The micrograph of Paramecium in Figure 2–12 on page 35 is a photomicrograph—that is, a micrograph made with light.

With the electron microscope, resolving power has been increased almost 1,000 times over that provided by the light microscope. This is achieved by using "illumination" consisting of electron beams instead of light rays. Areas in the specimen that permit the transmission of more electrons—"electron-transparent" regions—show up bright, and areas that absorb or deflect electrons—"electron-dense" regions—are dark. Electron microscopy at present, under the very best conditions, affords a resolving power of about 2 Å, roughly 500,000 times greater than that of the human eye. (A hydrogen atom is about 1 Å in diameter.)

Electrons, which have a very small mass, can pass through specimens only when the specimens are exceedingly thin. As a consequence, living matter has to be fixed, dehydrated, embedded in hard materials, and sliced by special cutting instruments before it can be examined by transmission electron microscopy. Moreover, most specimens are electron-transparent and have to be stained with heavy metals that bind differentially to different subcellular components. Thus only dead cells can be studied, and moreover, it is sometimes difficult to determine whether or not the specimens have been changed by the preparation methods.

Recently, electron microscopy has been expanded in scope by the development of the scanning electron microscope. In scanning electron microscopy, the electrons whose imprints are recorded come from the surface of the specimen. The electron beam is focused into a fine probe, which is rapidly passed back and forth over the specimen. Complete scanning from top to bottom usually takes a few seconds. As a result of the electron bombardment from the probe, the specimen emits low-energy secondary electrons. Variations in the surface of the specimen alter the number of secondary electrons emitted. Holes and fissures appear dark, and knobs and ridges are light. Electrons scattered from the surface plus secondary electrons are amplified and transmitted to a screen, which is scanned in synchrony with the electron probe. The series of micrographs of Paramecium and Didinium in this chapter were made with the scanning electron microscope.

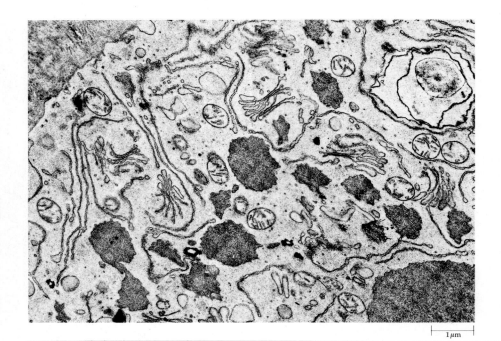

Electron micrograph of a portion of a cell from a barley root. In the upper left-hand corner is a thin double line, the cell membrane, outside of which is the cell wall. Numerous minute structures can be seen in the cytoplasm. Most of the smaller structures could not be resolved by a light microscope and, before the development of electron microscopy, were not known to exist.

1 μm

Model of a cell membrane, proposed by S. J. Singer and G. L. Nicolson. The membrane is composed of phospholipid molecules with their hydrophobic "tails" forming the inner layer, and of protein molecules (indicated in color). According to this model, the proteins vary from membrane to membrane, according to cell function, and also from place to place on the same membrane. Some are enzymes and some are probably involved in active transport.

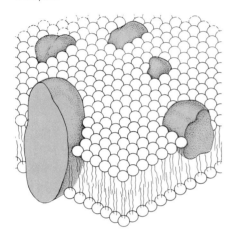

In the phospholipid bilayer are numerous globular proteins, some of which are embedded on one or the other surface and some of which extend through both layers (Figure 4-5). Some of the proteins are enzymes, some are carriers involved with the transport of molecules across the membrane, and others form a structural backbone. Proteins vary from cell to cell, depending on the cell's function, and also apparently from one area of a cell membrane to another. Moreover, they may move from place to place within the lipid bilayer.

Crossing the Membrane

In order to get into or out of a cell or into or out of a membrane-bound organelle, such as a chloroplast or a mitochondrion, substances have to pass through a cell membrane. The membrane is permeable to some substances, such as water, oxygen, and carbon dioxide, which move in and out by osmosis or diffusion. For example, carbon dioxide gas is constantly being produced as fuel molecules are oxidized for energy. As a result, there is more carbon dioxide inside the cell than out. Thus a diffusion gradient is maintained between the inside of the cell and the outside, and carbon dioxide moves out of the cell along this gradient. Conversely, oxygen gas is used up by the cell in the course of its internal activities, so oxygen present in air or water or blood tends to move into the cell by diffusion, again along a gradient.

Molecules may also move into and out of cells against a diffusion gradient. For instance, liver cells, which store sugar (see Figure 3-4, page 44), pick it up from the bloodstream even though the concentration of sugar is higher in the cell than in the bloodstream. Movement into and out of a cell against a diffusion gradient, as we noted in Chapter 2, is called *active transport*. Active transport can take place by means of carrier molecules, such as those shown in Figure 4-6.

A third type of transport into cells involves *phagocytosis* ("cell eating"). This process is diagrammed in Figure 4-7. It begins with the attachment of a solid particle to the cell membrane and is followed by the enveloping of the particle within a segment of the membrane. Thus the particle enters the cell within a

4-6

The "revolving door" model of active transport. (a) The molecule to be transported locks into the carrier molecule in much the same way a substrate locks into its enzyme. (b–d) The carrier molecule rotates, releasing the passenger molecule on the other side of the membrane, and (e, f) then returns to its former position. Sodium ions are pumped out of cells in this way.

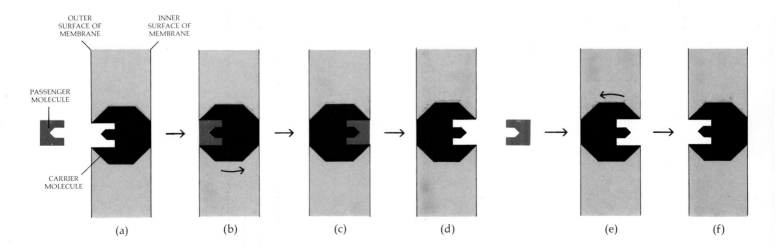

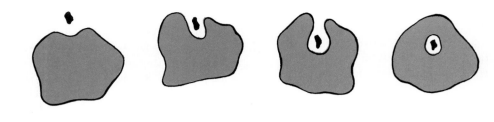

*Phagocytosis: Certain materials may induce
some types of membranes to form vacuoles
around them and so transport them into the
cytoplasm wrapped up in a membrane composed of a segment of the cell's outer membrane.*

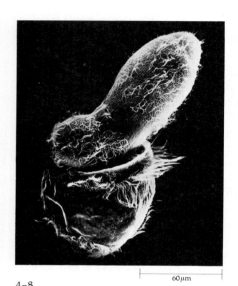

4-8

Paramecium *and* Didinium *(continued). Phagocytosis has begun (in this situation, folding helps).*

sac, or vacuole. It is digested within that sac. Amoebas engulf their prey in this
way, as shown in Figure 18–9 on page 213. Our own white blood cells, which
resemble amoebas, phagocytize bacteria and other invaders in phagocytic vacuoles. *Paramecium* feeds on bacteria in a similar fashion. In this case, the bacteria
enter the cell at only one relatively small area, the stoma, on the surface of the
Paramecium. Didinium also has a stoma, through which, in Figure 4–8, it is
attempting to force a *Paramecium.*

Pinocytosis ("cell drinking") resembles phagocytosis except that molecules in
solution, rather than solid particles, are taken into the cells.

Phagocytosis and pinocytosis can also work in reverse: A particle to be exported from the cell may be wrapped in a vacuole and transported to the cell
surface. Here the membrane of the vacuole fuses with the outer membrane of
the cell, expelling the vacuole's contents to the outside (Figure 4–9). This process is called *exocytosis.*

MANUFACTURE AND EXPORT BY CELLS

Cells continuously make proteins and other molecules. Some of these molecules—enzymes or hemoglobin, for instance—are used within the cell. Others,
such as collagen, digestive enzymes, hormones, or mucus, are released outside
the cell, sometimes carrying out their functions at a great distance, on a cellular
scale, from their source.

There is much that is not known about how cells can carry on so many
different activities at the same time, can produce enough of (and not too much
of) literally hundreds of different kinds of molecules simultaneously, and can
keep these processes separate from one another, all within a space of 10 to 30
micrometers. It is known, however, that proteins are assembled in the cytoplasm from amino acids. It is also known that this assembly process takes place,
in both prokaryotic and eukaryotic cells, on the ribosomes. (This process of
protein biosynthesis is described in more detail in Chapter 15.) All protein-
manufacturing cells contain ribosomes and, in general, the more proteins
(quantitatively speaking) a cell is making, the more ribosomes it has.

The way in which the ribosomes are distributed in the cell seems to have to
do with the way that the proteins are utilized. In cells that are making proteins
for their own use, such as embryonic cells, ribosomes tend to be distributed in
the cytoplasm. In cells that are making digestive enzymes or other proteins for
export, the ribosomes are found attached to a complex system of internal
membranes called the *endoplasmic reticulum.* Endoplasmic reticulum with ribosomes attached to it is known as rough endoplasmic reticulum (Figure 4–10).
Thus there is circumstantial evidence that endoplasmic reticulum with ribosomes on it is somehow involved both in producing proteins and in preparing
them for shipment out of the cell.

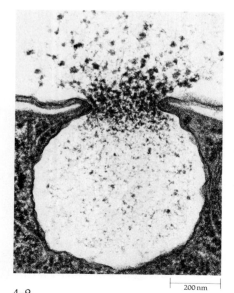

4-9

*Exocytosis. A secretory vesicle, formed
within the cell, discharges its contents. Note
that the membrane of the vesicle is fusing
with the cell membrane.*

Rough endoplasmic reticulum, which fills most of this micrograph, is a system of membranes that separates the cell into channels and compartments and provides surfaces on which chemical activities take place. The dense objects on the membrane surfaces are ribosomes. This cell is from a pancreas, an organ extremely active in the synthesis of digestive enzymes, which are "exported" to the upper intestine, where most digestion takes place. In the lower right-hand corner of the micrograph are one mitochondrion and, above it, a portion of another.

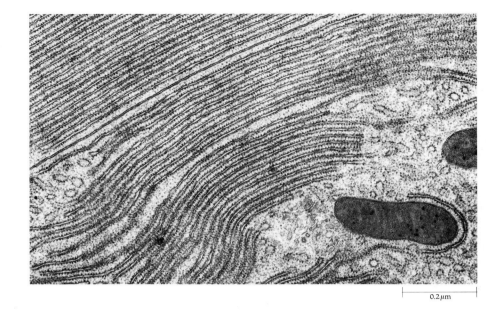

0.2 μm

This is not the only role of endoplasmic reticulum in the manufacturing activities of the cell. Membranes of the endoplasmic reticulum, like cell membranes, have proteins on and in them, and many of these proteins are enzymes. It seems likely that groups of enzymes involved in the same sort of activity are grouped together on the endoplasmic reticulum. Thus the endoplasmic reticulum provides an extended work surface for various cellular activities. Also, the endoplasmic reticulum divides the cell into compartments and so may help to separate the various chemical processes being carried out simultaneously by the cell.

Cells also contain smooth endoplasmic reticulum, that is, endoplasmic reticulum with no ribosomes on it. It is found largely in the form of tubules and plays a role in transporting substances from the cell's interior to the cell surface and in synthesizing steroid hormones.

Golgi Bodies

Golgi bodies are involved in the cell's export activities. Each Golgi body consists of a little group of flattened sacs composed of membranes stacked loosely on one another and surrounded by tubules and vesicles (very small vacuoles). Golgi bodies apparently serve as packaging centers, apparently especially for substances formed on the endoplasmic reticulum. Also, they are the sites of assembly for some complex molecules—for instance, combinations of sugars and proteins (glycoproteins)—that are found on the surfaces of cell membranes. In plant cells, they bring together the various components of plant cell walls. Golgi bodies are found in almost all eukaryotic cells. Animal cells usually contain 10 to 20 Golgi bodies, and plant cells may have several hundred.

Lysosomes

One type of small vesicle formed in the Golgi body is a *lysosome*. Lysosomes are essentially bags of hydrolytic (from "hydrolysis") enzymes enclosed in membranes that separate them from the rest of the cell. If the lysosomes break open,

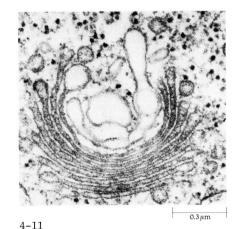

0.3 μm

4-11
Golgi bodies are also composed of special arrangements of membranes. Compounds formed within the cell are packaged within membrane-enclosed vesicles at the Golgi bodies. The vesicles then move to the cell surface and release their contents (as shown in Figure 4-9). In the case of plant cells, the contents of the vesicles are used in the construction of the plant cell walls.

4–12

Diagram illustrating the interaction of cyto-plasmic organelles in the synthesis of protein and in the packaging of macromolecules for export. Glycoproteins are combinations of sugars and proteins, and lipoproteins are combinations of fats and proteins. They are common components of cell walls and cell membranes.

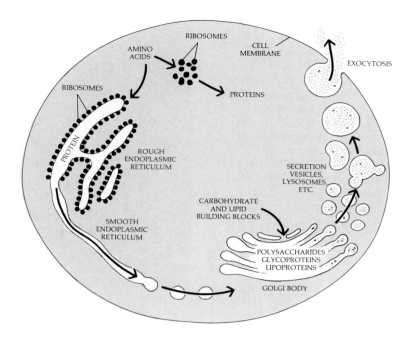

the cell itself is immediately destroyed, since the enzymes they carry are capable of breaking down all the major compounds found in a living cell.

White blood cells in the human body phagocytize bacteria. When this occurs, the lysosomes within the white cell fuse with the vacuoles containing the bacteria and release their hydrolytic enzymes. The bacteria are quickly digested. In similar fashion, the lysosomes of single-celled organisms, such as *Paramecium, Didinium,* and amoebas, fuse with the phagocytic vacuoles containing prey. Hydrolytic enzymes released by the lysosomes into the vacuoles digest the contents. Why the enzymes do not destroy the membranes of the lysosomes that carry them is a pertinent question yet to be answered.

Figure 4–12 summarizes the ways in which the rough and smooth endoplasmic reticulum and the Golgi body and its vesicles may interact to produce macromolecules for export.

GENERATING POWER

The activities of a cell require energy. Cells obtain their energy either directly from sunlight (autotrophs) or by breaking down (oxidizing) organic molecules made by other cells (heterotrophs).

Photosynthesis, the process by which light energy is transformed to chemical energy, requires special pigments, of which the most common is chlorophyll. For reasons which are not entirely understood, photosynthesis takes place only when the chlorophyll molecules are embedded in a membrane. In prokaryotes, the photosynthetic membranes are in the cytoplasm. In eukaryotes, they are organized in special organelles, the chloroplasts.

Cellular respiration is the process by which organic (carbon-containing) molecules are combined with oxygen, and carbon dioxide, water, and energy are produced. In prokaryotes, these processes take place on the cell membrane and in the cytoplasm. In eukaryotes, they are carried out in the mitochondria.

Plant cell walls consist mainly of cellulose fibers embedded in a matrix (the same construction principle used in fiberglass boats). As plant cells elongate, the original fibers become reoriented longitudinally, but new fibers are laid down in the opposite direction, maintaining the cross-ply construction.

4-14

Electron micrograph and diagram of two adjacent cell walls of vessel elements, the cells that form the tubes through which water is conducted in plants. You can see the middle lamella, the primary walls, and the layered secondary walls, deposited inside the primary wall. The cells are from the wood of a black locust tree. The protoplasts have died.

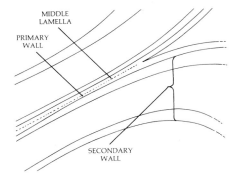

MIDDLE LAMELLA

PRIMARY WALL

SECONDARY WALL

Mitochondria are found in virtually all eukaryotic cells, including photosynthetic cells. Thus, the chloroplasts and mitochondria are the power generators of eukaryotic cells. Without the energy these organelles produce, most other cellular functions could not be carried out.

The processes of photosynthesis and cellular respiration will be described in more detail in Chapters 6 and 7.

THE CELL AS ARCHITECT

Cells build many structures. The cell wall of plants is an example. As plant cells divide, a thin layer of gluey material forms between them, which becomes the *middle lamella*. It is composed of pectins (the compounds that make jellies jell) and other polysaccharides. It holds the cells together. Next, under the middle lamella, the plant cell constructs its wall. The plant cell wall is composed of cellulose molecules wound together like wires in a cable and laid down in a softer matrix or bed. As you can see in Figure 3–5a on page 45, the molecules are laid down in sheets oriented at right angles to one another. (Persons familiar with building materials will note that the cellulose cell wall thus combines the structural features of both fiber glass and plywood.)

As the wall is constructed, small openings are left in it at crucial points, usually right opposite a gap in an adjacent cell. Some of the openings serve as passageways for plasmodesmata (singular, *plasmodesma*), which are fine strands of cytoplasm that connect plant cells together. These are believed to integrate the plant body by forming channels for the exchange of materials between cells.

In young plant cells, the wall is thin and flexible, permitting the cell to enlarge as it grows, but as the cell matures, it becomes more rigid. In some types of plant cells, such as those that form the woody tissue of tree trunks, a thicker, secondary cell wall is constructed. (The first wall is called the primary cell wall.) This wall is sometimes impregnated with a macromolecule called lignin, which has stiffening properties. In such cells, the living material of the cell often dies, leaving only the two cell walls.

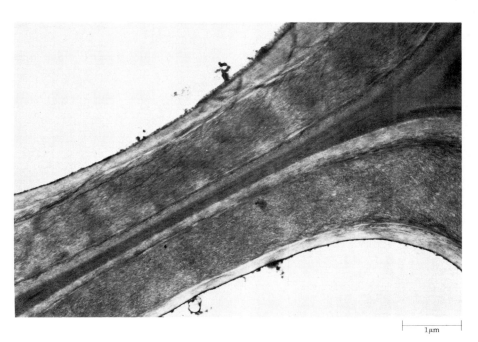

1 µm

Diagram of a portion of a microtubule. It is composed of globular proteins assembled in a helix.

Often these plant cell walls have specialized openings in them called pits. In some plants, water flows from one cell to another through these pits. In another type of water-conducting cell, the cell walls are constructed so they join together, end on end, to form continuous tubes. These water-conducting vessels are described in more detail in Chapter 20.

Cell wall materials are synthesized in the cytoplasm and are exported to the cell surface in the vesicles of Golgi bodies. Cellulose and other large molecules are assembled on the surface of the cell membrane.

Microtubules

Another type of structure designed and built by cells is the *microtubule* (Figure 4–15). Microtubules are composed of identical units of globular proteins. As these proteins are synthesized, they spontaneously form this tertiary structure (see Figure 3–17, page 52). Then the individual globular proteins come together, because of attractions among their R groups, to form a coil (a helix). These can assemble and disassemble rapidly and, as we shall see, are involved in cell movement. Their diameter is only 0.02 micrometer, and they are often very long.

HOW CELLS MOVE

Bacterial Flagella

Prokaryotes often have flagella that propel them through the water (Figure 4–16). Some have only a single flagellum, but others have many. Each flagellum may be several times longer than the cell itself. Bacterial flagella are very simple in construction; they are made up of one type of protein molecule, flagellin. Biochemical studies indicate that a bacterial flagellum is composed of three molecules of flagellin intertwined. The motion of the flagella is imparted to them by small bodies at the base of the flagellum (basal granules) to which the whiplike structure is attached.

A bacterial cell (Proteus mirabilis) *with multiple flagella.*

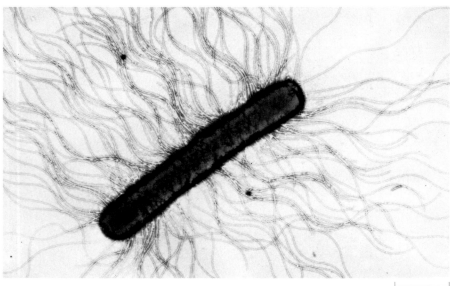

1 μm

Eukaryotic Cilia and Flagella

Cilia (from the Latin word for "eyelash") and flagella are two names for essentially the same structure in eukaryotes. (The names were given before the basic sameness was realized.) When they are shorter and occur in large numbers, the structures are more likely to be called cilia; when they are longer and fewer, they are usually called flagella. Thus we say that a _Paramecium_ or a _Didinium_ (Figure 4–17) has cilia, and _Chlamydomonas_ has two flagella. Lest you think this is simple, the numerous but long structures on _Trichonympha_ (Figure 3–5b) are called flagella.

One-celled organisms and also very small organisms, such as flatworms, are propelled by cilia. Many of the cells in our own bodies have identical structures. The motile power of the human sperm cell comes from its single powerful flagellum, or "tail."

Many of the cells that form the tissues of our bodies are also ciliated. These cilia do not move the cells, but rather serve to sweep substances across the cell surface. For example, cilia on the surface of cells of the respiratory tract beat upward, propelling bits of soot, dust, pollen, tobacco tar—whatever foreign substances we have inhaled either accidentally or on purpose—to the backs of our throats, where they can be removed by swallowing.

Only a few large groups of organisms—notably the angiosperms (flowering plants)—have no cilia or flagella in any cells.

All eukaryotic cilia and flagella have the same structure. The basic unit of this structure is the microtubule. In each cilium or flagellum, nine pairs of micro-

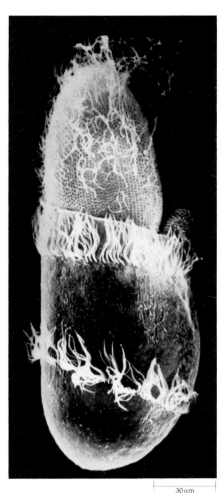

4–17

Paramecium _and_ Didinium _(continued). Both_ Paramecium _and_ Didinium _are characterized by multiple cilia (they are known, in fact, as ciliates, members of the order Ciliophora). One species of_ Paramecium _has 17,000 cilia on each cell. This_ Paramecium _is decreasing in volume; the_ Didinium _is squeezing water out of it as it engulfs it._

Table 4–2   _Comparison of Animal, Plant, and Prokaryotic Cells_

	PROKARYOTE	ANIMAL	PLANT
Cell membrane	Present	Present	Present
Cell wall	Present (noncellulose polysaccharide plus protein)	Absent	Present (cellulose)
Nucleus	No nuclear envelope	Surrounded by nuclear envelope	Surrounded by nuclear envelope
Chromosomes	Single, containing only DNA	Multiple, containing DNA and protein	Multiple, containing DNA and protein
Endoplasmic reticulum	Absent	Usually present	Usually present
Mitochondria	Absent	Present	Present
Plastids	Absent	Absent	Present in many cell types; chloroplasts in photosynthetic cells
Ribosomes	Present (smaller)	Present	Present
Golgi bodies	Absent	Present	Present
Lysosomes	Absent	Often present	Usually absent
Vacuoles	Absent	Small or absent	Usually large single vacuole in mature cell
9 + 2 cilia or flagella	Absent	Often present	Absent (in higher plants)
Centrioles	Absent	Present	Absent (in higher plants)

(a) *Diagram of a cilium with its basal body. All eukaryotic cilia and flagella, whether they are found on one-celled organisms or on the surfaces of cells within our own bodies, have this same basic structure, which con-*

sists of an outer ring of nine pairs of microtubules surrounding two additional microtubules in the center. The basal bodies from which they arise have nine outer triplets, with no microtubules in the center. (b)

Cross section of flagella. These are from Trichonympha, the one-celled organism shown in Figure 3–5b. All eukaryotic flagella and cilia look like this in cross section.

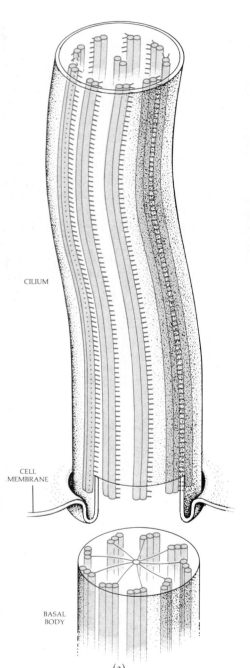

CILIUM

CELL
MEMBRANE

BASAL
BODY

(a)

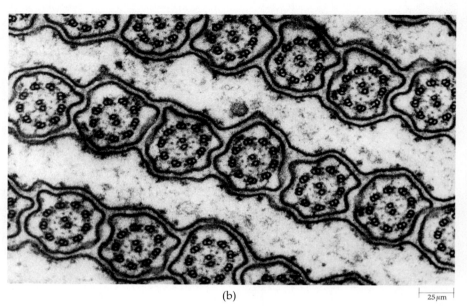

(b)

|— 25 μm

tubules form a ring that surrounds two additional microtubules in the center. In some electron micrographs, two little "arms" are visible on one of each pair of outer tubules; these are believed to be enzymes involved in converting chemical energy to kinetic energy—the movement of the cilium. If cilia or flagella are removed from cells, cut into little fragments, and placed in a suitable medium containing energy-rich chemicals, the fragments twitch. Such evidence indicates that, as is not the case with bacterial flagella, the machinery of movement of eukaryotic flagella is in the shaft of the flagellum itself.

The way in which eukaryotic flagella and cilia move is a matter of current interest and controversy. Some evidence suggests that the microtubules may slide along one another, thus causing the flagellum to bend. The direction in which it bends would be decided by which pairs of microtubules were interacting.*

Cilia and flagella arise from basal bodies, which are also made up of microtubules, although their number and arrangement are somewhat different, as you can see in Figure 4–18. The basal bodies are believed to keep the flagella supplied with fuel molecules and perhaps other substances as well.

The discovery of the complex internal structure of cilia and flagella, repeated over and over again throughout the living world, was one of the spectacular revelations of electron microscopy. For biologists, it is another glimpse down the long corridor of evolution, providing overwhelming evidence, once again, of the basic unity of living things.

---

* The molecular machinery responsible for the movements in a very different type of cell—the cells of skeletal muscles—is described on pages 319 to 322. You may want to read these pages now.

Animal cell, as interpreted from electron micrographs.

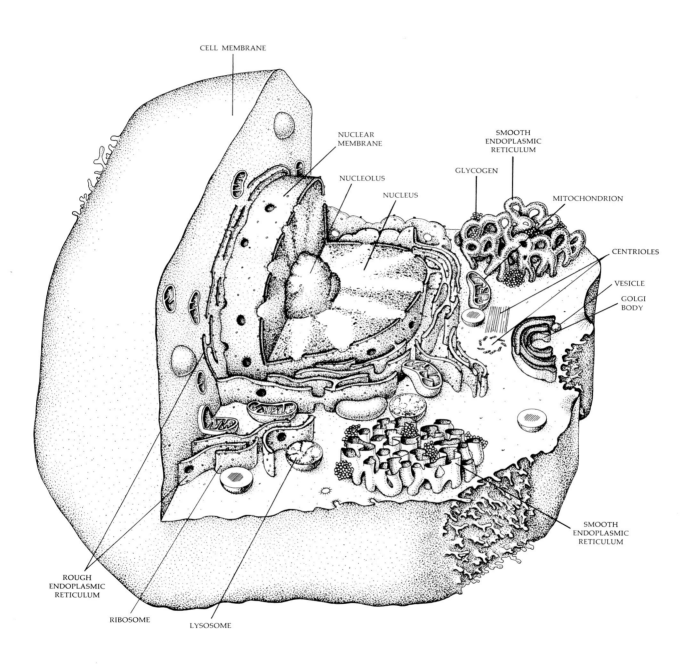

CELL MEMBRANE

NUCLEAR
MEMBRANE

SMOOTH
ENDOPLASMIC
RETICULUM

NUCLEOLUS

GLYCOGEN

NUCLEUS

MITOCHONDRION

CENTRIOLES

VESICLE

GOLGI
BODY

SMOOTH
ENDOPLASMIC
RETICULUM

ROUGH
ENDOPLASMIC
RETICULUM

RIBOSOME

LYSOSOME

This simple organism, called Pandorina, is made up of 32 cells, most of which are visible here, held together by a jellylike substance, whose outlines you can also see. Each of these cells is photosynthetic and can survive independent of the others. Each cell has two flagella, none of which is visible in this picture, probably because the flagella move too quickly. The flagella all point outward and beat in synchrony, so that the colony rolls through the water like a ball. To reproduce, each cell divides, producing a new cell inside, and then the parent colony breaks apart.

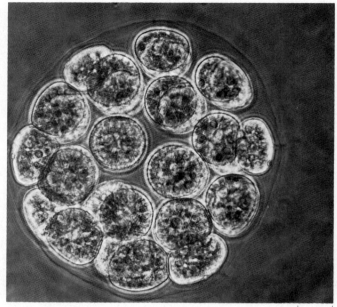

0.1 μm

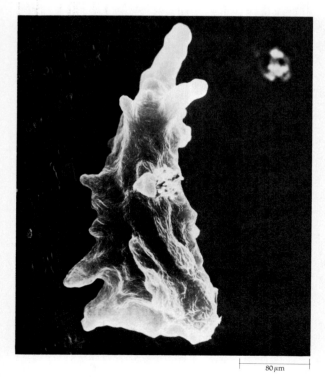

80 μm

Amoeba proteus, *a single-celled organism named for Proteus, a Greek god capable of changing his shape at will. Amoebas use their pseudopods both for moving and for capturing prey, such as the particle of debris in the foreground of this scanning electron micrograph.*

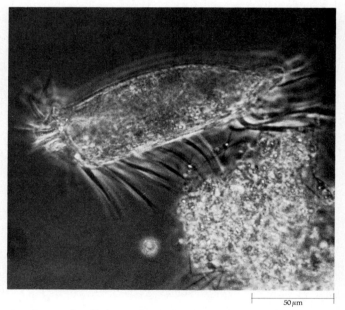

50 μm

Another single-celled organism, Euplotes patella, a ciliate. Its "legs" and other appendages are fused clumps of cilia. Their movements are highly coordinated, enabling Euplotes to move in a brisk and seemingly purposeful manner. Although sometimes referred to as "simple one-celled organisms," the protists in general, and the ciliates in particular, are probably the most complex of cells in terms of their many specialized structures and remarkable abilities.

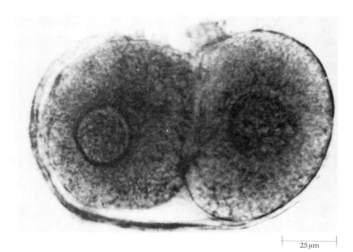

A human embryo at the two-celled stage. Within the cells, the nuclei are clearly visible. Each of these nuclei carries all the genetic information for every cell in the human organism.

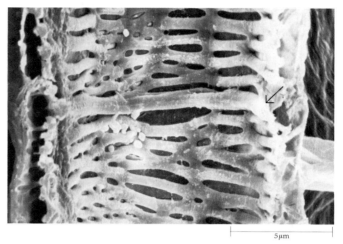

Like the red blood cells, these cells from the water-transporting vessels of a plant have been reduced at maturity to their essentials, in this case, the cell walls composed of cellulose and lignin. To obtain this micrograph, the vessel was cut lengthwise, so that half of the cylinder is seen. The view is from the inside. The thicker line indicated by the arrow is the result of the junction of the two cells. It is the remains of the two cell walls, which perforated as the cells matured, forming one continuous vessel.

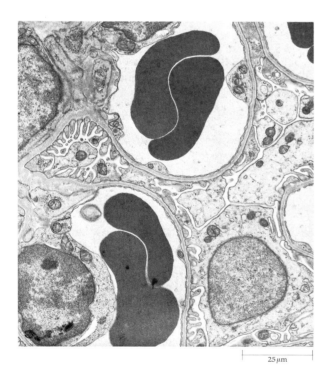

The four dark shapes shown here are red blood cells in the capillaries of a kidney. Unlike the other cells shown on these pages, red blood cells lack nuclei and mitochondria and other organelles (they lose them as they mature). These cells are essentially flexible bags of oxygen-bearing hemoglobin.

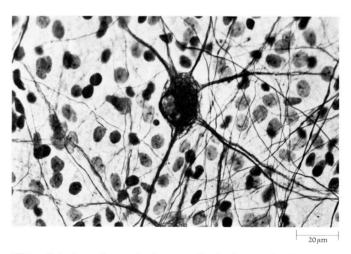

This cell is from the cerebral cortex. In the human brain, there are about 12 billion such nerve cells. The actions of these cells and the interconnections among them are responsible for consciousness, intelligence, dreams, and memory.

The cell nucleus fills the upper half of this electron micrograph. What you see here is the surface of the nuclear envelope. Clearly visible on this surface are pores, through which, it is believed, the nucleus and the cytoplasm are able to "communicate."

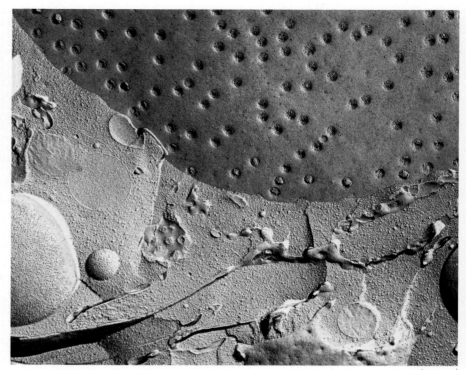

20 μm

THE CONTROL CENTER

Like all complex organizations, the cell requires a control center. In prokaryotic cells, this control center is a single, long molecule, the bacterial chromosome. In eukaryotic cells, the control center is in the form of a _nucleus_, which is a large, spherical body, usually the most prominent structure within the cell. The nucleus is surrounded by two membranes that together make up the nuclear envelope (Figure 4–20). Penetrating the surface of the envelope are nuclear pores. These pores apparently permit only certain large molecules to go through, in particular, molecules carrying "messages" from nucleus to cytoplasm.

The chromosomes, which are composed of DNA and protein, are found within the nucleus. When the cell is not dividing, the chromosomes are visible only as a tangle of fine threads, called _chromatin_. The most conspicuous body within the nucleus is the _nucleolus_. The nucleolus is the site at which major components of the ribosomes are manufactured.

The Functions of the Nucleus

The nucleus performs several crucial functions for the cell. First, stored in the chromosomes are all the blueprints for making the complex macromolecules on which the structures and activities of cells depend. Second, interacting with the cytoplasm, the nucleus helps to regulate the production and assembly of molecules. Third, the nucleus contains the hereditary information of the cell. The _Didinium_, having swallowed the _Paramecium_ (Figure 4–21), now is likely to divide, forming two new cells virtually identical to the parent cell, each of which will soon begin to stalk the _Paramecium_.

4-21
The end.

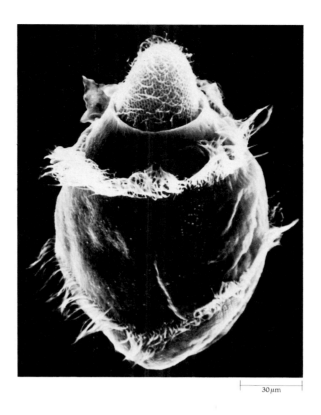

30 μm

SUMMARY

Cells are the basic units of biological structure and function. Most cells are between 10 and 30 micrometers in diameter. One limit on the size of cells is surface-to-volume ratio. Smaller units have a greater surface in proportion to their volume than large ones.

All cells are bounded by an outer membrane, the cell membrane. It is composed of two layers of phospholipid molecules, with globular proteins embedded in it. Materials pass through the membrane by diffusion, by active transport, and by phagocytosis, pinocytosis, and exocytosis.

Cells carry on a vast number of biochemical activities, including the biosynthesis of proteins and other molecules. Proteins are assembled on ribosomes, small cytoplasmic particles present in both prokaryotic and eukaryotic cells. The cytoplasm of eukaryotic cells is subdivided by a network of membranes known as the endoplasmic reticulum, which serves as a work surface for many of the cell's biochemical activities and also apparently as a system of channels for moving materials through the cell. Endoplasmic reticulum to which ribosomes are attached is known as rough endoplasmic reticulum. Cells that produce protein for export have extensive systems of rough endoplasmic reticulum.

Golgi bodies, also composed of membranes, are packaging centers for materials being moved into and out of the cell. They give rise to secretion vesicles and storage vesicles, such as lysosomes.

Lysosomes are sacs of digestive enzymes. They may fuse with other vesicles or vacuoles, such as food vacuoles, and digest their contents. Cells build their own structures. An example is the plant cell wall, composed of layers of cellulose molecules. Another example is the microtubule, made up of subunits of globular protein, coiled in a helix.

Cilia and flagella are hairlike structures found on the surface of many types of eukaryotic cells. They are associated with the movement of cells or the movement of materials along cell surfaces. They have a highly characteristic 9 + 2 structure, with nine pairs of microtubules forming a ring surrounding two central microtubules.

The nucleus of eukaryotic cells is surrounded by a double membrane, the nuclear envelope. The nucleus contains the hereditary information of the cell, and, interacting with the cytoplasm, helps to regulate the cell's ongoing activities.

QUESTIONS

1. Sketch a cell, either plant or animal. Include the principal organelles and label them.

2. What are the functions of the following: endoplasmic reticulum, Golgi body, ribosome, lysosome?

3. Sketch a cross section of a cilium.

4. Why is the secondary cell wall of a plant *inside* the primary cell wall? Where is the cell membrane in relation to the two cell walls?

5. Return to Chapter 2 and add approximate scale markers: Figures 2–1, 2–2, 2–3, 2–4, 2–9, and 2–12. Use a ruler and the scale mark at the bottom of each picture on pages 57 and 66 to determine (a) the diameter of a cilium, (b) the diameter of a microtubule within a cilium, (c) the thickness (roughly) of a cell membrane. (This is how it is done by microscopists.) Would a cilium be resolvable in a light microscope (that is, is its diameter more than 0.2 μm)?

6. On the basis of what you know of the functions of each of the structures in Table 4–2 (page 65), what components would you expect to find most prominently in each of the following cell types: muscle cells, sperm cells, green leaf cells, red blood cells, white blood cells?

7. Two brothers were under medical treatment for infertility. Microscopic examination of their semen showed that the sperm were immobile and that the little "arms" were missing from the microtubular arrays. The brothers also had chronic bronchitis and other respiratory difficulties. Can you explain why?

SUGGESTIONS FOR FURTHER READING

BAKER, J. J., and G. E. ALLEN: *Matter, Energy and Life: An Introduction for Biology Students*, 2d ed., Addison-Wesley Publishing Co., Inc., Reading, Mass., 1970.*

A book for students who have had no previous chemistry or physics, which deals with such topics as the structure of matter, the formation of molecules, the course and mechanism of chemical reactions, as well as with the chemistry of living systems.

JASTROW, ROBERT: *Red Giants and White Dwarfs*, Harper & Row, Publishers, Inc., New York, 1967.

A short, handsomely illustrated book covering "the evolution of stars, planets, and life."

* Available in paperback.

LEDBETTER, M. C., and KEITH R. PORTER: *Introduction to the Fine Structures of Plant Cells*, Springer-Verlag, New York, 1970.

An excellent atlas of electron micrographs of plant cells, with detailed explanations.

LEHNINGER, ALBERT L.: *Biochemistry*, 2d ed., Worth Publishers, Inc., New York, 1975.

This advanced text is outstanding both for its clarity and for its consistent focus on the living cell.

LOEWY, A. G., and P. SIEKEVITZ: *Cell Structure and Function*, 2d ed., Holt, Rinehart and Winston, Inc., New York, 1970.

An outstanding elementary text on cell structure and function.

OPARIN, A. I.: *The Origin of Life*, Dover Publications, Inc., New York, 1938.*

Oparin, a Russian biochemist, was the first to argue that life arose spontaneously in the oceans of the primitive earth. Although his concepts have been somewhat modified in detail, they form the basis for the present scientific theories on the origin of living things.

PONNAMPERUMA, CYRIL: *The Origins of Life*, E. P. Dutton & Co., Inc., New York, 1972.

A clear, authoritative, and up-to-date look at the origins of life on earth and the possibilities of life on other planets. Written by one of the leading investigators in this field, with abundant and imaginative illustrations.

PORTER, KEITH R., and MARY A. BONNEVILLE: *An Introduction to the Fine Structure of Cells and Tissues*, 3d ed., Lea & Febiger, Philadelphia, 1968.

An atlas of electron micrographs of animal cells; detailed commentaries accompany each. These are magnificent micrographs, and the commentaries describe not only what the picture shows but also the experimental foundations of knowledge of cell ultrastructures.

THOMAS, LEWIS: *The Lives of a Cell: Notes of a Biology Watcher*, Viking Press, Inc., New York, 1974.*

Thomas, a physician and medical researcher, reveals the extent to which science can tune our intellectual antennae, broaden our perceptions, and extend our appreciation of ourselves and of the world around us. Anyone who wants to refute the contention that science destroys human values need look no further than these short, sensitive essays.

WOLFE, STEPHEN L.: *Biology of the Cell*, Wadsworth Publishing Company, Inc., Belmont, Calif., 1972.

An outstanding synthesis of cell structure and function.

* Available in paperback.

SECTION 2

Cellular Energetics

5-1

The flow of energy in an ecological system. Radiant energy from the sun is transformed to chemical energy by the grasses of this African savanna. The zebras, eating the grasses, use this chemical energy for growth and convert some of it to kinetic energy, which may help to keep them from participating in another energy transfer involving the waiting lions. Here we have a biological example of the second law of thermodynamics. Of the radiant energy falling on the grass, less than 10 percent is converted to chemical energy. Less than 10 percent of the chemical energy stored in the grass is converted to chemical energy stored in the zebra, and less than 10 percent of the zebra's stored energy is transferred to its predator, the lion.

CHAPTER 5

The Flow of Energy

Life here on earth depends on the flow of energy from the thermonuclear reactions taking place at the heart of the sun. The amount of energy delivered by the sun is 13×10^{23} (the number 13 followed by 23 zeros) calories per year. It is a difficult quantity to imagine. For example, the amount of energy striking the earth every day is the equivalent of about a million Hiroshima-sized A-bombs.

About one-third of this solar energy is immediately reflected back out to space as light (as it is from the moon). Much of the remaining two-thirds is absorbed by the earth and converted to heat. Some of this absorbed heat energy serves to evaporate the waters of the ocean, producing the clouds that, in turn, produce the rainfall on the land. Solar energy, in combination with other factors, is also responsible for the movements of air and of water that help set patterns of climate over the surface of the earth.

A small fraction—less than one percent—of the solar energy reaching the earth becomes, through a series of operations performed by the cells of plants and animals, the energy that drives all the processes of life. Plant and animal cells change energy from one form to another, transforming the radiant energy from the sun into the chemical and mechanical energy used by everything that is alive (Figure 5–1).

In this chapter, we shall look at these processes in broad perspective. In the chapters that follow, the two principal and complementary stages of energy flow—photosynthesis and respiration—will be examined in greater detail.

THE LAWS OF THERMODYNAMICS

As we noted in the Introduction, certain laws, the laws of thermodynamics, apply to energy exchanges in both physical and biological systems. The *first law of thermodynamics* states: *Energy can be changed from one form to another, but it cannot be created or destroyed.* The total energy of any system plus that of its surroundings thus remains constant, despite its changes in form.

Light is a form of energy, as is electricity. Light can be changed to electrical energy, and electrical energy can be changed to light (for example, by letting it

flow through the tungsten wire in a light bulb). Energy can be stored in chemical bonds. When we burn carbohydrates, such as wood or paper, we release much of this energy in the form of heat. Warm-blooded animals "burn" (oxidize) sugars in maintaining their body temperature, converting chemical energy to heat energy, which is dissipated into the surrounding air or water.

Motion is another kind of energy. A gasoline engine, for example, produces energy in the form of motion. In order to do this, it must use up energy in another form—the chemical energy stored in the gasoline. By burning gasoline as fuel, the engine changes the chemical energy to heat and then changes the heat to mechanical movements.

Energy can be stored in various other forms. A boulder on a hill is an example of stored energy. The energy required to push the boulder up the hill is "stored" in the boulder as potential energy. When the boulder rolls downhill, the energy is released as kinetic (motion) energy (Figure 5–2).

Some useful energy is always converted to heat. When a boulder rolls downhill, its potential energy is converted to kinetic energy and also to heat energy due to friction. In a gasoline engine, about 75 percent of the energy originally present in the fuel is dissipated to the surroundings in the form of heat. Similarly, an electric motor converts only about 50 percent of the electrical energy into mechanical energy, the rest going into heat. Heat, you will recall, is simply the random motion of atoms or molecules. The higher the heat, the greater the random motion.

When we say that energy is "lost" as heat, what we actually mean is that it is no longer available to do work. Energy stored in a stick of dynamite can do work; that part of the energy that is released as heat at the time of the explosion is no longer available for work. The energy in the boulder at the top of the hill is also available to do work. If it is difficult to visualize a rolling boulder as having the capacity to do work, remember that boulders are not conventionally harnessed for this purpose. Substitute water, for which the first law of thermodynamics applies equally well. Suppose that, instead of the boulder, we transport water to the top of our imaginary hill and let it rush down again. We know from experience that the water could be used to turn a series of paddle wheels and that this machinery could be used to grind corn or do other work useful in human terms.

This brings us to the *second law of thermodynamics*. When stated in its simplest form, this law says that *in all natural processes, energy in a form to do work is eventually converted to heat energy, which is dissipated out into the surroundings.* Another way of stating this is that the disorder, or randomness, of every natural system in-

Some illustrations of the second law of thermodynamics. In nature, processes tend toward randomness, or disorder. Only an input of energy can reconstruct the initial state from the final state.

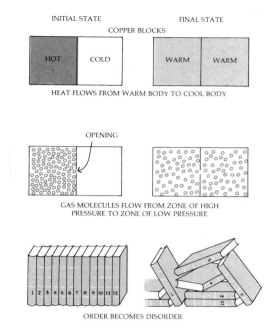

creases. In thermodynamics, this randomness is given the special name of *entropy* (Figure 5–3). This second law, unlike the first one, shows the direction that natural processes take. As a consequence, it is sometimes known as "time's arrow."

Now let us look at these two laws of thermodynamics as they apply to biological systems. The cell is a specialist in energy exchange, interconverting electrical energy, light energy, kinetic energy, and chemical energy and also shifting the energy from one type of chemical bond to another, more convenient form. Most of the attributes that we would select as characteristic of living things are forms of energy exchange. Living organisms constantly take in energy from outside sources and release it into the environment in a less useful form. The ultimate outside energy source is, of course, the sun. Life can continue to operate at this vast, perpetual deficit only because of the tremendous amount of solar energy that flows into the biosphere every minute of every day.

ENERGY AND THE ELECTRON

To understand how solar energy can be converted to chemical energy, we shall have to look once more at the structure of an atom. Atoms, as we pointed out in Chapter 2, are made up of a central nucleus that contains protons (positively charged particles) and an outer cloud of electrons, which are negatively charged particles moving in orbitals around the nucleus. Electrons are found at certain fixed distances from the nucleus, arranged in electron shells, which have different energy levels. Energy conversions in chemical systems involve the movement of electrons from one energy level to another.

To understand this a little better, return to the analogy of the boulder. A boulder sitting still on flat ground has no energy. If you push it up a hill, it gains

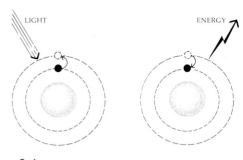

5-4

When an input of energy—such as light energy—boosts an electron to a higher energy level, the electron, like the boulder, possesses potential energy. This potential energy is released when the electron returns to its previous energy level.

potential energy. So long as it sits on the peak of the hill, it neither gains nor loses energy. If it is permitted to roll down the hill, some of the potential energy is converted into kinetic energy and the rest is converted into heat. Similarly, as we pointed out in Chapter 2, water on a hilltop has potential energy, unlike water at the bottom of the hill. Energy has been expended to get the water up there—whether you carried it, or an engine pumped it, or the sun's energy converted it to water vapor which eventually precipitated as snow or rain. If you trapped the water up there—in a bucket, say, or by a dam, or as snow—it would have potential, unreleased energy. If you released it, it could come down the hill and turn a waterwheel, converting some of its energy to another form. When it reached the bottom, all its potential energy would be spent.

An electron is like the boulder or the water on the hilltop in that an input of energy can raise it to a higher level. So long as it remains at this level, it possesses potential energy. When it returns to a lower level, its potential energy is released (Figure 5–4). The fall of electrons from higher to lower levels is the source of energy for biological work. Photosynthesis is a way of using light energy to push electrons to a higher level.

Oxidation-Reduction Reactions

The movement "uphill" or "downhill" of an electron often involves passing the electron from one atom or molecule to another atom or molecule. The passing of an electron from one molecule to another is known as an oxidation-reduction reaction. The loss of an electron is know as *oxidation,* and the compound that loses the electron is said to be oxidized. The reason electron loss is called oxidation is that often no electron loss will occur unless oxygen is available to accept the transferred electrons. As you know, you can stop a fire from burning by smothering it (that is, by cutting off its oxygen). Similarly, you can destroy the energy-yielding process of an animal by suffocating it, which cuts off the oxygen needed for cells to break down carbon compounds and convert their chemical energy to some other form of energy.

Reduction is, conversely, the gain of an electron. Oxidation and reduction take place simultaneously because an electron that is lost by one atom is accepted by another.

Often an electron travels in company with a proton—in short, as part of a hydrogen atom. In that case, oxidation involves the removal of the hydrogen ion (proton) and its electron from one substance, and reduction involves the transfer of both a hydrogen ion and an electron to another substance. The reduction of oxygen—the addition of hydrogen atoms—thus results in the formation of water. The reduction of carbon dioxide, which occurs in photosynthesis, can result in the formation of carbohydrates from carbon dioxide, hydrogen ions, and electrons. The oxidation of carbohydrate, whether it takes place in the cell or by burning wood or gas in a flame, yields carbon dioxide and water and releases the energy put into the molecule during photosynthesis. The water is formed when oxygen atoms accept the electrons (and hydrogen ions) removed from the carbohydrate.

As we noted previously, oxidation and reduction always take place simultaneously. However, if a reaction results in a net increase in energy, it is customarily referred to as a reduction reaction; for example, the chief result of photosynthesis is the *reduction* of carbon. Conversely, if there is a net decrease in chemical bond energy (with a release of energy as heat or light), the reaction is often referred to as an oxidation process. Sugar is *oxidized* to carbon dioxide and water.

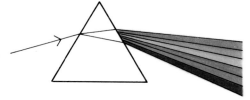

5-5

White light is actually a mixture of different colors, ranging from violet at one end of the spectrum to red at the other. It is separated into its component colors when it passes through a prism—"the celebrated phaenomena of colors," as Newton referred to it.

LIGHT AND LIFE

Almost 300 years ago, the English physicist Sir Isaac Newton (1642–1727) separated visible light into a spectrum of colors by letting it pass through a prism (Figure 5–5). Then by passing the light through a second prism, he recombined the colors, producing white light once again. By this experiment, Newton showed that white light is actually made up of a number of different colors, ranging from violet at one end of the spectrum to red at the other. Their separation is possible because light of different colors is bent at different angles in passing through the prism. Newton believed that light was a stream of particles (or, as he termed them, "corpuscles"), in part, because of its tendency to travel in a straight line.

In the nineteenth century, through the genius of James Clerk Maxwell (1831–1879), it came to be known that what we experience as light is in truth a very small part of a vast continuous spectrum of radiation, the electromagnetic spectrum. As Maxwell showed, all the radiations included in this spectrum travel in waves. The wavelengths—that is, the distances from one peak to the next—range from those of gamma rays, which are measured in angstroms, to those of low-frequency radio waves, which are measured in kilometers. Within the spectrum of visible light, red light has the longest wavelength, violet the shortest. Another feature that these radiations have in common is that, in a vacuum, they all travel at the same speed—300,000 kilometers per second.

By 1900, it had become clear, however, that the wave theory of light was not adequate. The key observation, a very simple one, was made in 1888: When a zinc plate is exposed to ultraviolet light, it acquires a positive charge. The metal, it was soon deduced, becomes positively charged because the light energy dislodges the electrons, forcing them out of the metal atoms. Subsequently, it was discovered that this photoelectric effect, as it is called, can be produced in all metals. Every metal has a critical wavelength for the effect; the light (visible or invisible) must be of that wavelength or a shorter wavelength for the effect to occur.

With some metals, such as sodium, potassium, and selenium, the critical wavelength is within the spectrum of visible light, and as a consequence, visible light striking the metal can set up a moving stream of electrons (such a stream is an electric current). The electric eyes that open doors for you at supermarkets or airlines terminals, burglar alarms, exposure meters, and television cameras all operate on this principle of turning light energy into electrical energy.

Wave or Particle?

Now here is the problem. The wave theory of light would lead you to predict that the brighter the light—that is, the stronger the beam—the greater the force with which the electrons would be dislodged. But as we have already seen, whether or not light can eject the electrons of a particular metal depends not on the brightness of the light but on its wavelength. A very weak beam of the critical wavelength or a shorter wavelength is effective, while a stronger beam of a longer wavelength is not. Furthermore, as was shown in 1902, increasing the brightness of the light increases the number of electrons dislodged but not the velocity at which they are ejected from the metal. To increase the velocity, one must use a shorter wavelength of light. Nor is it necessary for energy to be accumulated in the metal. With even a dim beam of a critical wavelength, an electron may be emitted the instant the light hits the metal.

To explain such phenomena, the particle theory of light was proposed by Albert Einstein in 1905. According to this theory, light is composed of particles of energy called *photons*. The energy of a photon is not the same for all kinds of light but is, in fact, inversely proportional to the wavelength—the longer the wavelength, the lower the energy. Photons of violet light, for example, have almost twice the energy of photons of red light, the longest visible wavelength.

The wave theory of light permits physicists to describe certain aspects of its behavior mathematically, and the photon theory permits another set of mathematical calculations and predictions. These two theories are no longer regarded as opposed to one another; rather, they are complementary, in the sense that both are required for a complete description of the phenomenon we know as light.

The Fitness of Light

Light, as Maxwell showed, is only a tiny band in a continuous spectrum. From the physicist's point of view, the difference between radiations we can see and radiations we cannot see—so dramatic to the human eye—is only a few nanometers of wavelength (Figure 5–6). Why does this particular small group of radiations, rather than some other, bathe our world in radiance, make the leaves grow and the flowers burst forth, cause the mating of fireflies and palolo worms, and when reflecting off the surface of the moon, excite the imagination of poets and lovers? Why is it that this tiny portion of the electromagnetic spectrum is responsible for vision, for the rhythmic, day-night regulation of many biological activities, for the bending of plants toward the light, and also for photosynthesis, on which all life depends? Is it an amazing coincidence that all these biological activities are dependent on these same wavelengths?

George Wald of Harvard, one of the greatest living experts on the subject of light and life, says no. He thinks that if life exists elsewhere in the universe, it is probably dependent on this same fragment of the vast spectrum. Wald bases this conjecture on two points. First, living things, as we have seen, are composed of large, complicated molecules held in special configurations and relationships to one another by hydrogen bonds and other weak bonds. Radiation of even slightly higher energies than the energy of violet light breaks these bonds and so disrupts the structure and function of the molecules. Radiations with wavelengths less than 200 nanometers drive electrons out of atoms. Light

5–6

Visible light is only a small portion of the vast electromagnetic spectrum. For the human eye, the visible spectrum ranges from violet light, which is made up of comparatively short rays, to red light, the longest visible rays.

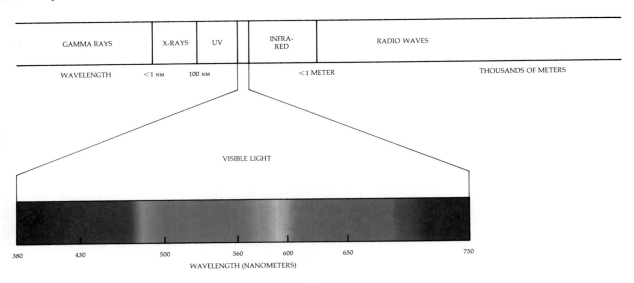

Chlorophyll a is a large molecule with a central core of magnesium held in a nitrogen-containing ring (a porphyrin ring). Attached to the ring is a long, insoluble carbon-hydrogen chain which serves to anchor the molecule in the internal membranes of the chloroplast. Chlorophyll b differs from chlorophyll a in having a CHO group in place of the encircled CH₃ group.

CHLOROPHYLL *a*

of wavelengths longer than those of the visible band is absorbed by water, which makes up the great bulk of all living things on earth. When it does reach molecules, its lower energies cause them to increase their motion (increasing heat) but do not trigger changes in their structure. Only those radiations within the range of visible light have the property of exciting molecules—that is, of raising electrons from one energy level to another—and so of producing biological changes.

The second reason for the visible band of the electromagnetic spectrum being "chosen" by living things is that it, above all, is what is available. Most of the radiation reaching the earth from the sun is within this range. Higher-energy wavelengths are screened out by the oxygen and ozone high in the atmosphere. Much infrared radiation is screened out by water vapor and carbon dioxide before it reaches the earth's surface.

This is an example of what has been termed "the fitness of the environment"; the suitability of the environment for life and that of life for the physical world are exquisitely interrelated. If they were not, life could not, of course, exist.

THE ROLE OF PIGMENTS

In order for light energy to be used by living systems it must first be absorbed. Pigments are substances that absorb light energy. They are always easy to recognize because they are colored. They have to be. A black pigment is black because it absorbs light of all wavelengths. Red pigment is red because it absorbs all wavelengths *except red;* that is, it reflects or transmits red. A red pigment viewed under blue light will appear black because it has nothing to reflect or transmit. The leaves of plants are green because chlorophyll molecules absorb violet, blue, and red light.

Chlorophyll *a,* the key compound in photosynthesis, contains a porphyrin ring (Figure 5–7). (A porphyrin ring is composed of four nitrogen-containing rings and their connecting carbon atoms.) In the center of the ring is an atom of magnesium. Attached to the porphyrin part of the molecule is a long hydrocarbon "tail." Chlorophyll *b* differs from chlorophyll *a* only in having a CHO group in the position at which chlorophyll *a* has a CH₃. In all photosynthetic cells (except the photosynthetic bacteria), chlorophyll *a* plays the primary role. Chlorophyll *b,* an accessory pigment, plays a secondary role, transferring energy to chlorophyll *a.*

PHOTOSYNTHESIS AND RESPIRATION

In photosynthesis, electrons are boosted uphill. The energy trapped in this reaction is then used to reduce carbon dioxide to building blocks from which sugars and other carbohydrates can be formed:

$$CO_2 + H_2O + energy \longrightarrow carbohydrate + O_2$$

This light-trapping process, as we noted previously, takes place in photosynthetic eukaryotes in specialized organelles known as chloroplasts.

The reduced carbon compounds, the carbohydrates, produced in the chloroplast may be used by the cell in which they are produced or they may be transported, usually in the form of sucrose, to other, nonphotosynthetic cells of the same plant. If the plant cells are eaten by a heterotroph, the sugars then become its source of chemical energy and if that heterotroph is, in turn, devoured, the stored chemical energy is passed along to that organism. Following

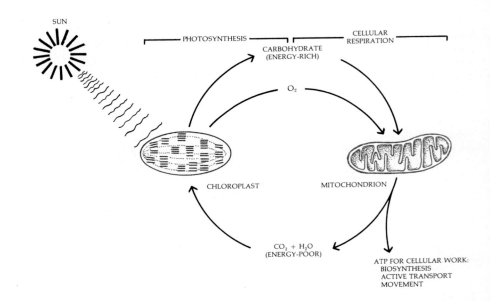

5–8
The flow of biological energy. The radiant energy of sunlight is produced by the fusion of hydrogen atoms to form helium. Chloroplasts, present in all photosynthetic eukaryotic cells, capture this energy and use it to convert water and carbon dioxide into carbohydrates, such as glucose, starch, and other foodstuff molecules. Oxygen is released into the air as a product of the photosynthetic reactions. Mitochondria, present in all eukaryotic cells, break down these carbohydrates and capture their stored energy in ATP molecules. This process, cellular respiration, consumes oxygen and produces carbon dioxide and water, completing the cycle.

the second law of thermodynamics, energy is dissipated in each of these transactions.

When the reduced carbon compounds formed by photosynthesis reach an energy-requiring cell, they are oxidized and their energy is released in the reaction known as cellular respiration:

$$\text{Carbohydrate} + O_2 \longrightarrow CO_2 + H_2O + \text{energy}$$

As you can readily see, this equation is exactly the reverse of the one that occurs in photosynthesis (Figure 5–8).

Similarly, when wood or fossil fuels are burned, carbon is oxidized and energy is released. However, in this case, the stored energy of sunlight is converted almost exclusively to heat energy. By contrast, when carbohydrates are broken down within a cell, a substantial proportion of the energy is recovered in a form in which it can do work for the cell. The molecule in which this energy is captured is known as adenosine triphosphate, abbreviated ATP.

ATP

Sugars and their derivatives are a storage form of energy and also a form in which energy is transferred from cell to cell and organism to organism. In that sense, they are like money in the bank. Before you can spend banked money to buy a hotdog or ride a bus, you have to cash a check and convert your "stored" money to another form, such as dollar bills or small change. ATP is equivalent to this immediately spendable form of money. Almost every energy-requiring transaction in the cell involves ATP, just as, alas, a large number of our daily transactions involve dollars and cents. For this reason, ATP is often referred to as the "universal currency" of the cell.

How does ATP perform this function? Again, to understand function we must look at structure. ATP is composed of three types of subunits. One of these is a compound known as adenine, which is categorized as a nitrogen base:

$$NH_2$$

ADENINE

Adenine is a compound we shall be mentioning frequently in Section 3 because in addition to its role as part of the ATP molecule, it is one of the principal components of the genetic material. (The use of adenine for these two quite different purposes is another example of the economy with which the cell operates.)

The second subunit of ATP is a five-carbon sugar, ribose, similar in basic design to the sugars described in Chapter 3:

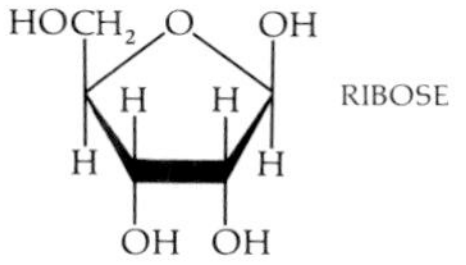

RIBOSE

The third subunit is phosphate, which is an atom of phosphorus combined with four oxygen atoms. Adenine plus the sugar plus one phosphate group make up adenosine monophosphate, AMP, a compound known as a *nucleotide*. Nucleotides are one of the basic and important many-purpose chemical combinations in the cell and will be discussed at greater length in Chapter 14, where the chemical structure of the genetic material is described.

In ATP, as the name implies, there are three phosphates:

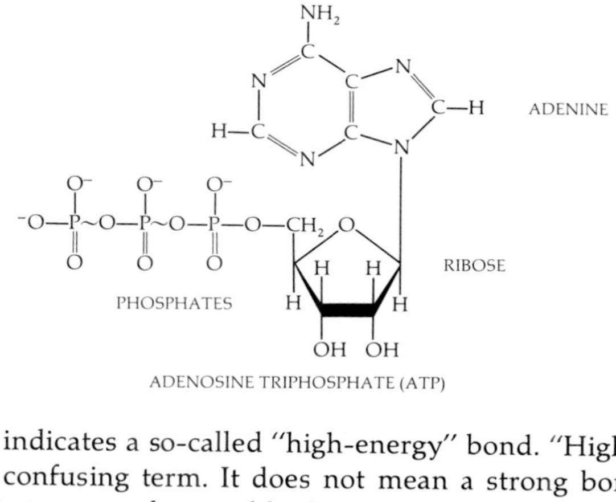

ADENINE

RIBOSE

PHOSPHATES

ADENOSINE TRIPHOSPHATE (ATP)

The symbol $\sim$ indicates a so-called "high-energy" bond. "High-energy bond" is a somewhat confusing term. It does not mean a strong bond, such as the covalent bond between carbon and hydrogen. A high-energy bond is one that yields its energy readily.

One of the principal characteristics of phosphate groups—and the reason for their playing an important role in chemical exchanges—is their capacity to form high-energy bonds. In ATP, two such bonds link the three phosphates together. Usually, in the course of cellular transactions, only one bond is broken. With the loss of one phosphate group, the ATP molecule becomes ADP, adenosine diphosphate, and about 7,000 calories of energy are released (per mole). The ADP molecule is then "recharged" with an input of 7,000 calories—regaining the third phosphate group and again becoming ATP—with the energy obtained from the oxidation of sugar or some other organic compound.

In eukaryotic cells, the transfer of energy from sugars and other carbohydrates to ATP takes place in a series of reactions that begin in the cytoplasm and are completed in the mitochondrion.

An Example of ATP Action

Let us look at a simple example of an energy exchange involving ATP. As we mentioned previously, sucrose is a disaccharide formed from the monosaccharides glucose and fructose. The formation of sucrose is an energy-requiring (endergonic) reaction. The energy for the reaction can be supplied by coupling the synthesis of sucrose to the removal of a phosphate group from the ATP molecule.

First, the terminal phosphate group of ATP is transferred to the glucose molecule.

$$ATP + glucose \longrightarrow glucose\ phosphate + ADP$$

In this reaction, some of the 7,000 calories available from the hydrolysis of ATP to ADP are conserved by the transfer of the phosphate group to the glucose molecule, which thus becomes "energized."

Next glucose phosphate reacts with fructose to form sucrose:

$$Glucose\ phosphate + fructose \longrightarrow sucrose + phosphate$$

In this second step, the phosphate group is released from the glucose and most of the energy made available by its release (energy originally derived from the ATP) is used to form the bond between glucose and fructose, producing sucrose. The free phosphate is then available, with an input of energy, to recharge an ADP molecule to ATP.

As we noted in Chapter 3, the formation of sucrose requires 5,500 calories per mole. The conversion of ATP to ADP + phosphate releases 7,000 calories per mole. Thus most of the energy in the phosphate bond has been effectively utilized to carry out work of the cell.

In the two chapters that follow, we shall present the processes of respiration and photosynthesis in greater detail. In the course of these presentations, we shall see how the life of the cell—so complex in its totality—is made up of series of small steps such as these, each carefully designed to make use, with great efficiency, of the skillfully stored energy of the sun.

SUMMARY

Living systems convert energy from one form to another in order to carry out the functions essential to their maintenance, growth, and reproduction. The sun is the original source of this energy.

The laws of thermodynamics govern exchanges of energy in biological systems. The first law states that energy can be transformed from one form to another but cannot be created or destroyed. The second law of thermodynamics states that energy in a form to do work is eventually converted to heat energy, which is dissipated out into the surroundings. In other words, the supply of useful energy is constantly diminishing and must be replaced.

The exchanges of energy in living cells usually involve the movement of electrons from one energy level to another. Such movements are known as oxidation-reduction reactions. An atom or molecule that loses electrons is oxidized; one that gains electrons is reduced.

Light energy is captured by the living world by means of pigments. The pigment involved in photosynthesis is chlorophyll. In photosynthetic eukaryotic cells, it is found in the internal membranes of the chloroplast.

In photosynthesis, the energy of light is used to reduce carbon, forming sugars and other carbohydrates from carbon dioxide and water:

$$CO_2 + H_2O + energy \longrightarrow carbohydrate + O_2$$

In cellular respiration, reduced carbon compounds are oxidized to carbon dioxide and water in a process that begins in the cytoplasm and is completed in the mitochondria:

$$Carbohydrate + O_2 \longrightarrow CO_2 + H_2O + energy$$

As the carbon compounds are broken down, a portion of their chemical energy is recovered in the form of ATP, which then is used to carry out the work of the cell.

QUESTIONS

1. What are the first and second laws of thermodynamics? Give examples of each.

2. In what way is the second law of thermodynamics relevant to present-day environmental problems?

3. Explain why the decrease in entropy that occurs in and is produced by living creatures (such as people building houses) does not violate the second law of thermodynamics.

CHAPTER 6

Photosynthesis: Harvesting the Sun

6-1

A cluster of cells of a green alga that lives in fresh water. Each cell is an individual, self-sufficient photosynthetic organism. The green color is due to chlorophyll, which, in these eukaryotic algae, is embedded in chloroplasts.

Photosynthesis, in eukaryotes, takes place only within highly specialized organelles, the chloroplasts. A single cell of a leaf may contain 40 to 50 chloroplasts, and a square millimeter, some 500,000. Chloroplasts may move within the leaf cell, orienting their surfaces, which are often lens-shaped, so that they catch the light.

Chloroplasts, as we noted previously, are surrounded by two membranes. Within the double membrane is a watery material, called the *stroma*, which is different in composition from the watery material that surrounds the organelles in the cytoplasm. In the light microscope under high power, it is possible to see little spots of green within the chloroplasts. These were called *grana* (grains) by the earlier microscopists, and this term is still in use. Under the electron microscope, however, it can be seen that the grana are part of an elaborate system of membranes called the *lamellae*. These membranes are organized in parallel pairs. The pairs of membranes are joined at each end, forming a disk-shaped structure, or *thylakoid* (from *thylakos,* the Greek word for "sac"). The grana are stacks of thylakoids.

Chlorophyll molecules are embedded in the lamellae. The exact way in which they are packaged is not known, but it is clearly important; isolated from its membranes, chlorophyll cannot carry out photosynthesis.

6-2

A chloroplast, in three dimensions. The light reactions take place in the internal membranes and the dark reactions in the stroma.

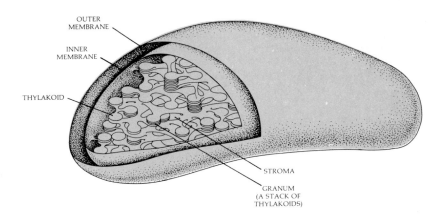

OUTER MEMBRANE

INNER MEMBRANE

THYLAKOID

STROMA

GRANUM (A STACK OF THYLAKOIDS)

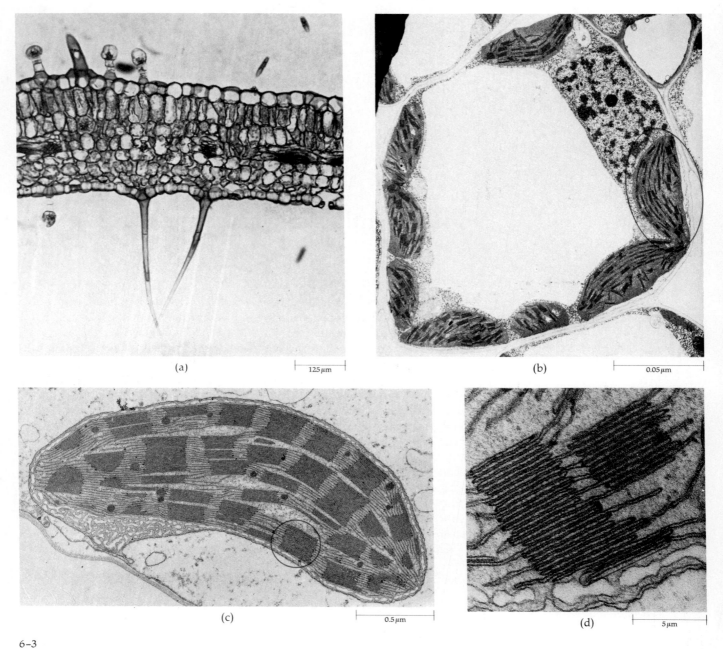

(a)   125 μm

(b)   0.05 μm

(c)   0.5 μm

(d)   5 μm

6–3

Journey into a chloroplast. (a) Photomicrograph of a leaf. The circle indicates one of the many photosynthetic cells. (b) Electron micrograph of a photosynthetic cell. The circle indicates a chloroplast. (Note that the chloroplasts are all near the surface of the cell. The center of the cell is filled with a large vacuole.) (c) A chloroplast. Chloroplasts, like mitochondria, are surrounded by a double membrane and have a system of internal membranes. The membranes are arranged in stacks, called grana. The circle indicates one of the many grana. (d) At still higher magnification, the grana can be seen to consist of flattened sacs, the thylakoids, composed of membranes. These membranes contain the chlorophylls and other pigments involved in photosynthesis.

PIGMENT SYSTEMS

The combination of chlorophyll and the other compounds involved in photosynthesis is known as a *pigment system*. In plants, pigment systems contain chlorophyll *a* and, in addition, chlorophyll *b* and carotene. Also packed in these pigment systems are the electron carriers, enzymes, and other molecules associated with the capture of light energy.

As we observed in the preceding chapter, pigments are compounds that absorb light. Even slight differences in the molecular structure can affect the pattern of light absorption, or *absorption spectrum*. Figure 6–4b, for example, shows the absorption spectra of chlorophylls *a* and *b*, which, as we noted previously, have very similar structures. Carotenoids (carotene) and carotene-like products, which are also found in chloroplasts, have different absorption spectra, as you can also see in Figure 6–4b. The combination of pigments makes it possible for the chloroplast to absorb light waves of a broader spectrum than could be absorbed by one type of molecule alone.

In the leaves of plants, there are two different kinds of pigment systems, known simply as Pigment System I and Pigment System II. Pigment System I contains a larger proportion of chlorophyll *a* to chlorophyll *b* than Pigment System II and also carotenoids of different types.

Each pigment system contains one reactive molecule, believed to be chlorophyll *a* packed in a special way in the chloroplast. Only this molecule can perform the crucial step of photosynthesis—the conversion of light energy to chemical energy.

In photosynthesis, electrons are ejected from the chlorophyll molecule, are captured and held at a high energy level, and are transferred downhill to a low energy level. To understand this process, we have to look at two types of energy-acceptor molecules. One of these is exemplified by NADP (nicotinamide adenine dinucleotide phosphate). NADP is composed of several subunits, including adenine (also present in ATP; see page 85).

The portion of the NADP molecule that accepts and releases electrons is shown in Figure 6–5. NADP in the reduced form (with two additional electrons) is designated $NADP_{red}$; NADP in the oxidized form (with the electrons removed) is $NADP_{ox}$.

The cytochromes are another type of electron-acceptor molecule (see Figure 6–6). Cytochromes contain an atom of iron, in a porphyrin ring similar to the porphyrin ring holding the magnesium atom in chlorophyll (see Figure 5–7).

6–4
(a) *The absorption spectrum of a pigment is measured by a spectrophotometer. This device directs a beam of light of each wavelength (at this moment, red) at the object to be analyzed and records what percentage of light of each wavelength is absorbed by the pigment sample as compared to a reference sample. Because the first mirror is lightly (half) silvered, half of the light is reflected and half is transmitted. The photoelectric cell is connected to an electronic device that automatically records the percentage absorbed at each wavelength.*
(b) *The absorption spectra of chlorophyll a, chlorophyll b, and the carotenoids found in chloroplasts of plants.*

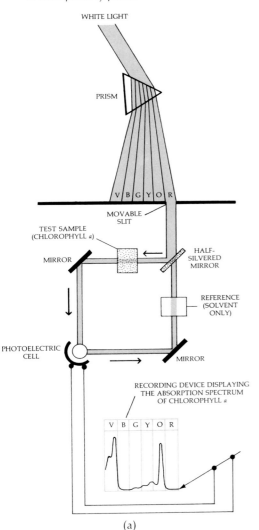

(a)

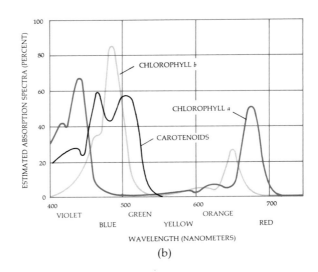

(b)

6-5

Oxidized (NADP_ox) and reduced (NADP_red) forms of the reactive portions of NADP, an electron acceptor. The colored dots indicate the two electrons added to NADP_ox, which then becomes NADP_red. The rest of the molecule, which consists of nucleotides of adenine and phosphates, does not change.

OXIDIZED FORM OF
NADP ($NADP_{ox}$)

REDUCED FORM OF
NADP ($NADP_{red}$)

6-6

Cytochromes are molecules in which an atom of iron (Fe) is held in a nitrogen-containing ring (a porphyrin ring). Cytochromes are involved in electron transfer. It is the iron within the molecule that actually combines with the electrons.

CYTOCHROME

The porphyrin ring is surrounded by a protein containing 100 or more amino acids. Each cytochrome differs in its protein chain and also in the energy level at which it holds the electrons. They thus work in sequence somewhat like a series of waterwheels. Water, as it flows from a point of higher water potential to one of lower water potential, releases energy which can be harnessed by waterwheels to do work. As electrons flow along the electron transport chain from a higher to a lower energy level, the cytochromes harness some of the released energy and, as we shall see, use some of it to make ATP from ADP and phosphate. Such a series of cytochromes, working in a sequence, is called an *electron transport chain.* Electron transport chains are present both in chloroplasts and in mitochondria.

PHOTOSYNTHESIS: LIGHT AND DARK REACTIONS

The reactions of photosynthesis, it is now known, take place in two stages. In the first stage, light energy is used to form ATP from ADP and to reduce (transfer electrons to) electron carrier molecules. These reactions require light and so are known as the *light reactions.* In the second stage of photosynthesis, the energy products of the first stage are used to reduce carbon from carbon dioxide to a simple sugar. Thus, in this second stage, the chemical energy of the electron carrier molecules is converted to forms suitable for transport and storage. At the same time, a carbon skeleton is formed on which other organic molecules can be built. The reactions of this second stage are known as the *dark reactions* because they do not require light. (However, they do not require darkness either; in most plants—except in the course of laboratory experiments—so-called "dark reactions" take place in the daytime and in the light.)

MODEL OF LIGHT REACTIONS

Figure 6-7 shows the current model of how the two systems work together in photosynthesis. Light energy first enters Pigment System II, where it is responsible for two important events: (1) electrons are boosted uphill from a reactive

92   Cellular Energetics

A summary of the light reactions of photosynthesis. Light energy trapped in the reactive molecule of Pigment System II boosts electrons uphill from chlorophyll to a primary electron acceptor. These electrons are replaced by electrons pulled away from water molecules, which then become protons (hydrogen atoms) and oxygen gas. The electrons are passed from the primary electron acceptor along an electron transport chain to a lower energy level, the reactive molecule of Pigment System I. As the electrons pass along the electron transport chain, some of their energy is packaged in the form of ATP. Light energy absorbed by Pigment System I boosts electrons to another primary electron acceptor. From this acceptor, they are passed via other electron carriers to $NADP_{ox}$ to form $NADP_{red}$. The electrons removed from Pigment System I are replaced by those from Pigment System II. ATP and $NADP_{red}$ represent the net gain from the light reactions.

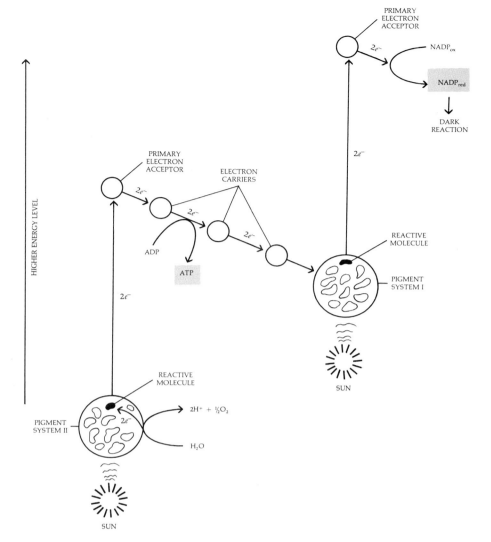

molecule, and (2) water is split into protons, electrons, and oxygen gas. As the electrons are removed, they are passed uphill at increasingly higher energy levels to a primary electron acceptor. (The electron acceptors in Figure 6–7 are indicated by circles.) The electrons then pass downhill along an electron transport chain to Pigment System I. In the course of this passage, ATP is formed from ADP.

In Pigment System I, light energy again boosts electrons from the reactive molecule to a primary electron acceptor. From this electron acceptor, they are passed downhill to a molecule of NADP.

The energy recovered from these two steps of photosynthesis is represented by the ATP molecule and the NADP $_{red}$. These molecules are the source of the chemical energy used to reduce carbon in the dark reactions.

Why are there two systems rather than one? The answer seems to lie in the problem of the electron "holes." Suppose for a moment that System I were isolated from System II. There would be no way to fill the electron "hole" left when the electron is boosted out of orbit around the reactive molecule in System I. This function is now carried out by the electrons boosted up from System II. And what fills the holes in System II? They are filled by electrons from the water molecules. When the electrons pass from the water molecule to the chlorophyll molecule (which has a greater electron affinity), the water molecule falls apart, releasing oxygen gas.

Thus all the pieces fit together.

Summary of Light Reactions

The reactions that we have just described are the "light reactions" of photosynthesis. In the course of these reactions, as we saw, light energy is converted to electrical energy—the flow of electrons—and the electrical energy is converted to chemical energy stored in the bonds of $NADP_{red}$ and ATP. In the words of Albert Szent-Györgyi: "What drives life is . . . a little electric current, kept up by the sunshine."

THE DARK REACTIONS

In the second stage of photosynthesis—the "dark reactions"—the energy generated by the light reactions is used to fix carbon, that is, to incorporate carbon into organic molecules. These reactions take place in the stroma of the chloroplasts.

The dark reactions involve a chemical cycle. Cycles are important in the chemistry of the cell because they permit the cell to use the same metabolic machinery over and over again. Each time the cycle is completed, the starting product is regenerated and so can be used again. (This is another example of the economy with which the cell operates.) The dark reactions of photosynthesis convert the chemical energy produced by the light reactions—the ATP and $NADP_{red}$ molecules—to forms more suitable for storage and transport. Moreover, they build the basic carbon structures from which all of the other organic molecules of living systems are produced.

The Calvin Cycle

The cycle of the dark reactions is named the Calvin cycle after its discoverer, Melvin Calvin of the University of California, Berkeley. In this cycle, the starting (and ending) compound is a five-carbon sugar with two phosphates attached, ribulose diphosphate (RuDP). Its structure is shown in Figure 6–8. The cycle begins when carbon dioxide enters the cycle and is incorporated into RuDP. The RuDP then immediately splits to form two molecules of phosphoglycerate (PGA). This reaction is catalyzed by a specific enzyme, RuDP carboxylase. Each step of the cycle is similarly regulated by a specific enzyme.

The complete cycle is diagramed in Figure 6–9. At each full turn of the cycle, a molecule of carbon dioxide enters the cycle, is reduced, and a molecule of RuDP is regenerated. Three revolutions of the cycle, with the introduction of

6–8

Calvin and his collaborators briefly exposed photosynthesizing algae to radioactive carbon dioxide ($^{14}CO_2$). They found that the radioactive carbon is first incorporated into ribulose diphosphate (RuDP), which then immediately splits to form two molecules of phosphoglycerate (PGA). The radioactive carbon atom, indicated in color, appears in one of the two molecules of PGA. This is the first step in the Calvin cycle.

Summary of the Calvin cycle. Three molecules of ribulose diphosphate (RuDP), a five-carbon compound, are combined with three molecules of carbon dioxide, yielding six molecules of phosphoglycerate, a three-carbon compound. These are converted to six molecules of glyceraldehyde phosphate. Five of these three-carbon molecules are combined and rearranged to form three five-carbon molecules of RuDP. The "extra" molecule of glyceraldehyde phosphate represents the net gain from the Calvin cycle. The energy that "drives" the Calvin cycle is in the form of ATP and NADP$_{red}$ produced by the light reactions.

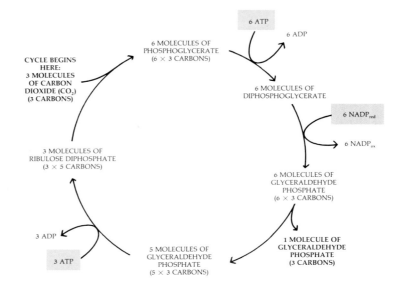

6-10

Glyceraldehyde phosphate molecule. This molecule is the direct product of the Calvin cycle. Two such molecules can combine to form glucose molecules by steps which are diagramed on page 100.

$$
\begin{array}{l}
CH_2-O-\textcircled{P} \\
| \\
CHOH \\
| \\
C=O \\
| \\
H
\end{array}
$$

GLYCERALDEHYDE
PHOSPHATE

three atoms of carbon, are necessary to produce one molecule of glyceraldehyde phosphate (Figure 6–10). The overall equation is

$$3RuDP + 3CO_2 + 9ATP + 6NADP_{red} \longrightarrow$$
$$3RuDP + \text{glyceraldehyde phosphate} + 9P_i + 9ADP + 6NADP_{ox}$$

Glyceraldehyde phosphate can then serve as the starting product for glucose, sucrose, cellulose, amino acids, or other compounds needed by the cell.

Four-Carbon Photosynthesis

It has recently been discovered that RuDP is not the only pathway in the dark reactions. Some plants first fix carbon dioxide in a four-carbon acid known as oxaloacetic acid; the enzyme involved in this reaction is PEP carboxylase (Figure 6–11. Such plants are known as C_4, or four-carbon plants, to distinguish them from the previously discussed plants, which are known as C_3, or three-carbon plants (because in these plants carbon is fixed first in the three-carbon compound PGA).

However, in C_4 plants, the carbon also enters the Calvin cycle. Once they are formed, the oxaloacetic acid molecules are oxidized and carbon dioxide is released. This carbon dioxide is then transferred to the RuDP of the Calvin cycle.

Why do C_4 plants employ such a seemingly cumbersome method of providing carbon dioxide to the Calvin cycle? The most important clue seems to be that C_4 plants evolved in the tropics. All are especially well adapted to high extremes of light, temperature, and dryness. Under such conditions, there is competition for carbon dioxide, and it has been demonstrated that C_4 plants use carbon dioxide more efficiently than do C_3 plants. This is due to the fact that the enzyme PEP carboxylase—the crucial enzyme in C_4 reactions—has a much greater affinity for carbon dioxide at low concentrations than the enzyme RuDP carboxylase of the Calvin cycle. The C_4 plant's two-cycle system of photosynthesis allows the plant to absorb carbon dioxide efficiently even at low concentrations and then to feed it into the Calvin cycle.

In C_4 plants, carbon dioxide is combined
with a compound known as PEP to yield the
four-carbon oxaloacetic acid. The reaction is
catalyzed by the enzyme PEP carboxylase.
The oxaloacetic acid is then broken down
again to CO_2, which enters the Calvin cycle.

$$\underset{\text{PEP}}{\begin{array}{c}\text{COOH} \\ | \\ \text{COPO}_3^{2-} \\ \| \\ \text{CH}_2 \end{array}} \xrightarrow[\text{CARBOXYLASE}]{\underset{\text{PEP}}{\text{CO}_2 + \text{H}_2\text{O}}} \underset{\begin{array}{c}\text{OXALOACETIC} \\ \text{ACID}\end{array}}{\begin{array}{c}\text{COOH} \\ | \\ \text{C=O} \\ | \\ \text{CH}_2 \\ | \\ \text{COOH} \end{array}}$$

Another advantage C_4 plants have over C_3 plants is that the C_4 photosynthesis can be carried out efficiently at much higher temperatures than C_3 photosynthesis. Consequently, C_4 plants flourish even at temperatures that are lethal to many C_3 species.

The list of plants known to utilize the four-carbon pathway has grown to over 100 genera, at least a dozen of which have both C_3 and C_4 species. This pathway undoubtedly has arisen many times independently in the course of evolution. Sugarcane, corn, and sorghum are among the best-known C_4 plants.

A familiar example of the competitive ability of C_4 plants is found in our lawns, which, in the cooler parts of the country at least, consist mainly of C_3 grasses such as Kentucky bluegrass. Crabgrass, which all too often overwhelms these dark green, fine-leaved grasses with patches of its yellowish green, broader leaves in summer, is a C_4 grass and so grows much more rapidly than the temperate C_3 grasses at high temperatures.

SUMMARY

Photosynthesis—the conversion of light energy to chemical energy—takes place within cellular organelles known as chloroplasts. These organelles are surrounded by two membranes. Inside the second membrane is a solution of organic compounds and ions known as the stroma, and a complex internal membrane system consisting of fused pairs of membranes that form sacs called thylakoids. The pigments and other molecules responsible for capturing light are located in and on these membranes.

Photosynthesis takes place in two stages: (1) the light reactions, in which light energy is captured by chlorophyll and converted to the chemical energy of ATP and $NADP_{red}$; and (2) the dark reactions, in which carbon atoms are reduced and carbohydrates formed.

Pigments are compounds that absorb light energy. The pigments involved in the light reactions include the chlorophylls and the carotenoids. Light absorbed by pigments boosts their electrons to a higher energy level. Because of the way the pigments are packed into the membranes, they are able to transfer this energy to reactive molecules, probably chlorophyll a molecules packed in a particular way. Two pigment systems have been identified in thylakoids.

The light reactions also involve electron acceptor molecules, such as NADP and cytochromes. Electrons pass from higher to lower energy levels as they pass from one acceptor molecule to another.

In the currently accepted model of the light reactions in photosynthesis, light energy first strikes Pigment System II, which contains several hundred molecules of chlorophyll a and chlorophyll b. Electrons are passed uphill to an electron acceptor from the reactive molecule. The electron holes are filled by electrons from molecules of water, which have split into protons and oxygen gas. The electrons then pass downhill to Pigment System I along an electron transport chain, in the course of which ATP is generated. Light absorbed in

Pigment System I results in the ejection of an electron from the reactive molecule of chlorophyll *a*. The resulting electron holes left in the reactive molecule of Pigment System I are filled by the electrons from Pigment System II. The electrons are accepted by the electron carrier molecule, NADP. The energy yield from the light reactions is represented by molecules of $NADP_{red}$ and ATP.

In the dark reactions, which take place in the stroma, the $NADP_{red}$ and ATP produced in the light reactions are used to reduce carbon dioxide to organic carbon. This is accomplished by means of the Calvin cycle. In the Calvin cycle, a molecule of carbon dioxide is combined with the starting material, a five-carbon sugar called ribulose diphosphate. At each turn of the cycle, one carbon atom enters the cycle. Three turns of the cycle produce a three-carbon molecule, glyceraldehyde phosphate. Two molecules of gylceraldehyde phosphate (six turns of the cycle) can combine to form a glucose molecule. At each turn of the cycle, RuDP is regenerated. The glyceraldehyde phosphate can also be used as starting material for other organic compounds needed by the cell. By this process, a flow of energy is created from the sun through living systems.

In C_4 plants, carbon dioxide is initially accepted by a compound known as PEP to yield the four-carbon oxaloacetic acid. The oxaloacetic acid is then oxidized, and carbon dioxide is tranferred to the RuDP of the Calvin cycle. At high temperatures and low CO_2 concentrations, C_4 plants are more efficient than C_3 plants.

QUESTIONS

1. Define the following terms: absorption spectrum, light reactions, dark reactions, and Calvin cycle.

2. Sketch a chloroplast and label its structures. Compare your sketch with the diagram on page 89.

3. Predict what colors of light might be most effective at stimulating plant growth. (Such is the principle of the special light bulbs used for plants.)

CHAPTER 7

Glycolysis and Respiration:
Fuel for the Living Cell

The oxidation of glucose (or other carbohydrates) is the process by which the energy of carbon-containing compounds—the products of photosynthesis—is transferred to ATP and so made available for the immediate energy requirements of cells. It is complicated in detail but simple in its overall design. It begins conventionally with the glucose molecule, the form in which organic carbon compounds usually reach the animal cell. This molecule is split, and the hydrogen atoms (that is, protons and electrons) are removed from the carbon atoms and combined with oxygen. The overall equation for this oxidation of glucose is:

Glucose + oxygen $\longrightarrow$ carbon dioxide + water + energy

or $C_6H_{12}O_6 + 6O_2 \longrightarrow 6CO_2 + 6H_2O + 686$ kilocalories

The energy released by this process is used to convert ADP to ATP.

7-1

Energy provided by oxidation of glucose and other organic compounds moves these dolphins through the water and the birds above them through the air. It also powers the many chemical activities taking place in their tissues and provides the heat that keeps their body temperatures well above that of the air and water around them.

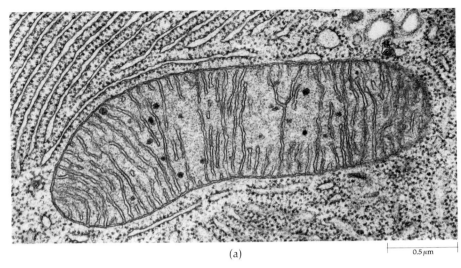

(a)

0.5 μm

7-2

(a) *Mitochondrion of a pancreatic cell. Mitochondria are typically found near the cellular structures that are performing the most work, in this case, rough endoplasmic reticulum.* (b) *The mitochondrion in three dimensions. The organelle is surrounded by two membranes. The inner membrane folds inward to make a series of shelves, or cristae. The enzymes and electron carriers involved in the final stage of cellular respiration are built into these internal membranes.*

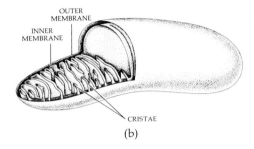

OUTER MEMBRANE

INNER MEMBRANE

CRISTAE

(b)

The breakdown of glucose takes place in two stages. The first is known as *glycolysis*. The second stage is *respiration*. Glycolysis takes place in the cytoplasm. In eukaryotic cells, respiration, which is the major source of ATP energy, takes place in the mitochondria (Figure 7–2).

As we mentioned previously, mitochondria are surrounded by two membranes. The outer one is smooth, and the inner one folds inward. The folds are called cristae. The more active a cell, the more numerous are both its mitochondria and the cristae within them. Within the inner compartment, surrounding the cristae, is a dense solution containing enzymes, water, phosphates, and other molecules involved in respiration. The mitochondrion is a self-contained chemical plant. The outer membrane lets most small molecules in or out freely, but the inner one permits the passage only of certain molecules and restrains the passage of others. Enzymes, cytochromes, and other compounds essential to the oxidation process are built into the internal membranes.

GLYCOLYSIS

Glycolysis is the first stage in the oxidation of glucose. In glycolysis, the six-carbon glucose molecule is split into two three-carbon molecules of a compound known as pyruvic acid (Figure 7–3). This is carried out in a series of reactions that release about 143 kilocalories. The next two pages describe the nine steps by which one glucose molecule is broken down (oxidized) into two pyruvic acid molecules.

7-3

In glycolysis, the six-carbon glucose molecule is split into two three-carbon molecules of a compound known as pyruvic acid.

GLUCOSE ⟶ 2 PYRUVIC ACID

Do not try to memorize these steps, but follow them closely. Notice how the carbon skeleton of the molecule is dismembered, step by step. Note especially the formation of ATP from ADP and of NAD_{red} from NAD_{ox}. (NAD, nicotinamide adenine dinucleotide, is an electron carrier that closely resembles NADP, differing from it by only one phosphate group.) ATP and NAD_{red} represent the cell's net gain from this energy transaction.

Step 1. The first steps in glycolysis require an input of energy. This activation energy is supplied by the hydrolysis of ATP to ADP. The terminal phosphate group is tranferred from an ATP molecule to the glucose molecule, to make glucose 6-phosphate. (This is also the first step, you may recall, in the biosynthesis of sucrose, page 86.) The combining of ATP with glucose to produce glucose 6-phosphate and ADP is an energy-yielding reaction, as we saw previously. Some of the energy released from the ATP is conserved in the chemical bond linking the phosphate to the sugar molecule. This reaction is catalyzed by a specific enzyme (hexokinase), and each of the reactions that follows is similarly regulated by a specific enzyme. (As we noted in Chapter 3, the names of enzymes characteristically end in "-ase.")

Step 2. The molecule is reorganized, again with the help of a particular enzyme. The six-sided ring characteristic of glucose becomes a five-sided fructose ring. As you know, glucose and fructose both have the same number of atoms—$C_6H_{12}O_6$—and differ only in the arrangements of these atoms.

Step 3. This step, which is similar to Step 1, results in the attachment of a phosphate to the first carbon of the fructose molecule, producing fructose 1,6-diphosphate, that is, fructose with phosphates in the 1 and 6 positions. Note that in the course of the reactions thus far two molecules of ATP have been converted to ADP and no energy has been recovered.

Step 4. The molecule is split into two three-carbon molecules. They are slightly different but as the glyceraldehyde phosphate is used up in subsequent reactions, the dihydoxyacetone phosphate is eventually converted to glyceraldehyde phosphate. Thus, all subsequent steps must be counted twice to account for the fate of one glucose molecule. With the completion of Step 4, the preparatory reactions, reactions that require an input of energy, are complete. If you follow Step 4 back to Step 1, you will see the pathway by which glyceraldehyde phosphate, the product of the Calvin cycle, is converted to glucose; this is an energy-yielding reaction.

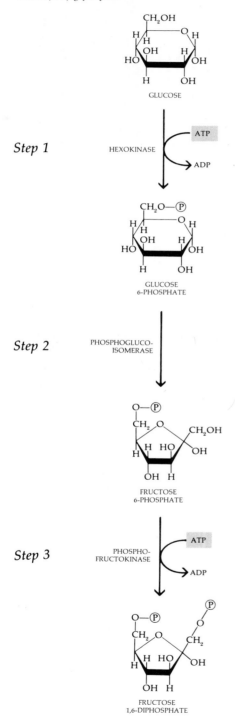

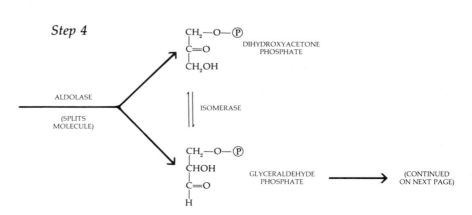

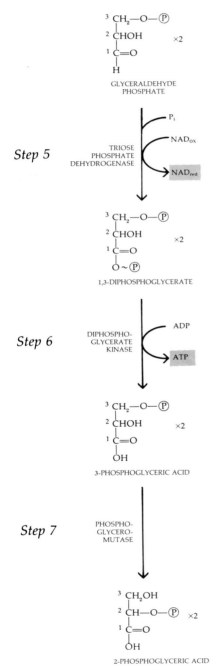

GLYCERALDEHYDE
PHOSPHATE

Step 5 — TRIOSE PHOSPHATE DEHYDROGENASE

1,3-DIPHOSPHOGLYCERATE

Step 6 — DIPHOSPHO-GLYCERATE KINASE

3-PHOSPHOGLYCERIC ACID

Step 7 — PHOSPHO-GLYCERO-MUTASE

2-PHOSPHOGLYCERIC ACID

Step 5. Glyceraldehyde phosphate molecules are oxidized—that is, hydrogen atoms with their electrons are removed—and NAD_{ox} becomes NAD_{red}. This is the first reaction from which the cell gains energy. Energy from this oxidation reaction is also used to attach phosphate groups to what is now the 1 position of each of the glyceraldehyde molecules. (The designation P_i indicates inorganic phosphate which is available in the cytoplasm.) Note that a "high-energy" bond is formed.

Step 6. The high-energy phosphate is released from the diphosphoglycerate molecule and used to recharge a molecule of ADP (a total of two molecules of ATP per molecule of glucose).

Step 7. The remaining phosphate group is transferred from the 3 position to the 2 position.

Step 8. In this step, a molecule of water is removed from the three-carbon compound, and a phosphate bond is formed.

Step 9. The phosphate is transferred to a molecule of ADP, forming another molecule of ATP (again, a total of two molecules of ATP per molecule of glucose).

Summary of Glycolysis

The complete sequence begins with one molecule of glucose. Energy is put into the sequence at Steps 1 and 3 by the transfer of a phosphate group from an ATP molecule—one at each step—to the sugar molecule. The six-carbon molecule splits at Step 4, and from this point onward, the sequence yields energy. At Step 5, a molecule of NAD_{ox} takes energy from the system and becomes NAD_{red}. At Steps 6 and 9, molecules of ADP take energy from the system, form additional phosphate bonds, and become ATP. Thus one glucose molecule is converted to two pyruvic acid molecules. The energy from the phosphate bonds of two ATP molecules is needed to initiate the glycolytic reaction, and two NAD_{red} are produced from two NAD_{ox} molecules and four ATP from four ADP molecules:

$$\text{Glucose} + 2\text{ATP} + 4\text{ADP} + 2NAD_{ox} \longrightarrow$$
$$2 \text{ pyruvic acid} + 2\text{ADP} + 4\text{ATP} + 2NAD_{red}$$

This series of reactions is carried out by virtually all living cells—from the simplest prokaryote to the eukaryotic cells of man.

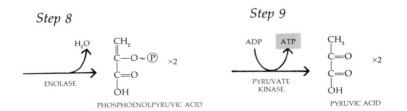

Step 8 — ENOLASE — PHOSPHOENOLPYRUVIC ACID

Step 9 — PYRUVATE KINASE — PYRUVIC ACID

ANAEROBIC FERMENTATION

Most of the rest of this chapter will concentrate on what happens to the pyruvic acid molecules in cells such as those of higher plants and animals in the presence of oxygen. The fact that glycolysis does not require oxygen, however, suggests that the glycolytic sequence evolved early, before free oxygen was present in the atmosphere, and that primitive one-celled organisms used glycolysis or something very much like it to extract energy from organic compounds which they absorbed from their watery surroundings. Moreover, certain modern cells are able to extract their energy from glycolysis alone if oxygen is not available to them. Yeast cells, for example, can live without oxygen. Under anaerobic (oxygenless) conditions, they convert glucose to pyruvic acid by the glycolysis sequence. The pyruvic acid so formed is then converted to alcohol by the steps shown in Figure 7–5. This process is a type of *anaerobic fermentation*.

When the glucose-filled juices of grapes and other fruit are extracted and stored in airtight kegs, the yeast cells, present on the skin of the fruit, turn the fruit juice to wine by converting glucose into alcohol. Yeast, like all living things, has a limited tolerance for alcohol, and when a certain concentration (about 12 percent) is reached, the yeast cells die and fermentation ceases.

Cells of higher animals do not form alcohol from pyruvic acid in the absence of oxygen, but rather produce another compound known as lactic acid (Figure 7–6). Lactic acid is produced, for example, in muscle cells during bursts of extra hard work, as by an athlete in a sprint. We breathe hard when we run fast, thereby increasing the supply of oxygen to the muscle cells, but the capacity of the muscles to do work is not limited by the amount of oxygen available to them at that moment. They can accumulate what is known as an oxygen debt by producing lactic acid from glucose (which is stored in the muscles in the form of glycogen). The accumulated lactic acid may be one of the substances that produces the sensations of muscle fatigue. After a sprint, we continue to breathe hard until we have paid off the oxygen debt and the accumulated lactic acid has been oxidized.

7–5

The steps by which pyruvic acid, formed by glycolysis, is converted to ethyl alcohol. In the first step, carbon dioxide is released. In the second, NAD_{red}, produced by glycolysis, is oxidized, and acetaldehyde is reduced. Most of the energy of the glucose remains in the alcohol, which is the end product of the sequence. The oxidation (without oxygen) of glucose to ethyl alcohol or to lactic acid (Figure 7–6) is called anaerobic fermentation.

PYRUVIC ACID (FROM GLYCOLYSIS)   ACETALDEHYDE   ETHYL ALCOHOL

7–6

In some biological systems, such as muscle cells, the end product of anaerobic fermentation is lactic acid. In the formation of lactic acid, NAD_{red} from glycolysis is oxidized and pyruvic acid is reduced. The NAD_{ox} molecules produced in this reaction and the one shown in Figure 7–5 are recycled in the glycolytic sequence.

PYRUVIC ACID (FROM GLYCOLYSIS)   LACTIC ACID

RESPIRATION

Some few organisms (anaerobes) that live without oxygen employ only anaerobic fermentation. In most cells, however, glycolysis is only a preparatory stage for cellular respiration, and it is in the latter stage that most of the cell's useful chemical energy is obtained.

Cellular respiration takes place in two stages: the Krebs cycle and the electron transport chain. Unlike glycolysis, they both require oxygen. In cellular respiration, pyruvic acid from glycolysis is oxidized to carbon dioxide and water, completing the breakdown on the glucose molecule.

The Krebs Cycle

Before entering the Krebs cycle, each of the three-carbon pyruvic acid molecules is oxidized. The third carbon is removed, carbon dioxide is formed, and one acetyl ($-COCH_3$) group is formed. In the course of this reaction, a molecule of NAD_{red} is produced from NAD_{ox}. These two-carbon groups are momentarily accepted by a compound known as _coenzyme A_ (a large molecule containing a nucleotide). The combination of the acetyl group and coenzyme A is known simply as acetyl CoA.

Fats and amino acids can also be converted to acetyl CoA and enter the respiratory sequence at this point.

In the Krebs cycle (Figure 7–8), the two-carbon acetyl group is combined with a four-carbon compound (oxaloacetic acid) to produce a six-carbon compound (citric acid). In the course of the cycle, two of the six carbons are oxidized to carbon dioxide, and oxaloacetic acid is regenerated. Each turn around the cycle uses up one acetyl group and regenerates a molecule of oxaloacetic acid, which is then ready to begin the sequence again. (Oxaloacetic acid, you will recall, is also the key molecule in C_4 photosynthesis.)

7–7

The three-carbon pyruvic acid molecule is oxidized to the two-carbon acetyl group, which is combined with coenzyme A to form acetyl CoA. The oxidation of the pyruvic acid molecule is coupled to the reduction of NAD_{ox}. Acetyl CoA enters the Krebs cycle.

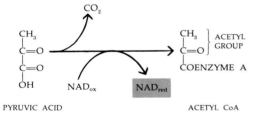

PYRUVIC ACID   ACETYL CoA

7–8

Summary of the Krebs cycle. In the course of the cycle, the carbons donated by the acetyl group are oxidized to carbon dioxide and the hydrogen atoms are passed to electron carriers. As in glycolysis, a specific enzyme is involved at each step. One molecule of ATP, three molecules of NAD_{red}, and one molecule of FP_{red} represent the energy yield of the cycle.

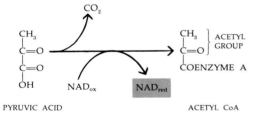

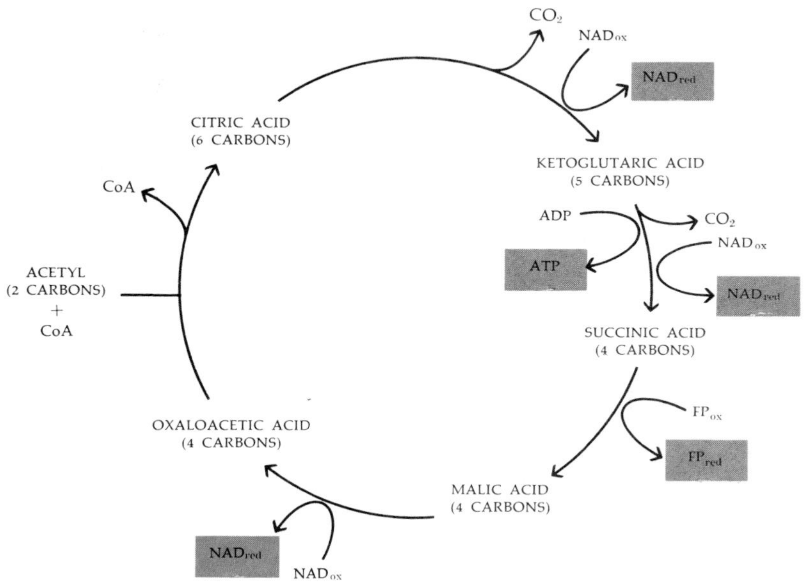

In the course of these steps, some of the energy released by the oxidation of the carbon atoms is used to convert ADP to ATP (one molecule per cycle), some is used to produce NAD_{red} from NAD_{ox} (three molecules per cycle), and some is used to reduce a second electron carrier, a flavoprotein, abbreviated FP (one molecule of FP_{red} from FP_{ox} per cycle).

$$Oxaloacetic\ acid + acetyl\ CoA + ADP + 3NAD_{ox} + FP_{ox} \longrightarrow$$
$$oxaloacetic\ acid + 2CO_2 + CoA + ATP + 3NAD_{red} + FP_{red}$$

Electron Transport

The glucose molecule is now completely oxidized. Some of its energy has been used to produce ATP from ADP. Most of it, however, still remains in electrons removed from the carbon atoms as they were oxidized and passed to the electron carriers NAD and FP. These electrons are at a high energy level. In the course of the electron transport chain, they are passed "downhill" and the energy released is used to form ATP molecules from ADP (Figure 7–9).

As electrons flow along the electron transport chain from a higher to a lower energy level, the cytochromes harness the released energy and use it to convert ADP to ATP. At the end of the chain, the electrons are accepted by oxygen and combine with protons (hydrogen ions) to produce water. Each time one pair of electrons passes from NAD_{red} to oxygen, three molecules of ATP are formed from ADP and phosphate. FP_{red} holds electrons at a slightly lower energy level than NAD_{red}. Each time a pair of electrons passes from FP_{red}, two molecules of ATP are formed.

We are now in a position to see how much of the energy originally present in the glucose molecule has been recovered in the form of ATP.

Glycolysis yielded two molecules of ATP directly and two molecules of NAD_{red}, for a total net gain of 8 ATP.

The conversion of pyruvic acid to acetyl CoA yields two molecules of NAD_{red} for each molecule of glucose and so produces six molecules of ATP.

The Krebs cycle yields, for each molecule of glucose, two molecules of ATP, six of NAD_{red}, and two of FP_{red}, or a total of 24 ATP.

7–9

Summary of glycolysis and respiration. Glucose is first broken down to pyruvic acid, with a yield of two ATP molecules and the reduction (dashed arrows) of two NAD molecules. Pyruvic acid is oxidized to acetyl CoA, and one molecule of NAD is reduced. (Note that this and subsequent reactions occur twice for each glucose molecule; this electron passage is indicated by solid arrows.) In the Krebs cycle, the acetyl group is oxidized and the electron acceptors, NAD and FP, are reduced. NAD and FP then transfer their electrons to the series of cytochromes that make up the electron transport chain. As the cytochromes pass the electrons "downhill," the energy released is used to make ATP from ADP.

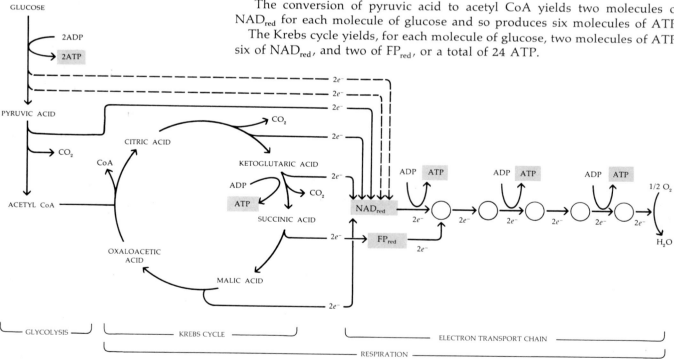

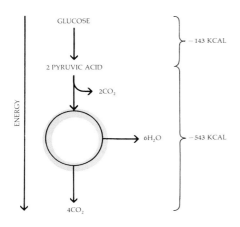

GLUCOSE

−143 KCAL

2 PYRUVIC ACID

2CO₂

6H₂O

−543 KCAL

ENERGY

4CO₂

7–10

Energy changes in the oxidation of glucose. The complete respiratory sequence (glucose + 6O₂ → 6CO₂ + 6H₂O) proceeds with an energy drop of 686 kilocalories. Of this, about 40 percent (266 kilocalories) is conserved in 38 ATP molecules. In anaerobic glycolysis, by contrast, only 2 ATP molecules are produced, representing only about 2 percent of the available energy of glucose.

As the balance sheet, Table 7–1, shows, the complete yield from a single molecule of glucose is 38 molecules of ATP. Note that all but two of the 38 molecules of ATP have come from reactions taking place in the mitochondrion, and all but four involve the oxidation of NAD_{red} or FP_{red} along the electron transport chain.

In the course of glycolysis 686 kilocalories are released per mole (180 grams) of glucose. Almost 40 percent of this, about 266 (7 × 38) kilocalories, has been captured in the "high-energy" bonds of the 38 ATP molecules.

Table 7–1 *Summary of Energy Yield from One Molecule of Glucose*

Glycolysis:	2 ATP 2 NAD_{red} ⟶ 6 ATP }	⟶ 8 ATP
Respiration:		
Pyruvic acid ⟶ acetyl CoA:	1 NAD_{red} ⟶ 3 ATP (× 2)	⟶ 6 ATP
Krebs cycle:	1 ATP 3 NAD_{red} ⟶ 9 ATP } 1 FP_{red} ⟶ 2 ATP	⟶ 24 ATP

These ATP molecules are, in turn, the immediate energy source for most of the ongoing activities of the cell (see Figure 7–11). As the energy of the ATP molecules is expended, the flow of energy through living systems that began in the chloroplast of the photosynthetic cell is completed.

PHOTOSYNTHESIS AND THE ACCUMULATION OF OXYGEN

The first photosynthetic organism probably appeared almost 3 billion years ago. Before the evolution of photosynthesis, the physical characteristics of the planet itself were by far the most powerful forces in shaping the course of natural selection. But with the evolution of photosynthesis, organisms began to change the face of the planet, and they have continued to do so, at an ever-increasing rate, up to the present day.

The atmosphere in which the first cells evolved probably consisted largely, as we noted previously, of ammonia and methane and other carbon gases. Gradually, methane and ammonia were broken down, and their hydrogen atoms, which are very light, were released into space as hydrogen gas (H_2). These earliest forms of life were, of course, adjusted to living in an environment without free oxygen, and, in fact, oxygen would have been poisonous to them. Their energy probably came from glycolysis. Anaerobic glycolysis (page 102) might have resulted in the accumulation of carbon dioxide in the atmosphere. Thus carbon dioxide and nitrogen replaced methane and ammonia as significant components of the atmosphere.

Then, it is hypothesized, there slowly evolved photosynthetic

organisms that used carbon dioxide as their carbon source and released oxygen, as do most modern photosynthetic forms. As these photosynthetic organisms multiplied, free oxygen began to accumulate. In response to the new selection pressures of these changing conditions, cell species arose for which oxygen was not a poison but a requirement of existence.

As we have seen, oxygen-consuming organisms have an advantage over those that do not use oxygen because a higher yield of energy can be extracted per molecule from the oxidation of carbon-containing compounds than from anaerobic fermentation processes, in which fuel molecules are not broken down completely. Energy resulting from the use of oxygen by cells made possible the development of increasingly active, increasingly complex organisms. Without oxygen, the higher forms of life that now exist on earth could not have evolved.

The planet continues to be dependent on photosynthesis for its oxygen. In fact, all the oxygen of the biosphere is renewed about every 2,000 years by photosynthetic cells. And the carbon dioxide produced in the course of respiration is the raw material on which photosynthesis depends.

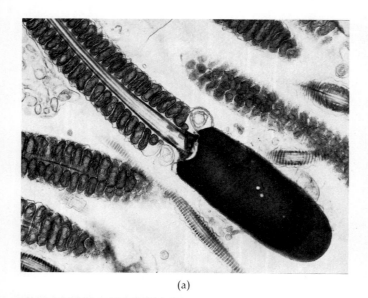

(a)

(b)

(c)

7-11
Some uses of ATP energy. (a) The rapid movements of the sperm cell, like all movements of cilia and flagella, are powered by ATP. Note the chain of mitochondria surrounding the tail of the sperm. (b) Muscular energy, the energy being expended by this corn snake, also requires ATP. The prey is a white-footed mouse. (c) ATP energy is also expended in bioluminescence, as shown in the firefly. These insects, which are actually beetles, use bioluminescence for sexual signaling (see page 468). Bioluminescence, chemical energy, is transferred back to its original source, light energy.

SUMMARY

The oxidation of glucose is a chief source of energy in most cells. As the glucose is broken down in a series of small enzymatic steps, the energy in the molecule is packaged in the form of "high-energy" bonds in molecules of ATP.

The first phase in the breakdown of glucose is glycolysis, in which the six-carbon glucose molecule is split into two three-carbon molecules of pyruvic acid, and two new molecules of ATP and two of NAD_{red} are formed. This reaction takes place in the cytoplasm of the cell.

The second phase in the breakdown of glucose and other fuel molecules is respiration. It requires oxygen and, in eukaryotic cells, takes place in the mitochondria. It occurs in two stages: the Krebs cycle and the electron transport chain.

In the course of respiration, the three-carbon pyruvic acid molecules are broken down to two-carbon acetyl groups, which then enter the Krebs cycle. In the Krebs cycle, the two-carbon acetyl group is broken apart in a series of reactions to carbon dioxide. In the course of the oxidation of each acetyl group, four electron acceptors (three NAD and one FP) are reduced, and another molecule of ATP is formed.

The final stage of breakdown of the fuel molecule is the electron transport chain, which involves a series of electron carriers and enzymes embedded in the inner membranes of the mitochondrion. Along this series of electron carriers, the high-energy electrons accepted by NAD_{red} and FP_{red} during the Krebs cycle pass downhill to oxygen. Each time a pair of electrons passes down the electron transport chain, ATP molecules are formed from ADP and phosphate. In the course of the breakdown of the glucose molecule, 38 molecules of ATP are formed, most of them in the mitochondrion.

QUESTIONS

1. Distinguish between the following: NAD_{ox}/NAD_{red}, glycolysis/fermentation, Calvin cycle/Krebs cycle.

2. In general terms, describe the events that take place during cellular respiration. Compare your description with the chart on page 104.

3. If oxygen-breathing organisms are so much more successful, in evolutionary terms, than anaerobes, why should there be any anaerobes left on this planet? Why shouldn't they all have become extinct long ago?

SUGGESTIONS FOR FURTHER READING

CONANT, JAMES BRYANT (ed.): *Harvard Case Histories in Experimental Science*, vol. 2, Harvard University Press, Cambridge, Mass., 1964.

Case #5, Plants and the Atmosphere, edited by Leonard K. Nash, describes the early work on photosynthesis, presented often in the words of the investigators themselves. The narrative illuminates the historical context in which the discoveries were made.

LEHNINGER, ALBERT L.: *Bioenergetics: The Molecular Basis of Biological Energy Transformations*, 2d ed., W. A. Benjamin, Inc., New York, 1971.

A brief, well-written account of the flow of energy in living systems.

MOROWITZ, H. J.: *Energy Flow in Biology*, Academic Press, Inc., New York, 1968.

A stimulating analysis of the relationship between energy flow and biological organization.

RABINOWITCH, EUGENE, and GOVINDJEE: *Photosynthesis*, John Wiley & Sons, Inc., New York, 1969.*

A lucid introduction, suitable for undergraduate students, to the processes of photosynthesis and to related physical and chemical concepts, such as entropy and free energy.

* Available in paperback.

SECTION 3   Genetics

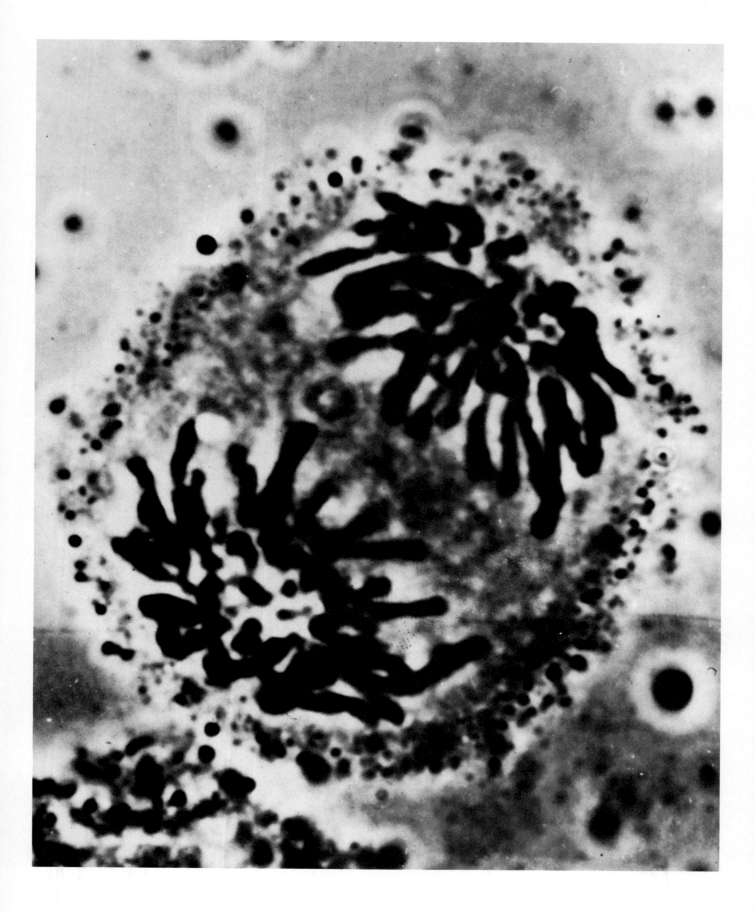

CHAPTER 8

The Nucleus and Mitosis

Genetics is the branch of biology that deals with heredity, the process by which traits are passed from parents to offspring. As we saw in Chapter 1, the capacity to reproduce is one of the special characteristics of living matter, and for more than 3 billion years, living things have been reproducing themselves, generation after generation, with each parent passing on to its offspring all the biological instructions necessary for the offspring to develop into the same kind of organism as the parent. A primary question of genetics is how this hereditary information is transmitted by the parent to its young. Or to put it more simply, why do cats have kittens, dogs have puppies, and hens have chicks? Why do you resemble your father or mother? What is the basis of this hereditary information? How is it stored between generations? How is it passed from one generation to another?

Since every organism—whether a *Paramecium,* an oak, or a human—starts as a single cell, the hereditary information must be carried somewhere, somehow in that single cell. More than a century ago, biologists began to suspect that the key to the mystery of heredity was to be found within the cell's nucleus, and as it turned out, they were quite right.

THE CELL'S NUCLEUS

The nucleus of the eukaryotic cell is a prominent, roughly spherical body. Unlike the cytoplasm, which contains many distinct structures, the nucleus contains little that can be seen with any kind of microscope. All that is usually visible is its outer membranes, a tangle of threadlike material, and one or more spherical bodies, the nucleoli (singular, <u>nucleolus</u>). Yet despite its simple appearance, the nucleus plays a crucial role in the life of the organism, both as the repository of the hereditary information and as the control center of the individual cell.

Some Experimental Evidence

One of the most important early observations of the role of the nucleus in the life of the cell was made about a hundred years ago by a German embryologist, Oscar Hertwig, observing the eggs and sperm of a sea urchin. Sea urchins produce eggs and sperm in great numbers. The eggs are relatively large—about

8-1

A human cell divides. The dark bodies are chromosomes, the carriers of the genetic information. The chromosomes have duplicated and moved apart. Each set of chromosomes is an exact copy of the other. Thus the two new cells, the daughter cells, will contain the same hereditary material.

Electron micrograph of the nucleus of the alga Chlamydomonas. *The dark body in the center of the nucleus is the nucleolus. The nucleolus is the site of production of major components of ribosomes; you can see partially formed ribosomes in its center. Notice also the nuclear envelope with its many pores, two of which are indicated by arrows. To the left of the nucleus is a portion of a chloroplast containing starch grains. A Golgi body is below the nucleus to the left, and mitochondria are visible to the right.*

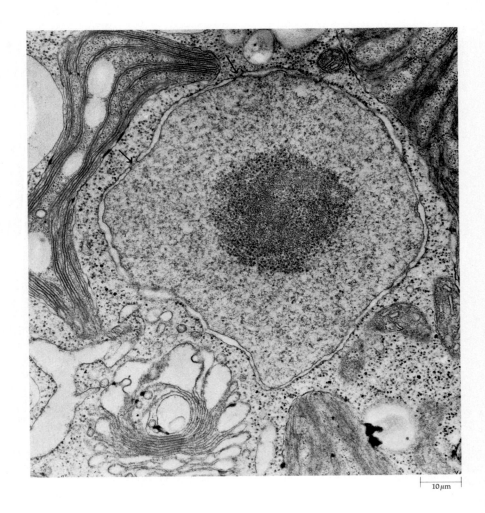

10 μm

as large as a human ovum (see Figure 9–1), which they closely resemble—and so are easy to observe. They are fertilized in the open water, rather than internally, as in the case with land-dwelling vertebrates such as ourselves. Watching the eggs being fertilized under his microscope, Hertwig observed that only a single sperm cell was required and that when this sperm cell penetrated the egg, its nucleus was released and fused with the nucleus of the egg. This observation, confirmed by other scientists and in other kinds of organisms, was important in establishing the fact that the nucleus is the carrier of heredity: the only link between father and offspring is the nucleus of the sperm.

Since Hertwig's time, a number of experiments have been performed concerning the role of the nucleus in the cell. In one simple experiment, the nucleus was removed from an amoeba by microsurgery. The protist stopped dividing and, in a few days, it died. If, however, a nucleus from another amoeba was reimplanted within 24 hours after the original one was removed, the cell survived and divided normally.

In the early 1930s, Joachim Hammerling studied the comparative roles of the nucleus and the cytoplasm by taking advantage of some unusual properties of the marine alga *Acetabularia.* The body of *Acetabularia* consists of a single, huge cell 2 to 5 centimeters in height. Individuals have a cap, a stalk, and a "foot," all of which are differentiated portions of the single cell. If the cap is removed, the

8-3

(a, b) *One species of* Acetabularia *has an umbrella-shaped cap, and another has a ragged, petal-like cap. If the caps are removed, others form, similar in appearance to the amputated one. However, if the nucleus (which is in the "foot" of the cell) is removed at the same time as the cap and a new nucleus from another species is transplanted, the cap (c) which forms will have a structure with characteristics of both species. If this cap is removed, the next cap (d) that grows will be characteristic of the cell which donated the nucleus, not of the cell which donated the cytoplasm.*

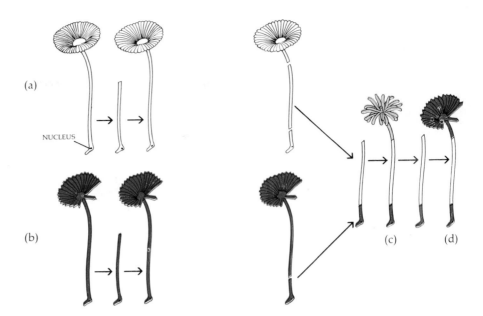

cell will rapidly regenerate a new one. Different species of *Acetabularia* have different kinds of caps. *Acetabularia mediterranea*, for example, has a compact umbrella-shaped cap and *Acetabularia crenulata* has a cap of petal-like structures.

Hammerling took the "foot," which contains the nucleus, from a cell of *A. crenulata** and grafted it onto a cell of *A. mediterranea*, from which he had first removed the "foot" and the cap. The cap that then formed had a shape intermediate between the two. When this cap was removed, the next cap that formed was completely characteristic of *A. crenulata* (Figure 8-3).

Hammerling interpreted these results as meaning that certain cap-determining substances are produced under the direction of the nucleus. These cap-determining substances accumulate in the cytoplasm, which is why the first cap that formed after nuclear transplantation was of an intermediate type. By the time the second cap formed, however, the cap-determining substances already in the cytoplasm had been exhausted, and the form of the cap was completely under the control of the new nucleus.

Conclusion: The Functions of the Nucleus

We can see from these experiments that the nucleus performs two crucial functions for the cell. First, it carries the hereditary information for the cell, the instructions that determine whether a particular organism will develop to become a *Paramecium*, an oak, or a human—and not just any *Paramecium*, oak, or human, but one which resembles the parent or parents of that particular, unique organism. Second, as Hammerling's work indicated, the nucleus exerts a continuing influence over the ongoing activities of the cell, ensuring that the complex molecules that cells require are synthesized in the number and of the kind needed.

---

*According to scientific convention it is not necessary to spell out the generic name of an organism more than once in the same discussion.

Diagram of a chromosome at the beginning of mitosis. The chromosomal material has replicated during interphase so that each chromosome now consists of two identical parts, called chromatids. The kinetochore is the area of attachment of the two chromatids.

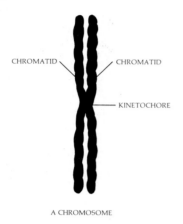

CHROMATID   CHROMATID

KINETOCHORE

A CHROMOSOME

8–5

If you look carefully, you can see that each of these chromosomes actually consists of two longitudinal halves. The kinetochore is the constricted area visible in each. Kinetochores are often, but not always, near the center of the chromosome.

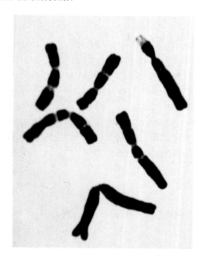

CELL DIVISION

At the time of cell division, the nucleus changes its appearance. The threadlike material, the chromatin, condenses into a group of structures, the chromosomes,* which appear rodlike under the light microscope. Because of the behavior of these bodies at the time of cell division, it was long suspected that they were the carriers of the hereditary instructions, a fact which has now been proved by studies we shall describe later in this section.

Cell division in eukaryotes consists of two overlapping stages, *mitosis* and *cytokinesis*. In mitosis, the nuclear membrane breaks down; the chromosomes, which have previously been replicated, are divided equally; and two new nuclei are formed. In cytokinesis, the cytoplasm of the parent cell is divided and packaged into two parts, each containing one of the nuclei.

Prior to mitosis, the chromosomal material is replicated, so that at the beginning of mitosis each chromosome consists of two identical sister chromosomes called *chromatids* joined together at a *kinetochore* (Figure 8–4). During mitosis, the chromatids are separated, and each of the two new, "daughter" cells receives one complete set of the chromosomes that were present in the parent cell. Cytokineses is not as precise as chromosome division. The various cytoplasmic organelles are apportioned more or less equally between the daughter cells.

Usually, at the end of cell division, as a result of cytokinesis, the two daughter cells will be of about the same size and will have about the same numbers and kinds of organelles. As a result of mitosis, they will contain *exactly* the same chromosomal material.

MITOSIS

The process of mitosis is conventionally divided into four phases: prophase, metaphase, anaphase, and telophase; of these, prophase is usually much the longest. If a mitotic division takes 10 minutes (which is about the minimum time required), during about six of these minutes the cell will be in prophase. The schematic drawings that follow show mitosis as it takes place in an animal cell.

Between divisions, the cell is in interphase. Although little can be seen in the nucleus during interphase, and it is not considered one of the stages of mitosis, it is an essential preparation for it, since it is during this period that replication of the chromosome occurs.

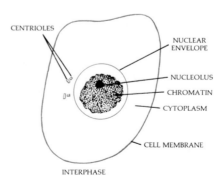

CENTRIOLES

NUCLEAR ENVELOPE

NUCLEOLUS

CHROMATIN

CYTOPLASM

CELL MEMBRANE

INTERPHASE

* Chromosomes ("colored bodies") were given this name by the early microscopists because they took up certain kinds of stains used in preparing specimens for microscopy and so were first visualized as brightly colored.

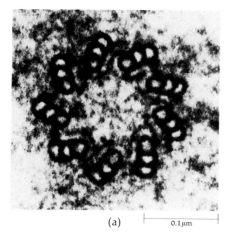

(a)   0.1 μm

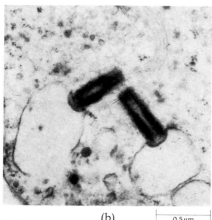

8–6   (b)   0.5 μm

(a) *Centrioles are structurally identical to basal bodies. It is hypothesized that basal bodies organize microtubules into cilia and flagella and that, similarly, centrioles organize the microtubules of spindle fibers. However, plant cells, which do not have centrioles, are able to form spindles.*

(b) *Mother-daughter centriole pair. Centrioles are also usually—but not always—formed from preexisting centrioles, with the daughter appearing at right angles to the mother centriole.*

In early prophase, the chromatin condenses, and the individual chromosomes begin to become visible. At this time each chromosome* consists of two duplicate chromatids pressed closely together longitudinally and connected at the kinetochore (Figure 8–5). Two centriole pairs are visible at one side of the nucleus, outside the nuclear envelope. Each pair consists of one mature centriole and a smaller daughter centriole lying at right angles to it (Figure 8–6).

During prophase, the centriole pairs move apart. Between the centriole pairs, forming as they separate (or perhaps separating as they form), are spindle fibers made up of microtubules and other proteins. Additional fibers, known collectively as the aster, radiate outward from the centrioles. The nucleolus usually disperses, disappearing from view. The nuclear envelope breaks down, marking the end of prophase.

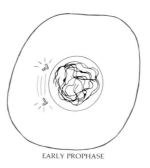

EARLY PROPHASE

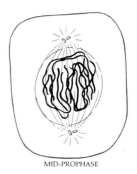

MID-PROPHASE

By the end of prophase, the centriole pairs are at opposite ends of the cell, and the members of each pair are of equal size. The spindle is fully formed. It is a three-dimensional football-shaped structure consisting of three groups of protein fibers: (1) astral rays, (2) continuous fibers reaching from pole to pole, and (3) shorter fibers that are attached to the kinetochores of the individual chromosomes (Figure 8–7). Since spindle fibers become neither thicker nor thinner during mitosis, it appears that they lengthen when new material is added at one end and shorten when it is removed.

During early metaphase the chromosomes move back and forth within the spindle, apparently maneuvered by the spindle fibers, as if they were being tugged first toward one pole and then the other. Finally they become arranged precisely at the midplane (equator) of the cell. This marks the end of metaphase.

EARLY METAPHASE

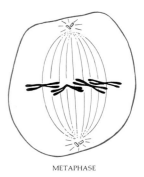

METAPHASE

* In counting chromosomes, you may not know whether to count a chromosome that has replicated but has not divided as 1 or 2. Such a chromosome is counted as 1. The trick is to count kinetochores.

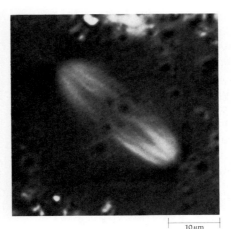

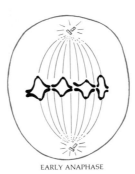 

EARLY ANAPHASE   LATE ANAPHASE

At the beginning of anaphase, each kinetochore divides, and the chromatids (now chromosomes) separate. The two duplicate chromosomes pull apart slowly, apparently drawn toward the poles by the spindle fibers. The kinetochores move first, while the "arms" of the chromosomes drag behind.

During anaphase, the identical sets of chromosomes move toward the opposite poles of the spindle.

8-7
This unusual micrograph of a spindle in an egg cell of a marine worm emphasizes the spindle's three-dimensional qualities. The bright streaks are the spindle fibers. The chromosomes appear as indistinct gray bodies near the equator of the spindle. Spindle fibers are made up of microtubules and other proteins. They attach to chromosomes at their kinetochores.

By the beginning of telophase, the chromosomes have reached the opposite poles and cytokinesis has begun. The spindle disperses. During this period a new daughter centriole arises from each mature centriole, so that each new cell has two centriole pairs.

EARLY TELOPHASE

During late telophase, the nuclear envelopes re-form, and the chromosomes once more become diffuse.

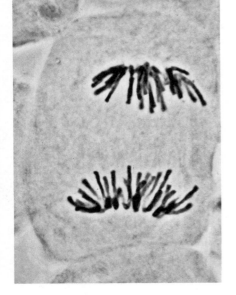

8-8
Early telophase in an onion cell.

LATE TELOPHASE

8-9

*Mitosis in the embryonic cells of a white-
fish. (a) Prophase. The chromosomes have become visible, the nuclear envelope has broken
down, and the spindle apparatus has
formed. (b) Metaphase. The chromosomes,
guided by the spindle fibers, are lined up at
the equator of the cell. Some of them appear
to have begun to separate. (c) Anaphase.
The two sets of chromosomes are moving
apart. (d) Telophase. The chromosomes are
completely separated, the spindle apparatus
is disappearing, and a new cell membrane is
forming which will complete the separation
of the two daughter cells.*

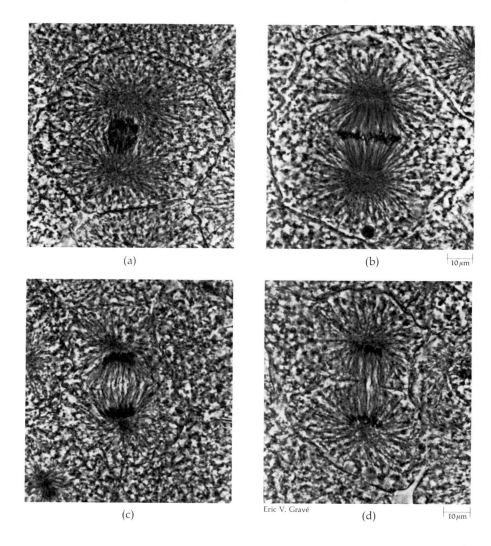

(a)

(b)

10 μm

(c)

Eric V. Gravé

(d)

10 μm

8-10

*Cells in mitosis in the tip of an onion root.
The root will continue to grow for the life
of the plant.*

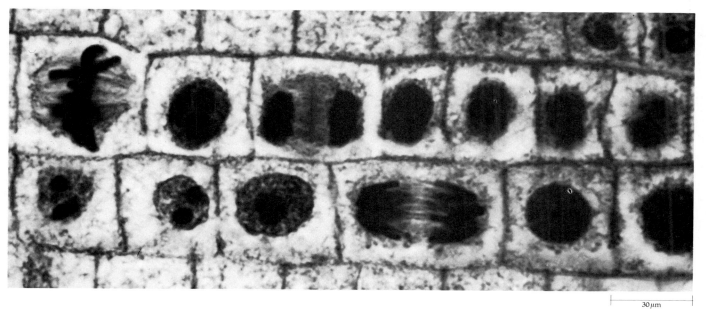

30 μm

In plants, the final separation of the two cells is effected by the formation of a structure known as a cell plate. Vesicles appear across the equatorial plane of the cell and gradually fuse, forming a flat membrane-bound space, the cell plate, which extends outward until it reaches the wall of the dividing cell. The dark forms on either side of the cell plate are the chromosomes.

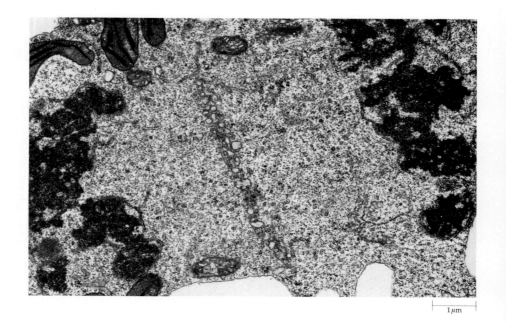

1 μm

CYTOKINESIS

Cytokinesis usually begins during telophase of mitosis. It differs in some respects in plant and animal cells. In animal cells, during late anaphase or early telophase, the cell membrane begins to pinch in along the circumference of the cell in the area where the equator of the spindle used to be. At first a furrow appears on the surface, and this gradually deepens into a groove. Eventually the connection between the daughter cells dwindles to a slender thread, which soon parts.

In plant cells, the cytoplasm is divided by the formation of a *cell plate*, which is formed from a line of vesicles produced from Golgi bodies (Figure 8–11). The vesicles form in the interior of the cell and move outward. They eventually fuse to form a flat, membrane-bound space. As more vesicles fuse, the edges of the growing plate fuse with the membrane of the cell. In this way, space is established between the daughter cells, completing their separation. Cellulose is laid down against the membranes, forming the cell walls, and the space itself eventually becomes impregnated with pectins and becomes the middle lamella (see page 63).

When cell division is complete, two daughter cells are produced, indistinguishable from one another and from the parent cell. The precise, ordered events of mitosis ensure the continuity of the genetic material and the preservation of the information that it carries.

THE CELL CYCLE

Dividing cells pass through a regular sequence of events, known as the cell cycle (Figure 8–12). The S (synthesis) phase of the cell cycle is the period during which the genetic material (DNA) is duplicated. G (gap) phases precede and follow the S phase. The G_1 period occurs after mitosis and precedes the S phase; the G_2 period follows the S phase and occurs before mitosis. The G and S phases together constitute interphase. Mitosis occupies the rest of the cell cycle.

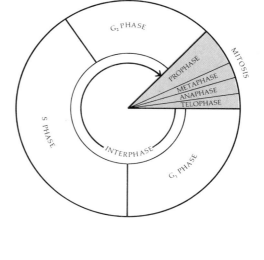

The cell cycle. Dividing cells go through four principal phases, including the phases of mitosis (shown in color) and the S phase, during which the chromosomes are duplicated. Separating mitosis and the S phase are two G phases. The first of these (G_1) is a period of general growth and replication of cytoplasmic organelles. During the second (G_2), structures directly associated with mitosis, such as the spindle fibers, are synthesized. After the G_2 phase comes mitosis, which is, in turn, divided into four phases.

Completion of the cycle requires varying periods of time, depending on both the type of cell and external factors, such as temperature or available nutrients.

The G_1 phase is principally a period of growth of the cytoplasmic material, including all the various organelles. Also, during this G_1 period, according to current hypotheses, substances are synthesized that either inhibit or stimulate the S stage and the rest of the cycle, thus determining whether or not cell division will occur. During the G_2 phase, structures involved directly with mitosis, such as the spindle fibers, are synthesized.

Some cells pass through successive cell cycles repeatedly. This group includes the one-celled organisms and certain cells in growth centers of both plants and animals. Some highly specialized cells, such as nerve cells, no longer replicate once they are mature. A third group of cells retains the capacity to divide but does so only under special circumstances. Cells in the human liver, for example, do not ordinarily divide, but if a portion of the liver is removed surgically, the remaining cells (even if only about a third of the total remain) continue to replicate themselves until the liver reaches its former size. Then they stop, but no one knows why. Further knowledge of the control mechanisms involved would not only be interesting biologically but might be of great importance in the control of cancer. Cancer cells differ from normal cells largely in that they keep on dividing at the expense of normal tissues.

THE FUNCTION OF MITOSIS

Mitosis is the means by which organisms grow. Each of us, for example, began as a single cell—the fertilized egg—which, by the process of mitosis, developed into a large, complex multicellular organism. Nor does mitosis necessarily stop when growth stops. Plants and some animals continue to grow throughout their life spans. Even in organisms whose growth is more or less fixed at maturity—such as ourselves—the cells of some tissues divide continuously. Notable examples are the outer layers of skin and the inner layers of the intestines; old cells are shed constantly from these surfaces and new ones take their place.

In one-celled organisms, mitosis is a means of reproduction—the way in which one organism becomes two organisms. Even some large organisms reproduce by mitosis. Sea anemones or sponges, for example, may break apart to form new sea anemones or sponges; plants may produce roots or runners from which new individuals arise; or, as in *Hydra* (Figure 8–13), a new individual may

8-13

In this multicellular animal, a Hydra, you can see a new organism developing on the body of the parent. It will grow tentacles and begin feeding while still attached but will eventually break away and begin to live an independent existence. It will, of course, be genetically identical to its parent.

0.5 mm

develop on the body of the parent. Reproduction that involves only mitosis is known as asexual reproduction. It is also sometimes called vegetative reproduction (page 286), since it is particularly common among plants. Asexual reproduction confers a sort of immortality, since the individual in this way actually continues to exist as long as its offspring survive.

SUMMARY

The nucleus is the most prominent structure within most cells. It is surrounded by a double membrane, the nuclear envelope. Within it are the nucleolus and the chromosomes, the bodies that contain the genetic information of the cell.

Cell division includes division of the nucleus (mitosis), during which the chromosomes are apportioned between the two daughter cells, and division of the cytoplasm (cytokinesis). When the cell is in interphase (not dividing), the chromosomes are visible only as thin strands of threadlike material. During this period, if mitosis is to take place, the chromosomal material is replicated. During mitosis, the chromosomal material condenses and each chromosome appears as two identical chromatids, held together at the kinetochore. At the end of mitosis, the chromatids separate, and each chromatid, now a daughter chromosome, moves to one of the daughter cells. In this way, the chromosomes are equally divided between the two new cells. Each daughter cell thus contains the same genetic material as that of the mother cell.

Mitosis is the means by which multicellular organisms grow and by which they replace damaged or wornout tissues. For one-celled organisms and for some many-celled organisms it is also a form of reproduction.

QUESTIONS

1. Define the following terms: chromatid, chromosome, kinetochore, nucleus, nucloelus, centriole, spindle.

2. Identify the stages of mitosis visible in Figure 8–10.

3. Describe what is happening in each of the drawings below.

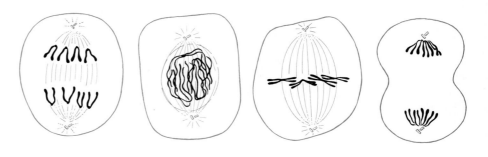

CHAPTER 9

Meiosis and Sexual Reproduction

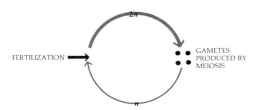

9-1

A human ovum surrounded by sperm cells. Note the great difference in size between the two types of cells. Each, however, contributes the same number of chromosomes to the fertilized egg from which the new individual will develop.

Sexual reproduction is reproduction by two parents, each of which contributes genetically to the new offspring. It always involves two events: *fertilization* and *meiosis*. Fertilization is the means by which the different genetic contributions of the two parents are brought together to form the new genetic identity (genome) of the offspring. Meiosis is a special kind of nuclear division which evolved from and uses much of the same machinery as mitosis. However, as you will see, it differs from it in some important respects.

MEIOSIS

To understand meiosis, we must look again at the chromosomes. Every organism has a chromosome number characteristic of its particular species.* A mosquito has 6 chromosomes per cell; a cabbage, 18; corn, 20; *Hydra*, 30; a sunflower, 34; a cat, 38; a man, 46; a plum, 48; a dog, 78; and a goldfish, 94.

However, the sex cells—eggs and sperm—have exactly half the number of chromosomes as are characteristic of the somatic (body) cells of an organism. The number of chromosomes in the sex cells is referred to as the haploid ("single") number, and the number in the somatic cells as the diploid ("double") number. For brevity, the haploid number is designated n and the diploid number $2n$. In man, for example, $n = 23$ and $2n = 46$. When a sperm fertilizes an egg, the two haploid nuclei fuse and the diploid number is restored (Figure 9-2). The diploid cell produced by the fusion of two gametes is known as a *zygote*. (Cells that have more than one double set of chromosomes, such as those in higher plants, are known as polyploid.)

---

* A notable exception is found in higher plants, in which many cells, for reasons which are not known, tend to have more than one set of chromosomes.

9-2

Sexual reproduction is characterized by two events: the coming together of the sex cells (fertilization) and meiosis. Following meiosis, the number of the chromosomes is single, or haploid (n). Following fertilization, the number is double, or diploid (2n).

FERTILIZATION   GAMETES PRODUCED BY MEIOSIS

121   MEIOSIS AND SEXUAL REPRODUCTION

Fertilization and meiosis occur at different points in the life cycle of different organisms. (a) In some one-celled organisms, meiosis occurs immediately after fertilization, and most of the life cycle is spent in the haploid state (signified by the thin line). (b) In plants, fertilization and meiosis are separated, and the organism characteristically has a diploid phase and a haploid phase. (c) In animals, meiosis is immediately followed by fertilization. As a consequence, during most of the life cycle the organism is diploid (signified by the thick line).

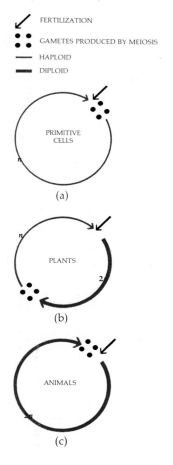

(a)

(b)

(c)

9–4

The life cycle of Chlamydomonas *is of the type shown in Figure 9–3a. The organism is haploid for most of its life cycle (thin arrows). Fertilization, the fusing of cells of different mating strains (indicated here by + and −), temporarily produces the diploid state (thick arrows). The diploid zygote divides meiotically, forming four new haploid cells, which will probably divide by mitosis (asexually) but may enter another sexual cycle.*

In every diploid cell, half of the chromosomes are derived from the chromosomes that originally came from one parent, and half are derived from those of the other. Any particular chromosome from one parent has a partner, originally from the other parent, which resembles it in size and shape and also, as we shall see, in the type of hereditary information it contains. These pairs of chromosomes are known as *homologues*.

In the special kind of nuclear division called meiosis, the diploid number of chromosomes is reduced to the haploid number. If this reduction did not occur, the chromosome number would double each generation. Moreover, as we shall see, meiosis is in itself a source of new genetic combinations.

MEIOSIS AND THE LIFE CYCLE

Meiosis occurs at different times during the life cycle of different organisms (Figure 9–3). In the protist *Chlamydomonas* it occurs following fusion of the mating cells (Figure 9–4); the cells are ordinarily haploid, and meiosis restores the haploid number. In plants, such as ferns, a haploid phase typically alternates with a diploid phase (Figure 9–5). The common and conspicuous form of a fern is the sporophyte, the diploid organism. By meiosis, fern sporophytes produce spores, usually on the undersides of their leaves. These spores have only the haploid number of chromosomes. They germinate to form much smaller plants (gametophytes), often only a few cell layers thick. In these plants, all the cells are haploid. The small, haploid plants produce gametes, which fuse and then develop into a new, diploid sporophyte. This process, in which the haploid stage is followed by the diploid stage and again by the haploid stage, is known as *alternation of generations*. As we shall see, alternation of generations occurs, although not always in the same form, in all plants.

Human beings have the typical animal life cycle, in which meiosis immediately precedes fertilization and virtually all of the life cycle is spent in the diploid state (Figure 9–6).

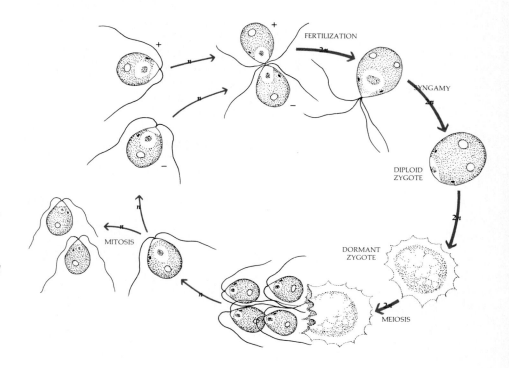

9-5

The life cycle of a fern is of the type shown in Figure 9-3b. Following meiosis, spores, which are haploid, are produced in the sporangia and shed (extreme right). The spores develop into haploid gametophytes. In many species, the gametophytes are only one layer of cells thick and are somewhat heart-shaped, as shown here (bottom). From the lower surface of the gametophyte, filaments, the rhizoids, extend downward into the soil.

On the lower surface of the gametophyte are borne the flask-shaped archegonia, which enclose the egg cells, and the antheridia, which enclose the sperm. When the sperm are mature and there is an adequate supply of water, the antheridia burst, and the sperm cells, which have numerous flagella, swim to the archegonia and fertilize the eggs. From the fertilized (2n) egg, the 2n sporophyte grows out of the archegonium within the gametophyte. After the young sporophyte becomes rooted in the soil, the gametophyte disintegrates. When it becomes mature, the sporophyte develops sporangia, in which meiosis occurs, and the cycle begins again.

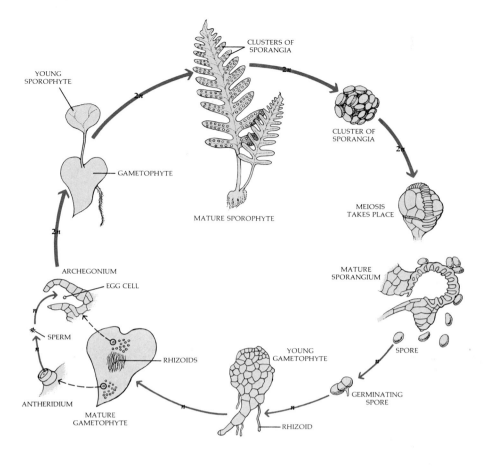

9-6

The life cycle of Homo sapiens. Gametes, which are haploid, are produced by meiosis. At fertilization, they fuse, restoring the diploid number in the fertilized egg. The fertilized egg develops into a mature man or woman, which again produces haploid gametes. As is the case with most other animals, the life cycle is almost entirely diploid, the only exception being the gametes. This is the type of life cycle diagrammed in Figure 9-3c.

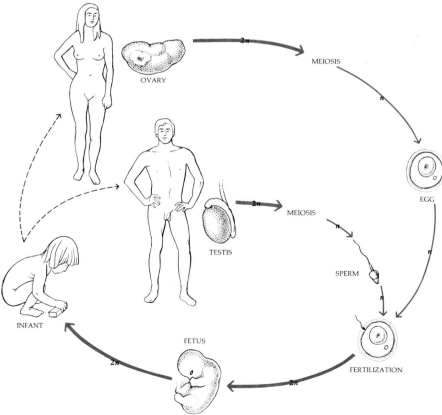

(a) *During prophase of meiosis, chromosomes become arranged in homologous pairs. Each homologous pair consists of four chromatids and is therefore also known as a tetrad (from the Greek tetra, meaning "four"). (b) Crossing over. During meiosis I, homologous chromosomes pair up to form tetrads, each consisting of four chromatids. Homologues connect at crossover points, where exchange of genetic material—crossing over—takes place. In this way, genetic material is redistributed among homologous chromosomes.*

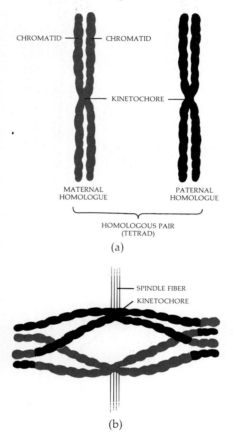

(a)

(b)

The events that take place during meiosis somewhat resemble those that take place during mitosis, and meiosis, which is believed to have evolved from mitosis, uses much of the same cellular machinery. But the differences are important. First, and most obvious, whereas mitosis takes place in one stage, meiosis takes place in two stages, involving two successive divisions and resulting in four nuclei instead of two. Second, mitosis produces two nuclei that are identical to each other and to the parent cells from which they are formed (identical both in chromosome number and in the composition of the individual chromosomes). Meiosis, by contrast, results in four nuclei that are not identical to each other and that have only half the number of chromosomes present in the parent cell.

In the following discussion, we shall describe meiosis in a plant cell that has eight chromosomes ($2n = 8$). Four of the eight chromosomes were inherited from one parent and four from the other parent. For each chromosome from one parent there is a homologous chromosome, or homologue, from the other parent—a chromosome of similar size and shape which contains the other parent's genetic instructions for the same set of hereditary characteristics, flower color, for instance.

During interphase, the chromosomes are replicated, so that by the beginning of meiosis each chromosome consists of two identical chromatids held together at the kinetochore, as shown in Figure 9–7a. In the first prophase of meiosis, the chromosomes come into view and the nuclear membrane begins to dissolve.

During the first prophase, prophase I, the homologous chromosomes come together in pairs. Since each chromosome consists of two identical chromatids, the pairing of the homologous chromosomes actually involves four chromatids. Crossing over, the interchange of segments of one chromosome with corresponding segments from its homologue, takes place at this time (Figure 9–7b). Such exchanges of chromosomal material are important sources of variation in the hereditary material. (We shall discuss crossing over in more detail in Chapter 12.)

If you remember the arrangement of the homologous chromosomes at this stage of meiosis, you will be able to remember all the subsequent events with little difficulty.

Toward the end of prophase, the spindle apparatus is formed, the nucleolus disperses, and the nuclear envelope breaks down.

In metaphase I, the four homologous pairs line up along the equatorial plane of the cell. Each pair consists of four chromatids.

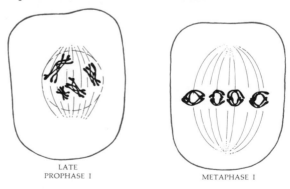

LATE
PROPHASE I

METAPHASE I

At anaphase I, the homologous chromosomes, each consisting of two chromatids, separate, apparently maneuvered by the spindle fibers attached to the kinetochores. The homologues, which seem to have a strong affinity for one

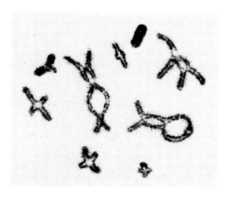

9–8

Late prophase I. All four chromatids can be seen in most of the tetrads. The spindle fibers are not visible.

another, pull slowly away from each other. The kinetochores do not divide, as they did in mitosis, and so the two chromatids of each chromosome do not separate.

By the end of the first meiotic division, the homologues have separated. The cell enters interphase. The chromosomes disappear from view and nuclear membranes may re-form.

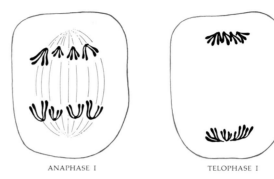

ANAPHASE I   TELOPHASE I

Each nucleus now contains only half the number of chromosomes of the original cell: these chromosomes are probably different from any one that was present in the original cell because of exchanges taking place during crossing over. This interphase differs from the interphase preceding mitosis and from the interphase preceding the first meiotic division in one crucial respect: no duplication of the chromosomal material takes place.

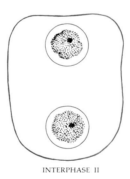

INTERPHASE II

The second meiotic division resembles mitosis. At the beginning of the second meiotic division, the chromosomes condense again. There are four in each nucleus (the haploid number), and they are still in the form of chromatids held together at the kinetochore.

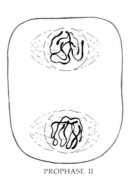

PROPHASE II

9–9

The end of spore formation in the royal fern
(Osmunda regalis).

During prophase II, the nuclear envelopes, if present, dissolve, and the spindle fibers begin to appear. During the second metaphase, the four chromosomes in each nucleus line up on the equatorial plane. At anaphase II, as in anaphase of mitosis, the chromosomes separate, and each chromatid (now a single chromosome) moves toward one of the poles.

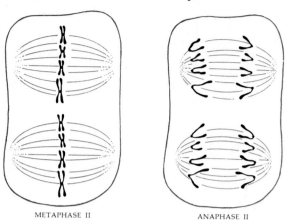

METAPHASE II   ANAPHASE II

During telophase, a nuclear membrane forms around each set of chromosomes. There are now four in all, each containing the haploid number of chromosomes.

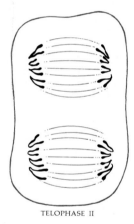

TELOPHASE II

Cytokinesis proceeds as it does following mitosis as cell membranes and cell walls form, dividing the cytoplasm. The cells begin to differentiate into gametes (in animals) or spores (in plants).

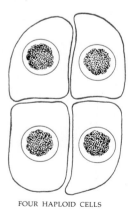

FOUR HAPLOID CELLS

MEIOSIS IN THE HUMAN SPECIES

In our own species, meiosis takes place in the reproductive organs, the testes of the male and the ovary of the female. In the male, a cell known as a primary spermatocyte undergoes the two divisions of meiosis to produce four spermatids, which then differentiate into sperm (Figure 9–10). The ejaculate of a normal human male contains 300 million sperm cells.

In the human female, the meiotic divisions are equal in terms of the genetic material, but unequal in the way the cytoplasm is apportioned (Figure 9–11). One egg cell (ovum) is produced and two or three polar bodies, which contain the other postmeiotic nuclei. These then disintegrate. As a result of this unequal division, the ovum is well supplied with nutrients for the developing embryo. In Figure 9–1, two polar bodies can be seen in the upper part of the micrograph.

9–10

The series of changes resulting in the formation of sperm cells begins with the growth of spermatogonia into large cells known as primary spermatocytes. At the first meiotic division, each primary spermatocyte divides into two haploid secondary spermatocytes. The second meiotic division results in the formation of four haploid spermatids. The spermatids differentiate into functional sperm.

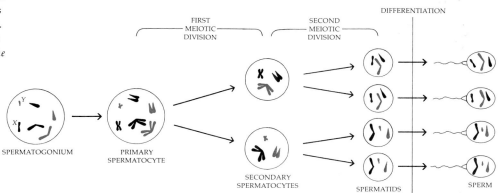

9–11

Formation of the ovum. A primary oocyte undergoes a meiotic division to produce a secondary oocyte and a polar body. This first meiotic division begins in the human female during the third month of fetal development and ends at ovulation, which may take place 50 years later. The second meiotic division, which produces the egg cell and a second polar body, does not take place until after fertilization. The first polar body may also divide.

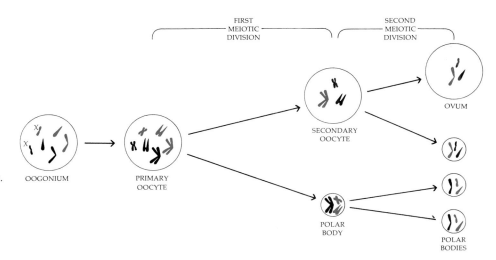

A comparison of mitosis and meiosis. (In these examples, 2n = 6.)

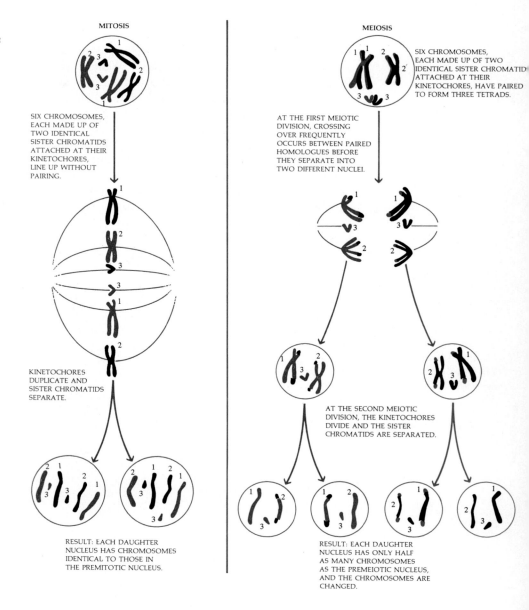

MITOSIS

SIX CHROMOSOMES, EACH MADE UP OF TWO IDENTICAL SISTER CHROMATIDS ATTACHED AT THEIR KINETOCHORES, LINE UP WITHOUT PAIRING.

KINETOCHORES DUPLICATE AND SISTER CHROMATIDS SEPARATE.

RESULT: EACH DAUGHTER NUCLEUS HAS CHROMOSOMES IDENTICAL TO THOSE IN THE PREMITOTIC NUCLEUS.

MEIOSIS

SIX CHROMOSOMES, EACH MADE UP OF TWO IDENTICAL SISTER CHROMATIDS ATTACHED AT THEIR KINETOCHORES, HAVE PAIRED TO FORM THREE TETRADS.

AT THE FIRST MEIOTIC DIVISION, CROSSING OVER FREQUENTLY OCCURS BETWEEN PAIRED HOMOLOGUES BEFORE THEY SEPARATE INTO TWO DIFFERENT NUCLEI.

AT THE SECOND MEIOTIC DIVISION, THE KINETOCHORES DIVIDE AND THE SISTER CHROMATIDS ARE SEPARATED.

RESULT: EACH DAUGHTER NUCLEUS HAS ONLY HALF AS MANY CHROMOSOMES AS THE PREMEIOTIC NUCLEUS, AND THE CHROMOSOMES ARE CHANGED.

THE FUNCTION OF SEXUAL REPRODUCTION

Sexual reproduction demands a tremendous expenditure of energy on the part of the organisms involved, both in the production of gametes, most of which are wasted, especially in the male, and in courtship and other strategies to ensure fertilization. Moreover, as we have seen, organisms can reproduce very adequately without either fertilization or meiosis. Why, then, does sexual reproduction occur?

The answer that biologists offer to this question is that organisms that produce a variety of offspring are more likely to have surviving offspring than organisms that produce only offspring identical to themselves. These variations are the raw material on which natural selection can act; thus organisms that reproduce sexually are more likely to evolve. The fact that all higher animals—the most complex and advanced organisms, evolutionarily speaking—reproduce sexually supports this point of view. We shall explore this subject further in Section 7.

SUMMARY

Sexual reproduction involves a special kind of nuclear division called meiosis. Meiosis is the process by which the genetic material is reassorted and cells are produced which have the haploid chromosome number ($1n$). The other principal component of sexual reproduction is fertilization, the coming together of haploid cells, which restores the diploid number ($2n$). There are characteristic differences among major groups of organisms as to where in the life cycle these events take place.

At the start of meiosis, the chromosomes arrange themselves in pairs. The members of the pairs are known as homologues. One homologue of each pair is of maternal origin and one is of paternal origin. Each homologue consists of two identical chromatids. Early in meiosis, crossing over occurs between homologues, resulting in exchanges of genetic material.

In the first stage of meiosis, the homologues are separated. Two nuclei are produced, each with a haploid number of chromosomes, each consisting of two chromatids. The cell enters interphase, but the chromosomal material is not duplicated. In the second stage of meiosis, the chromatids separate as in mitosis. When the two new nuclei divide, four haploid cells result.

Meiosis results in cells that are different from the parent cell both in number of chromosomes and (because of crossing over) in chromosome composition. Thus they provide a source of variation and so make it possible for evolution to take place.

QUESTIONS

1. Why are there no centrioles in the drawings?

2. Draw a diagram of a cell with eight chromosomes ($n = 4$) at meiotic prophase I. Label each pair of chromosomes differently (e.g., one pair labeled A^1 and A^2, another B^1 and B^2, etc.)

3. Diagram the possible gametes resulting from a single meiotic division in a plant with six chromosomes ($n = 3$).

4. What is happening at each step in the illustrations which appear below?

CHAPTER 10

From an Abbey Garden:
The Beginning of Genetics

10-1

Gregor Mendel, in experiments carried out in a monastery garden, showed that hereditary determinants are carried as separate units from generation to generation. Mendel's discoveries explained how inherited variations can persist for generation after generation.

In the preceding chapter, we saw that the nucleus carries the information that is passed from parent to offspring and that controls cellular structure and function. Each time a cell divides, this information is passed to daughter cells by complex movements of bodies known as chromosomes, which bear this information. But what do we mean when we speak of "hereditary information"? What is passed from parent to child? What shape and form does it take?

At about the same time that Hertwig was observing the fusion of sperm and egg and Darwin was preparing *The Origin of Species* for publication, Gregor Mendel was conducting the first experiments that provided useful answers to these questions. His work, carried out in a quiet monastery garden and ignored until after his death, marks the beginning of modern genetics.

THE CONCEPT OF THE GENE

By Mendel's time, breeding experiments with domestic plants and animals had shown that both parents contribute to the characteristics of the offspring and that these contributions are carried in the male and female sex cells, the sperm and the eggs.

Mendel's great contribution was to demonstrate that inherited characters are carried as discrete units, which are parceled out in different ways, or *reassorted*, in each generation. These discrete units eventually came to be known as *genes*.

Mendel was successful where his predecessors had failed largely for two reasons. First, he planned his experiments carefully and imaginatively, choosing for study definite and measurable hereditary differences. Second, he was one of the first to apply mathematics to the study of biology. Even though his mathematics was simple, the idea that exact rules could be applied to the study of living organisms was startlingly new. (Mendel had studied physics, which may well be why he was prepared to think in terms of "laws.") He organized his data in such a way that his results could be evaluated simply and objectively and could be checked and duplicated by other scientists, as they eventually were.

A FIT MATERIAL

For his experiments in heredity, Mendel chose the common garden pea. It was a good choice. The plants were commercially available and easy to cultivate.

130   Genetics

10-2

In a flower, pollen develops in the anther and the egg cells in the ovule. Pollination occurs when pollen grains, trapped on the stigma, germinate and grow down to the ovule, where they release sperm cells. The fertilized egg develops within the ovule, and ovule and embryo form the peas (the seeds).

Pollination in most species involves the pollen from one plant (often carried by an insect) being caught on the stigma of another plant. This is termed cross-pollination.

In the pea flower, however, the stigma and anthers are completely enclosed by petals, and the pea, unlike most, does not open until after fertilization has taken place. Thus the plant normally self-pollinates. In his crossbreeding experiments, Mendel pried open the bud before the pollen matured and removed the anthers with tweezers. Then he artificially pollinated the flower by dusting the stigma with pollen collected from other plants.

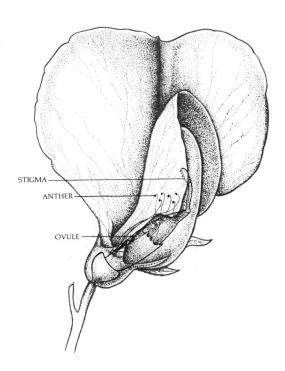

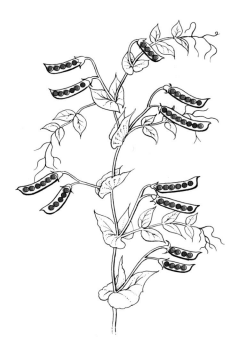

10-3

Garden pea plant with seeds (peas) in the approximate ratio of three yellow seeds for each green seed, as observed by Mendel.

Different varieties had clearly different characteristics, which "bred true," reappearing in crop after crop, and although the plants could be crossbred experimentally, accidental crossbreeding (as, for example, by insects) would not be likely to confuse the experimental results (Figure 10-2). As Mendel said in his original paper, "The value and utility of any experiment are determined by the fitness of the material to the purpose for which it is used."

Mendel started with 32 different types of pea plants, which he studied for two years before he began his experiments. As a result of these observations, he selected for study seven traits that appeared as conspicuously different characteristics in different types of plants. One variety of plant, for example, always produced yellow peas, or seeds, and another always produced green ones. In one variety, the seeds, when dried, had a wrinkled appearance; in another variety, they were smooth. These seven traits and their alternatives are listed in Table 10-1.

Mendel then performed crosses among the different types of pea plants. For instance, he used the pollen from a white flower to fertilize a red flower and the pollen from a plant that bred true for yellow seeds to fertilize the flower from a

Table 10-1   *Mendel's Pea-Plant Experiment*

| | ORIGINAL CROSSES | | F_2 GENERATION | | |
TRAIT	DOMINANT × RECESSIVE		DOMINANT	RECESSIVE	TOTAL
Seed form	Round	× Wrinkled	5,474	1,850	7,324
Seed color	Yellow	× Green	6,022	2,001	8,023
Flower position	Axial	× Terminal	651	207	858
Flower color	Red	× White	705	224	929
Pod form	Inflated	× Constricted	882	299	1,181
Pod color	Green	× Yellow	428	152	580
Stem length	Tall	× Dwarf	787	277	1,064

A pea plant homozygous for red flowers is represented as WW in genetic shorthand. The gene for red flowers is designated W because of a convention by which geneticists, in indicating a pair of alleles, use the first letter of the less common form (white). The capital indicates the dominant, the lowercase the recessive. A WW plant can produce egg cells or sperm cells with only a red-flower (W) gene. The female symbol ♀ indicates that this flower contributed the egg cells, or female gametes.

A white pea plant (ww) can produce egg cells or sperm cells with only a white-flower (w) gene. The male symbol ♂ indicates that this flower contributed the sperm cells, or male gametes.

When a w sperm cell fertilizes a W egg cell, the result is a Ww pea plant, which, since the W gene is dominant, will produce red flowers. However, this Ww plant can produce egg cells or sperm cells with either a W or a w allele.

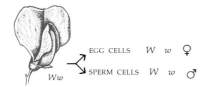

And so, if the plant self-pollinates, four possible crosses can occur:

♀ W × ♂ W ——→ red flowers
♀ W × ♂ w ——→ red flowers
♀ w × ♂ W ——→ red flowers
♀ w × ♂ w ——→ white flowers

These results are summarized in Figure 10-5.

green-seeded plant. When he had performed these experimental crosses between the different varieties, he found that in every case in the first generation (now known as F_1 in biological shorthand), one of the alternate traits disappeared completely without a sign. For example, all the offspring of the cross between yellow-seeded plants and green-seeded plants were as yellow-seeded as the yellow-seeded parent. This trait (yellow seeds) and the other such traits of the F_1 generation Mendel called _dominant;_ the traits that disappeared in the first generation (such as green seeds) he called _recessive._

The interesting question was: What had happened to the recessive trait—the wrinkledness of the seed or the greenness of its color—that had been passed on so faithfully for generations by the parent stock? Mendel let the pea plant itself carry out the next stage of the experiment by permitting the F_1 to self-pollinate. The recessive traits reappeared in the second (F_2) generation. In Table 10-1 are the results of Mendel's actual counts. Before you read any further, look at these figures and see if you can detect any relationships among them. Some interpreters of Mendel's work believe that he first formulated his theory and then performed the experiments to test it. Others believe that he examined the data first and then worked out a hypothesis to explain them. In either case, these numbers were the basis for Mendel's first "law."

SEGREGATION

Did you notice that the dominant and recessive traits appear in the second, or F_2, generation in ratios of about 3:1? How do the recessives disappear so completely and then appear again, and always in such constant proportions? It was in answering this question that Mendel made his greatest contribution. He saw that the appearance and disappearance of traits and their constant proportions could be explained only if hereditary characters are always determined by discrete factors. Moreover, he reasoned, these factors must occur in the offspring as pairs—one inherited from each parent—that are separated again when the sex cells are formed, producing two kinds of gametes, with one factor of the pair in each.

This hypothesis is known as Mendel's first law, or the _principle of segregation._

The two genetic factors in a pair might be the same, in which case the self-pollinating plant would breed true. Or the two factors might be different; such different, or alternative, forms of the same gene came to be known as _alleles._ Yellow-seededness and green-seededness, for instance, are determined by alleles, different forms of the gene (factor) for seed color.

When the alleles of a gene pair are the same, the organism is said to be _homozygous_ for that particular trait; when the alleles are different, the organism is _heterozygous_ for that trait.

When gametes are formed, genes are passed on to them; but each gamete contains only one of two possible alleles. When two gametes combine in the fertilized egg, the genes occur in matched pairs again. One allele may be dominant over another allele; in this case, the organism will appear as if it had only this gene. This outward appearance is known as its _phenotype._ However, in its genetic makeup, or _genotype,_ each allele still exists independently and as a discrete unit even though it is not visible in the phenotype, and the recessive allele will separate from its dominant partner when gametes are again formed. Only if two recessive alleles come together—one from the female gamete and one from the male—will the phenotype then show the recessive trait.

When pea plants homozygous for red flowers are crossed with pea plants having white flowers, only pea plants with red flowers are produced, although each plant in the F_1 generation will carry a gene for red and a gene for white. Figure 10–5 shows what happens in the F_2 generation if the F_1 generation self-pollinates. Notice that the result would be the same if the individuals were cross-fertilized with others of the same genotype, which is the way these experiments are performed with plants that are not self-pollinating.

In order to test his hypothesis (diagramed in Figure 10–5), Mendel performed an additional experiment. He crossed white-flowering plants with white-flowering plants and confirmed that they bred true—that he got only white-flowering plants. Next he crossed one of his F_1 hybrids, the result of a cross between red- and white-flowering plants, with a true-breeding white-flowering plant. To the outside observer, it would appear as if Mendel were simply repeating his first experiment, crossing plants having red flowers with plants having white flowers. But he knew that if his hypothesis were correct, his results would be different from those of his first experiment. In fact, he actually predicted the results of such a cross before he made it. Can you? The easiest way is to diagram

10–5

A cross between a pea plant with two dominant genes for red flowers (WW) and one with two recessive genes for white flowers (ww). The phenotype of the offspring in the F_1 generation is red, but note that the genotype is Ww. The F_1 heterozygote self-pollinates, producing four kinds of gametes, ♀ W, ♂ W, ♀ w, and ♂ w, in equal proportions. The W and w sperm cells and eggs combine randomly to form, on the average, ¼ WW (red), ½ Ww (red), and ¼ ww (white) offspring. It is this underlying 1:2:1 genotypic ratio that accounts for the genotypic ratio of 3 dominants (red) to 1 recessive (white). The distribution of traits in the F_2 generation is shown by a Punnett square, named after the English geneticist who first used this sort of checkerboard diagram for the analysis of genetic traits.

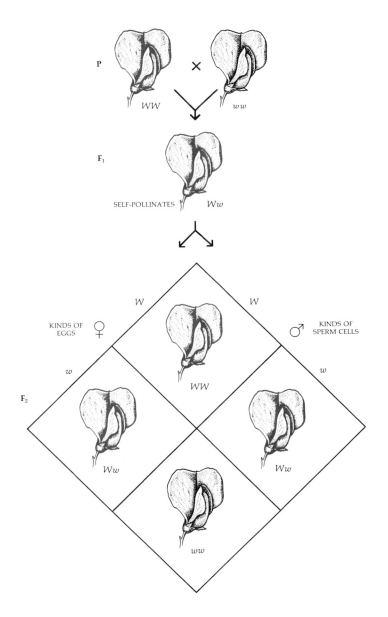

A testcross. In order for a pea flower to be white, the plant must be homozygous for the recessive gene (ww). But a red pea flower can come from a plant with either a Ww or a WW genotype. How could you tell such plants apart? Mendel solved this problem by breeding such plants with homozygous recessives. As shown here, a phenotypic ratio in the F$_1$ generation of 2 red to 2 white indicates a heterozygous red-flowering parent. What would have been the result if the plant being tested had been homozygous for red flowers?

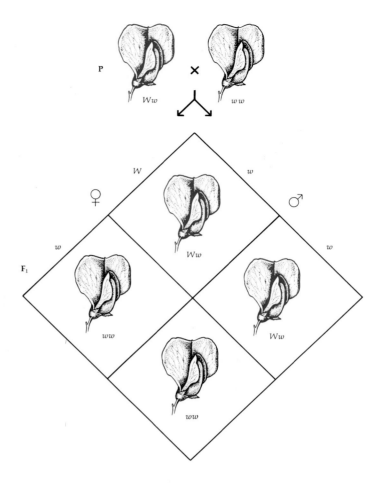

it, as in Figure 10-6. This experiment, which reveals the genotypes, is known as a testcross. A testcross is an experimental cross between an individual with the dominant phenotype for a given trait with another individual that is homozygous recessive for the trait. The resulting ratio of offspring will indicate whether the individual with the dominant phenotype was homozygous or heterozygous for the trait being studied. In the testcross shown in Figure 10-6, the cross revealed that the genotype of the individual being tested was *Ww* rather than *WW*.

INDEPENDENT ASSORTMENT

In a second series of experiments, Mendel studied crosses between pea plants that differed simultaneously in two characteristics; for example, one parent plant had peas that were round and yellow, and the other had peas that were wrinkled and green. The round and yellow traits, you will recall (see Table 10-1), are both dominant, and the wrinkled and green are recessive. As you would expect, all the seeds of the F$_1$ generation were round and yellow. When the F$_1$ seeds were planted and the flowers allowed to self-pollinate, 556 seeds were produced. Of these, 315 showed the two dominant characteristics, round and yellow, but only 32 combined the recessive traits, green and wrinkled. All the rest of the seeds were unlike either parent; 101 were wrinkled and yellow, and 108 were round and green. Totally new combinations of characteristics had appeared. This experiment did not contradict Mendel's previous results. Round and wrinkled still appeared in the same 3:1 proportion (423 round to 133 wrinkled), and so did yellow and green (416 to 140 green). But the round and the yellow traits and the wrinkled and the green ones, which had originally combined in one plant, behaved as if they were entirely independent of one another. From this, Mendel formulated his second law, the *principle of independent assortment*, which states that pairs of alleles (genes) assort independently from each other when two or more pairs of alleles are involved in a cross.

Figure 10-7 diagrams Mendel's results and shows why, in a cross involving two pairs of alleles, each pair with one dominant and one recessive allele, the ratio of distribution will be, on the average, 9:3:3:1 (with 9 representing the proportion of F_2 progeny that will show the two dominant traits, 1 the proportion that will show the two recessive traits, and 3 and 3 the proportions of the two alternative combinations of dominants and recessives). This is true whether one of the original parents is homozygous for both recessive traits and the other homozygous for both dominant ones, as in the experiment just described ($RRYY \times rryy$), or each parent is homozygous for one recessive and one dominant trait ($rrYY \times RRyy$). You can demonstrate this last point by drawing a Punnett square.

10-7

One of the experiments from which Mendel derived his principle of independent assortment. A plant homozygous for round (RR) and yellow (YY) peas is crossed with a plant having wrinkled (rr) and green (yy) peas. The F_1 generation are all round and yellow, but notice how the traits will, on the average, appear in the F_2 generation. Of the 16 kinds of offspring, 9 show the two dominant traits (RY), 3 show one combination of dominant and recessive (Ry), 3 show the alternate combination (rY), and 1 shows the two recessives (ry). This 9:3:3:1 distribution is always the expected result from a cross involving two pairs of independent dominant-recessive alleles. (The letters R and Y are used because round and yellow are the less common forms in nature.)

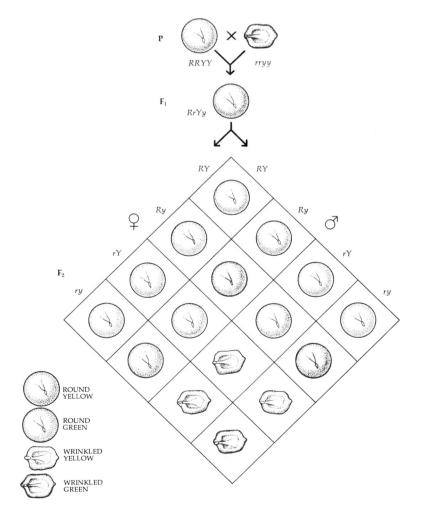

In applying statistics to the study of genetics, Mendel was stating that the laws of chance apply to biology as they do to the physical sciences. Toss an imaginary coin. The chance that it will turn up heads is fifty-fifty, or $\frac{1}{2}$. The chance that it will turn up tails is also fifty-fifty, or $\frac{1}{2}$. Now toss two imaginary coins. The chance that one will turn up heads is again $\frac{1}{2}$. The chance that the second will turn up heads is also $\frac{1}{2}$. The chance that both will turn up heads is $\frac{1}{2} \times \frac{1}{2}$, or $\frac{1}{4}$.

The probability of two independent events occurring together is simply the probability of one occurring alone multiplied by the probability of the other occurring alone. The chance of both turning up tails is similarly $\frac{1}{2} \times \frac{1}{2}$. The chance of the first turning up tails and the second turning up heads is $\frac{1}{2} \times \frac{1}{2}$, and the chance of the second turning up tails and the first heads is $\frac{1}{2} \times \frac{1}{2}$. We can diagram this in a Punnett square: The square indicates that the combination in each square has an equal chance of occurring. It was undoubtedly the observation that one-fourth of the offspring in the F_2 generation showed the recessive phenotype that indicated to Mendel that he was dealing with a simple case of the laws of probability.

If there were three coins involved, the probability of any given combination would be simply the product of all three fifty-fifty possibilities: $\frac{1}{2} \times \frac{1}{2} \times \frac{1}{2}$, or $\frac{1}{8}$. Similarly, with four coins, the probability of any given combination is $\frac{1}{2} \times \frac{1}{2} \times \frac{1}{2} \times \frac{1}{2}$, or $\frac{1}{16}$. The Punnett square on page 135 expresses the probability (or chance) of each of any one of four possible combinations.

If you toss two coins 4 times, it is unlikely that you will get the precise results diagramed below. However, if you toss two coins 100 times, you will come close to the results predicted in the Punnett square, and if you toss two coins 1,000 times, you will be very close indeed. As Mendel knew, the relationship between dominants and recessives might well not have held true if he had been dealing with a small sample. The larger the sample, the more closely it will conform to results predicted by the laws of chance.

INFLUENCE OF MENDEL

Mendel's experiments were first reported in 1865 before a small group of people at a meeting of the Brünn Natural History Society. None of them, apparently, understood what Mendel was talking about. (So do not feel dismayed if you have to read this chapter again.) But his paper was published the following year in the *Proceedings* of the Society, a journal which was circulated to libraries all over Europe. In spite of this, his work was ignored for 35 years, during most of which he devoted himself to the administrative duties of an abbot, and he received no scientific recognition until after his death. (He was, to use DuPraw's phrase, an odd traveler whose tale could not be made to fit.)

It was not until 1900, 35 years after Mendel first reported his work, that biology was finally prepared to accept his findings. Within a single year, his paper was independently rediscovered by three scientists working in three different European countries. Each of them had done similar experiments and was searching the scientific literature to seek confirmation of his results. And each found, in Mendel's brilliant analysis, that much of their work had been anticipated.

GENES AND CHROMOSOMES

During the 35 years that Mendel's work remained in obscurity, great improvements were made in microscopy and, as a consequence, in the study of the structure of the cell (cytology). It was during this period that chromosomes were first discovered and their movements at meiosis and mitosis, described in the preceding chapters, were observed and recorded.

Two years after the rediscovery of Mendel, in 1902, Walter Sutton, then a graduate student at Columbia University, was observing meiosis in the production of sperm cells in grasshoppers and was struck by the parallels between what he was seeing and the laws of Mendel: Chromosomes come in pairs (Figure 10–8), so do Mendelian factors (genes). Chromosome pairs (homologues) separate when the gametes are formed; so do genes. And genes and chromosomes come together again in pairs in the offspring. On the basis of these parallels, Sutton proposed that the factors observed by Mendel were carried on the chromosomes.

10–8

Chromosomes from a diploid cell of a grasshopper. Note that even though these chromosomes are not paired, it is possible to pick out some of the homologues. The observation that chromosomes come in homologous pairs was one of Sutton's clues to the meaning of meiosis.

Many characteristics in man are governed by simple Mendelian traits. One such characteristic is tongue rolling. The ability to roll one's tongue is a simple dominant. Can you roll your tongue? Can your parents? What is your genotype for tongue rolling? If you cannot roll your tongue, does this mean that neither of your parents can? This may seem to be a very trivial sort of characteristic, one that neither natural selection nor society would favor, but oddly enough, very small differences such as this may have considerable importance over an evolutionary time span.

Of more immediate consequence are a number of congenital diseases that are the result of the coming together of recessive genes. One such disease is sickle cell anemia. The sickling gene is a recessive; in the heterozygote, it is usually detectable only by laboratory tests. In persons homozygous for the sickling gene, however, a large proportion of the red blood cells "sickle"—that is, form a sickle shape—and then clog the small capillaries, causing often painful blood clots and depriving vital organs of their full supply of blood. This produces intermittent illness and often a shortened life span. About 4 percent of the population in some tropical regions in Africa are born with sickle cell anemia, and almost half of the members of some African tribes are known to carry the recessive gene.

Another characteristic in man that follows a simple Mendelian pattern of inheritance is the ABO blood groups, described on page 373.

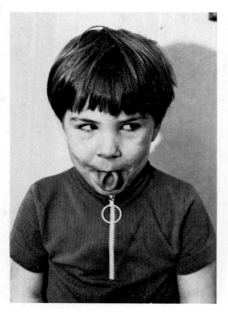

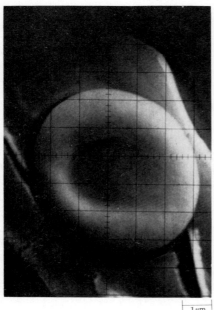

1 μm

1 μm

The tongue-rolling ability is transmitted by a dominant gene. Seven out of ten people have this ability. Do you? You can perform a simple experiment in genetics by checking to see if your parents and your brothers and sisters can do it.

Scanning electron micrographs of normal (left) and sickle (right) red blood cells.

What about Mendel's law of independent assortment in relation to the movement of the chromosomes at meiosis? This law states that members of different pairs of alleles assort independently. We can see that this also can be true if—and this is an important point—the genes are on different chromosomes, as shown in Figure 10–9.

Sutton's hypothesis was widely accepted and eventually proved. And genetics was off to a racing start.

10-9

The chromosome distributions in Mendel's cross of round yellow and wrinkled green peas, according to Sutton's hypothesis. Although the pea has 14 chromosomes (n = 7), only 4 are shown here, the two carrying the genes for round or wrinkled and the two carrying the genes for yellow or green. (This is analogous to what Mendel did when he selected the two traits to study.) As you can *see, one parent is homozygous for the recessives, one for the dominants. Therefore, the only gametes they can produce are RY and ry. (Remember, R now stands not just for the trait but for the chromosome carrying the trait, as do the other letters.) The F$_1$ generation, therefore, must be Rr and Yy. When a mother cell of this generation undergoes meiosis, R is separated from r and Y from y* *when the respective homologues separate at anaphase 1. Four different types of n egg cells are possible, as the diagram reminds us, and also four different types of pollen nuclei. These can combine in 4 × 4, or 16, different ways; the ways in which they can combine are illustrated in the Punnett square at the bottom of the figure.*

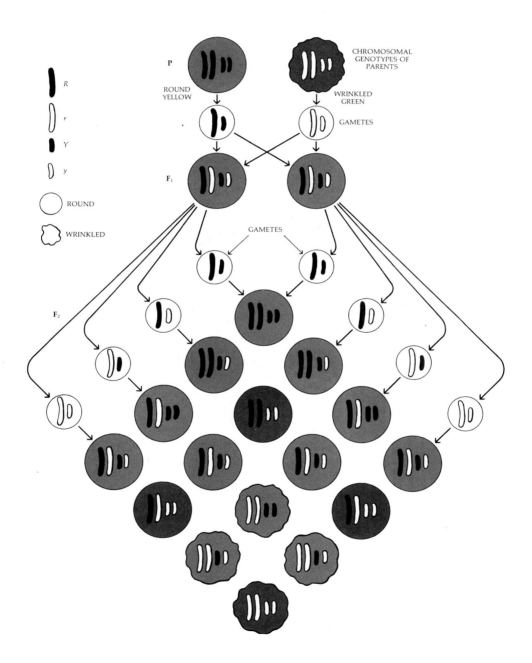

SUMMARY

What is the nature of the hereditary information that is passed from cell to cell and generation to generation? Gregor Mendel provided the first answers to this question when he showed that hereditary traits were passed from parent to offspring as distinct units.

According to Mendel's principle of segregation, these discrete factors (later called genes) appear in pairs in organisms, one of each pair inherited from each parent. Such alternative forms of the same gene are known as alleles. When gametes are formed, alleles are separated.

The genetic makeup of an organism is known as its genotype. The appearance, or outward characteristics, of an organism, is its phenotype. Both alleles may be alike (homozygosity), or they may be different (heterozygosity). Although both alleles are present in the genotype, only one may be detected in the phenotype. The allele that is expressed in the phenotype is the dominant gene; the one that is concealed in the phenotype is the recessive gene. When two organisms that are heterozygous for different alleles of the same gene are crossed, the ratio of dominant to recessive in the phenotype is 3:1.

Mendel's other great principle, the principle of independent assortment, applies to the behavior of two or more genes. This law states that the members of each pair of alleles segregate independently (if, as it was shown later, the genes are on different chromosomes). In crosses involving two independent pairs of alleles, the expected phenotypic ratio is 9:3:3:1.

The importance of Mendel's work, first published in 1865, was not recognized until it was rediscovered in 1900. Shortly thereafter, the movement of the chromosomes at meiosis was correlated with Mendel's laws, and an important hypothesis was formed. According to this hypothesis, the Mendelian units, or genes, are located on the chromosomes and are distributed with them.

QUESTIONS

1. Distinguish between the following terms: gene/allele, dominant/recessive, homozygous/heterozygous.

2. Name Mendel's two "laws" and explain each, using hypothetical human examples.

3. Assume that brown eyes are dominant over blue eyes (as they usually are). What is the phenotype of a man with two brown-eyed parents, supposing that one parent is heterozygous for the trait and the other parent homozygous? What are his possible genotypes?

4. If such a man marries a woman with blues eyes, what are the possibilities for eye color among their children? Suppose they have one with blue eyes; what would you then know about his genotype? Explain your results by drawing a Punnett square.

5. Suppose this couple had four brown-eyed children. What would you then know about the father's genotype?

6. PKU, phenylketonuria, is a disease caused by the coming together of two recessive genes. It causes mental deficiency. If two healthy parents have a child with PKU, what are their genotypes with respect to PKU? What are their chances of having another child with the same disease?

CHAPTER 11

More about Genes

11-1

Hugo de Vries is shown standing next to Amorphophallus titanum, *which has the largest flower cluster of any of the flowering plants. De Vries, a Dutch botanist, was the first to recognize the nature of mutations and their role in hereditary processes.*

MUTATIONS AND EVOLUTION

One of Mendel's rediscoverers, a Dutch botanist named Hugo de Vries, was studying genetics in the evening primrose. Heredity in the primrose, he found, was generally orderly and predictable, as in the garden pea, but occasionally a characteristic appeared that was not present in either parent or indeed anywhere in the lineage of that particular plant. De Vries hypothesized that this characteristic came about as the result of a change in a gene and that the trait embodied in the changed gene was then passed along like any other hereditary trait. De Vries spoke of this hereditary change as a *mutation* and of the organism that carried it as a *mutant.**

During this same period (the early 1900s), biologists were attempting to bring together the new developments in genetics and Darwin's theory of evolution. Mendel's work answered the troublesome question of how genetic traits persist, skipping generations, disappearing and reappearing. It also provided for small variations among populations, with each individual representing a new mixture of traits. But it was hard to see how larger changes could come about—how new species could come into being. This question was made more difficult by the fact that the age of the earth was at that time believed to be much shorter than we know it to be today, and so there was much less time for evolution to have taken place.

De Vries and other geneticists seized upon mutations—changes by "leaps and jumps," to quote de Vries—as the principal factor in evolution, and this "leaping" theory of evolution persisted for some time. As one of its supporters, T. H. Morgan, stated, "Evolution is not a war of all against all, but it is largely a creation of new types for unoccupied or poorly occupied places in nature."

Eventually Darwin's original theory was proved right: Natural selection acting on small variations is the essential process in evolution. Natural selection acts on the whole organism—the phenotype—and not usually on individual traits. Mutation, however, is the source of new genes. These new genes are

* It is one of the ironies of history that only about 2 of some 2,000 changes in the evening primrose observed by de Vries were actually mutations. The rest were due to new combinations of genes rather than to actual changes—mutations—in any particular gene. However, de Vries' definition of a mutant and his recognition of the importance of the concept of mutation are valid, although most of his examples are not.

worked into the genotype, being brought into new combinations and relationships by the process of meiosis.

In 1927, H. J. Muller found that treatment of gametes by x-rays greatly increases the rate of appearance of mutations. It was soon discovered that other radiations, such as ultraviolet light, and also some chemicals can act as *mutagens*, agents that produce mutations. Studies of mutations have shown that whether they are deliberately produced by radiations and chemicals or are "spontaneous,"* they are usually detrimental to the organism, and in fact, many of them are lethal. This is not surprising. If one were to change a word at random in a Shakespearean sonnet or a wire at random in a television set, an improvement would be unlikely, and the results might well be disastrous. It is, in part, because of the effects on the sex cells—effects whose consequences will be concealed until the next generation—that many are concerned about the increase in radiation exposure caused by increased use of nuclear energy or by unwarranted or careless use of x-rays. (Radiation exposure can also produce harmful changes in somatic cells which may result in cancer—as shown, for example, by the increased incidence of leukemia among the "survivors" at Hiroshima.)

INCOMPLETE DOMINANCE

During the decade that followed the discovery of Mendel's work, many studies were carried out which, while confirming his work in principle, showed that the action of genes is more complex than it had at first appeared to be. An important example is *incomplete dominance* (also called codominance).

Dominant and recessive traits are not always so clear-cut as in the pea plant. Some traits do appear to blend. For instance, a cross between a red snapdragon and a white snapdragon produces heterozygotes that are pink (Figure 11–2). But when members of this generation self-pollinate, the traits begin to sort themselves out once again. As we shall see, what occurs in the snapdragon and in similar crosses is actually a result of the combined effect of gene products.

BROADENING THE CONCEPT OF THE GENE

Mendel's experiments seemed to suggest that each gene affects a single characteristic in a one-to-one relationship. It was soon discovered, however, that a single gene sometimes affects many traits in an organism (*pleiotropy*), and that, conversely, a single trait is often affected by many genes (*polygenic inheritance*).

Pleiotropy

One of the first investigators to demonstrate pleiotropy was Theodosius Dobzhansky. Dobzhansky arbitrarily selected 12 mutant female fruit flies, each with a single different mutation that changed a specific characteristic, such as eye or body color or wing shape. In each of these flies, he examined the shape of the spermatheca, a special sperm-storing organ associated with the female reproductive tract. Ten out of the twelve mutants showed a variation from normal in the size and shape of this particular organ, although each mutant fly was thought to differ from the norm in only a single obvious characteristic. Thus it was apparent that the genes affected more than one characteristic of the flies.

11–2

A cross between a red snapdragon and a white snapdragon. This looks very much like the cross between a red- and a white-flowering pea plant shown in Figure 10–5, but there is a significant difference. Although the Mendelian ratio of 1:2:1 appears in the F₂ generation, the heterozygote is pink rather than red.

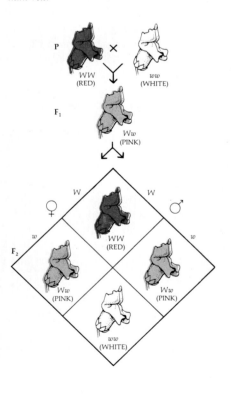

* "Spontaneous," in this sense, means that we do not know why they happen.

GENOTYPE AND PHENOTYPE

There is a crucial difference between genotype and phenotype. From the moment of its conception, every organism is acted upon by the environment, and the expression of any gene is always the result of the interaction of gene and environment. To take a simple, familiar example, a seedling may have the genetic capacity to be green, to flower, and to fruit, but it will never turn green if it is kept in the dark, and it may not flower and fruit unless certain precise environmental requirements are met. Himalayan rabbits are all white if they are raised at high temperatures (above 35°C). However, rabbits of the same genotype, when raised at room temperature, have black ears, forepaws, noses, and tails.

In humans, also, similar genotypes are expressed quite differently in different environments. A simple example is found in height, which, as we have noted, is influenced by a group of genetic factors. This influence is indicated by the fact that tall parents generally have tall children and short parents tend to have short children. On the other hand, adult height is clearly also influenced by environmental factors in childhood, such as diet, sunshine, and incidence of disease. The population of the United States grew taller with each generation for four generations, presumably as a result of general improvements in the standard of living and in the average diet (this trend has just come to an end; grown children in this country are no longer, on the average, taller than their parents).

As René Dubos reminds us,* not only are children growing taller, but they are growing faster; a boy now reaches full height at about 19 years, but 50 years ago, maximum stature was not usually attained until age 29. Of more social consequence is the fact that puberty is also being reached earlier. In Norway, for example, the mean age of the onset of menstruation has fallen from 17 in 1850 to 13 in 1960. Historical evidence indicates, however, that also in Imperial Rome and Western Europe in Shakespeare's time, teen-agers reached puberty at an early age (Juliet, remember, was not yet 14). Apparently, the slowing of the growth and maturation rate of the general population was a

The characteristic color pattern of the Siamese cat is a result of an interaction between genotype and environment. Because of a mutation in the gene affecting coat color, the gene is only expressed in the cooler, more peripheral areas of the body, such as the tips of the ears and tail, the muzzle, and the paws. As is often the case with animals with mutations resulting in reduced pigmentation (white tigers, pearl minks, and albino rats), Siamese cats are also frequently cross-eyed.

consequence of the industrial revolution and increasing urbanization, which reduced the standards of nutrition and health. Thus we have a situation in which the genetic potential has remained apparently unaltered over the centuries, but the phenotypic expression has undergone fluctuations. A side effect of this change in phenotypic expression is that teen-agers are now reaching physical maturity early in a society in which childhood and dependency have become greatly prolonged.

* René Dubos, So Human an Animal, Charles Scribner's Sons, New York, 1968.

Sometime later, another investigator, Hans Gruneberg, approaching the problem from the other direction, studied a whole complex of congenital deformities in rats, including thickened ribs, a narrowing of the tracheal passage, a loss of elasticity of the lungs, hypertrophy of the heart, blocked nostrils, a blunt snout, and needless to say, a greatly increased mortality. All these changes, he was able to demonstrate, were caused by a single mutation, that is, a mutation involving only one gene. This particular gene produces a protein involved in the formation of cartilage, and since cartilage is one of the most common structural substances of the body, the widespread effects of such a gene are not difficult to understand. In fact, it is very likely that Mendel's allele for wrinkled, for example, affected other characteristics of the pea.

The frizzle trait in fowl (Figure 11–3) is another example of pleiotropy.

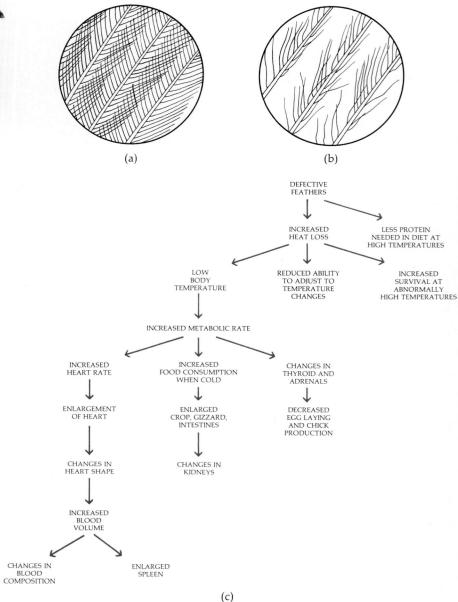

11-3

The frizzle trait in fowl is an example of pleiotropy, the capacity of a gene to produce a variety of phenotypic effects. "Frizzle" is manifested primarily by differences in the feathers. (a) Under low magnification, feathers from normal birds show a closely interwebbed structure. (b) Frizzle feathers are weak and stringy and provide poor insulation. Some of the consequences of the manifestation of this single gene are shown in (c). Notice that the frizzle trait is a liability at low temperatures but may increase survival at high temperatures.

Polygenic Inheritance

A trait affected by a number of genes does not show a distinctive clear-cut difference between groups—such as the differences tabulated by Mendel—but rather shows a gradation of small differences, which is known as continuous variation. If you make a chart of genetic differences among individuals that are affected by a number of genes, you get a curve such as that shown in Figure 11-4.

Fifty years ago, the average height in the United States was less but the shape of the curve was the same; in other words, the great majority fell within the middle range and the extremes in height were represented by only a few individuals. Some of these height variations are produced by environmental

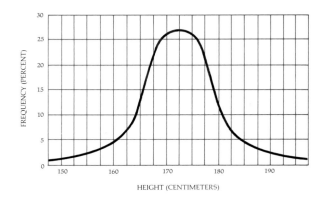

factors, such as diet, but even if all the men in a population were maintained from birth on the same type of diet, there would still be a continuous variation in height in the population. This is due to genetic differences in hormone production, bone formation, and numerous other factors. A number of other characteristics in man, notably skin color, are controlled by polygenic inheritance.

Table 11-1 illustrates a simple example of polygenic inheritance, color in wheat kernels, which is controlled by two pairs of genes. Human skin color is under a similar kind of genetic control.

Table 11-1 *The Genetic Control of Color in Wheat Kernels*

PARENTS:	$R_1R_1R_2R_2 \times r_1r_1r_2r_2$ (DARK RED) (WHITE)		
F_1:	$R_1r_1R_2r_2$ (MEDIUM RED)		
F_2:	Genotype		Phenotype
1	$R_1R_1R_2R_2$		Dark red
2 } 4 2	$R_1R_1R_2r_2$ $R_1r_1R_2R_2$		Medium-dark red Medium-dark red
4 } 6 1 1	$R_1r_1R_2r_2$ $R_1R_1r_2r_2$ $r_1r_1R_2R_2$		Medium red Medium red Medium red
2 } 4 2	$R_1r_1r_2r_2$ $r_1r_1R_2r_2$		Light red Light red
1	$r_1r_1r_2r_2$		White

15 red to 1 white

SUMMARY

Mutations are abrupt changes in genotype. Together with the genetic recombinations that occur at fertilization and meiosis, they are the source of variations necessary for biological evolution.

Incomplete dominance is a situation in which the effects of both alleles are apparent in the heterozygote.

Genes commonly affect more than one characteristic; this property of the gene is known as pleiotropy.

Many characteristics are under the control of a number of separate genes. This phenomenon is called polygenic inheritance. Traits under the control of a number of genes typically show continuous variation, as represented by a bell-shaped curve.

Genes determine only potential capacities. Interactions of the genotype and the environment determine the phenotype.

QUESTIONS

1. The so-called "blue" (really gray) Andalusian variety of chicken is produced by a cross between the black and white varieties. What color chickens (and in what proportions) would you expect if you crossed two blues? If you crossed a blue and a black?

2. Skin color in one strain of mice is determined by five different genes. The colors range from almost white to dark brown. Would it be possible for any pair of mice to produce offspring darker or lighter than either parent?

3. You and a geneticist are looking at a mahogany-colored Ayrshire cow with a newly born red calf. You wonder if it is male or female, and the geneticist says it is obvious from the color which sex the calf is. He explains that in Ayrshires the genotype AA is mahogany and aa is red, but the genotype Aa is mahogany in males and red in females. What is he trying to tell you—that is, what sex is the calf?

CHAPTER 12

White-Eyed and Other Fruit Flies

FEMALE   MALE

12–1

The fruit fly and its chromosomes. This species has only four pairs of chromosomes, a fact that simplified Morgan's experiments. Six of the chromosomes (three pairs) are autosomes (including the two small spherical chromosomes in the center) and two are sex chromosomes.

Early in the 1900s, Thomas Hunt Morgan began a study of genetics at Columbia University, founding what was to be the most important laboratory in the field for several decades. By a remarkable combination of foresight and good fortune, he selected as his experimental material the fruit fly *Drosophila.*

Drosophila means "lover of dew," although actually this useful little fly is not attracted by dew but feeds on the fermenting yeast which it finds on rotting fruit. The fruit fly was a likely choice for a geneticist since it is easy to breed and maintain. These tiny flies, each only about 3 millimeters long, produce a new generation every two weeks. Each female lays hundreds of eggs at a time, and an entire population can be kept in a half-pint bottle, as they were in Morgan's laboratory.

The little fruit fly proved to be a "fit material" for a wide variety of genetic investigations. In the decades that followed, *Drosophila* was to become famous as the biologist's principal tool in studying animal genetics.*

THE SEX CHROMOSOMES

Fruit flies have only four pairs of chromosomes (Figure 12–1), a feature that turned out to be particularly useful. Three pairs are structurally the same in both sexes, but the fourth pairs are different; it is this fourth pair that determines the sex of the fly. In the female fly, the two sex chromosomes, as they are called, are structurally the same. These are, by convention, termed X chromosomes, and so the female is designated XX. The sex chromosomes of the male consist of one X chromosome, which is the same as the female X chromosome, and one Y chromosome. Thus males in the fruit fly (and also in mammals) are designated XY. However, there are some animal groups, such as birds, in which the females are XY and the males XX.

In XY males, when the sex cells are formed by meiosis, half of the sperm carry an X chromosome and half carry a Y chromosome. All the gametes formed by a female carry the X. The sex of the offspring is determined by

---

* Geneticists have often used for their experiments such "insignificant" little plants and animals—Mendel's pea plants, for instance, or Hertwig's sea urchins—organisms that seem to occupy very unimportant and out-of-the-way places in the natural order. Underlying this approach is the geneticist's assumption that genetic principles are universal, applying equally to all living things.

12–2

How the sperm cell determines the sex of human offspring. (a) At meiosis, every egg cell receives an X chromosome from the mother. (b) A sperm cell may receive either an X chromosome or a Y chromosome. (c) If a sperm cell carrying an X chromosome fertilizes the egg, the offspring will be female (XX). (d) If a sperm cell carrying a Y chromosome fertilizes the egg, the offspring will be male (XY).

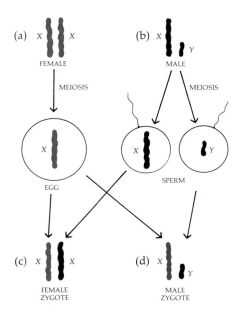

whether the female gamete carrying the X chromosome is fertilized by a male gamete carrying an X chromosome or a male gamete carrying a Y chromosome (Figure 12–2). Since equal numbers are produced, there is theoretically an even chance of male and female.*

GENES AND THE SEX CHROMOSOMES

One of the first mutants to turn up in Morgan's laboratory was a male fly with white eyes. (Red is the usual color of a fruit fly's eyes.) The white-eyed male was mated to a red-eyed female, and all the offspring had red eyes. Then Morgan crossbred the offspring, just as Mendel had in his pea experiments, and this is what he got:

Red-eyed females	2,459
White-eyed females	0
Red-eyed males	1,011
White-eyed males	782

Why were there no white-eyed females? Perhaps the gene for white eyes was carried only on the Y chromosome and not on the X. To test this hypothesis, he crossed the original white-eyed male with one of the F_1 females. And the result was:

Red-eyed females	129
White-eyed females	88
Red-eyed males	132
White-eyed males	86

* In actual fact, the ratio of human male to human female births is about 106 to 100. The reason for this is not known, but it has been suggested that because the Y chromosome is slightly lighter than the X chromosome, the male-determining sperm may have an advantage in getting to the egg.

Offspring of a cross between a white-eyed female fruit fly and a red-eyed male fruit fly, illustrating what happens when a recessive gene, indicated by the white dot, is carried on an X chromosome. The F₁ females, with one X chromosome from the mother and one from the father, are heterozygous (Ww) and so will be red-eyed. But the F₁ males, with their single chromosome received from the mother carrying the recessive (w) trait, will be white-eyed because the Y chromosome carries no gene for eye color. Thus the recessive allele on the X chromosome inherited from the mother will be expressed.

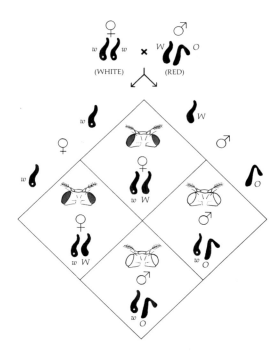

Morgan and his co-workers examined these figures and came to a second conclusion (the right one this time): the gene for eye color is carried only on the X chromosome. (In fact, as it was later shown, the Y chromosome carries almost no genetic information at all.) The white allele is recessive. Thus a heterozygous female would never have white eyes—which is why there are no white-eyed females in the F₁ generation. However, a male that received an X chromosome carrying the allele for white eyes would always be white-eyed since no other allele would be present (Figure 12–3).

Further experimental crosses proved Morgan's hypothesis to be right. They also showed that white-eyed fruit flies are more likely to die before they hatch than red-eyed fruit flies, which explains the lower-than-expected numbers in the F₁ generation and the testcross.

These experiments introduced the concept of sex-linked traits. As we shall see, such traits are important in human genetics. They also established what Sutton had hypothesized some years before: Genes *are* on chromosomes.

LINKAGE

Mendel, you will remember, showed that certain pairs of alleles, such as round and wrinkled, assort independently of other pairs, such as yellow and green. However, as we noted previously, genes follow Mendel's rule of independent assortment only if they are on different chromosomes. The genes on any one chromosome are said to be in *linkage groups.*

As increasing numbers of mutants were found in Columbia University's *Drosophila* population, the mutations began to fall into four linkage groups, in accord with the four pairs of chromosomes visible in the cells. Indeed, in all organisms which have been studied in sufficient genetic detail, the number of linkage groups and the number of pairs of chromosomes have been the same.

CALICO CATS, BARR BODIES, AND THE LYON HYPOTHESIS

A dark spot of chromatin—called a Barr body—can be seen at the outer edge of the nucleus of female mammalian cells in interphase. According to the Lyon hypothesis—named after Mary Lyon, who hypothesized it—this dark spot is an inactivated X chromosome. Early in embryonic life, according to Lyon, this inactivation occurs in one or the other X chromosome in each cell of the female mammal (except for those cells from which egg cells will form). Thus all the somatic cells of female mammals are not identical but are one of two types, depending on which of the X chromosomes is active.

Calico cats (also sometimes known as tortoise-shell cats) have coats that are both black and yellow (actually an orange-yellow). They are also almost always female. In cats, the alleles for black or yellow coat color are carried on the X chromosome, so calico cats neatly fit the Lyon hypothesis. There are, however, occasional male calicos. These are suspected of having an extra X chromosome, a supposition supported by the observation that they are almost always sterile.

CROSSING OVER

Large-scale studies of linkage groups soon revealed some unexpected difficulties. For instance, most fruit flies have gray bodies and long wings. These individuals, which show the common, characteristic features of the population, are known as "wild-type" flies. When wild-type flies were bred with mutant fruit flies having black bodies and short wings (both recessive traits), all the progeny had gray bodies and long wings, as would be expected. Then the F_1 generation was inbred. Two outcomes seemed possible:

1. The two recessives would be assorted independently and would appear in the 9:3:3:1 ratio, indicating that they were on different chromosomes.
2. The two recessives would be linked. In this case, 75 percent of the flies would be gray with long wings, and 25 percent, homozygous for both recessives, would be black with short wings.

In the case of these particular traits, the results most closely resembled outcome 2, but they did not conform exactly. In a few of the offspring, the traits seemed to assort independently, not together; that is, some few flies appeared that were gray with short wings, and some that were black with long wings. How could this be? Somehow the genes had moved out of their linkage groups.

To find out what was happening, Morgan tried a testcross, breeding one of the F_1 generation with a homozygous recessive. If black and gray, long and short assorted independently—that is, if they were on different chromosomes—25 percent of the offspring of this cross should be black with long wings, 25 percent gray with long wings, 25 percent black with short wings, and 25 percent gray with short wings. On the other hand, if the two traits (color and wing size) were on the same chromosome and so moved together, half of the offspring of the testcross should be gray with long wings and half should be black with short wings. But actually, as it turned out, over and over, in counts of hundreds of fruit flies resulting from such crosses, 41.5 percent were gray with long wings, 41.5 percent were black with short wings, 8.5 percent were gray with short wings, and another 8.5 percent were black with long wings.

Morgan was convinced by this time that genes are located on chromosomes. It now seemed clear that the two traits were located on the same chromosome since they did not show up in the 25:25:25:25 percentage ratios of separately assorted genes. The only way in which the observed figures could be explained was if one assumed (1) that the two genes were on one chromosome and their two alleles on the homologous chromosome, and (2) that sometimes genes could be exchanged between homologous chromosomes.

As we noted in Chapter 9, it now has been established that exchange of portions of the chromosome—crossing over—takes place at the beginning of meiosis. As you can see in Figure 12–4, the chromatids of homologous chromosomes stick together during the first meiotic prophase, forming what are known as chiasmata. Presumably crossing over occurs at the chiasmata.

12–4

Homologous chromosomes at the beginning of meiosis in a grasshopper. All four chromatids are visible. Crossing over—exchanges of genetic material—has probably occurred at the point at which these chromatids intersect. The arrows indicate the kinetochores.

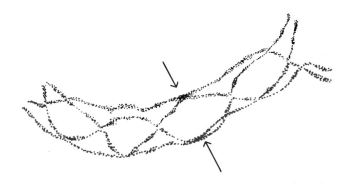

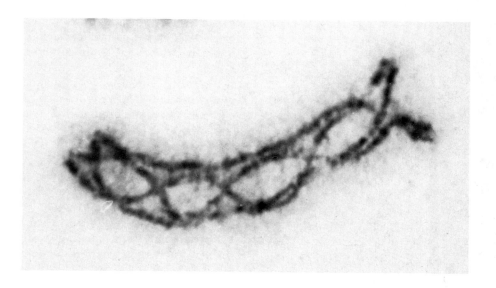

Crossing over takes place when breaks occur in chromatids at the beginning of meiosis, when the chromosomes are paired, and the broken end of each chromatid joins with the chromatid of a homologous chromosome. In this way, alleles are exchanged between chromosomes. The white circles symbolize kinetochores.

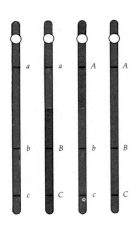

12–6

A portion of the genetic map of Drosophila melanogaster, *showing the positions (loci) of some of the genes on chromosome 2 and the distances between them, as calculated by the frequency of crossovers. As you can see, more than one gene may affect a single characteristic, such as eye color.*

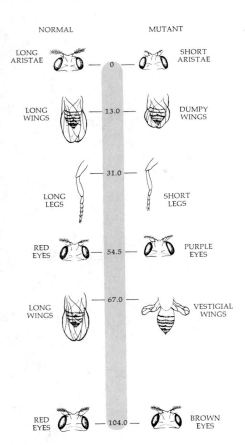

NORMAL		MUTANT
LONG ARISTAE	0	SHORT ARISTAE
LONG WINGS	13.0	DUMPY WINGS
LONG LEGS	31.0	SHORT LEGS
RED EYES	54.5	PURPLE EYES
LONG WINGS	67.0	VESTIGIAL WINGS
RED EYES	104.0	BROWN EYES

"MAPPING" THE CHROMOSOME

With the discovery of crossovers, it began to seem clear not only that the genes are carried on the chromosomes, as Sutton had hypothesized, but that they must be arranged in a definite, fixed linear array along the length of the chromosome. Furthermore, all the alleles must be at corresponding sites on homologous chromosomes (the site of a gene on a chromosome is called its locus). If this were not true, exchange of sections of chromosomes could not possibly result in an exact exchange of alleles.

As other traits were studied, it became clear that the percentage of separations, or crossovers, between any two genes, such as gray body and long wing, was different from the percentage of crossovers between two other genes, such as gray body and long leg. In addition, as Morgan's experiments had shown, these percentages were very fixed and predictable. It occurred to A. H. Sturtevant, one of the many brilliant young geneticists attracted to Morgan's laboratory during these golden days of *Drosophila* genetics, that the percentage of crossovers probably had something to do with the distances between the genes, or in other words, with their spacing along the chromosome. (You can see in Figure 12–5, for example, that in a crossover, the chances of a strand breaking and recombining with its homologous strand somewhere between *B* and *C* is less likely than this happening somewhere between *A* and *C*. This simple concept opened the way to the "mapping" of chromosomes.)

Sturtevant postulated (1) that genes are arranged in a linear series on chromosomes; (2) that genes which are close together will be separated by crossing over less frequently than genes which are farther apart; and (3) that it should therefore be possible, by determining the frequencies of crossovers, to plot the sequence of the genes along the chromosome and the relative distances between them. (The distances cannot be absolute because breaking and rejoining of chromosome segments may be more likely to occur at some sites on the chromosome than on others.) In 1913, Sturtevant began a series of crossover studies in fruit flies. As a standard unit of measure, he arbitrarily took the distance that would give (on the average) one crossover per 100 fertilized eggs. Thus genes with 10 percent crossover would be 10 units apart; those with 8 percent crossover would be 8 units apart. By this method, he and other geneticists constructed genetic maps locating a wide variety of genes and their mutants in *Drosophila* (Figure 12–6).

However, there is still no explanation of how homologous chromosomes break in exactly corresponding sets and so exchange equal amounts of genetic material at crossing over. Presumably, increased knowledge of the structure of the chromosome (Chapter 16) will shed new light on this question.

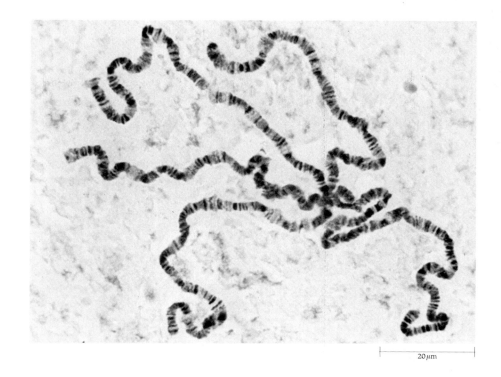

12-7
Chromosomes from the salivary gland of a Drosophila *larva. These chromosomes are 100 times larger than the chromosomes in ordinary body cells, and their details are therefore much easier to see. (This micrograph was taken with a light microscope, not an electron microscope.) Because of the distinctive banding patterns, it is possible in some cases to assign genes to specific loci on particular chromosomes.*

20 µm

GIANT CHROMOSOMES

In *Drosophila*, as in many other insects, certain cells do not divide during the larval stage of the insect. In such cells, however, the chromosomes continue to replicate, over and over again, but since they do not separate after replication, they simply become larger and larger. In 1933, such chromosomes were reported in the salivary glands of the larvae (immature forms) of *Drosophila*. As you can see in Figure 12–7, these giant chromosomes are marked by very distinctive dark and light bands. The bands are areas in which the chromatin is more tightly packed. Changes in these banding patterns were found to correlate with observed genetic changes in the flies. In addition, crossovers and breaks could actually be *seen;* thus further confirmation of Sturtevant's brilliant hypothesis came 20 years later from an unexpected source.

SUMMARY

The fruit fly *Drosophila*, because it is easy to breed and maintain, has been used in a wide variety of genetic studies. *Drosophila* has four pairs of chromosomes; three pairs are structurally the same in both sexes, but the fourth pairs, the sex chromosomes, are different. In fruit flies, as in most species, the two sex chromosomes are XX in the females and XY in the males.

At the time of meiosis, the sex chromosomes are segregated. Each egg cell will receive an X chromosome, but half the sperm cells will receive X chromosomes and half will receive Y chromosomes. Thus it is the sperm cell that determines the sex of the embryo.

In the early 1900s, experiments with mutations in the fruit fly showed that certain characteristics are sex-linked, that is, are carried on the sex chromosomes. Recessive genes carried on the X chromosomes (the Y chromosome contains almost no genetic information) appear in the phenotype far more often in males than in females; a female heterozygous for a sex-linked characteristic will show the dominant trait, whereas a single recessive gene in the male, if carried on the X chromosome, will result in a recessive phenotype since no allele is present.

Some genes assort independently in breeding experiments, and others tend to remain together. Genes that tend to travel together are said to be in the same linkage groups.

Genes are sometimes exchanged between homologous chromatids at meiosis. Such crossovers could take place only if (1) the genes are arranged in a fixed linear array along the length of the chromosomes, and (2) the genes are at corresponding sites (loci) on homologous chromosomes. On the basis of these assumptions, chromosome maps, showing the relative positions of gene loci along *Drosophila* chromosomes, were developed from crossover data provided by breeding experiments.

QUESTIONS

1. Draw a diagram similar to Figure 12–2 indicating sex determination in a robin.

2. A couple has three girls. What are the chances that the next child will be a boy?

3. Suppose you would like to have a family consisting of two girls and a boy. What are your chances, assuming you have no children now? If you already have one boy, what are your chances of completing your family as planned?

4. Construct a pedigree, similar to that shown in Figure 12–3, for a female calico cat, assuming she mates with a yellow male.

5. In a series of breeding experiments, a linkage group composed of genes *A*, *B*, *C*, *D*, and *E* was found to show the following crossover frequencies:

		Gene					
		A	*B*	*C*	*D*	*E*	
	A	—	8	12	4	1	
	B	8	—	4	12	9	Crossovers
Gene	*C*	12	4	—	16	13	per 100
	D	4	12	16	—	3	fertilized eggs
	E	1	9	13	3	—	

Using Sturtevant's standard unit of measure, "map" the chromosome.

CHAPTER 13

Sex Linkage and Human Karyotypes

SEX-LINKED CHARACTERISTICS

Color Blindness

As in *Drosophila,* the Y chromosome of a human male carries much less genetic information than the X chromosome. Genes for color vision, for example, are carried on the X chromosome in humans but not on the Y chromosome. Color blindness is produced by a recessive allele of the normal gene. The allele for complete color vision is dominant; a woman with one X chromosome with the normal allele and one X chromosome with the allele for color blindness will have normal color vision. If she transmits the X chromosome with the recessive allele to a daughter, the daughter also will have normal color vision if she receives a normal X chromosome from her father (that is, if he is not color-blind). If, however, the X chromosome with recessive allele is transmitted from mother to son, he will be color-blind since, lacking a second X chromosome, he

13-1

The normal diploid chromosome number of a human being is 46, 22 pairs of autosomes and the 2 sex chromosomes. The arrangement here is called a karyotype. The autosomes are grouped by size (A, B, C, etc.), and then the probable homologues are paired. A normal woman has two X chromosomes and a normal man, shown here, an X and a Y.

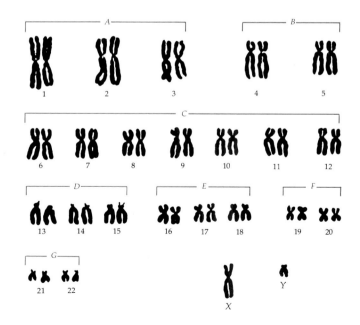

Color blindness in humans is carried by a recessive allele on the X chromosome. In the chart, the mother has inherited one normal and one defective allele. The normal allele will be dominant, and she will have normal color vision. However, half her eggs (on the average) will carry the defective allele and half will carry the normal allele—and it is a matter of chance which kind is fertilized. Since her husband's Y chromosome, the one that determines a son rather than a daughter, carries no gene for color discrimination, the single gene the wife contributes (even though it is a recessive gene) will determine whether or not the son is color-blind. Therefore, half her sons (on the average) will be color-blind. Assuming that her children marry individuals with normal X chromosomes, the expected distribution of the trait among her grandchildren will be as shown on the chart.

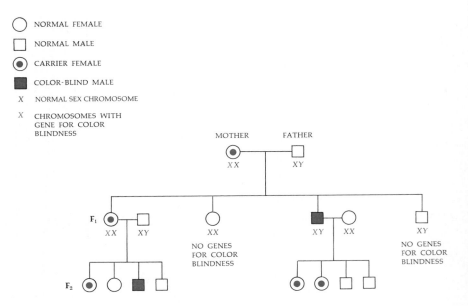

13-3

Gregory Efimovitch Rasputin (1871–1916). Nicholas II and Alexandra of Russia fell under the sinister influence of Rasputin because of the hemophilia of their son Alexis, on whom Rasputin exerted apparently mystical healing powers. The gene for hemophilia had been inherited from Queen Victoria (see Figure 13–4). Alexis did not die of his disease, but was executed with other members of the royal family in 1918.

has only the recessive allele (Figure 13–2). Nonsexual characteristics, such as color blindness, that are controlled by alleles on the X chromosome are said to be _sex-linked_.

Hemophilia

A classic example of a recessive inheritance transmitted on the X chromosome is the hemophilia that has afflicted some royal families of Europe since the nineteenth century. Hemophilia is a group of diseases in which the blood does not clot normally. In some kinds of hemophilia, even minor injuries carry the risk of the patient's bleeding to death. Queen Victoria was probably the original carrier in the family. Because none of her forebears or collateral relatives was affected, we conclude that the mutation may have occurred on an X chromosome in one of her parents or in the cell line from which her own eggs were formed. Prince Albert, Victoria's consort, could not have been responsible; male-to-male inheritance of the disease is impossible. (Why?) One of her sons, Leopold, Duke of Albany, died of hemophilia at the age of 31. At least two of Victoria's daughters were carriers, since a number of *their* descendants were hemophiliacs. And so, through various intermarriages, the disease spread from throne to throne across Europe (Figure 13–4). In the Tsarevitch, son of the last Tsar of Russia, the gene for hemophilia inherited from Victoria had considerable political consequences.

HUMAN CHROMOSOMAL ABNORMALITIES

From time to time, usually because of "mistakes" at the time of meiosis, homologues may not separate. In this case, one of the sex cells has one too many chromosomes and the other one too few. This phenomenon is known as _nondisjunction_. The cell with one too few cannot produce a viable embryo, but the one with one too many sometimes can. The result is an individual with an extra chromosome in every cell of his or her body.

13-4

As this chart shows, Queen Victoria was the original carrier of the hemophilia that has afflicted male members of the royal families of Europe since the nineteenth century. The British royal family escaped the disease because King Edward VII, and consequently all his progeny, did not inherit the defective gene.

UNAFFECTED MALES

AFFECTED MALES

MALES WITH UNKNOWN PHENOTYPE

FEMALES NOT KNOWN TO BE CARRIERS

FEMALES KNOWN TO BE CARRIERS

FEMALES NOT ADEQUATELY PROGENY TESTED

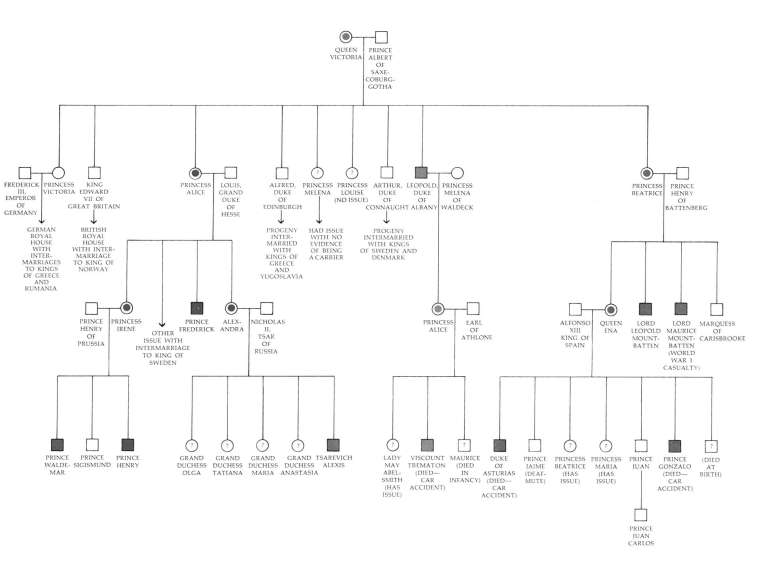

13-5

The frequencies of births of infants with Down's syndrome in relation to the ages of the mothers. The number of cases shown for each age group represents the occurrence of mongolism in every 2,240 births by mothers in that group. As you can see, the risk of having a child with Down's syndrome increases rapidly after the age of 40.

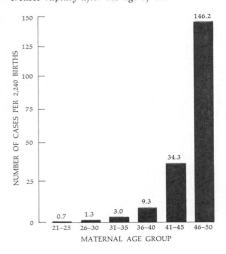

The presence of additional chromosomes often produces widespread abnormalities. Many such infants are stillborn or die soon after birth. Among those who survive, many are mentally deficient. In fact, studies of abnormalities in human chromosome number among living subjects are often carried out on patients in mental hospitals. These patients frequently have abnormalities of the heart and other organs as well.

Down's Syndrome

One of the most familiar chromosomal abnormalities is the form of mental deficiency known as mongolism. This name derives from the characteristic appearance of the eyefold in these people, which makes them look "foreign," or mongoloid, to Europeans. Mongolism usually involves more than one defect and so is more precisely referred to as a syndrome, a group of disorders that occur together. In medical terminology it is usually referred to as Down's syndrome, after the physician who first described it. The syndrome includes, in most cases, not only mental deficiency but a short, stocky body type with a thick neck and, often, abnormalities of other organs, especially the heart.

Down's syndrome and a number of other defects involving gross abnormalities of chromosomes are more likely to occur among infants born to older women (Figure 13–5). The reasons for this are not known, but the formation of the egg cells is well under way in the human female before she is born, so the increasing incidence of abnormalities may be correlated in some way with the aging of the mother's reproductive cells.

The most common cause of the genetic abnormality that produces Down's syndrome is nondisjunction involving chromosome 21. This results in an extra chromosome 21 (Figure 13–6) in the cells of the defective child.

Down's syndrome may also be the result of an abnormality in the chromosomes of one of the parents. In the case of Down's syndrome, the abnormality involved is _translocation_. Translocation occurs when a portion of a chromosome is broken off and becomes attached to another chromosome. The patient with translocation mongolism usually has a third chromosome 21 (or, at least, most of it) attached to a larger chromosome, such as 15. (Thus, in both cases, the patient has three chromosome 21s, or their equivalent.)

13-6

The karyotype of a male patient with Down's syndrome caused by nondisjunction. Note that there are three chromosomes 21.

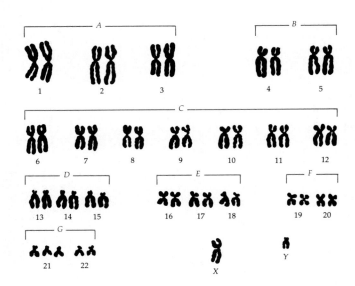

Transmission of translocation mongolism. The father, top row, has normal pairs of chromosomes 21 and 15, and each of his sperm cells will contain a normal 21 and a normal 15. The mother (more frequently, although not always, the translocation carrier) has one normal 15, one normal 21, and a translocation 15/21. She herself appears normal, but her chromosomes cannot pair normally at meiosis. There are six possibilities for the offspring of these parents: the infant will (1) die before birth (three of the six possibilities), (2) be mongoloid, (3) be a translocation carrier like the mother, or (4) be normal. Tests for the chromosomal abnormality can be made in prospective parents and in the fetus before birth.

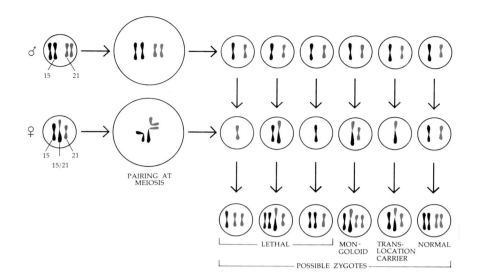

When cases of Down's syndrome due to translocation are studied, it is usually found that one parent, although phenotypically normal, has only 45 separate chromosomes—one chromosome being composed of most of chromosomes 15 and 21 joined together. The possible genetic makeups of the offspring of this parent are diagramed in Figure 13–7. Three out of the six possible combinations are lethal. One of the remaining three will produce Down's syndrome, one will be normal, and one will be a carrier.

Thus parents who have had one abnormal child are faced with the terrible decision of whether or not to risk having another infant. Now there are special clinics throughout the country which can help them in this decision. In these clinics, skin cells from stillborn or abnormal infants are cultivated in a test tube for examination at the time of mitosis, when the chromosomes become visible. If an infant is found to have an abnormal karyotype, the skin cells of the parents can be similarly tested. If the parents show an abnormality, they are warned that they are likely to transmit it to future infants through their gametes.

If the karyotypes of both parents are normal, they do not run a greater-than-average risk for the mother's age group of having another congenitally defective child.

It is now possible to detect the presence of extra chromosomes in the fetus when it is 16 weeks old. This diagnosis can be made from a sample of the fluid surrounding the fetus. This fluid contains cells shed from the surface of the body. If a chromosomal defect is found, the parents may elect to abort the fetus.

Abnormalities in the Sex Chromosomes

Nondisjunction may also produce individuals with extra sex chromosomes. An XY combination in the twenty-third pair, as you know, is associated with maleness, but so is XXY and XXXY and even XXXXY. These males are usually sexually underdeveloped and sterile, however. XXX combinations sometimes produce normal females, but many of the XXX women and all XO women (women with only one X chromosome) are sterile.

Many individuals with sex chromosome abnormalities are mentally retarded. More recently it has been discovered that white males with an extra Y chromosome (XYY) are found in a higher percentage in institutions for mentally

PREPARATION OF A KARYOTYPE

Chromosome typing for the identification of hereditary defects is being carried out at an increasing number of genetic counseling centers throughout the United States. The result of the procedure is known as a karyotype. The chromosomes shown in a karyotype are actually chromatid pairs held together by their kinetochores. White blood cells in the process of dividing have *been interrupted at metaphase by the addition of colchicine, which prevents the subsequent steps of mitosis from taking place. After treating and staining, the chromosomes are photographed, enlarged, cut out, and arranged according to size. Certain abnormalities, such as an extra chromosome or piece of a chromosome, can be detected.*

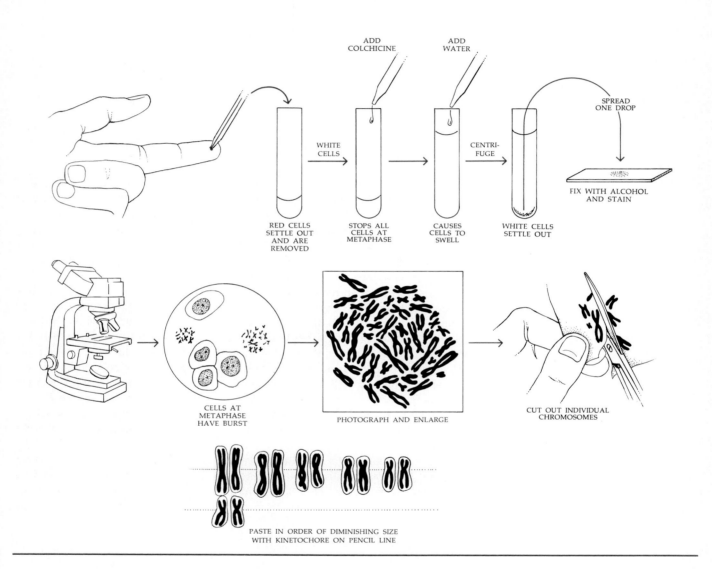

ADD COLCHICINE

ADD WATER

SPREAD ONE DROP

WHITE CELLS

CENTRIFUGE

FIX WITH ALCOHOL AND STAIN

RED CELLS SETTLE OUT AND ARE REMOVED

STOPS ALL CELLS AT METAPHASE

CAUSES CELLS TO SWELL

WHITE CELLS SETTLE OUT

CELLS AT METAPHASE HAVE BURST

PHOTOGRAPH AND ENLARGE

CUT OUT INDIVIDUAL CHROMOSOMES

PASTE IN ORDER OF DIMINISHING SIZE WITH KINETOCHORE ON PENCIL LINE

160   Genetics

disturbed or criminally insane than among males studied at random. (In the general population, on the basis of the rather limited studies done, about 1 in 1,000 males is XYY. Among the institutionalized males, the rate is about 2 per 100.) Although some XYY males are of normal intelligence and show no signs of mental disturbance, the XYY genotype is associated with below-average intelligence, tallness, and severe acne. At one time it was believed that men with the extra Y chromosome were likely to be more aggressive; a recent study in Denmark has not supported this hypothesis, however.

SUMMARY

Humans have two sex chromosomes (X and Y) and 22 pairs of autosomes. For study, the chromosomes are arranged in a karyotype.

In humans (and all other mammals), the Y chromosome determines maleness. The Y chromosome carries fewer genes than the X chromosome. As a consequence, if a man receives an X chromosome carrying recessive alleles, these alleles (which in the female would be dominated by the normal alleles) will usually be expressed. Characteristics resulting from the expression of such genes are said to be sex-linked or X-linked. Color blindness and hemophilia are examples of sex-linked characteristics which are carried by females but seen chiefly among males.

Nondisjunction is the failure of two homologues to separate at the time of meiosis. As a consequence of nondisjunction, children may be born with an extra chromosome. The presence of such an extra chromosome can produce abnormalities such as Down's syndrome.

QUESTIONS

1. Under what conditions would color blindness be found in a woman? If she married a man who was not color-blind, would her sons be color-blind? Her daughters?

2. Individuals with Down's syndrome frequently must be institutionalized for their entire lives, often at state expense. Do you believe that the government would be justified in undertaking measures to reduce the number of these births? What measures?

CHAPTER 14

The Double Helix

The work discussed in the previous chapters of this section is often referred to as "classical genetics." In some ways, this is an unfortunate term because "classical" carries a sense of something that happened a long time ago and that is interesting today only for historical reasons. Classical genetics provides the framework for all modern genetics, and classical studies continue today. As we shall see in Section 7, they are the basis for population genetics, one of the most dynamic of the modern branches of biology.

Classical genetics showed that genes are carried on chromosomes, that they come in alternate forms (alleles), that alleles occupy corresponding positions (loci) on homologous chromosomes, and that chromosomes with their alleles are reassorted at meiosis. Classical geneticists discovered that segments of chromatids are exchanged during meiosis, and they made use of these exchanges to construct chromosome maps. They introduced the concept of mutations and of interactions of genes with other genes, as in polygenic

14-1

Modern genetics has been concerned with the chemical basis of heredity. The genetic information has been shown to be contained in a large, complex molecule known as deoxyribonucleic acid (DNA). This electron micrograph shows the genetic material of a bacteriophage, a virus that attacks bacterial cells. It is in the form of one long, continuous molecule of DNA. In the center of the micrograph is the outer coat of the virus, made of protein, from which the DNA has been released.

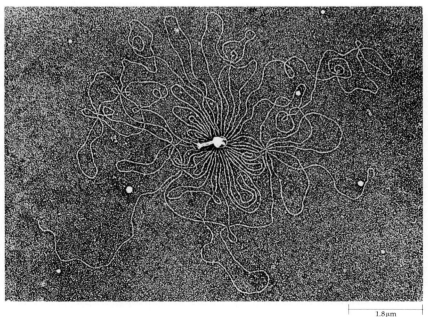

1.8μm

inheritance, to produce the phenotype. All this work—and more—was accomplished in less than half a century.

A turning point in genetics came when scientists began to turn their attention to the question of how it was possible for these little lumps of matter—the chromosomes—to be the bearers of what they had come to realize must be an enormous amount of complex information. The chromosomes, like all the rest of the living cell, are composed of atoms arranged into molecules. Some scientists, a number of them quite eminent in the field of genetics, thought it would be impossible to understand the complexities of genetics in terms of the structure of "lifeless" chemicals. (Although they did not use the term to describe themselves, they were, in fact, the vitalists of the twentieth century.) Others thought that if the chemical structure of the chromosomes was understood, we could then come to understand how chromosomes could function as the bearer of the genetic information. The work of the latter group is now termed "molecular genetics."

The first question was a simple one: What are chromosomes made of? Chemical analysis shows that the chromosomes of eukaryotes are made up of protein and, in smaller amounts, a chemical known as deoxyribonucleic acid, now usefully abbreviated as DNA. So the scientists who believed that the key to understanding heredity was to be found in the chemistry of the gene were faced with a second, more difficult question: Are genes protein or DNA?

PROTEIN OR DNA?

In the 1940s and early 1950s, when this question was first being seriously considered, proteins seemed to be the more promising alternative. Biochemists had just begun to work out the structure of proteins. They had discovered that enzymes are proteins, that protein molecules are large and complex, and that the number of structurally different proteins present in the living world—that is, proteins differing in the number and sequence of their component amino acids—is enormous. Therefore, it was possible to see how proteins, with their 20 amino acids arranged in varying sequences, might spell out a "language of life," as our own language is spelled out using only the 26 letters of the alphabet. The various amino acid arrangements might, in other words, be biological "sentences," dictating the directions for the many activities of the cell, with each different protein specific for one particular activity.

Some important research focused further attention on proteins and heredity. George Beadle and Edward Tatum, who later received the Nobel Prize for their research, demonstrated that a mutation involving a single gene of the mold *Neurospora* could result in a change in one of the enzymes produced by the cells

14–2

Three enzymatic reactions involved in the biosynthesis of the amino acid arginine. Beadle and Tatum postulated that each enzyme is produced by a separate gene. To test this hypothesis, they sought mutant strains of Neurospora *that could not carry out one particular biochemical step, such as the production of citrulline from ornithine, or arginine from citrulline. Such mutants could be analyzed by observing whether they could or could not grow in a particular medium. For example, a mold that could not perform step 2 could satisfy its arginine requirements with arginine or citrulline, but not with ornithine.*

14-3

How Beadle and Tatum tested the mutants of Neurospora. *By these experiments, they were able to show that a change in a single gene results in a change in a single enzyme. (a) Spores are removed from fruiting bodies of* Neurospora *and (b) each spore is transferred to an enriched medium, containing all* Neurospora *normally needs for growth plus supplementary amino acids. (c) A fragment of the mold is tested for growth in the minimal medium. If no growth is observed on the minimal medium, it may mean that a mutation has occurred that renders this mutant incapable of making a particular amino acid, and so tests are continued. (d) Subcultures of mold that grow on the enriched medium but not on the minimal one are tested for their ability to grow in minimal media supplemented with only one of the amino acids. As in the example shown here, a mold that has lost its capacity to synthesize the amino acid proline is unable to survive in a medium that lacks that amino acid. Further tests are then made to discover, in each case, which enzymatic step has been impaired.*

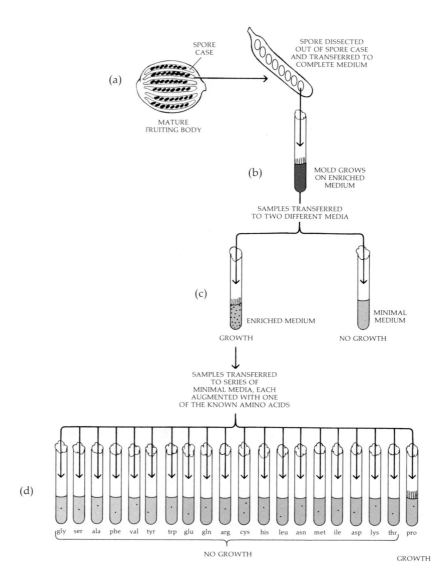

(Figure 14–2). (*Neurospora,* a common mold of bread, is haploid, like *Chlamydomonas,* so a change in a single gene is not masked by the effects of an allele.) On the basis of their experiments, summarized in Figure 14–3, Beadle and Tatum formulated the one-gene–one-enzyme theory. This theory states that genes control a cell's activities by controlling enzymes, with each gene responsible for one particular enzyme. This theory was later generalized to "one gene, one protein" and became the basis for much subsequent research.

The Structure of Hemoglobin

Linus Pauling, also a Nobel laureate, was one of the first to see some of the implications of these new ideas. Hemoglobin, the substance in vertebrate red blood cells that carries oxygen, is a protein made up of chains of amino acids. Perhaps, Pauling reasoned, human diseases, such as sickle cell anemia, involving red blood cells can be traced to a variation from normal in the protein structure of the hemoglobin molecule—a typographical error, so to speak, in

An example of the remarkable precision of the "language" of proteins. Portions of the beta chains of the hemoglobin A (normal) molecule and the hemoglobin S (sickle cell) molecule are shown. The entire structural difference between the normal molecule and the sickle cell molecule (literally, a life-and-death difference) consists of this single substitution of one valine for one glutamic acid.

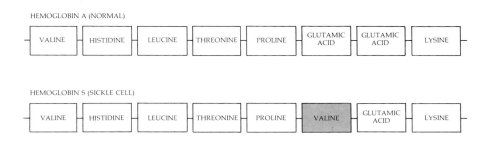

HEMOGLOBIN A (NORMAL)

| VALINE | HISTIDINE | LEUCINE | THREONINE | PROLINE | GLUTAMIC ACID | GLUTAMIC ACID | LYSINE |

HEMOGLOBIN S (SICKLE CELL)

| VALINE | HISTIDINE | LEUCINE | THREONINE | PROLINE | VALINE | GLUTAMIC ACID | LYSINE |

one important sentence in the hypothetical language of the cells. He took samples of hemoglobin from normal persons, from persons with sickle cell anemia, and from persons heterozygous for the trait. To study the differences in these proteins, he used a process known as electrophoresis, in which organic molecules are dissolved in a solution and exposed to a weak electric current. Very small differences, even in very large molecules, may be reflected in the electric charges of the molecules, and the molecules will move differently in the electric field. The normal person, he found, makes one sort of hemoglobin, the person with sickle cell anemia makes a slightly different sort, and the person who carries one copy of the recessive gene for sickling makes both types of hemoglobin.

A few years later, it was found that the actual difference between the normal and the sickle hemoglobin molecules lies in 2 of the molecule's 600 amino acids. The hemoglobin molecule is composed of four polypeptide chains—two identical alpha chains and two identical beta chains—each made up of about 150 amino acids. In a precise location in each beta chain, one glutamic acid is replaced by one valine (Figure 14-4). The language of proteins is fantastically precise, and the consequences of even a small error or small variation may be great.

So it is easy to see why many prominent investigators, particularly those who had been studying proteins, believed that the genes themselves were proteins, that the chromosomes contained master models of all the proteins that would be required by the cell, and that enzymes and other proteins active in cellular life were copied from these master models. This was a logical hypothesis—but, as it turned out, it was wrong.

THE DNA HYPOTHESIS

The Transforming Factor

To trace the beginning of the other hypothesis—the one that proved to be right—it will be necessary to go back to 1928 and pick up an important thread in modern biological history. In that year, an experiment was performed which seemed at the time very remote from either biochemistry or genetics. Frederick Griffith, a public-health bacteriologist, was studying the possibility of developing a vaccine against pneumococci, the bacterial cells that cause one kind of pneumonia. The pneumonia-causing strains, he found, are all surrounded by special capsules, or sheaths, composed mainly of polysaccharides (Figure 14-5). The existence of the sheath and its composition are both genetically determined—that is, they are inherited properties of the bacteria. Griffith thought that injections of heat-killed disease-causing bacteria or of live, closely related

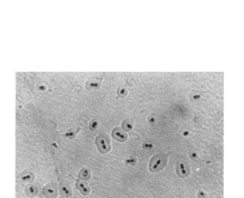

(a) $\vdash\!\!-\!\!-\!\!\dashv$ 0.2 μm

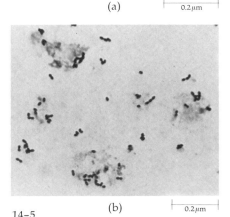

(b) $\vdash\!\!-\!\!\dashv$ 0.2 μm

14-5

(a) Encapsulated and (b) nonencapsulated forms of pneumococci. The encapsulated form, which is resistant to phagocytosis by white blood cells, produces pneumonia; the nonencapsulated form is harmless.

14-6

Discovery of the transforming factor, a substance that can transmit genetic characteristics from one cell to another, resulted from studies of pneumococci bacteria that cause pneumonia in mice and men. One strain of these bacteria has capsules; another does not. The capacity to make capsules and cause disease is an inherited characteristic, passed from one bacterial generation to another as the cells divide. (a) Injection into mice of encapsulated pneumococci killed the mice. (b) The nonencapsulated strain was harmless. (c) If the encapsulated strain was heat-killed before injection, it too was harmless. (d) If, however, heat-killed encapsulated bacteria were mixed with live nonencapsulated bacteria and the mixture was injected into mice, the mice died. (e) Blood samples from the dead mice revealed live encapsulated pneumococci. Something had been transferred from the dead bacteria to the live ones that endowed them with the capacity to make capsules and cause pneumonia. This "something" was later isolated and found to be DNA.

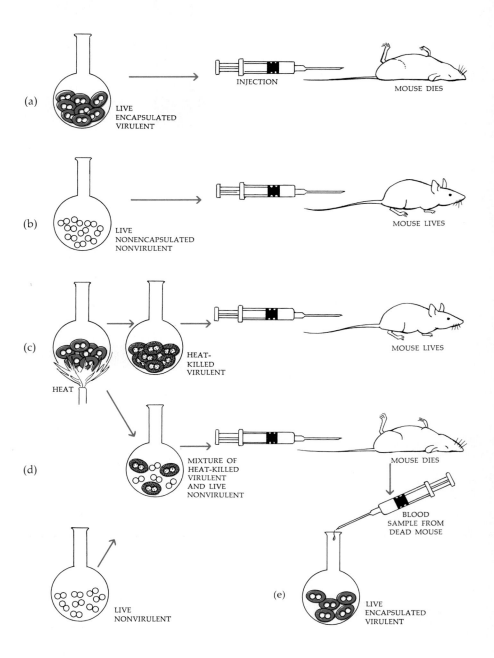

but harmless strains—ones without capsules—might immunize mice to live virulent pneumococci. As a result of his experiments, which are outlined in Figure 14-6, Griffith found that something can be passed from dead to live bacteria that changes their hereditary characteristics. This phenomenon was known as *transformation* and the "something" that caused it was called the *transforming factor.*

In 1943, O. T. Avery and his co-workers at the Rockefeller Institute demonstrated that the transforming factor is DNA. Subsequent experiments showed that a variety of genetic traits can be passed from members of one strain of bacterial cells to those of another, similar colony by means of isolated DNA.

DNA had first been isolated by a German chemist named Friedrich Miescher in 1869—in the same remarkable decade in which Darwin published *The Origin*

14-7

(a) *A nucleotide is made up of three different parts: a nitrogen base, a sugar, and a phosphate.* (b) *Each nucleotide in DNA contains one of the four possible nitrogen bases, a deoxyribose sugar, and a phosphate.*

(a)   BASIC STRUCTURE OF A NUCLEOTIDE:

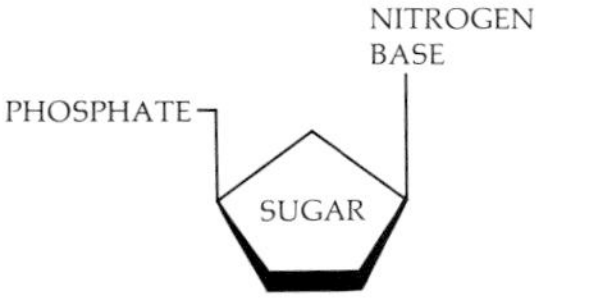

(b)   THE FOUR KINDS OF NUCLEOTIDES THAT MAKE UP THE DNA MOLECULE:

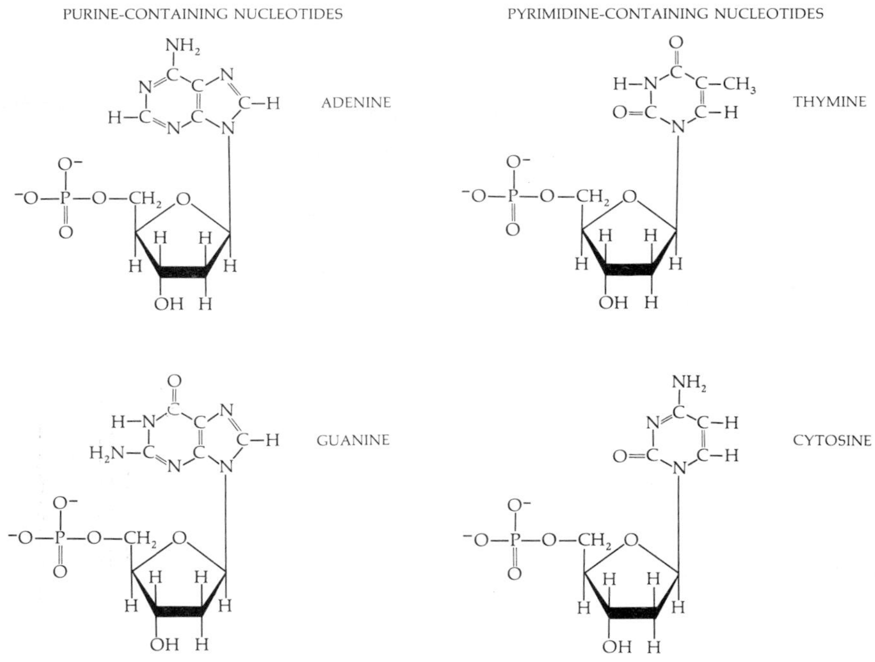

PURINE-CONTAINING NUCLEOTIDES   PYRIMIDINE-CONTAINING NUCLEOTIDES

of *Species* and Mendel presented his results to an audience of 40 at the Natural History Society in Brünn. The substance Miescher isolated was white, sugary, slightly acid, and contained phosphorus. Since he found it only in the nuclei of cells, he called it nucleic acid. This name was later amended to deoxyribonucleic acid, to distinguish it from a similar chemical also found in cells, ribonucleic acid.

By the time of Avery's discovery, it was known that DNA is made up of nucleotides (Figure 14–7). Each nucleotide consists of a nitrogen-containing compound, a deoxyribose sugar, and a phosphate. The nitrogen-containing compounds are of two kinds: purines, which have two rings, and pyrimidines, which have one ring. There are two kinds of purines found in DNA, adenine (A) and guanine (G), and two kinds of pyrimidines, cytosine (C) and thymine (T). So DNA is made up of four types of nucleotides, differing only in terms of their nitrogen-containing purine or pyrimidine.

Avery's results offered evidence for DNA as the genetic material, but his discovery was slow to gain full recognition. This was partly because bacteria, which are, of course, prokaryotes, were considered "lower" and "different" and partly because the DNA molecule—made up of only four components—seemed too simple for the enormously complex task of carrying the hereditary information.

The "Bacteria Eaters"

A second, crucial experiment was made with an even "lower" organism, a virus that attacks bacteria. These viruses, known as *bacteriophages* ("bacteria eaters"), were originally chosen for study because they reproduce at such a phenomenal rate. Moreover, the bacteria they attack are the familiar *Escherichia coli,* the most common bacteria found in the healthy human intestine. Bacteriophages invade bacterial cells, multiply within them, and then escape—usually bursting (lysing) the cell. The entire infection cycle can take place in as brief a period as 20 minutes, during which time several hundred virus progeny can be produced from a single virus.

Another interesting thing about bacteriophages is that they consist only of DNA and protein (see Figure 14–1), the two leading contenders for the role as carriers of the genetic material. The question of which type of molecule carries the viral genes—the hereditary information by which new viral particles are made—was answered in 1952 by Alfred D. Hershey, working with Martha Chase. Their series of experiments is summarized in Figure 14–8. Remember that protein contains sulfur and no phosphorus and DNA contains phosphorus and no sulfur; you will then see why this experiment showed that only the DNA of the bacteriophages was involved in the replication process and that the protein could not be the hereditary material.

Electron micrographs later confirmed that this type of bacteriophage attaches to the bacterial cell wall by its tail and injects its DNA into the cell, leaving the empty protein coating (the "ghost") on the outside (Figure 14–9). In short, the protein is just a container for the DNA, which carries all the hereditary information of the virus and directs the synthesis both of new DNA and of new protein for the viral progeny.

14–8

A summary of the Hershey-Chase experiments demonstrating that DNA is the hereditary material of a virus.

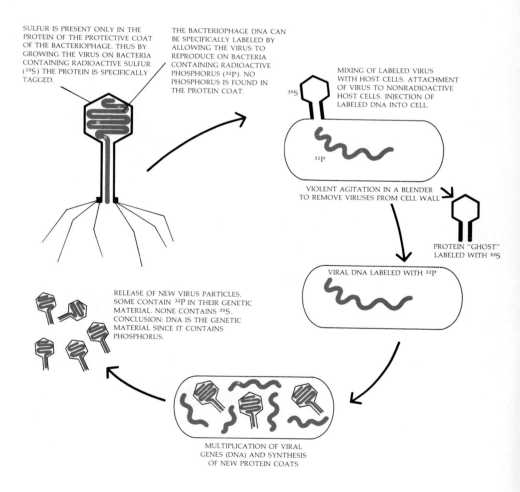

SULFUR IS PRESENT ONLY IN THE PROTEIN OF THE PROTECTIVE COAT OF THE BACTERIOPHAGE. THUS BY GROWING THE VIRUS ON BACTERIA CONTAINING RADIOACTIVE SULFUR (^{35}S) THE PROTEIN IS SPECIFICALLY TAGGED.

THE BACTERIOPHAGE DNA CAN BE SPECIFICALLY LABELED BY ALLOWING THE VIRUS TO REPRODUCE ON BACTERIA CONTAINING RADIOACTIVE PHOSPHORUS (^{32}P). NO PHOSPHORUS IS FOUND IN THE PROTEIN COAT.

MIXING OF LABELED VIRUS WITH HOST CELLS. ATTACHMENT OF VIRUS TO NONRADIOACTIVE HOST CELLS. INJECTION OF LABELED DNA INTO CELL.

^{35}S

^{32}P

VIOLENT AGITATION IN A BLENDER TO REMOVE VIRUSES FROM CELL WALL

PROTEIN "GHOST" LABELED WITH ^{35}S

VIRAL DNA LABELED WITH ^{32}P

RELEASE OF NEW VIRUS PARTICLES. SOME CONTAIN ^{32}P IN THEIR GENETIC MATERIAL. NONE CONTAINS ^{35}S. CONCLUSION: DNA IS THE GENETIC MATERIAL SINCE IT CONTAINS PHOSPHORUS.

MULTIPLICATION OF VIRAL GENES (DNA) AND SYNTHESIS OF NEW PROTEIN COATS

Further Evidence for DNA

The role of DNA in transformation and in viral infection formed very convincing evidence that DNA is the genetic material. Two other lines of work also helped to lend weight to the argument. Alfred Mirsky and others, in a long series of careful studies, showed that, in general, the somatic cells of any given species contain equal amounts of DNA. The principal exceptions are the gametes, which regularly contain just half as much DNA as other cells.

A second important series of contributions was made by Erwin Chargaff. Chargaff analyzed the purine and pyrimidine content of the DNA of many different kinds of living things and found that the nitrogen bases do not occur in exact proportions. DNA molecules, he found, can have a great deal of variety in composition. However, the proportions of the four nitrogen bases are the same in all cells of a given species, even though they vary from one species to another. In other words, the bases, according to Chargaff's analysis, do not occur in a regular sequence—ATCG, ATCG, for instance—because if they did, the proportions would all be the same. Rather, they are arranged irregularly—ATCG, TTAC, for instance. Therefore, these variations could very well provide a "language" in which the instructions controlling cell growth could be written. Some of Chargaff's results are reproduced in Table 14-1. Can you, by examining these figures, notice anything interesting about the proportions of purines and pyrimidines?

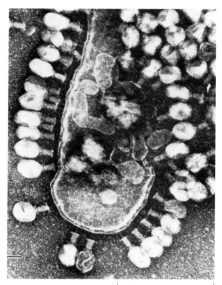

14-9

Electron micrograph of bacteriophages attacking a cell of Escherichia coli. *The viruses are attached to the bacterial cell by their tails. Some of the viruses have injected their DNA into the cell, as revealed by their empty heads. A complete cycle of virus infection takes only about 20 minutes. At the end of that period, several hundred new virus particles are released from the cell.*

Table 14-1   *Proportions of the Nitrogen Bases in the DNA of Several Species*

SOURCE	PURINES		PYRIMIDINES	
	ADENINE	GUANINE	CYTOSINE	THYMINE
Man	30.9%	19.9%	19.8%	29.4%
Sheep	29.3	20.7	20.8	29.2
Salmon	29.7	20.8	20.4	29.1
Sea urchin	32.8	17.7	17.3	32.1
Wheat germ	28.1	21.8	22.7	27.4
Escherichia coli	26.0	24.9	25.2	23.9

THE WATSON-CRICK MODEL

In the early 1950s, a young American scientist, James D. Watson, went to Cambridge, England, on a research fellowship to study problems of molecular structure. There, at the Cavendish Laboratory, he met physicist Francis Crick. Both were interested in DNA, and they soon began to work together to solve the problem of its molecular structure. They did not do experiments in the usual sense but rather undertook to examine all the data about DNA and unify them into a meaningful whole.

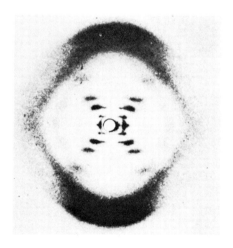

14-10

X-ray diffraction photograph of DNA taken by Rosalind Franklin in the laboratories of Maurice Wilkins, who shared the Nobel Prize with Watson and Crick. The reflections crossing in the middle indicate that the molecule is a helix. The heavy dark regions at the top and bottom are due to the closely stacked bases perpendicular to the axis of the helix.

The Known Data

By the time Watson and Crick began their studies, quite a lot of information on the subject had already accumulated. It was known that the DNA molecule is very large and also that it is long and thin. It was known that DNA contains nucleotides, each consisting of either one purine or one pyrimidine plus a deoxyribose sugar and a phosphate group. X-ray studies (Figure 14-10) showed that the molecule takes the form of a helix (coil). In 1950, Pauling had shown that proteins sometimes take this form (see page 51) and that the helical structure is maintained by hydrogen bonding between successive turns in the helix.

The double-stranded helical structure of DNA, as first presented in 1953 by Watson and Crick. The framework of the helix is composed of the sugar-phosphate units of the nucleotides. The rungs are formed by the four nitrogen bases adenine and guanine (the purines) and thymine and cytosine (the pyrimidines). Each rung consists of two bases. Knowledge of the distances between the atoms, determined from x-ray diffraction pictures, was crucial in establishing the structure of the molecule.

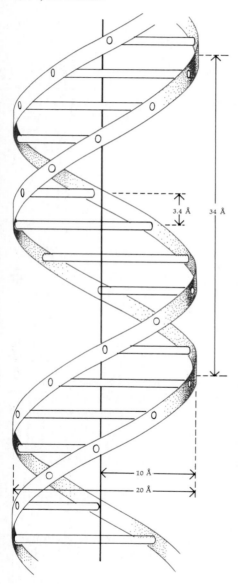

3.4 Å   34 Å

10 Å

20 Å

Pauling had suggested that the structure of DNA might also be based on a single helical framework. And, finally, there were the Chargaff data (Table 14–1), which indicated that the ratio of DNA nucleotides containing thymine to those containing adenine is approximately 1:1 and that the ratio of nucleotides containing guanine to those containing cytosine is also approximately 1:1.

Building the Model

From these data, Watson and Crick attempted to construct a model of DNA that would fit the known facts. They were very conscious of the biological role of DNA. In order to carry such a vast amount of information, the molecules would have to be heterogeneous and varied. Also, there had to be some way for them to replicate readily and with great precision in order that faithful copies could be passed from cell to cell and from parent to offspring, generation after generation.

Let us see what the Watson-Crick model looked like. If you take a ladder and twist it into the shape of a helix, keeping the rungs perpendicular, this would form a crude model of the molecule (Figure 14–11). The two longitudinal railings are made up of alternating sugar and phosphate molecules. The rungs of the ladder are formed by the nitrogenous bases—adenine (A), thymine (T), guanine (G), and cytosine (C)—one base bonded to each sugar-phosphate and two bases forming each rung. The paired bases meet across the helix and are joined together by hydrogen bonds, the relatively weak, common, and very important chemical bonds that Pauling had demonstrated in his studies of the structures of proteins. (Hydrogen bonds are described in Chapter 2.)

The distance between the two sides, or railings, according to x-ray measurements, is 20 angstroms. Two purines in combination would take up more than 20 angstroms, and two pyrimidines would not reach all the way across. But if a purine paired in each case with a pyrimidine, there would be a perfect fit. The paired bases—the "rungs" of the ladder—would therefore always be purine-pyrimidine combinations.

As Watson and Crick worked their way through these data, they assembled actual tin-and-wire models of the molecule (see page 41), seeing where each piece would fit into the three-dimensional puzzle. First, they noticed, the nucleotides along any one strand of the double helix could be assembled in any order: ATGCGTA-CATTGCCA, and so on (Figure 14–12). Since a DNA molecule may be several hundred nucleotides long, there is a possibility for great variety. The meaning of this variety, the molecular heterogeneity of DNA, will be explored more thoroughly in the next chapter.

DNA Replication

The most exciting discovery came, however, when they set out to construct the matching strand. They encountered an interesting and important restriction. Not only could purines not pair with purines and pyrimidines not pair with pyrimidines, but because of the configurations of the molecules, adenine could pair only with thymine and guanine only with cytosine. (Look at Table 14–1 again and see how well these physicochemical requirements confirm Chargaff's data.)

This fact about the structure of the DNA molecule immediately suggested the method by which it reproduces itself. At the time of chromosome replication, the molecule opens up, bit by bit, the bases separating at the hydrogen bonds. The two strands separate, and new strands form along each old one,

14–12

(a) *The structure of a portion of one strand of a DNA molecule. Each nucleotide consists of a sugar, a phosphate group, and a purine or pyrimidine base. The sugar of each nucleotide is linked by a phosphate group to the sugar of an adjacent nucleotide. The se-* *quence of nucleotides varies from one molecule to another. In the figure, the order of nucleotides is TTCAG. (b) The double-stranded structure of a portion of the DNA molecule. The strands are held together by hydrogen bonds between the bases. Because* *of bonding requirements, adenine can pair only with thymine and guanine only with cytosine. Thus the order of bases along one strand determines the order of bases along the other.*

(a)

(b)

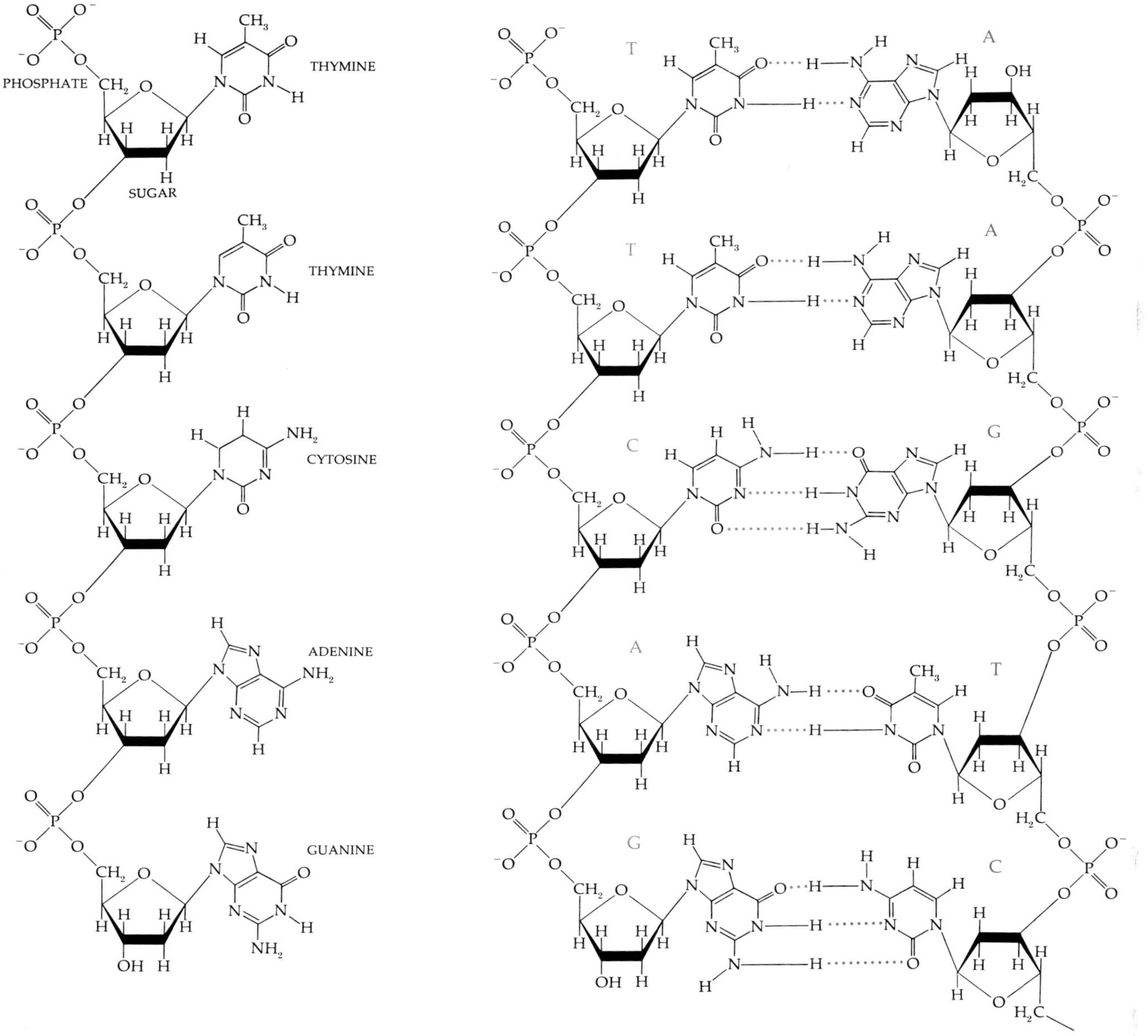

Then there is the question, what would have happened if Watson and I had not put forward the DNA structure? This is "iffy" history which I am told is not in good repute with historians, though if a historian cannot give plausible answers to such questions I do not see what historical analysis is about. If Watson had been killed by a tennis ball I am reasonably sure I would not have solved the structure alone, but who would? Olby has recently addressed himself to this question. Watson and I always thought that Linus Pauling would be bound to have another shot at the structure once he had seen the King's College x-ray data, but he has recently stated that even though he immediately liked our structure it took him a little time to decide finally that his own was wrong. Without our model he might never have done so. Rosalind Franklin was only two steps away from the solution. She needed to realise that the two chains must run in opposite directions and that the bases, in their correct tautomeric forms, were paired together. She was, however, on the point of leaving King's College and DNA, to work instead on TMV with Bernal. Maurice Wilkins had announced to us, just before he knew of our structure, that he was going to work full time on the problem. Our persistent propaganda for model building had also had its effect (we had previously lent them our jigs to build models but they had not used them) and he proposed to give it a try. I doubt myself whether the discovery of the structure could have been delayed for more than two or three years.

There is a more general argument, however, recently proposed by Gunther Stent and supported by such a sophisticated thinker as Medawar. This is that if Watson and I had not discovered the structure, instead of being revealed with a flourish it would have trickled out and that its impact would have been far less. For this sort of reason Stent had argued that a scientific discovery is more akin to a work of art than is generally admitted. Style, he argues, is as important as content.

I am not completely convinced by this argument, at least in this case. Rather than believe that Watson and Crick made the DNA structure, I would rather stress that the structure made Watson and Crick. After all, I was almost totally unknown at the time and Watson was regarded, in most circles, as too bright to be really sound. But what I think is overlooked in such arguments is the intrinsic beauty of the DNA double helix. It is the molecule which has style, quite as much as scientists. The genetic code was not revealed all in one go but it did not lack for

Watson and Crick with one of their models of DNA.

impact once it had been pieced together. I doubt if it made all that difference that it was Columbus who discovered America. What mattered much more was that people and money were available to exploit the discovery when it was made. It it this aspect of the history of the DNA structure which I think demands attention, rather than the personal elements in the act of discovery, however interesting they may be as an object lesson (good or bad) to other workers.

Francis Crick: "The Double Helix: A Personal View," Nature, **248**: 766–769, 1974.

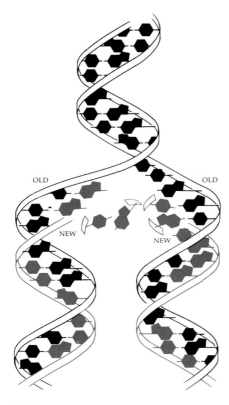

14-13

The DNA molecule shown here is in the process of replication, separating down the middle as its base units separate at the hydrogen bonds. (For clarity, the bases are shown out of plane.) Each of the original strands then serves as a template along which a new, complementary strand forms from nucleotides available in the cell.

using the raw materials in the cell. If a T is present on the old strand, only an A can fit into place with the new strand; a G will pair only with a C; and so on. In this way, each strand forms a copy of its original partner strand, and two exact replicas of the molecule are produced (Figure 14-13). The age-old question of how hereditary information is duplicated and passed on for generation after generation had, in principle, been answered.

DNA as the Carrier of Genetic Information

The Watson-Crick model also shows clearly how the DNA molecule is able to carry the genetic information. The information is coded in the sequence of the bases, and *any* sequence of the four pairs (AT, TA, CG, GC) is possible. Since the number of paired bases ranges from about 10,000 for the simplest known virus up to an estimated 10 billion in the 46 chromosomes of man, the possible variations are astronomical. The DNA from a human cell—which, if extended in a single thread, would be almost 2 meters long—contains genetic instructions that, if spelled out in English, would require some 600,000 printed pages averaging 500 words each, or a library of about a thousand books. Obviously, the DNA structure can well explain the endless diversity among living things.

Once the structure of DNA was revealed, there was no longer any serious question as to its genetic role.

SUMMARY

Proteins are large and complex molecules, with great specificity of structure. Even small differences in structure can produce large differences in function (as in the case of hemoglobin). Protein structure is controlled by genes—"one gene, one protein."

DNA (deoxyribonucleic acid) is the genetic material of the cell. Investigations showing that the transforming factor in bacteria and the carrier of the genetic information in bacteriophages are both DNA provided some of the first evidence for this hypothesis.

Further support for the genetic role of DNA came from two more findings: (1) Almost all tissue cells of any given species contain equal amounts of DNA. (2) The proportions of nitrogen bases are the same in the DNA of all cells of a given species, but they vary in different species.

The structure of DNA was reported in 1953 by Watson and Crick. The DNA molecule, they found, is a double-stranded helix, shaped somewhat like a twisted ladder. The two sides of the ladder are composed of repeating groups of a phosphate and a five-carbon sugar. The "rungs" are made up of paired bases, one purine base pairing with one pyrimidine base. There are four bases—adenine (A), guanine (G), thymine (T), and cytosine (C)—and A can pair only with T, and G only with C. The four bases are the four "letters" used to spell out the genetic message. The paired bases are joined by hydrogen bonds.

The DNA molecule is self-replicating. The two strands come apart down the middle, breaking at the hydrogen bonds, and each strand forms a new complementary strand from nucleotides available in the cell.

On the basis of this structure, as revealed by Watson and Crick, the role of DNA as the carrier and transmitter of the genetic information became widely accepted.

QUESTIONS

1. One of the chief arguments for proteins being the genetic material is that proteins are heterogeneous in structure. Explain why the genetic material must have this property. What feature in the Watson-Crick DNA model was important in this respect?

2. The bread mold with which Beadle and Tatum worked is haploid. Would haploid organisms follow Mendel's rules? Explain. Why would this feature make it easier and faster to conduct genetic experiments?

3. What are the steps by which Griffith demonstrated the existence of the transforming principle? Can you think of any implications of Griffith's discovery for modern medicine?

4. Describe the Hershey-Chase experiment and the conclusions drawn from it.

5. Suppose you are talking to someone who has never heard of DNA. How would you support an argument that DNA is the genetic material? List at least five of the strong points in such an argument.

CHAPTER 15

Breaking the Code

Like most important scientific discoveries, the Watson-Crick model raised more questions than it answered.

It was now known that genes are made of DNA and that the products of genes are specific proteins. When it had been thought that genes *were* proteins, biologists had hypothesized that the "gene proteins" formed the models or the molds for the cellular proteins, the basic structural matter of life. During this period, scientists began talking about templates. Templates are patterns, or guides, and the word "template" is usually associated with the metalwork patterns used in industry. By extension, the word came to be applied to a biological molecule which, by its shape, directs or molds the structure of another molecule. Each old strand of DNA, for example, serves as a template for the formation of its new partner strand during replication of the DNA molecule.

For a time after the structure of DNA was elucidated, biochemists struggled with the problem of how DNA could also be a template for the formation of protein molecules. Several ingenious schemes were proposed, but it was simply not possible to get a satisfactory physiochemical "fit." The relationship between DNA and protein was apparently a more complicated one. If the proteins, with their 20 amino acids, were the "language of life," to extend the metaphor of the early 1950s, the DNA molecule, with its four nitrogen bases, could be envisioned as a sort of code for this language. So the term "genetic code" came into being.

THE CODON

As it turned out, the idea of a "code of life" was useful not only as a dramatic metaphor but also as a working analogy. Scientists seeking to understand how the DNA so artfully stored in the nucleus could dictate the sequences of amino acids in the quite dissimilar structures of protein molecules approached the problem by the methods used by cryptographers in deciphering codes. Twenty amino acids are generally found in proteins, and there are four different nucleotides in nucleic acids. If each nucleotide "coded" one amino acid, only four could be provided for. If two nucleotides specified one amino acid, there could be a maximum number, using all possible arrangements, of 4^2, or 16—still not quite enough. Therefore, at least three nucleotides must specify each amino acid, following the code analogy. This would provide for 4^3, or 64, possible

15-1

Chemically, RNA is very similar to DNA, but there are two differences in its chemical groups. One difference is in the sugar component; instead of deoxyribose, RNA contains ribose, which has an additional oxygen atom. The other difference is that instead of thymine, RNA contains the closely related pyrimidine uracil (U). (A third, and very important, difference between the two is that most RNA does not possess a regular helical structure and is usually single-stranded.)

DEOXYRIBOSE RIBOSE

THYMINE URACIL

combinations. Thus it appeared that a group of three nucleotides would code for an amino acid; such a group came to be known as a <u>codon</u>. The postulate of a three-nucleotide codon was widely and immediately adopted as a working hypothesis, although it was not actually proved until a decade after the Watson-Crick discovery. Proof depended on answering yet another question: How is the code translated?

15-2

The beginning of the process of protein biosynthesis is the formation of messenger RNA (mRNA). At the chromosome, a strand of RNA is copied from a sequence of DNA. The new mRNA is formed by the same base-pairing mechanism by which new DNA is synthesized. The strand of mRNA is a "negative print" of the sequence of nucleotides in the DNA. Transfer RNA and ribosomal RNA are also formed from chromosomal DNA by the same copying process.

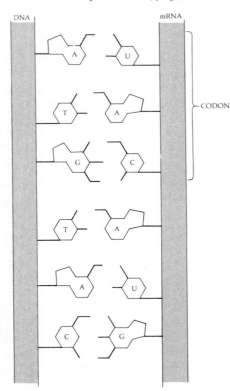

DNA   mRNA

CODON

THE RNAs

One of the principal clues used in answering this question was that cells which are making large amounts of protein are rich in ribonucleic acid (RNA). RNA differs from DNA in only a few respects. The sugar, or ribose, component of the molecule contains one more atom of oxygen (*deoxy* simply means "minus one oxygen"), and in place of thymine (T), RNA has another pyrimidine, uracil (U) (Figure 15-1). Perhaps more significant, since this part of the story has to do with the function of RNA, the RNA molecule is only rarely found in a fully double-stranded form. As a consequence, its properties and so its activities are quite different from those of DNA.

The role of RNA molecules in "translating" the code was studied by breaking apart *Escherichia coli* cells, separating their contents into various fractions, and seeing which fractions and (finally) which components of which fractions were essential for protein biosynthesis in a test tube. It was found that the machinery is complex, involving, among other things, three forms of RNA.

Another requirement for protein synthesis, it was found, is the presence of ribosomes, a fact which had already been suspected, because electron micrographs had shown that cells making large amounts of protein are rich in ribosomes. Ribosomes are made of protein and RNA, and this form of RNA came to be called <u>ribosomal RNA</u> (Figures 15-3 and 15-4). The genes that code for ribosomal RNA are in the nucleolus (see Figure 8-2).

One kind of RNA that was found to be involved in protein synthesis occurs in the form of long (from a few hundred to 10,000 nucleotides) single strands. This type, for reasons which will soon become clear, is known as *messenger RNA* (mRNA). Messenger RNA is formed along a DNA template (Figure 15-2).

The third type of RNA needed for protein synthesis is called *transfer RNA* (tRNA). These molecules are relatively short and partially coiled (Figure 15-5).

In addition, protein synthesis requires, as you would expect, amino acids, certain enzymes, and ATP. Now, with this rather large cast of characters in mind, let us look at what happens.

15–3

Diagram of a ribosome from a bacterial cell. As you can see, it consists of two subunits, one larger than the other, and each composed of specific RNA and protein molecules.

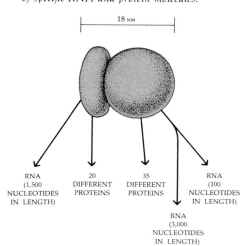

15–4

Polyribosomes. Each polyribosome is a group of ribosomes "reading" the same mRNA strand. These polyribosomes are from yeast cells.

15–5

(a) Structure of a tRNA molecule. These molecules consist of about 80 nucleotides linked together in a single chain. One end of the chain always terminates in a guanine nucleotide, and the other in a CCA sequence. The amino acid is linked to the tRNA at the CCA end. The other nucleotides vary according to the particular tRNA. All tRNA molecules appear to have the configuration shown here; in some, however, there is an additional loop. The "cloverleaf" is, in addition, twisted in a three-dimensional structure. Some of the bases are hydrogen-bonded to one another, as in the DNA molecule. The unpaired bases at the bottom of the diagram (indicated in color) serve as the anticodon and "plug in" the molecule to an mRNA codon. (b) Relationship between a specific mRNA codon and its corresponding tRNA anticodon.

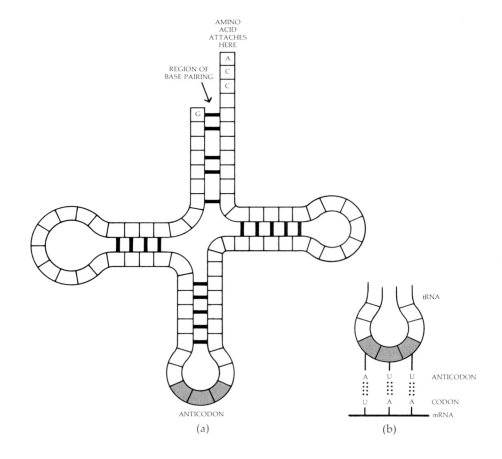

Protein biosynthesis begins when a strand of mRNA, with the help of a particular enzyme, RNA polymerase, forms against a segment of one strand of the DNA helix. (There is evidence that the DNA uncoils somewhat and a short section of the double helix opens to permit the synthesis of RNA.) The mRNA forms along the DNA strand according to the same base-pairing rules as those that govern the formation of a DNA strand, except that in mRNA, uracil substitutes for thymine. Because of the copying mechanism, the mRNA strand when completed carries an inverse copy of the DNA message. This stage of protein biosynthesis—the formation of a coded strand of RNA—is known as *transcription* (Figure 15–6).

Before the mRNA leaves the nucleus of the eukaryotic cell, an additional segment of RNA is usually added to the molecule. It is a sequence of about 200 nucleotides of adenine. Its function is unknown, although it has been hypothesized to be involved in the next step, which is the movement of the mRNA across the nuclear envelope. (Bacterial cells have no nuclear envelopes and their mRNAs have no added sequences of this sort, which is suggestive evidence in support of this hypothesis.)

The mRNA then moves into the cytoplasm. Here are the amino acids, special enzymes, ATP molecules, ribosomes, and molecules of tRNA (see Figure 15–5a). There are at least as many kinds of tRNA as there are kinds of amino acids found in proteins and all are formed along a segment of the DNA molecule, just as mRNA is. Each type of tRNA attaches by one end to a particular amino acid; these attachments each involve a special enzyme and a molecule of ATP.

Once in the cytoplasm, the mRNA molecule attaches to a ribosome. (The way in which the molecule attaches is a subject of current investigation.)

At the point at which the mRNA molecule is exposed on the surface of the ribosome, a molecule of tRNA, with its particular amino acid in tow, zeroes into position. The tRNA molecule finds its proper place by means of a nucleotide triplet—called an *anticodon*—which pairs with the nucleotide triplet (the codon) on the mRNA molecule (see Figure 15–5b).

As the ribosome moves along the mRNA strand, the next tRNA molecule with its amino acid moves into place (Figure 15–7). The energy that held the tRNA molecule to the amino acid is now utilized to forge the peptide link between the two amino acids. The first tRNA molecule then detaches itself from the mRNA molecule, leaving the dipeptide (two amino acids) in place, and the tRNA becomes available once more. These tRNA molecules apparently can be used over and over again. This stage in the biosynthesis of proteins—the assembly of amino acids in specific sequences—is known as *translation*.

15–6

Transcription. This electron micrograph shows a portion of DNA from the nucleolus of an amphibian egg cell. The fine fibrils are RNA molecules that have formed along the DNA strand. The arrows indicate the direction in which the transcription is proceeding. (How could you tell even if the arrows were not present?)

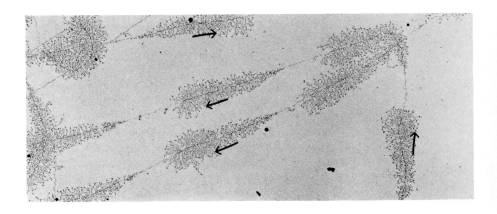

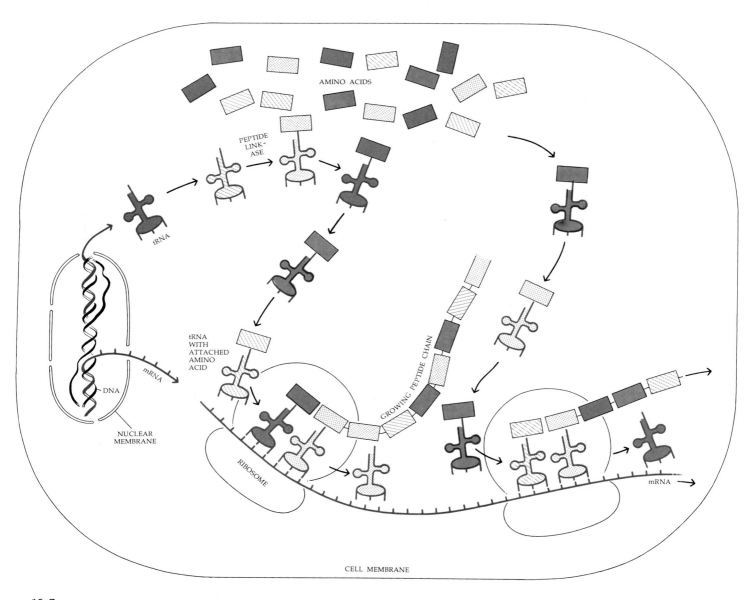

15-7

How a protein is made. At least 20 different kinds of tRNA molecules are formed on the DNA in the nucleus of the cell. These molecules are so structured that each can be attached (by a special enzyme) at one end to a specific amino acid. Each carries somewhere in the molecule an anticodon which fits an mRNA codon for that particular amino acid. The process of protein biosynth-esis begins when an mRNA strand is formed on the DNA template in the nucleus and travels to the cytoplasm. A ribosome attaches to the strand, and, at the point of attachment, the matching tRNA molecule, with its amino acid, plugs in momentarily to the codon in the mRNA. As the mRNA strand moves along the ribosome, a tRNA bound to its particular amino acid fits into place and the first tRNA molecule is released, leaving behind its amino acid. As the process continues, the amino acids are brought into line one by one, following the exact order laid down by the DNA code, and are formed into a protein chain, which may be anywhere from fifty to hundreds of amino acids long.

In the figure, the following labels appear: AMINO ACIDS, PEPTIDE LINK-ASE, tRNA, tRNA WITH ATTACHED AMINO ACID, DNA, mRNA, NUCLEAR MEMBRANE, RIBOSOME, GROWING PEPTIDE CHAIN, CELL MEMBRANE, mRNA

The existence of mRNA was postulated in 1961 by the French scientists François Jacob and Jacques Monod. Almost immediately, Marshall Nirenberg of the United States Public Health Service set out to test the mRNA hypothesis. He added several crude extracts of RNA from a variety of cell sources into extracts of *Escherichia coli*—that is, material which contained amino acids, ribosomes, ATP, and tRNA extracted from *E. coli* cells—and found that they all stimulated protein synthesis in the *E. coli* extract. In other words, the *E. coli* material started producing protein molecules even when the RNA "orders" it received were from a "complete stranger."

Perhaps if *E. coli* could read a foreign message and translate it into protein, Nirenberg reasoned, it could read a totally synthetic message, one dictated by the scientists themselves. A man-made RNA was available; Severo Ochoa of New York University had developed a method for linking together ribonucleotides into a long strand of RNA. The trouble with the method, from Nirenberg's point of view, was that there was no way to control the order in which an assortment of ribonucleotides would be assembled. For Nirenberg's purposes, the order was of the utmost importance. He wanted to know the exact contents of any message that he dictated.

A simple solution for this seemingly perplexing problem suddenly presented itself: Use an RNA molecule which consisted of only one ribonucleotide repeated over and over again. Nirenberg and his associate, Heinrich Matthaei, selected the ribonucleotide containing uracil. They then prepared 20 different test tubes, each containing cellular extracts of *E. coli* with ribosomes, tRNA, ATP, the necessary enzymes, and 20 different amino acids. In each test tube, one kind of amino acid, and only one, carried a radioactive label. The synthetic poly-U, as it was called, was added to each test tube. In 19 of the test tubes, nothing detectable occurred, but in the twentieth one, the one in which the radioactive amino acid was phenylalanine, the investigators were able to detect newly formed, radioactive polypeptide chains. When the polypeptide was analyzed, it was found to consist only of phenylalanines, one after another. Nirenberg and Matthaei had dictated the message "uracil-uracil-uracil-uracil-uracil-uracil-uracil-uracil-uracil . . . ," and a clear answer had come back, "phenylalanine-phenylalanine-phenylalanine. . . ."

It was not long before tentative codes were worked out for all the amino acids, using synthetic mRNA. A synthetic polynucleotide made up entirely of adenine (poly-A), for instance, makes a peptide chain composed entirely of lysine. When methods were worked out for controlling the order of the nucleotides in the synthetic RNA, it was possible to determine the rest of the codons (Figure 15–8).

Since more than 60 combinations code for 20 amino acids, you can see that there are a number of "synonyms" among the codons. Characteristically, these synonyms almost always differ only in the third nucleotide, leading to the speculation that the first two may be sufficient to hold the tRNA in most instances.

Some of the biological implications of these findings are strikingly clear. Consider mutations, for example. Mutations involve changes in the nucleotides of the DNA. These changes, once they occur, are faithfully replicated and passed on from cell to cell, generation after generation. Let us take another look at sickle cell anemia in the light of Figure 15–8. Normal hemoglobin contains glutamic acid; sickle cell hemoglobin contains valine (see Figure 14–4). GAA or GAG specifies glutamic acid; GUA or GUG specifies valine. So the difference

The genetic code, consisting of 64 triplet combinations (codons) and their corresponding amino acids (see page 49). Since 61 triplets code 20 amino acids, there are "synonyms," as many as six for leucine, for example. Most of the synonyms, as you can see, differ only in the third nucleotide. Of the 64 codons, only 61 specify particular amino acids. The other three codons are stop signals which cause the chain to terminate. The code is shown here as it would appear in the mRNA molecule. How would you determine the corresponding DNA codons?

SECOND LETTER

		U	C	A	G	
FIRST LETTER	U	UUU } phe / UUC / UUA } leu / UUG	UCU } ser / UCC / UCA / UCG	UAU } tyr / UAC / UAA stop / UAG stop	UGU } cys / UGC / UGA stop / UGG trp	U / C / A / G
	C	CUU } leu / CUC / CUA / CUG	CCU } pro / CCC / CCA / CCG	CAU } his / CAC / CAA } gln / CAG	CGU } arg / CGC / CGA / CGG	U / C / A / G
	A	AUU } ile / AUC / AUA / AUG met	ACU } thr / ACC / ACA / ACG	AAU } asn / AAC / AAA } lys / AAG	AGU } ser / AGC / AGA } arg / AGG	U / C / A / G
	G	GUU } val / GUC / GUA / GUG	GCU } ala / GCC / GCA / GCG	GAU } asp / GAC / GAA } glu / GAG	GGU } gly / GGC / GGA / GGG	U / C / A / G

THIRD LETTER

between the two hemoglobins lies in the replacement of one adenine by uracil in an RNA molecule, which, since it dictates a protein that contains more than 150 amino acids, must contain more than 450 bases. In other words, the tremendous functional difference between the two hemoglobins can be traced to a single "misprint" in more than 450 nucleotides.

SUMMARY

Genetic information is coded in the molecules of DNA, and these, in turn, determine the sequence of amino acids in molecules of protein. One gene contains the information needed to specify the complete sequence of one polypeptide chain.

The way in which the gene directs the production of a protein, according to current theory, is as follows: Messenger RNA (mRNA) molecules are copied from a strand of chromosomal DNA. The mRNA forms along one of the strands of DNA, following the principles of base pairing first suggested by Watson and Crick, and therefore is complementary to it. This process is known as transcription.

The mRNA strand leaves the cell's nucleus and attaches to a ribosome. At a point where the strand of mRNA is in contact with the ribosome, small molecules of RNA, known as transfer RNA (tRNA), which serve as adapters between the mRNA and the amino acids, attach temporarily to the mRNA strand. Their sites of attachment are determined by groups of three nucleotides—codons—on the mRNA molecule. The corresponding sites on the tRNA molecules are called anticodons. The bonding between codons and anticodons takes place by the same base-pairing principle as that which holds together the two strands of the double helix of DNA. Each tRNA molecule carries the specific amino acid that matches the mRNA codon into which the tRNA fits. Thus, following the sequence dictated by the DNA, the amino acid units are brought into line one by one and are formed into a polypeptide chain. This process is known as translation.

The codons for all the amino acids have now been identified.

1. Explain the term "genetic code." In what ways is it a useful analogy?

2. Describe the structure and function of each of the three types of RNA involved in protein biosynthesis.

3. Define and distinguish transcription and translation.

4. In a hypothetical segment of DNA, the sequence of bases is AAGTTTGGT-TACTTG. What would be the sequence of bases in an mRNA strand transcribed from this DNA segment? What would be the amino acids coded by the mRNA? Does it matter where the mRNA starts reading? Show why.

5. Fill in the missing letters in the following:

<u>T</u> <u>G</u> <u>T</u>	_ _ _	_ _ _	⎫
_ _ _	<u>C</u> _ _	_ _ _	⎬ DNA
<u>U</u> _ _	_ <u>C</u> <u>A</u>	_ _ _	mRNA codon
_ _ _	_ _ _	<u>G</u> <u>C</u> <u>A</u>	tRNA anticodon

What amino acids will each triplet code for?

CHAPTER 16

The Structure of the Chromosome and Regulation of Gene Expression

Subsequent experiments have shown that, as Nirenberg's work indicated, the genetic code is universal—the same for *Escherichia coli* as *Homo sapiens* and for all other organisms. (This fact is striking evidence that all living things are descended from a common ancestor.) However, more recent work in molecular genetics has emphasized that, despite their basic similarities, there are important differences between the chromosomes of bacteria (with which much of the original work was done) and those of eukaryotes (which are now receiving more attention).

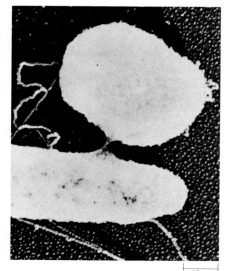

16–1

Electron micrograph of two conjugating E. coli *bacteria. A slender connecting bridge has been formed between an elongated donor cell and a rotund recipient. DNA enters the recipient cell through this bridge.*

$\overline{\quad 15\mu m \quad}$

THE BACTERIAL CHROMOSOME

The bacterial chromosome is a single, circular loop of DNA with little or no protein attached to it. Not all of a bacteria's genes are in its chromosome. In addition to the chromosomes, some bacteria also have from one to twenty much smaller circular DNA molecules, called *plasmids,* which replicate in synchrony with the cell. Plasmids sometimes become part of the larger DNA molecule, as we shall describe.

The Chromosome and Conjugation

Bacteria mate (bacterial mating is called *conjugation*). When this event occurs, a donor cell passes some or all of its chromosome to another cell (Figure 16–1). Mating can be detected when recipient cells suddenly acquire a new characteristic, such as resistance to penicillin.

Like many other phenomena, conjugation has been studied most extensively in *E. coli.* In *E. coli,* cells are either donors or recipients of genetic material. Whether or not a cell can function as a donor depends on its possession of a particular plasmid known as an F (fertility) factor. (A donor is thus known as F^+ and a recipient as F^-.) Like some other plasmids, the F factor can exist independently in the cell or can be part of the chromosome.

The F-factor plasmid can be passed from one cell to another. When this happens, the recipient cell becomes a donor cell.

Transfer of a part of the chromosome occurs only when the F factor is part of the *E. coli* chromosome. The F factor, like other genes, has its own position on the chromosome in relation to the other bacterial genes.

During conjugation, the chromosome opens and begins to replicate at the point of attachment of the F factor. As it replicates, the free end of the chromosome moves into the F⁻ cell (Figure 16–2). Usually only part of the chromosome is transferred. This fragment can then replace part of the chromosome of the F⁻ cell.

The conjugation process takes about 90 minutes (at 37°C). By separating the cells at successively later stages in the process (by whirling them at high speed in an electric blender), it is possible to determine the sequence in which the various genes enter one recipient cell. Thus, a map can be constructed of the chromosome (Figure 16–3).

16-2

(a) In bacterial conjugation, a replica of the F (fertility) factor may be transferred from a donor (F⁺) cell to a recipient (F⁻) cell. The recipient then becomes a donor. (b) When the F factor is part of the chromosome (such cells are referred to as HFr, for high frequency of recombination), the chromosome itself, or a portion of it, enters the recipient cell. That portion may become part of the recipient's chromosome, replicating with it. The recipient cell usually remains an F⁻ cell.

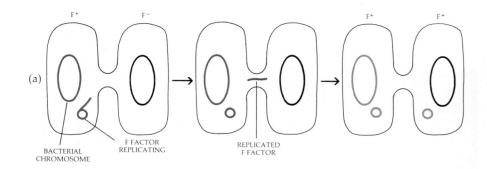

(a)

BACTERIAL CHROMOSOME   F FACTOR REPLICATING   REPLICATED F FACTOR

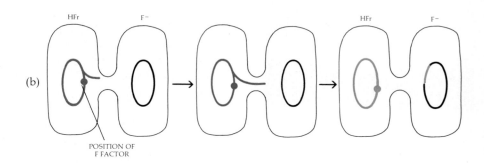

(b)

POSITION OF F FACTOR

16-3

In bacterial conjugation, genes are transferred from one cell to another in a definite order and at a constant rate. By employing mating strains with a variety of different mutations, it was possible to map the bacterial chromosomes simply by observing the order in which genes entered the recipient cell. The letters on the chromosome represent genetic markers.

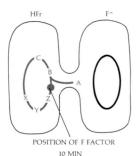

POSITION OF F FACTOR
10 MIN

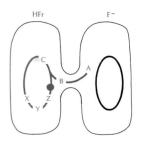

20 MIN

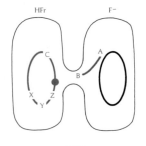

30 MIN

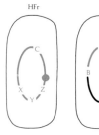

60 MIN

For many years, transfers of information among bacterial cells have been regarded mainly as laboratory phenomena, chiefly of interest to research scientists engaged in studies such as gene mapping. However, it is now clear that such transfers also occur frequently in nature among certain groups of bacteria.

In a large bacterial population, there may be a few cells which, as a result of mutation, are resistant to a particular drug—streptomycin, for example. If the bacterial population is exposed to streptomycin—as when a patient is being treated with a drug—susceptible cells are destroyed and the resistant ones remain. These multiply and produce a population of bacterial cells made up entirely of resistant individuals. This is one reason why responsible physicians advocate restraint in the use of antibiotics. (A second reason is that with each treatment with antibiotics, the chances of the patient's developing an allergic reaction to that drug increases.)

It is now known that bacteria can become drug-resistant by the transfer of plasmids from resistant to nonresistant cells. Like F factors, these plasmids are readily passed from cell to cell. Under experimental conditions, 100 percent of a population of sensitive cells can become resistant within an hour after being mixed with suitable resistant bacteria. Such transfers were first discovered in a strain of Shigella, a bacterium that causes

dysentery. Further studies showed that not only can Shigella transfer drug resistance to other Shigella organisms but also the innocuous E. coli can transfer resistance factors to this and other, unrelated groups of bacteria. Infectious resistance is now found among an increasing number of types of bacteria, including those that cause typhoid fever, gastroenteritis, plague, and undulant fever as well as dysentery.

These findings cast considerable doubt on the advisability of the current widespread use of antibiotics. For example, antibiotics are ineffective against viruses, yet many physicians (sometimes at their patients' insistence) routinely prescribe them for the common cold, a virus-caused disease. Antibiotics are routinely added to animal feed because they promote growth. In members of one farm family whose chickens were being fed antibiotics, 36 percent of the bacteria present in their intestines were resistant to three or more antibiotic compounds compared with 6 percent of a neighboring control family.

The problem is, of course, that if these practices continue, an increasingly larger percentage of pathogenic bacteria will be resistant to any treatment now available for them. Thus, by our own lack of wisdom, we are threatening to negate some of the greatest medical advances of our century.

16–4

Splitting of lactose (milk sugar) to galactose and glucose requires the enzyme beta-galactosidase. Beta-galactosidase is an inducible enzyme; that is, its production is regulated by an inducer—in this case, lactose.

Control of Gene Expression in Bacteria

The chromosome of *E. coli* is calculated, on the basis of its weight, to consist of about 4 million nucleotide pairs. If you estimate that the polypeptide chains of proteins average 300 amino acids, you can calculate that the chromosome could consist of more than 4,000 genes. In fact, as many as 3,000 different proteins have been isolated from *E. coli* cells. Yet, at any one time, a relatively small proportion of the cell's genes are active.

For example, cells of *E. coli* supplied with the disaccharide lactose need the enzyme beta-galactosidase to split the disaccharide into glucose and galactose before they can use it as an energy source (Figure 16–4). If lactose is not present, there is only about one molecule of this enzyme per cell. In cells growing on lactose, however, approximately 3,000 molecules of beta-galactosidase are present in every normal *E. coli* cell. This represents about 3 percent of all the

According to the operon theory, the synthesis of proteins is regulated by interactions involving either a repressor and inducer or a repressor and corepressor. (a) In the case of inducible enzymes, such as beta-galactosidase, the repressor molecule is active until it combines with the inducer (in this case, lactose). It is then inactivated, and so the genes in the operon are no longer repressed. (b) In the case of a repressible enzyme, the repressor is not active until it combines with the corepressor. Thus, in the absence of the corepressor, the genes in the operon are active.

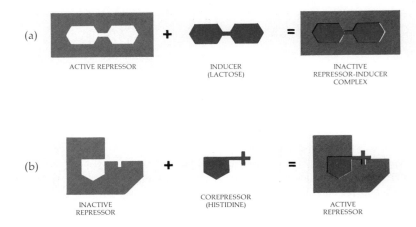

(a)

ACTIVE REPRESSOR   +   INDUCER (LACTOSE)   =   INACTIVE REPRESSOR-INDUCER COMPLEX

(b)

INACTIVE REPRESSOR   +   COREPRESSOR (HISTIDINE)   =   ACTIVE REPRESSOR

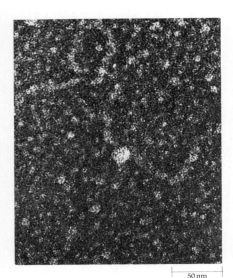

50 nm

16–6

A repressor protein (the white spherical form below the center of the micrograph) attached to the operator of the lactose operon.

protein in the cell. When the galactosidase is needed, it is produced from new mRNA. Substances such as lactose that increase the amount of enzyme produced by a cell are known as *inducers,* and the enzymes they influence are known as inducible enzymes (Figure 16–5a).

Some substances act to repress enzyme production. For example, *E. coli* can make all its own amino acids. If a particular amino acid—histidine, for example—is present in the medium, however, the cell will then stop making all of the enzymes associated with the biosynthesis of histidine. Such enzymes are known as repressible enzymes, and the substances that inhibit their production are known as *corepressors* (Figure 16–5b).

The Operon

To explain how bacterial cells regulate enzyme biosynthesis, the French scientists François Jacob and Jacques Monod developed the hypothesis of the *operon.* An operon is a group of structural genes aligned along a single segment of DNA. In the beta-galactosidase system, the operon includes the gene that codes for beta-galactosidase and two other genes, which also code for enzymes involved in lactose metabolism. These genes are adjacent to one another and are transcribed consecutively, one after the other along a single strand of DNA, forming a single mRNA molecule. Adjacent to the operon, as shown in Figure 16–7a, are (1) the *promoter,* a short segment of DNA to which RNA polymerase attaches at the initiation of transcription, and (2) the *operator,* which serves as the on-off switch for the operon.

Whether the operon is on or off is controlled by yet another gene, the *regulator.* The regulator codes for a protein, known as the *repressor,* which apparently binds to the operator. When the repressor is bound to the operator, no mRNA is produced by the operon (Figure 16–7b).

The repressor is controlled by another "signal" compound. In the case of inducible enzymes, this compound is the inducer. The inducer binds with the repressor molecule and changes its shape so that it can no longer attach itself to the DNA. In the absence of the repressor, mRNA molecules are formed along the structural genes, and from these molecules proteins are produced. When the supply of the inducer is used up, the regulator once again assumes control, and mRNA production and protein formation cease. In the beta-galactosidase system, the inducer is lactose. The lactose binds with the repressor and inactivates it, thus permitting the operon to make enzymes.

16-7

(a) *An operon is a group of structural genes that code for proteins, often enzymes that work sequentially in a particular enzyme sequence. Adjacent to the operon on the bacterial chromosome is a segment of DNA that contains the promoter and the operator. The operator slightly overlaps the promoter. The promoter is the site at which RNA polymerase attaches; this is the enzyme responsible for the transcription of mRNA. Another gene involved in operon function is the regulator. The regulator gene, which is not necessarily adjacent to the other genes, produces a regulatory protein known as a repressor. Some operons are inducible, others repressible.*

(b) *In operons activated by inducers, the regulatory gene codes for a protein that binds to DNA at the operator and so prevents the RNA polymerase from initiating transcription. The inducer counteracts the effect of the repressor, probably by binding with it and changing its shape. Thus, when the inducer is present, the repressor can no longer attach to the operator, and synthesis of mRNA proceeds.*

(c) *In operons regulated by corepressors, the repressor can bind to the operator only when it is combined with a corepressor. Thus transcription proceeds until corepressor is produced.*

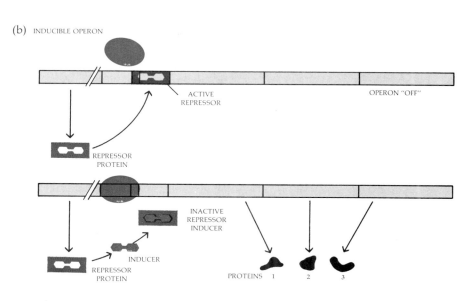

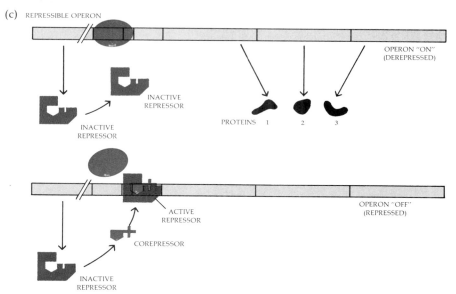

In the case of repressible enzymes, the "signal" compound is a corepressor. The repressor is active only when bound with the corepressor (Figure 16-7b). In further confirmation of the operon hypothesis, mapping studies show that genes for enzymes with related activities occur in clusters. The operon system is now widely accepted as the principal control mechanism in bacteria.

THE EUKARYOTIC CHROMOSOME

The eukaryotic chromosome is far more complex in structure. Unlike the bacterial chromosome, it is more than half protein. These proteins are believed to play a major role in maintaining the structure of the chromosomes and in the events that take place during mitosis and meiosis. They are also believed to be involved in regulating gene expression in eukaryotic cells.

The problem of gene expression is much different in a multicellular eukaryote from what it is in *E. coli.* A multicellular eukaryote usually starts life as a fertilized egg. The egg cell undergoes multiple mitotic divisions, producing many cells, and at some stage these cells begin to differentiate, becoming muscle cells, bone cells, blood cells, intestinal cells, and so forth. Each different cell type, as it differentiates, begins to produce characteristically different proteins that distinguish if from other types of cells. This is nicely illustrated by the mammalian red blood cell, which first produces one type of fetal hemoglobin, then a second type, and then, sometime after the birth of the organism, begins producing the alpha and beta chains characteristic of the adult. Thus the genes are expressed as a carefully controlled sequence, one after the other. The DNA segments that code for all these hemoglobin molecules are expressed only in the red blood cells.

Why is it that other types of cells do not produce hemoglobin? Perhaps the DNA coding for hemoglobin has been lost as these cells differentiate. Experimental evidence indicates, however, that this is not the case. For example, J. B. Gurdon at Oxford University in England removed nuclei from the intestinal cells of tadpoles and placed them into frogs' egg cells in which the nuceli had been destroyed. Some of the egg cells then developed into normal frogs (Figure 16-8). Therefore, the nucleus of a differentiated cell—the intestinal cell—still contained all the genetic information originally present in the egg cell and needed for the cellular activities of every cell of the body.

In other words, the DNA segments that code for hemoglobin are present in skin cells and heart cells and liver cells and nerve cells and, indeed, every one of the more than 100 different types of cells in the body. And the DNA sequence that codes for insulin is present not only in special cells of the pancreas but also in all the other cells. Following this argument further, one is rapidly led to conclude that in any cell in any multicellular organism, most of the DNA has been selectively silenced.

There is considerable evidence that relates the way that the DNA and protein are packed together into the chromosome with the expression of the genes. First, RNA is transcribed only when the nucleus is in interphase, never when the chromatin is tightly coiled as it is during mitosis.

A second example is found in studies of the giant chromosomes of insects (see page 153). At various stages of larval growth in insects, it is possible to observe diffuse thickenings, or "puffs," in various regions of these chromosomes. The puffs are separated strands of DNA, and studies with radioactive isotopes indicate that these puffs are sites of rapid RNA synthesis. When ecdysone, a hormone that produces molting in the insects, is injected, the puffs

*In experiments by J. B. Gurdon, the nucleus
was removed from an intestinal cell of a tadpole and implanted into an egg cell in which
the nucleus had been destroyed. In many
cases, the egg developed normally, indicating
that the intestinal cell nucleus contained all
the information required for all the cells of
the organism.*

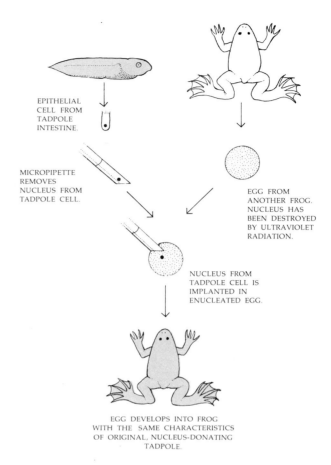

EPITHELIAL
CELL FROM
TADPOLE
INTESTINE.

MICROPIPETTE
REMOVES
NUCLEUS FROM
TADPOLE CELL.

EGG FROM
ANOTHER FROG.
NUCLEUS HAS
BEEN DESTROYED
BY ULTRAVIOLET
RADIATION.

NUCLEUS FROM
TADPOLE CELL IS
IMPLANTED IN
ENUCLEATED EGG.

EGG DEVELOPS INTO FROG
WITH THE SAME CHARACTERISTICS
OF ORIGINAL, NUCLEUS-DONATING
TADPOLE.

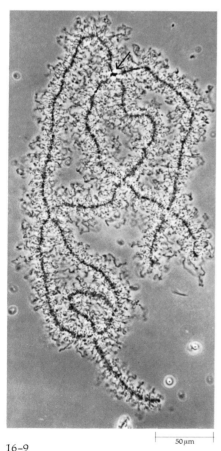

16–9

*Two homologous lampbrush chromosomes
held together by three chiasmata and a kinetochore (indicated by the arrow).*

occur in a definite sequence which can be related to the developmental stage of
the animal. For example, in one species of *Drosophila,* ecdysone initiates three
new puffs and causes increases in 18 other puffs within 20 minutes after it is
injected; during this same period, 12 other puffs decrease in size. After four to
six hours, five additional puffs can be seen. The looping out of the DNA occurs
before RNA synthesis is initiated. The mechanism of this unraveling, or spinning out, of the chromosomal DNA is not known.

Somewhat similar observations have been made with a third type of chromosomal material, lampbrush chromosomes, so-called because they resemble
the brushes used to clean kerosene lamps (Figure 16–9). They are found in the
nuclei of egg cells of fish, birds, reptiles, and amphibians. These cells are
extremely actively engaged in the RNA and protein syntheses required for the
rapid growth of the mature egg.

As you can see in Figure 16–9, each chromosome has a number of lateral
loops, which branch out from the main axis. These loops appear to be reeled in
and out from the body of the chromosome. Each of these loops consists of a thin
fiber. If this thin fiber is exposed to an enzyme that digests away the associated
protein, all that remains is a thin filament about the diameter of a single DNA
helix. These loops, like the chromosome puffs in giant chromosomes, are also
the sites of active RNA synthesis.

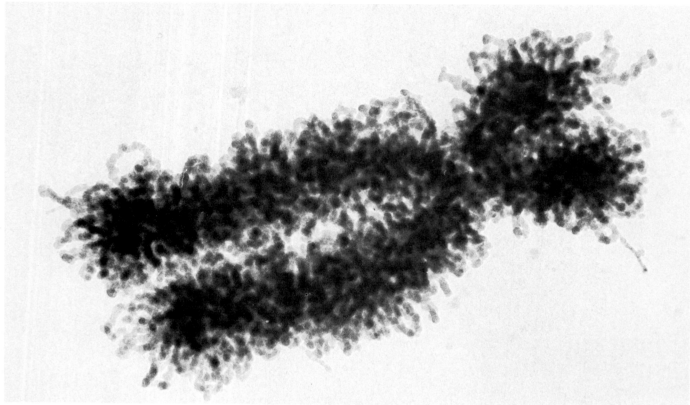

E. J. DuPraw

16–10

Human chromosome 12. Note that at the tips, the fibers do not form free ends but loop back into the chromosome. This feature, found in all eukaryotic chromosomes studied, suggests that the chromosome contains only a single molecule of DNA.

Thus, several different lines of evidence suggest that the tight coiling of DNA prevents the transcription of mRNA and that, conversely, the unfolding of DNA is related to mRNA transcription.

DNA of the Eukaryotic Chromosome

DNA is an "exquisitely thin filament," in the words of E. J. DuPraw, who calculates that a length sufficient to reach from the earth to the sun would weigh only half a gram. The DNA of each chromosome is believed to be in the form of a single molecule. In a human chromosome, each of these molecules is believed to be from 3 to 4 centimeters long. Thus each human cell contains between 1 and 2 meters of DNA, and the entire human body contains some 25 billion kilometers of DNA double helix.

The technical problems of investigating a molecule as large as the DNA of a eukaryotic cell are enormous. Two methods are currently being used. First, because of the base-pairing properties of DNA, it is possible to compare structures of different DNA molecules and different sequences on the same molecule without actually identifying them. If DNA is heated gently, the hydrogen bonds holding the two strands together are broken and the strands separate. (This process is called denaturation.) When the solution is cooled slightly, the bonds re-form. If, for example, you take DNA from one species, label it with radioactive phosphorus (^{32}P) (as in the Hershey-Chase experiment, page 168), heat it, and mix it with denatured DNA of another species, the extent to which the DNAs from the two different species combine can be measured quite accurately (Figure 16–11). This provides an estimate of the similarity and dif-

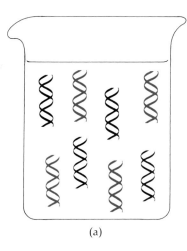

(a)

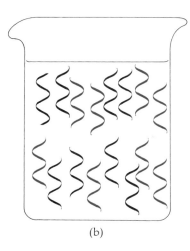

(b)

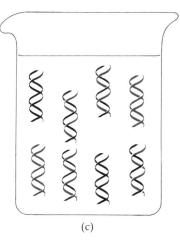

(c)

16–11

(a) *Two types of DNA are mixed in a beaker. One (indicated in color) is labeled with radioactive phosphorus. (b) When heated, the double-stranded DNA separates into two single strands. (c) When cooled, the DNA strands recombine in double helices. The double-stranded molecules are then separated from the single-stranded (by weight) and, the amount of radioactivity in the newly formed duplexes is analyzed. The extent to which radioactive and nonradioactive strands are able to form duplexes in a given length of time is an indicator of the similarities of the DNAs.*

ferences between the two samples. Similarly, the combination of a labeled mRNA molecule with a DNA molecule makes it possible to detect whether or not the mRNA was transcribed from that DNA.

Studies of this sort have revealed that the DNA of the eukaryotic cell is composed of three distinct kinds of sequences. Some are very short (about 100 base pairs long) and highly repetitive; in a crab, for instance, the sequence is A-T-A-T-A-T-A. . . . Such sequences may be repeated as many as 10^7 times per cell and may make up as much as 10 percent of all the DNA. This first type is known as satellite DNA. Satellite DNA, it is now known, occurs at the region of the kinetochore. This finding, plus the very homogeneous composition of satellite DNA, indicates it plays a structural role.

The second type of DNA is also made up of repeated sequences, but these are of a larger size, containing some several thousand base pairs. The number of such large-sized repeated sequences is estimated at anywhere from 100 to several thousand copies a cell. These longer repetitive segments are hypothesized to play a regulatory role, although the evidence yet is slight.

The third type of DNA appears as unique sequences—single copies—and makes up perhaps 50 percent of the total DNA.

One of the big surprises of this decade's investigations into the molecular biology of the gene is the discovery that only half of the DNA carries the genetic message and that the rest is highly repetitious and relatively simple.

Restriction Enzymes

Recently enzymes have been discovered that cleave DNA at specific sites. One, for instance, known as Eco RI (isolated from a plasmid-containing *E. coli*), cleaves DNA only at the sequence GAATTC. (These enzymes, which are known as restrictive enzymes, apparently have the function of recognizing and destroying DNA that is foreign to the cell.) The analysis of large protein molecules was made possible in large part by the availability of such specific enzymes—concerned with digestion—that hydrolyzed the peptide bonds at specified points in the polypeptide chain, yielding relatively small, uniform fragments for analysis. Restriction enzymes, it is believed, will serve as similar probes for DNA structure. They are also being used in a host of quite different investigations, generally described by the controversial term "genetic engineering," which will be described in the next chapter.

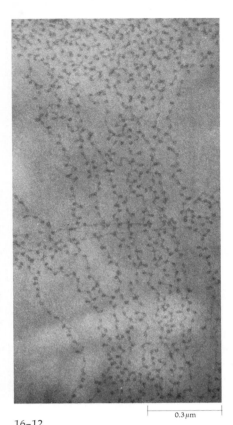

16-12

Chromatin from the red blood cell of a chicken. The distance between beads is about 14 nanometers, and the diameter of each bead is about 7 nanometers. Each bead is composed of about 200 base pairs of DNA and an assembly of histone molecules. Some more recent evidence suggests that the spaces between the beads are artifacts and that some unfolding may have taken place when the molecules were being prepared for electron microscopy. Future preparations may show all beads and no string.

Proteins of the Eukaryotic Chromosome

There are two types of proteins in the chromosome. One group is composed of large molecules that are varied and heterogeneous and, for these reasons, are believed to play a regulatory role. They are difficult to analyze and, as yet, little is known about them.

Histones

The second is a group of relatively small molecules known as histones; they are positively charged (basic) and so are attracted to the negatively charged (acidic) DNA. They are always present in chromatin, very similar from species to species, and are synthesized in large numbers during the S phase of the cell cycle (see page 118) when DNA is also being synthesized. At one time, because of the obvious intimate relationship between DNA and the histone proteins, they were believed to be the regulatory proteins. However, because of their lack of heterogeneity (and therefore lack of specificity), they are now believed to be important and universal structural elements of chromatin.

Much recent attention has been focused on the way in which DNA and histones might fit together. Until recently, it had been widely accepted that the DNA double helix was coated in some way with histones and twisted on itself in a large single helix, which investigators called the supercoil. Most recently the supercoil has been superseded and a wholly new, wholly unexpected structure has taken its place. This structure, which is based on electron micrographs (Figure 16-12) and is supported by biochemical evidence, resembles beads on a string. The beads are tight clusters of DNA and histones, and the string is short stretches of DNA. When a fragment of DNA is tied up in a bead, it is about one-sixth the length it would be if it were fully extended. Presumably the beads and string are further folded in some way.

When Watson and Crick deduced the structure of DNA, the meaning of the structure—in terms of the replication of DNA—was immediately obvious to them. The structure of chromatin at this moment is, functionally speaking, quite mysterious, but nonetheless intriguing.

SUMMARY

There are major structural differences between the prokaryotic and the eukaryotic chromosome, and these differences appear to be related to the differences in the problems of gene regulation in the two types of organisms.

The bacterial chromosome is a single, circular loop of DNA. In addition, a bacterial cell contains a number of much smaller, also circular molecules of DNA called plasmids.

In the course of bacterial conjugation (a form of mating), the DNA of a bacterial cell replicates, and as it replicates, some or all of the chromosome is passed to a recipient cell. Donor cells are characterized by the presence of the F (fertility) factor. The F factor is a plasmid and, like many other plasmids, it can exist independently or as part of a chromosome. An isolated F factor can be passed from one conjugating cell to another; the recipient cell then becomes a donor cell. When the F factor is integrated in the chromosome, the chromosome can break at that point and replicate, and all or part of one of the daughter chromosomes may be passed to a recipient cell.

A principal means of regulation of gene expression in bacteria is the operon system. This system is made up of a promoter, the site at which formation of the mRNA begins; an operator, which is the site of regulation; and the operon, one or more structural genes that code for polypeptide chains. The operator is under the influence of another gene, known as the regulator. The regulator codes for a protein, known as a repressor, which attaches to the DNA molecule at the operator and blocks mRNA transcription. The repressor does not act alone but rather in conjunction with another compound, often an end product of the enzyme sequence coded for by the operon. This compound may be an inducer, in which case it inactivates the repressor, or it may be a corepressor, in which case it makes it possible for the repressor to function.

The eukaryotic chromosome, which is far more complex in structure, consists of DNA complexed with proteins. Each chromosome, according to present evidence, consists of a single, very long DNA double helix. Three types of DNA sequences have been discovered. In one type, the sequences are short and highly repetitive. This type of DNA is found in the region of the kinetochore and is believed to serve a structural function. In the second type, the sequences are long and highly repetitive. In the third type, the sequences are moderately long, heterogeneous, and nonrepetitive. Only this third type, which comprises about 50 percent of the total DNA, is believed to contain the genetic message.

In chromatin, the DNA is combined tightly with a group of very simple, basic (positively charged) proteins known as histones. Recent evidence indicates that the histone-DNA complexes resemble beads on a string.

QUESTIONS

1. Identify the following terms: repressible enzyme, operator gene, plasmid, conjugation, F factor.

2. Compounds can be used by cells in two different ways. One type is broken down (usually as an energy source). Another is used as a building block for a larger molecule. Which type of compound would you expect to function as a corepressor? As an inducer? Do the examples given in the text conform to these expectations?

3. Gene expression theoretically can be regulated at the level of transcription, translation, or activation of the protein (see page 53). What would be the advantages of each, in terms of the cell? Under what circumstances might one type be more useful than another? Which is the more economical?

4. In what way does bacterial mating resemble fertilization? How does it differ from it?

5. As bacterial mating indicates, it is possible to separate the production of new genetic material from reproduction. Why do you think these two processes are combined in eukaryotic cells?

CHAPTER 17

Matters Arising

The discovery of the molecular biology of the gene clearly ranks among the great scientific achievements of all time. The recognition of the significance of this new knowledge has led to public concern as to the ways in which it might be utilized. Comparisons have been made between these advances and those that led, little more than a generation ago, to the explosion of the bomb over Hiroshima. Science fiction, following the ever-successful tradition of Dr. Frankenstein, has created more modern monsters, ranging from virulent microorganisms to multiple copies of Adolf Hitler. Equally fictional are the speculations of some science writers and, alas, some scientists, ranging from those who predict the swift demise of the human species to others, less alarmed but sometimes more alarming, who see visions of a brave new world achieved by the new technology. In this chapter, building on some of the basic scientific data of the previous chapters, we shall try to present these controversies in such a way that you can form your own judgments.

GENETIC ENGINEERING

Genetic engineering has come to mean the substitution of new genes for old. We have already seen two examples of how such substitutions may take place: transformation in bacteria, which, you will recall, provided one of the first clues to the nature of the genetic material, and bacterial conjugation, described in the previous chapter. We shall describe a third type now, because it is pertinent to this discussion.

Transduction

Transduction is the transfer of genes from one cell to another by means of viruses. It has been observed with bacterial viruses, which, as we noted in Chapter 14, are composed of nucleic acid wrapped in a protein coat. Transduction occurs when some of the nucleic acid of the host cell becomes wrapped in the protein coat of a virus and is carried to a new host cell.

Two forms of transduction are known: general and restricted. General transduction occurs when fragments of the DNA of a bacterial cell, broken down in the course of infection, are caught up in a bacteriophage head and are carried to a new host cell. These particles carry little or no bacteriophage DNA and so they cannot set up an active infection in the recipient cell. Instead, the portion of the bacterial chromosome carried into the new host may become a part of the host chromosome.

Restricted transduction occurs only with certain types of bacterial viruses known as lysogenic bacteriophages. Such bacteriophages do not necessarily set up an infective cycle when they enter a host cell but may instead exist as a molecule of naked DNA—in other words, a plasmid. Like the F factor and some other plasmids, they may become part of the cell chromosome (Figure 17–1). They are then called prophages or proviruses.

From time to time the relationship between a bacterial cell and its resident virus may shift, and the virus breaks loose from the chromosome and sets up an infective cycle. Exposure to x-rays, ultraviolet, or certain chemicals (all generally the same agents that cause mutations) increases the rate of activation of viruses in host bacteria. Bacteria harboring prophages are known as "lysogenic" bacteria. When a lysogenic bacterium releases its viruses, the viruses infect other nearby bacterial cells.

When lysogenic viruses leave a host bacterial cell chromosome, they sometimes take a fragment of the chromosome with them (Figure 17–2). This fragment is replicated with them through any infectious cycle. If one of these viruses becomes a part of a chromosome of a new host, the genes from the previous host may be inserted into the new host's chromosome and become a part of its genetic equipment. For instance, the bacteriophage lambda can carry the bacterial genes that produce the enzymes concerned with the breakdown of

17–1

When certain types of bacterial viruses infect bacterial cells, one of two events may occur. (a) The virus DNA may enter the cell and set up an infection, such as we described in Chapter 14, or (b) the DNA of the virus may simply lie latent in the cell. In the latter case it may become part of the bacterial chromosome, replicating with it. From time to time, such a virus becomes activated and sets up a new infective cycle. Bacteria harboring such viruses are known as lysogenic.

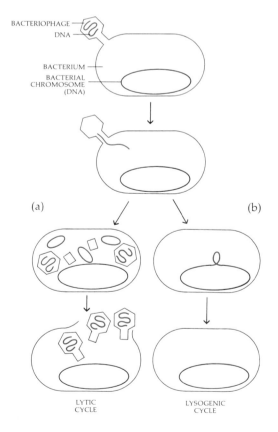

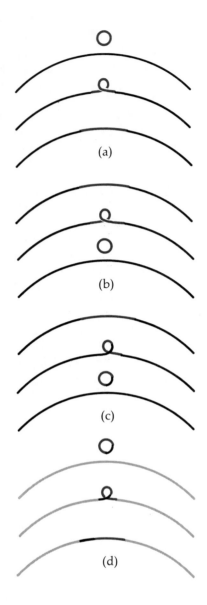

17-2
(a) *Like the F factor, a virus may become part of a bacterial chromosome.* (b) *As a result of x-rays, chemical treatment, or "spontaneously," the virus may break loose from the chromosome, setting up a new infectious cycle.* (c) *Sometimes the virus takes a fragment of adjacent host chromosome with it. This fragment may be large enough to include several host genes.* (d) *When and if the virus combines with the chromosome of a new host cell, the cell acquires the new genetic information. This process is known as transduction.*

galactose—in other words, the galactose operon. When a lambda provirus carrying the galactose genes becomes a provirus in a bacterial cell that cannot synthesize one or more of these enzymes, the infected cell gains the capacity to utilize galactose for growth.

It has long been suspected that certain viruses can enter into relationships with eukaryotic cells similar to those relationships between the lysogenic bacteria and their viruses. Herpes simplex, the virus that causes fever blisters, is an example. If you suffer from fever blisters or know someone who does, you know that they tend to break out repeatedly in the same person and that they characteristically occur in times of stress, such as when that person has a fever or eats certain food or gets a sunburn (ultraviolet exposure). It is believed that certain cells constantly harbor the herpes simplex virus, that it multiplies when the cells multiply, and that it causes cell damage—the fever blister—only when it is activated.

New DNA Combinations

Transformation, bacterial conjugation, and transduction are not new discoveries. There are, however, new developments that make it possible to manipulate these phenomena to some extent. One of these is the discovery of a group of enzymes known as DNA ligases that mend breaks in DNA molecules. The second is the isolation of various DNA restriction enzymes, which we mentioned briefly in the previous chapter. Some of these enzymes cleave the molecule symmetrically, cutting through both strands at the same site. Others cleave the molecule asymmetrically, cutting through the two strands a few bases apart. This can have surprising consequences. Look again at the sequence at which Eco RI cleaves the molecule: GAATTC. Now visualize the double-stranded molecule. It would be:

$$\ldots GAATTC \ldots$$
$$\ldots CTTAAG \ldots$$

Eco RI cleaves each strand between the guanine and the adenine:

$$\ldots G\overset{\downarrow}{A}ATTC \ldots$$
$$\ldots CTTAA\underset{\uparrow}{G} \ldots$$

What is left, as you will readily see, is:

$$\ldots G \qquad\qquad AATTC \ldots$$
$$\text{and}$$
$$\ldots CTTAA \qquad\qquad G \ldots$$

The protruding ends, . . . TTAA and AATT . . . , will thus join again with one another and also—and this is most important—with any other segment of DNA that has been cleaved by the same enzyme. Thus it is now possible to splice together segments of DNA from, apparently, a limitless variety of sources. (Such DNA molecules have been called chimeras after the mythological beast that was part lion, part goat, and part snake.)

Stanley Cohen of Stanford University Medical School and a group of coworkers have isolated a small plasmid of *E. coli*, designated pSC101, which makes the bacteria resistant to the antibiotic tetracycline. pSC101 has only one

17-3

An enzyme (Eco RI) has been found that opens DNA molecules at specific sites, leaving "sticky" ends exposed. These ends, consisting of TTAA and AATT sequences can rejoin or can join with any other fragment of DNA that has been cleaved by the same enzyme. Thus it is possible to splice together DNA from different sources, such as inserting a foreign gene into a bacterial plasmid, as shown here.

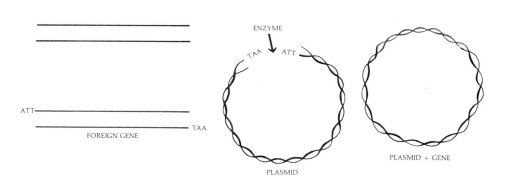

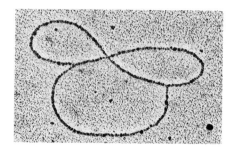

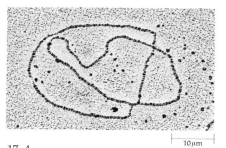

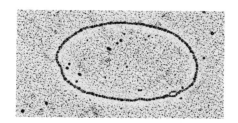

17-4

A plasmid replicating. At almost 5:00 in the chromosome in the upper micrograph, you can see the two DNA strands beginning to separate (each consisting, as you will remember, of one old strand and one new one). In the center micrograph, the replication process is more than half completed, and in the lower, the two strands are almost at the point of separation.

GAATTC sequence in the entire molecule and, as a consequence, is cleaved at only one spot by Eco RI. They have shown that the insertion of an additional segment of a DNA molecule into the plasmid (as shown in Figure 17-3) does not affect the uptake of the plasmid by *E. coli* or its capacity to make the recipient cells tetracycline-resistant. Moreover, the plasmid replicates when the cell does.

Thus, at present, biochemical geneticists have available two different types of hosts for new genes, bacteriophages and bacterial plasmids, and a means to introduce the DNA into the hosts. One present limit on the new technology is the difficulty in isolating specific genes for transfer. At present, the most practical available technique is to expose large DNA molecules to Eco RI or some other restriction enzyme and cleave them into smaller segments of manageable size but unknown function. (This process is known as "shotgunning.")

The Possibilities

What can be done with this new technology? Some researchers, including Robert Sinsheimer of the California Institute of Technology, have suggested that viruses might be tailor-made to carry particular genes to particular cells and so to cure diseases such as PKU or sickle cell anemia. Others have pointed out that if, for example, the DNA segments that code for the polypeptide chains of insulin were inserted into a bacterial plasmid, one could have a low cost, virtually limitless source of insulin, or of any other biological protein useful in medical treatment. One group is more intrigued with the possibility of being able to isolate a eukaryotic gene and study it outside of the enormously complex eukaryotic cell and in the well-understood and relatively simple *E. coli* host. A particularly important practical application might be the production of agriculturally important plants with combinations of new characteristics—such as the capacity both to carry out C_4 photosynthesis (page 95) and to fix nitrogen (page 515). Some investigators are working with four strains of the bacteria *Pseudomonas*, each of which has a capacity to break down a component of fuel oil, and are attempting to combine them genetically into one super oil-eater that, in powdered form, can simply be sprinkled on oil spills.

However, despite the possibilities and the enthusiasm they engendered, the hazards are too great to ignore. *Escherichia coli* lives in the digestive tract of man and of many other animals. Ideally, these studies should be made with a rare prokaryote, whose spread could be closely controlled. But *E. coli* is so well understood—probably the most extensively studied of any organism—that the scientists are understandably reluctant to abandon it. (Also, the notion of a rare prokaryote is a contradiction in terms; all prokaryotes multiply with great

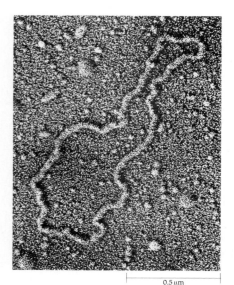

17-5

pSC101, a plasmid of E. coli *that confers resistance to tetracycline.*

rapidity.) What would be the consequences of creating a new race of *E. coli* with the capacities to produce insulin or growth hormone or some other biologically potent protein? (A hazard made more remote, we should add, by the fact that eukaryotic DNA does not seem to code for complete proteins in prokaryotic cells.) The fact that many plasmids confer drug resistance also bothers critics of the program. (pSC101 is so common in nature that its use is not considered dangerous, but others might well be.) Moreover, it is all very well to create a super oil-eater in the service of mankind, but even working only with prokaryotic genes might it not be possible to produce—either accidentally or intentionally—an infectious organism of unheard-of virulence and unbreakable resistance to known antibiotic treatment? Also, because of shotgunning, there is no way to predict what genes or new combinations of genes might be spread abroad.

In 1974, in a totally unprecedented move, a group of molecular biologists—all of whom were working actively in these investigations—called for a temporary ban on all experiments involving DNA recombinations, and for almost two years the work was virtually at a standstill.

In June 1976, the National Institutes of Health issued guidelines governing research on recombinant DNA. Some kinds of experiments are prohibited, including recombinations that increase the virulence of a disease-causing organism or that might transfer drug resistance to an organism that does not acquire it naturally. Also, conditions are specified under which various categories of experiments may be carried out. These are designed to protect both laboratory personnel and the general public. For example, the *E. coli* used in those experiments will be "disabled"; that is, they will lack the capacity to synthesize their cell wall, for example, or to make thymine, and so will be able to survive only in an enriched laboratory medium, like the *Neurospora* strains of Beadle and Tatum.

Undoubtedly, the guidelines, which were under debate for more than a year, will be revised as experience increases.

CLONING

The fact that each cell carries all of an organism's genes suggests that whole organisms can be generated from a single somatic cell. This form of genetic manipulation, which is known as *cloning*, has also been the subject of recent public attention.

A clone is a group of organisms that have been produced from a single parent by a series of mitotic divisions and that are (with the exception of possible mutations) genetically identical. Clones are found in nature only among one-celled organisms and in a few invertebrates and plants that reproduce asexually. It has been discovered, however, that it is possible to produce a whole carrot from an isolated mature carrot cell, and, as we noted in the previous chapter, nuclei removed from a somatic cell, in the frog, direct the development of an egg cell. More recent work has indicated that cloning may be extremely valuable in the propagation of new plant types. It could be used for instance to cultivate large numbers of artificial, ordinarily infertile hybrids or to produce whole plants from single cells into which new genes have been inserted.

Its application to man, although more publicized, is less clear. One possibility that has been mentioned is the production of organs for transplant. There are two main obstacles to the widespread use of transplanted organs to replace diseased ones. One is the lack of suitable replacement organs, which need to be healthy, relatively young, and undamaged. The other, and at present more

QUESTIONS DISCUSSED AT A MEETING OF JURISTS AND
PHYSICIANS ON "ETHICAL ISSUES IN GENETIC
COUNSELING AND THE USE OF GENETIC KNOWLEDGE"

Is a genetic counselor's responsibility first to the couple involved, or to society?

What constitutes a "defective" fetus?

When a mother being tested for mongolism in her unborn child turns out negative in that respect but the test shows the child has an extra Y chromosome—which might predispose him to antisocial behavior as an adult—should the doctor tell her? And if he does, what should she do?

Does an unborn baby have "rights"—including even the right not to be conceived (by cloning) as an identical twin of many others, or of his father? Can a parent ethically give "consent" on behalf of an unborn child—for gene manipulation, for example?

What will be the effect on society, the family, and the individual himself of being able to choose in advance which sex a child will be?

Mass prenatal screening can detect some 130 biochemical abnormalities. Who should do such screening? Which defects should have priority in the search? Should it be voluntary or compulsory?

Who should have access to the information from prenatal tests? How should it be used?

Could a child with genetic defects arising from a rubella infection early in pregnancy sue his mother's obstetrician—or his parents—for "wrongful life" in failing to abort him?

What public policy can and should be worked out to prevent abuses of current and future advances in genetics?

serious, obstacle is that an animal body reacts more or less strongly against any cells that are not genetically identical to its own cells, rejecting the transplant. For this reason, tissues can be exchanged readily only between identical twins. Currently, transplants are done whenever possible between genetically similar persons, often members of the immediate family. Also, drugs are given to the patient to reduce his reactions to the foreign tissue, but since these drugs reduce his reactions to *all* foreign substances, they greatly increase the dangers of infection. One possible way to overcome these two obstacles might be the production of new organs from the patient's own cells growing in tissue culture; such organs, clearly, would be acceptable to his body.

Before such a feat could even be attempted, far more would have to be known about the control of development and differentiation. Needless to say, the cultivation of human replacement tissues is, at the moment, a very remote possibility. It has also been suggested—although not, so far, by scientists in the field—that in this way a family could "reincarnate" a loved one simply by removing a cell from his body, a society could reproduce its leaders in politics, sciences, and the arts, and a government could produce large numbers of people of a proved useful type. This leads us to a further area of controversy, which will be discussed in the following paragraphs.

NATURE AND NURTURE

A little more than half a century ago, a leading American psychologist, John Watson, threw out a challenge: "Give me a dozen healthy infants, well-formed, and my own specified world to bring them up in, and I'll guarantee to take any one at random and train him to become any type of specialist I might select—doctor, lawyer, merchant, chief, and yes, even beggarman and thief, regardless of his talents, penchants, tendencies, abilities, vocations, and the race of his ancestors."*

*J. B. Watson, *Behaviorism*, Peoples Institute, New York, 1925, p. 82.

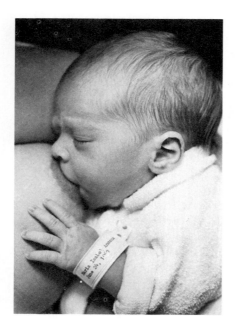

17–6
Very few patterns of human behavior are clearly based solely on instinct. One such, of obvious survival value, is the suckling response, present in infants at birth.

The pendulum now seems to have swung entirely in the other direction, and no longer, in the popular view, is an infant completely plastic, capable of being molded into any shape by his environment. Rather, a member of the human species is seen as an untidy collection of behavioral souvenirs, acquired during his ascendancy from the apes (or perhaps even on the long route from the amoeba).

The reason for this change in attitude has little to do with studies actually carried out in man but is rather a product of a new and extremely interesting discipline known as ethology. Ethology is the study of animal behavior under natural conditions, and ethologists are particularly concerned with species differences in behavior and their evolutionary origin (as contrasted, for example, with studying a solitary pigeon shut up in a box).

In animals, some patterns of behavior are clearly inherited. A spider, for instance, although raised in solitary confinement, can build an extremely complex web on its first try. Among vertebrates, many species of fish and birds have elaborate courtship and territory-defending rituals, whose basic patterns are part of their genetic equipment, that is, instinctive rather than learned. Cross-breeding experiments in animals with different behavior patterns sometimes reveal the extent of the genetic component. Lovebirds of one species, for instance, cut long strips of paper bark or leaves with which to build their nests. The bird then grasps one end of the strip in its bill, tucks the strip into the feathers of the lower back, and so carries it off to its nest. Another species cut materials in the same way but carry the strips back to the nest in their bills. Hybrids between the two species always try to tuck the nesting material into their feathers but invariably fail to do so. Sometimes the tucking movements are incomplete, or sometimes the bird will hold the strip by the middle or the end, will fail to let go at the right moment, or will tuck it into some inappropriate part of the body. The F_1 hybrids, which were studied over a three-year period, never entirely gave up trying the tucking movements, but by the end of the period, the futile activities related to this method of carrying had decreased considerably.

Mammals also show inherited behavior. For example, the first time a flying squirrel raised in a bare laboratory cage isolated from other squirrels is given nuts or nutlike objects, it will "bury" them in the bare floor, making scratching movements as if to dig out the earth, pushing the nuts down into the "hole" with its nose, covering them over with imaginary earth, and stamping on them.

Wild rats are much more aggressive in their behavior than the white rats used in laboratories, which are generally docile and often affectionate toward their keepers. The offspring of wild rats, even though raised in a laboratory, are also aggressive. (These wild animals have adrenal glands about twice as large as those of their gentle cousins, which appears to provide a physiological basis for their behavior.)

Is man naturally aggressive? Does he inherit genes for territoriality? Are some races more intelligent than others? At this time, very, very little is known about these questions. Schizophrenia has recently been shown to have a strong genetic basis, as indicated by studies of identical twins separated at birth (by being adopted into different families). These studies revealed that, if one twin developed schizophrenia, the other was far more likely to be schizophrenic than an unrelated person of the same age. Some forms of mental deficiency have been associated with chromosomal abnormalities, as previously noted, and with genetic "errors," as in cases of PKU. Identical twins separated at birth have been found to have IQs more like one another than like other members of the household in which they are raised. However, recent studies purporting to show significant differences in IQ between races or ethnic groups have been under attack, largely because of environmental inequities or the cultural bias of the test itself. Even using standard IQ tests, the overlaps between any two groups are large enough to invalidate any predictions concerning any individual.

Beyond such scraps of evidence as we have cited, there are very few data. The nature of one's nature is, of course, an interesting subject and one in which anyone can participate, since, even if special knowledge is required, it is not available. Consequently, at present, arguments about these issues must be regarded as social and ethical rather than scientific.

SUMMARY

Controversial subjects in the field of genetics include genetic engineering, cloning, and hereditary influences on human behavior and intelligence.

Genetic engineering is used to describe the introduction of new genetic material into a cell. In nature, DNA is passed between prokaryotic cells by transformation, conjugation, and transduction. The possibilities of exploiting one or more of these forms of genetic exchange has been increased by the discovery of enzymes that cleave the DNA molecule in such a way that new segments can be inserted into an existing bacterial or viral chromosome or into a plasmid.

Cloning is the production of multicellular organisms from a single somatic cell. Its most immediate applications appear to be in the production of new plant varieties.

There are very few data available on the existence of hereditary influences on human behavior or intelligence.

QUESTIONS

1. Describe the possible outcomes of infection of a bacterial cell by a bacterial virus.

2. Three types of plasmid were mentioned in this chapter. Describe them.

3. Obviously experiments using a plasmid that confers resistance to an antibiotic has disadvantages. However, it also has an important advantage. What is it?

4. How would you answer the questions in the boxed essay on page 199? Who do you think should determine such policies?

SUGGESTIONS FOR FURTHER READING

JACOB, FRANÇOIS: *The Logic of Life: A History of Heredity,* Pantheon Books, New York, 1973.

Jacob's principal theme concerns the changes in the way man has looked at the nature of living beings. These changes, which are part of man's total intellectual history, determine both the pace and direction of scientific investigation. The opening chapters are particularly brilliant.

LURIA, S. E.: *Life: The Unfinished Experiment,* Charles Scribner's Sons, New York, 1973.

An interpretation of biology written for the layman by a Nobel laureate. A National Book Award winner.

PETERS, JAMES A. (ed.): *Classic Papers in Genetics*, Prentice-Hall, Inc., Englewood Cliffs, N.J., 1959.*

Includes papers by most of the scientists responsible for the important developments in genetics: Mendel, Sutton, Morgan, Beadle and Tatum, Watson and Crick, Benzer, etc. You should find this book very interesting; the authors are surprisingly readable, and the papers give a feeling of immediacy that no account can achieve.

STRICKBERGER, MONROE W.: *Genetics*, 3d ed., The Macmillan Company, New York, 1975.

An up-to-date and cohesive account of the science, this book is much broader in coverage than most texts at this level.

WATSON, J. D.: *The Double Helix*, Atheneum Publishers, New York, 1968.*

"Making out" in molecular biology. A brash and lively book about how to become a Nobel laureate.

WATSON, J. D.: *Molecular Biology of the Gene*, 3d ed., W. A. Benjamin, Inc., Menlo Park, Calif., 1976.*

For the student who wants to go more deeply into the questions of molecular biology, this is a detailed and authoritative account.

*Available in paperback.

PART II   Organisms

SECTION 4   The World of Living Things

18–1
*A confrontation between representatives of
two prominent phyla. On the left, a tropical
grasshopper; on the right, an insect-eating
marsupial (a mouse opossum) from South
America. The grasshopper held its ground,
and, after weighing the situation carefully,
the mouse opossum withdrew.*

CHAPTER 18

The Diversity of Life

In this part of the book we are going to be concerned with organisms, the living things that populate the world around us. The chief focus will be on the higher plants, particularly the flowering plants, and on the higher animals, particularly humans. These are the subjects of Sections 5 and 6, respectively. Here we are going to introduce these organisms by examining the modern representatives of some of their predecessors and, as far as we are able, by tracing their evolutionary history.

THE CLASSIFICATION OF ORGANISMS

Most people are concerned with only very limited areas of the natural world. Gauchos, the cowboys of Argentina, who are famous for their horsemanship, have some 200 names for different colors of horses but divide all the plants known to them into four groups: *pasto,* or fodder; *paja,* bedding; *cardo,* wood; and *yuyos,* everything else.

Most of us are like the gauchos. Once beyond the range of common plants and animals, and perhaps a few uncommon ones that are of special interest to us, we usually run out of names. Biologists, however, face the problem of systematically investigating, identifying, and exchanging information about well over a million different kinds of organisms. In order to do this, they must have a system for naming all these organisms and for grouping them together in orderly and logical ways. The problems of developing such a system are immensely complicated. Some 700,000 different kinds of insects, for instance, have been named and classified to date.

In the eighteenth century, the binomial ("two-name") system became generally accepted, and this system is still in use today. The binomial consists of two parts, the name of the genus and that of the species. The wolf, for example, belongs to the genus *Canis* (the Latin word for "dog"), and its scientific name is *Canis lupus,* which distinguishes it from such near relations as the coyote *(Canis latrans)* and the domestic dog *(Canis familiaris).* Every individual is thus a member of a particular *species,* which, in turn, belongs to a larger group, a *genus* (plural, genera).

According to the modern system of classification, genera are grouped into *families,* families into *orders,* orders into *classes,* classes into *phyla* (singular, phylum), and phyla into *kingdoms.* Notice, in Table 18–1, how much information

(a)

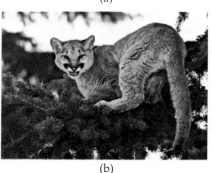

(b)

(c)

18–2

Three members of the genus Felis: (a) *an ocelot,* Felis pardalis; (b) *a cougar,* Felis concolor; *and* (c) *a domestic cat,* Felis cattus.

Table 18-1 *Biological Classifications*

Red Maple

CATEGORY	NAME	CHARACTERISTICS
Kingdom	Plantae	Multicellular, primarily terrestrial photosynthetic organism with cellulose cell walls and chlorophylls *a* and *b* and carotenes
Sub-kingdom	Embryophyta	Plants forming embryos
Phylum	Tracheophyta	Vascular plants
Sub-phylum	Sperma-tophytina	Seed plants, generally with large leaves with a complex vascular pattern
Class	Angiospermae	Flowering plants, seed enclosed in ovary
Subclass	Dicotyledoneae	Embryo with two seed leaves (cotyledons)
Order	Sapindales	Soapberry order; trees and shrubs
Family	Aceraceae	Maple family; trees of temperate regions
Genus	*Acer*	Maple and box elder
Species	*rubrum*	Red maple

Man

CATEGORY	NAME	CHARACTERISTICS
Kingdom	Animalia	Multicellular; requiring organic substances as energy source
Phylum	Chordata	Having a dorsal hollow nerve cord, a notochord, and gills at some stage of life cycle
Sub-phylum	Vertebrata	Spinal cord enclosed in a vertebral column, brain enclosed in a skull
Class	Mammalia	Young nourished by milk glands; possessing hair; body cavity divided by a muscular diaphragm; red blood cells without nuclei; endothermic ("warmblooded")
Order	Primates	Hands and feet with at least four mobile digits, the thumb and large toe usually opposable; usually have only one young, helpless at birth
Family	Hominidae	Two-legged, with hands and feet differently specialized; all digits with flattened nails; no tail
Genus	*Homo*	Completely bipedal after infancy; arms shorter than legs; thumb large; big toe not opposable; face small; incisors and canines small; large brain; long childhood
Species	*sapiens*	High forehead; small jaw; prominent chin; small molars; sparse body hair

you have about an organism once you know its classification.

In this text, we recognize five kingdoms of organisms: Monera, Protista, Fungi, Plantae, and Animalia. In doing so, we follow a system first proposed by R. H. Whittaker.* It is not, however, the only system in modern use. In fact, until recently, most biologists recognized only two kingdoms: Plantae and Animalia. The monerans, most protistans, and all the fungi were considered plants. The newer classifications are based both on new information and on shifts in emphases. Ideally, one would wish that the different categories reflected degrees of evolutionary relationship. In this regard, as we shall see, some of these groupings are much more satisfactory than others. (A more complete classification of organisms is found in Appendix C.)

* *Science,* **163**:150–160, 1969.

18–3

Representatives of the five kingdoms. (a) Monerans. Blue-green algae. Colonies of these cells, which are prokaryotes, are common in freshwater habitats. (b) Protists. Vorticella, like most protists, is a single cell. It attaches itself to a substrate by a long stalk. A contractile fiber, visible in this micrograph, runs through the stalk. (c) Fungi. An inky cap mushroom. Fungi are characterized by a multicellular underground network and also spores and sporangia which, in mushrooms, are borne in these familiar structures. (d) Plants. Marsh marigolds. The flowers of angiosperms, attractive to pollinators, are among the principal reasons for their evolutionary success. (e) Animals. A luna moth. A nervous system with a variety of sense organs is a chief characteristic of the animal kingdom.

(a)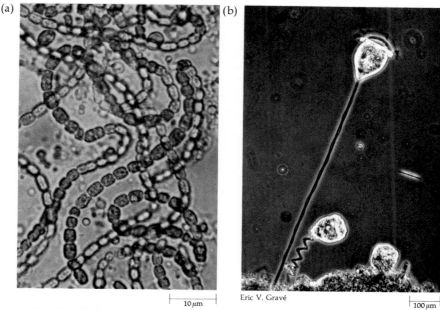

(b)

Eric V. Gravé

10 μm   100 μm

(c)

(e)

(d)

The kingdom Monera includes the only living prokaryotes, the bacteria and the blue-green algae. As we noted in Section 1, prokaryotes are by far the simplest of all cells. Their genetic material is in the form of a long, continuous molecule of DNA. The DNA is not combined with protein, it is not surrounded by a special membrane, and there is no mitosis or meiosis. The cytoplasm contains ribosomes but no complex membrane-bound organelles. The cell is surrounded by a cell membrane and a cell wall. This wall, which is a combination of a unique polysaccharide and protein, is different in composition from the cell walls of any other organism. (This unusual cell-wall composition, shared by the bacteria and the blue-green algae, was one of the clues to the recognition of the close evolutionary ties between these two groups.)

BACTERIA

Bacteria are not only the oldest but also the most abundant group of organisms in the world. They can live in places and under conditions that support no other forms of life. They have been found in the icy wastes of Antarctica, in the near-boiling waters of natural hot springs, and in the dark depths of the ocean. All else failing, many kinds can take the form of hard, resistant spores, which may lie dormant for years until conditions become more favorable for growth.

Bacteria obtain energy in a great variety of ways. Some are autotrophs. Among the autotrophs are photosynthetic bacteria which, like green plants, capture light energy in specialized pigments (bacteriochlorophyll or chlorobium chlorophyll). However, unlike algae and plants, photosynthetic bacteria lyse hydrogen sulfide (H_2S) and other compounds rather than water and do not release oxygen gas. Other autotrophic bacteria are chemoautotrophic; these obtain their energy from the oxidation of inorganic molecules such as certain compounds of nitrogen, sulfur, and iron.

18–4

Examples of three classes of bacteria. (a) The myxobacteria are distinguished by their slime tracks. Also, some types form fruiting bodies, such as those in this micrograph of Stigmatella aurantiaca. (b) The spirochetes are spiral bacteria up to 500 micrometers long, which is an enormous size for bacteria. They move by means of a rapid spinning or whirling of the cell around its long axis. Treponema pallidum, two of which are shown here, is the causative agent of syphilis. (c) Rickettsia are the smallest known cells. Typhus is caused by Rickettsia typhi, shown here. It is spread from rats to man by fleas. It can then be transmitted by body lice, and, under crowded conditions, large numbers of people can be infected in a very short time. More human lives have been taken by rickettsial diseases than by any other infection except malaria. At the siege of Granada in 1789, 17,000 Spanish soldiers were killed by typhus, 3,000 in combat. In the Thirty Years War, the Napoleonic campaigns, and the Serbian Campaign during World War I, typhus was also the decisive factor.

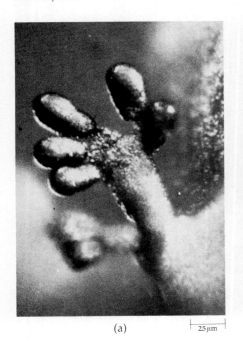

(a)   25 μm

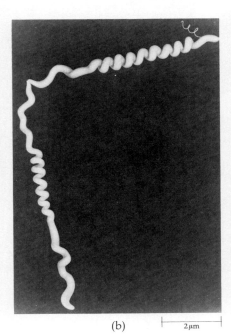

(b)   2 μm

(c)   1 μm

Despite the relative simplicity of bacterial cells, they are able to respond to stimuli. Some photosynthetic bacteria exhibit phototaxis—moving toward a light by differential motion of their flagella. Escherichia coli and other flagellated bacteria similarly exhibit chemotactic (chemical-sensing) responses that are made possible by changes in the direction of flagellar rotation. When presented simultaneously with both attractant and repellent, they respond to whichever one is present in the more effective concentration. Such a response indicates that these bacteria probably have a processing mechanism that compares opposing signals from different chemoreceptors, sums these signals up, and then communicates the sum to the flagella. The entire system has been shown to be controlled by proteins in the cell membranes.

Recently, scientists at the University of California, working with a strain of Salmonella, were able to show that the cells could move up a very slight concentration gradient of a nutrient amino acid (serine). How are the bacteria able to distinguish an area of lower concentration from an area of only very slightly higher concentration? There seem to be only two possible answers to this question. One is that they sense the gradient instantaneously by comparing the concentration of serine at their "head" with that at their "tail." This possibility has been rejected because the bacteria are so extremely small and the differences in concentration are too minute and too variable. The second hypothesis, supported by preliminary data, is that they are able to compare concentrations at many body lengths apart. In other words, information about the concentration at one time and place is stored by the bacteria and compared with information received at another time and place—a primitive form of memory.

18–5

Preparing a rice paddy in Bali, Indonesia. The blue-green algae on the surface of the water provide nitrogen for the rice plants.

Most bacteria are heterotrophs, and it is this group that is of the greatest importance to man. Some of the heterotrophs parasitize other organisms. These are the disease-causing (pathogenic) bacteria. These forms often cause illness by the production of specific poisons ("toxins"). However, by far the largest numbers of heterotrophic bacteria live on dead organic matter. These bacteria are of the utmost importance to man and all other living things because they are the principal decomposers of the biosphere. Through the action of the bacteria (and the fungi, which are also decomposers), materials incorporated into the bodies of once-living organisms are released and made available for successive generations of living things. We shall discuss this important subject of recycling in greater detail in Chapter 36.

BLUE-GREEN ALGAE

Blue-green algae resemble bacteria in their small size, in their simple, prokaryotic organization, and in the unusual composition of their cell walls. Some biologists believe that they should be classified as another type of photosynthetic bacteria.

The blue-green algae resemble eukaryotic algae and also plants in their photosynthesis: In all these groups, the principal photosynthetic pigment is chlorophyll *a*, and photosynthesis involves the lysis (splitting) of the water molecule (pages 92 to 93). The chlorophyll molecules and other photosynthetic pigments of the blue-green algae are embedded in membranes, as are the pigment systems of eukaryotic cells. However, these membranes are not enclosed in a special membrane-bound organelle (the chloroplast) as is the case with the eukaryotic cells.

Blue-green algae grow for the most part in fresh water. They are sometimes found as isolated cells, but more often they form clusters, threads, or chains. Blue-green algae are among the few microorganisms able to fix nitrogen (that is, to incorporate atmospheric nitrogen into organic compounds) and are important contributors of nitrogen to rice cultivation.

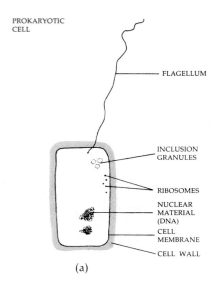

PROKARYOTIC
CELL

FLAGELLUM

INCLUSION
GRANULES

RIBOSOMES

NUCLEAR
MATERIAL
(DNA)

CELL
MEMBRANE

CELL WALL

(a)

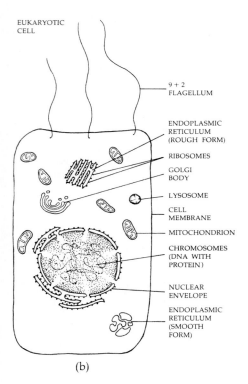

EUKARYOTIC
CELL

9 + 2
FLAGELLUM

ENDOPLASMIC
RETICULUM
(ROUGH FORM)

RIBOSOMES

GOLGI
BODY

LYSOSOME

CELL
MEMBRANE

MITOCHONDRION

CHROMOSOMES
(DNA WITH
PROTEIN)

NUCLEAR
ENVELOPE

ENDOPLASMIC
RETICULUM
(SMOOTH
FORM)

(b)

18–6

Comparison between (a) a prokaryotic cell and (b) a eukaryotic cell. The eukaryote contains a membrane-bound nucleus; complex membrane-bound organelles, such as mitochondria and chloroplasts; endoplasmic reticulum; and Golgi bodies. Its chromosomes are a complex of DNA and special proteins. Flagella, when present, have a characteristic 9 + 2 structure. All these features are lacking in prokaryotes.

THE VIRUSES

Viruses, because of their simplicity, are traditionally studied with the bacteria and blue-green algae. Some biologists believe that the earliest forms of life were viruslike; others contend that viruses should not be regarded as living organisms at all but rather as parts of cells that have set up a partially independent existence. S. E. Luria, who was awarded the Nobel Prize for his work with bacterial viruses, has called them "bits of heredity looking for a chromosome." (As we noted in Chapter 17, some viruses do indeed "find" a chromosome.)

Viruses are parasites that can multiply only within a living cell. All viruses consist of nucleic acid—either DNA or RNA—and protein. The protein forms an outer coat, which protects the nucleic acid and determines what sort of cell the virus can parasitize. Usually particular viruses attack only particular cells: an influenza virus invades the lining of the respiratory tract, a polio virus attacks intestinal cells and sometimes nerve cells, and so forth.

Either before the virus enters the cell or just after (depending on the type of virus), the viral nucleic acid slips out of its protein coat. In the case of the DNA viruses, the DNA of the virus replicates and also codes for messenger RNA. The mRNA, in turn, produces enzymes and coat protein needed by the virus, which uses raw materials (such as amino acids and nucleotides) and cellular machinery (such as the ribosomes and transfer RNA) of the host cell. In the case of the RNA viruses, the nucleic acid both replicates and serves directly as messenger RNA, producing the enzymes and coat proteins needed by the virus. The end product of either type of infection is hundreds or often thousands of new viral particles produced and assembled within the infected cell, which is often broken apart and destroyed as the particles are released. Characteristically, it is this lysing of the host cell that causes the symptoms associated with a virus infection.

THE PROTISTS

The kingdom Protista includes all eukaryotic one-celled heterotrophs and also all eukaryotic algae, some of which are multicellular. Among protists, the DNA of the nucleus is combined with protein, organized into chromosomes, and enveloped in membranes, the nuclear envelope. Protists have mitochondria, and photosynthetic protists have a chloroplast or chloroplasts as well. Many have cilia or flagella with the characteristic 9 + 2 structure (page 66). Many have a sexual cycle—such as that of *Chlamydomonas* (page 122) for example—and in those that do not, biologists tend to believe that some form of sexual reproduction once existed and was lost.

There are nine phyla of Protista (Table 18–2), seven of which are algae. Biologists generally agree (1) that the protists represent a number of quite different evolutionary lines, and (2) that all other eukaryotic organisms—fungi, plants, and animals—originated from primitive protists.

ORIGIN OF THE PROTISTS

The step from the prokaryotes (the monerans) to the first eukaryotes (the protists) was one of the big evolutionary transitions, perhaps second only to the origin of life. The question of how it came about is a matter of current and lively discussion. One interesting hypothesis is that larger, more complex cells evolved as a result of certain prokaryotes taking up residence inside other cells.

18–7

(a) *Adenovirus, one of the many viruses that cause colds in humans. This virus is an icosahedron. Each of its 20 sides is an equilateral triangle composed of identical protein subunits. Many viruses, and also Buckminster Fuller's geodesic domes, are constructed on this principle. There are 252 subunits in all. Within the icosahedron is a core of double-stranded DNA. (b) A model of the adenovirus, made up of 252 tennis balls. (c, d) Electron micrograph and diagram of influenza virus. The virus is composed of a core of RNA surrounded by a lipoprotein envelope through which protrude stubby protein spikes. The influenza virus, for reasons which are not understood, mutates frequently. The changes in its nucleic acid alter its proteins and, hence, previously formed antibodies no longer "recognize" it. New strains of influenza virus are likely to arise more rapidly than new vaccines can be produced to combat them. (e, f) Electron micrograph and diagram of a bacteriophage, showing the many different structural components of the protein coat. The DNA of the virus codes for all these structural proteins.*

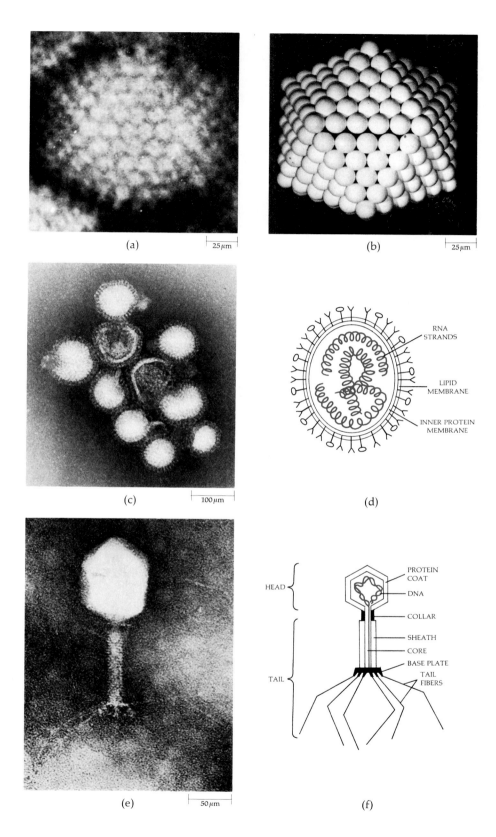

(a)   25 μm

(b)   25 μm

(c)   100 μm

(d)

RNA STRANDS
LIPID MEMBRANE
INNER PROTEIN MEMBRANE

(e)   50 μm

(f)

HEAD

PROTEIN COAT
DNA
COLLAR
SHEATH
CORE
BASE PLATE
TAIL FIBERS

TAIL

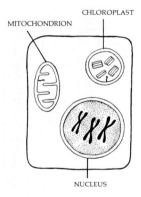

CHLOROPLAST

MITOCHONDRION

NUCLEUS

18–8

A hypothesis for the evolution of the eu-
karyotic cell. Invaginating (inward-turning)
cell membranes enclose duplicates of a pro-
karyotic chromosome within double mem-
branes. The structures so formed then, in the
course of evolution, slowly take on separate
specialized functions, becoming the eukaryotic
nucleus, mitochondrion, and chloroplast.

About 2.5 billion years ago, oxygen gas began slowly to accumulate in the atmosphere (page 105). Those bacteria that were able to convert to the use of oxygen in ATP production gained a strong advantage, and so such forms began to prosper and increase. Some of these evolved into modern forms of bacteria. Others, according to the theory, became symbionts within larger cells and evolved into mitochondria. Among the evidence that supports this latter idea is the following: mitochondria contain their own DNA, and this DNA is present in a single, continuous molecule, like the DNA of bacteria; many of the same enzymes contained in the cell membranes of bacteria are found in mitochondrial membranes; the ribosomes of mitochondria resemble those of bacteria both in their size (they are smaller than those of eukaryotic cells) and in some details of their chemical composition; mitochondria appear to be produced only by other mitochondria, which divide within their host cell. (However, to make the situation more complex, the DNA of mitochondria does not code for all the mitochondrial proteins. Host-cell DNA is also required for mitochondrial structures.)

We know little about those original cells in which these bacteria first set up housekeeping—or, indeed, if they actually existed. But if they did exist, they probably had no means of using oxygen for cellular respiration and so were dependent entirely on glycolysis as an energy source, which, as we have seen, is relatively inefficient. Cells with respiratory assistants would have been more efficient than those lacking them and so would have outnumbered them and crowded them out.

In an analogous fashion, photosynthetic prokaryotes ingested by larger, nonphotosynthetic cells are believed to be the forerunners of chloroplasts. By this symbiosis ("living together"), the smaller cells gained nutrients and protection, and the larger cells were given a new energy source.

This hypothesis accounts for the presence in eukaryotic cells of the complex organelles not found in the far simpler prokaryotes. It gains support from the fact that many modern organisms contain symbiotic algae.

An alternative hypothesis is that eukaryotes developed from prokaryotes by a process of the infolding (invagination) of cellular membranes (Figure 18–8). This explanation is attractive because it accounts for the presence of the nuclear envelope as well as other principal features of eukaryotes.

Table 18–2   *The Kingdom Protista*

Phylum Protozoa	Unicellular heterotrophs, including flagellates (mastigophores), amoebalike organisms (sarcodines), ciliates, and sporozoans
Phylum Euglenophyta	*Euglena* and related algae, all unicellular, mostly freshwater
Phylum Chrysophyta	Diatoms and related algae, all unicellular, mostly marine
Phylum Pyrrophyta	Dinoflagellates and related algae, all unicellular, marine and freshwater
Phylum Xanthophyta	Yellow-green algae, all unicellular
Phylum Chlorophyta	Green algae, including unicellular, colonial, and multicellular forms; mostly freshwater
Phylum Phaeophyta	Brown algae, including the kelps, all multicellular, almost all marine.
Phylum Rhodophyta	Red algae, all multicellular, mostly marine
Phylum Gymnomycota	Slime molds

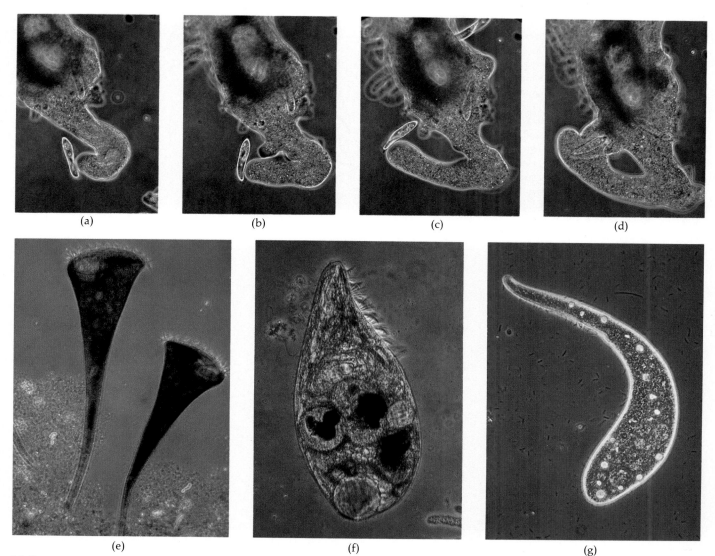

18–9

(a–d) *The giant amoeba* Chaos chaos *captures a* Paramecium. *Although seemingly disorganized,* Chaos chaos *is able to sense its prey, move toward it, and send out a pseudopod ("false foot") of the right size and shape to envelop it.* (e) Stentor, *a ciliate. Extended, the cell looks like a trumpet, crowned with a wreath of membranelles. These beat in rhythm, creating a powerful vortex that draws edible particles up to and into the funnel-like groove leading to the cytostome ("mouth"). Elastic protein threads, myonemes, run the length of the body. When contracted,* Stentor *is an almost perfect sphere.* (f) Blepharisma, *a ciliate, with food visible in several vacuoles.* (g) Dileptus, *a ciliate.*

TYPES OF PROTISTS

Protozoa

The Protozoa—"first animals"—are a collection of varied and interesting organisms, all of which are unicellular and all of which are heterotrophs. There are four main classes in the phylum Protozoa: Mastigophora, Sarcodina, Ciliophora, and Sporozoa.

The Mastigophora are characterized by their flagella; most have one or two, but some have many. Most are free-living, but some are parasitic, including *Trychonympha* (Figure 3–5b) and *Trypanosome gambiense,* the flagellate that causes African sleeping sickness.

The Sarcodina includes the amoebas and amoebalike organisms. They have no cell wall and generally move and feed by the formation of pseudopodia, which are extensions of the cytoplasm (Figure 18–9a–d).

The Ciliophora, or ciliates, are the most specialized and complicated of the Protozoa and indeed probably the most complex of all cells. They are characterized by cilia, all of which have a 9 + 2 structure. (Cilia and flagella differ

18-10

Life cycle of Plasmodium vivax, *the sporozoan that causes malaria in man. The cycle begins (a) when a female Anopheles mosquito bites a person with malaria and, along with his blood, sucks up gametes (b) of the protozoan. In the mosquito's body, the gametes unite (c) and form a zygote (d). From the zygotes, multinucleate structures called oocysts develop (e), which, within a few days, divide into thousands of very small, spindle-shaped cells, sporozoites (f), which then migrate to the mosquito's salivary glands. When the mosquito bites another victim (g), she infects him with the sporozoites. These first enter liver cells (h), where they undergo multiple division (i). The products of these divisions (merozoites) enter the red blood cells (j), where again they divide repeatedly (k). They break out of the blood cells (l) at regular intervals of about 48 or 72 hours, producing the recurring episodes of fever characteristic of the disease. After a period of asexual reproduction, some of these merozoites become gametes and, if they are ingested by a mosquito at this stage, the cycle begins anew.*

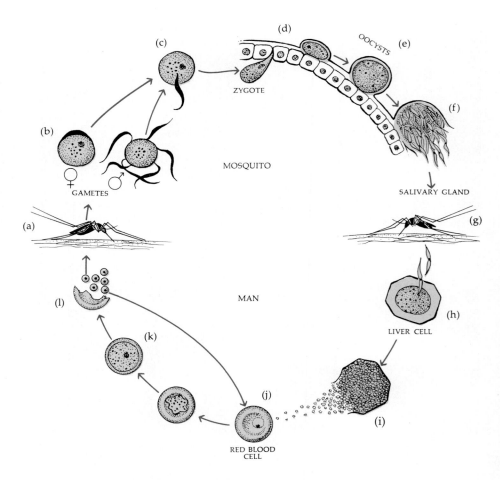

principally in their length; cilia are shorter and also they are usually more numerous. If their basic structure had been known at the time they were first observed, they probably would not have been given two names.) *Paramecium* and *Didinium*, featured in Chapter 4, are ciliates, as is *Stentor* (Figure 18–9).

The Sporozoa are all parasites. The most infamous are the members of the genus *Plasmodium* that cause malaria, which is, on a worldwide basis, still a principal cause of death among humans. The complex life cycle of a *Plasmodium* is shown in Figure 18–10.

Algae

The seven phyla of algae listed in Table 18–2 resemble one another in that they are all made up principally of photosynthetic organisms (though several phyla include a few forms that are clearly related although not photosynthetic). They all have chlorophyll *a* as their principal photosynthetic pigment, and they are unicellular or have relatively simple multicellular forms, as compared with the plants. The seven phyla can be distinguished from one another in a variety of ways, such as their accessory photosynthetic pigments, the presence or absence of a cell wall, and the composition of the cell wall, if present. Accessory pigments that mask the green of the chlorophylls are responsible for the different colors of the members of the various phyla.

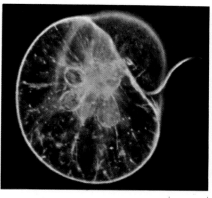

18-11　　　(a)　　　|—100 μm—|

Some prominent members of the plankton community. (a) Noctiluca scintillans, *the "shining light of the night," is bioluminescent, as its name implies. Yellow-brown diatoms that have been ingested can be seen inside the cell. (b) A diatom, showing the characteristically intricately marked shell, which contains silicone. (c) A dinoflagellate (Ceratium tripos). These organisms often have thick cellulose walls and bizarre shapes. You can see one flagellum in motion.*

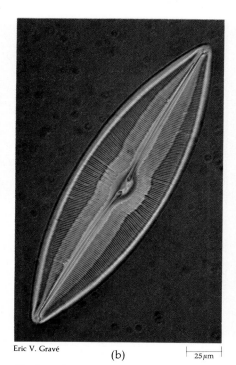

Eric V. Gravé　　　(b)　　　|—25 μm—|

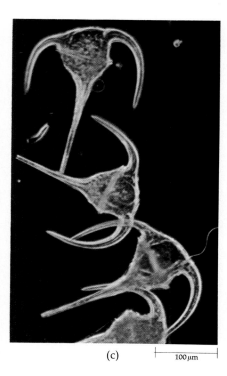

(c)　　　|—100 μm—|

18-12

Ulva, or sea lettuce, in which the cells divide both longitudinally and laterally, with a single division in the third plane, producing a broad plant body two cells thick.

Plankton

The one-celled algae float upon the surface of the seas or on bodies of fresh water. Such free-floating populations are known as *plankton*, from *planktos*, the Greek word for "wandering." The photosynthetic members of the plankton community—sometimes called *phytoplankton*—carry out most of the photosynthesis that takes place in the ocean and are the food source for most of the ocean's inhabitants (Figure 18–11).

Seaweeds

The larger, multicellular algae make up the seaweeds. These include the Rhodophyta (red algae), Phaeophyta (brown algae), and some members of the Chlorophyta (green algae). These forms are found near the shores, where the waters are much richer in minerals (such as nitrogen and phosphorus) washed down from the land. The shortage of such minerals in the open ocean is the limiting factor in any modern plans to farm the seas.

These forms that evolved along the shores were under different evolutionary pressures than the planktonic algae. Chief factors in this environment are the waves and tides, and one important adaptation found among the seaweeds is *holdfasts*, special anchoring structures that adhere to the rocks. The upper portions of the seaweed characteristically are thin and spread out; they produce enough sugars to nourish the other parts of the organism, which may be a meter or more below the surface of the water and overshadowed by the upper portions of the algae. Some of the large red algae and brown algae have specialized conducting tissues that transport the products of photosynthesis downward.

Not only did these marine algae, the seaweeds, take on shapes, but they also developed specialized colors. Water filters out the light, removing first the longer wavelengths, the reds and oranges; at the deeper levels, only a faint blue light penetrates. The red algae capture the energy of these blue rays. The brown algae (the brown results from a mixture of green and orange pigments) absorb blue-green lights, and the green algae, whose principal pigments are the chlo-

(a)

(b)

18-13

(a) *A brown alga,* Laminaria, *showing hold-fasts, stipes, and blades.* (b) *Unlike the brown algae, most red algae are made up of filaments. The branched filaments of this red alga are hooked, enabling it to cling to other seaweeds.*

rophylls, make fullest use of the longer-waved red light. As a consequence of the evolution of specialized pigments, the various seaweeds came to use the entire spectrum of light as well as the living space along the rocky shore.

Chlorophyta: Ancestors of the Land Plants

The Chlorophyta (green algae), of which *Chlamydomonas* (page 20) is an example, are of particular interest to biologists because they are believed to be the group from which the plants arose. Evidence for this relationship is based on a comparison between modern chlorophytes and modern plants: like the plants, the green algae contain chlorophylls *a* and *b* and carotenoids as their photosynthetic pigments; they accumulate their food as starch; and their cell walls are composed of cellulose. Many species of green algae are colonial or multicellular. (In colonial organisms, the cells making up the colony may be interconnected, but there is no division of labor among them. In a multicellular organism, different cells are specialized for different functions.) Green algae have achieved multicellularity in several different ways. Some multicellular representatives of the green algae are shown in Figure 18-14.

All multicellular green algae have a life cycle that includes *alternation of generations,* which is, as we described in Chapter 9, an alternation between a diploid form that reproduces asexually (by spores) and a haploid form that reproduces sexually (by gametes). The diploid, spore-producing form is known as the *sporophyte,* and the structures in which spores are produced are known as *sporangia.* The haploid, gamete-producing form is known as the *gametophyte,* and the structures in which gametes are produced are known as *gametangia.* In some algae, the gametes are identical in appearance; such gametes are known as isogametes. In other algae, one gamete is larger than the other; the larger gamete is the egg cell and the smaller the sperm. Gametangia in which eggs are produced are called *archegonia,* and those in which sperm are produced are called *antheridia.* The life cycle of *Ulva,* a green alga that produces isogametes, is shown in Figure 18-15. This reproductive pattern is another link between the green algae and plants, in all of which some form of alternation of generations is found.

(a)   10mm

(b)   25μm

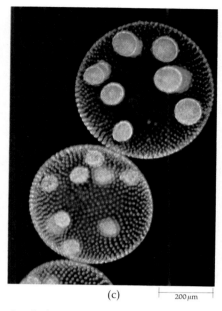

(c)   200μm

18-14

It is possible to identify several evolutionary lines among the Chlorophyta, each starting presumably with a single-celled organism resembling Chlamydomonas and each representing a different trend toward multicellularity. (a) Valonia, which is about the size of a hen's egg, with many nuclei but no parti-

tions separating them. (b) Spirogyra, in which the cells all elongate and then are divided by transverse cross walls so that the cells in the plant body are strung together in long, fine filaments, in which the chloroplasts form spirals that look like strips of green tape within each cell. (c) Volvox, in which

hundreds (sometimes thousands) of Chlamydomonas-like cells are united by tiny strands of protoplasm into a hollow sphere. New daughter colonies can be seen forming inside the mother colony. Ulva (Figure 18-12) represents a fourth trend toward multicellularity.

18-15

In the sea lettuce, Ulva, we can see the reproductive pattern known as alternation of generations, in which one generation produces spores, the other gametes. The haploid (n) gametophyte produces haploid isogametes, and the gametes fuse to form a diploid (2n) zygote. A sporophyte, a plant in which all the cells are diploid, develops from the zygote. This plant produces haploid spores by meiosis. The haploid spores develop into haploid gametophytes, and the cycle begins again.

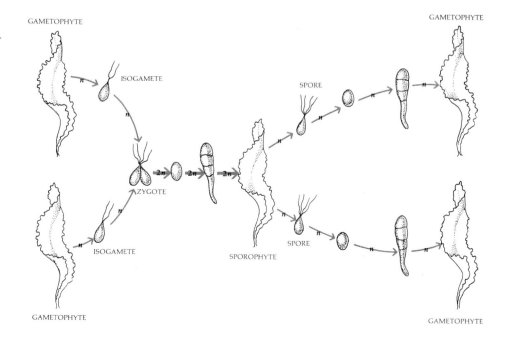

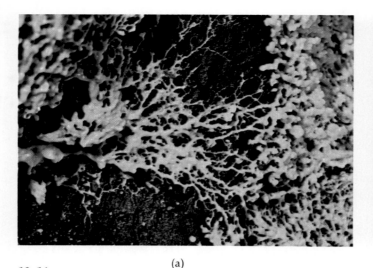

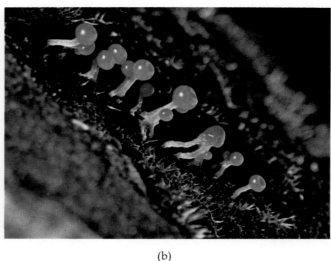

18–16

(a) *Plasmodium of a slime mold. Such a plasmodium can pass through a piece of silk or filter paper and come out the other side apparently unchanged.* (b) *Sporangia of a plasmodial slime mold on a rotting log.*

Slime Molds

The slime molds (phylum Gymnomycota) are a group of curious organisms that resemble amoebas. However, during some or all of their life cycle, the slime molds are multicellular or, at least, multinucleate (having many nuclei but not partitioned by cell membranes).

Slime molds reproduce by means of spores. A spore, by definition, is a cell that is capable of developing into an adult organism without fusion with another cell; thus it contrasts with a gamete, which must unite with another gamete to form a zygote, which then develops into an adult individual (page 121). The spores of slime molds, like the spores of many other organisms, are produced by meiosis in specialized structures known as sporangia (Figure 18–16). Sporangia often elevate the spores, which facilitates their dispersal by air currents. (It will be necessary for you to remember these new terms, spore and sporangia, for they will recur throughout this chapter.)

THE FUNGI

The fungi are a group of organisms so unlike any others that although they were long classified with the plants, it has come to seem appropriate to assign them to a separate kingdom. Except for some one-celled forms, such as the yeasts, the fungi are basically composed of masses of filaments. A fungal filament is called a *hypha*, and all the hyphae of a single organism are collectively called a *mycelium*. The walls of the hyphae usually contain chitin, a polysaccharide that is never found in plants. (It is, however, the principal component of the exoskeletons—the hard outer coverings—of insects and other arthropods.) Protoplasm flows within the mycelium. The complex reproductive structures of fungi, such as mushrooms, are composed of tightly packed hyphae.

All fungi are heterotrophs. They obtain food by absorbing organic compounds, sometimes digesting the compounds first by means of enzymes they secrete into the food mass. Consequently, they live in soil, in water, or in some other medium containing organic substances. Growth is their only form of mobility (except for sex cells or spores, which may travel through air or water).

Some fungi are of great economic importance in the production of wine, beer, bread (yeasts), and cheeses, such as Roquefort and Camembert, and anti-

ERGOTISM, THE WITCHES OF SALEM, AND LSD

Many fungi are parasitic on higher plants. The disease of plants called ergot is caused by Claviceps purpurea, *a parasite of rye and other grasses. Although ergot seldom causes serious damage to the crop of rye, it is dangerous because a small amount mixed with rye grains is enough to cause severe illness among domestic animals or among the people who eat bread made with this flour. Ergotism, the toxic condition caused by eating grain infected with ergot, occurred frequently during the Middle Ages, when it was known as St. Anthony's fire. In one epidemic in 944, more than 40,000 people died. In 1722, ergotism struck down the cavalry of Czar Peter the Great on the eve of battle for the conquest of Turkey and thus changed the course of history.*

Ergotism is often accompanied by psychotic delusions. Recent evidence suggests that the adolescent girls who denounced the "witches" of Salem, Massachusetts, were victims of ergotism. As recently as 1951, there was an outbreak in a small French village in which thirty people became temporarily insane, imagining that they were pursued by demons and snakes; five of the villagers died.

Ergot, which causes muscles to contract and blood vessels to constrict, is used in medicine. It is also the initial source for the psychedelic drug lysergic acid diethylamide (LSD).

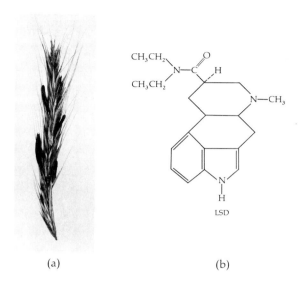

(a)

(b)

(a) The purple-black, hard resting structures associated with ergot seen among the spikelets of rye. (b) The chemical structure of LSD, lysergic acid diethylamide, derived from the fungus.

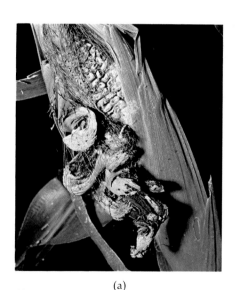

(a)

(b)

(c)

(d)

18–17

(a) Corn smut, a common fungus disease of corn. The black, dusty-looking masses are spores. (b) A morel. These (and truffles) are among the most prized of the edible fungi. (c) Sulfur shelf fungi, which grow on decaying wood. (d) Amanita muscaria. Members of this genus include the most beautiful and also the most poisonous of the mushrooms. The ring near the top of the stalk is one of the identifying characteristics of the group. One mushroom has been picked to show the gills, on which the sexual spores are formed.

biotics, including penicillin. Fungi are the cause of many important plant diseases, including potato blight, which caused the great potato famines in Ireland, and downy mildew of grapes, which threatened the entire French wine industry during the latter part of the nineteenth century. Fungi also cause a number of diseases of man, such as ringworm and thrush, and are important as destroyers of food and clothing (especially cotton and leather).

Together with the bacteria, fungi are the decomposers of the world, and as we shall see in Section 8, their activities are as vital to the continued survival of other forms of life as are those of the food producers.

REPRODUCTION IN THE FUNGI

Fungi reproduce both asexually and sexually. Asexual reproduction takes place either by the fragmentation of the mycelium (with each fragment becoming a new individual) or by the production of spores. In some of the fungi, spores are produced in sporangia. The bright colors and powdery textures associated with many particular types of molds are the colors and textures of spores and sporangia, which are often elevated above the mycelium on sporangiophores.

Sexual reproduction is often initiated by the coming together of hyphae of different mating strains. Either before or after they come in contact, the touching tips of the hyphae may develop into specialized reproductive structures (the gametangia). The gametangia fuse, the nuclei unite, forming a zygote, and meiosis takes place. Reproduction in one type of fungus, the black bread mold, is shown in Figure 18-18.

18–18

Asexual and sexual reproduction in the common black bread mold Rhizopus. *The mold consists of branched hyphae, including rhizoids (rootlike structures), which anchor the mycelium; stolons, which run above ground; and sporangiophores. At maturity, the fragile walls of the sporangia disintegrate, releasing the asexual spores, which are carried away by air currents. Under suitable conditions of warmth and moisture, the spores germinate, giving rise to new masses of hyphae. Sexual reproduction occurs when two hyphae from different mating strains come together, forming gametangia, which fuse to form a thick-walled, resistant zygote. After a period of dormancy, the zygote undergoes meiosis and germinates, producing a new sporangium.*

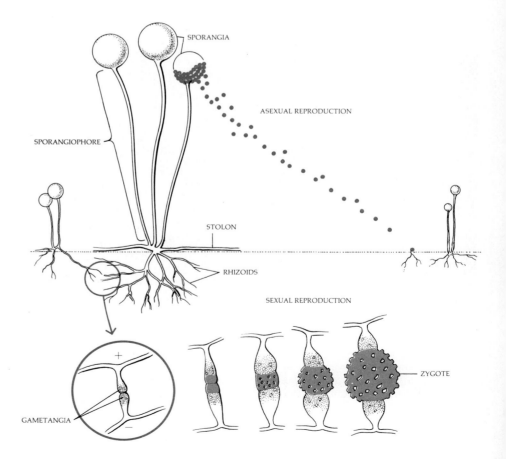

18–19

(a) *Lichen growing on a dead cedar in a North Carolina salt marsh.* (b) *British soldier lichen (Cladonia). Each soldier (so called because of the scarlet color) is about 3 millimeters tall.*

(a)

(b)

THE LICHENS

A lichen is a specific combination of a fungus and an alga. The organisms resulting from these symbiotic combinations are completely different from either the alga or the fungus growing alone, as are the physiological conditions under which the lichen can survive. The lichens are widespread in nature, and they are often the first colonists of bare rocky areas. Lichens do not need an organic food source, as do their component fungi, and unlike free-living algae they can remain alive even when very desiccated (dried out). Because they absorb substances from rainwater, they are particularly susceptible and sensitive to airborne toxic compounds; thus, the presence or absence of lichens is a sensitive index of air pollution.

Lichens reproduce most commonly by the breaking off of fragments containing both fungal hyphae and algae. New lichens are formed, according to some recent research, by the capture of an alga by a fungus. Sometimes the alga is destroyed by the fungus; if it survives, a lichen is produced.

THE PLANTS

The common ancestor of all the modern plants seems to have been a relatively complex multicellular green alga that invaded the land some 500 million years ago. It may have been "prepared" for its invasion of land by its differentiation into a holdfast and an upper photosynthetic area, such as are seen among the modern coastal seaweeds. (This "preparation" is known as *preadaptation*. Preadaptation is the occurrence in an organism of a structure or function that, in the course of evolution, in changing environments takes on a new role.)

Once this ancestral plant had made the transition to land, some new evolutionary developments took place. These must have occurred early in the evolutionary history of plants since most modern plants, even though very diverse, share them.

These adaptations include a waxy coating, the cuticle, which coats the above-ground parts of the plant body and retards water evaporation, and stomata, the specialized openings in leaves and green stems through which carbon dioxide enters during photosynthesis. Another adaptation is the surrounding of the gametangia and sporangia by protective layers of cells that keep the reproductive cells from drying out. Finally, as we noted previously, all plants have a reproductive cycle that involves alternation of generations.

In the process of photosynthesis, plants and algae use the energy of the sun to convert carbon dioxide and water into carbohydrates. These carbohydrates are then oxidized either by the plants themselves, by heterotrophs that eat the plants, by decomposing, or, less frequently, by burning (as in a forest fire). In the course of oxidation, carbon dioxide is again formed. The amount of carbon dioxide involved in photosynthesis on an annual basis is about 100 billion metric tons. The amount returned as carbon dioxide as a result of oxidation of these living materials is about the same, differing only by 1 part in 10,000. This very slight imbalance is caused by the burying of organisms in sediment or mud under conditions in which oxygen is excluded and decay is only partial. This accumulation of partially decayed material is known as peat. The peat may eventually become covered with sedimentary rock, placing it under pressure. Depending on time, temperature, and other factors, peat may become compressed into soft or hard coal, petroleum, or natural gas—the so-called fossil fuels.

During certain periods in the earth's history, the rate of fossil fuel formation was greater than at other times. One such period was the Carboniferous, some 300 million years ago. The lands were low, covered by shallow seas or swamps, and, in what are now the temperate regions, conditions were favorable for growth year round. The principal large plants were ferns, tree-sized lycophytes and horsetails, and seed ferns and other primitive gymnosperms; their fossils are found near the large coal seams of this period.

Although man has burned peat and coal for domestic uses for several thousand years, fossil fuels have been mined and consumed on a large scale only since about 1900. The growth of industries, of cities, and of the human population closely parallels the great increase in fuel consumption. Although fossil fuel is still being formed, the rate of formation is so slow that for all practical purposes, supplies are not renewable and so are finite. As the resources become depleted, the cost in energy of extracting the fuel will increase until the process is no longer economically attractive, and at that time, alternative sources will be vigorously sought. Many have been suggested, including the direct use of solar energy, harnessing wind and flowing water (actually indirect uses of solar energy), the surge of the tides, or tapping the heat energy beneath the surface of the earth. The most plausible new energy source is nuclear power, which, like the energy of fossil fuels, has its source in reactions that took place long ago in the heart of the sun. Many scientists, however, are highly concerned about the biological costs of this both new and ancient energy source.

18–20

A bryophyte, a haircap moss with spore capsules. The lower green structures are the female gametophytes, which are haploid (n); the stalks and capsules are the diploid (2n) sporophytes.

THE BRYOPHYTES

The two major groups of plants, the *bryophytes* and the *tracheophytes*, separated early in evolutionary history, probably more than 400 million years ago. The bryophytes, which include the mosses, liverworts, and hornworts, are plants that did not develop elaborate conducting systems. They are comparatively simple in their structure and are relatively small, usually less than 15 centimeters in height. They have rhizoids rather than roots and absorb moisture through leaflike structures as well as from the ground. The terms "root" and "leaf" are reserved for organs with vascular tissues. As you would imagine, they are most abundant in moist areas. In the bryophytes, unlike any tracheophytes, the gametophyte (the haploid form) is larger than the sporophyte (Figure 18–20). The sperm are released when sufficient moisture is present, enabling them to swim (by flagella) to the archegonium, to which they are attracted chemically. Fusion of egg and sperm takes place within the archegonium.

Inside the archegonium, the zygote develops into a sporophyte, which remains attached to the gametophyte and is often nutritionally dependent on it. Typically, the sporophyte has a single, large sporangium (often in the form of a capsule) elevated on a stalk, from which the spores are discharged.

Asexual reproduction, often by fragmentation, is also common.

THE TRACHEOPHYTES

The tracheophytes, or vascular plants, which dominate the modern landscape, are characterized by their efficient systems for the transport of water and

18-21

(a) *In the club mosses,* Lycopodium, *the sporangia are borne on specialized leaves, sporophylls, which are aggregated into a cone at the apex of the branches, as shown in this club moss, a running ground pine. The airborne, waxy spores give rise to small, independent, subterranean gametophytes. The sperm, which are biflagellated, swim to the archegonium, and the plant embryo develops there. (b) The horsetails, of which there is only one living genus (Equisetum), are easily recognized by their jointed, finely ribbed stems, which contain silicon. At each node, there is a circle of small, scalelike leaves. Spore-bearing leaves (sporophylls) are clustered into a cone at the apex of the stem. The gametophytes are independent, and the sperm are coiled, with numerous flagella. (c) An oak fern. Among ferns, the leaf, or frond, is divided into leaflets, or pinnae. Spores are born in sporangia, either on the underside of leaves or on separate stalks.*

(a)

(b)

(c)

sugars. Modern tracheophytes include the club mosses, horsetails, ferns, gymnosperms, and angiosperms.

Trends among Vascular Plants

Better Conducting Systems

Among the vascular plants, there have been three marked evolutionary trends. The first has been the development of increasingly efficient conducting systems. One conducting system, the xylem, transports water from the roots to the leaves, often more than 100 meters. The other conducting system, the phloem, conducts sucrose and other products of photosynthesis from the leaves to the nonphotosynthetic cells of the plants. The structure and function of the vascular system in modern angiosperms will be described in Chapter 20.

Reduction of the Gametophyte

The second pronounced trend has been the reduction in importance of the gametophyte generation. Among the ferns, as we saw in Chapter 9, the gametophyte is separate from the sporophyte but is much smaller. In most species of fern, there is only one type of gametophyte, which is bisexual and produces both male and female gametes. In some few, however, there are male and female gametophytes.

Among the gymnosperms, there are two types of gametophytes, one male and one female, both of which are entirely heterotrophic, being dependent on the parent plant for their nutrition and development. The female gametophyte, in fact, never leaves the protection of the sporophyte, and the egg cell is fertilized there by sperm from the male gametophyte, the pollen grain (Figure 18–22).

In the angiosperms, the most recently evolved of the vascular plants, the gametophytes are microscopic. The female gametophyte, or embryo sac, con-

18-22

(a) *Male cones of jack pine (Pinus banksiana) shedding pollen. The pollen grains are male gametophytes, which complete their maturation when they reach the ovules, embedded in the female cones. There they release sperm cells that fuse with the egg cells on the scales of the female cone. (b) Female cone. The female gametophyte develops on the base of a scale of the cone and is fertilized there. When the seed (ovule plus embryo) is mature, it drops from the cone.*

(a)   (b)

tains only eight haploid nuclei, and the male gametophyte, the pollen grain, is a single cell with three nuclei. And yet the ancestral pattern persists, and it is not possible to understand reproduction among this most important group of modern plants without knowledge of its legacy from the past.

The Seed

The third major development among the vascular plants, and perhaps the most important to survival on land, is the seed. The seed is a complex structure in which the embryo sporophyte is contained within a protective outer covering, the seed coat (Figure 18-23). The seed coat, which is derived from tissues of the parent sporophyte, protects the immature plant from drying out while it remains dormant, sometimes for years, until conditions are favorable for its germination.

Seeds became important toward the close of the Paleozoic era, which ended some 230 million years ago. Earlier in this era, in the Carboniferous period, the temperature was warm and the lands were low, covered by shallow seas or great swamps. Water was plentiful, and there was little seasonal variation in temperature. This was the period in which our coal deposits were formed from the masses of vegetation in the swamps. At the close of the Carboniferous period, there were worldwide changes of climate, with widespread glaciers and droughts. The seed was in existence at this time; according to the fossil record, some of the fernlike plants and even some of the club mosses had seedlike structures. But it was not until the close of the Permian period, when the land became colder and drier, that plants that had seeds gained a major evolutionary advantage and the seed plants became the dominant plants of the land.

The Seed Plants

The two major modern groups of plants are both seed plants: the *gymnosperms*, "naked seed" plants, and *angiosperms* (from the Greek word *angio*, meaning vessel—literally a seed borne in a vessel). The largest and most familiar group of gymnosperms are the conifers, "cone-bearers," in which the female gametophyte develops on the scale of a cone and is fertilized there. The life cycle of a familiar conifer, a pine, is shown in Figure 18-24.

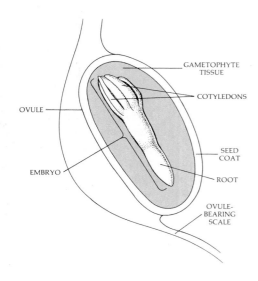

18-23

Pine seed. The ovule has hardened into a seed coat, enclosing the female gametophyte and the embryo, which now consists of an embryonic root and a number of embryonic leaves, the cotyledons. When the seed germinates, the root will emerge from the seed coat and penetrate the soil. When the root takes in water, the tightly packed cotyledons will elongate and swell with the moisture, rising above ground on the lengthening stem and forcing off the seed coat.

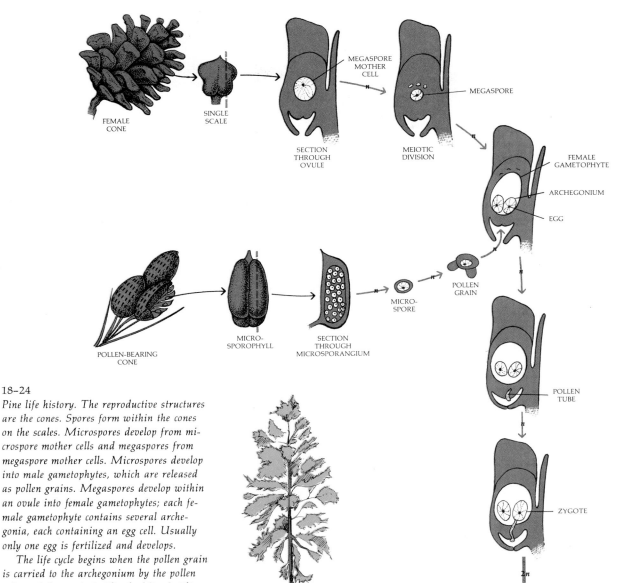

FEMALE
CONE

SINGLE
SCALE

SECTION
THROUGH
OVULE

MEGASPORE
MOTHER
CELL

MEIOTIC
DIVISION

MEGASPORE

FEMALE
GAMETOPHYTE

ARCHEGONIUM

EGG

POLLEN-BEARING
CONE

MICRO-
SPOROPHYLL

SECTION
THROUGH
MICROSPORANGIUM

MICRO-
SPORE

POLLEN
GRAIN

POLLEN
TUBE

ZYGOTE

CONE
SCALE

FEMALE
GAMETOPHYTE

EMBRYO

SEED
COAT

MATURE SEED
ON CONE SCALE

SEEDLING
(SPOROPHYTE)

18–24

Pine life history. The reproductive structures are the cones. Spores form within the cones on the scales. Microspores develop from microspore mother cells and megaspores from megaspore mother cells. Microspores develop into male gametophytes, which are released as pollen grains. Megaspores develop within an ovule into female gametophytes; each female gametophyte contains several archegonia, each containing an egg cell. Usually only one egg is fertilized and develops.

The life cycle begins when the pollen grain is carried to the archegonium by the pollen tube and the egg is fertilized. After fertilization, the ovule and its contents become the seed; the seed contains the developing plant embryo and surrounding tissue derived from the gametophyte, which nourishes the embryo. As the seed matures, the cone opens, releasing the winged seeds, which germinate, producing the sporophyte. Both types of cones develop on the sporophyte.

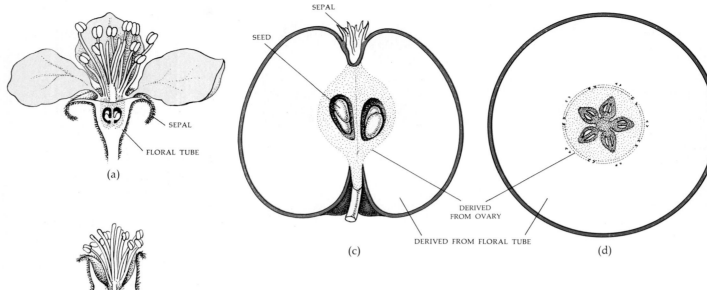

SEED

SEPAL

SEPAL

FLORAL TUBE

DERIVED FROM OVARY

DERIVED FROM FLORAL TUBE

(a)

(b)

(c)

(d)

18–25

Development and structure of the apple.
(a) Flower of apple. (b) Older flower, after
petals have fallen. (c, d) Longitudinal and
cross sections of the mature fruit. The core
of the apple is the ripened ovary wall. The
fleshy edible part develops from the floral
tube.

Among the angiosperms, the female gametophyte develops within the flower, where it is fertilized. The mature seed is not exposed, as it is in the gymnosperms, but is enclosed in a carpel, which is part of the flower. The carpel (and sometimes other parts of the flower as well) and the seed become the fruit (Figure 18-25). Thus the angiosperm seed has an additional covering that encloses and protects it and often aids in its dispersal (Figure 18-26).

THE ROLE OF PLANTS

The only forms of life on land that do not depend on plants for their existence are a few kinds of microscopic autotrophic prokaryotes. For all other living things, the chloroplast of the plant cell is the "needle's eye" through which the sun's energy is channeled into the biosphere. Even those animals that eat only other animals—the carnivores—could not exist if their prey, or their prey's prey, had not been nourished by plants.

Moreover, plants are the channels by which many of the simple inorganic substances vital to life enter the biosphere. Carbon is taken from the carbon dioxide of the atmosphere and "fixed" into organic compounds during the process of photosynthesis. Elements such as nitrogen and sulfur are taken from the soil in the form of simple inorganic compounds and incorporated into proteins, vitamins, and other essential organic compounds within green plant cells. Animals cannot make these organic compounds from inorganic materials and so are entirely dependent on plants for these compounds as well as for their source of energy.

The plants on which the human species depends are almost exclusively angiosperms. Among the angiosperms, the cereal grasses, such as wheat, rice, and corn, are of particular importance because their dry fruits (grains) are relatively easy to store and are high in calories and in protein. The ability to stockpile food in the form of grain (and also in the form of domesticated animals that ate the grasses and the grain) was an essential component in the agricultural revolution of some 10,000 or 12,000 years ago. Without the angiosperms, it is unlikely that modern civilization could have come into being.

18-26

Angiosperms are characterized by fruits. A fruit is a mature ovary, including the seed or seeds and, often, accessory parts of the flower. The fruit aids in dispersal of the seed. Some fruits are borne on the wind, some are carried from one place to another by animals, some float on water, and some are even forcibly ejected by the parent plant. (a) This chipmunk is holding 93 ragweed seeds in his cheek pouches. (b) In milkweed, the fruit bursts open when it is ripe, releasing seeds with tufts of silky hair that aid in their dispersal.

(a)   (b)

THE ANIMALS

Animals are many-celled heterotrophs. They depend directly or indirectly for their nourishment on land plants or algae. Most digest their food in an internal cavity, and most store food as glycogen or fat. Their cells do not have walls. Most move by means of contractile cells (muscle cells) containing characteristic proteins. Reproduction is usually sexual. The higher animals—the arthropods and the vertebrates—are the most complex of all organisms, with many kinds of specialized tissues, including elaborate sensory and neuromotor mechanisms not found in any of the other kingdoms. (Higher, in this context, means more recently evolved.)

For most of us, animal means mammal, and mammals are, in fact, the chief focus of attention in Section 6. However, the mammals, or even the vertebrates as a whole, represent only a small fraction of the animal kingdom. More than 90 percent of the different species of animals are invertebrates—that is, animals without backbones—and most of these are insects.

ORIGINS OF THE ANIMALIA

Animals, like plants, presumably had their origins among the protists, although in the case of animals we have fewer clues about which particular group of protists most closely resemble the ancestral ones. By the Cambrian period, which ended some 500 million years ago, many of the present-day types of invertebrates were already in existence. Although numerous efforts have been made over the past hundred years to trace the evolutionary relationships of these earlier forms, the Precambrian fossil record is inadequate to verify any hypotheses about the chronological order of their appearance. Most of the evidence for charts such as that shown in Figure 18-44, on page 238, comes from studies of modern forms.

There are about 30 phyla of invertebrates, each distinguished from all the others by a particular type of body plan. Of these, we are going to discuss only a few, concentrating on the most "successful," as judged on the basis of number of surviving species and individuals.

SPONGES: PHYLUM PORIFERA

Sponges, the most primitive of the modern animals, represent a level of organization somewhere between a colony of cells and a true multicellular organism. As with colonial forms, if the individual cells of the sponge are separated, such as by pressing the tissue through a fine cloth, and then mixed together again, the cells will reassemble, resuming their previous form.

In contrast to true colonial forms, however, the cells of the sponge are to a certain extent differentiated and specialized. The different cell types include feeding cells (the flagellated collar cells, or choanocytes); epithelial cells, some of which contain contractile fibers; and amoebalike cells (archeocytes), which serve several functions, including that of carrying digested food from the collar cells to the epithelial cells and other nonfeeding types. Other cells produce the stiffening structures, either inorganic spicules or organic fibers, which form the skeleton, the only part of the sponge that remains when it is dried and cleaned.

Because all the digestive processes of sponges are carried out within single cells, even a giant sponge—and some stand taller than a man—can eat nothing larger than microscopic particles, obtained by filtering the water that is driven through the body of the sponge by the flagella of the collar cells. Sponges are well suited to a slow life on the ocean floor. Sunlight is not necessary to their existence, and they have been found at depths that support few other living creatures. However, they are generally regarded as representing an evolutionary dead end; that is, no other groups were derived from them.

18-27
A cluster of sponges.

18-28

The body of a simple sponge is dotted with tiny pores, from which the phylum derives its name (Porifera, or "pore bearer"). Water containing food particles is drawn into the internal cavity of the sponge through these pores and is exhaled out the osculum. The water is moved by the beating of the flagella protruding from the collars of choanocytes and by the sucking effect of flow of the local currents across the osculum. Each collar is made up of about 20 filaments, each of which is retractile. The lashing of the flagella directs a current of water through the filaments. Minute particles are filtered out and cling to one or more filaments and are then drawn into the cell and digested. The digested food is then shared by diffusion with other sponge cells. A sponge filters as much as a ton of water to gain an ounce of body weight.

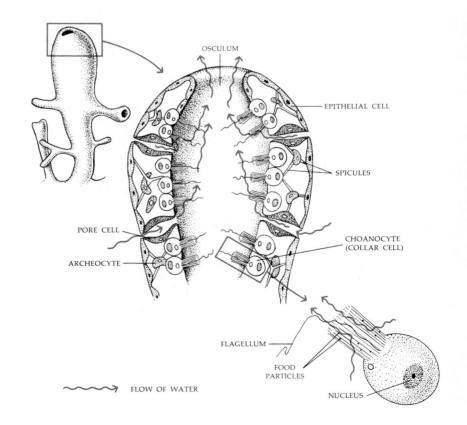

OSCULUM

EPITHELIAL CELL

SPICULES

PORE CELL

ARCHEOCYTE

CHOANOCYTE
(COLLAR CELL)

FLAGELLUM

FOOD
PARTICLES

NUCLEUS

FLOW OF WATER

Table 18-3 *Major Physical and Biological Events in Geologic Time*

MILLIONS OF YEARS AGO	ERA	PERIOD	EPOCH	LIFE FORMS	CLIMATES AND MAJOR PHYSICAL EVENTS
	CENOZOIC	Quaternary	Recent Pleistocene	Age of man. Planetary spread of *Homo sapiens;* extinction of many large mammals. Deserts on large scale.	Fluctuating cold to mild. Four glacial advances and retreats (Ice Age); uplift of Sierra Nevada.
1½–7		Tertiary	Pliocene	Large carnivores. First known appearance of hominids (manlike primates).	Cooler. Continued uplift and mountain building, with widespread extinction of many species.
7–26			Miocene	Whales, apes, grazing mammals. Spread of grasslands as forests contract.	Moderate uplift of Rockies.
26–38			Oligocene	Large, browsing mammals. Apes appear.	Rise of Alps and Himalayas. Lands generally low. Volcanoes in Rockies.
38–53			Eocene	Primitive horses, tiny camels, modern and giant types of birds.	Mild to very tropical. Many lakes in western North America.
53–65			Paleocene	First known primitive primates and carnivores.	Mild to cool. Wide, shallow continental seas largely disappear.
65–136	MESOZOIC	Cretaceous		Extinction of dinosaurs. Marsupials, insectivores, and angiosperms become abundant.	Lands low and extensive. Last widespread oceans. Elevation of Rockies at end of period.
136–195		Jurassic		Dinosaurs' zenith. Flying reptiles, small mammals. Birds appear. Gymnosperms, especially cycads, and ferns.	Mild. Continents low. Large areas in Europe covered by seas. Mountains rise from Alaska to Mexico.
195–225		Triassic		First dinosaurs. Primitive mammals appear. Forests of gymnosperms and ferns.	Continents mountainous. Large areas arid. Eruptions in eastern North America. Appalachians uplifted and broken into basins.
225–280	PALEOZOIC	Permian		Reptiles evolve. Origin of conifers and possible origin of angiosperms; earlier forest types wane.	Extensive glaciation in southern hemisphere. Appalachians formed by end of Paleozoic; most of seas drain from continent.
280–345		Carboniferous Pennsylvanian Mississippian		Age of amphibians. First reptiles. Variety of insects. Sharks abundant. Great swamps, forests of ferns, gymnosperms, and horsetails.	Warm. Lands low, covered by shallow seas or great coal swamps. Mountain building in eastern U.S., Texas, Colorado. Moist, equable climate, conditions like those in temperate or subtropical zones, little seasonal variation, root patterns indicate water plentiful.
345–395		Devonian		Age of fish. Amphibians appear. Shellfish abundant. Lunged fish. Rise of land plants. Extinction of primitive vascular plants. Origin of modern subclasses of vascular plants.	Europe mountainous with arid basins. Mountains and volcanoes in eastern U.S. and Canada. Rest of North America low and flat. Sea covers most of land.
395–440		Silurian		Earliest vascular plants. Rise of fish and reef-building corals. Shell-forming sea animals abundant. Modern groups of algae and fungi.	Mild. Continents generally flat. Again flooded. Mountain building in Europe.
440–500		Ordovician		First primitive fish. Invertebrates dominant. Invasion of land by plants?	Mild. Shallow seas, continents low; sea covers U.S. Limestone deposits; microscopic plant life thriving.
500–600		Cambrian		Age of marine invertebrates. Shell animals.	Mild. Extensive seas. Seas spill over continents.
	PRECAMBRIAN			Earliest known fossils.	Dry and cold to warm and moist. Planet cools. Formation of earth's crust. Extensive mountain building. Shallow seas.

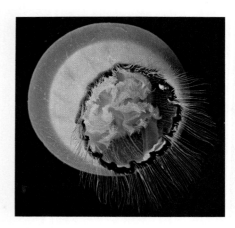

18-29

The moon jellyfish, Aurelia limbata, *is about a foot in diameter. Like other jellyfish, it moves through the water by contracting and relaxing a ring of muscles encircling the mouth.*

COELENTERATES: PHYLUM COELENTERATA

The coelenterates are a large group of aquatic animals—including jellyfish, *Hydra*, corals, and sea anemones—which are characterized by a hollow body made up of two layers of tissue. Between these two layers is a jellylike filling, most conspicuous in the jellyfish. The interior cavity of the body is known as the *coelenteron*, from which the coelenterates get their name. There are two basic body plans: polyp and medusa (Figure 18-30a).

The life cycle of coelenterates characteristically includes both polyp and medusa and also an immature larval form, known as the planula, which is a small, free-swimming ciliated organism. In the typical life cycle, the planula settles to give rise to a polyp, which reproduces asexually and may form extensive colonies. From the polyp, young jellyfish (medusas) may bud off. These are sexually reproducing adult forms that give rise to planulas again (Figure 18-31). All adult forms are radially symmetrical, which means that their body parts are arranged around an axis, like spokes around the hub of a wheel.

In both polyps and medusas, food is captured by means of tentacles and pushed into the coelenteron, which opens wide and engulfs it. Within the coelenteron the food is digested to some extent by enzymes secreted by the cells lining the cavity. Digestion is then completed within food vacuoles in the cells lining the coelenteron. The indigestible remains are then ejected by the same opening through which the food enters. The coelenterates are able to eat almost any prey they can stuff into their remarkably expandable body cavities.

18-30

(a) *Among coelenterates, there are two basic body plans: the vase-shaped polyp (left) and the bowl-shaped medusa (right). The coelenteron, characteristic of the phylum, is a digestive cavity with a single opening. The coelenterate body has two tissue layers, ectoderm and endoderm, with gelatinous mesoglea between them. (b) Cnidoblasts, specialized cells located in the tentacles and body wall, are a distinguishing feature of coelenterates. The interior of the cnidoblast is filled by a nematocyst, which consists of a capsule containing a coiled tube, as shown on the left. A trigger on the cnidoblast, responding to chemical or mechanical stimuli, causes the tube to shoot out, as shown on the right. The capsule is forced open and the tube turns inside out, exploding to the outside. The cnidoblast cannot be "reloaded"; it is absorbed and a new cell grows to take its place.*

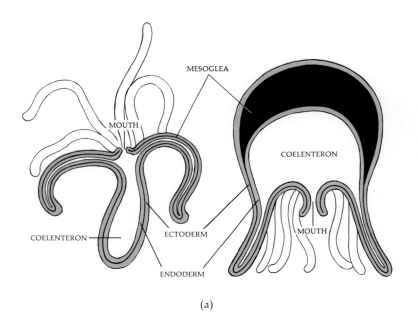

MESOGLEA

MOUTH

COELENTERON

COELENTERON

ECTODERM

MOUTH

ENDODERM

(a)

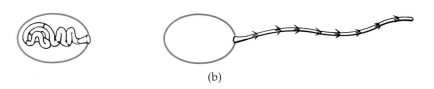

(b)

The life cycle of Aurelia. *Sperm and egg cells are released from adult medusas into the surrounding water. Fertilization takes place, and the resulting zygote develops first into a hollow sphere of cells, the blastula, and then*

elongates and becomes a ciliated larva called a planula. The planula eventually settles to the bottom, attaches by one end to some object, and develops a mouth and tentacles at the other end, thus transforming into the

polyp stage. The body of the polyp grows and, as it grows, begins to form medusas, stacked upside down like saucers, which bud off, one by one, and grow into full-sized jellyfish.

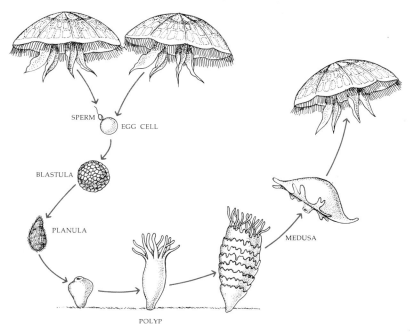

SPERM   EGG CELL

BLASTULA

PLANULA

MEDUSA

POLYP

18-32

The sea anemone is a coelenterate in which the medusa phase has been lost. It feeds by tentacles, moving food into its coelenteron, which is divided longitudinally by partitions.

Coelenterates have primitive nervous systems. The medusa form—exemplified by the jellyfish—has two rings of nerve cells that circle the margin of the bell, whereas the polyp form—as in *Hydra* (page 119)—has a continuous network of nerve cells just below the surface of the epidermis, which links the body into a functional whole and makes possible coordinated movements. Highly specialized stinging cells, cnidoblasts, are found only among the coelenterates and are used for defense and to capture prey (Figure 18-30b).

The most ecologically important members of this group are the coral builders, which are responsible for the formation of great land masses in the sea where ordinarily no land would exist. The 2,000-kilometer-long Great Barrier Reef off the northeast shores of Australia and the Caroline Islands in the Pacific is an example of such a coral-created land mass. A coral reef is composed primarily of the accumulated limestone skeletons of coral coelenterates, covered by a thin crust occupied by the living colonial animals.

Many biologists believe that a primitive coelenterate is to be found on the evolutionary pathway leading from the protists to the higher forms. A principal clue in this evolutionary puzzle, they point out, is the ciliated larva. It is easy to imagine a gradual transition between such a form and the simplest of the flatworms, the next big step in the order of complexity. As shown in Figure 18-44, on page 238, a form corresponding to the planula of the jellyfish and other modern coelenterates may have been the starting point for wormlike forms in the kingdom Animalia.

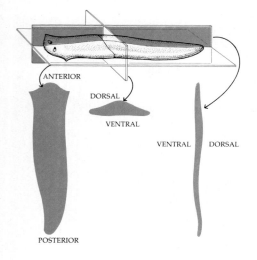

In a bilaterally symmetrical organism, such as the planarian shown here, the right and left halves of the body are mirror images of one another. The upper and lower (or back and front) surfaces are known as dorsal and ventral. The end that goes first is termed anterior and the rear, posterior.

ANTERIOR

DORSAL

VENTRAL

VENTRAL   DORSAL

POSTERIOR

FLATWORMS: PHYLUM PLATYHELMINTHES

The flatworms are the simplest animals, in terms of body plan, to show *bilateral symmetry* (Figure 18–33). In bilaterally symmetrical animals, the body plan is organized along a longitudinal axis, with the right half an approximate mirror image of the left half. A bilaterally symmetrical animal can move more efficiently than a radially symmetrical one (which is better adapted to a sedentary existence). It also has a top and a bottom, or in more precise terms (applicable even when it is turned upside down or, as in the case of humans, standing upright), a dorsal and a ventral surface. Like most bilateral organisms, the flatworm also has a distinct "headness" and "tailness." Apparently, when one end goes first, it is advantageous to collect the sensory cells into that end. With the aggregation of sensory cells, there came a concomitant gathering of nerve cells; this gathering is a forerunner of the brain.

Flatworms have the three distinct tissue layers—ectoderm, mesoderm, and endoderm—characteristic of all animals above the coelenterate level of organization. Moreover, not only are their tissues specialized for various functions, but two or more types of tissue cells may combine to form organs. Thus while coelenterates are largely limited to the tissue level of organization, flatworms can be said to have gained the organ level of complexity.

The free-living flatworms, of which the freshwater planarians are familiar examples, suck bits of dead animals into their highly branched digestive cavities, where they are digested by the cells lining the cavity. The indigestible residue is ejected through the mouth. Two light-sensitive spots at the anterior end resemble a pair of crossed eyes, and two projections on either side of the head, which resemble ears, are areas sensitive to chemical stimuli such as emanate from meat or other food. Within the mesoderm is a fairly complex musculature, which enables the animal, when disturbed, to move rapidly with a sort of loping motion, resulting from wavelike contractions of muscles. Planarians have an extensive excretory system and complex reproductive organs. The phylum also includes the parasitic flukes and tapeworms, which often have complex life cycles involving several hosts in succession.

18–34

Example of a flatworm: the freshwater planarian. The planarian, which is carnivorous, feeds by means of its extensible pharynx. (a) The nervous system is indicated in color. Note that some of the fibers have been aggregated into two cords, one on each side of the body, and there is a cluster of nerve cells in the head, the beginnings of a brain. The ocelli are light-sensitive areas. (b) Planarian, like other flatworms but unlike coelenterates, has three layers of body tissues. (c) A carnivore, the planarian feeds by means of its extensible pharynx.

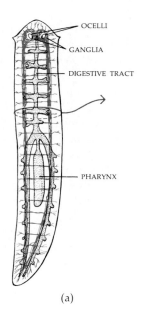

OCELLI

GANGLIA

DIGESTIVE TRACT

PHARYNX

(a)

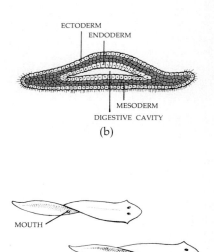

ECTODERM

ENDODERM

MESODERM

DIGESTIVE CAVITY

(b)

MOUTH

PHARYNX

(c)

RIBBON WORMS: PHYLUM RHYNCHOCOELA

The ribbon worms, although they constitute a small phylum, are of special interest to biologists attempting to reconstruct the evolution of the invertebrates because they appear to be closely related to the flatworms, with an important difference: they have a one-way digestive tract beginning with a mouth and ending with an anus. This is a far more efficient arrangement than the one-opening digestive system of the coelenterates and flatworms; in the one-way tract, food moves assembly-line fashion, with the consequent possibility of specialization of various segments of the tract for different stages of digestion.

This phylum is called Rhynchocoela ("beak" plus "hollow") because these worms are characterized by a long, retractile, slime-covered tube (proboscis), which lies in a special cavity. The proboscis, sometimes armed with a barb, seizes prey and draws it to the mouth, where it is engulfed.

ROUNDWORMS: PHYLUM NEMATODA

The number of species of roundworms has been variously estimated as low as 10,000 and as high as 400,000 to 500,000. Roundworms range from microscopic forms to parasites 2 meters long, but they are all structurally so similar that they are considered to constitute only one phylum, the nematodes.

Nematodes have a three-layered body plan and a tubular gut with a mouth and an anus. They are unsegmented and are covered by a thick, continuous cuticle, which they must shed as they grow, like arthropods. They also have a pseudocoelom, as shown in Figure 18-39, page 235. Most are free-living microscopic forms. It has been estimated that a spadeful of good garden soil usually contains about a million nematodes. Some are parasites; all species of plants and animals are parasitized by at least one species of nematodes. Man is host to about fifty species.

18–35
(a) *A planarian. The two eyes are actually light-sensitive patches (ocelli). A planarian can see about as well as you can with your eyes closed. (b) A ribbon worm in an embarrassing situation. Ribbon worms range in length from less than 2 centimeters to 30 meters in length and through virtually all the colors of the spectrum. All members of the species, however, have flat, thin (seldom more than ½ centimeter thick) bodies, a mesoderm, a mouth-to-anus digestive tract, and a long, muscular tube that can be thrust out to grasp prey. (c) A soil-dwelling roundworm (a nematode). This particular organism is being attacked by a predatory fungus, which traps the worm by means of small adhesive knobs. Once captured, fungal hyphae grow into the body of the worm and digest it.*

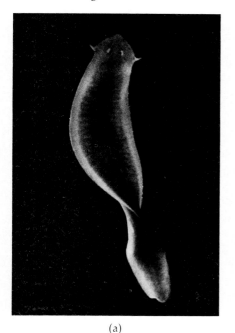

(a)

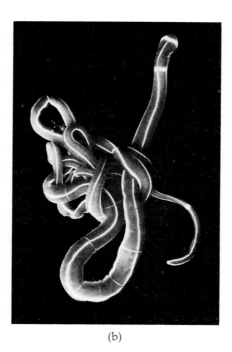

(b)

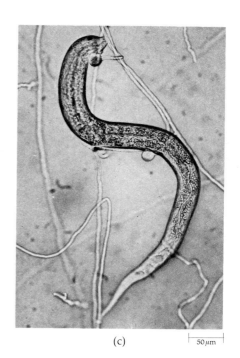

(c)

50 μm

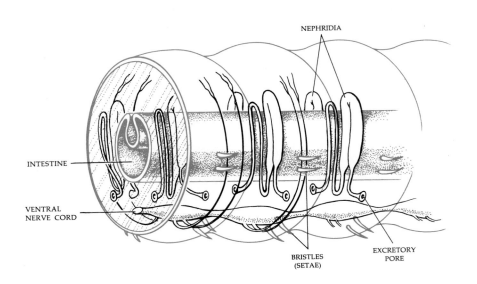

NEPHRIDIA

INTESTINE

VENTRAL
NERVE CORD

BRISTLES
(SETAE)

EXCRETORY
PORE

18-36

Three segments of the earthworm, an annelid. On each segment are four pairs of bristles, which are extended and retracted by special muscles. These anchor one part of the body while the other part moves forward. Two excretory tubes, or nephridia, are in each segment (except the first three and the last). Each nephridium really occupies two segments since it opens externally by a pore in one segment and internally by a ciliated funnel in the segment immediately in front of it. The intestine, nephridia, and other internal organs are suspended in the large coelom, which also serves as a hydraulic skeleton.

18-37

This annelid, a polychaete worm, unlike the more familiar earthworm, has a well-differentiated head with sensory appendages and lateral parapodia ("side feet") with many setae.

SEGMENTED WORMS: PHYLUM ANNELIDA

This phylum includes almost 7,000 different species of marine, freshwater, and soil worms, including the familiar earthworm. The term annelid means "ringed" and refers to the most distinctive feature of this group, which is the division of the body into segments, not only by rings on the outside but by partitions on the inside (Figure 18-36). This segmented pattern is found in a modified form in higher animals, too, such as dragonflies, millipedes, and lobsters, which are thought to have evolved from ancestors that probably gave rise also to modern annelids. Although the earthworm is probably the most familiar of the annelids, it is atypical because of its much-reduced appendages.

The annelids have a three-layered body plan, a tubular gut, and a well-developed, closed circulatory system which transports oxygen (diffused through the skin) and food molecules (from the gut) to all parts of the body (Figure 18-38). The excretory system is made up of specialized paired tubules, *nephridia*, which occur in each section of the body except the head. Annelids have a nervous system and a number of special sense cells, including touch cells, taste receptors, light-sensitive cells, and cells concerned with the detection of moisture.

The annelids also contain *coeloms* (Figure 18-39); they are the first group of animals discussed so far in which this important evolutionary development is seen. In the flatworms, the mesoderm is packed solid with muscle and other tissues, but in the annelids there is a fluid-filled cavity, the coelom, in this middle layer. (Note that the term "coelom," although it sounds similar to "coelenteron" and comes from the same Greek root, meaning "cavity," refers quite specifically to a cavity *within* the mesoderm, whereas the coelenteron is a digestive cavity lined by endoderm.) The gut is suspended within this cavity by mesenteries, made of double layers of mesoderm. The fluid in the coelom constitutes a sort of hydraulic skeleton for the earthworm, supporting the soft tissues in the same way water keeps a fire hose distended.

Most of the higher invertebrates, and all of the vertebrates as well, have a body plan that includes the coelom. Within the coelom, organ systems can bend, twist, and fold back on themselves, increasing their functional surface areas and filling, emptying, and sliding past one another, surrounded by lubricating coelomic fluid. Examples of such organ systems are the human lung, constantly expanding and contracting in the chest cavity, and the 6 meters or more of coiled human intestine, both of which occupy the coelom.

18–38

The digestive tract and circulatory system of an earthworm. (a) The mouth leads into a muscular pharynx, which sucks in decaying vegetation and other material. These are stored in the crop and ground up in the gizzard with the help of soil particles. The rest of the tract is a long intestine (gut) in which food is digested and absorbed. (b) The circulatory system of the earthworm is made up of longitudinal vessels running the entire length of the animal, one dorsal and several ventral. Smaller vessels (the parietal vessels) in each segment collect the blood from the tissues and feed it into the muscular dorsal vessel, through which it is pumped forward. In the anterior segments are five pairs of hearts—muscular pumping areas in the blood vessels—whose irregular contractions force the blood downward through the parietal vessels to the ventral vessels, from which it returns to the posterior segments. The arrows indicate the direction of blood flow.

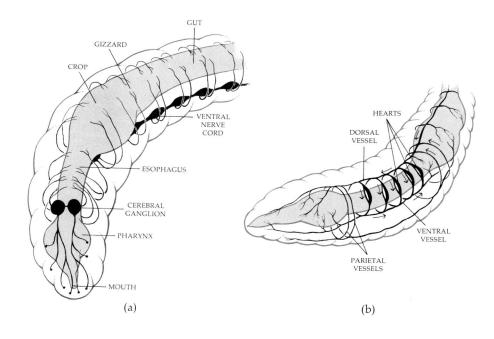

(a)

(b)

18–39

Basic body plans of the animal phyla, as shown in cross section. (a) A body which consists of only two tissue layers is characteristic of coelenterates. (b) Flatworms and ribbon worms have three-layered bodies, with the layers closely packed on one another. (c) Nematodes have three-layered bodies with a pseudocoelom between the endoderm and mesoderm. (d) Annelids and most other animals, including vertebrates, have bodies that are three-layered with a cavity, the coelom, within the middle layer (mesoderm). The mesodermal mesenteries suspend the gut within the body wall.

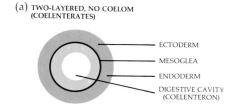

(a) TWO-LAYERED, NO COELOM
(COELENTERATES)

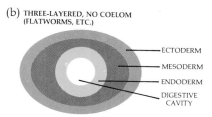

(b) THREE-LAYERED, NO COELOM
(FLATWORMS, ETC.)

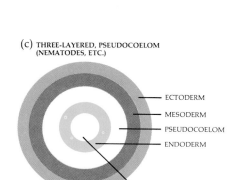

(c) THREE-LAYERED, PSEUDOCOELOM
(NEMATODES, ETC.)

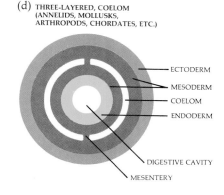

(d) THREE-LAYERED, COELOM
(ANNELIDS, MOLLUSKS, ARTHROPODS, CHORDATES, ETC.)

MOLLUSKS: PHYLUM MOLLUSCA

The mollusks constitute one of the largest phyla of animals, both in numbers of species and in numbers of individuals. They are characterized by soft bodies within a hard, calcium-containing shell, although in some forms the shell has been lost in the course of evolution, as in slugs and octopuses, or greatly reduced in size, as in squids. There are three major classes of mollusks: (1) the gastropods, such as the snails, which generally have one shell; (2) the bivalves, including the clams, oysters, and mussels, which have two shells joined by a hinge ligament; and (3) the cephalopods, the most active and most intelligent of the mollusks, including the cuttlefish, squids, and octopuses.

Although the mollusks are diverse in size and shape, they all have the same fundamental body plan. There are three distinct body zones: a head-foot, which contains both the sensory and the motor organs; a "hump," or visceral mass,

18-40

Mollusks are characterized by soft bodies composed of a head-foot, a visceral mass, and a mantle, which can secrete a shell. They breathe by gills, except for the land snails, in which the mantle cavity has been modified for air breathing. The hypothetical primitive mollusk is shown in (a). The three major modern groups are the bivalves, such

as the clam (b), which are generally sedentary and feed by filtering water currents, created by beating cilia, through large gills; the gastropods, exemplified by the snail (c), in which the visceral mass has become coiled upward and the gut turned back so that mouth, anus, and gills all share the same small aperture in the mantle; and the cepha-

lopods, such as the squid (d), in which the head is modified into a circle of tentacles and part of the head-foot forms a tubelike siphon through which water can be forcibly expelled, providing for locomotion by jet propulsion. The arrows indicate the direction of water movement.

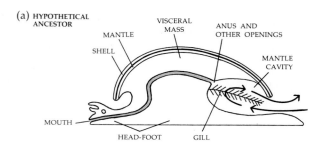

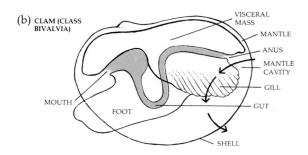

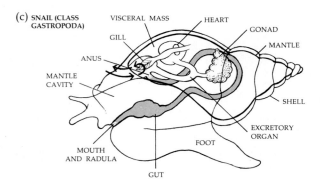

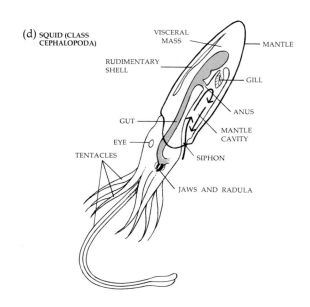

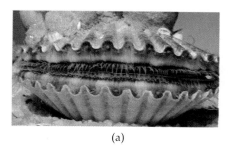

(a)

(b)

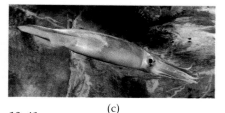

(c)

18–41

(a) *A bivalve, a blue-eyed scallop. Its eyes, visible among its tentacles, have lenses and so are presumably able to form images.*
(b) *A land-dwelling gastropod. The shell, which is secreted by the mantle and grows*

(d)

as the soft body grows, covers and protects the visceral mass. The head contains sensory organs, including two eyes at the tips of the longer tentacles. (c) A squid. Jet-propelled, the squid is moving to the left.

Note that its siphon is directed to point backward. (d) Octopus macropus. Note its well-developed eyes and the suckers under the arms.

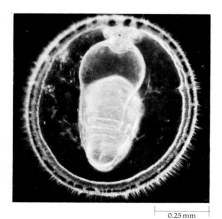

|—— 0.25 mm ——|

18–42

The trochophore larva. Although their adult forms are very different, certain annelids, mollusks, and arthropods have larvae of this type. This particular larva will develop into a polychaete worm.

which contains the organs of digestion, excretion, and reproduction; and a mantle, which hangs over and enfolds the visceral mass and which secretes the shell. The mantle cavity, a space between the mantle and the "hump," houses the gills, and the digestive, excretory, and reproductive systems discharge into it.

A characteristic organ, found only in this phylum, is the radula. This tooth-bearing tonguelike organ is variously used by different kinds of mollusks to scrape off algae, to drill holes in the shells of barnacles or other mollusks, or to aid in the ingestion of prey animals. It is present in all mollusks except the bivalves.

The molluscan circulatory system, which includes a chambered heart, is the most efficient circulatory system of the invertebrates. The nervous systems of the gastropods and bivalves are simple, but among the cephalopods, the octopus, in particular, has a highly developed nervous system and a brain and eyes that rival those of the vertebrates in complexity, although different in design.

Although the annelids and the mollusks are quite different in their basic body plans, there are some evolutionary links between them. One of these is the trochophore larva (Figure 18-42).

ARTHROPODS: PHYLUM ARTHROPODA

18-43
South American katydid, showing the hard, many-jointed exoskeleton characteristic of arthropods. The slits in each foreleg are the insect's ears.

Phylum Arthropoda—the joint-legged animals—is divided into three major classes, the arachnids, the crustaceans, and the insects, and several minor ones, which include centipedes, millipedes, and horseshoe crabs. The identifying characteristics of the three major classes are summarized in Table 18-4.

Arthropoda is by far the largest phylum of animals, with more than 765,000 species classified to date. There are more species of insects than of all other animals combined; of beetles alone, some 275,000 have been classified. Some experts estimate that, including the unclassified forms, there are more than 10 million living insect species.

The characteristic features of the arthropods are their segmented organization, similar to that of the annelids, from which they almost certainly evolved, and their hard outer covering, or *exoskeleton*, containing chitin. In the evolution of the arthropods, the various segments of the body became specialized in different ways. The most anterior segments form the head, the next constitute the thorax (head and thorax segments are sometimes fused into a cephalo-thorax), and the final segments are the abdomen (Figure 18-45).

In some of the arthropods, such as the centipedes, which have a pair of legs on almost every segment of the clearly wormlike body, the appendages are uniform in size and structure, but more typically, arthropods have highly differentiated appendages, specialized for walking, swimming, feeling, feeding, chewing, biting, and other such functions, depending on the species.

The exoskeleton deters predators and, in the land forms, protects the animals

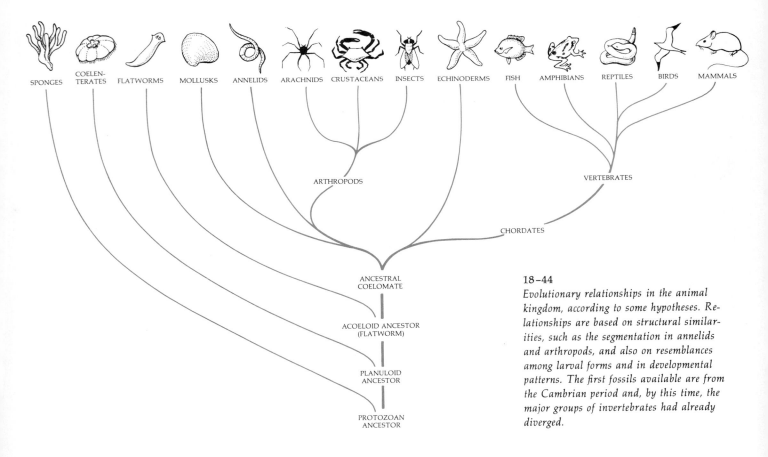

18-44
Evolutionary relationships in the animal kingdom, according to some hypotheses. Relationships are based on structural similarities, such as the segmentation in annelids and arthropods, and also on resemblances among larval forms and in developmental patterns. The first fossils available are from the Cambrian period and, by this time, the major groups of invertebrates had already diverged.

18-45

(a) *In the grasshopper, an insect, the head consists of six fused segments which have appendages specialized for biting and chewing. Each of the three segments of the thorax carries a pair of legs (three pairs in all), and two of them carry wings (in the grasshopper, the forewings are hardened as protective covers). The spiracles in the abdomen open into a network of chitin-lined tubules, the tracheae, through which air circulates to various tissues of the body. This sort of tubular breathing system is found only among insects and some other land-dwelling arthropods. (b) Mouthparts of a grasshopper. The mandibles are crushing jaws. The labium and the labrum are the lower and upper lip. The maxillae move food into the mouth.*

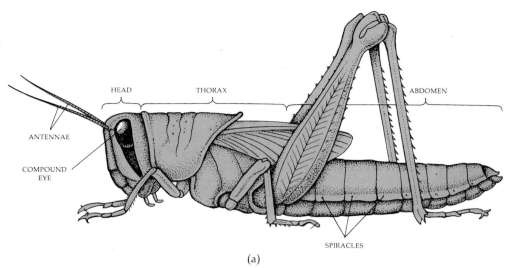

(a)

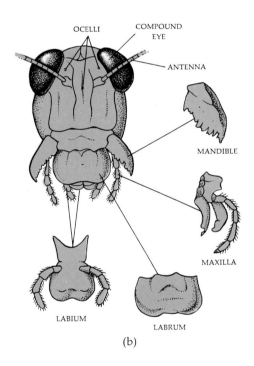

(b)

from desiccation. It is many-jointed and is attached to the musculature so that it moves with the animal, rather than restricting the animal to moving inside it, as do the mollusk within their shells. However, it covers the animal completely and does not grow; as a consequence, arthropods must molt, a process by which the old exoskeleton is discarded and a new and larger one is formed (page 45). At molting time, the epidermis secretes an enzyme that dissolves the inner layer of the exoskeleton, and a new skeleton, not yet hardened, is formed beneath the old one. The animal wriggles out from the old skeleton, which splits open. After emerging, the arthropod expands rapidly by taking up air or water, stretching the new exoskeleton before it hardens.

All arthropods are segmented, a characteristic that suggests that they share a common ancestry with the annelids. In the higher arthropods, the segments have become fused, but the basic segmented pattern is still clearly evident in the adult forms of more primitive arthropods—such as centipedes and millipedes—and the immature stages of many (witness the caterpillar).

The anterior section of the arthropod gut is lined with chitin and serves largely to grind food into smaller particles. Large digestive glands secrete enzymes into the midsection, where digestion takes place. Arthropods have a complex endocrine (hormone-producing) system, which plays a major role in molting. Their circulatory system is an open system, in which the blood flows from the blood vessels into blood sinuses, or cavities. The heart is a tube with paired openings suspended within a blood cavity which pumps the blood into the sinuses.

The insects and some arachnids have unusual respiratory machinery consisting of a system of chitin-lined air ducts (tracheae) that pipe air directly into various parts of the body. Air flow is regulated by the opening and closing of special pores (spiracles) on the exoskeleton. Terrestrial arthropods that do not have tracheae have book lungs, so called because they are constructed in a series of leaflike plates. Both tracheae and book lungs are found only in this phylum. Excretion in terrestrial forms is by means of tubes (called Malpighian tubules) attached to and emptying into the hindgut. Malpighian tubules are also an exclusive arthropod characteristic.

The nervous system of the arthropods is complex, making possible the intricate and finely tuned movements involved in activities such as flight, mating in midair, and building webs and hives. Arthropods have a number of extremely

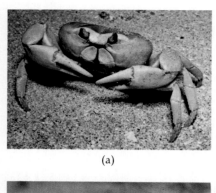

(a)

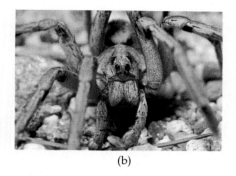

(b)

(c)

(d)

(e)

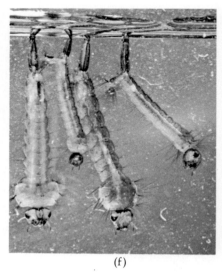

(f)

18-46

Some arthropods. (a) A crab, a familiar crustacean. Note the compound eyes, the fused cephalothorax, and the many appendages. (b) A wolf spider. These common spiders have eight eyes, four small ones in a row beneath four large ones. The chelicerae are visible beneath the mouth. Ducts from a pair of poison glands lead through the chelicerae, which are used for capturing and paralyzing prey. (c) A stag beetle, so called because of the antlerlike mandibles characteristic of the pugnacious males. (d) A tiger beetle. The hairs on the legs are touch receptors. (e) A lady bug, also a beetle. (f) Mosquito larvae. Aquatic, they hang by their tails on the underside of the water and extend respiratory tubes above the water's surface.

(a)

(b)

(c)

(d)

18-47

Developmental stages of a spicebush swallowtail butterfly: (a) the egg; (b) caterpillar, the larval form; (c) pupa, the resting form in which metamorphosis takes place; (d) a butterfly, the adult, reproductive form.

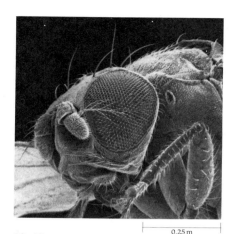

18-48

Scanning electron micrograph of the compound eye of Drosophila. *Although insect eyes cannot change focus, they can observe objects as close as a millimeter from the lens, a useful adaptation for an insect.*

sensitive sensory organs, such as the compound eye (Figure 18-48) found among crustaceans and insects.

Arthropods, especially insects, such as ants, honeybees, and termites, live in highly organized societies. The social organization of honeybees is described in Chapter 38.

The enormous success of the insects appears to be related to the high degree of specialization of the individual species. Insects are usually so selective about where they live and how they eat that many different species can live noncompetitively in a very small area—such as on a single plant or in or on one small animal.

Most insects go through definite developmental stages. In some species the infant, although sexually immature, looks like a small copy of the adult; it grows larger by a series of molts until it reaches full size. In others, such as the grasshopper, the newly hatched young is wingless and somewhat different in proportions from the adult, but it is otherwise similar.

Almost 90 percent of all insects, however, undergo a complete metamorphosis, so that the adult is entirely different from the immature form. The immature eating and feeding forms may all correctly be referred to as *larvae*, although they are also commonly known as caterpillars, grubs, or maggots, depending on the species. Following the larval period, the insect undergoing complete metamorphosis enters an outwardly quiescent pupal stage, in which extensive remodeling of the organism occurs. The adult insect emerges from the pupa. Thus the insect that undergoes complete metamorphosis exists in four different forms in the course of its life history: the egg, the larva, the pupa, and the adult. The larvae and adults are so different that they usually do not compete for food or other resources, another example of the extreme specialization found in the insect world.

Table 18-4   *Major Classes of the Phylum Arthropoda*

	NO. OF SPECIES	BODY PARTS	HEAD APPENDAGES	LEGS	WINGS	EYES	RESPIRATION	DISTRIBUTION
Subphylum Chelicerata								
Class Arachnida (Scorpions, spiders, daddy-longlegs, mites, ticks)	35,000	Two (cephalothorax, abdomen)	Chelicerae, pedipalps, no antennae	4 pairs	None	Simple	Book lungs, tracheae, or both	Terrestrial, some aquatic (secondarily)
Subphylum Mandibulata								
Class Crustacea (Lobsters, crabs, barnacles, *Daphnia*, copepods)	30,000	Two (cephalothorax, abdomen) or three (head, thorax, abdomen)	Antennae (2 pairs), mandibles, maxillae (2 pairs)	Numerous (on both thorax and abdomen)	None	Usually compound	Gills	Marine and freshwater, some terrestrial
Class Insecta	700,000	Three (head, thorax, abdomen)	Antennae (1 pair), mandibles, maxillae (2 pairs)	3 pairs (on thorax only)	2 pairs, 1 pair, or none	Compound and simple	Tracheae	Terrestrial, some aquatic (secondarily)

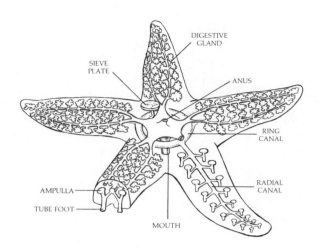

The water vascular system of the starfish is its means of locomotion. Water enters through minute openings in the sieve plate and is drawn, by ciliary action, down a tube to the ring canal. Five radial canals, one for each arm, connect the ring canal with many pairs of tube feet, which are hollow, thin-walled cylinders ending in suckers. Each tube foot connects with a rounded muscular sac, the ampulla. When the ampulla contracts, the water in it, prevented by a valve from flowing back into the radial canal, is forced under pressure into the tube foot. This stiffens the tube, making it rigid enough to walk on and extends the foot until it attaches to the substrate by its sucker. The longitudinal muscles of the foot then contract, forcing the water back into the ampulla, creating suction that holds the foot to the surface. If the tube feet are planted on a hard surface, such as a rock or a clamshell, the collection of tubes will exert enough force by suction to pull the starfish forward or to open the clam.

ECHINODERMS: PHYLUM ECHINODERMATA

The echinoderms ("prickly skins") include the sea urchins, sand dollars, sea cucumbers, and sea lilies. Adult echinoderms are radially symmetrical, like the coelenterates. The most familiar of the echinoderms are the starfish, which have the spiny skin from which the phylum derives its name. Most starfish have five arms, but some have multiples of five, up to as many as twenty. Other members of the phylum also have bodies arranged in fives or multiples of five. The water vascular system (Figure 18–49) provides suction for the clinging and pulling activities of the tube feet, with which many members of the phylum cling, move, and attack prey.

Although the adults are radially symmetrical, the larvae of the members of this phylum are bilaterally organized; it is believed that the radial symmetry is a late, secondary development in the evolution of the group. In the development of these larvae, the first opening of the alimentary canal of the embryo becomes the anus, while the mouth breaks through secondarily at the other end of the larval digestive tract. Animals with such a pattern of early development are known as deuterostomes ("second the mouth"), in contrast to the protostomes such as mollusks, annelids, and arthropods, in which the mouth develops first. The primitive chordates from which the vertebrates arose also are deuterostomes. For this reason, among others, the vertebrates are believed to have a closer evolutionary relationship with this group than with the other invertebrate phyla, as shown in Figure 18-44 on page 238.

CHORDATES: PHYLUM CHORDATA

Mammals and other vertebrates belong to a large phylum of animals comprising the chordates. *Amphioxus*, a lancelet, is a good example of a chordate (Figure 18-51). The lancelet is a small, sliver-shaped, semitransparent animal found in shallow marine waters all over the warmer parts of the world. Although it can swim very efficiently, it spends most of its time buried in the sandy bottom, with only its mouth protruding above the surface. This animal has three features that identify it as a member of the chordate phylum. The first is the *notochord*, a skeletal rod which extends the length of the body and serves as a firm but flexible axis. The notochord is a structural support. Because of it, *Amphioxus* can swim with strong undulatory motions that move it through the water with a speed unattainable by the flatworms or aquatic annelids.

The second chordate characteristic is the nerve cord, a hollow tube that runs beneath the dorsal surface of the animal above the notochord. (The principal

Sand dollar. Note the five-segmented body plan. Like other echinoderms, the sand dollar has a water vascular system and tube feet. The mouth is on the lower surface.

Amphioxus, a lancelet, exemplifies three distinctive chordate characteristics: (1) a notochord, the dorsal rod that extends the length of the body; (2) a dorsal, tubular nerve cord; and (3) pharyngeal gill slits.

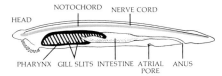

nerve cord in the invertebrates, by contrast, is almost always near the ventral surface.)

The third characteristic is a pharynx with gill slits. The pharyngeal gill slits become highly developed in fishes, in which they serve a respiratory function, and traces of them even remain in the human embryo. In *Amphioxus*, they serve primarily for collecting food. The cilia around the mouth and at the opening of the pharynx pull in a steady current of water, which passes through the pharyngeal slits into a chamber known as the atrium and then exits through the atrial pore. Food particles are collected in the sievelike pharynx.

The chordates are believed to have arisen from a group of organisms that resembled the modern tunicates. Tunicates, in their larval form, have chordate characteristics (Figure 18–52).

The Principal Chordates: The Vertebrates

The vertebrates constitute the largest and most familiar group of chordates. All vertebrates have a backbone, or vertebral column, as their structural axis; this is a flexible bony support that develops around the notochord, supplanting it entirely in most species. The backbone is made up of bony elements, the vertebrae, which encircle the nerve cord along the length of the spine. The brain is similarly enclosed and is protected by bony skull plates. Between the vertebrae are cartilaginous disks, which give the vertebral column its flexibility. Associated with the vertebrae are a series of muscle segments, the myotomes, by which sections of the vertebral column can be moved separately. This segmented pattern persists in the embryonic forms of higher vertebrates but is largely lost in the course of development.

One of the great advantages of this bony endoskeleton, as compared with the exoskeletons of the invertebrates, is that it is composed of living tissue that can grow with the animal. In the developing vertebrate embryo, the skeleton is largely cartilaginous, the bone gradually replacing cartilage in the course of

18–52

Two stages in the life of a tunicate. At the left is the "tadpole" larva; the adult form is shown on the right. After a brief free-swimming existence, the larva settles to the bottom and attaches at the anterior end. Metamorphosis then begins. The larval tail, with the notochord and dorsal nerve cord, disappears, and the animal's entire body is turned 180°. The mouth is carried backward to open at the end opposite that of attachment, and all the other internal organs are also rotated back. As some biologists reconstruct the past, tunicate larvae wriggled up the rivers, where they gave rise, after many generations and millions of years, to the ancestral vertebrates.

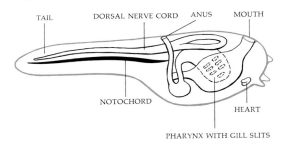

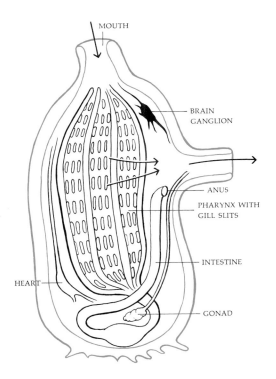

18-53

X-ray of an elk fetus. The bones have been stained to show them more clearly. The skeleton is still largely cartilaginous. Notice the legs, for example: only the dark areas are bone. As animals with endoskeletons mature, bone gradually replaces cartilage. Thus, the skeleton grows as the animal does, rather than having to be replaced, as is the case with an exoskeleton.

maturation (Figure 18-53). The growing tips of the bones characteristically remain cartilaginous until the animal reaches its full size. In addition to its powers of growth, bone can also repair itself, unlike the lifeless tissue of a clamshell, for example.

There are seven classes of vertebrates: the fish (comprising three classes), the amphibians, the reptiles, the birds, and the mammals.

Fish

The first fish were jawless and had a strong notochord running the length of their bodies. Today these jawless fish (Class Agnatha), once a large and diverse group, are represented only by the hagfish and the lampreys, which are the most primitive living vertebrates. They have a notochord throughout their lives, like *Amphioxus,* and a cartilaginous skeleton. The sharks (including the dogfish) and the skates, the second major class of fish, also have a completely cartilaginous skeleton. Their skin is covered with small pointed teeth (denticles), which resemble vertebrate teeth structurally and give the skin the texture of coarse sandpaper.

The third major class is the osteichthyans, the fish with bony skeletons. This group includes the trout, bass, salmon, perch, and many others—most of the familiar freshwater and saltwater fish.

According to present interpretations of the evidence, the bony fish went through most of their evolution in fresh water and spread to the seas at a much later period. The fresh water in which they lived was shallow and often stagnant (devoid of oxygen). Some primitive osteichthyans had air-filled swim bladders that served as flotation chambers. Under the pressures of competition for oxygen, these swim bladders, in some species, evolved into simple lungs that served as accessory organs to the gills.*

Lunged fish were the most common fish in the later Devonian period. Some of these lunged fish, ancestors of modern lungfish, would bury themselves in the mud during periods when the water dried up completely. In others, skeletal supports evolved that served to prop up the thorax of the fish. These fish could gulp air even when their bodies were not supported by water. Eventually, according to this hypothesis, these lobe-finned fish gained the capacity to waddle to a new body of water. Thus the transition to land began as an attempt to remain in the water.

* Lungs and gills and their comparative functions will be described in more detail in Chapter 27.

18-54

Ray fish, like skates and sharks, are primitive cartilaginous fish, having existed in their present form for about 350 million years. Their flattened body is an adaptation to bottom living.

Amphibians

Amphibians descended from the air-breathing, lunged fish. Modern amphibians include frogs and toads (which are tailless as adults) and salamanders (which have tails throughout their lives). They can readily be distinguished from the reptiles by their thin, scaleless skins, which serve as accessory breathing organs. (Frogs also have lungs, into which they gulp air, but salamanders breathe entirely through their skins and the mucous membranes of their throats.) Because water evaporates rapidly through these skins, the animals can readily die of desiccation in a dry environment.

Most frogs in cold climates have two life stages (hence their name, from *amphi* and *bios*, meaning "two lives"). The eggs are laid in water and are fertilized externally. They hatch into gilled larvae (tadpoles). The tadpoles later develop into adults which lose their gills and develop lungs. The adults live out of the water, at least in the summer. However, there are many variations on this theme. Some of the American salamanders fertilize their eggs on land; the males deposit sperm packets that are picked up by the females. Many amphibians are now known to skip the free-living larval stage. The eggs, which may be laid on land, in a hollow log or cupped leaf, or even carried by the parent, hatch into miniature versions of the adult. Thus the modern amphibians are highly evolved in their own way—as far removed from their ancestral forms—as man is in his way.

18-55
Many frogs are clearly fishlike in their larval (tadpole) stages. As adults, they require water to reproduce and their moist skins are an important accessory breathing organ. Frogs are carnivores. They catch insects with a flick of their long tongues, which are attached at the front of their mouths and which have a sticky, flypaper-like surface.

Reptiles

As you will recall, the vascular plants freed themselves from the water by the development of the seed. Analogously, the vertebrates became truly terrestrial with the evolution in the reptiles of the amniote egg, an egg that carries its own

18-56
Reptiles lay eggs in which the embryos can develop on land. These newly hatched black snakes are sunning themselves, a behavior characteristic of exotherms—animals that take in heat from the environment. Black snakes, like all other snakes, are carnivorous, with the prey devoured live and whole.

HOMOLOGY AND ANALOGY

Homologous structures arise from a common evolutionary origin and do not necessarily involve similar functions. For example, in the course of adaptive change, the fundamentally similar forearms of different vertebrates have become quite different in overall shape and in function. The accompanying illustration shows the foreleg of a crocodile, the wing of a bird, the flipper of a whale, the leg of a horse, the wing of a bat, and the arm of a man. The crocodile is placed first because it is closest to the ancestral type—the form from which all the others arose. In the drawing, the bones are color-keyed to indicate that they are

homologous structures—the "same" structure, modified by evolution for different purposes. Note that in the horse, the two bones of the forearm are fused.

Analogy, by contrast, refers to structures that are superficially similar but have an entirely different evolutionary background—the wing of a bird and the wing of an insect, for example, or the spine of a cactus (a modified leaf) and the thorn of a rose (a modified branch).

Classification systems in biology attempt to reflect homology and not analogy.

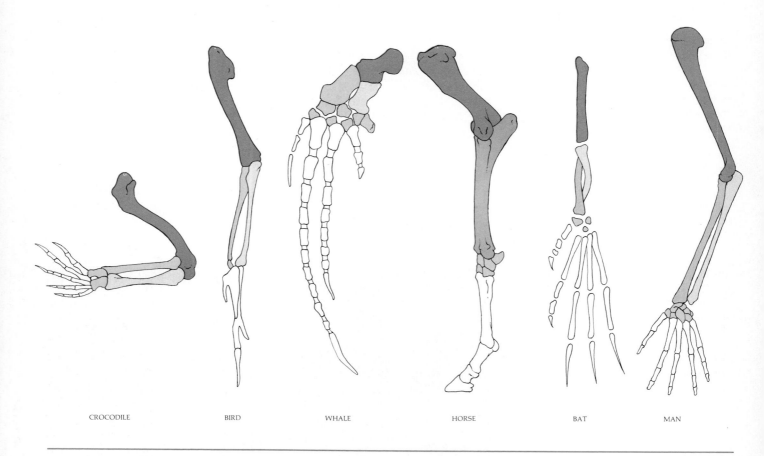

CROCODILE   BIRD   WHALE   HORSE   BAT   MAN

water supply and so can survive on land. The reptilian egg, which is much like the familiar hen's egg in basic design, contains a large yolk, the food supply for the developing embryo; abundant albumin; and a water supply. A membrane, the amnion, surrounds the developing embryo in a liquid-filled space that substitutes for the ancestral pond. The gilled stage is passed in a shelled egg or in the maternal oviduct or uterus. In mammals also, although their eggs typically develop internally, the embryos are enclosed in water within an enveloping membrane and pass through their gilled stages before birth.

Reptiles are characteristically four-legged, although the legs are absent in most snakes and some lizards. In keeping with their terrestrial existence, reptiles have a dry skin usually covered with protective scales. Modern reptiles, of which there are 5,400 species, include lizards, snakes, turtles, and crocodiles.

Until recently the dinosaurs, which are clearly of reptilian descent, were assumed to have been cold-blooded (exothermic). Several lines of evidence indicate, however, that at least some groups of dinosaurs were warm-blooded (endothermic).

Birds

Birds developed from one class of reptiles, and in their skeletal structure, birds are essentially reptiles highly specialized for flight (Figure 18–57). Their bodies are lightened by air sacs and also by having hollow bones. The oceanic frigate, a seagoing bird with a wingspread of more than 2 meters, has a skeleton that weighs only 110 grams (about 4 ounces). The most massive bone in the bird skeleton is the breastbone, or sternum, which provides the keel for the attachment of the huge muscles that operate the wings. Flying birds have jettisoned all extra weight; the female's reproductive system has been trimmed down to a single ovary, and even this becomes large enough to be functional only in the mating season.

Birds have feathers, which is one of the taxonomic distinctions, and they maintain a high and constant body temperature, which distinguishes them from most of the modern reptiles (although a few, such as leatherback turtles, show some degree of endothermy). In modern birds, feathers serve both to aid in flight and as insulation. (Only animals that are endothermic require insulation; animals that warm their bodies by exposure to the environment would find insulation a disadvantage.) Birds also have scales, a reminder of their reptilian ancestry. They are further characterized by young that are born at an immature stage and usually require a long period of parental care.

18–57

(a) *The oldest known fossil bird,* Archaeopteryx, *dates from the middle Jurassic period, about 150 million years ago. It still has many reptilian characteristics. The teeth and the long, jointed tail are not found in modern birds. The clearly evident feathers may have been related as much to endothermy as to flight. (b) A modern bird, a marabou stork from Uganda, preparing for takeoff.*

(a)

(b)

Mammals

Mammals also descended from a group of extinct reptiles now believed to be endothermic and perhaps fur-bearing. Characteristics that distinguish mammals from most other living vertebrates are: (1) mammals have hair (rather than scales or feathers); (2) they nurse their young; and (3) they are endothermic and maintain a high and relatively constant body temperature.

In nearly all mammalian species, the young are born alive, as they are in some fish and reptiles, which retain the eggs in their bodies until they hatch. Some very primitive mammals, however, such as the duckbilled platypus, lay shelled eggs but nurse their young after hatching. The *marsupials*, which include the opossum and the kangaroo, also bear their young alive, but they differ from the major group of mammals in that the infants are born at a tiny and immature stage and are often kept in a special protective pouch in which they suckle and continue their development (Figure 18–58). Most of the familiar mammals are *placentals*, so called because they have an efficient nutritive connection, the placenta, between the uterus and the embryo. As a result, the young develop to a much more advanced stage before birth. Thus the young are afforded protection during their most vulnerable period. The earliest placentals were small, shy, and probably nocturnal; they undoubtedly lived mostly on insects, grubs, worms, and eggs, devoting much of their energies to avoiding the carnivorous dinosaurs.

Among the living placentals, there are only a few major evolutionary lines, which are summarized in Figure 18–60.

(a)

(b)

18–58

(a) *Marsupial infants are born at an immature stage and continue their development attached to a nipple in a special protective pouch of the mother. This tiny kangaroo accidentally became dislodged from its mother's pouch. As you can see, it is still attached to the nipple. After the picture was taken, the baby was restored to the pouch, with no apparent ill effects from its premature introduction to the outside world. (b) Opossum infants spend about two weeks in the womb and about three months in their mother's pouch, passively attached to a teat. A newborn opossum is about the size of a honeybee.*

18–59

An assortment of mammals. (a) Lagomorphs, such as the jack rabbit shown here, have two pairs of upper incisors, whereas rodents, such as lemmings (b), have only one. In rodents, the teeth grow continually. Carnivores, such as the Weddell seal (c) and the lion (d), are adapted to hunt and kill for food. The lion's prey is a perissodactyl, an odd-toed ungulate. (e) Hippopotamuses, which are artiodactyls (two-toed ungulates), graze by night and spend most of the daylight hours resting in the water. (f) Elephants, the Proboscidea, are the largest land mammals living today; some reach a weight of 7.5 metric tons.

(a)

(b)

(c)

(d)

(e)

(f)

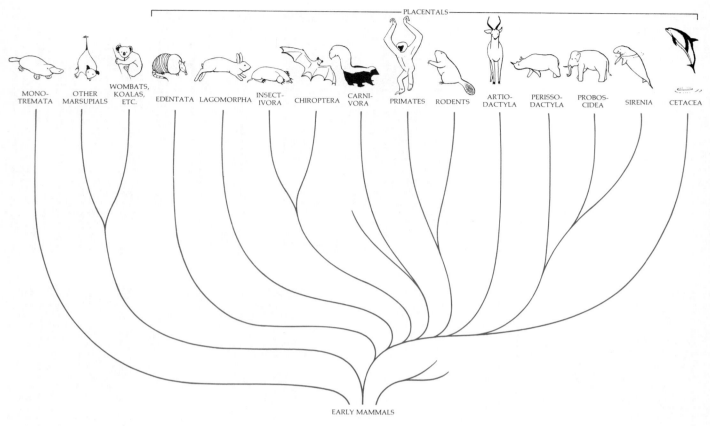

PLACENTALS

MONO-TREMATA · OTHER MARSUPIALS · WOMBATS, KOALAS, ETC. · EDENTATA · LAGOMORPHA · INSECT-IVORA · CHIROPTERA · CARNI-VORA · PRIMATES · RODENTS · ARTIO-DACTYLA · PERISSO-DACTYLA · PROBOS-CIDEA · SIRENIA · CETACEA

EARLY MAMMALS

18–60
Evolution of the mammals.

The primates, the order to which man belongs, are placental mammals characterized by three kinds of teeth (canines, incisors, molars), opposable first digits (thumbs and usually toes), two pectoral mammae, expanded cerebral cortex, and a tendency toward single births. Man is distinguished from the other primates by his upright posture, his long legs and short arms, his high forehead and small jaw, and his lack of body hair.

Among all the mammals, man is one of the least specialized. He is an omnivore. Unlike the Carnivora, which are meat eaters, and the several orders of herbivores, he eats a wide variety of fruits, vegetables, and other animals. His hands closely resemble those of a reptile, in contrast to the highly specialized forelimbs developed by, for example, the whales, bats, and horses. His sensory organs are crude compared with those of insects or of many other mammals. Man has, however, one area of extreme specialization: the brain. Because of his brain, man is unique among all the other animals in his capacity to reason, to speak, to plan, to learn, and so, to some extent, to control his own future on this planet.

SUMMARY

Organisms are named according to a binomial ("two-word") system, which designates the genus and species of the organism. Genera, in turn, are grouped into families, families into orders, orders into classes, classes into phyla, and phyla into kingdoms. Ideally, these groupings reflect the evolutionary history of the organism. Five kingdoms are recognized in this text: Monera, Protista, Fungi, Plantae, and Animalia.

The Monera comprises all the living prokaryotes: the bacteria and the blue-green algae. Prokaryotes are small, relatively simple cells without a membrane-bound nucleus or organelles. Their genetic material is in the form of a DNA molecule not combined with protein. They have a cell wall of unique composition. The bacteria are a diverse group of organisms, including autotrophs (both chemoautotrophs and photosynthetic autotrophs) and heterotrophs. Among the heterotrophs are the decomposers, which are ecologically of great importance in the recycling of organic matter, and the pathogenic (disease-causing) bacteria. All blue-green algae are photosynthetic. A few species of bacteria and blue-green algae are important in nitrogen fixation.

Viruses are parasites that can multiply only within a living cell, often only one particular type of cell. A virus consists essentially of nucleic acid (either RNA or DNA) enclosed in an outer coat of protein. Viruses do not clearly belong to any of the five kingdoms of organisms, and, in fact, some biologists do not consider them to be living organisms.

The other four kingdoms of organisms are all composed of eukaryotes. In eukaryotic cells, DNA, the genetic material, is complexed with proteins and organized into chromosomes which are contained within a nucleus bound by a double membrane, the nuclear envelope. Eukaryotic cells also have membrane-bound organelles, including mitochondria and, in the case of photosynthetic cells, chloroplasts. They have cilia and flagella with a characteristic 9 + 2 structure.

The kingdom Protista is a highly diverse group, comprising unicellular eukaryotic heterotrophs, the Protozoa; the slime molds; and seven phyla of algae, both unicellular and simple multicellular forms. Unicellular algae are the photosynthetic components of the plankton, the food source for many marine organisms; brown algae, red algae, and some green algae are the common marine seaweeds. Some multicellular algae are differentiated into a blade ("leaf"), stipe ("stem"), and holdfast, which anchors them to rocks or some other substrate.

The green algae, which resemble plants in many characteristics, are believed to be the group from which the plants arose. These resemblances include: (1) having chlorophylls *a* and *b* and carotene as the photosynthetic pigments; (2) having cellulose cell walls; and (3) having a life cycle in which a haploid form, the gametophyte, alternates with a diploid form, the sporophyte (alternation of generations).

Fungi are heterotrophs and, with a few exceptions, are simple multicellular organisms. They absorb organic materials from the substrate on or in which they live. Most have cell walls that contain chitin. The body of a multicellular fungus is a mycelium, which is made up of a mass of filaments, called hyphae. Many fungi reproduce by spores that are often enclosed in sporangia. Fungi are ecologically important as decomposers. They are also economically useful in the production of foodstuffs and antibiotics. They are the causes of a number of serious plant diseases as well as some diseases in animals.

Lichens are symbiotic combinations of fungi and algae that are able to adapt to conditions—such as lack of water and of organic matter—under which neither form could survive alone.

Plants are multicellular, photosynthetic organisms adapted for life on land. All have chlorophylls *a* and *b* and carotene as their photosynthetic pigments, and the membranes in which these pigments are embedded are organized in chloroplasts. All plants have cellulose cell walls. All have a waxy outer coating, the cuticle; stomata and multicellular reproductive structures; and a life cycle with alternation of generations. There are two major groups: the bryophytes and the tracheophytes.

The bryophytes are relatively small plants lacking elaborate vascular systems; they include the mosses, liverworts, and hornworts. In the bryophytes the gametophyte (haploid) generation is the larger, dominant form.

The tracheophytes are the vascular plants. The largest groups of these are the gymnosperms and the angiosperms. The gymnosperms are "naked seed" plants in which the ovules and seeds are exposed. The angiosperms are the flowering plants in which the seed (the mature ovule and its contents) is enclosed in an outer protective layer, formed from parts of the flower. The fruit—the seed and its outer covering layers—is often dispersed by animals. Land animals are dependent directly or indirectly on plants as their source of food energy.

Animals are multicellular heterotrophs that depend directly or indirectly on plants or algae as their energy source. Almost all digest their food in an internal cavity. Most are motile. Reproduction is usually sexual. Most—more than 90 percent—are invertebrates, animals without backbones.

Major phyla of invertebrates include: (1) Porifera (sponges), which have a unique body plan, simple cellular organization, and feed by choanocytes; (2) Coelenterata, characterized by radial symmetry, a two-layered body with mesoglea, a coelenteron, and cnidoblasts; (3) Platyhelminthes (flatworms), with a three-layered body; (4) Rhynchocoela (ribbon worms), with a three-layered body plan, one-way digestive tract, and retractile proboscis; (5) Nematoda (roundworms), a very large group characterized by a three-layered body, a one-way digestive tract, and a pseudocoelom; (6) Annelida, in which the body is divided externally and internally into segments and which have a coelom and a simple circulatory system; (7) Mollusca, characterized by a head, a muscular foot (variously modified), and a mantle that secretes shell; (8) Arthropoda, segmented animals with a hard, jointed exoskeleton, the largest phylum in the animal kingdom, including arachnids, crustaceans, and insects; and (9) Echinodermata, with radial symmetry, a spiny exoskeleton, and a water vascular system.

The vertebrates are the principal members of the phylum Chordata. They are distinguished from the invertebrates by a bony internal skeleton. The major classes of vertebrates include fish, amphibians, reptiles, birds, and mammals.

QUESTIONS

1. Name the principal differences between prokaryotes and eukaryotes.

2. Before the structure of any virus was known, Crick and Watson predicted that the protein coats of viruses would prove to be made up of a large number of identical subunits. Can you explain the basis of their prediction?

3. Distinguish among the following: sporangia, sporangiophore, sporophyte.

4. Define alternation of generations, using as your example the pine life cycle.

5. Diagram the life cycle of *Ulva* using a heavy line for the $2n$ phase and a thin line for the n phase.

6. How and why would an organizm "lose" a sexual cycle? What would be the advantage from the point of view of evolution?

7. Obviously in order for a characteristic to be selected for, in evolutionary terms, a character must be advantageous at the time selection is taking place. How, then, can you explain preadaptation?

8. How would you distinguish an insect from an arachnid? An insect from a crustacean?

9. Name the identifying characteristics of the phylum Chordata and indicate the functional significance of each.

10. Birds invest a large amount of time and energy in caring for eggs and young, a feature that would appear to be disadvantageous. Can you think of an advantage to birds of having the young produced at such an immature stage?

11. There are disadvantages to humans in being bipedal; running on two legs is certainly slower and more tiring than running on four, for instance. What are the compensating advantages? (We should warn you that the answer to this question is not fully agreed upon by the authorities.)

SUGGESTIONS FOR FURTHER READING

AHMADJIAN, V.: *The Lichen Symbiosis,* Blaisdell Publishing Company, Waltham, Mass., 1967.

In this small but fascinating book, Ahmadjian reviews the nature of the relationship between the fungal and algal components of a lichen.

ALEXOPOULOS, C. J.: *Introductory Mycology,* 2d ed., John Wiley & Sons, Inc., New York, 1962.

A general, brief introduction to the fungi.

BARNES, ROBERT D.: *Invertebrate Zoology,* 3d ed., W. B. Saunders Company, Philadelphia, 1974.

One of the best general introductions to protozoans and invertebrates.

BONNER, J. T.: *The Cellular Slime Molds,* 2d ed., Princeton University Press, Princeton, N.J., 1968.

A record of experimental work with a small but fascinating group of organisms.

BUCHSBAUM, RALPH, and LORUS J. MILNE: *The Lower Animals: Living Invertebrates of the World,* Doubleday & Company, Inc., Garden City, N.Y., 1962.

A collection of handsome photos of the invertebrates accompanied by a text prepared by two noted zoologists but directed toward the general reader.

CORNER, E. J. H.: *The Life of Plants,* Mentor Books, New American Library, Inc., New York, 1968.*

A renowned botanist with a flair for poetic prose describes the evolution of plant life, telling how plants modify their structures and functions to meet the challenge of a new environment as they invade the shore and spread across the land.

CURTIS, HELENA: *The Marvelous Animals,* Natural History Press, Garden City, N.Y., 1968.

An introduction to protozoans.

DAWSON, E. Y.: *Marine Botany: An Introduction,* Holt, Rinehart and Winston, Inc., New York, 1966.

A short, lively text that covers seaweeds, marine bacteria, fungi, phytoplankton, and sea grasses.

* Available in paperback.

DESMOND, ADRIAN J.: *The Hot-Blooded Dinosaurs: A Revolution in Paleontology*, The Dial Press, Inc., New York, 1976.

Desmond, a science historian, describes the development of evolutionary theories, particularly as they were influenced by the discovery of dinosaur fossils and presents a lively review of the evidence that is leading an increasing number of paleontologists to the conclusion that dinosaurs were endothermic. (If you are not interested in the history of paleontology, you may want to begin in the middle.) The text is well written and the illustrations are wonderful.

EVANS, HOWARD E.: *Life on a Little-Known Planet*, E. P. Dutton & Co., Inc., New York, 1968.

Professor Evans is Curator in the Department of Entomology of Harvard's Museum of Comparative Zoology and also the author of many popular articles and books on insects. This book profits from his wide knowledge, clarity, and humor.

HICKMAN, CLEVELAND P.: *Biology of the Invertebrates*, 2d ed., The C. V. Mosby Company, St. Louis, 1973.

Probably the best general text on the invertebrates.

KLOTS, ALEXANDER B., and ELSIE B. KLOTS: *Living Insects of the World*, Doubleday & Company, Inc., Garden City, N.Y., 1962.

A spectacular gallery of insect photos. The text is informal but informative, written by experts for laymen.

LARGE, E. C.: *The Advance of the Fungi*, Dover Publications, Inc., New York, 1962.*

A fascinating popular account of the closely interwoven histories of fungi and man, first published in 1940.

MARGULIS, LYNN: *Origin of Eukaryotic Cells*, Yale University Press, New Haven, Conn., 1970.

A fascinating discourse on the origin of eukaryotic cells by serial symbiotic events, beautifully illustrated and well reasoned.

PICKET-HEAPS, J. D.: *Green Algae: Structure, Reproduction and Evolution in Selected Genera*, Sinauer Associates, Inc., Sunderland, Mass., 1975.

A beautifully illustrated book providing much insight into the variety of form and function in the cells of the green algae.

RAVEN, PETER H., RAY F. EVERT, and HELENA CURTIS: *Biology of Plants*, 2d ed., Worth Publishers, Inc., New York, 1976.

This general botany text contains an excellent presentation of the evolution of plants and related organisms.

ROMER, ALFRED: *The Vertebrate Story*, 4th ed., The University of Chicago Press, Chicago, 1959.

The history of vertebrate evolution, written by an expert but as readable as a novel.

ROSEBURY, THEODOR: *Life on Man*, The Viking Press, Inc., New York, 1969.

An account of our resident microbes and our attitudes toward them—"from Aristophanes to Lenny Bruce."

RUSSELL-HUNTER, W. D.: *A Biology of Higher Invertebrates*, The Macmillan Company, New York, 1968.*

Short, authoritative accounts, with emphasis on function.

RUSSELL-HUNTER, W. D.: *A Biology of Lower Invertebrates*, The Macmillan Company, New York, 1968.*

* Available in paperback.

SLEIGH, M.: *The Biology of Protozoa*, Edward M. Arnold, Publishers, Ltd., London, 1973.*

A general biology of the Protozoa, including chapters on structure, metabolism, reproduction, and ecology, with many excellent illustrations.

SMITH, A. H.: *The Mushroom Hunter's Field Guide*, 2d ed., The University of Michigan Press, Ann Arbor, Mich., 1966.

A clear, concise, well-illustrated guide to edible mushrooms, enlivened with good advice and pertinent anecdotes.

STANIER, R. Y., M. DOUDOROFF, and E. ADELBERG: *The Microbial World*, 3d ed., Prentice-Hall, Inc., Englewood Cliffs, N.J., 1970.

An introduction to the biology of microorganisms, with special emphasis on the properties of bacteria.

WELLS, M. J.: *Brain and Behavior in Cephalopods*, Stanford University Press, Stanford, Calif., 1962.

Experimental analyses of behavior in the octopus and squid.

* Available in paperback.

SECTION 5   Plants and the Land

CHAPTER 19

Introducing the Land Plant: The Leaf and Photosynthesis

For most of earth's history, the land was bare. A billion years ago, seaweeds may have clung to the shores at low tide and perhaps some gray-green lichens patched a few inland rocks, but had anyone been there to observe it, the earth's surface would generally have appeared as barren and forbidding as the surface of Mars is today. According to the fossil records, plants first began to invade the land a mere half billion years ago, and not until then did the earth truly come to life. As a film of green spread from the edges of the waters, other forms of life, the heterotrophs, were able to follow. The shapes of these new forms and the ways in which they lived were determined by the plant life that preceded them because, in freeing themselves from the water, they became increasingly dependent on plants not only for their food—their chemical energy—but also for their nesting, hiding, stalking, and breeding places.

In all communities except those created by man, the character of the plants still determines the character of the animals and other forms of life that inhabit that particular area. Even man, who has seemingly freed himself from the life of the land and even, on occasion, from the surface of the earth, is still dependent on the photosynthetic events that take place in the green leaves of plants.

The modern plant can best be understood in terms of its long evolutionary history and, in particular, its transition to land. In this section, we shall focus our attention on the largest and most abundant group of plants—the angiosperms. The angiosperms are characterized by specialized organs—flowers—in which sexual reproduction takes place and in which the seed is formed and develops into a fruit. There are about 235,000 species of angiosperms, divided into two large subclasses, the dicots (170,000 species) and the monocots (65,000 species). A summary of the differences between the two groups can be found on page 265.

Figure 19–2 shows, in a diagrammatic form, the plant body of a common and familiar angiosperm, a geranium. This plant, like other vascular plants, is characterized by a root system that anchors the plant in the ground and collects water and minerals from the soil; a stem or trunk that raises the photosynthetic parts of the plant toward the sun; and organs highly specialized for light capture and photosynthesis, the leaves. These are all interconnected by a complicated and efficient system for the transport of food, water, and minerals. Such a system is the distinguishing feature of all vascular plants, a group that also includes the conifers and other gymnosperms and the ferns (see page 223).

These characteristics—leaves, stems, and roots—are all adaptations to a photosynthetic life on land.

19–1

One of our most successful angiosperms, showing two of the characteristics of this large group. One is the flower, in which the seed develops; the other is the fruit, which aids in dispersal of the seed. The fruit in this case is a plumelike structure that is caught by the wind.

19–2

The body plan of a complex land plant. The above-ground structures constitute the shoot, consisting of the stem, the leaves, whose primary function is photosynthesis, and the flowers, the reproductive organs. Leaves appear at regions on the stem known as nodes. The portions of the stem between nodes are called internodes. The below-ground structures, the roots, supply water and minerals to the plant body.

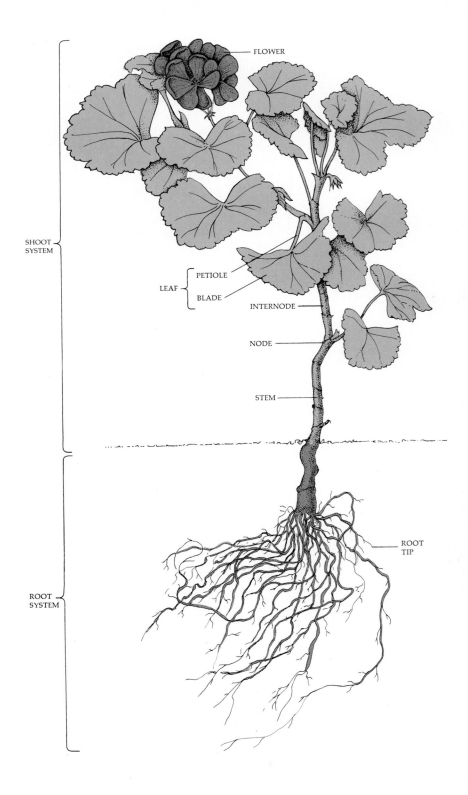

SHOOT SYSTEM

FLOWER

LEAF — PETIOLE / BLADE

INTERNODE

NODE

STEM

ROOT SYSTEM

ROOT TIP

THE LEAF AND PHOTOSYNTHESIS

Let us look first at the cells of the leaf that carry out this work of photosynthesis. The photosynthetic cells are of a general type known as *parenchyma cells;* they are the cells from which all other cells of the plant body have evolved. In the leaf (Figure 19–3), parenchyma cells make up the tissue called *mesophyll,* "middle

19–3
Diagram of the interior of a leaf.

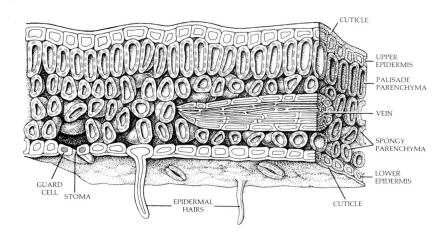

CUTICLE

UPPER
EPIDERMIS

PALISADE
PARENCHYMA

VEIN

SPONGY
PARENCHYMA

LOWER
EPIDERMIS

CUTICLE

GUARD
CELL

STOMA

EPIDERMAL
HAIRS

19–4
Veins in the leaf of a maple. Continuous with the vascular tissue of the stem and root, they branch and divide into finer and finer bundles, reaching a short distance from every photosynthetic cell.

leaf." There are two types of parenchyma cells in the mesophyll: palisade parenchyma, column-shaped cells near the upper surface of the leaf, and spongy parenchyma, which underlie the palisade cells.

One can best understand the anatomy of the leaf—and indeed, of the entire plant body—if one compares these photosynthetic cells with the single-celled algae shown on page 88. The point is that they are remarkably similar. Their principal cell organelles are chloroplasts, and the chloroplasts, like those of other green algae and of all plants, contain chlorophylls *a* and *b* and carotene. They are surrounded by a cell membrane outside of which is a cell wall composed mostly of cellulose. In addition to sunlight, these cells require water, carbon dioxide, and a few minerals. The algal cells can obtain these from the water in which they live. The photosynthetic cells of plants, however, require a complex life support system; the plant body is, in effect, that life support system.

The blade of the leaf is a very thin, very broad light-collecting surface whose primary function is the exposure of the photosynthetic cells to sunlight. In the leaf, the photosynthetic cells are sandwiched between two layers of small, stiff-walled epidermal cells that are transparent and so permit the free passage of light to the photosynthetic parenchyma. The outer surfaces of the epidermal cells are coated with a waxy layer, the *cuticle*, which retards the escape of water from the interior of the leaf.

The supply of water to the photosynthetic cells is the most crucial problem that plants had to solve, evolutionarily speaking, in their transition to land. The success of the various groups of modern plants is in direct proportion to the efficiency of their water transport systems. Water is brought into the leaf through the vascular tissues; the vascular tissues of the leaf are called *veins* (Figure 19–4). Sugars, produced by photosynthesis, are removed from the leaf by the veins and transported to other parts of the plant body. (We shall discuss the structure and function of the vascular tissues in more detail in the following chapter.) The vascular tissues of the leaf pass through the petiole (leaf stalk) and are continuous with the vascular bundles of the stem and root. Within the leaf, water moves from the xylem cells—the water-conducting cells of the vascular tissues—into the spaces surrounding the spongy parenchyma. As we noted, its escape from the leaf is retarded by the waxy cuticle of the epidermal cells.

The leaf, as we have described it so far, appears to be a waterproof, airproof package in which photosynthetic cells are held aloft and advantageously displaced to sunlight. However—and this is a serious difficulty—photosynthesis also requires carbon dioxide, and the source of carbon dioxide for land plants is in the air. A perfect solution would be a system for getting carbon dioxide into the photosynthetic cells without permitting any loss of water from the mesophyll.

The imperfect process of evolution has produced a compromise in the form of _stomata_ (singular, stoma, from the Greek word for mouth). Stomata are very small openings, or pores, in the leaf epidermis (Figure 19–5); they are bordered by guard cells that open and close them, thus regulating the escape of water and the exchange of gases. Ninety percent of the water that leaves the plant body escapes through the stomata. (The rest is lost through the cuticle.)

Figure 19–6 shows the mechanism of stomatal movement. The movement is controlled by the turgor pressure of the guard cells. (Turgor is described on page 35.) When the water available to the leaf drops below a certain point, the turgor pressure of the guard cells drops and the stomata close, thus conserving the remaining water.

Factors other than water loss also affect the stomata, however. In many species, for example, the stomata close regularly in the evening and open in the morning, even though there may be no changes in water available to the plant. In most species, an increase in carbon dioxide concentration in the mesophyll causes the stomata to close. Temperatures higher than 30° to 35°C can cause stomatal closing in some species. Many plants in hot climates close their stomata regularly at midday.

In a number of succulents, the stomata are closed all day long, opening only at night. Such plants include cacti and the pineapple, and members of the stonecrop family, Crassulaceae. At night, when the stomata are open, the plants take in carbon dioxide and convert it to four-carbon acids. During the day, the carbon dioxide is released from the acids and used immediately in photosynthesis. This pathway of photosynthesis, known as CAM (Crassulacean acid metabolism), resembles, in principle, the C_4 photosynthesis described on page 95.

The way in which various factors such as light, temperature, and carbon dioxide concentration affect the turgor pressure of guard cells is not known.

19–5

(a) _Portion of the leaf of a spiderwort_ (Tradescantia). _Two stomata are visible. The one at the left is partly open and the other is closed._ (b) _View of an opened stoma of cucumber._

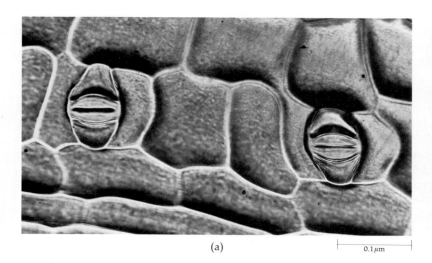

(a)   0.1 μm

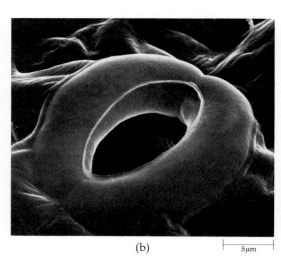

(b)   5 μm

Mechanism of stomatal movement. Each stoma is flanked by two guard cells that open the stoma when they are turgid and collapse and close it when they lose turgor. In many species, the guard cells have thickened walls adjacent to the stomatal opening. As turgor pressure increases, the thinner parts of the cell wall are stretched more than the thicker parts, causing the cells to bow out and the stoma to open. A surface view is shown in (a) and a cross section in (b).

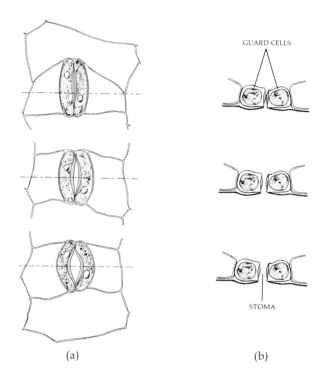

(a)   (b)

The fact that guard cells have chloroplasts and other epidermal cells do not may be a clue.

Stomata are commonly most abundant on the undersurface of leaves. They may be very numerous; for example, in tobacco leaves, there are 12,000 stomata per square centimeter of leaf surface.

SOME SPECIAL ADAPTATIONS

Leaves come in all shapes and sizes, ranging from broad fronds to tiny scales. Some of these variations can be correlated with the environments in which the plants live. Small, leathery leaves are often associated with harsh, dry climates, where the light-capturing surface has to be sacrificed for water conservation. This trend reaches its extreme in the desert cacti, which have no leaves at all. In these plants, photosynthesis takes place in the fleshy stems. Conversely, large leaves with broad surfaces are often found in plants that grow under the canopy in a tropical rain forest, where water is plentiful but where there is intense competition for light.

In some plants, leaves are modified as spines, which are hard, dry, and nonphotosynthetic. (The terms spine and thorn are often used interchangeably; however, thorns are technically modified branches.) In others, such as the garden pea, leaves are modified as tendrils. In many plants, leaves are succulent—that is, they are modified for water storage. Others are adapted for food storage. A bulb, such as the onion, is a large bud consisting of a short stem with many modified leaves attached to it. The "head" of a cabbage also consists of a stem bearing numerous thick, overlapping leaves. In some plants, the petioles become thick and fleshy: celery and rhubarb are two familiar examples.

19–7
The four seasons in a deciduous forest in Illinois. In such forests, the trees produce their leaves early in spring and begin to manufacture food; they lose them again in the autumn and enter an essentially dormant condition, thus passing the unfavorable growing conditions of winter.

MONOCOTS AND DICOTS

The angiosperms are divided into two broad groups: the dicots and the monocots. The names refer to the fact that the embryo in the dicots has two cotyledons ("seed leaves") and in the monocots has one. The veins of dicot leaves are usually fanlike or *featherlike; those of monocot leaves are usually parallel. In the dicots, the vascular tissues are arranged around a central core in the stem; in monocots, they are scattered. Dicots characteristically have taproots, and monocot roots are often fibrous.*

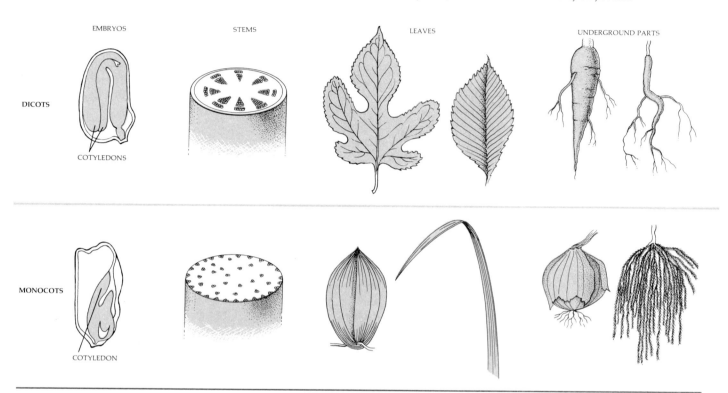

EMBRYOS   STEMS   LEAVES   UNDERGROUND PARTS

DICOTS

COTYLEDONS

MONOCOTS

COTYLEDON

THE ABSCISSION OF LEAVES

As we noted, most of the water lost by a plant escapes through the leaves. Angiosperms first evolved in the tropics. The first angiosperms were *perennials*, plants in which vegetative structures survive year-round and from year to year. Many groups of early angiosperms adapted to seasonal water shortages by becoming *deciduous*; that is, they dropped their leaves during periods of drought. Because of this adaptation, the plants were able to spread northward, where seasonal water shortages result from the cold.

The fall of the leaf comes about when special enzymes break up the middle lamellas (see page 63) in a layer of weak, thin-walled cells in the petiole. These cells are called the *abscission layer* (Figure 19–8). As the leaf ages, the cells separate. Before the leaf falls, a protective layer, the leaf scar, develops.

The dropping of leaves in deciduous trees comes about not as a direct reaction to cold or drought. Rather, in response to an environmental cue—the relative length of day and night—the plant produces chemicals that cause leaf abscission. The way this and other such responses come about is the subject of Chapter 22.

As the leaf dies, new chlorophyll can no longer be formed and the green color gradually disappears, unmasking the yellows and oranges of the carotenoids

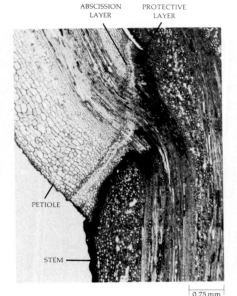

ABSCISSION LAYER   PROTECTIVE LAYER

PETIOLE

STEM

0.75 mm

19-8
Abscission layer in a maple leaf, as seen in a longitudinal section through the base of the petiole.

and producing the blaze of colors characteristic of the deciduous forest in the fall. (The red and purple colors seen in some fall leaves are due to another group of pigments, the anthocyanins. These pigments, which are in solution in the vacuoles, form when the temperature drops.)

Another adaptation to seasonal water shortages is in the life cycle of the *annual plants,* in which the leaves and all other vegetative parts die at the end of the growing season, leaving only the dormant seed, which then resumes growth after the period of water shortage is past. In another group of plants, known as *biennials,* the period from seed formation to flower spans two growing seasons. The first season of growth ends with the formation of a root, a stem, and a rosette of leaves. In the second growing season, stem elongation, flowering, and fruiting take place. Many common food plants are biennials, including cabbage and other members of the cabbage family (see page 445), the sugar beet, and parsley.

SUMMARY

The vascular plant is characterized by a root system, which, in the great majority of species, anchors the plant to the ground and collects water and minerals from the soil; a stem or trunk, which contains vascular and supporting tissues; and leaves.

Leaves are organs specialized for light capture and photosynthesis. The photosynthetic cells of leaves are parenchyma cells. The mesophyll of the leaf is composed of palisade parenchyma and spongy parenchyma. It is enclosed by a layer of epidermal cells that are covered by a waxy, water-resistant cuticle. Bundles of vascular tissue bring water to the leaf cells. The vascular tissues of leaves are called veins.

Stomata, minute pores in the epidermis, regulate the exchange of gases in the leaf. Most of the water that leaves the plant body escapes through the stomata. The opening and closing of the guard cells bordering the stomata are regulated by turgor pressure of the cells.

In some species, leaves are modified as spines or tendrils or for water or food storage. Some perennial plants (plants in which vegetative parts persist year-round and from year to year) have adapted to periods of seasonal drought by becoming deciduous. Other plants have adapted by completing the growing cycle—seed to seed—in a single season (annuals) or in two seasons (biennials).

QUESTIONS

1. Sketch the interior of a leaf, labeling all its principal parts. Compare your drawing with the one on page 261.

2. Superimpose on this sketch arrows showing the movement of water, of oxygen, and of carbon dioxide.

3. In what ways are C_4 photosynthesis and CAM photosynthesis similar? In what way are they different?

4. How is it physically possible for an increase in turgor of the guard cells to open the stomata? Consider what would happen if you partly inflated a cylindrical balloon, applied a strip of adhesive tape along its length, and then inflated it further. What does the experiment suggest about the role of wall thickenings of the guard cells?

CHAPTER 20

Roots, Stems, and Transport Systems

Roots, as we noted in the previous chapter, are specialized for anchoring the plant and for taking up water and minerals essential to the cells that are doing the photosynthetic work of the plant. Characteristically, the root system makes up more than half of the plant body. The lateral spread of tree roots is usually greater than the spread of the crown of the tree. In a study made on a four-month-old rye plant, the total surface area of the root system, including root hairs, was found to be 639 square meters, 130 times the surface area of the leaves and stem.

ANATOMY OF THE ROOT

The internal structure of the root is comparatively simple. There are three concentric layers: the *epidermis*, the *cortex*, and the *vascular cyclinder* (Figure 20–1).

The Epidermis

The epidermal cells of the root, which enclose the entire surface of the root, absorb the water and minerals from the soil. As you would expect, they lack the cuticle found on the surface of the epidermal cells of the leaf. They are also characterized by fine threadlike structures, known as root hairs (Figure 20–2). Root hairs are not appendages, but are slender extensions of the epidermal cells themselves; the nucleus of the epidermal cell is often found within the root hair.

In the study of the rye plant previously mentioned, the roots were estimated to have some 14 billion root hairs; placed end to end, they would have extended more than 10,000 kilometers. Most of the water and minerals that enter the root are taken up through these delicate outgrowths of the epidermal cells.

The Cortex

As you can see in Figure 20–1, the cortex occupies by far the greatest area of the young root. The cells of the cortex are parenchyma cells, like those of the leaf; however, as you would also expect, they lack chloroplasts. They store starch and other organic substances. The cortical cells are connected with one another by plasmodesmata (see page 63). The tissue of the cortex contains many air

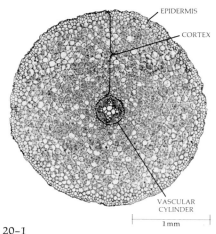

EPIDERMIS

CORTEX

VASCULAR CYLINDER

1 mm

20–1
Root of a buttercup in cross section.

20-2

Root hairs of radish seedling. Most of the uptake of water and minerals takes place through the root hairs, which form just behind the growing tip of the root.

spaces. Oxygen-containing air enters these spaces through the epidermal cells and is used by the cortical cells in respiration.

Unlike the rest of the cortex, the cells of the innermost layer, the *endodermis*, are compact and have no spaces between them. Each endodermal cell is characterized by the presence of a *Casparian strip*, which forms a waxy band within the cell wall. The strip is continuous and is not permeable to water. Therefore, water and dissolved substances, which pass freely around the other cortical cells and through their cell walls, must pass through the cell membranes of endodermal cells. As you will recall (page 59), water, oxygen, and carbon dioxide pass freely through cell membranes, but many ions and other substances do not. Therefore, the membranes of the endodermal cells regulate the passage of such substances into the vascular tissues of the root.

The Vascular Cylinder

The vascular cylinder of the root consists of the vascular (conducting) tissues surrounded by one or more layers of cells, the *pericycle*, which completely surrounds the vascular tissues. Branch roots (also called lateral roots) arise from the pericycle. In most species, the vascular tissues are grouped in a solid cylinder, as shown in Figure 20-1. In some, however, they form a hollow cylinder around a pith (Figure 20-4). Figure 20-5 shows the details of the vascular cylinder of a buttercup.

20-3

Diagrammatic cross section of a root, showing the two pathways of uptake of water and minerals. Along pathway A, water moves by osmosis and solutes by active transport through the cellular membranes of a series of

living cells. Along pathway B, water flows through the cell walls and along their surfaces, and the solutes flow with the water or by diffusion. Notice the location of the Casparian strip and how it blocks off pathway

B all around the vascular cylinder of the root. In order to pass the Casparian strip, the solutes must be transported through the cell membranes of the endodermal cells, as in pathway A.

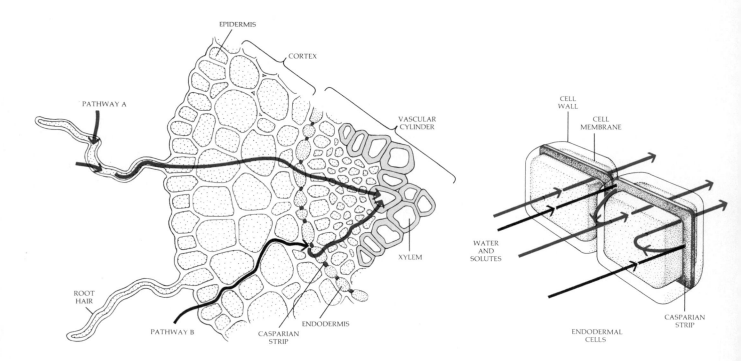

20-4

Cross section of the root of a corn plant, showing the vascular cylinder enclosing the pith. Part of a branch root can be seen at the lower right.

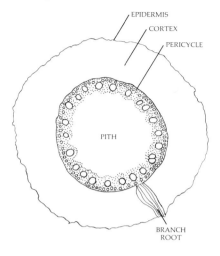

EPIDERMIS
CORTEX
PERICYCLE
PITH
BRANCH ROOT

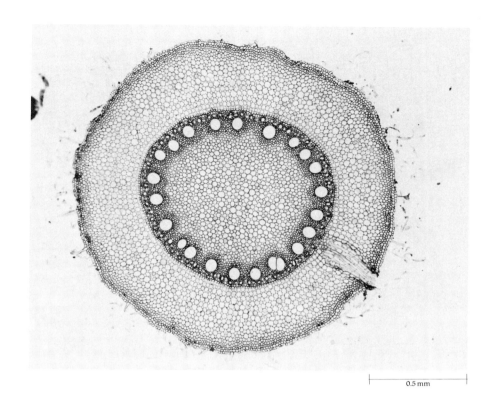

0.5 mm

20-5

Details of the vascular cylinder shown in Figure 20-1. The endodermis is considered part of the cortex. The outermost layer of the vascular cylinder is the pericycle, from which branch roots arise. The conducting tissues, the xylem and the phloem, are within the pericycle.

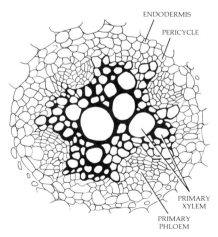

ENDODERMIS
PERICYCLE
PRIMARY XYLEM
PRIMARY PHLOEM

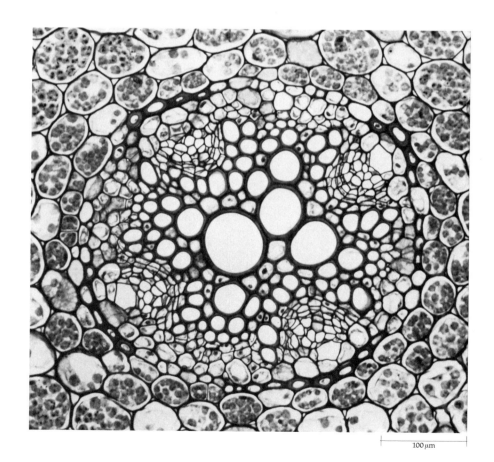

100 μm

(a)   (b)   (c)

20–6

Types of root systems. (a) Taproot of dandelion (a dicot). (b) Fibrous root characteristic of grasses (monocots). (c) Prop roots of corn, also a monocot.

TYPES OF ROOTS

The first root of a plant, which originates in the embryo, is called the primary root. In dicots, this root develops into a *taproot* and gives rise to secondary roots or branch roots (Figure 20–6a). In monocots, the primary root is usually short-lived and the final root system develops from the base of the stem; such roots are called *adventitious roots*. These adventitious roots and their branches develop into a fibrous root system (Figure 20–6b). Aerial roots are adventitious roots produced from above-ground structures. The aerial roots of some plants serve as prop roots, such as those found in corn (Figure 20–6c). Many tropical trees have prop roots, such as the red mangrove and the banyan tree. In some species that live in swamps, where the soil is low in oxygen, air roots are believed not only to anchor the plant but also to supply the root with the oxygen needed for respiration.

THE ANATOMY OF THE STEM

The stem holds the leaves up to the light and provides for the transportation of substances to and from the leaves. The outer surface of a young green stem, like that of the leaf and the root, is made up of epidermal cells. Like the leaf, it is covered with a waxy cuticle and contains stomata.

Ground Tissue

The bulk of the tissue of a young stem is known as the *ground tissue*. Like the mesophyll of the leaf, ground tissue is composed mostly of parenchyma cells. The turgor of these cells provides the chief support for young green stems. Parenchyma cells also accumulate and store food, usually in the form of starch, and in some plants, such as cacti and other succulents, they are specialized for storing water.

Stems also have supporting tissues formed from collenchyma and sclerenchyma cells. Collenchyma cells differ from the thin-walled parenchyma cells of the stem in having primary walls* that are unevenly thickened (Figure 20–7a). Collenchyma cells are often located just inside the epidermis.

Sclerenchyma cells are of two types: fibers and sclereids (Figure 20–7b and c). Fibers, which are elongated, somewhat elastic cells, typically are grouped in strands or bundles arranged in patterns characteristic of the plant. They are often associated with the vascular tissues. Plant fibers have long been useful to man in such forms as flax, hemp, jute, and raffia. Sclereids, which are variable in form, are also present in stems and, in addition, in seeds, nuts, and fruit stones, where layers of sclereids form the hard outer coverings. Sclerenchyma cells differ from collenchyma in three respects: (1) they have secondary walls; (2) the walls often contain lignin, a complex macromolecule that impregnates the cellulose and toughens and hardens it; and (3) they are sometimes dead at maturity—that is, without living protoplasts.

* Primary and secondary plant cell walls are described on page 63.

20–7

Some types of cells found in the ground tissue of stems. (a) Collenchyma cells. Their irregularly thickened cellulose walls contain pectin and large amounts of water and they are plastic, so they permit growth while providing support. (b) Fibers. Their specialization is thickened, often lignified cell walls that give them strength and rigidity. Many, but not all, are dead at maturity, like those shown here. (c) Sclereids. These cells, which have very thick lignified walls, are often found in seeds and fruit, as well as in stems. These sclereids are from a pear; they are what gives the fruit its characteristic gritty texture. This cell is a living cell with slender branches of cytoplasm extending through the cell walls. These terminate in plasmodesmata.

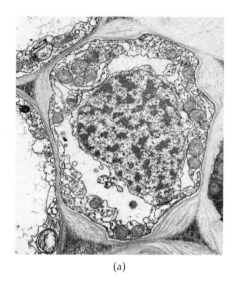

(a)

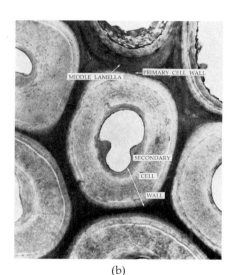

(b)

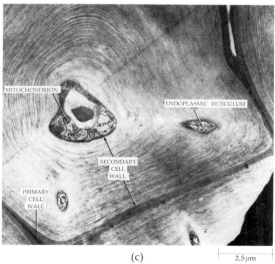

(c)   2.5 μm

Vascular Tissues

The vascular tissues of the plant pass through this ground tissue. In dicots, the vascular tissues form either a continuous cylinder (Figure 20–8a) or a series of separate, but interconnected bundles arranged in a hollow cylinder (Figure 20–8b). The ground tissue lying within the cylinder is known as the _pith_, and the ground tissue outside the cylinder is the cortex. In monocots, numerous vascular bundles are scattered through the ground tissue, which is not separated into pith and cortex (Figure 20–8c).

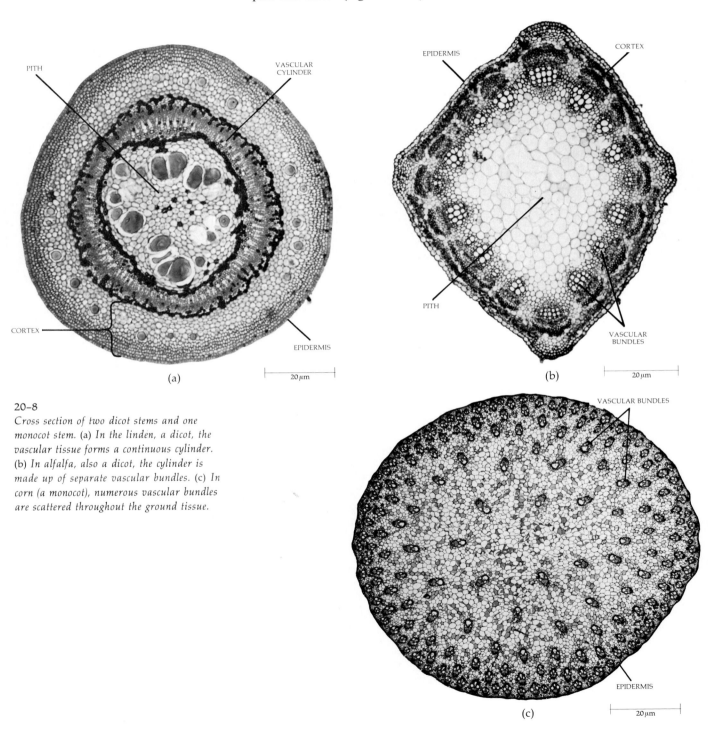

20–8
Cross section of two dicot stems and one monocot stem. (a) In the linden, a dicot, the vascular tissue forms a continuous cylinder. (b) In alfalfa, also a dicot, the cylinder is made up of separate vascular bundles. (c) In corn (a monocot), numerous vascular bundles are scattered throughout the ground tissue.

(a)

(b)

20–9
(a) *The tendrils of grape vines are modified stems.* (b) *Thorns are modified branches. In this photograph of a hawthorn, you can see that the thorns arise in the axils of the leaves.*

SPECIAL ADAPTATIONS OF THE STEM

The stems of some climbing plants coil themselves around the structures on which they are growing. Others produce modified branches in the form of tendrils. (As we saw in the previous chapter, tendrils may also be modified leaves.) The tendrils of English ivy, grape, and Virginia creeper are all modified stems.

Runners, such as those found in most varieties of strawberry, are long, slender stems that grow along the surface of the soil. Rhizomes are underground stems which often, as in the grasses, produce stems bearing leaves and flowers.

ADAPTATIONS FOR FOOD STORAGE

In many plants, roots or stems are adapted for storage. Such plants have often been selected and cultivated to produce vegetables for human consumption. (A vegetable, strictly speaking, is a nonreproductive plant part.) Beets, carrots, and sweet potatoes are modified, edible roots. White potatoes are tubers, which are modified rhizomes (stems).

TRANSPORT SYSTEMS: WATER AND MINERALS

As we have seen, roots, stems, and leaves are interconnected by a continuous system of vascular tissues. In the rest of this chapter, we are going to look more closely at these tissues and their functions, beginning first with the *xylem*, the tissue that conducts water and minerals through the roots to the rest of the plant body.

20–10

Tracheids and vessel elements are the conducting cells of the xylem in angiosperms. (a) Tracheids are a more primitive and less efficient type of conducting cell. Water passing from one tracheid to another passes through pits. Pits are not perforations but areas in which there is no secondary cell wall. Water

moving from one tracheid to another passes through two primary cell walls and the middle lamella. Vessel elements differ from tracheids in that the primary walls of vessel elements are perforated at the end where they are joined with other vessel elements. (b) There may be numerous small perforations

in adjoining walls of vessel elements, or (c) the adjoining walls may break down completely as the cells mature, forming a single opening. Vessel elements are also characteristically shorter and wider than tracheids and their adjoining walls are less oblique.

(a)

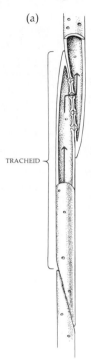

TRACHEID

(b)

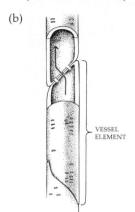

VESSEL
ELEMENT

(c)

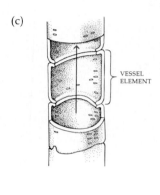

VESSEL
ELEMENT

Xylem is composed of several types of cells: parenchyma, which store food and water; fibers, which provide stiffening and strength; and, in angiosperms, *tracheids* and *vessel elements* (Figures 20–10 and 20–11). Both tracheids and vessel elements have thick secondary walls containing lignin. Also, both are dead at maturity. Tracheids are long, thin cells that overlap one another on their tapered ends. These overlapping surfaces contain thin areas, pits, where no secondary wall has been deposited. Water passes from one tracheid to the next through these pits. Vessels differ from tracheids in that their end walls either contain multiple perforations or are broken down entirely. Thus, the cells form a continuous *vessel*, which is a much more effective conduit than a series of tracheids. (Gymnosperms have only tracheids.)

Water Movement

As we noted in Chapter 19, plants lose water from the stomata during gas exchange. Largely as a consequence of this process (which is the major route of water loss), the quantity of water passing through a plant is enormous—far greater than that used by an animal of comparable weight. An animal requires less water because a great deal of its water recirculates through its body over and over again, in the form, in vertebrates, of blood plasma. In plants, more than 90 percent of the water that enters the roots is given off into the air as water vapor. A single corn plant needs 160 to 200 liters of water—as much as 200 kilograms—from seed to harvest, and an acre of corn requires 2 million liters, or more than half a million gallons, of water a season. The loss of water vapor from the plant body is known as *transpiration.*

Water enters the body of most plants almost entirely through the roots. During periods of rapid transpiration, water may be removed from around the roots so quickly that the soil in the vicinity of the roots becomes depleted; water will then move by diffusion toward the roots through the soil. (For a review of diffusion and osmosis, see pages 32 to 34. An understanding of these processes is necessary for understanding the rest of this chapter.)

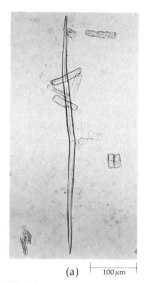   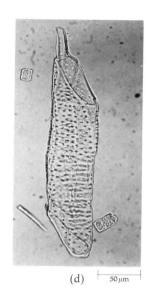 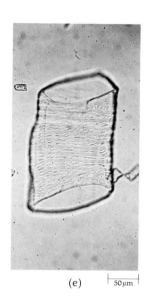

(a) |—100 μm| (b) |—50 μm| (c) |25 μm| (d) |—50 μm| (e) |—50 μm|

20-11

Cell types in the xylem of an oak: (a) a fiber with some parenchyma cells; (b) a tracheid *(the spots in the wall are pits); (c) a narrow vessel element; (d, e) wide vessel elements.* *Both fibers and vessel elements evolved from primitive tracheids.*

20-12

Guttation droplets on the edge of a wild strawberry leaf. Guttation, the loss of liquid water, is a result of root pressure. The water escapes through specialized pores located near the ends of the principal veins of the leaf. Guttation, which is restricted to small plants, usually occurs at night when the air is moist.

Root cells, like other living parts of the plant, contain a higher concentration of salts and minerals than does soil water, and so water from the soil enters the roots by osmosis. Osmotic pressure is sufficient to move water a short distance up the stem; for example, the phenomenon known as guttation (Figure 20–12) is a consequence of osmosis. But how can water reach 20 meters high to the top of an oak tree, travel three stories up the stem of a vine, or move 100 meters up in a tall redwood? Active transport cannot be the answer, because vessel elements are dead cells.

One important clue is the observation that during times when the most rapid transpiration is taking place—which is, of course, when the flow of water up the stem must be the greatest—xylem pressures are characteristically negative (less than atmospheric pressure). The existence of negative pressure can be demonstrated readily. If you peel a piece of bark from a transpiring tree and make a cut in the xylem, no sap runs out. In fact, if you place a drop of water on the cut, the drop will be drawn in.

What is the pulling force? It is not simple suction, as the negative pressure might indicate. Suction simply removes air from a system so that the water (or other liquid) is pushed up by atmospheric pressure. But atmospheric pressure is only enough to raise water (against no resistance) about 11 meters at sea level, and many trees are much taller than 11 meters.

According to the now generally accepted theory, the explanation is to be found not in the properties of the plant but in the remarkable properties of water, to which the plant has become exquisitely adapted. As we pointed out in Chapter 2, in every water molecule, the pair of hydrogen atoms is linked to a single oxygen atom. The hydrogen atoms are also held to the oxygen atoms of the nearest water molecules by hydrogen bonds. This secondary attraction can produce a tensile strength of as much as 140 kilograms per square centimeter (2,000 pounds per square inch) in a thin column of water. In the leaf, water evaporates, molecule by molecule, from the cell walls as a consequence of the lower water potential of the intercellular spaces in the leaves. The water potential of the leaf cell falls, and water from the vessels or tracheids moves, molecule

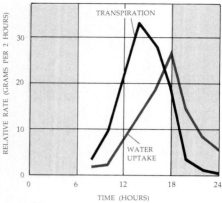

20-13
Measurements in ash trees show that a rise in water uptake follows a rise in transpiration. These data suggest that the loss of water generates forces for its uptake.

by molecule, into the leaf cell. But each molecule in the xylem vessel is linked to other molecules in the vessel, and they, in turn, are linked to others, forming one long, narrow, continuous strand of water reaching right down to a root tip. As the molecule of water moves into the leaf cell, it tugs the next molecule along behind it.

Because the diameter of the vessels is very small and because the water molecules adhere to the walls, even as they are cohering to one another, gas bubbles, which could rupture the column, do not usually form. The pulling action, molecule by molecule, causes the negative pressure observed in the xylem. The technical term for a negative pressure is *tension*, and this theory of water movement is known as the *cohesion-tension theory*.

The energy for the evaporation of water molecules—and thus for the movement of water and minerals through the plant body—is supplied not by the plant but directly by the sun (Figure 20-14).

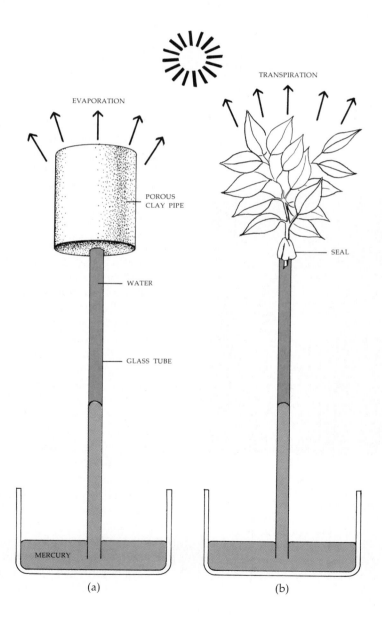

20-14
(a) *A simple model that illustrates the cohesion-tension theory. A piece of porous clay pipe, closed at both ends, is filled with water and attached to the end of a long, narrow glass tube also filled with water. The water-filled tube is placed with its lower end below the surface of a volume of mercury contained in a beaker. As water molecules evaporate from the pores in the pot, they are replaced by water "pulled up" through the narrow glass tube in a continuous column. As the water evaporates, mercury rises in the tube to replace it. (b) Transpiration from plant leaves results in sufficient water loss to create a similar negative pressure.*

"FUNGUS-ROOTS": MYCORRHIZAE

In recent years, it has been found that a particular kind of association between fungi and the roots of plants plays a crucial role in mineral nutrition. If seedlings of many forest trees are grown in nutrient solutions and then transplanted to prairie and other grassland soils, they fail to grow and eventually die from malnutrition, despite the fact that soil analysis shows that there are abundant nutrients in the soil. If a small amount (0.1 percent by volume) of forest soil containing fungi is added to the soil around the roots of the seedlings, however, they will grow promptly and normally. The restoration of normal growth is caused by the functioning of mycorrhizae ("fungus-roots"), which are symbiotic associations between roots and fungi.

In mycorrhizal associations, only the cortex of the root is invaded by the fungus, which sometimes forms a sheath around the root. Roots with mycorrhizae usually lack root hairs, the role of water and mineral uptake from the soil evidently being assumed by the fungus.

The exact relationship between roots and fungi is not known. Apparently the roots secrete sugars, amino acids, and possibly some other organic substances that are used by the fungus.

Although the evidence is just now accumulating, it seems that the chief role of the fungi in such relationships is the disintegration of soil materials and the absorption and transport of the released soil nutrients to the plant.

For example, if the mycorrhizal fungi in nursery soils are killed by fumigation, many plants are unable to utilize phosphorus, even when it is present in the form of easily soluble compounds. In nature, the growth of seedlings often seems to be limited by the amount of phosphorus available in the seed, until a mycorrhizal association can be set up. These symbiotic fungi are so essential to forest trees that one might consider them as a part of the tree's root system rather than as independent inhabitants of the soil.

A study of the fossils of early vascular plants has revealed that mycorrhizae were as frequent then as they are in modern vascular plants. This has led to the very interesting suggestion that the evolution of mycorrhizal associations may have been the critical step allowing the plants to make the transition to the bare and relatively sterile soils of the then unoccupied land.

Effects of mycorrhizae on tree nutrition. Nine-month-old seedlings of white pine were grown for two months in a sterile nutrient solution and then transplanted to prairie soil. The seedlings on the left were transplanted directly. The seedlings on the right were placed for two weeks in forest soil before being transplanted to the prairie.

Minerals

Minerals generally are present in the soil and move into the plant with the water. The need of plants for elements such as nitrogen or magnesium can be deduced from an analysis of the molecules of which plant cells are composed. Other requirements are determined by studying the capacity of plants to grow in distilled water to which small amounts of various minerals are added. This sounds easier than it is. Sometimes, it has been found, a substance—chlorine, for example—is needed in such small amounts that it is almost impossible to set up experimental conditions that exclude it, and so it is difficult to prove that its absence is lethal.

As a result of such studies, six elements that plants require in relatively large amounts (macronutrients) and seven that are needed in smaller quantities (micronutrients, or trace elements) have been identified (see Table 20–1).

The greatest requirement is for nitrogen. Although nitrogen is the most abundant element in air, most plants cannot use gaseous nitrogen and are dependent upon ammonium (NH_4^+) and nitrogen oxides—mostly nitrates (NO_3^-)—from the soil. Plant cells reduce the nitrates to ammonium, and the ammonium is then combined with carbon-containing compounds to form amino acids, nucleotides, chlorophyll, and other nitrogen-containing compounds.

Table 20–1 *Summary of Mineral Elements Required by Plants*

ELEMENT	FORM IN WHICH ABSORBED	APPROXIMATE CONCENTRATION IN WHOLE PLANT (AS % OF DRY WEIGHT)	SOME FUNCTIONS
Macronutrients			
Nitrogen	NO_3^- (or NH_4^+)	1–3%	Component of amino acids, proteins, nucleotides, nucleic acids, chlorophyll, and coenzymes
Potassium	K^+	0.3–6%	Involved in amino acid and protein synthesis; activator of many enzymes
Calcium	Ca^{2+}	0.1–3.5%	Component of cell walls; decreases cell permeability
Phosphorus	H_2PO_4 or HPO_4^{2-}	0.05–1.0%	Component of "high-energy" phosphate compounds (ATP and ADP), nucleic acids, sugars, and phospholipids
Magnesium	Mg^{2+}	0.05–0.7%	Component of the chlorophyll molecule; activator of many enzymes
Sulfur	SO_4^{2-}	0.05–1.5%	Component of some amino acids and of coenzyme A
Micronutrients			
Iron	Fe^{2+}, Fe^{3+}	10–1,500 parts per million (ppm)	Involved in chlorophyll synthesis; component of cytochromes
Chlorine	Cl^-	100–10,000 ppm	Probably essential in photosynthesis in the reactions in which oxygen is produced
Copper	Cu^{2+}	2–75 ppm	Activator of some enzymes
Manganese	Mn^{2+}	5–1,500 ppm	Activator of some enzymes
Zinc	Zn^{2+}	3–150 ppm	Activator of many enzymes
Molybdenum	MoO_4^{2-}	0.1–5.0 ppm	Required for nitrogen metabolism
Boron	BO^{3-} or $B_4O_7^{2-}$ (borate or tetraborate)	2–75 ppm	Influences Ca^{2+} utilization
Elements Essential to Some Plants or Organisms			
Cobalt	Co^{2+}	Trace	Required by nitrogen-fixing microorganisms
Sodium	Na^+	Trace	Required by some desert and salt-marsh species and may be required by all plants that utilize C_4 photosynthesis

20–15

Most plants obtain their nitrogen in the form of inorganic salts taken up with the soil water, but for some unusual species, such as the sundew shown here, insect bodies are the chief nitrogen source. The plant has a rosette of 8 to 15 club-shaped leaves, each about 2 centimeters long and not quite a centimeter wide at the tip. The upper exposed surface of the leaf is covered with hairlike structures, the tentacles, each about 3 millimeters long, enlarged at the tip. A crystal droplet of sticky secretion surrounds each enlargement. These droplets trap insects or other comparably small animals, and then the tentacles secrete enzymes that digest the prey.

Animals can make some amino acids from ammonium and organic carbon compounds, but they lack the biosynthetic pathways to make others—the so-called "essential amino acids." These they must obtain—as they obtain their carbohydrates—either directly or secondarily from plants.

You might anticipate that organisms make use of what is most readily available, as indeed they seem to have done when life originated from elements in the gases of the primitive atmosphere. But the table reveals some findings that you might not expect. Sodium, for instance, which is one of the most abundant of the elements, is not required at all by the great majority of plant species. The fact that plants did not find a use for sodium appears even stranger when you consider that sodium is vital to the function of animals. In the seas, where both plants and animals seem to have had their origins, sodium is the most abundant element and is far more readily available than potassium, which it closely resembles in its essential properties. Similarly, although silicon and aluminum are almost always present in large amounts in soils, few plants require silicon and none requires aluminum. On the other hand, all plants need molybdenum, which is relatively rare.

As we noted earlier, plant cells are able to select particular elements from the soil and reject others because of the Casparian strip, which blocks free passage into the vascular system.

TRANSPORT SYSTEMS: SUGARS AND THE PHLOEM

The photosynthetic cells of the plant, which are most abundant in the leaf, provide the food energy for all of the other cells of the plant. The cells lacking chloroplasts are heterotrophs, just as all animal cells are. Organic compounds, mostly sugars, are conducted to these cells through the *phloem*. (It is customary to think of xylem as transporting water up and phloem as transporting sugars down, but if you think of the various shapes of plants you can see that water must also often be transported laterally, as along a rhizome, or even down, as to the branches of a weeping willow. Conversely, sugars must often go upward, as into a flower or fruit.)

In angiosperms, the conducting cells of the phloem are *sieve-tube elements*. A *sieve tube* is a vertical column of sieve-tube elements joined by their end walls. These end walls, called *sieve plates,* have openings, or pores, leading from one sieve-tube element to the next (Figure 20–16).

20–16

In angiosperms, the conducting elements of the phloem are sieve tubes, made up of individual cells, the sieve-tube elements. These cells, which lack nuclei at maturity, are always found in close association with companion cells, which do have nuclei. Sieve-tube elements are joined by a sieve plate. The light-colored (electron-transparent) material near the pore is callose, a polysaccharide characteristically associated with sieve plates, especially in old or injured cells. Callose plugs the pores when a sieve cell is injured, thus preventing leakage. Immediately to the left of the upper sieve-tube element, identifiable by its dense cytoplasm, is a portion of a companion cell. A parenchyma cell is to the right.

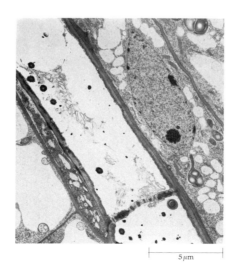

5 μm

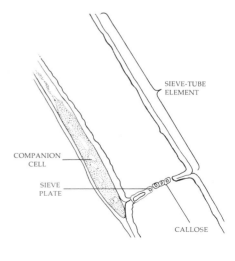

SIEVE-TUBE ELEMENT

COMPANION CELL

SIEVE PLATE

CALLOSE

Sieve-tube elements are alive at maturity; their nuclei and many of their organelles disintegrate, however, as the sieve-tube elements mature. The remaining cytoplasm, consisting largely of a watery fluid, forms a film along the longitudinal surface of the cell. Sieve-tube elements are always associated with *companion cells*, which are thought to provide nuclear functions and energy for the sieve tubes. A sieve tube can function only if its cell membrane is intact, and it is likely that the companion cell helps to maintain the membrane. Phloem also contains parenchyma cells, which store food and water, and often supporting fibers.

Translocation

The process by which the products of photosynthesis are transported to other tissues is known as *translocation*. The way in which translocation takes place is still a matter of controversy. It is generally agreed, however, that the sieve-tube elements, and perhaps parenchyma and companion cells as well, must be alive for translocation to take place. It has also been established that substances can move in opposite directions at the same time through the phloem, although it is not clear whether they move both ways at the same time in the same sieve tube.

Translocation does not take place by simple diffusion; too much sugar is moved too fast. Rates of up to 100 centimeters per hour are common, whereas diffusion from a 10 percent solution would take place at less than 2 centimeters an hour.

Some investigators hypothesize that translocation takes place by *cyclosis*, which is a regular, circular streaming of the cytoplasm observed in many plant cells. One attractive feature of this hypothesis is that it explains how substances can move in two directions at the same time. However, cyclosis has never been observed in mature sieve-tube elements. Also, cyclosis, like diffusion, is too slow a process to account for the known rate of movement.

One investigative technique that has proved useful in these studies is the exposure of plants to radioactive carbon dioxide ($^{14}CO_2$). The sugars made from this carbon dioxide are thus labeled with a radioactive tag, and so their passage through the plant can be traced. Another involves the use of aphids for taking samples of sieve-tube cytoplasm (Figure 20–17).

20–17

Assistance by aphids. (a) Aphids are very small insects that feed on plants, sucking out their juices; you probably have seen them on rose bushes. The aphid drives its sharp mouthparts, or stylets, like a hypodermic needle between the epidermal cells and then, (b) as the micrograph reveals, taps the contents of a single sieve-tube element. If the aphid is anesthetized, it is possible to sever the stylet and leave it undisturbed in the cell. The fluid will continue to exude through the stylet from the sieve tube for several days, and pure samples of the fluid flowing through the sieve tube can be collected for analysis without damaging the sieve tube or interfering with its function.

(a) 1 mm

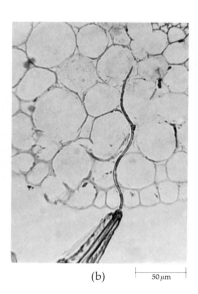

(b) 50 μm

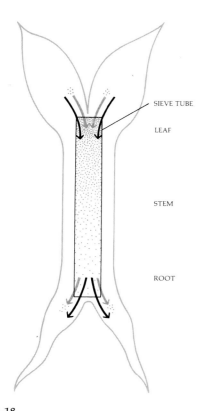

20-18

Model of pressure-flow mechanism. Sugar (colored arrows) enters a sieve tube of a leaf by active transport. As a consequence of the increased concentration of sugar, water (black arrows) enters the sieve tube by osmosis. Sugar, required by the root cells, is removed from the phloem by active transport, and the sugar concentration of the sieve tube falls. As a consequence of the lowered sugar concentration, water leaves the sieve tube by osmosis. Because of the active secretion of sugar into the sieve tube at one end and the active absorption of sugar from the sieve tube at the other end, a flow of sugar solution takes place along the tube.

SIEVE TUBE

LEAF

STEM

ROOT

The Pressure-Flow Mechanism

The most widely accepted hypothesis for translocation is the *pressure-flow mechanism*. According to this model, sugars move through the sieve tube along concentration gradients.

Sugar manufactured in the photosynthetic cells of the leaf is actively secreted—"pumped"—into the sieve tubes of the veins by neighboring cells, such as companion cells. This decreases the water potential in the sieve tube and causes water to move into the sieve tube from the xylem by osmosis. At the root tips or other nutrient-requiring tissue, the sugar is removed from the sieve tube, again as the result of the expenditure of energy by neighboring cells (probably the companion cells). The water molecules follow the sugar molecules out, again by osmosis. Thus the water flows in at one end of the sieve tube and out at the other, and between these two points, the water and its solutes, including sugar, move passively by bulk flow (Figure 20–18).

SUMMARY

Roots are organs specialized for anchoring the plant in the soil, for absorbing water and minerals from the soil, and for transporting them to other parts of the body. Anatomically, the root consists of three concentric layers: the epidermis, the cortex, and the vascular cylinder. The epidermis is the protective and absorbing tissue. Much of the absorption takes place through fine extensions of epidermal cells known as root hairs. The cortex consists largely of parenchyma cells adapted to the storage of water and sugars. The innermost layer of the cortex is the endodermis. Endodermal cells are characterized by Casparian strips, which inhibit the movement of water and solutes into the vascular tissues. Because of the Casparian strips, plants are able to regulate their uptake of minerals. The vascular cylinder consists of xylem and phloem, surrounded by a layer of pericycle, from which branch roots arise.

Stems hold the photosynthetic organs—the leaves—up to the light, conduct water to the photosynthetic cells, and transport sugars from them. The outer surface of a green stem is made up of epidermal cells. The bulk of the stem is ground tissue, which may be divided into an outer cylinder (the cortex) and an inner core (the pith). The ground tissue is largely composed of parenchyma cells but also may contain collenchyma cells and sclerenchyma (fibers and sclereids).

The tissue through which water is transported is the xylem. In angiosperms, the xylem is composed of parenchyma cells, fibers, vessel elements, and tracheids. Water transport takes place through the vessel elements, which form vessels, and the tracheids. Both of these cell types are dead at maturity. According to the cohesion-tension theory, water moves through vessels and tracheids under negative (less than atmospheric) pressure. Because the molecules of water cling together (cohere), a continuous column of water molecules is pulled, molecule by molecule, from the root as water evaporates into the air spaces of the leaf. The loss of water vapor from the leaf and other plant parts is known as transpiration.

Minerals, including nitrogen (in the form of nitrate and ammonium ions), enter the plant body in solution in the water from the soil.

Sugars move from the photosynthetic cells of the plant to other cells of the plant body by way of the phloem. In angiosperms, phloem consists of sieve-tube elements, companion cells, parenchyma cells, and sometimes fibers. The conducting cells are the sieve-tube elements and, unlike the conducting cells of the xylem, these cells are alive at maturity. According to the pressure-flow hypothesis, sugars are pumped into the sieve tubes in the leaf by active transport and are pumped out of them in various parts of the plant body where they are needed for energy. These pumping actions create a difference in water potential along the sieve tube, and thus water and the sugars dissolved in it move by bulk flow along the sieve tube.

QUESTIONS

1. Define the following terms: parenchyma, xylem, collenchyma, pericycle, phloem, rhizome, fiber, endodermis, tracheid.

2. Sketch a cross section of a root, identifying the tissue layers.

3. Sketch a cross section of a dicot stem, identifying epidermis, cortex, phloem, xylem, and pith. Compare your sketch with the micrographs on page 272.

4. Diagram the movement of water in (a) tracheids, (b) a vessel, (c) a sieve tube.

5. Which of the properties of water discussed in Chapter 2 are important to phenomena described in this chapter? How have plants adapted to these properties?

6. Most plants cannot live in areas in which there is a high salt concentration, such as salt marshes. Explain.

7. Consider a tree transpiring most rapidly at midday and an investigator with a sensitive instrument for measuring changes in the diameter of the trunk. If water is pulled up from the top (cohesion-tension theory), what changes in diameter should be observed from night to day? (The change was, in fact, one of the early data in support of the theory.)

CHAPTER 21

CHAPTER 21

Reproduction and Development

With plants, as with animals, a new individual comes into being with the fusion of sperm and egg. In the angiosperms, this process comes about as a result of an organ which, by happy coincidence, is extraordinarily beautiful to the human eye: the flower. Unlike the reproductive organs of animals, which are permanent structures that develop in the embryo, flowers are transitory, developing and being discarded seasonally. Each consists of four sets of floral appendages, which grow in spirals or whorls. Each of these floral parts, evolutionarily speaking, is a modified leaf.

The outermost parts of the flower are the *sepals*, which are commonly green and obviously leaflike in structure. The sepals, collectively known as the *calyx*, enclose and protect the flower bud. Next are the *petals*, collectively called the *corolla*; these are also usually leaf-shaped but are often brightly colored. They advertise the presence of the flower among the green leaves, attracting insects or other animals that visit flowers for their nectar—a sugary liquid secreted that attracts them—or for other edible substances. These animals, as they forage for food, carry pollen from flower to flower.

21-1
Primitive flowers are believed to have been radially symmetrical, with numerous separate floral parts, such as this magnolia.

Structure of a flower. A flower that possesses both stamens and carpel, such as this flower, is known as a perfect flower. Pollen grains have been deposited on the sticky surface of the stigma and one is growing down the style to an ovule.

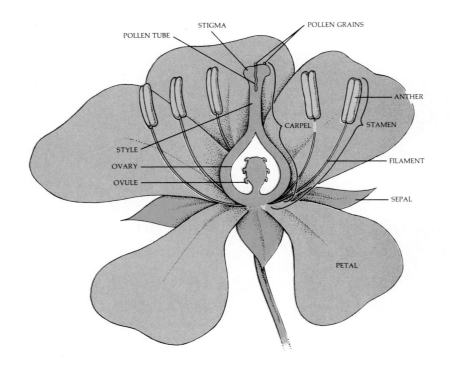

21–3

Corn is monoecious; that is, it has separate male and female flowers but both are borne on the same plant. The tassels are male (pollen-producing) flowers. Each thread of corn silk is the combined stigma and style of a female flower.

Within the corolla are the *stamens*. Each stamen consists of a single elongated stalk, the *filament*, and at the end of the filament, the *anther*. Then pollen grains are male gametophytes.* They are released from the anther often in large numbers, usually through narrow slits or pores.

The centermost appendages of the flower are the *carpels*, which contain the female gametophytes. Typically, a carpel consists of a *stigma*, which is a sticky surface specialized to receive the pollen; a slender stalk, the *style*, down which the pollen tube grows; and a hollow base, the *ovary*. Within the ovary are the *ovules*. Each ovule encloses a female gametophyte with a single egg cell. When the egg cell is fertilized, the ovule develops into a seed.

In some species, flowers are either male (staminate) or female (carpellate). Male and female flowers may be present on the same plant, as in corn, squash, oaks, and birches, or on different plants, such as the tree of heaven (*Ailanthus*), holly, the date palm, and the American mistletoe. Species in which separate male and female flowers are borne on the same plant are said to be monoecious ("in one house"), and species in which the male and female flowers are on separate plants are known as dioecious ("in two houses").

EVOLUTION OF THE FLOWER

As we noted in Chapter 18, the angiosperms are believed to have evolved from the gymnosperms. Gymnosperms, like angiosperms, are vascular plants, although their vascular systems are less highly developed. However, unlike the angiosperms, they do not produce a flower and their seeds are not enclosed in fruit (hence the name "naked seed").

The gymnosperms from which the angiosperms evolved were probably wind-pollinated, as are modern gymnosperms. And as is the case with the

* For a review of the concept of alternation of generations, see page 122.

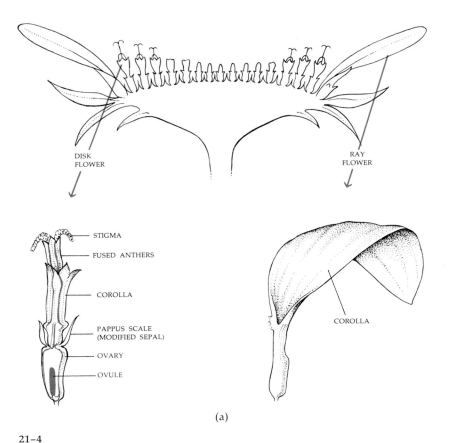

DISK FLOWER

RAY FLOWER

STIGMA

FUSED ANTHERS

COROLLA

PAPPUS SCALE (MODIFIED SEPAL)

OVARY

OVULE

COROLLA

(a)

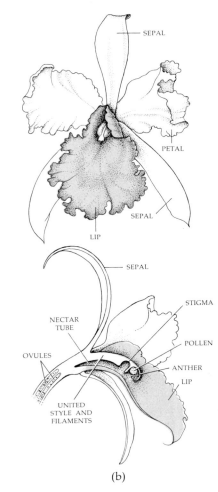

SEPAL

PETAL

SEPAL

LIP

SEPAL

STIGMA

NECTAR TUBE

POLLEN

OVULES

ANTHER

LIP

UNITED STYLE AND FILAMENTS

(b)

21–4

Diagrams and examples of two large families of flowers, composites (family Asteraceae) and orchids (family Orchidaceae).
(a) *The organization of the head of a composite. The individual flowers are subordinated to the overall effect of the head, which acts as a large single flower in attracting insects. The internal structure of the disk flower is indicated in color.* (b) *The parts of an orchid flower. The lip is a modified petal that serves as a landing platform for insects.* (c) *Ox-eye daisies* (Chrysanthemum leucanthemum), *representative composites. The ray flower is often sterile.*
(d) Epipactus gigantea, *stream orchids.*

The Orchidaceae, with about 20,000 species, is the largest family of flowering plants; composites are the second largest family, with some 13,000 species. Orchids are monocots; composites are dicots.

(c)

(d)

285 REPRODUCTION AND DEVELOPMENT

VEGETATIVE REPRODUCTION

Unlike the higher animals, among which reproduction is almost exclusively sexual, many plants reproduce both sexually and asexually. In fact, asexual reproduction is often referred to as vegetative reproduction, even when it occurs among animals. Organisms produced by asexual reproduction are genetically identical to their single parent, whereas organisms produced by sexual reproduction, which involves meiosis and fertilization, are different from both parents.

There are many forms of asexual reproduction among plants. One of the most familiar occurs by means of horizontal stems growing either above ground (runners) or below ground (rhizomes). Strawberries are a common example of plants that propagate by runners, as are spider plants (walking anthericum), commonly grown as hanging plants. Plants that reproduce by rhizomes include potatoes, many flowering garden perennials, such as lilies-of-the-valley and irises, and the sod-forming grasses of lawns and pastures. Both runners and rhizomes develop adventitious roots.

Many members of the lily family, which includes onions and tulips as well as lilies, reproduce asexually, by bulbs. Some species of plants with arching stems, such as raspberries, develop new roots from stem tips that touch the soil, and new plants may form if the stem is subsequently broken, separating it from the parent plant. The leaves of some plants, such as African violets, develop adventitious roots when they are detached from the parent plant and may give rise to new individuals in that way. Some species of Kalanchoe produce plantlets in the margins of leaves which later drop to the ground and develop into separate plants. If a dandelion is injured or broken near its upper portion, a callus forms that plugs the wound, and eventually two to five new plants grow from this callus tissue. Thus both suburban householders and grazing animals have highly beneficial effects on the dandelion population.

Tiny plantlets growing along the leaf margins of Kalanchoe.

The capacity of many species of plants to reproduce asexually has been exploited by man in developing domestic varieties of plants for food or ornamental use. Such plants are, of course, genetically identical to the parent stock, and so vegetative reproduction is a way of preserving uniformity. Many plants are reproduced by stem cuttings, which simply involves sticking young stems in the ground and protecting them from drying air until adventitious roots appear. Rooting can often be facilitated by hormone treatment (see Chapter 22). Another artificial form of plant propagation is grafting, in which a stem cutting is attached to the main stem of a rooted, woody plant. Most fruit trees and roses are propagated in this way.

Many economically important plants are sterile and can only be propagated vegetatively; these include pineapples, bananas, seedless grapes, navel oranges, and numerous ornamental plants.

modern gymnosperms, the female gametophytes probably exuded droplets of sticky sap in which pollen grains were caught and drawn into the ovule. Insects, probably beetles, feeding on the leaves and other plant parts must have come across the protein-rich pollen grains and the sticky, sugary droplets. As they began returning regularly to these newfound food supplies, they inadvertently carried pollen from plant to plant.

Beetle pollination must have been more efficient than wind pollination for some species because, clearly, selection began to favor plants with insect pollinators. The more attractive the plants were to the beetles, the more frequently they would be visited and the more seeds they would produce. Any chance variations that made the visits more frequent or that made pollination more efficient thus offered immediate advantages; more seeds would be formed, and more offspring would survive. Plants began to produce nectar and nectaries, structures which hold the sweet syrup. White or brightly colored flowers evolved that called attention to the nectar and other food supplies. The carpel,

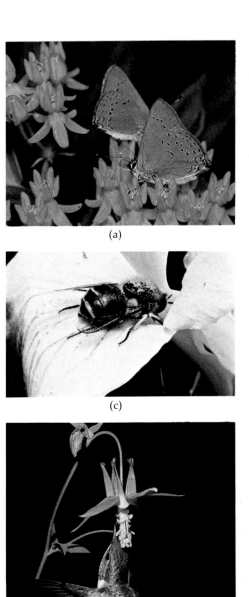

(a)

(b)

(c)

(d)

(e)

(f)

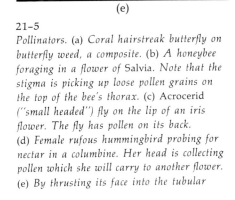

(g)

21-5

Pollinators. (a) Coral hairstreak butterfly on butterfly weed, a composite. (b) A honeybee foraging in a flower of Salvia. Note that the stigma is picking up loose pollen grains on the top of the bee's thorax. (c) Acrocerid ("small headed") fly on the lip of an iris flower. The fly has pollen on its back. (d) Female rufous hummingbird probing for nectar in a columbine. Her head is collecting pollen which she will carry to another flower. (e) By thrusting its face into the tubular corolla of an organ-pipe cactus flower, a Leptonycteris bat is able to lap up nectar with its long, bristly tongue. Pollen grains clinging to its face and neck are transferred to the next flower visited by the bat. (f) Long-horn beetle eating thistle pollen. (g) Syrphid fly receiving pollen from the anther (which is touching its back) of an Epipactus gigantea orchid. This species is the only pollinator for this flower.

Pollen grains. The walls of the pollen grain protect the male gametophyte on the journey between the anther and the stigma. These outer surfaces, which are remarkably tough and resistant, are often elaborately sculptured. As you can see, the pollen grains of different species are distinctly different: (a) morning glory (spiny pollen grains such as these are common among composites) and (b) a horse chestnut (each grain contains three slitlike pores, a characteristic of many dicots).

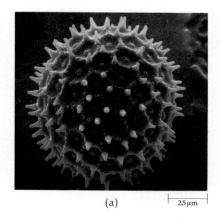

(a)   25 μm

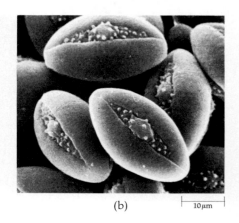

(b)   10 μm

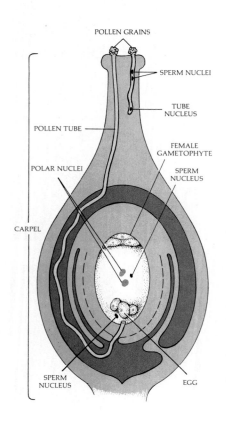

21–7

Fertilization in angiosperms. The pollen tube of the male gametophyte, or pollen grain, grows down through the style and enters the ovule, in which the female gametophyte has developed to a seven-cell stage. One of the sperm nuclei unites with the egg cell, forming the zygote. The other sperm nucleus fuses with the two polar nuclei that are present in a single large cell (which in the drawing fills most of the ovule). From the resulting triploid (3n) cell, the endosperm will develop. The carpel shown here contains a single ovule.

originally a leaf-shaped structure, became folded on itself, enclosing and protecting the ovule from hungry pollinators. By the beginning of the Cenozoic era, some 65 million years ago, the first bees, wasps, butterflies, and moths had appeared. These are long-tongued insects for which flowers are often the only source of nutrition. From this time onward, flowers and insects had a profound influence on one another's history, each shaping the other as they evolved together.

A flower that attracts only a few kinds of animal visitors and attracts them regularly has an advantage over flowers visited by more promiscuous pollinators: much less of its pollen is liable to be lost on a plant of another species. Similarly, it is advantageous to the insect to have a private food supply, relatively inaccessible to others. Most of the distinctive features of modern flowers are special adaptations that encourage constancy of particular pollinators. The varied colors and odors are "brand names" for guiding pollinators. The diverse shapes such as deep nectaries and complex landing platforms that are found, for example, in orchids, snapdragons, and irises, represent ways of excluding indiscriminate pollinators.

POLLINATION

By the time a pollen grain is released from its parent flower, it characteristically consists of three haploid nuclei (two sperm nuclei and a so-called tube nucleus), a small amount of dense cytoplasm, and a tough wall.

Pollen is commonly produced in great quantities; the probability of any particular pollen grain reaching the stigma of an appropriate flower is very small. The pollen grain contains its own nutrients and has so tough an outer coating that intact grains have been found in peat bogs thousands of years old.

In lower plants, you will recall, there is a distinct cycle of alternation of generations in which the sporophyte produces spores that produce gametophytes and the gametophytes produce gametes, which unite and so begin a new sporophyte generation. In the course of plant evolution, the gametophyte stage has been steadily reduced and, in the angiosperms, all that remains of the male gametophyte is the tough, tiny pollen grain. The sperm cells are the gametes.

Once on the stigma, the pollen grain germinates, and a pollen tube grows down through the style into an ovule. The ovule holds the female gametophyte, which also has become reduced in size in the course of evolution. Typically, it consists of seven cells, with a total of eight haploid nuclei. One of the smaller cells, containing a single haploid nucleus, is the egg.

One sperm nucleus moves down the pollen tube and unites with the egg. This fertilized cell, the zygote, develops into the embryo plant. The second sperm nucleus unites with the two polar nuclei (so called because they move to the center from each end, or pole, of the female gametophyte). From the 3*n* cell produced, a specialized tissue called the *endosperm* develops, which completely surrounds and nourishes the embryo. Thus, as a result of pollination, two extraordinary events take place. One is double fertilization, which is the fusion of egg and sperm plus the simultaneous fusion of the second male gamete with the polar nuclei. The other is triple fusion, the coming together of three nuclei to make a 3*n* tissue. These occur in all the natural world only among the flowering plants.

THE EMBRYO

The zygote (the fertilized egg cell) divides mitotically, and as the embryo grows, its cells begin to *differentiate*—become different from one another—and the embryo begins to take on a characteristic shape, a process known as *morphogenesis*.

Growth Areas

In the earliest stages of embryonic growth, cell division takes place throughout the body of the infant plant. As the embryo grows older, however, the addition of new cells becomes gradually restricted to certain parts of the plant body: the *apical meristems* of the root and the shoot. These are composed of cells that are physiologically young and able to continue to divide. During the rest of the life of the plant, all the primary growth—which chiefly involves the elongation of the plant body—originates in these meristems.

The existence of such meristematic areas, which add to the plant body throughout the life of the plant, is one of the principal differences between plants and animals. Plants continue to grow during their entire life span. Growth in plants is the counterpart, to some extent, of mobility in animals. By growth, for example, a plant modifies its relationship with the environment, turning toward the light and extending its roots, and by producing flowers, it seeks a mate.

The Seed and the Fruit

As the embryo develops, the petals and stamens of the parent flower fall away, and the wall of the ovary develops into the fruit. The ovule (which comes from the maternal sporophyte) becomes the seed, with the 3*n* endosperm and the young sporophyte growing within it. In a peach, for instance, which contains only one ovule in each ovary, the skin, the succulent edible portion of the fruit, and the stone are three distinctive layers of the wall of the mature ovary (the base of the carpel). The almond-shaped structure within the stone is the seed. In a pea, the pod is the mature ovary wall and the peas are the seeds (the mature ovules and their contents). A raspberry is an aggregate of many fruits from a single flower, each fruit containing a single seed and each formed from a separate carpel.

When the seed matures, the embryo enters a period of dormancy. The seed coat, which develops from the outer layers of the ovule, thickens and hardens, and the seed, which includes the embryo and its stored food, falls from the parent plant. Two of the many mechanisms for seed dispersal are shown on page 227.

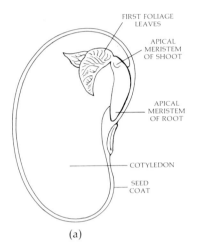

FIRST FOLIAGE LEAVES

APICAL MERISTEM OF SHOOT

APICAL MERISTEM OF ROOT

COTYLEDON

SEED COAT

(a)

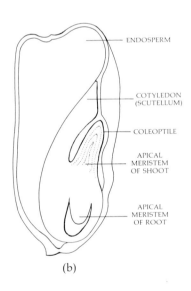

ENDOSPERM

COTYLEDON (SCUTELLUM)

COLEOPTILE

APICAL MERISTEM OF SHOOT

APICAL MERISTEM OF ROOT

(b)

21-8

Seeds. (a) In dicots such as the common bean, the endosperm is digested as the embryo grows and the food reserve is stored in the fleshy cotyledons. (b) In corn and other monocots, the single cotyledon, known as the scutellum, absorbs food reserves from the endosperm.

Grains are the small, one-seeded fruits of grasses. Because they are dry, they can be stored by man as food. The collecting and storing of grains from wild grasses was believed to be an important impetus to the agricultural revolution of some 10,000 years ago (see page 579). Modern man is heavily dependent on cultivated wheat, rice, corn, rye, and other grains. In many countries, they constitute the chief food resource.

The fruit of wheat, sometimes known as the kernel, is made up of the embryo, the endosperm, and the surrounding seed coats. More than 80 percent of the bulk of the wheat kernel and 70 to 75 percent of its protein are in the endosperm. White flour is made from the endosperm. Wheat germ, the embryo, forms about 3 percent of the kernel. It is usually removed as wheat is processed because it contains oil, which makes the grain more likely to spoil. Bran is the seed coat plus the aleurone layer (outer part of endosperm); it constitutes about 14 percent of the kernel. The bran is also removed when wheat is milled to make white flour. Actually, the bran somewhat decreases the caloric value of the wheat kernel. Because it is mostly cellulose, bran cannot be digested by man and tends to speed the passage of food through the intestinal tract, resulting in decreased absorption. The wheat germ and the bran, which contain most of the vitamins, are sometimes used for human consumption but more often are fed to livestock.

Wheat is about 9 to 14 percent protein, most of which, as we noted, is contained in the endosperm. Its protein value is diminished, however, by its lack of certain essential amino acids, notably lysine.

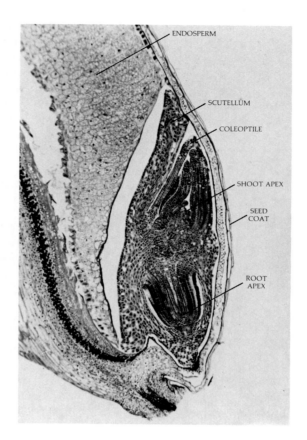

Longitudinal section of a wheat kernel.

SEED DORMANCY

The seeds of most wild plants undergo a period of dormancy before they will germinate (begin to grow). This ensures that the seed will "wait" at least until the next favorable growth period. In a few cases, seeds have remained dormant and yet viable—with the embryo in a state of suspended animation—for hundreds of years.

The seed coat apparently plays a major role in maintaining dormancy. In some species, the seed coat seems to act primarily as a mechanical barrier, preventing the entry of water and gases, without which growth is not possible. In these cases, growth is promoted by the seed coat being worn away in various ways—such as being washed by rainfall, abraded by sand or soil, burned away by a forest fire, or partially digested as it passes through the digestive tract of a bird or other animal. In other species, dormancy seems to be maintained chiefly by the presence of chemical inhibitors in the seed coat or in the embryo. These inhibitors undergo chemical changes in response to various environmental factors, such as light or prolonged cold or a sudden rise in temperature, which degrade them or neutralize their effects, or, if present in the seed coat, they may be washed or eroded away. Eventually, the embryo is released.

PRIMARY GROWTH

During dormancy, the seed contains very little moisture (only about 5 to 10 percent of its total weight). Dormancy ends with a massive entry of water (imbibition) into the seed. The seed coat ruptures and the embryo sporophyte emerges.

Primary growth begins immediately. It involves the formation and elongation of stems and roots, the expansion of leaves, and the differentiation of the conduction tissues and other specialized tissues of the young shoot and root. All primary growth originates in the apical meristems of the shoot and root.

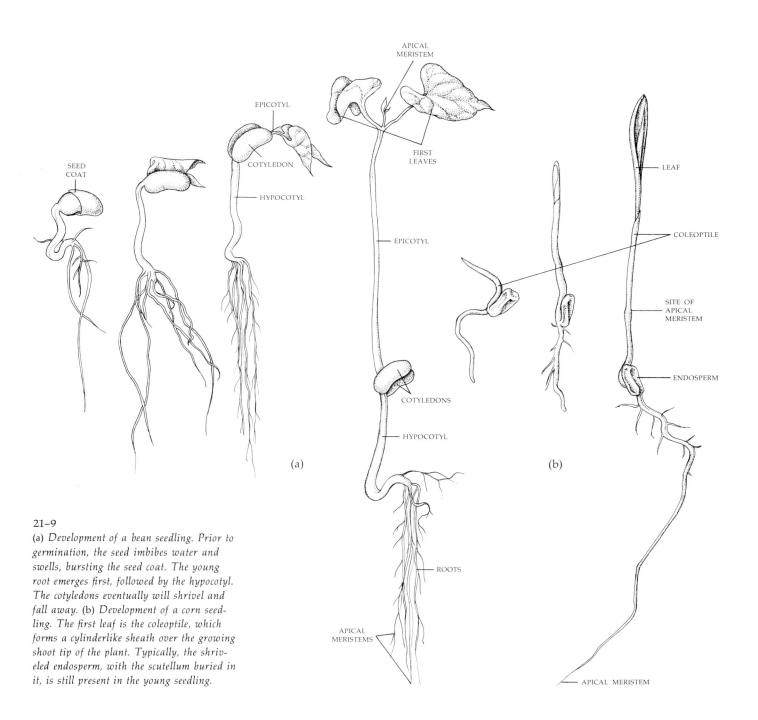

21-9

(a) Development of a bean seedling. Prior to germination, the seed imbibes water and swells, bursting the seed coat. The young root emerges first, followed by the hypocotyl. The cotyledons eventually will shrivel and fall away. (b) Development of a corn seedling. The first leaf is the coleoptile, which forms a cylinderlike sheath over the growing shoot tip of the plant. Typically, the shriveled endosperm, with the scutellum buried in it, is still present in the young seedling.

21–10

The developmental regions of a dicot root. New cells are produced by the division of cells within the meristem. The cells above the meristem undergo a characteristic series of changes as the distance increases between them and the root tip. First, there is a maximum rate of cell division, followed by cell elongation, which accounts for most of the lengthening of the root. As the cells elongate, they differentiate into various specialized tissues of the root. The protoderm becomes the epidermis, the ground meristem becomes the cortex, including the endodermis, and the procambium becomes the pericycle, xylem, and phloem. Some of the cells produced in the apical meristem form the protective root cap.

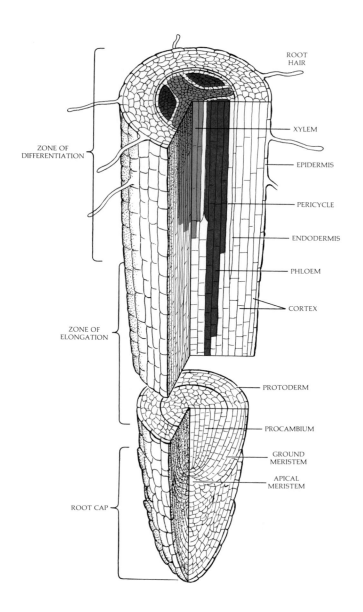

The Root

The first part of the embryo to break through the seed coat, in nearly all seed plants, is the embryonic root. Figure 21–10 diagrams the growing zones of the root of a dicot. At the very tip is the root cap, which protects the apical meristem as the root tip is pushed through the soil. The cells of the root cap wear away and are constantly replaced by new cells from the meristem. The meristem is the source of new cells. All the other cells in the root are the progeny of these relatively few meristematic cells, which divide continuously. Some of the daughter cells remain in the meristem. Others become cells of the root cap and others form the complex tissues of the root. Just above the point where cell division ceases, the cells gradually elongate, growing to 10 or more times their previous length, often within the span of a few hours.

As the cells elongate, they begin to differentiate, forming first the conducting cells of the phloem and then of the xylem. Farther up the root, the endodermis takes shape. Within the endodermis, the pericycle forms. At about this same level of the root, the epidermal cells differentiate and begin to extend root hairs into the crevices between the soil grains.

This same basic pattern of growth is seen in the first root of a seedling and in the growing root tips of a tall tree.

The Shoot

The shoot includes all the above-ground parts of the plant. The organization of the developing shoot tip is somewhat similar to that seen in the root: first, a zone in which most of the cell division takes place; next, a zone of cell elongation; and finally, a zone of differentiation. These zones are not as distinct in the shoot as they are in the root, however, because of the regular occurrence of nodes and their appendages.

As in the root, the outermost layer of cells develops into the epidermis. In the shoot, these cells have a waxy cuticle. Other cells differentiate to form ground tissue and primary vascular tissues. The pattern of development is more complicated, however, than in the root tip, since the apical meristem of the shoot is the source of tissues that give rise to new leaves, branches, and flowers.

Figures 21–11 shows the shoot tip of a lilac. Here you can see the apical meristem, which is very small, and the beginnings—primordia—of leaves. As you can see, leaves are formed in an orderly sequence at the shoot tip. The vascular tissue begins to differentiate in the leaf primordia, eventually becoming part of the general vascular system that connects the plant from root to leaf tip. As the internodes elongate, the young leaves become separated so that the leaf pairs clustered so tightly together around the apex in Figure 21–11 will eventually be spaced out along the stem of the plant.

As the nodes are separated by elongation of the internodes, small apical meristems (buds) are left in the axils of the leaves. These buds may remain dormant or they may give rise to branches, runners, rhizomes, tubers, or flowers.

21–11

Longitudinal section of the shoot tip of a lilac (Syringa vulgaris). At the tip of the stem is the apical meristem, the central zone of cell division. The leaf primordia, from which new leaves will form, originate along the sides of the shoot apex. In this picture, two leaf primordia can be seen arising from the apical meristem. Successively older leaf primordia had formed on both sides. The developing stem below the apical meristem and the lateral buds on either side are also regions of active cell division.

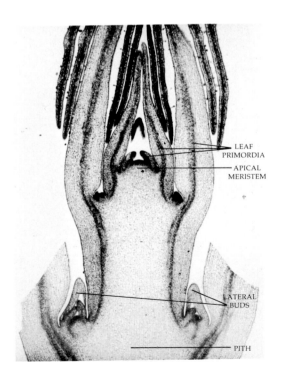

LEAF PRIMORDIA

APICAL MERISTEM

LATERAL BUDS

PITH

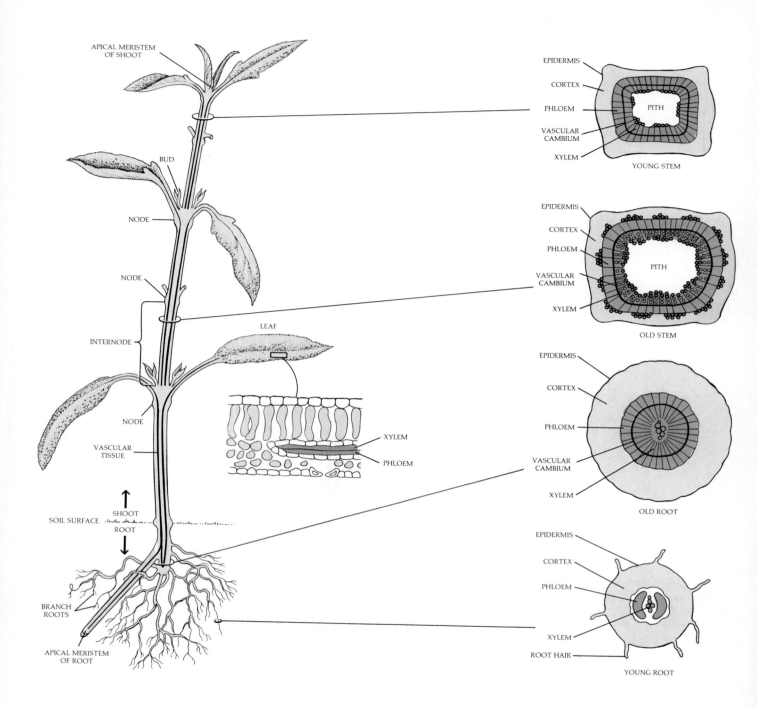

APICAL MERISTEM
OF SHOOT

BUD

NODE

NODE

INTERNODE

LEAF

NODE

VASCULAR
TISSUE

SHOOT

SOIL SURFACE

ROOT

BRANCH
ROOTS

APICAL MERISTEM
OF ROOT

XYLEM

PHLOEM

EPIDERMIS
CORTEX
PHLOEM
PITH
VASCULAR
CAMBIUM
XYLEM
YOUNG STEM

EPIDERMIS
CORTEX
PHLOEM
PITH
VASCULAR
CAMBIUM
XYLEM
OLD STEM

EPIDERMIS
CORTEX
PHLOEM
VASCULAR
CAMBIUM
XYLEM
OLD ROOT

EPIDERMIS
CORTEX
PHLOEM
XYLEM
ROOT HAIR
YOUNG ROOT

21–12
Diagram of a sage plant (Salvia officinalis),
showing the principal organs and tissues of
the plant body.

SECONDARY GROWTH

Secondary growth is the process by which the thickness of trunks, stems, branches, and roots is increased in those parts of the plant body where primary growth is no longer taking place (Figure 21–13). Secondary growth is most evident in woody plants. Unlike the primary tissues, the tissues produced by secondary growth are not derived from the apical meristems. They are the result of the production of new cells by the vascular cambium and cork cambium, which are called *lateral meristems.*

The *vascular cambium* is a thin, cylindrical sheath of cells between the xylem and the phloem. In plants with secondary growth, the cambium cells divide continually during the growing season, adding new xylem cells—that is, secondary xylem—on the outside of the primary xylem, and secondary phloem on the inside of the primary phloem. Some daughter cells remain as a cylinder of undifferentiated cambium. By this continuous formation of layers of xylem and phloem, tree trunks increase their diameter as they increase their height.

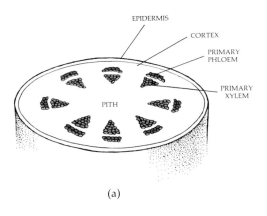

(a)

21–13

(a) *Stem of a dicot before the onset of secondary growth.* (b) *Beginnings of secondary growth. Secondary xylem and secondary phloem are produced by the vascular cambium, a meristematic tissue formed late in primary growth. As the trunk increases in diameter, the epidermis is stretched and torn, which apparently triggers the formation of the cork cambium, from which cork is formed, replacing the epidermis.* (c) *Cross section of a three-year-old stem, showing annual growth rings. Rays are strands of living cells that transport nutrients and water laterally (across the trunk). On the perimeter of the outermost growth ring of xylem is the vascular cambium, encircled by a band of secondary phloem. The primary phloem and also the cortex will eventually disappear. In an older stem, the thin cylinder of active secondary phloem is immediately adjacent to the vascular cambium.*

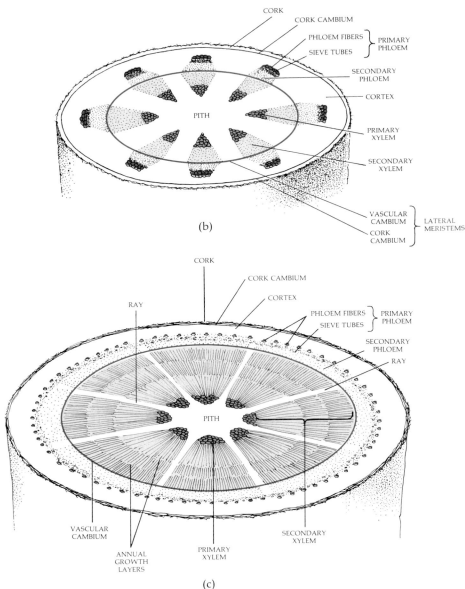

(b)

(c)

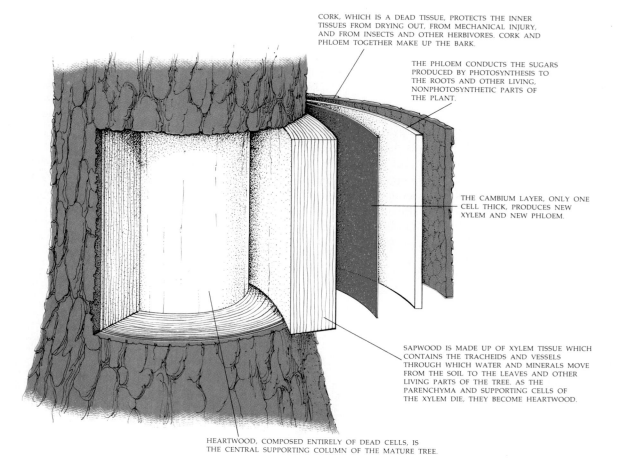

CORK, WHICH IS A DEAD TISSUE, PROTECTS THE INNER TISSUES FROM DRYING OUT, FROM MECHANICAL INJURY, AND FROM INSECTS AND OTHER HERBIVORES. CORK AND PHLOEM TOGETHER MAKE UP THE BARK.

THE PHLOEM CONDUCTS THE SUGARS PRODUCED BY PHOTOSYNTHESIS TO THE ROOTS AND OTHER LIVING, NONPHOTOSYNTHETIC PARTS OF THE PLANT.

THE CAMBIUM LAYER, ONLY ONE CELL THICK, PRODUCES NEW XYLEM AND NEW PHLOEM.

SAPWOOD IS MADE UP OF XYLEM TISSUE WHICH CONTAINS THE TRACHEIDS AND VESSELS THROUGH WHICH WATER AND MINERALS MOVE FROM THE SOIL TO THE LEAVES AND OTHER LIVING PARTS OF THE TREE. AS THE PARENCHYMA AND SUPPORTING CELLS OF THE XYLEM DIE, THEY BECOME HEARTWOOD.

HEARTWOOD, COMPOSED ENTIRELY OF DEAD CELLS, IS THE CENTRAL SUPPORTING COLUMN OF THE MATURE TREE.

21–14

A tree trunk showing the relationships of the successive concentric layers. The heartwood is composed entirely of dead xylem cells. The cortex, which is outside the phloem in a green stem, is sloughed off in the process of secondary growth.

As the tree grows older, the living cells of the xylem in the center of the trunk die, and the vessels cease to function. This nonliving wood is called heartwood, as distinct from sapwood, which consists of living cells and functional vessels.

As the girth of stems and roots increases by secondary growth, the epidermis becomes stretched and torn. In response to this tearing process, a new type of cambium, known as <u>cork cambium</u>, forms. From this cork cambium, cork, which is a dead tissue at maturity, is produced.

As trees grow, season after season, the new xylem may form visible growth layers, or rings. These visible growth rings are the result of a difference in the density of the wood produced early in the growing season—when the cells are larger and have thinner walls—and late in the season. The age of such trees can be estimated by counting the number of growth rings in a section near its base.

SUMMARY

The flower is the reproductive structure of the angiosperms. The anthers of the flower produce the pollen grain, the male gametophyte; each anther is attached to a slender filament. The entire structure, anther plus filament, is known as the stamen. The carpel typically consists of the stigma, an area on which the pollen

Table 21-1 *Summary of Main Cell Types in Angiosperms*

CELL TYPE	LOCATION	CHARACTERISTICS	FUNCTION
Apical meristem	Apices of shoots and flowers; in root, beneath root cap	Many-sided, small, thin-walled cells; nucleus large; vacuoles usually small	Origin of primary meristematic tissues and leaf primordia
Epidermis	Surface of entire plant	Flattened, variable in shape, overlaid by cuticle in shoot; includes guard cells	Boundary between plant body and external environment
Parenchyma	Everywhere, usually dominant in pith, cortex, mesophyll, phelloderm	Many-sided, usually thin-walled; abundant air spaces between cells	Photosynthesis, storage, differentiation into other cell types
Collenchyma	Stands in cortex of stem and in leaves	Elongate, with irregularly thickened primary walls	Support for young stems and leaves
Sclereid	Scattered in parenchyma tissue; seed coats	Irregular; massive secondary wall; often dead at maturity	Produces hard texture
Fiber	Primary and secondary xylem and phloem	Very long, narrow cell, with thick cell walls; often dead at maturity	Support
Tracheid	Xylem	Elongated, with secondary walls containing pits; dead at maturity	Conduction of water and solutes
Vessel element	Interconnected series (= vessels) in xylem	Similar to tracheid, but broader, with end walls perforated; dead at maturity	Conduction of water and solutes
Sieve-tube element	Phloem, usually connected with companion cells (parenchyma), forming interconnected series (= sieve tube)	Elongate, cylindrical, with specialized sieve plates; nucleus lacking at maturity	Conduction of sugars
Vascular cambium	Lateral, between phloem and xylem tissues	Elongate fusiform (spindle-shaped) and short-ray initials	Produces secondary xylem and secondary phloem
Cork (phellem)	Surface of stems and roots with secondary growth	Flattened cells, compactly arranged; protoplast absent at maturity, the cells often air-filled	Restricts gas exchange and water loss

grains germinate, a slender style, and at its base, the ovary. The ovary contains one or more ovules. Within each ovule, the female gametophyte containing the egg cell develops. The ovule later becomes the seed, and the ovary becomes the fruit. The various shapes and colors of flowers evolved under selection pressures for more efficient pollinating mechanisms.

In flowering plants, a new life cycle begins when a pollen grain germinates upon the stigma of a flower of the same species, sending a pollen tube down the style and into an ovule. Within the ovule is the female gametophyte, which usually consists of seven cells with a total of eight haploid nuclei, including the egg nucleus. One of the three haploid nuclei in the pollen grain (a sperm nucleus) unites with the egg cell nucleus, and the other sperm nucleus unites with the two polar nuclei of the female gametophyte. The plant embryo develops from the first union, and the nutritive endosperm, which is a triploid ($3n$) tissue, from the second. These phenomena of double fertilization and triple fusion are found only among the flowering plants.

The embryo has two apical meristems—the apical meristem of the shoot and the apical meristem of the root—and either one or two cotyledons ("seed leaves"). From the endosperm, the cotyledons absorb nutrients which nourish the growing embryo. Early in embryonic life, cell division becomes confined to limited areas in the plant body: the apical meristems of the root and the shoot.

The seed consists of the embryo and the seed coat—that is, the ovule and its contents. It also often contains a nutrient tissue, the endosperm. The petals, stamens, and other floral parts of the parent plant may fall away as the ovary ripens into fruit and the seed forms. The embryo commonly enters a dormant period.

When the seed germinates, growth of the root and shoot proceeds from the apical meristems of the embryo. Certain cells within the meristems divide continuously. Some of these cells remain meristematic, continuing to divide, while others elongate and then differentiate, forming, according to their position, various specialized cells, including those of the xylem and the phloem. In the root, the apical meristem forms a root cap, which protects the root tip as it is pushed through the soil. Root hairs, which are the principal pathways of absorption by the roots, are outgrowths of the epidermal cells.

Leaf primordia arise from the apical meristem of the shoot. The young leaves are separated as the internodes elongate. Buds (apical meristems) in the axils of the leaves may give rise to branches or flowers.

Secondary growth is the process by which the woody plants increase their girth. Such growth arises primarily from the vascular cambium, a sheath of meristematic tissue completely surrounding the xylem and completely surrounded by phloem. The cambium cells divide during the growing season, adding new xylem cells (secondary xylem) outside the primary xylem and new phloem cells (secondary phloem) on the inside of the primary phloem. As the trunk increases in girth, the epidermis is eventually ruptured and destroyed and is replaced by cork.

QUESTIONS

1. Define the following terms: secondary growth, rhizome, pollen grain, endosperm, apical meristem, bud, primary growth, lateral meristem.

2. Sketch a flower. Give the function of each floral part.

3. Sketch a dicot embryo at the time of seed release. Identify each part in terms of the future development of the plant body.

4. Sketch a tree trunk with secondary growth. Compare your sketch with Figure 21–13c.

5. Suppose you carve your initials a meter above the ground on the trunk of a mature tree that is growing vertically at an average rate of 10 centimeters a year. How high will your initials be at the end of 2 years? At the end of 20 years?

6. What plant part are you eating when you eat the following: carrots, eggplant, lima beans, corn on the cob, string beans, celery, tomatoes, spinach, peanuts, brussel sprouts, artichokes, asparagus?

CHAPTER 22

Plant Hormones and Plant Responses

22–1

The Darwins' experiment. (a) Light striking a growing shoot tip (such as the tip of this oat seedling) causes it to bend toward the light. (b) Placing an opaque cover over the tip of the seedling inhibits this bending response, but (c) an opaque collar placed below the tip does not. These experiments indicate that something produced in the tip of the seedling and transmitted down the stem causes the bending.

We saw in the previous chapter that as a plant grows it does far more than simply increase its mass and volume. It differentiates, forming a variety of cells, tissues, and organs, and undergoes morphogenesis, taking on the shape characteristic of the adult sporophyte. Moreover, many of its activities are finely tuned to its environment and to the changing pattern of the seasons. Many of the details of how these processes are regulated are not known, but it has become clear that plant development and growth depend on the interplay of a number of internal and external factors.

Chief among the internal factors are the plant hormones. Hormones, by definition, are substances that are produced in one tissue and transported to another, where they exert highly specific effects. Typically they are active in very small quantities. In the shoot of a pineapple plant, for example, only 6 micrograms of auxin, a common growth hormone, are found per kilogram of plant material. One enterprising plant physiologist calculated that the weight of the hormone in relation to that of the shoot is comparable to the weight of a needle in a 22-ton haystack. The term "hormone" comes from the Greek word meaning "to excite." It is now clear, however, that many hormones have inhibitory influences. So, rather than thinking of hormones as stimulators, it is perhaps more useful to consider them as chemical messengers. But this term, too, needs qualification. As we shall see, the response to the particular "message" depends not only on its content but upon how it is "read" by its recipient.

THE AUXINS

The first plant hormones to be isolated were the auxins. The effects of these hormones were observed by Charles Darwin and his son Francis and first reported in *The Power of Movement in Plants,* published in 1881. The Darwins were studying the bending toward light (phototropism) of grass seedlings. They noted that the bending takes place below the tip, in the lower part of the shoot. Then they showed that if they covered just the terminal portion of the shoot of the seedling with a cylinder of metal foil or a hollow tube of glass blackened with India ink and exposed the plant to a light coming from the side, the characteristic bending of the shoot did not occur. If, however, the tip was enclosed in a transparent glass tube, bending occurred normally. Bending also occurred normally when the lightproof cylinder was placed below the tip (Figure 22–1). "We must therefore conclude," they stated, "that when seedlings are

Went's experiment. (a) He cut the coleoptile tips from oat seedlings and placed them on a slice of agar. After about an hour, he cut the agar in small blocks and (b) placed each block, off-center, on a decapitated seedling (the leaf is pulled up). (c) The seedlings bent away from the side on which the block was placed.

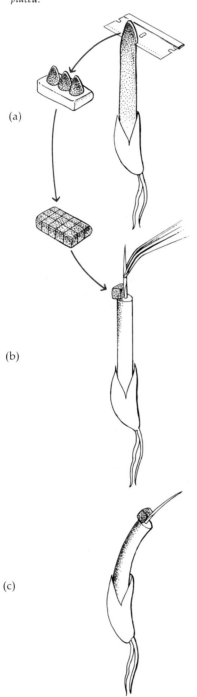

(a)

(b)

(c)

freely exposed to a lateral light some influence is transmitted from the upper to the lower part, causing the material to bend."

In 1926, the Dutch plant physiologist Frits W. Went succeeded in separating this "influence" from the plants that produced it. Went cut off the leaf (coleoptile) tips from a number of oat seedlings. He placed the tips on a slice of agar (a gelatinlike substance), with their cut surfaces in contact with the agar, and left them there for about an hour. He then cut the agar into small blocks and placed a block off-center on each stump of the decapitated plants, which were kept in the dark during the entire experiment. Within one hour, he observed a distinct bending *away* from the side on which the agar block was placed (Figure 22–2).

Agar blocks that had not been exposed to a coleoptile tip produced either no bending or only a slight bending toward the side on which the block had been placed. Agar blocks that had been exposed to a section of coleoptile lower on the shoot produced no physiological effect.

Went interpreted these experiments as showing that the coleoptile tip exerted its effects by means of a chemical stimulus (in short, a hormone) rather than a physical stimulus, such as an electrical impulse. This chemical stimulus came to be known as auxin, a term coined by Went from the Greek word *auxein*, "to increase." Several different substances with auxin activity have now been isolated from plant tissues, and others have been synthesized in the laboratory.

The phototropism observed by the Darwins results from the fact that under the influence of light, auxins migrate from the light side to the dark side of the tip. The cells on the dark side, having more auxin, elongate more rapidly than those on the light side, causing the plant to bend toward the light. This response has high survival value for young plants (Figure 22–3).

Other Auxin Actions

Auxins have a number of other effects in the growing plant. Auxins produced in the meristem produce elongation of the shoot. This effect is produced by increasing the plasticity of the cell wall. The cell then takes up more water and elongates.

In small concentrations, auxins stimulate the growth of adventitious roots; rooting preparations used by gardeners contain auxin. In large concentrations, they inhibit the growth of the main roots, which may explain why seedlings can orient themselves in the ground (Figure 22–4).

Although auxins stimulate growth in apical buds, in some species they inhibit the growth of lateral buds. If you pinch off the growing tip (meristem) of the stem of the houseplant *Coleus*, for example, the lateral buds begin to grow vigorously, producing a plant with a bushier, more compact body. If you treat the "eyes" of a potato with auxin, they will be inhibited from sprouting and so can be stored longer. The "eyes" are actually lateral buds. The formation of the abscission layer (page 265) has been correlated with diminished production of auxin in the leaf. Auxins are also involved in the growth of fruits (Figure 22–5).

The mechanism or mechanisms by which one chemical can produce so many varied and even opposite effects is not known. In fact, exposing a plant tissue to a hormone has been compared to putting a coin in a vending machine. You may get your morning newspaper, a candy bar, or four minutes of country and western music. It depends not so much on the coin as on the machine in which you put it. Similarly, the effects of auxins and other plant hormones depend not only on the chemical but also on the target tissues and the chemical environment in which these tissues find themselves.

22-3

An example of phototropism. Here bean seedlings turn toward the sun.

22-4

Why shoots grow upward and roots downward. (a) When a seedling is perpendicular, the auxin produced in the tip is distributed evenly and the plant grows erect. (b) However, when a seedling is placed on its side, auxin is transported laterally and accumulates in cells on the lower side of the shoot; (c) *these cells then grow more rapidly and so the shoot grows upward. It has been hypothesized that auxin similarly accumulates in the cells on the lower side of the root, causing the cells to grow more slowly, so that the root tip turns down.*

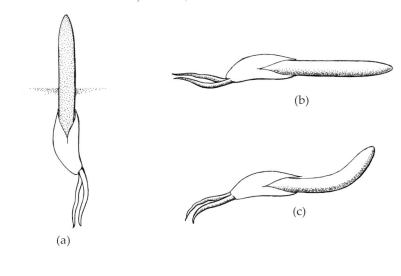

(a)

(b)

(c)

22-5

(a) Normal strawberry and (b) strawberry from which all the seeds were removed. (c) Strawberry in which three horizontal rows of seeds were left. (d) Growth in strawberry induced by one developing seed. (e) Growth induced by three developing seeds. If a paste containing auxin is applied to the strawberry from which the seeds have been removed, the strawberry develops normally.

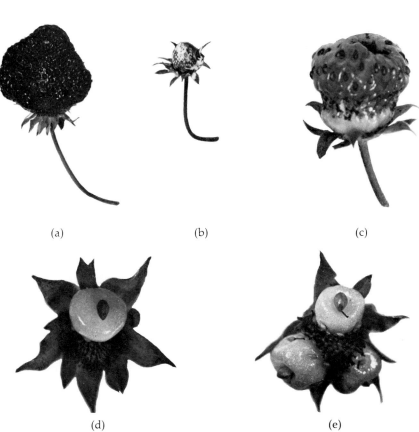

(a)   (b)   (c)

(d)   (e)

THE GIBBERELLINS

The gibberellins were first discovered by a Japanese scientist who was studying a disease of rice plants, "foolish seedling disease." The diseased plants grew rapidly but were spindly and tended to fall over. The cause of the symptoms, the scientist found, was a chemical produced by a fungus, *Gibberella fujikuroi*, which parasitized the seedlings. The substance, which he named gibberellin, was subsequently isolated not only from the fungus but from many species of plants.

The most remarkable results are seen when gibberellins are applied to certain plants that, because of a single mutation, have lost the capacity to synthesize gibberellins themselves. Such plants are genetic dwarfs. Under gibberellin treatment, these dwarfs become indistinguishable from normal tall plants.

Some plants—lettuce and cabbage are common examples—first grow as rosettes; the leaves develop but the internodes do not elongate until just before flowering. At that time the stem elongates rapidly, a phenomenon known as bolting. In the case of cabbage and other biennials, bolting and flowering do not normally occur until after a period of cold. However, in cabbage and many other plants, bolting and flowering can be caused by gibberellin (Figure 22-6).

Gibberellins have also been shown to play a role in the germination of certain seeds. In grass seeds, there is a specialized layer of cells, the aleurone layer, just inside the seed coat. These cells are rich in protein. When the seeds imbibe water, just prior to germination, the embryo produces gibberellin, which diffuses to the aleurone layer. In response to the gibberellin, the aleurone cells produce enzymes that hydrolyze the starch in the endosperm, converting it to soluble sugars which the embryo can use (Figure 22-7). In this way the embryo itself calls forth the substances needed for its survival and growth at the time they are required.

22-6
Bolting and flowering in cabbage following gibberellin treatment. The plant on the left was not treated.

22-7
An experiment investigating the action of gibberellin in barley seeds. Forty-eight hours before the picture was taken, each of these seeds was cut in half and the embryo removed. The seed at the bottom was treated with plain water, the seed in the center was treated with a solution of 1 part per billion of gibberellin, and the seed at the top was treated with 100 parts per billion of gibberellin. As you can see, digestion of the starchy storage tissue has begun to take place in the seeds treated with gibberellin.

BIOLOGICAL CLOCKS

Many plants open their leaves in the daytime and fold them at night. As long ago as 1729 a French scientist noticed that these movements continue even when plants are kept in continuous dim light. It is now generally agreed that such movements are endogenous—that is, they arise within the plant itself. The mechanism by which they are controlled is known as a biological clock. The nature of the clock, however, is not known.

Biological clocks are believed to play an essential role in many aspects of plant and animal physiology. For instance, insects are more active in the early evening hours. Bats that feed on insects begin to fly each evening just when the insects are most available. Moreover, caged bats under controlled and constant laboratory conditions continue to show this sort of activity about every 24 hours. (The "about" is important since it indicates that the response is not to some environmental cue.)

Some plants secrete nectar or perfume at certain specific times of the day. As a result, insects—which have their own biological clocks—become accustomed to visiting these flowers at these times, thereby ensuring maximum rewards for both the insects and the flowers.

The ability to tell time also appears to be involved in a number of complex and fascinating phenomena such as the extraordinary navigatory ability of migrating birds and turtles.

Man, too, is tied to his own endogenous rhythms. It has long been known that he is more likely to be born between 3 and 4 A.M. and also to die in these same early morning hours. Body temperature fluctuates about 1° during the course of the day, usually reaching a high about 4 P.M. and a low about 4 A.M. Alcohol tolerance is greatest, fortunately, at 5 P.M. Many people have an automatic internal alarm system that wakes them at the same hour every morning—whether they want to or not. Hormone secretion, heart rate, blood pressure, and urinary excretion of potassium, sodium, and calcium in man all vary according to a circadian rhythm. A recent study by the Federal Aviation Agency showed that pilots flying from one time zone to another—from New York to Europe, for instance—exhibit "jet lag," a general decrease in mental alertness and ability to concentrate and an increase in decision time and physiological reaction time. Measurements of circadian rhythms show that the body may be "out of sync" for as much as a week after such a flight. This brings into question present policies of diplomats speeding to foreign capitals at times of international crisis or of troops being air-transported into combat.

(a)

Richard F. Trump, OMIKRON

(b)

Leaves of the wood sorrel, day (a) and night (b). One hypothesis concerning the function of such "sleep" movements is that they protect the leaves from absorbing moonlight on bright nights, thus protecting photoperiodic phenomena. Another theory, proposed by Darwin almost 100 years ago, is that the folding protects against heat loss from the leaves by night.

22-8

Two buds forming on undifferentiated tissue from a geranium following treatment with both an auxin and a cytokinin. Callus from some types of plants will continue to grow as undifferentiated tissue, or roots, or buds, depending on the relative proportions of auxins and cytokinins.

22-9

Examples of each of three main groups of plant hormones: (a) an auxin (indoleacetic acid, or IAA); (b) gibberellic acid (GA₃), a gibberellin, one of more than 20 that have been isolated from plants and fungi, and (c) a cytokinin, kinetin. Other members of each group show slight structural differences between one another and often, slight differences in activity.

INDOLEACETIC ACID
(IAA)

(a)

GIBBERELLIC ACID (GA₃)

(b)

KINETIN

(c)

CYTOKININS AND AUXIN INTERACTIONS

Cytokinins were first found in coconut milk, which is a liquid endosperm. Their principal action is to promote cell division, hence their name (from the term "cytokinesis").

Studies of interactions involving auxin and cytokinins are helping physiologists understand how plant hormones work to produce the total growth pattern of the plant. Apparently, the undifferentiated plant cell—such as the meristematic cell—has two courses open to it: Either it can enlarge, divide, enlarge, and divide again, or it can elongate without cell division. The cell that divides repeatedly remains essentially undifferentiated, or embryonic, whereas the elongating cell tends to differentiate and become specialized. In studies of tobacco stem tissues, the addition of an auxin to the tissue culture produced rapid cell expansion, so that giant cells were formed. A cytokinin alone had little or no effect. Auxin plus cytokinin resulted in rapid cell division, so that large numbers of relatively tiny cells were formed.

By slight alterations in the relative concentrations of auxin and cytokinin, investigators have been able to affect the development of undifferentiated cells growing in tissue culture. When a high concentration of auxin is present, undifferentiated tissue gives rise to organized roots. With higher concentrations of cytokinins, buds appear (Figure 22–8).

However, lest you think this is simple, we shall describe another tissue culture study, in which tuber tissue of the Jerusalem artichoke was used. In this study, it was shown that a third substance, the calcium ion, can modify the action of the auxin-cytokinin combination. Auxin plus low concentrations of cytokinin was shown to favor cell enlargement, but as calcium ion was added to the culture, there was a steady shift in the growth pattern from cell enlargement to cell division. High concentrations of calcium prevent the cell wall from expanding, and at such concentrations the cell switches course and divides. Thus, not only do hormones modify the effects of hormones, but these combined effects are, in turn, modified, by nonhormonal factors, such as calcium and, undoubtedly, many others.

PLANTS IN TEST TUBES

As long ago as the 1930s, scientists developed techniques for the growing of plant cells in test tubes. Tiny fragments of meristem are implanted under sterile conditions in a medium containing minerals and various combinations of organic compounds. Under these conditions, the meristematic cells proliferate to form clumps of undifferentiated cells. Subsequently, it was discovered that, by adjusting the hormone balance in the medium, it was possible to make these cells differentiate and grow into mature plants. Using these techniques, hundreds or even thousands of subcultures of meristematic tissue can be produced in a relatively short time and in a short space. Then, by alternating the hormone balance, each of these can be turned into a small but perfect plant. This technique has already been employed in the culture of exotic orchids and is being extended to other hard-to-grow varieties.

Within the last decade, a group of Japanese scientists, using similar techniques, was able to grow isolated protoplasts—plant cells without cell walls—in the laboratory. These protoplasts can be grown as isolated cells, like cultures of bacteria. Alternatively, if they are grown in a suitable medium, they will regenerate cell walls, multiply, and differentiate into whole plants.

These cells are being used in a variety of ways. For instance, it has been found that they can be used in rapid screening tests to determine resistance to infectious diseases or to detect nutritional requirements. In this way, a scientist not only can work much more rapidly, but can do assays with millions of cells growing in a very small space, as compared to the far fewer number of plants that can be grown in a field or a greenhouse. Moreover, it has been found possible to fuse protoplasts from two different species of plant and create a hybrid. So far this has only been accomplished in the case of a hybrid of tobacco that had already been produced by crossbreeding. (Producing a plant whose characteristics were known was a necessary preliminary test for the method.) However, this initial success has, of course, raised the hopes of producing entirely new superplants—organisms that might be capable, for instance, of simultaneously carrying out C_4 photosynthesis and fixing nitrogen. So far, fusion of protoplasts from cells as distantly related as corn and soybeans has been successful, but these hybrid cells have not been induced to develop into plants.

OTHER HORMONES

Soon after the discovery of the growth-promoting hormones, plant physiologists began to speculate that hormones that inhibited growth would be found, since it is clearly advantageous to the plant not to grow at certain times and in certain seasons. Not long afterward, an inhibitory hormone was isolated from dormant buds. Subsequently, the same hormone was discovered in leaves, where it was found to promote abscission (see page 265). It is called abscisic acid.

Ethylene is an unusual hormone in that it is a gas, a simple hydrocarbon. Its effects have been known for a long time. In the early 1900s, many fruit growers made a practice of improving the color of citrus fruits by "curing" them in a room with a kerosene stove. (Long before this, the Chinese used to ripen fruits in rooms where incense was being burned.) It was long believed that it was the heat that ripened the fruits. Ambitious fruit growers, who went to the expense of installing more modern heating equipment, found to their sorrow that this was not the case. As experiments showed, it is actually the incomplete combustion products of the kerosene that are responsible. The most active gas was identified as ethylene. As little as 1 part per million of ethylene in the air will speed the ripening process. Subsequently, it was found that ethylene is produced by plants, as well as by kerosene stoves, that it appears just before and also during fruit ripening, and that it is responsible for a number of the changes in color, texture, and chemical composition that take place as fruits mature.

Experiments that indicate the existence of a flowering hormone. (a) When certain plants, such as the cocklebur shown here, are exposed to an appropriate light cycle, those with leaves flower and those without leaves do not. When even one-eighth of a leaf remains on a plant, flowering occurs, and the illumination of a single leaf—not necessarily the whole plant—suffices. These experiments indicate that a chemical originating in the leaves causes the plant to flower. (b) This conclusion is supported by experiments on branched plants. Exposure of one branch to the light induces flowering on the other branch as well, even when only a portion of a leaf is present on the lighted branch. (c) When two plants are grafted together, exposure of one of the plants to the light cycle induces flowering in both the lighted plant and the grafted one. This effect occurs even when a piece of paper is inserted across the graft.

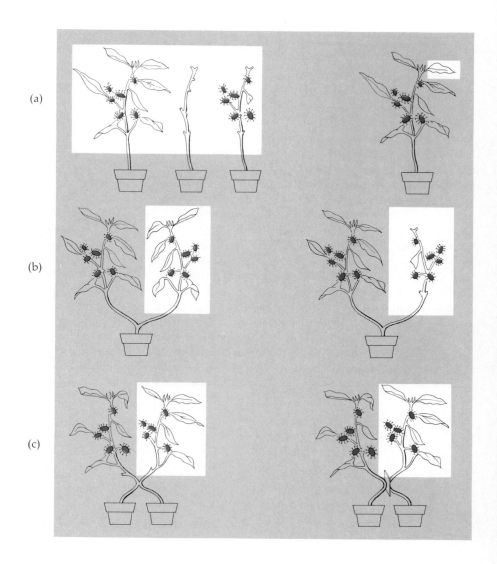

(a)

(b)

(c)

A third hormone which has been postulated to exist, but never isolated, is a flowering hormone. Experiments that indicate the existence of such a hormone are outlined in Figure 22–10.

PLANTS AND PHOTOPERIODISM

In many regions of the biosphere, the most important environmental changes affecting plants (and indeed, land organisms, in general) are those that result from the changing seasons. Plants are able to accommodate themselves to these changes because of their capacity to sense and even to anticipate the yearly calendar of events—the first frost, the spring rains, long dry periods, long growing spells, and even the time that nearby plants of the same species will be in flower. For many plants, all of these determinations are made in the same way: by measuring the relative periods of light and darkness. This phenomenon is known as *photoperiodism.*

Photoperiodism and Flowering

The effects of photoperiodism on flowering are particularly striking (Figure 22–11). Plants are of three general types: day-neutral, short-day, and long-day. Day-neutral plants flower without regard to day length. Short-day plants flower in early spring or fall; they must have a light period shorter than a critical length—for instance, the cocklebur flowers when exposed to $15\frac{1}{2}$ hours or *less* of light. Other short-day plants are poinsettias, strawberries, primroses, ragweed, and some chrysanthemums.

Long-day plants, which flower chiefly in the summer, will flower only if the light periods are *longer* than a critical length. Spinach, potatoes, clover, henbane, and lettuce are examples of long-day plants.

The discovery of photoperiodism explained some puzzling data about the distribution of common plants. Why, for example, is there no ragweed in northern Maine? The answer, they found, is that ragweed starts producing flowers when the day is less than $14\frac{1}{2}$ hours long. The long summer days do not shorten to $14\frac{1}{2}$ hours in northern Maine until August, and then there is not enough time for ragweed seed to mature before the frost. For similar reasons, spinach cannot produce seeds in the tropics. Spinach needs 14 hours of light a day for a period of at least two weeks in order to flower, and days are not this long in the tropics.

22–11

(a) *The relative length of day and night determines when plants flower. The four curves depict the annual change in day length in four North American cities at four different latitudes. The black lines indicate the effective photoperiod of three different short-day plants. The cocklebur, for instance, requires $15\frac{1}{2}$ hours or less of light. In Miami, it can flower as soon as it matures, but in Winnipeg the buds do not appear until early in August, so late that frost usually kills the plants before the seed is mature. (b) Relationship between day length and the developmental cycle of plants in the temperate zone.*

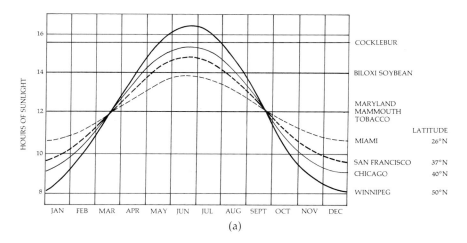

(a)

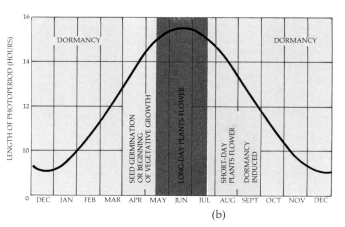

(b)

22-12

The cocklebur, a short-day plant, was an important tool for experimental studies on photoperiodism.

Note that cocklebur and spinach will both bloom if exposed to 14 hours of daylight, yet one is designated as short-day and one as long-day. The important factor is not the absolute length of the photoperiod but rather whether it is longer or shorter than a particular critical interval for that variety. And in some varieties, 5 or 10 minutes' difference in exposure can determine whether or not a plant will flower.

These early field studies on photoperiodism in plants were carried out by scientists at the U.S. Department of Agriculture, known as the Beltsville group, for the small town in Maryland where they carried out their studies.

Measuring the Dark

Subsequently, other investigators, Karl C. Hamner and James Bonner, began a laboratory study of photoperiodism, again using the cocklebur as an experimental tool. As we mentioned previously, the cocklebur is a short-day plant, requiring 16 hours or less of light per 24-hour cycle to flower. It is particularly useful for experimental purposes because a single exposure under laboratory conditions to a short-day cycle will induce flowering two weeks later, even if the plant is immediately returned to long-day conditions. Also, as we saw in Figure 22–10, it can withstand a good deal of rough treatment.

In the course of these studies, in which they tested a variety of experimental conditions, the investigators made a crucial and totally unexpected discovery. If the period of darkness is interrupted by as little as a one-minute exposure to a 25-watt bulb, flowering does not occur. Interruption of the light period by darkness has no effect whatsoever on flowering. Subsequent experiments with other short-day plants showed that they, too, required periods not of uninterrupted light but of uninterrupted darkness (Figure 22–13).

On the basis of the findings of the Beltsville group, commercial growers of chrysanthemums had found that they could hold back blooming in the short-

22-13

Experiments on photoperiodism showed that plants measure the darkness rather than the light. Short-day plants flower only when the darkness exceeds some critical value. Thus, the cocklebur, for instance, will flower on 8 hours of light and 16 hours of darkness. If the 16-hour period of darkness is interrupted even very briefly, as shown on the right, the plant will not flower. The long-day plant, on the other hand, which will not flower on 16 hours of darkness, will flower if the darkness is interrupted. Long-day plants flower only when the darkness is less than some critical value.

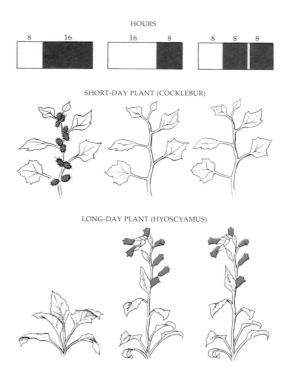

HOURS

SHORT-DAY PLANT (COCKLEBUR)

LONG-DAY PLANT (HYOSCYAMUS)

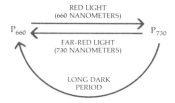

$$P_{660} \xrightleftharpoons[\text{FAR-RED LIGHT}]{\text{RED LIGHT}} P_{730}$$

RED LIGHT
(660 NANOMETERS)

FAR-RED LIGHT
(730 NANOMETERS)

LONG DARK
PERIOD

22–14

P_{660} changes to P_{730}, the active form, when exposed to red light, which is present in sunlight. P_{730} reverts to P_{660} when exposed to far-red light. In darkness, P_{730} slowly reverts to P_{660}.

day plants by extending the daylight with artificial light. Now, on the basis of the new experiments by Hamner and Bonner, they were able to economize on their electric bill and still have flowers for late-season football games.

What about long-day plants? They also measure darkness. A long-day plant that will flower if it is kept in a laboratory in which there is light for 16 hours and dark for 8 hours will also flower on 8 hours of light and 16 hours of dark if the dark is interrupted by even a brief exposure of light.

Photoperiodism and Phytochrome

Following up on the clues provided by the Hamner and Bonner experiments, the Beltsville group was able to detect and eventually to isolate the pigment involved in photoperiodism. This pigment, which they called *phytochrome*, exists in two different forms: one is P_{660} and one is P_{730}. The numbers refer to the wavelengths of light the two forms absorb; 660 nanometers is red light and 730 nanometers is far-red. P_{660} absorbs red light and is converted to P_{730}, which is the active form. This conversion takes place in daylight or in incandescent light; in both of these types of light, red wavelengths predominate over far-red. When P_{730} absorbs far-red light, it is converted back to P_{660}. The P_{730} to P_{660} conversion can also take place in the dark, which is how it usually occurs in nature (Figure 22–14). P_{730}, the active form of the pigment, promotes flowering in long-day plants and inhibits flowering in short-day plants.

Other Phytochrome Responses

Many types of small seeds, such as lettuce, germinate only when they are in loose soil, near the surface. Otherwise the seedling would have little chance of reaching the light. Red light, a sign that sunlight is present, stimulates and phytochrome mediates the response. Similarly, phytochrome is involved in the early development of seedlings. When a seedling develops in the dark, as it normally does underground, the stem elongates rapidly, pushing the shoot up through the soil layers. Any seedling grown in the dark will be elongated and spindly with small leaves (Figure 22–15). It will also be almost colorless, because

22–15

Dark-grown seedlings, such as the ones on the right, are thin and pale with longer internodes and smaller leaves than normal seedlings, such as those on the left. This group of characteristics, known as etiolation, has survival value for the seedling because it increases its chances of reaching light before its stored energy supplies are used up.

Richard F. Trump, OMIKRON

the chloroplasts do not synthesize chlorophyll until they are exposed to light. Such a seedling is said to be etiolated. When the seedling tip reaches the light, normal growth begins. Phytochrome is involved in the switching of etiolated to normal growth. If a dark-grown bean seedling is exposed to only one minute of red (660 nanometers) light, it will respond with normal growth. If however, the exposure to red light is followed by a one-minute exposure to far-red light, etiolated growth continues.

The way in which phytochrome acts is not known. One recent suggestion is that it alters the permeability of the cell membrane, permitting other substances to enter the cell, or perhaps, inhibiting their entry, and that these substances, which probably include hormones, regulate the cell's activities.

TOUCH RESPONSES IN PLANTS

Plants respond to touch, often in surprising ways. One of the most common examples is seen in tendrils, which wrap around any object with which they come in contact (Figure 22–16) and so enable the plant to cling and climb. The response can be rapid; a tendril may wrap around a support in less than a minute. Cells touching the support shorten slightly and those on the other side elongate. There is some evidence that auxin plays a role in this response.

A more spectacular touch response is seen in the sensitive plant, *Mimosa pudica*, in which the leaflets and sometimes entire leaves droop suddenly when touched (Figure 22–17). This response is a result of a sudden change in turgor pressure. There is some controversy about its survival value to the plant. *Mimosa pudica* often grows in dry, exposed, windy areas. Strong winds may shake the leaves enough to make them fold up, so conserving water. Another suggestion is that the wilting response makes the plant unattractive to large herbivores. Finally, it is conceivable that the folding response startles insects; there have been claims that other "nonsensitive" species of *Mimosa* growing near *Mimosa pudica* show more evidence of attack by chewing insects.

The triggering of turgor changes by touch is also involved in the capture of prey by the carnivorous Venus flytrap *(Dionaea muscipila).* The leaves of the

22–16
Tendrils of Smilax *(greenbrier). Twisting is caused by varying growth rates on different sides of the tendril.*

22–17
Sensitive plant (Mimosa pudica). *(a) Normal position of leaves and leaflets. (b, c) Successive responses to touch.*

(a)

(b)

(c)

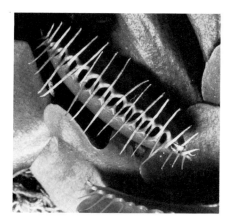

22–18
Touch responses in the Venus flytrap.

22–19
(a) *Diagram of a sundew tentacle, with electrodes in place. (b) A record of electric impulses produced by positioning a fruit fly so that its feet stroked the head of a tentacle. The recording ended when the contact was broken as the tentacle bent.*

Venus flytrap are hinged in the middle, and each leaf half is equipped with three sensitive hairs. When an insect walks on one of these leaves, attracted by the nectar on the leaf surface, it brushes against the hairs, triggering the traplike closing of the leaf. The toothed edges mesh, the leaf halves gradually squeeze closed, and the insect is pressed against digestive glands on the inner surface of the trap.

The trapping mechanism is so specialized that it can distinguish between living prey and inanimate objects, such as pebbles and small sticks, that fall on the leaf by chance: The leaf will not close unless two of its hairs are touched in succession or one hair is touched twice.

Studies have recently been made by investigators at Washington University in St. Louis on how the sundew traps insects. As you saw in Figure 20–15, the club-shaped leaves of the plant are covered with tiny tentacles. A sticky droplet surrounding the tip of each tentacle attracts insects. When an insect is caught on the tip of a tentacle, the tentacle bends in rapidly, carrying the prey to the center of the leaf, where it is digested by enzymes. Microelectrodes placed in the tentacles (Figure 22–19) have revealed that the rapid response is due to an electric impulse that moves down the tentacle. This impulse is similar to the nerve impulse in animals (see page 354).

The investigators predict that electric signals will be found to coordinate a variety of functions in plants.

SUMMARY

Plants respond to factors both in their internal and external environments. Such responses enable the plant to develop normally and to remain in touch with external conditions. Hormones are important factors in plant response. A hormone is a chemical produced in particular tissues of an organism and carried to other tissues of the organism, where it exerts specific effects. Characteristically, it is active in extremely small amounts.

Auxins are hormones that are produced principally in rapidly dividing tissues, such as coleoptile tips and apical meristems. They cause lengthening of the shoot and the coleoptile, chiefly by promoting cell elongation. They often inhibit growth in lateral buds, thus restricting growth principally to the apex of the plant. The same quantity of auxin that promotes growth in the stem inhibits growth in the main root system. Auxins promote the initiation of branch roots and adventitious roots; however, they retard abscission in leaves and fruits. In fruits, auxin produced by seeds stimulates growth of the ovary wall. The capacity of auxins to produce such varied effects appears to result from different responses of the various target tissues.

(a)

SECRETION DROP

HEAD

NECK

STALK

BASE

RECORDING ELECTRODES:
HEAD

UPPER STALK

LOWER STALK

(b)

FLY INTRODUCED

50 MV

20 SEC

The gibberellins were first isolated from a parasitic fungus that causes abnormal growth in rice seedlings. They were subsequently found to be natural growth hormones also present in plants. Gibberellins induce bolting and flowering in cabbage. In dwarf plants, application of gibberellins restores normal growth. Gibberellins are also involved in seed germination in grasses. They stimulate the production of hydrolyzing enzymes that act on the stored starch of the endosperm, converting it to sugar, which nourishes the embryo.

The cytokinins, a third class of growth hormone, promote cell division. It is possible by altering the concentrations of auxins and cytokinins to alter patterns of growth in undifferentiated plant tissue in tissue culture.

Abscisic acid, a growth-inhibiting hormone, induces dormancy in vegetative buds and accelerates abscission. Ethylene is a gas produced by the incomplete combustion of hydrocarbons; it also emanates from fruits during the ripening process. Ethylene promotes the ripening of fruits and is now considered a natural growth regulator. There is evidence for the existence of a flowering hormone (or hormones), but it has not yet been isolated.

Plants respond to a number of environmental stimuli. Photoperiodism is the response of organisms to changing periods of light and darkness in the 24-hour day. Such a response controls the onset of flowering in many plants. Some plants will flower only when the periods of light exceed a critical length. Such plants are known as long-day plants. Other plants, short-day plants, flower only when the periods of light are less than some critical period. Day-neutral plants flower regardless of photoperiods. Interruption of the dark phase of the photoperiod, even by a brief exposure to light, can serve to reverse the photoperiodic effects, indicating that the dark period rather than the light period is critical.

Phytochrome, a pigment commonly present in small amounts in the tissues of higher plants, is the receptor molecule for transitions between light and darkness. The pigment can exist in two forms, P_{660} and P_{730}. P_{660} absorbs red light (which is also ordinarily present in white, or visible, light) of a wavelength of 660 nanometers and is thereby converted to P_{730}, which absorbs far-red light of a 730-nanometer wavelength. With ordinary day-night sequences, P_{730} is converted back to P_{660} over a period of hours in the dark; it can also be converted to P_{660} by exposure to far-red light. P_{730} is the active form of the pigment; it promotes flowering in long-day plants and inhibits flowering in short-day plants. It promotes germination in lettuce seeds and normal growth in seedings. Its mechanism of action is unknown.

Some species of plants respond to touch. Examples include the winding of tendrils, the collapse of the leaves of the sensitive plant, the triggering of the carnivorous Venus flytrap, and the bending of the tentacles of the sundew. The latter has been shown to be a result of an electric impulse.

QUESTIONS

1. Define the following terms: phototropism, photoperiodism, hormone, long-day plant.

2. Name the five types of hormones that have been identified chemically and briefly describe the effects of each.

3. Describe Went's experiments and conclusions that can be drawn from them.

4. Plants must, in some manner, synchronize their activities with the seasons. Of the clues which they might use, day (or night) length seems to have been selected. What might be the advantage of using photoperiod rather than, say, temperature as a season detector?

5. Traveling from north to south, one can find varieties of the same species with different photoperiodic requirements. How would you expect them to differ?

SUGGESTIONS FOR FURTHER READING

GALSTON, A. W.: *The Green Plant,* Prentice-Hall, Inc., Englewood Cliffs, N.J., 1968.*

   A convenient and concise summary of plant growth and development.

GREULACH, VICTOR: *Plant Structure and Function,* The Macmillan Company, New York, 1973.

   A short, lucid introduction to botany, with emphasis on physiology.

LEDBETTER, M. C., and KEITH PORTER: *Introduction to the Fine Structure of Plant Cells,* Springer-Verlag, New York, 1970.

   An excellent atlas of electron micrographs of plant cells, with a detailed explanation of each.

RAVEN, PETER, RAY EVERT, and HELENA CURTIS: *Biology of Plants,* 2d ed., Worth Publishers, Inc., New York, 1975.

   An up-to-date and handsomely illustrated general botany text.

RAY, PETER M.: *The Living Plant,* 2d ed., Holt, Rinehart and Winston, Inc., New York, 1971.*

   An outstanding, short text.

RICHARDSON, MICHAEL: *Translocation in Plants,* St. Martin's Press, Inc., New York, 1968.*

   An extremely useful, brief review of experimental work on the movement of water in plants.

SALISBURY, FRANK B., and CLEON ROSS: *Plant Physiology,* Wadsworth Publishing Co., Inc., Belmont, Calif., 1969.

   A good, modern plant physiology text for more advanced students.

TORREY, JOHN G.: *Development in Flowering Plants,* The Macmillan Company, New York, 1967.*

   How flowering plants develop, with emphasis on the underlying physiological processes.

* Available in paperback.

SECTION 6   Human Physiology

23–1
Genus, Homo; *species,* sapiens. *An immature form.*

CHAPTER 23

The Anatomy of Man

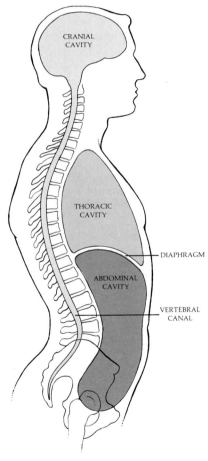

CRANIAL CAVITY

THORACIC CAVITY

DIAPHRAGM

ABDOMINAL CAVITY

VERTEBRAL CANAL

23-2

The two principal body cavities of verte-brates are the dorsal cavity and the ventral cavity. The dorsal cavity includes the verte-bral canal and the cranial cavity. The ven-tral cavity is, in effect, the coelom, also found in invertebrates. In mammals, a muscular diaphragm divides the coelom into the tho-racic cavity and the abdominal cavity.

In this section of the book we shall examine man (a term that includes, equally, woman) as an example of an animal organism. Let us begin more or less where Chapter 18 left off, with the emergence of the vertebrates.

It is safe to assume that all of us reading this book are vertebrates. As such, we have bony, articulated (jointed), internal skeletons (endoskeletons) that support the soft tissues of our bodies. Our spinal cords are surrounded by bones, the vertebrae, and our brains are enclosed in protective casings, our skulls. We move our skeletons by means of muscles, which are attached either directly to the bones of our skeletons or to tendons attached to the bones.

As in other vertebrates, and most invertebrates as well, our bodies contain a coelom (see page 234). Our coelom is divided into two parts, the thoracic cavity and the abdominal cavity, which are separated by a sheet of muscle, the dia-phragm (Figure 23–2). The thoracic cavity contains the heart, the lungs, and the upper portion of the digestive tract; the abdominal cavity contains the stomach, intestines, liver, and other organs concerned with digestion, excretion, and reproduction. The lower part of the abdominal cavity, although it is not struc-turally separated from the upper part, is often referred to as the pelvic cavity.

Another cavity, the dorsal cavity, is divided into the cranial cavity, which is surrounded by the skull and contains the brain, and the vertebral canal, which is surrounded by the vertebrae and contains the spinal cord.

CELLS AND TISSUES

The human body, like that of all other higher organisms, is made up of a variety of different cells. These cells are organized into *tissues*, groups of cells similar in structure and function.

Although experts can distinguish more than a hundred different types of cells in the human body, they are customarily classified in terms of only four tissue types: (1) epithelial, (2) connective, (3) muscle, and (4) nerve.

Epithelial Tissue

Epithelial tissue consists of a continuous sheet of cells covering an internal or external surface. One surface is always attached to an underlying layer of a fibrous polysaccharide material produced by the epithelial cells themselves.

The three types of epithelial cells that cover the inner and outer surfaces of the body. Squamous cells, which usually perform a protective function, make up the outer layers of the skin and the lining of the mouth and other mucous membranes. There are usually several layers of these flat cells piled on top of one another. Cuboidal and columnar cells, in addition to lining various passageways, perform much of the chemical work of the body.

SQUAMOUS

CUBOIDAL

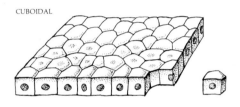

COLUMNAR

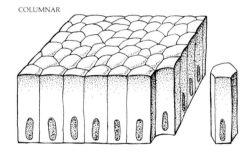

23-4

(a) Collagen, which is produced by connective tissue cells, makes up about one-third of the protein in the human body. Collagen is composed of long polypeptide molecules which assemble spontaneously to form the cablelike fibrils shown here. (b) Section of a bone cell (osteocyte), surrounded by calcified extracellular ground substance (bone). The large nucleus occupies the upper half of the cell. Cytoplasmic processes extend through the bone.

This layer is known as the basement membrane. Epithelium is classified according to the shape of the individual cells as squamous, cuboidal, or columnar (Figure 23-3). It may consist of only a single layer of cells (simple epithelium) or several layers (stratified epithelium).

Epithelium forms the covering of the whole body, where it provides a protective covering and contains various sensory nerve endings. It also forms the lining membranes of internal organs, cavities, and passageways; hence, as a moment's reflection will reveal, everything that goes into and out of the body must pass through epithelial cells.

The epithelium of the body cavities and passageways frequently contains specialized cells that secrete mucus, which lubricates the surfaces. _Glands_ are special types of epithelial tissue whose cells produce specific substances, such as perspiration, saliva, hormones, or digestive enzymes. Glandular epithelium is composed of cuboidal or columnar epithelial cells.

Connective Tissue

Connective tissue binds together and supports epithelial, muscle, and nerve tissues. Unlike epithelial tissue, connective tissue has relatively few cells, widely separated from one another by large amounts of intercellular substances. The intercellular substances include connecting and supporting fibers, such as collagen, a major component of skin, tendons, ligaments, and bones (Figure 23-4a). Other intercellular substances are elastic fibers, which are often found in the walls of hollow, distensible organs, such as the stomach and the uterus, and reticular fibers, which form networks inside solid organs. All of these fibers are embedded in a ground substance that is more or less fluid and amorphous (without any shape or form).

Bone, like other connective tissues, consists of cells, fibers, and ground substance (Figure 23-4b). It is unlike the others in that its extracellular components are calcified. Bone tissue is amazingly strong and light; our bones make up only about 18 percent of our weight.

Blood and lymph are connective tissues in which the ground substance is plasma. The characteristic cells of blood and lymph are described in Chapter 26.

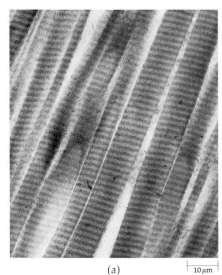

(a)   ⊢ 10 μm ⊣

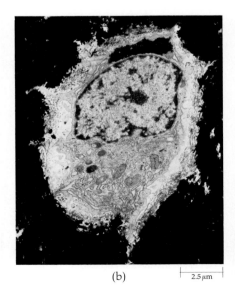

(b)   ⊢ 2.5 μm ⊣

Photomicrograph of skeletal muscle fibers, showing striated pattern. The arrow indicates a small blood vessel, a capillary. Within it, you can see red blood cells, which carry oxygen to the muscle fibers, each of which is a single cell.

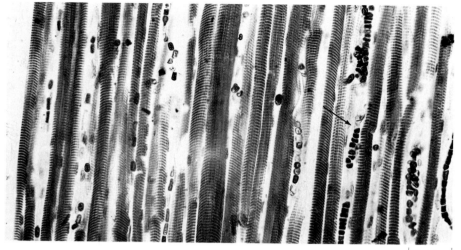

150 μm

Muscles attached to bone move the vertebrate endoskeleton. They characteristically work in antagonistic pairs, with one slowly relaxing as the other contracts. For example, when you move your hand toward your shoulder, as shown here, the biceps contracts and the triceps relaxes. When you move your hand down again, the triceps contracts, while the biceps relaxes. The muscles that move the skeleton, such as those diagramed here, are known as striated muscles.

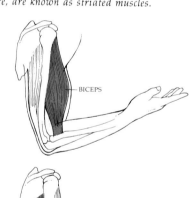

—BICEPS

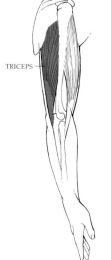

TRICEPS—

Muscle Tissue

About 40 percent of our body's weight is made up of muscle tissue. There are two general types, smooth muscle and striated muscle, so called because of its striped appearance (Figure 23-5). Smooth muscle surrounds the walls of the internal organs, such as the digestive tract and uterus, and is not generally under voluntary control. Striated muscle is the type of muscle that moves the skeleton. A special type of striated muscle, cardiac muscle, makes up the wall of the heart.

In this chapter, we shall focus on skeletal muscle because it is the subject of much current research, and much of what we know about the other forms of muscle is inferred from what has been learned about skeletal muscle. Also, studies of the machinery of skeletal muscle, involving as they do both electron microscopy and biochemistry, have brought scientists tantalizingly close to visualizing the precise role of individual molecules.

Muscle Action

Skeletal muscles, like all muscles, act by contracting. A skeletal muscle is typically attached to two or more bones, either directly or by means of the tough strands of connective tissue known as tendons. Some of these tendons, such as those that connect the fingers and their muscles, may be very long. When the muscle contracts, the bones move around a joint, which is held together by ligaments and contains a lubricating fluid. Most of the skeletal muscles of the body work in antagonistic pairs, one muscle flexing, or bending, the joint, the other extending, or straightening, it (Figure 23-6).

A muscle, such as the biceps, consists of bundles of muscle fibers—often hundreds of thousands of fibers—held together by connective tissue. Each fiber is a single cell with many nuclei. These fibers are very large cells—50 to 100 micrometers in diameter and, often, several centimeters long. They are surrounded by an outer cell membrane that has been given the special name of sarcolemma.

Embedded in the cytoplasm of each muscle cell, that is, each fiber, are some 1,000 to 2,000 smaller structural units, the *myofibrils*. They run parallel to the length of the cell. Each myofibril is, in turn, composed of units called *sarcomeres*. The repetition of these units gives the muscle its characteristic striped pattern.

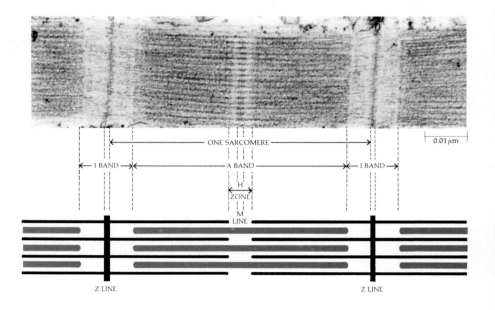

23-7
Electron micrograph and diagram of a sarco-mere, the contractile unit of muscle. Each sarcomere is composed of an array of thick and thin protein filaments arranged longitu-dinally. The Z line is where thin (actin) fila-ments from adjoining sarcomeres interweave. The A band is formed by thick filaments, and the I band by thin filaments. The H zone is that part of the A band which has no thin filaments. The thick filaments are interconnected and held in place at the M line. Muscle contraction involves the sliding of the thin filaments between the thick ones.

Figure 23–7 shows a sarcomere as seen in a longitudinal section of muscle. As the diagram shows, each sarcomere is composed of two types of filaments running parallel to one another. The thicker filaments in the central portion of the sarcomere are composed primarily of a protein known as *myosin;* the thinner filaments are made primarily of *actin,* also a protein. In cross section, each thick filament can be seen to be surrounded by six thin filaments.

Muscles function by contracting. From the cell physiologist's point of view, the exciting feature of the sarcomere is that it is the contractile mechanism. When muscle is stimulated, the thin (actin) filaments of the sarcomere slide past the thick (myosin) filaments. Since the thin filaments are anchored into each Z line, this causes each sarcomere to shorten, and thus the myofibril as a whole contracts. According to the current hypothesis, cross bridges between the thick and thin filaments form, break, and re-form rapidly, as one filament "walks" along the other (Figure 23–8).

Actin and Myosin

The actin strands in muscle are composed, it has been found, of many smaller globular subunits assembled in a long chain. As shown in Figure 23–8e, each thin filament is composed primarily of two such actin chains wound around each other.

The myosin molecule has the longest protein chains known, each one consisting of some 1,800 amino acid units. Each long protein chain has a globular structure at one end. The myosin molecule consists of two of these chains wound around each other, with the globular "heads" free. The thick filaments, in turn, are composed of bundles of myosin molecules. These globular heads apparently have two crucial functions: They are the binding sites which link the actin and myosin molecules, and they also act as enzymes to split ATP to ADP, thus providing the energy for muscle contraction.

To sum up, when a muscle fiber is stimulated, the thin (actin) filaments slide

23-8

(a) *Skeletal muscle is composed of individual muscle cells, the muscle fibers. These are cylindrical cells, often many centimeters long, with numerous nuclei. (b) Each muscle fiber is made up of many cylindrical subunits, the myofibrils. These are rods of contractile proteins that run from one end of the fiber to the other. (c) The myofibril is divided into segments, sarcomeres, by thin, dark partitions, the Z lines. The Z lines of adjacent myofibrils are in line with each other and appear to run from myofibril to myofibril* across the fiber, giving the muscle cell its striated appearance. (d) Each sarcomere is made up of thick and thin filaments. Chemical analysis shows that the thick filaments consist of bundles of a protein called myosin. Each individual myosin molecule is composed of two protein chains wound in a helix; the end of each chain is folded into a globular structure. (e) Each thin filament consists of two actin strands coiled about one another in a helical chain. Each strand is composed* of globular actin molecules. (f) According to current hypothesis, the globular myosin heads protruding from the thick filaments serve as hooks or levers, attaching to the actin molecules of the thin filament, pulling them toward each other, shortening the sarcomere and contracting the myofibril. When the myofibrils contract, the fiber contracts, and when enough fibers contract, the entire muscle shortens, producing skeletal movements.*

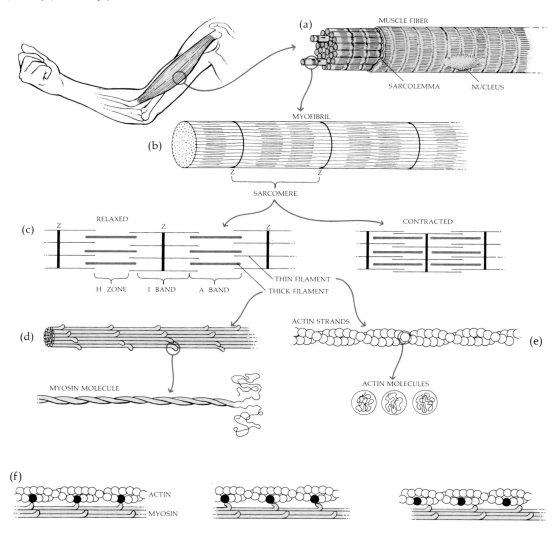

between the thick (myosin) filaments, hydrolyzing ATP to ADP in the process. The sarcomeres contract and the fibers contract. If enough fibers contract, a whole muscle contracts, and the organism or part of the organism moves. Thus the intricate machinery of the sarcomere is, in essence, a mechanism for the conversion of chemical energy into kinetic energy, the energy of motion.

23-9

Electron micrograph and diagram of smooth muscle from the lining of a trachea (windpipe) of a bat. Smooth muscle is made up of long, spindle-shaped cells containing contractile proteins. Portions of a number of these can be seen in the picture. Part of the nucleus of one cell, distinguishable by its dark color and granular texture, can be seen in the left half of the micrograph. Very thin fibrils run lengthwise through the cells and make up the bulk of the cytoplasm.

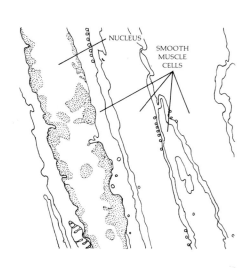

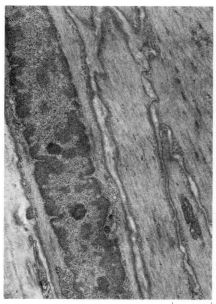

2.5 μm

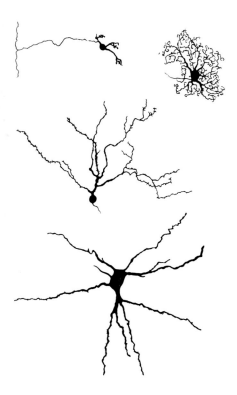

23-10

Four neurons that carry messages within the central nervous system. These drawings were made from tissues treated with a silver stain, which selectively stains the nerve cell membrane. The stain travels along the many intricately branching nerve fibers, tracing out their delicate patterns.

Smooth Muscle

Smooth muscle, unlike striated muscle, consists of cells with single nuclei (Figure 23-9). These cells contain numerous fibrils that run lengthwise through the cell but are not arranged in any discernible pattern. Like striated muscle, smooth muscle cells contain large amounts of actin and myosin, which are presumed, but not proved, to be in the fibrils and to play a role in the contraction of smooth muscle. Smooth muscle also requires ATP for contraction.

Nerve Tissue

The fourth major tissue type is nerve tissue. Nerve cells receive and transmit signals from either the external or the internal environment. They also process and store this information and are responsible for the complex functions of consciousness, memory, and thought. The basic units of nervous tissue are the neurons, or nerve cells, which make up about 10 percent of nervous tissue. The other 90 percent is made up of glial cells, which support the neurons physically and also nourish them.

A neuron consists typically of a cell body, which contains the nucleus and most of the metabolic machinery of the cell, and one or more long extensions, the nerve fibers (Figure 23-10). Along these fibers, nerve impulses are transmitted from one part of the body to another. These cells are the most morphologically spectacular of the body's cells. Nerve cells may reach astonishing lengths. For example, the fiber of a single motor neuron—a nerve cell that activates muscle—may extend from the spinal cord down the whole length of the leg right to the toe. Or a sensory neuron—one that transmits sensations to the brain—based near the spinal cord may send one fiber down to the toe and another fiber up the entire length of the spinal cord to terminate in the lower part of the brain. In a man, such a cell might be close to 2 meters long (5 meters in a giraffe).

A nerve consists of numerous nerve fibers and blood vessels bound together by connective tissue.

23–11

The knee-jerk reflex, a very simple example of nerve-cell action. When the ligament below the knee cap is tapped sharply, an impluse is carried by a sensory neuron to the spinal cord, where it is transmitted to a motor neuron. The impulse is carried by the motor neuron to the appropriate muscle, which contracts.

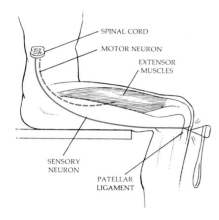

The Reflex Arc

One of the simplest examples of nerve cell function is the reflex arc. In vertebrates, it always involves two types of neurons: a sensory neuron, which carries impulses from a receptor cell to the central nervous system (the spinal cord and brain), and a motor neuron, which carries impulses from the central nervous system to some other part of the body. It also frequently involves one or more interneurons in the central nervous system, which relay impulses between the motor and sensory neurons.

In a simple reflex action, such as that illustrated in Figure 23–11, the stimulus is received by the sensory neuron and is transmitted to the spinal cord. Here the sensory neuron activates a motor neuron. The motor neuron stimulates the muscle, which contracts, producing the familiar "knee jerk.'"

In Chapter 25, we shall describe the structure of nerve cells in more detail, in particular the way in which they transmit the nerve impulse.

ORGANS AND ORGAN SYSTEMS

Organs are composed of functionally related groups of tissue. The stomach, for example, is made up of layers of glandular epithelium (the stomach lining), connective tissue, nerves, blood vessels, and smooth muscle. Organs, in turn, generally form part of *organ systems*. The stomach, for instance, is part of the digestive system. One of the major organ systems, the urinary system, is shown in Figure 23–12.

23–12

The urinary system is one of the major organ systems of the body. Fluids are processed in the kidneys, and waste products and water—the urine—pass along a pair of tubes, the ureters, to the bladder, where they are stored. Urine leaves the body by way of the urethra, which, in the male mammal, also serves as the passageway for semen.

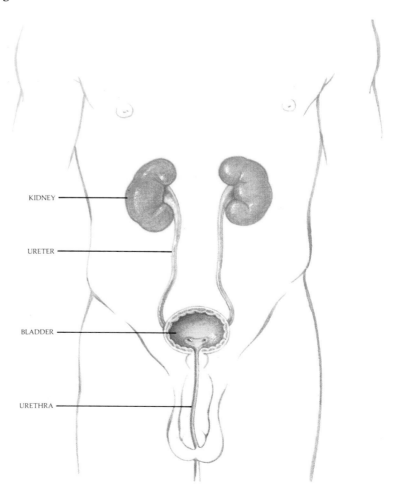

Thus, as we have seen, the human body is composed of trillions of cells of more than a hundred kinds, grouped together in many different structural and functional units. Moreover, each cell is also working for itself, breaking down glucose, making ATP, building and maintaining membranes and organelles, producing enzymes and other protein molecules, dividing or not dividing. Yet in order for the organism as a whole to survive, all of these many cells, tissues, and organs have to function as a whole.

For example, do you recall what it feels like to be in an absolute rage? The physical characteristics of a rage result from a simultaneous discharge of many nerve fibers. Certain of these cause the blood vessels in the skin and intestinal tract to contract; this contraction increases the return of blood to the heart, raising the blood pressure and sending more blood to the muscles. The heart beats both faster and stronger, and the respiratory rate increases. The pupils dilate. The muscles underlying the hair follicles in the skin contract; this is probably a legacy from our furry forebears, which looked larger and more ferocious with their hair standing on end. The rhythmic movement of the intestines stops, and sphincters, the muscles that close the intestines and the bladder, relax; these reactions inhibit digestive operations, but the relaxing of the sphincters may also have the decidedly unuseful consequence of causing involuntary defecation or urination. Hormones are produced which cause the release of large quantities of sugar from the liver into the bloodstream; this sugar provides an extra energy source for the muscles. The adrenal glands pour out epinephrine (adrenaline). As a consequence, the body as a whole is prepared for appropriate action—or, at least, action that would have been appropriate at some stage in our cultural evolution.

However, obviously, response to the external environment is only a small part of the organism's need for integration of its many individual functional units. Equally complex and vital, although less obvious, activities are required for the constant regulation of the internal environment. This maintenance of a constant internal environment is known as *homeostasis,* and as we noted in Chapter 1, homeostasis—"staying the same"—is one of the chief characteristics of living organisms. Evolution may, in fact, be viewed as a steady increase in the capacity for homeostasis. It is by regulating the internal environment that organisms have been able to free themselves to an increasing degree from the external environment, moving from salt water to fresh, and from the water to the land and air.

Homeostasis is characteristic even of single cells and very simple organisms. In an organism as complex as man, it involves constant monitoring and regulation of a large number of different factors, including oxygen and carbon dioxide, glucose and other nutrients, hormones, ions, and other organic and inorganic substances. Despite changes in the external environment and the organism's responses to it, the concentrations of these substances in the body fluids remain relatively unchanged.

Virtually every cell, tissue, and organ in the body contributes in some way to this remarkable stability. The liver acts as a metabolic factory, removing organic molecules or adding them as they are needed. Oxygen for ATP production may be used rapidly, as during muscle exertion, or slowly, as during sleep, but the rates at which the lungs take in oxygen and the heart pumps blood are regulated so that the supply in the bloodstream available to the individual cells remains constant. The kidney processes and removes just the right amount of wastes, water, and salts, and so on, down the long list of organs, tissues, and cells. Therefore, although we examine the various organ systems of the body one by one—as it is conventional to do—it is important to remember that all are acting in concert and that the function of each is affected by all the others. In Chapter 25, we shall discuss the two major control agencies—the nervous system and the endocrine system—in more detail.

SUMMARY

We are vertebrates. As such, we have a bony, jointed supporting endoskeleton, a vertebral column that encloses our spinal cords, and a skull. Our bodies contain a coelom that is divided by a muscular diaphragm into two compartments, the abdominal cavity and the thoracic cavity. Our bodies are made up of cells of four principal tissue types: epithelial tissue, connective tissue, muscle, and nerve. Epithelium serves as a covering or a lining for the other tissues of the body. It also secretes mucus, perspiration, saliva, hormones, and digestive enzymes. Connective tissue strengthens and supports the other tissues of the body. It is characterized by intercellular substances. Muscle is composed of assemblies of contractile fibers. It converts chemical energy to kinetic energy. Nerve tissue carries signals from one part of the body to another. The reflex arc is a simple example of neural function.

Tissues are groups of cells that are structurally and functionally similar. Various types of tissues are grouped in different ways to form organs, and organs form organ systems.

The various cells, tissues, and organs of the body operate in an integrated manner so that the organism is capable of responding as a whole to its external environment and of regulating its internal environment (homeostasis).

QUESTIONS

1. Define the following terms: coelom, gland, organ, intercellular substances, muscle fiber.

2. Name the four principal types of tissue and give examples of each.

3. What are the three types of epithelial tissue?

4. Diagram a sarcomere, relaxed, and label your drawing.

5. Name the principal components of a reflex action.

6. Can you interpret this electron micrograph? (You may be able to find the clue you need on page 320.)

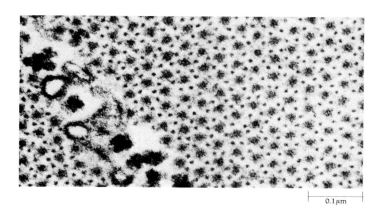

0.1 μm

CHAPTER 24

Reproduction and Development

Life began for each of us with the fusion of two remarkable cells, a sperm and an ovum. In the first part of this chapter, we shall describe the formation of these cells and the events leading to their encounter. In the last part of the chapter, we shall examine the stages by which the cell resulting from this fusion, the fertilized egg, develops into the many cells and tissues and organs that make up the human body.

SPERM AND THE MALE REPRODUCTIVE SYSTEM

Normal males from adolescence until old age produce an average of several hundred million sperm a day. These cells are produced in the *testes* (singular,

24–1

Diagram of the human male reproductive tract, showing the penis and scrotum before (dashed lines) and during erection. Sperm cells formed in the seminiferous tubules of the testis enter the vas deferens, which connects with a duct from the seminal vesicle and then passes through the prostate gland. The sperm cells are mixed with fluids, mostly from the seminal vesicles and prostate gland. The resulting mixture, the semen, is released through the urethra of the penis. The urethra is also the passageway for urine, which is stored in the bladder.

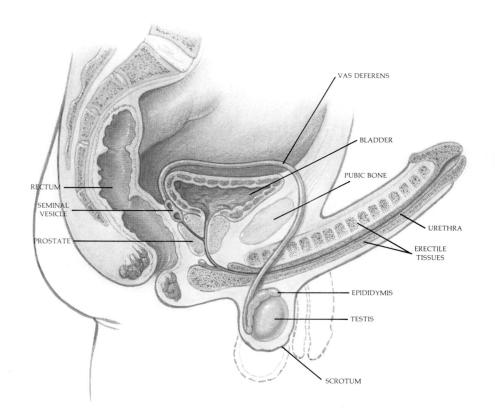

VAS DEFERENS

BLADDER

PUBIC BONE

RECTUM

SEMINAL VESICLE

PROSTATE

URETHRA

ERECTILE TISSUES

EPIDIDYMIS

TESTIS

SCROTUM

24-2

(a) *The testis is made of tightly packed coils of seminiferous tubules containing sperm cells in various stages of development. (b) As shown in the diagram of a cross section of one tubule, spermatogonia develop into primary spermatocytes. These divide (first meiotic division) into secondary spermatocytes. Each secondary spermatocyte then divides (second meiotic division) into two equal-sized spermatids. These differentiate into functional sperm cells. This process takes eight to nine weeks. During this period, the Sertoli cells support and nourish the developing sperm. (c) Diagram of a human sperm cell. The mature cell consists primarily of the nucleus, carrying the "payload" of tightly condensed DNA and associated protein, the very powerful tail (flagellum), and mitochondria, which provide the power for sperm movement. The acrosome is a specialized lysosome (see page 61).*

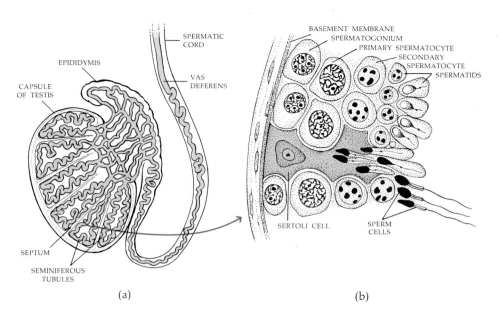

(a)   (b)

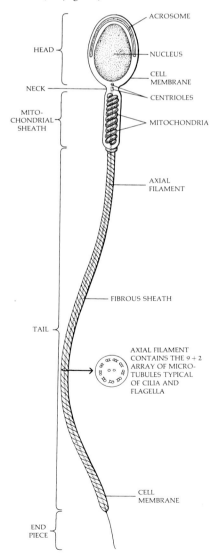

(c)

testis), which are the male gonads (gamete-producing organs). Each human testis contains about 125 meters of tightly coiled seminiferous ("seed-bearing") tubules. Cells in these tubules, the spermatocytes, divide meiotically,* each producing four haploid cells, the spermatids. These cells then differentiate into the highly specialized sperm cells. The stages in the development of the sperm cells are diagramed in Figure 9–10 on page 127.

Pathway of the Sperm

From the testis the sperm move to the epididymis, where the process of maturation continues. From the epididymis, the sperm are propelled to the vas deferens, a further extension of the tubular system, which leads from the testis into the abdominal cavity. Sperm are stored in the epididymis and the first portion of the vas deferens. Vasectomy, an increasingly popular method of birth control, involves tying off the vas deferens just above the testis, as shown in Figure 24–3.

Within the abdominal cavity, the vas deferens loops around the bladder, where it merges with the duct of the seminal vesicle. The vas deferens from each testis then enters the prostate gland and merges with the urethra, which extends the length of the penis. The penis serves both for the excretion of urine and the ejaculation of sperm.

Erection of the penis is an adaptation that ensures deposit of sperm in the female reproductive tract. It occurs as a consequence of an increased flow of blood that fills the spongy, erectile tissues of the penis (Figure 24–4). The blood flow is controlled by nerve fibers to the blood vessels of the erectile tissues. Ejaculation of sperm from the penis is a reflex similar to the simple reflex described in Chapter 23. Stimulation of the penis, such as may be produced by repeated thrusting in the vagina, results in the contraction of smooth muscle in the walls of the vas deferens, causing the sperm to empty into the urethra. Other muscles forcibly expel the sperm from the penis. These muscular spasms resulting in ejaculation account for many of the sensations associated with orgasm.

* For a review of meiosis, see pages 124 to 126.

During a vasectomy, the vas deferens on each side is severed and tied off, preventing the release of sperm from the testes. The

sperm cells are reabsorbed by the body, and the semen is normal except for the absence of sperm. This is a safe and almost painless

procedure that does not require a general anesthetic or hospitalization. Its chief drawback is that it is generally not reversible.

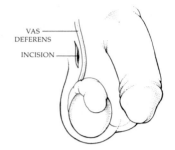

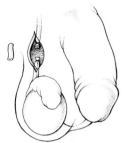

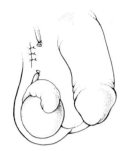

24-4

A cross section of a human penis. The penis is formed of three cylindrical masses of spongy erectile tissue that contain a large number of small spaces, each about the size of a pinhead. Erection of the penis is caused by dilation of the blood vessels carrying blood to the spongy tissues.

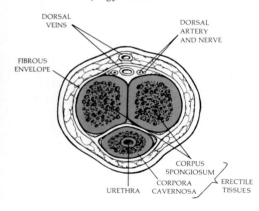

As the sperm cells move through the vas deferens, fluids (about 3 milliliters per ejaculation) are added from the seminal vesicles and prostate gland. This seminal fluid, which contains fructose, nourishes the cells and suspends them, giving them additional motility. It also helps buffer the sperm against the acidic vaginal secretions. Sperm and fluid together form the semen.

We are going to continue to follow the course of the sperm, but let us digress briefly, first, to look at some of the control mechanisms involved in spermatogenesis (sperm formation) and related activities, and second, to look at the anatomy of the female reproductive tract and the development of the ovum.

The Role of Hormones

In addition to producing the sperm cells, the testes are also the source of male hormones, known collectively as *androgens*. Androgens are produced by cells, known as interstitial cells, found in the connective tissue surrounding the seminiferous tubules. The principal androgen is testosterone.

Androgens produced in the male fetus are important in the development of the external sex organs. When the human male is about 10 years old, renewed androgen production is associated with the enlargement of the penis and testes and also the prostate, the seminal vesicles, and other accessory organs.

Androgens also affect other parts of the body not directly involved in the production and delivery of sperm. In the human male, these effects include growth of the larynx and an accompanying deepening of the voice, muscle development, skeletal size, and distribution of body hair. Androgens stimulate the biosynthesis of proteins, and therefore of muscle tissue, and also the formation of red blood cells. They stimulate sweat glands, whose secretion attracts bacteria and so produces the body odors associated with sweat after puberty. They may cause the oil-secreting glands of the skin to become overactive, resulting in acne. Characteristics associated with sexual development but not directly involved with reproduction are known as *secondary sex characteristics* (Figure 24–5).

In animals other than man, the androgens are responsible for the lion's mane, the powerful musculature and fiery disposition of the stallion, the cock's comb and spurs, the bright plumage of many adult male birds, and a variety of behavior patterns, such as the scent marking of dogs, the courting behavior of sage grouse, and various forms of aggression toward other males found in a great many vertebrate species.

24–5
Secondary sex characteristics among vertebrate males.

Regulation of Hormone Production

The production of testosterone is regulated by hormones produced by the pituitary gland. The pituitary, which is about the size and shape of a kidney bean, is located just beneath and is controlled by an area of the brain known as the hypothalamus. Under the influence of the hypothalamus, the pituitary produces a gonadotropin (a gonadotropic, or gonad-stimulating, hormone) called luteinizing hormone, or LH.

LH acts on the interstitial cells to stimulate their output of testosterone. The increased concentrations of testosterone, in turn, inhibit the release of LH by the pituitary (Figure 24–6). This type of regulatory system, in which the system is shut off when the products of its operation reach a certain level, is known as negative feedback. The most familiar example of a negative feedback system is a thermostat that turns off a furnace when the temperature rises above the desired level. Many examples of negative feedback are known, both in biological systems and in those engineered by man.

Another pituitary gonadotropin, also under the control of the hypothalamus, is follicle-stimulating hormone, or FSH. FSH acts on the seminiferous tubules, stimulating the development of sperm cells.

Production of the male sex hormone testosterone is controlled by a negative feedback system. The hypothalamus, a brain center, regulates the pituitary gland's production of the gonadotropin LH. LH stimulates production of testosterone by the interstitial cells of the testes. As testosterone increases to a certain level of concentration in the blood, it inhibits the production of LH. Decreased levels of LH result in decreased production of testosterone. FSH, another gonadotropin secreted by the pituitary under the influence of the hypothalamus, stimulates sperm production.

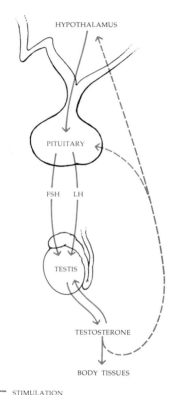

HYPOTHALAMUS

PITUITARY

FSH   LH

TESTIS

TESTOSTERONE

BODY TISSUES

———— STIMULATION
- - - - INHIBITION

Environmental stimuli can affect the production of testosterone. Studies on bulls have shown, for example, that after a bull sees a cow, the concentration of LH in his bloodstream increases as much as seventeenfold. Human testosterone production has also been shown to fluctuate in response to environmental stimuli. In some animals, production is strictly seasonal. For example, in many species of deer testosterone is almost undetectable in adult males during the period from spring to midsummer, while new antlers are growing. As soon as testosterone concentrations begin to rise, the antlers die, losing their nerve and blood supply and their "velvet" cover. While their antlers are hard and insensitive, stags have pushing contests in which they lock antlers. Hierarchy within a group is determined by antler size. In early fall, testosterone levels begin to soar, the stag's voice "changes," his neck and shoulder muscles thicken. A period of intense sexual activity, known as rutting, begins. Each stag sets out to corral a harem, which he defends until rutting season is over. Then he returns to the "stag line," his antlers drop off, and the testosterone in his blood becomes once more undetectable.

THE OVUM AND THE FEMALE REPRODUCTIVE SYSTEM

The female gonads are the ovaries, each a solid mass of cells about 3 centimeters long. The oocytes from which the eggs (ova) develop are in the outer layer of cells. The primary oocytes begin to form about the third month of fetal development. By the time of birth, the two ovaries contain some 2 million primary oocytes, which have reached prophase of the first meiotic division. By the time of puberty, the number of primary oocytes has dwindled to 80,000, all still in the first prophase. Then, responding to some unknown stimulus, the pituitary begins its production of the gonadotropins LH and FSH, the ovary begins to produce the hormones estrogen and progesterone, and the menstrual cycle and *ovulation* (the release of the egg cell from the ovary) begin.

The onset of the menstrual cycle marks the beginning of puberty in the human female. The average age of puberty is 13½ years, but the range is very wide. The increased production of female sex hormones induces the development of secondary sex characteristics, such as pubic and axillary (underarm) hair and enlargement of hips and breasts.

Under the influence of hormones, the first meiotic division of a primary oocyte resumes, resulting in a secondary oocyte and a polar body. The process is completed about the time of ovulation. For the next 30 to 40 years, the oocytes mature, usually one at a time, about every 28 days. Therefore, as many as 50 years may elapse between the beginning and the end of the first meiotic division in a particular oocyte. The second meiotic division, resulting in production of an ovum and a polar body from the secondary oocyte, takes place only after fertilization.

24–7

The chemical structures of testosterone, estrogen, and progesterone. Note that the hormones differ only very slightly chemically, in contrast to the great differences in their physiological effects—another example of the extreme specificity of biochemical actions. All of these belong to a group of chemicals known as steroids, characterized by the four-ring structure shown here.

TESTOSTERONE

ESTROGEN

PROGESTERONE

24-8

The female reproductive organs. Notice that the uterus lies at right angles to the vagina. This is one of the consequences of the bipedalism and upright position of Homo sapiens *and one of the reasons that childbirth is more difficult for the human female than for other mammals.*

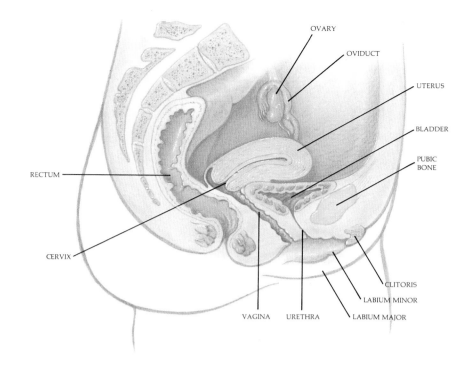

24-9

Oocytes develop near the surface of the ovary within follicles. After an oocyte is discharged from a follicle (ovulation), the remaining cells of the ruptured follicle give rise to the corpus luteum. If the oocyte is not fertilized, the corpus luteum is reabsorbed in two to three weeks. If the oocyte is fertilized, the corpus luteum persists, producing progesterone, which prepares the uterus for the embryo.

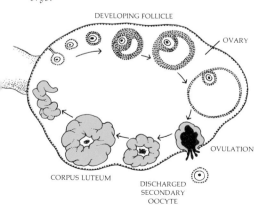

The maturation of the ovum involves not only meiosis but an increase in size, reflecting the accumulation of stored food reserves and metabolic machinery, such as ribosomes, messenger RNA, and enzymes required for the early stages of development. As you can see in Figure 9–11 on page 127, oocytes do not divide into equal-sized cells, as spermatocytes do. Instead, one very large cell (100 micrometers in diameter in humans) is produced, and the other nuclei are, in effect, discarded.

The cells surrounding the oocyte in the ovary are known as follicle cells, and oocyte and follicle cells are called the *ovarian follicle*. The follicle cells nourish the developing egg cell and also secrete estrogen, which initiates the buildup of the lining of the uterus (the endometrium). After the egg cell has been discharged from the ovary, the follicle cells undergo changes, forming the *corpus luteum* ("yellow body"). The series of steps involved in the maturation and release of an oocyte are shown in Figure 24–9. After ovulation, the cells of the corpus luteum secrete estrogen, although in decreased amounts, and also a second hormone, progesterone. Progesterone further prepares the endometrium for pregnancy and stimulates the milk-forming ducts of the breast. If pregnancy occurs, the corpus luteum persists. If pregnancy does not occur, the corpus luteum degenerates within 14 days, estrogen and progesterone levels fall, and the lining of the uterus is discarded. This recurring pattern of events is known as the *menstrual cycle* (Figure 24–10).

The released oocyte is drawn into the oviduct (also called the Fallopian tube) by the beating of cilia on long fingerlike projections at the opening of the tube (Figure 24–11). (This mechanism is so effective that women who have only one ovary and only one oviduct—and these on opposite sides of the body—have become pregnant.) The egg cell then moves slowly down the oviduct, propelled by the cilia that line the tube. The journey from the oviduct to the uterus takes about 5 days. An oocyte lives, however, only about 24 hours after it is ejected from the follicle, unless it is fertilized. So fertilization, if it occurs, generally takes place in an oviduct.

24-10

Diagram of events taking place during the menstrual cycle. The cycle begins with the first day of menstrual flow, which is caused by the shedding of the endometrium, the lining of the uterine wall. The increase of FSH during the first week promotes the growth of the ovarian follicle and its secretion of estrogen. Under the influence of estrogen, the endometrium regrows. A sharp increase of LH from the pituitary about midcycle stimulates the release of the egg cell (ovulation). (It is not known what role, if any, is played by the simultaneous increase in FSH.) Following ovulation, LH and FSH levels drop. The follicle is converted to the corpus luteum, which secretes estrogen and also progesterone. Progesterone further stimulates the endometrium, preparing it for implantation. If pregnancy does not occur, the corpus luteum degenerates, the production of progesterone and estrogen falls, the endometrium begins to slough off, FSH and LH concentrations increase once more, and the cycle begins anew.

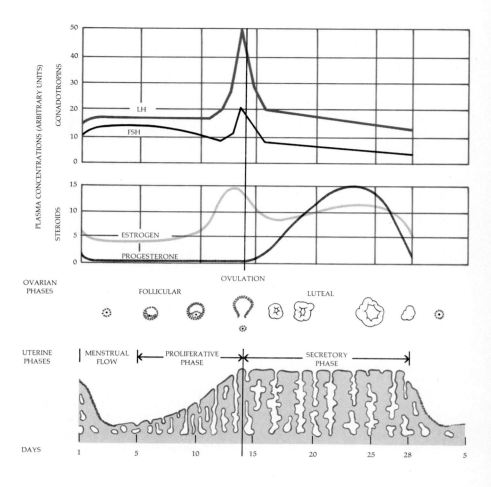

24-11

Fertilization of the egg by sperm. Once a month in the nonpregnant female of reproductive age, an oocyte breaks loose from the ovary and is swept into one of the oviducts. Fertilization, when it occurs, generally takes place within an oviduct, after which the fertilized egg becomes implanted in the lining of the uterus. Muscular movements of the oviduct, plus the beating of the cilia that line it, propel the egg cell down the tube toward the uterus. If the egg cell is not fertilized, it dies, usually within 12 to 24 hours. A sperm cell has an average life of 48 hours within the female reproductive tract.

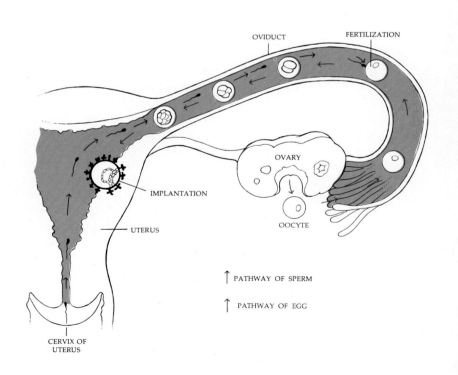

Table 24-1 *Major Mammalian Gonadotropic and Sex Hormones*

HORMONE	PRINCIPAL SOURCE	PRINCIPAL EFFECTS	CONTROL
FSH	Pituitary	Stimulates sperm production; stimulates growth of ovarian follicle, stimulates estrogen production	Hypothalamus
LH	Pituitary	Stimulates testosterone production; stimulates release of egg cell, stimulates progesterone production	Hypothalamus
Testosterone	Testes	Produces and maintains male sex characteristics, stimulates sperm production	LH
Estrogen	Ovary (follicle, corpus luteum), placenta	Produces and maintains female sex characteristics, thickens lining of uterus	FSH
Progesterone	Ovary (corpus luteum), placenta	Prepares uterine lining for pregnancy, inhibits uterine movements, promotes development of milk ducts	LH

Uterus, Vagina, and Vulva

The uterus is a hollow, muscular, pear-shaped organ about 7.5 centimeters long and 5 centimeters wide. It is lined by the endometrium. The opening of the uterus is the cervix, through which the sperm pass on their way toward the oocyte and through which the fetus emerges at the time of birth. The vagina is a muscular tube about 7.5 centimeters long that leads from the cervix of the uterus to the outside of the body. It is the receptive organ for the penis and also the birth canal. Its opening is between the urethra, the tube leading from the bladder, and the anus.

The external genital organs of the female, collectively known as the vulva, include the clitoris, which corresponds to the penis in the male (in the early embryo, the structures are identical), and the labia. The labia (singular, labium) are folds of skin. The labia majora are fleshy and, in the adult, covered with pubic hair; they enclose and protect the underlying, more delicate structures. The labia minora are thin and membranous.

Orgasm in the Female

Under the influence of a variety of stimuli, the clitoris, the labia, and other tissues in the pelvic region of the female become engorged and distended with blood, as does the penis of the male. This process is somewhat slower in women than in men, largely because the valves trapping the blood in the female sexual structures are not as efficient as in the male. The distension of the tissues is accompanied by the secretion into the vagina of a fluid that lubricates the walls of the vagina.

Orgasm in the female, as in the male, is the consequence of rhythmic muscular contractions, followed by expulsion of the blood trapped in the engorged tissues. Homologous muscles produce orgasm in the two sexes, but in the female there is no ejaculation of fluid through the urethra, as occurs in the male. Orgasm in the female is not necessary for conception.

Table 24-2 *Methods of Contraception Currently Available*

METHOD	MODE OF ACTION	EFFECTIVENESS (PREG-NANCIES PER 100 WOMEN PER YEAR)	ACTION NEEDED AT TIME OF INTERCOURSE	REQUIRES INSTRUCTION IN USE	POSSIBLE UNDESIRABLE EFFECTS
Vasectomy	Prevents release of sperm	0	None	No	Usually produces irreversible sterility
Tubal ligation	Prevents passage of egg cell to uterus	0	None	No	Usually produces irreversible sterility
"The pill" (estrogen and progesterone)	Prevents follicle maturation and ovulation	0–10	None	Yes, timing	Early—some water retention, breast tenderness, nausea Late—possible blood clots, hypertension
Intrauterine device (coil, loop, IUD)	Possibly prevents implantation	1–5	None	No	Menstrual dis-comfort, possible displacement or loss of device, possible uterine infection
"Minipill" (progesterone alone)	Probably prevents sperm from entering uterus	1–10	None	Yes, timing	?
"Morning-after pill"	Arrests pregnancy, probably by preventing implantation, 50x normal dose of estrogen	?	None	Yes, timing	Breast swelling, nausea, water retention, cancer (?)
Condom (worn by male)	Prevents sperm from entering vagina	3–10	Yes, male must put on after erection	Not usually	Some loss of sensation in male
Diaphragm with spermicidal jelly	Prevents sperm from entering uterus, jelly kills sperm	3–17	Yes, insertion before intercourse	Yes, must be inserted correctly each time	None known
Vaginal foam, jelly alone	Spermicidal, mechanical barrier to sperm	3–22	Yes, requires application before intercourse	Yes, must use within 30 minutes of intercourse; leave in at least 6 hours after	None usually, may irritate
Withdrawal	Removes penis from vagina before ejaculation	9–25	Yes, withdrawal	No	Frustration in some
Rhythm	Abstinence during probable time of ovulation	13–21	None	Yes, must know when to abstain	Requires abstinence during part of cycle
Douche	Washes out sperm that are still in the vagina	?–40	Yes, immediately after	No	None

CONTRACEPTIVE TECHNIQUES

A variety of contraceptive techniques is now available for couples who wish to prevent or defer pregnancy. In Table 24-2, they are rated in order of effectiveness, in terms of average number of pregnancies per year among the women of child-bearing age using the techniques. (Using no contraceptive methods, 90 percent of child-bearing age women having regular sexual intercourse become pregnant within a year.) In most cases, two figures are given for effectiveness. The first, lower, figure is an "ideal" figure, obtainable when the method is used consistently and correctly. The second figure is an average figure, reflecting actual experience.

HUMAN DEVELOPMENT

Fertilization and Implantation

About 300 to 400 million sperm cells are released into the vagina at intercourse. Of these, several hundred thousand, swimming at the rate of about 2 to 3 centimeters an hour, make their way up the oviducts, moving against the beating of the cilia that line the tubes. Sperm cells can survive about 48 hours in the female reproductive tract.

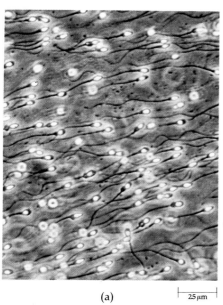

(a) |— 25 µm —|

24-12

(a) *Human sperm. About 300 to 400 million sperm cells are present in the ejaculate of an average, healthy adult male. (b) A human egg. The dark strands within the nucleus are chromosomes. A polar body is at the lower right. (a, Fritz Goro, Time-Life Picture Agency; b, Roberts Rugh and Landrum B. Shettles, M.D., From Conception to Birth: The Drama of Life's Beginnings, Harper & Row, Publishers, Inc., New York, 1971.)*

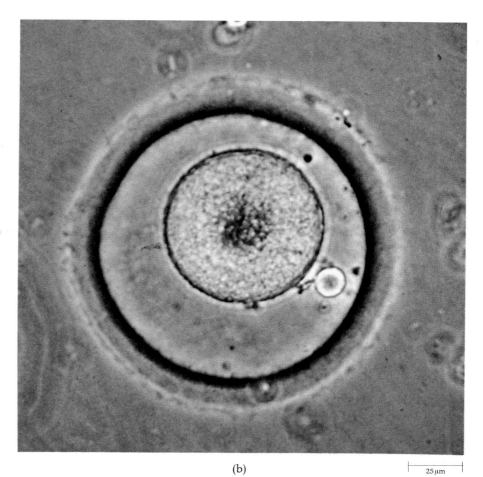

(b) |— 25 µm —|

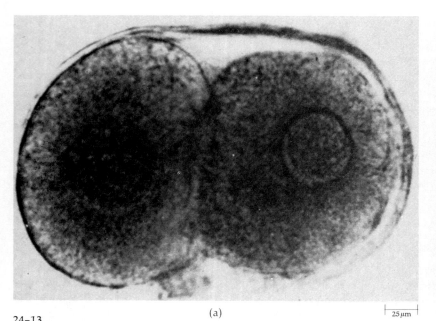

24-13

(a) *A human egg at the two-celled stage. The cells are still surrounded by an outer mem-* *brane. (b) The cells continue to divide, but, since the mass of the embryo does not in-* *crease, they are still easily contained within the same membrane.*

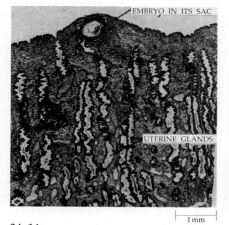

EMBRYO IN ITS SAC

UTERINE GLANDS

1 mm

24-14

Implantation. The tiny embryo invades the lining of the uterus within a week after fertilization. Subsequently, the placenta begins to form; this organ is the source of hormones that help to maintain pregnancy. Implantation usually occurs 3 to 4 days after the young embryo reaches the uterus.

Although only one sperm cell fertilizes the ovum, a large number must reach it for fertilization to occur (see Figure 9–1 on page 121). Apparently, the sperm cells release enzymes that lyse the follicle cells surrounding the ovum. The membrane of one sperm cell fuses with the membrane of the egg, and the sperm cell's contents are emptied into the egg's cytoplasm. The two nuclei meet within the egg cell and merge. This is the most important event in an individual's life. In fact, this event determines whether or not that particular individual—that genetic identity—will ever exist.

After fertilization, the egg continues its passage down the oviduct, where the first cell divisions take place. At about 36 hours after fertilization, the fertilized egg divides to form two cells; at 60 hours, the two cells divide to form four cells. At three days, the four cells divide to form eight. As the cells divide further, an inner space is developed, so that a hollow ball of cells is formed. At this stage, the ball of cells is called a *blastula*. One side of the ball of cells, which is thicker than the other side, will develop into the embryo itself, and the remaining cells will develop into the membranes that enclose the embryo. Although the embryo soon consists of many cells, it is not significantly larger than the single egg cell from which it originated.

At about the sixth day, the embryo makes contact with the tissues of the uterus. As a result of chemical interactions of the tiny mass of cells and the rich uterine lining, the outer epithelium of the lining breaks down and the embryo becomes embedded (implanted) in the nourishing tissue (Figure 24–14). As the embryo descends into the tissues of the uterus, it becomes surrounded by ruptured blood vessels and the nutrient-filled blood escaping from them. By this stage, the embryo is developing extraembryonic membranes analogous to the membranes surrounding the chick embryo (see the boxed essay on the next page). The outermost of these membranes, the chorion, develops fingerlike

THE AMNIOTE EGG

As we noted in Chapter 18, one of the most significant evolutionary steps among the Animalia was the development of the amniote egg, which contains its own water supply and so can be laid on land. In the amniote egg, the embryo is enveloped in a liquid-filled membrane, the amnion.

The amniote egg actually includes four membranes, known collectively as the extraembryonic membranes. Each develops from the embryo itself. The figures show the development of these membranes in the chick egg, the most familiar example of the amniotic egg.

In early development, body folds of the embryo begin to separate from the underlying yolk. One membrane, the yolk sac, grows around and almost completely encloses the yolk. A second, the allantois, arises as an outgrowth of the rear of the gut. The third and fourth are elevated over the embryo by a folding process during which the membrane is doubled. When the folds fuse, two separate membranes are formed. The inner one is the amnion, and the outer one is the chorion. The chorion eventually fuses with the allantois to form the chorioallantoic membrane, which, in the later stages of development, encloses embryo, yolk, and all the other structures.

Although only a few mammals lay eggs (these are the monotremes, such as the duckbilled platypus), the mammalian fetus develops within a system of membranes which closely resemble those found in the amniote egg. Even the yolk sac is still present, enclosing a space where the yolk "used to be." Among the placental mammals, the outermost embryonic membrane, the chorion, has become specialized to form part of the placenta, through which the embryo receives its nourishment.

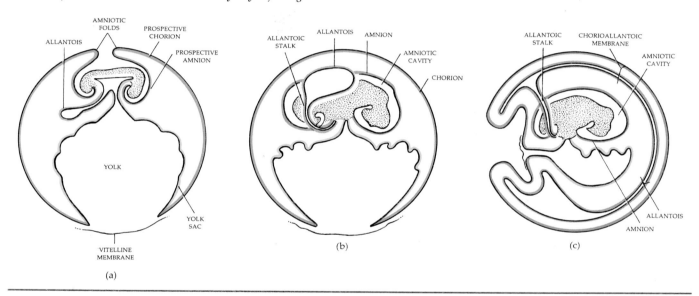

(a)   (b)   (c)

projections that further invade the mother's uterine tissues. Blood vessels develop in these tissues that connect with the blood vessels of the embryo as they develop. Ultimately, the placenta forms.

The placenta is a spongy tissue through which oxygen, food molecules, and wastes are exchanged between mother and embryo. It is formed from a maternal tissue, the endometrium, as well as from the chorion of the fetus, and has a rich blood supply from both. Within the placenta, the capillaries of the fetal and maternal circulatory systems intertwine but are not directly connected. Molecules, including food and oxygen, diffuse from the maternal bloodstream through the placental tissue and into the blood vessels that carry them into the fetus. Similarly, carbon dioxide and other waste products from the fetus are picked up from the placenta by the maternal bloodstream and carried away for disposal through the mother's lungs and kidneys.

24-15

Levels of estrogen, progesterone, and chorionic gonadotropin excreted in the urine during pregnancy. Urinary excretion rates indicate the concentrations of these hormones in the blood.

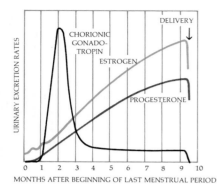

24-16

As the embryo develops, two ridges form and enlarge. The groove between them is called the neural groove. From the neural groove, the dorsal nerve cord—one of the "trade-marks" of the chordates—develops. These two folds soon close. This embryo is 18 days old.

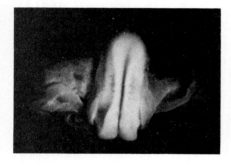

The placenta is also a source of hormones. Early in pregnancy, it produces a gonadotropin known as chorionic gonadotropin (Figure 24-15). Pregnancy tests involve the detection of this hormone in blood or urine. Chorionic gonadotropin stimulates the corpus luteum to continue its production of estrogen and progesterone. Later in pregnancy, the placenta itself produces estrogen and progesterone in large amounts. As the embryo grows larger, it remains attached to the placenta by the long umbilical cord, which permits it to float freely in its sac of amniotic fluid.

Completing the First Trimester

During the second week of its life, the embryo grows to 1.5 millimeters in length and its major body axis begins to develop. (In this and subsequent measurements, the fetus is measured from crown to rump.) As the body elongates, it can be seen to be divided into segments, known as somites.

During the third week, the embryo grows to 2.3 millimeters long and most of its major organ systems begin to form: the neural groove (Figure 24-16), which is the beginning of the central nervous system (spinal cord and brain), the first organ system to develop; the heart and blood vessels; the primitive gut; and the muscle rudiments.

By 21 days, the eyes begin to form. Also, by this time, about 100 cells have been set aside (in the yolk sac) as germ cells, from which the ova or sperm cells of the individual will eventually develop. By 24 days, the very rudimentary heart, still only a tube, begins to flutter and then to pulsate. From this time on, the heart will not stop its 100,000 beats per day until the death of the individual.

By the end of the first month, the embryo is 5 millimeters (about 0.2 inch) in length and has increased its mass 7,000 times. The neural groove has closed, and the embryo is now C-shaped (Figure 24-17). At this stage, it can be clearly seen that the tissues lateral to the notochord are arranged in paired somites.

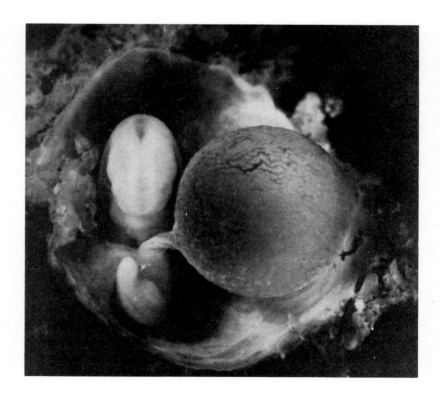

24-17

A human embryo at 28 days. The balloon-like structure is the yolk sac. The embryo is curved toward you. At the top you can see the bulge of the rudimentary brain. The "seam" where the neural ridges closed is still visible.

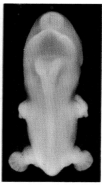

24-18

A human embryo at 40 days, front and back views. Notice the spinal cord, brain, and paddlelike feet. The embryo is now about 16 millimeters long, little more than half an inch. (Roberts Rugh and Landrum B. Shettles, M.D., From Conception to Birth: The Drama of Life's Beginnings, Harper & Row, Publishers, Inc., New York, 1971.)

Each embryo has 40 pairs of somites, from which muscles, bones, and connective tissues will develop. This segmentation of the muscles persists in the adult forms of lower vertebrates—fish, particularly—but not in the higher, terrestrial vertebrates. The heart, even as it beats, develops from a simple contracting tube to a four-chambered vessel.

During the second month, the embryo increases in mass about 500 times. By the end of this period, it weighs about 1 gram ($\frac{1}{30}$ ounce), slightly less than the weight of an aspirin tablet, and is about $2\frac{1}{2}$ centimeters long. Despite its small size, it is almost human-looking, and from this time on it is generally referred to as a fetus (Figure 24–19). Its head is still relatively large, because of the early and rapid development of the brain, but the head size will continue to be reduced in proportion to body size throughout gestation (and throughout childhood as well). Arms, legs, elbows, knees, fingers, and toes are all forming during this time, and as another reminder of our ancestry, there is a temporary tail. The tail reaches its greatest length in the second month and then gradually begins to disappear. The primitive reproductive organs have begun to form by this time. The liver now constitutes about 10 percent of the body of the fetus and is its main blood-forming organ.

By the end of the second month, the major steps in organ development are more or less complete and the embryo appears quite human in its external form. In fact, the rest of development is mostly concerned with growth and the maturation of physiological processes.

The first two months are the most sensitive period as far as the possible influence of external factors is concerned. For example, when the arms and legs are mere rudiments (fourth and fifth weeks), a number of substances can upset the normal course of events and result in limb abnormalities. An increasing number of drugs have been linked to congenital deformities, and pregnant women and their doctors have become far more cautious about the use of any

24-19

Embryo at 2 months. The yolk sac is now smaller in comparison to the embryo but still persists. The umbilical cord, connecting the embryo to the placenta, contains both veins and arteries. Portions of the skeleton are visible. (Rugh and Shettles, op. cit.)

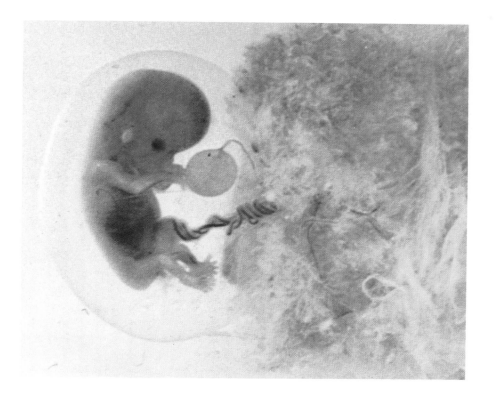

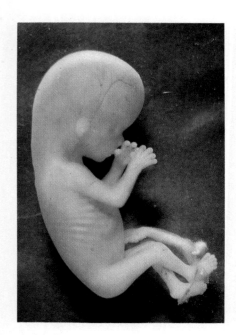

24-20

Another 2-month-old human embryo. Note the development of the skeleton and of the hands and feet. (Rugh and Shettles, op. cit.)

kind of medication during these critical first months. Similarly, exposure to x-rays at doses that would not affect an adult or even an older fetus may produce permanent abnormalities.

Infections may also affect the development of the embryo. Rubella (German measles) is a very mild disease in children and adults. Yet when contracted by the mother during the fourth through the twelfth weeks of pregnancy, it can have damaging effects on the formation of the heart, the lens of the eye, the inner ear, and the brain, depending on exactly when the infection occurs in relation to embryonic development.

During the third month, the fetus begins to move its arms and kick its legs, and the mother may become aware of its movements. Reflexes, such as the startle reflex and (by the end of the third month) sucking, first appear at this time. Its face becomes expressive; the fetus can squint, frown, or look surprised. Its respiratory organs are fairly well formed by this time but, of course, are not yet functional. The external sexual organs begin to develop.

By the end of this month, the fetus is about 9 centimeters long from the top of its head to its buttocks and weighs about 15 grams (½ ounce). It can swallow and occasionally does swallow some of the fluid that surrounds it in the amniotic sac. The finger, palm, and toe prints are now so well developed that they can be clearly distinguished by ordinary fingerprinting methods. The kidneys and other structures of the urinary system develop rapidly during this period, although waste products are still disposed of through the placenta. By the end of this period— the first trimester of development—all the major organ systems are laid down.

24-21

Human fetus at 2 months, 1 week, now almost 4 centimeters long. The eyes have lenses but are covered by lids that will fuse during the third month and remain closed for the next three months. (Rugh and Shettles, op. cit.)

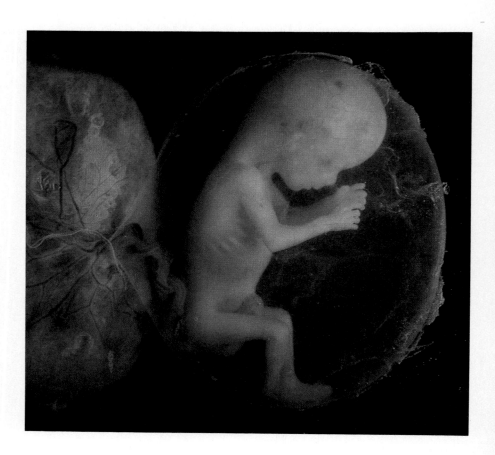

Fetus at 16 weeks. Blood vessels are visible through the translucent skin. The hands and feet are well formed; even the fingernails are clearly visible. The fetus is now about 13 centimeters long and fills the uterus, which is expanding as the fetus grows. (Rugh and Shettles, op. cit.)

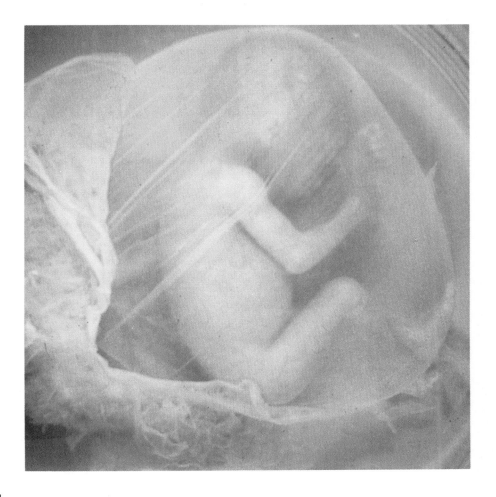

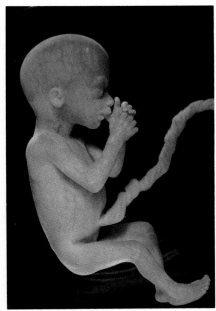

24–23

A 17-week-old fetus, sucking its thumb. (Rugh and Shettles, op. cit.)

The Second Trimester

During the fourth month, movements of the fetus become obvious to the mother. Its bony skeleton is forming and can be visualized by x-rays. The body is becoming covered with a protective cheesy coating. The four-month-old fetus is about 14 centimeters long and weighs about 115 grams (4 ounces).

By the end of the fifth month, the fetus has grown to almost 20 centimeters and now weighs 250 grams. It has acquired hair on its head, and its body is covered with a fuzzy, soft hair called the lanugo, from the Latin word for "down." Its heart, which beats between 120 and 160 times per minute, can be heard with a stethoscope. The five-month-old fetus is already discarding some of its cells and replacing them with new ones, a process that will continue throughout its lifetime. At this stage, the placenta covers about 50 percent of the uterus.

During the sixth month, the fetus has a sitting height of 30 to 36 centimeters and weighs about 680 grams. By the end of the sixth month, it could survive outside the mother's body, although probably only with respiratory assistance in an incubator. Its skin is red and wrinkled, and the cheesy body covering, which helps protect the fetus against abrasions, is now abundant. Reflexes are more vigorous. In the intestines is a pasty green mass of dead cells and bile, known as meconium, which will remain there until birth.

Final Trimester

During the final trimester, the fetus increases greatly in size and weight. In fact, it doubles in size just during the last two months. During this period, many new nerve tracts are forming and new brain cells are being produced at a very rapid rate. By the seventh month, brain waves can be recorded, through the abdomen of the mother, from the cerebral cortex of the fetus. Some recent research indicates that the protein intake of the mother is important during this period if the child is to have full development of its brain. Alcohol consumed by the mother during this trimester is concentrated by and can lead to tissue damage in the fetus.

During the last month of pregnancy, the growth rate of the baby begins to slow down. (If it continued at the same rate, the child would weigh 200 pounds by its first birthday.) The placenta begins to regress and becomes tough and fibrous.

24–24

A human fetus, shortly before birth, showing the protective membranes surrounding it and the uterine tissues. The cervical plug is composed largely of mucus. It develops under the influence of progesterone and serves to exclude bacteria and other infectious agents from the uterus. In 95 percent of all births, the fetus is in this head-down position.

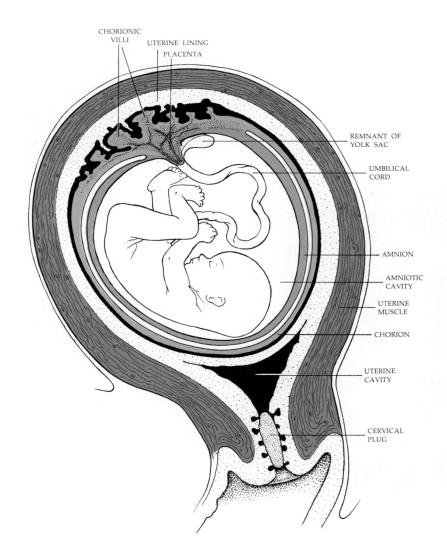

CHORIONIC VILLI

UTERINE LINING

PLACENTA

REMNANT OF YOLK SAC

UMBILICAL CORD

AMNION

AMNIOTIC CAVITY

UTERINE MUSCLE

CHORION

UTERINE CAVITY

CERVICAL PLUG

24-25

Birth of a baby. (a) Mother in delivery room just before the beginning of the second stage of labor. (b) A strong contraction. (c) The baby appears. Visible on the top of the baby's head is a suction device (much safer than forceps) that has been placed there to aid in maneuvering the infant into position for delivery. (d) Doctor holding newborn infant.

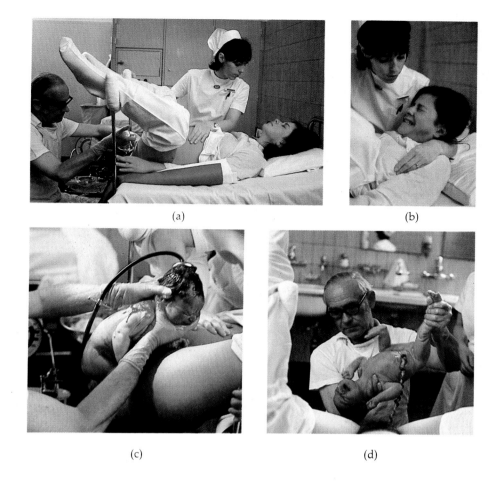

(a)

(b)

(c)

(d)

Birth

The date of birth is calculated as about 266 days after conception or 280 days after the beginning of the last regular menstrual period, but only some 75 percent of babies are born within two weeks of the scheduled time. Labor is divided into three stages: dilation, expulsion, and placental stages. Dilation, which lasts from 2 to 16 hours (it is longer with the first baby than with subsequent births), begins with the onset of contractions of the uterus and ends with the full dilation, or opening, of the cervix. At the beginning of the stage, uterine contractions occur at intervals of about 15 to 20 minutes and are relatively mild. By the end of the dilation stage, contractions are stronger and occur about every 1 to 2 minutes. At this point, the cervix is dilated to about 10 centimeters in diameter. Rupture of the amniotic sac, with the expulsion of fluids, usually occurs during this stage.

The second, or expulsion, stage lasts 2 to 60 minutes. It begins with the full dilation of the cervix and the appearance of the head in the cervix, called crowning. Contractions at this stage last from 50 to 90 seconds and are 1 or 2 minutes apart.

The third, or placental, stage begins immediately after the baby is born. It involves some contractions of the uterus and the expelling from the vagina of fluid, blood, and finally the placenta (also called the afterbirth) with the umbilical cord attached. The infant has begun his or her separate existence.

SUMMARY

The male and female gametes, sperm and ova, are formed by meiosis in the gonads (the testes and the ovaries). Sperm cells are produced in the seminiferous tubules of the testes. These sperm cells enter the epididymis, a tightly coiled tubule overlying the testis. The epididymis is continuous with the vas deferens, which leads through the abdominal cavity, around the bladder, and into the prostate. Within the prostate, each vas deferens merges with the urethra, which leads through the penis.

The penis is composed largely of spongy erectile tissue that can become engorged with blood, enlarging and hardening it. At the time of ejaculation, sperm are propelled along the vas deferens by contractions of a surrounding coat of smooth muscle. Secretions from the seminal vesicles and the prostate are added to the sperm as they pass to the urethra. Semen is expelled from the urethra by muscular contractions.

Production of sperm and the development of male secondary sex characteristics are under the control of hormones, including testosterone (an androgen) and two gonadotropins, luteinizing hormone (LH) and follicle-stimulating hormone (FSH). LH acts on the interstitial cells, located between the seminiferous tubules, to stimulate the production of testosterone. FSH and testosterone stimulate the production of sperm. The gonadotropins are produced by the pituitary gland under the regulation of the hypothalamus, a brain center. Production of LH is inhibited by the presence of testosterone through a negative feedback system.

The female gamete-producing organs are the ovaries. The primary oocytes develop within nests of cells called follicles. The first meiotic division of the oocyte begins in the female fetus and is completed at ovulation. The second meiotic division is completed at fertilization. On the average, one oocyte is released every 28 days; it travels down the oviduct to the uterus. Fertilization of the egg, if it occurs, usually takes place in an oviduct. Later, it normally becomes implanted in the lining (the endometrium) of the uterus. If the oocyte is not fertilized, it degenerates and the endometrial lining is shed (menstruation).

The production of ova, the menstrual cycle, and development at puberty of the uterus, vulva, and breasts and other secondary sex characteristics of the female are controlled by estrogen and progesterone, both steroid hormones, and the gonadotropins LH and FSH. Estrogen is produced by the ovarian follicles before ovulation. After ovulation, the corpus luteum, which forms from the emptied follicle, produces both estrogen and progesterone. Progesterone and estrogen combined stimulate the growth of the endometrium.

Fertilization occurs in the oviduct, and the early stages of development take place there. During these stages, there is a large increase in the number of cells but little or no increase in the total size of the embryo. Implantation in the endometrium occurs about a week after fertilization. Subsequently, extraembryonic membranes combine with tissues of the endometrium to form the placenta, through which the fetus receives its food and oxygen and excretes carbon dioxide and other waste products. The fetus is attached to the placenta by the umbilical cord.

During the first three months, although the fetus remains very small (less than 8 centimeters in length), all the major organ systems are established and externally the fetus takes on a human appearance. This is the most critical period of development for the production of abnormalities.

During the second trimester, movements of the fetus can be felt by the mother, the heartbeat can be heard with a stethoscope placed on the mother's abdomen, and the skeleton, although still cartilaginous, can be visualized by x-rays. By the end of six months, survival outside the uterus is possible, although an incubator would probably be required.

During the third trimester, there is a great increase in size and weight. New brain cells are formed rapidly during this period.

Birth occurs, on the average, 266 days after conception. Labor consists of a period of dilation and contraction, during which the cervix opens and the muscular wall of the uterus contracts with increasing force; expulsion, during which the baby is expelled from the uterus; and the placental stage, during which the placenta, blood, and other tissues are expelled.

QUESTIONS

1. Define the following: secondary spermatocyte, Sertoli cell, corpus luteum, FSH, vagina, oviduct, implantation, amnion.

2. Diagram the feedback regulation of house temperature by a thermostat.

3. What would constitute the semen of a man who had had a vasectomy?

4. During which days in the menstrual cycle is a woman most likely to become pregnant? (Include data on longevity of eggs and sperm in making this calculation.) Why, in your opinion, is use of the calendar rhythm method of birth control so ineffective?

CHAPTER 25

Integration and Coordination

The nervous system and the endocrine system, the hormone-secreting glands and their products, are the two great communication networks of the body. They operate continuously to integrate and control the body's activities. In this chapter, we are going to examine these vast communication networks in order to provide a background for understanding the control of such functions as circulation, respiration, digestion, and excretion, which are the subjects of the chapters that follow. The discussion of the brain and of some of the current research on brain function is reserved for the final chapter.

THE NERVOUS SYSTEM

In the preceding chapter, and also in Chapter 21, you have seen some of the ways in which hormones affect particular cells in the organism, stimulating or inhibiting their growth or other activities. Functionally, the nervous system differs from the endocrine system chiefly in its capacity for rapid response—a nerve impulse can travel through the body in a matter of milliseconds. Hormones, however, move at a somewhat slower rate (by way of the bloodstream) and, characteristically, elicit a slower response. Plants, which are very slow living organisms, rely largely on an elaborate interplay of hormones to coordinate their activities. Animals, in general, are characterized by the presence of networks of specialized nerve cells. Mammals, birds, octopuses, and many insects, in particular, have highly developed and highly tuned nervous systems that monitor internal and external changes and initiate responses to them.

NEURONS

The functional unit of the nervous system, as we noted in Chapter 23, is the neuron. A neuron has three basic parts: the cell body itself, which contains the nucleus; the _dendrites_, which receive stimuli and transmit impulses to the cell body (the cell body itself may also receive stimuli); and the _axon_, which transmits impulses away from the dendrites and cell body (Figure 25–1). Axons and dendrites are commonly referred to as nerve fibers. A single neuron may have many dendrites, but it usually has only one axon, although the axon may be branched. Vertebrate nerve fibers are often enveloped in a myelin sheath formed by Schwann cells (Figure 25–2). Myelinated fibers transmit messages more rapidly than fibers without a myelin sheath.

25-1

A motor neuron, a nerve cell that transmits signals to muscles. The stimulus is received at any point on the naked nerve surface, but usually by the dendrites. The nerve impulse travels along the axon, which is insulated by a myelin sheath composed of Schwann cells. The nodes of Ranvier are gaps in the axon sheath that occur at the junctions of adjacent Schwann cells. The cell body of a motor neuron always lies within the central nervous system. The axon of such a cell in a large vertebrate may be several meters long.

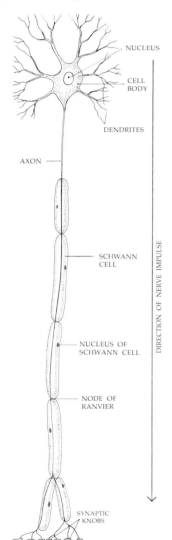

NUCLEUS

CELL BODY

DENDRITES

AXON

DIRECTION OF NERVE IMPULSE

SCHWANN CELL

NUCLEUS OF SCHWANN CELL

NODE OF RANVIER

SYNAPTIC KNOBS

25-2

(a) Part of a cross section of a nerve. You can see some of the connective tissue surrounding the nerve at the bottom of the micrograph, and numerous nerve fibers are visible. The dark areas are myelin sheaths, which insulate the fibers. They are produced by specialized cells known as Schwann cells. The fibers are bound together by connective tissue. (b) Formation of a myelin sheath. As the Schwann cell grows, it extends itself around and around the axon and gradually extrudes its cytoplasm from between the layers. In the micrograph the cytoplasms and nuclei of Schwann cells can be seen on the outer edges of some of the myelin sheaths.

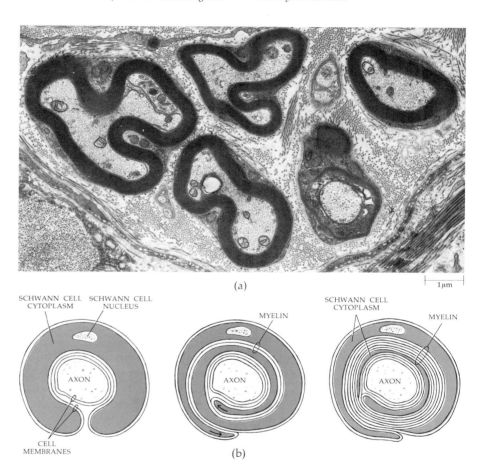

(a)

1 μm

SCHWANN CELL CYTOPLASM   SCHWANN CELL NUCLEUS

MYELIN

SCHWANN CELL CYTOPLASM

MYELIN

AXON

AXON

AXON

CELL MEMBRANES

(b)

As we mentioned in Chapter 23, there are three types of neurons in the nervous system: sensory neurons (also called afferent neurons), interneurons, and motor neurons (also called efferent neurons). *Sensory neurons* are activated by sensory stimuli. These stimuli may be received by naked nerve endings; some of the pain receptors in the human skin appear to be of this sort. More often the sensory receptors are specialized cells (Figure 25–3), often within organs, such as the eye or the ear. The impulse is carried along nerve fibers (dendrites) of the sensory neuron toward its cell body, which lies just outside the spinal cord. The impulse is then transmitted along the axon of the sensory neuron, which enters the spinal cord and, in turn, transmits the impulse to an *interneuron*. The impulse may then be transmitted to other regions of the spinal cord or brain, or, in the case of a simple reflex, to the dendrites or cell body of a *motor neuron* and transmitted along its axon to an effector, which may be either a gland or muscle (Figure 25–4). These events are summarized in Figure 25–5.

Receptors found in skin. Free nerve endings are pain receptors, end-bulbs of Krause are thought to be receptors for cold, Meissner's corpuscles may be receptors for superficial touch, *and the onion-shaped Pacinian corpuscles are for pressure. Nerve fibers are shown in color. In the Pacinian corpuscle, the best studied of the four, a single* myelinated nerve fiber (dendrite) enters the corpuscle, which is composed of many concentric layers of connective tissue. Pressure on these outer layers stimulates the dendrite.

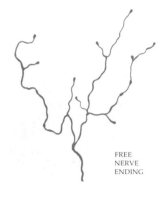

FREE
NERVE
ENDING

END-BULB
OF KRAUSE

MEISSNER'S
CORPUSCLE

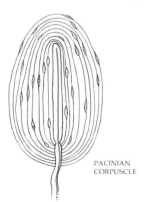

PACINIAN
CORPUSCLE

In the course of evolution, relatively few changes have occurred in the structure of the individual nerve cells, which are remarkably the same throughout the animal kingdom, but vast changes have occurred in the number and arrangement of such cells and in the uses made of them. There are about 12 billion neurons in the human body, the great majority of which are interneurons.

Impulses are transmitted from one neuron to another across specialized junctions called *synapses*. Nerves, as we noted earlier, are composed of many nerve fibers from many neurons—usually hundreds and sometimes thousands. Each fiber is capable of transmitting separate messages, like the wires in a telephone cable. The fibers themselves do not contain nuclei; the Schwann cells encircling them are nucleated, however.

The cell bodies of neurons are often found in clusters. Clusters of cell bodies within the brain and spinal cord are called *nuclei*: clusters of cell bodies outside the brain and spinal cord are called *ganglia* (singular, ganglion).

THE CENTRAL AND PERIPHERAL NERVOUS SYSTEMS

The brain and spinal cord constitute the *central nervous system*. In vertebrates the central nervous system is encased in bony supporting and protecting structures, the vertebral, or spinal, column and the skull.

The nerves that pass from the brain and spinal cord to the muscles, glands, and sensory receptors of the body constitute the *peripheral nervous system*. (Note, however, that fibers of the peripheral nervous system—either dendrites or axons—extend into the central nervous system.) Peripheral nerves are classified as *cranial nerves*, nerves connecting directly to the brain (such as the facial nerves), and *spinal nerves*, those that connect with the spinal cord.

Figure 25-6 shows a section of the human spinal cord. As you can see, it is divided into gray matter and white matter. The *gray matter* contains interneurons, the cell bodies of some motor neurons, and glial cells. As we mentioned in Chapter 23, the glial cells, which make up 90 percent of the nervous system, nourish and physically support the neurons. Glial cells are especially abundant in the brain and spinal cord.

The axon of a motor neuron divides into branches, each forming a neuromuscular junction with a different muscle fiber. The motor neuron and the numerous muscle fibers it innervates are known as a motor unit. Stimulation of a motor neuron stimulates all of the fibers in that motor unit. Within a given muscle, fibers of different motor units are intermingled.

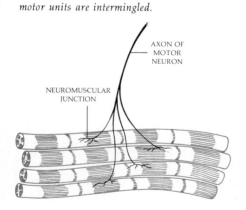

AXON OF
MOTOR
NEURON

NEUROMUSCULAR
JUNCTION

25-5

Diagram showing the functions of the different types of neurons. Sensory neurons transmit information from sensory receptors to interneurons, which transmit signals to other interneurons or to motor neurons. Depending on the nature and intensity of the original stimulus, a motor neuron may be activated. In this case, an impulse is transmitted along the fiber (axon) of the motor neuron to an effector. The cell body of the sensory neuron is in the ganglion.

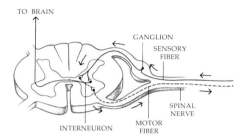

The *white matter* contains nerve fibers (axons) carrying information from various parts of the body to the brain (ascending fibers) and those relaying instructions from the brain (descending fibers). Fibers carrying similar types of information are grouped together in fiber tracts.

Pairs of spinal nerves connect with the spinal cord through spaces between the vertebrae. Each of these pairs innervates (supplies with nerves) the muscles, glands, and sensory receptors of a different and distinct area of the body. In mammals there are 31 such pairs.

The motor and sensory components of the spinal nerves separate from each other before they connect with the spinal cord. The sensory fibers feed into the dorsal (back) side of the cord; they either synapse immediately with interneurons or motor neurons as in a simple reflex, or they turn and ascend toward the brain. The cell bodies of these neurons are in the dorsal root ganglia outside the spinal cord. Motor fibers emerge from the spinal cord on the ventral (front) side. The cell bodies of these fibers are in the spinal cord, where they synapse with interneurons in the spinal cord, as shown in Figure 25–5, or with fibers descending from the brain. Most peripheral nerves contain fibers of both sensory and motor neurons.

Divisions of the Peripheral Nervous System: Somatic and Autonomic

As you can see in Table 25–1, there are two divisions of the peripheral nervous system: the somatic and autonomic. The *somatic nervous system* includes both motor and sensory neurons. The *autonomic nervous system* is entirely a motor system, consisting of the neurons that control cardiac muscle, glands, and smooth muscle (the type of muscle found in the walls of blood vessels and in the digestive, respiratory, and reproductive tracts). The autonomic nervous system is generally categorized as an "involuntary" system, in contrast to the "voluntary" somatic system, which controls the muscles that we can move at will, that is, the skeletal muscles.

25-6

A segment of the human spinal cord. Each spinal nerve divides into two fiber bundles, the sensory root and the motor root, at the vertebral column. The sensory bundle connects with the cord dorsally; the cell bodies of the sensory neurons are in the dorsal root ganglia. The motor bundle connects ventrally with the spinal cord. The sympathetic ganglia are part of the autonomic nervous system (see next page). The butterfly-shaped gray matter within the spinal cord is composed of groups of cell bodies, and the surrounding white matter consists of ascending and descending tracts of fibers.

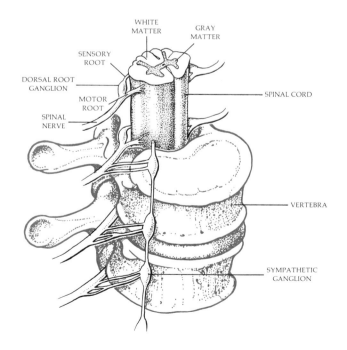

You will readily recognize that the distinction here between "voluntary" and "involuntary" is not clear-cut. Skeletal muscles often move involuntarily, as in a reflex action, and it is reported that some persons, particularly practitioners of yoga, have learned to control their rate of heartbeat and the contractions of some smooth muscle.

Anatomically, the motor neurons of the somatic system are distinct and entirely separate from those of the autonomic system, although fibers of both types may be carried in the same nerve. The cell bodies of the motor neurons of the somatic system are located within the central nervous system, with long nerve fibers (axons) running uninterruptedly all the way to the skeletal muscle. The motor fibers of the autonomic nervous system also originate in cell bodies inside the central nervous system, but they do not travel all the way to their target organ, or effector. Instead, motor neurons of the autonomic system form a synapse with a second neuron in a ganglion outside the central nervous system. This neuron innervates the target muscle or gland. These postganglionic fibers, as they are called, constitute a characteristic difference between the autonomic and somatic nervous systems. The principal differences between these two systems are summarized in Table 25-2.

Divisions of the Autonomic Nervous System: Sympathetic and Parasympathetic

The autonomic nervous system itself has two divisions, the sympathetic and the parasympathetic, which are anatomically and functionally distinct. The *parasympathetic nervous system* consists of nerve fibers from the brain and from the lower region of the spinal cord. The *sympathetic nervous system* originates in the thoracic and lumbar areas of the spinal cord. In the parasympathetic system, the point of synapse is near or in the target organ, whereas in the sympathetic system, many synapses are in a chain of ganglia running parallel to the spinal cord. They also differ in the chemical transmitter substance they release. Most postganglionic sympathetic nerve endings release epinephrine or norepinephrine, the same chemicals that are released by the adrenal gland (described later in this chapter). All parasympathetic endings release a chemical called acetylcholine. The role of these chemical transmitters is described on pages 355 to 356.

As you can see in Figure 25-7, most of the major visceral organs of the body are innervated by fibers from both the sympathetic system and the parasympathetic system. The effects of the two systems are often antagonistic to each other, simulation of one system increasing the activity of the muscle or gland while stimulation of the other inhibits it. In contrast, nerves of the somatic nervous system cannot inhibit skeletal muscle cells; they can only excite or not excite them.

Table 25-2 *Differences between the Somatic and Autonomic Nervous Systems*

	POINT OF SYNAPSE	EFFECTOR	EFFECT
Somatic	None outside central nervous system	Only skeletal muscle	Always excitation
Autonomic	Ganglia outside central nervous system	Smooth or cardiac muscle, or gland	Excitation or inhibition

25-7

The autonomic nervous system, consisting of the sympathetic and the parasympathetic systems. The presynaptic fibers of the parasympathetic system exit from the base of the brain and from the sacral region of the spinal cord and synapse at or near the target organs. The sympathetic system originates in the thoracic and lumbar regions; presynaptic fibers of the sympathetic system synapse in the sympathetic chain or in other ganglia, such as the celiac ganglion, part of the solar plexus. Most, but not all, internal organs are innervated by both systems, which usually function in opposition to each other. In general, the sympathetic system produces the effect of exciting organs involved in fight or flight reactions, and the parasympathetic system stimulates more tranquil functions, such as digestion.

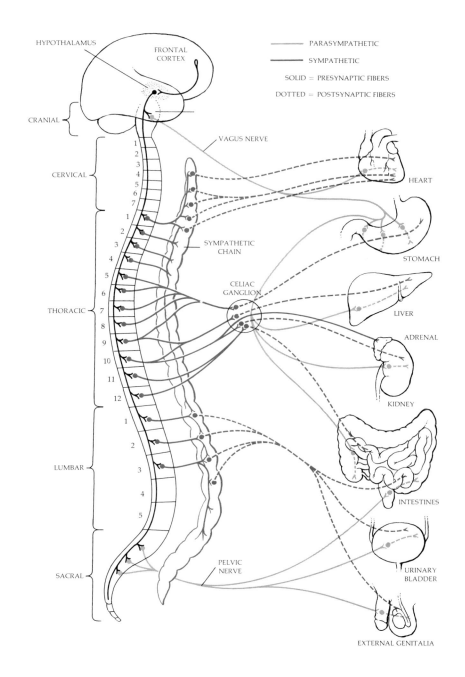

Table 25-3   *Divisions of the Autonomic Nervous System*

DIVISIONS	CNS CONNECTIONS	POINT OF SYNAPSE	CHEMICAL TRANSMITTER	EFFECTS
Sympathetic	Thoracic and lumbar regions of spinal cord	Ganglia near spinal cord	Epinephrine or norepinephrine	Helps organism to cope with external environment
Parasympathetic	Brain stem and sacral (tail) area of spinal cord	Ganglia near target organ	Acetylcholine	Supports restorative, resting body functions

SOME INVERTEBRATE NERVOUS SYSTEMS

In the coelenterate Hydra, nerve cells form a diffuse continuous network. The neurons of the network receive information from sensory receptor cells, which can be found among the epithelial cells on the inner (feeding) and outer surfaces of the animal. The neurons stimulate epithelial muscular cells which cause movements in the body wall. Hydra also contains independent effectors such as cnidoblasts (page 230), which act as both receptors and effectors.

The nervous system of the planarian is more highly organized than that of Hydra. Some of the nerve net is condensed into two cords, and there are more neurons at the anterior end of the body.

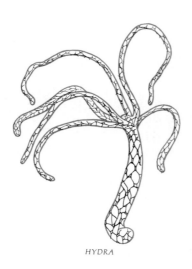

HYDRA

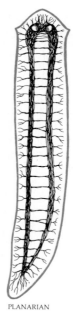

PLANARIAN

In the earthworm, the two conducting channels characteristic of more primitive organisms have come together in a double nerve cord, which runs down the ventral surface of the body. The nerve cord forks just below the pharynx and the two forks meet again, terminating in two ganglia that are large enough to be called the brain.

Arthropod nervous systems are characterized by a chain of ganglia interconnected by a bundle of nerve fibers that run along the ventral surface. As a consequence of this type of nervous system, many quite complicated arthropod activities—such as complex movements of the finely articulated appendages—can be carried out at a local level. Arthropods, especially the insects, are also characterized by a variety of highly specialized receptors.

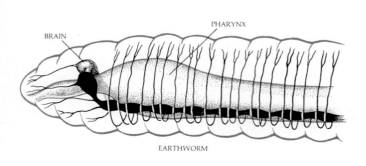

BRAIN

PHARYNX

EARTHWORM

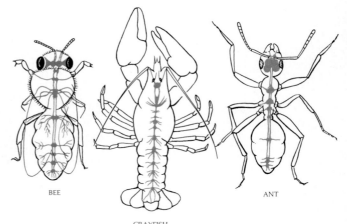

BEE

CRAYFISH

ANT

The changes that occur in a state of rage, which we described at the end of Chapter 23, are initiated by the simultaneous discharge of nerves of the sympathetic system preparing our bodies for "fight or flight." The parasympathetic system, on the other hand, is more involved with the restorative activities of the body, that is, with rest and rumination. Parasympathetic stimulation slows down the heartbeat and increases the muscular movements of the intestine and the secretions of the salivary gland. Most large organs, such as the heart, are under the control of both sympathetic and parasympathetic nerves, which work in close cooperation. Some tissues, such as the hair muscles, the small blood vessels, and the sweat glands, are under sympathetic control alone.

TRANSMISSION OF THE NERVE IMPULSE

As we noted previously, communication in the nervous system takes place along the nerve fibers of the neurons. In order to understand how the nerve impulse travels along an axon, we need to look more closely at the axon itself, particularly at the properties of its cell membrane.*

The Nerve Impulse

In all cells, there is an electric potential between the inside and the outside of the cell membrane. This potential is produced by differences in the concentrations of ions with plus and minus charges trapped on opposite sides of the cell membrane. An electric potential is a form of potential energy, like a boulder on the top of a hill or water behind a dam. In an electric wire, the potential energy is converted to electrical energy when electrons run along the wire. In a living organism (and also in a storage battery), the electric current consists not of electrons but of moving ions. The force with which the charged particles—electrons or ions—move, which is analogous to the force of water running downhill, is measured in volts.

The electric potential of the membrane at rest is about 70 millivolts (a millivolt is 1/1,000 of a volt). Figure 25-8 shows how this _resting potential_—the electric potential of the membrane at rest—is measured and what happens if the membrane of the axon is stimulated. In nature, this stimulus is a signal from a sensory receptor or from another neuron. In the laboratory, this effect is achieved by a brief electrical stimulus to the outside of the membrane. As you can see in the diagram and also on the oscilloscope, the inner surface of the membrane then becomes momentarily positive in relation to the outer surface of the membrane. This change in electric potential is called depolarization and the wave of depolarization is known as the _action potential_. The action potential is about 120 millivolts.

The remarkable feature of the action potential is that when the axon is stimulated once at any point, the action potential travels the entire length of the fiber with undiminished voltage. This action potential—the wave of depolarization that moves along the membrane of the fiber—is the _nerve impulse_. It is the form in which the nervous system carries communications.

---

* The studies on how the nerve impulse is transmitted—which won a Nobel Prize for A. L. Hodgkin and A. F. Huxley—were carried out on the axon and therefore we refer primarily to the axon in this discussion; however, nerve impulses travel along the long dendrites of sensory neurons in the same way.

25–8

The electrical potential of the membrane of the axon is measured by microelectrodes connected to an oscilloscope. When both electrodes are outside the membrane, no potential is recorded. When one electrode penetrates the membrane, the oscilloscope shows that the interior is negative with respect to the exterior and that the difference between the two is about 70 millivolts. This is the resting potential. When the axon is stimulated and a nerve impulse passes along it, the oscilloscope shows a brief reversal of polarity—that is, the interior becomes positive in relation to the exterior. This reversal in polarity is the action potential.

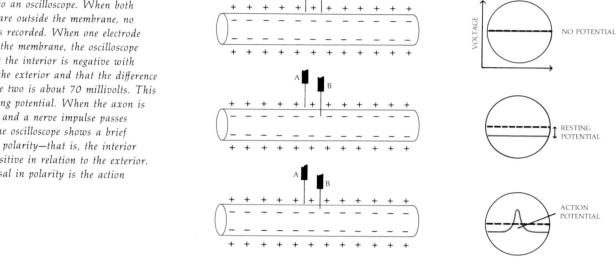

EVENTS AT AXON   EVENTS AT OSCILLOSCOPE

NO POTENTIAL

RESTING POTENTIAL

ACTION POTENTIAL

25–9

(a) *Resting axon, with membrane polarized.*

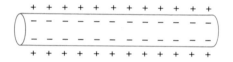

(b) *Action potential. A portion of the membrane becomes momentarily permeable to Na⁺ ions. Na⁺ ions rush in, and the membrane is depolarized. The small arrows indicate the movement of positive charge along the inside of the membrane.*

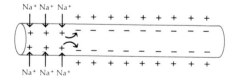

The Action Potential and the Sodium Pump

As we noted earlier, the electric potential is produced by differences in the distribution of ions on either side of the membrane. According to the present widely held hypothesis of neuron action, this difference is the result of the activity of a carrier molecule on the membrane (see Figure 4–6 on page 59) known as the sodium-potassium pump. This molecule pumps out sodium ions (Na^+) and pumps in potassium ions (K^+). As a consequence, the concentration of K^+ ions is about 30 times higher inside the axon than in the fluid outside, and the concentration of Na^+ ions is about 10 times higher outside the axon than inside. In its resting state, the membrane is impermeable to Na^+ ions and relatively permeable to K^+; as a consequence, the Na^+ ions are kept on the outside of the axon and the K^+ ions tend to leak out, moving down the steep diffusion gradient. It is this leaking out of the K^+ ions that makes the inside of the membrane slightly negative, creating the resting potential.

When the membrane is stimulated, it suddenly becomes highly permeable to Na^+. The Na^+ rushes in down the concentration gradient. This influx of positively charged ions momentarily reverses the polarity of the membrane so that it becomes positive on the inside and negative on the outside, producing the action potential. This change in permeability lasts for only about 1.5 milliseconds. Then the membrane regains its previous impermeability to Na^+, the K^+ ions renew their outward movement, and the resting potential is restored. (The actual number of ions involved, like the voltage, is very small. Only one in every 100,000 K^+ ions within a cell need diffuse out to restore the resting potential, and only a very few Na^+ ions need enter to cause the depolarization of the membrane.)

Now let us return to the very important question of why the wave of depolarization continues to move along the axon. As you probably know, electrically charged particles with the same charge repel one another, and particles with opposite charges attract one another. The fatty membrane of the axon acts as insulation, however, which is why, when the axon is in the resting state, it is possible to have positively charged particles on the outside and negatively charged particles on the inside of the membrane (or vice versa).

Following the action potential, Na⁺ ions are once more pumped out and the resting potential is restored. However, the movement of positive charges along the inside of the membrane has begun to depolarize the adjacent segment of membrane. As a consequence, the Na⁺ ions will enter there, forming a new action potential.

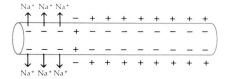

At the peak of the action potential, when the inside of the membrane at the active region is positive, positively charged ions move from this region to adjacent areas inside the axon, which are still negative, so the adjacent area, in turn, becomes depolarized. Na⁺ ions rush in across the momentarily depolarized membrane, creating a new action potential and also depolarizing another, adjacent area of the membrane. As a consequence of this renewal process, an axon, which would be a very poor conductor of an ordinary electric current, is capable of transmitting a nerve impulse often over a considerable distance with absolutely undiminished strength.

The transmission of the nerve impulse is an all-or-nothing reaction. The size of the action potential is limited by the concentration of ions on either side of the membrane, which does not vary to any appreciable extent. As a result, every time the nerve cell is stimulated, the action potential is the same. The message transmitted by a neuron thus depends not on the magnitude of action potentials but rather on their frequency and duration (Figure 25–11b). It is, in effect, a Morse code with dashes but no dots.

The Synapse

As we noted previously, a signal travels from one nerve cell to another across a specialized junction known as a synapse. The neurons do not touch at the synapse; there is a space between them known as the synaptic gap. Transmission across most synapses in mammals is by chemical means. In synaptic knobs at the end of the axon are numerous small vesicles, visible in electron micrographs (Figure 25–12), which contain a chemical transmitter substance. Arrival of the action potentials at the axon terminal causes these vesicles to empty their contents into the synaptic gap. The transmitter substance crosses the gap and combines with receptor molecules on the membrane of the postsynaptic cell, changing the permeability of the membrane. A number of these chemical transmitters have been tentatively identified. Two principal ones are acetylcholine and epinephrine.

Unlike the nerve impulse along the fiber—which is an all-or-nothing proposition—signals transmitted by chemicals across a synaptic junction can modulate one another. A single neuron may receive signals from hundreds of synapses and, based on its processing of all the signals, it will or will not transmit the impulse. Synapses are therefore relay and control points that are extremely important in the functioning of the nervous system.

25-11

(a) *Nerve impulses can be monitored by relatively simple electronic recording instruments. The impulses from any one neuron are all the same, that is, each impulse has the same duration and voltage as any other.* (b) *However, the frequency—the number of impulses per unit time—varies, depending on the intensity of and duration of the stimulus. This recording from a single neuron of a cat shows the change in the nerve code when the neuron is stimulated. In this case, the stimulus was a flash of red light, indicated by the bar. (The slight differences in the height of the action potentials are recording artifacts.)*

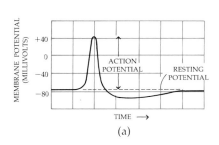

(a)

(b)

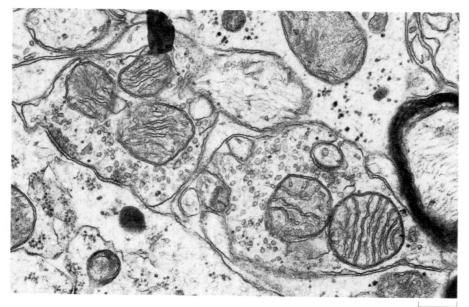

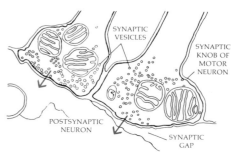

SYNAPTIC
VESICLES

SYNAPTIC
KNOB OF
MOTOR
NEURON

POSTSYNAPTIC
NEURON

SYNAPTIC
GAP

0.05 μm

25–12

Electron micrograph and diagram of a synaptic area in the spinal cord of a bat. In the micrograph, part of the cell body of one neuron occupies the left-hand corner of the picture. Two of the synaptic knobs of a motor neuron fill most of the rest of the micrograph. These contain numerous synaptic vesicles, which appear in the picture as small, grayish granules. At the synaptic junction, the membranes of the two cells are intact and clearly separated from one another by an intercellular space, the synaptic gap. The vesicles, triggered by the nerve impulse, release their chemical contents into the gap. The chemicals stimulate the adjacent (postsynaptic) neuron and thus relay signals from one nerve cell to another, as indicated by the arrows. The outlines of the axons above the synaptic knobs were not included in the extremely thin slice of tissue prepared for this electron micrograph.

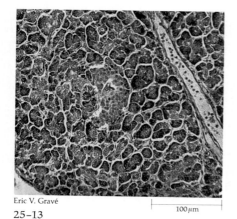

Eric V. Gravé

25–13

A cross section of pancreatic tissue. The pancreas is both an exocrine and an endocrine gland. The group of small cells near the center of the micrograph are islet cells, which secrete the hormones insulin and glucagon into the bloodstream. The surrounding cells produce digestive enzymes, which are carried through the pancreatic duct to the small intestine. The canal-like structure traversing the micrograph is a branch of the duct.

100 μm

The chemicals that transmit signals across nerve junctions are rapidly destroyed by specific enzymes after their release or are taken up again by the synaptic knob. Such removal of the transmitter puts a halt to its effect. This in itself is an essential feature in the control of the activities of the nervous system. The activity of many drugs, such as LSD and psilocybin, is thought to depend on their interference with chemical transmission across synapses (see Table 30–2 on page 417).

THE ENDOCRINE SYSTEM

Glands, which are epithelial tissues specialized for secretion, are classified as exocrine or endocrine. *Exocrine glands* secrete their products into ducts; examples are the sweat glands of the human skin and digestive glands. *Endocrine glands* are glands that secrete their products into the bloodstream (or, more precisely, into the extracellular space from which the substances diffuse into the bloodstream). They are sometimes referred to as "ductless glands." "Endocrine" is generally, though not quite precisely, used as a synonym for "hormone-secreting," and endocrinology means the study of hormones.

In the following pages we shall discuss some of the principal endocrine glands of mammals and the hormones they secrete. In the previous chapter, we described the gonads and the hormones they produce. The hormones involved in digestion, such as those produced by the pancreas, and in kidney function will be taken up in later chapters.

Table 25-4. *Some of the Principal Endocrine Glands of Vertebrates and the Hormones They Produce*

GLAND	HORMONE	PRINCIPAL ACTION	MECHANISM CONTROLLING SECRETION	CHEMICAL COMPOSITION
Pituitary, anterior lobe	Thyroid-stimulating hormone (TSH)	Stimulates thyroid	Thyroxine in blood; hypothalamic releasing hormone	Glycoprotein
	Follicle-stimulating hormone (FSH)	Stimulates ovarian follicle, spermatogenesis	Estrogen in blood; hypothalamic releasing hormone	Glycoprotein
	Luteinizing hormone (LH)	Stimulates interstitial cells in male, corpus luteum and ovulation in female	Testosterone or progesterone in blood; hypothalamic releasing hormone	Glycoprotein
	Adrenocorticotropic hormone (ACTH)	Stimulates adrenal cortex	Adrenal cortical hormone in blood; hypothalamic releasing hormone	Protein
	Growth hormone	Stimulates bone and muscle growth, inhibits oxidation of glucose, promotes breakdown of fatty acids	Hypothalamic releasing hormone	Protein
	Prolactin	Stimulates milk production and secretion in "prepared" gland	Hypothalamic inhibiting hormone	Protein
Thyroid	Thyroxine, other thyroxinelike hormones	Stimulate and maintain metabolic activities	TSH	Iodinated amino acids
	Calcitonin (thyrocalcitonin)	Inhibits release of calcium from bone	Concentration of calcium in blood	Peptide (32 amino acids)
Parathyroid	Parathyroid hormone (parathormone)	Stimulates release of calcium from bone, promotes calcium uptake from gastrointestinal tract, inhibits calcium excretion	Concentration of calcium in blood	Protein
Ovary, follicle	Estrogens	Develop and maintain sex characteristics in females, initiate buildup of endometrium	FSH	Steroids
Ovary, corpus luteum	Progesterone and estrogens	Promote continued growth of endometrium	LH	Steroids
Testis	Testosterone	Supports spermatogenesis, develops and maintains sex characteristics of males	LH	Steroid
Hypothalamus (via posterior pituitary lobe)	Oxytocin	Stimulates uterine contractions, milk ejection	Nervous system	Peptide (9 amino acids)
	Antidiuretic hormone (vasopressin)	Controls water excretion	Osmotic concentration of blood; nervous system	Peptide (9 amino acids)
Adrenal cortex	Cortisol, other cortisol-like hormones	Affect carbohydrate, protein, and lipid metabolism	ACTH	Steroids
	Aldosterone	Affects salt and water balance	Renin from kidney, K^+ ions in blood	Steroid
Adrenal medulla	Epinephrine and norepinephrine	Increase blood sugar, dilate some blood vessels, increase rate of heartbeat	Nervous system	Catecholamine
Pancreas	Insulin	Lowers blood sugar, increases storage of glycogen	Concentration of glucose in blood	Protein
	Glucagon	Stimulates breakdown of glycogen to glucose in the liver	Concentration of glucose and amino acids in blood	Protein

Some of the hormone-producing (endocrine) organs. The pituitary releases hormones that, in turn, regulate the hormone secretions of the sex glands, the thyroid, and the adrenal cortex (the outer layer of the adrenal gland). The pituitary is itself under the regulatory control of an area in the brain known as the hypothalamus, indicated in color. The hypothalamus thus is the major link between the two integrating and control centers.

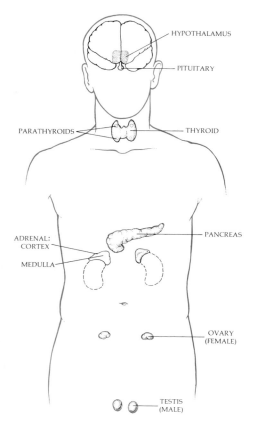

THE PITUITARY GLAND

As you can see in Table 25–4, the pituitary gland produces six different hormones. One of these is growth hormone, sometimes called somatotropin. It promotes the growth of bone and muscle. As is the case with most of the hormones, growth hormone is best known, both to scientists and laymen, by the effects caused by too much or too little hormone activity. If there is a deficit in somatotropin production in childhood, a midget results, the so-called "pituitary dwarf." An excess of somatotropin results in a giant; most circus giants are the result of an excess of growth hormone. Excessive growth hormone in the adult does not lead to giantism, since not all adult tissues respond to growth hormone, but to acromegaly, an increase in the size of the jaw and the hands and feet, adult tissues that are still sensitive to the effects of growth hormone.

Prolactin, also produced by the anterior pituitary, stimulates secretion of milk in mammals. Milk secretion begins in the mother shortly after delivery, as a result of the hormonal changes that take place following the expulsion of the fetus from the uterus. As long as the infant continues to nurse, the impulses produced by the suckling of the breast are transmitted by way of the central nervous system to the pituitary, causing it to produce prolactin. The prolactin, in turn, acts upon the breast to maintain the production of milk. Once suckling ceases, the synthesis and release of prolactin stops and so does milk production. Thus supply is regulated by demand. In some birds, prolactin stimulates nesting behavior and the production of crop milk.

The other four hormones secreted by the anterior pituitary are tropic hormones, hormones that act upon other glands to regulate their secretions. Two of these tropic hormones, the gonadotropins FSH and LH, which act upon the gonads, were discussed in the previous chapter. TSH, the thyroid-stimulating hormone, stimulates cells in the thyroid gland to increase their productivity of thyroxine, the thyroid hormone. Like the gonadotropins, TSH is regulated by negative feedback, being sensitive to the concentration of thyroxine in the blood. ACTH, adrenocorticotropic hormone, has a similar feedback relationship with cortisol.

THE HYPOTHALAMUS

The pituitary is under the influence of the hypothalamus and also, by way of the hypothalamus, is affected by other brain centers. The hypothalamus is also itself the source of two hormones, oxytocin and antidiuretic hormone, or ADH. (ADH is sometimes called vasopressin because it increases blood pressure in many vertebrates; it has no such effect in man, however, except in very high doses.) Oxytocin accelerates childbirth by acting on the smooth muscles of the uterus and promoting contractions. ADH, as we shall see in Chapter 29, regulates the excretion of water by the kidneys. These hormones are secreted by the hypothalamus and are stored in the posterior lobe of the pituitary from which they are released.

Figure 25–15 summarizes the negative feedback system linking the hypothalamus and pituitary with the thyroid, adrenal cortex, and gonads. With the discovery of the close working relationship between the pituitary and the hypothalamus, it became clear that the nervous and endocrine systems are not dual methods of control but are, in fact, a single regulatory system integrating the organism's activities and promoting its homeostasis.

25–15

The production of many major hormones is regulated by a complex negative feedback system involving the pituitary and the hypothalamus. The hypothalamus stimulates the pituitary to secrete tropic hormones, and these in turn stimulate the secretion of hormones from the thyroid, adrenal cortex, and gonads (the testes or the ovaries). When the hormones produced by these target glands reach a certain concentration in the blood, the hypothalamus stops stimulating the pituitary, the pituitary stops producing hormones, and production of hormones by the target glands also stops. By way of the hypothalamus, which receives information from many other parts of the brain, hormone production is regulated also in response to changes in the external and internal environments.

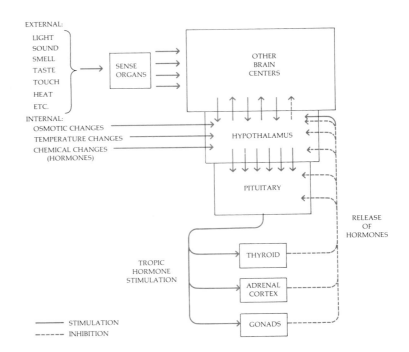

EXTERNAL:
LIGHT
SOUND
SMELL
TASTE
TOUCH
HEAT
ETC.

INTERNAL:
OSMOTIC CHANGES
TEMPERATURE CHANGES
CHEMICAL CHANGES
(HORMONES)

SENSE ORGANS

OTHER BRAIN CENTERS

HYPOTHALAMUS

PITUITARY

RELEASE OF HORMONES

TROPIC HORMONE STIMULATION

THYROID

ADRENAL CORTEX

GONADS

——— STIMULATION
------ INHIBITION

ADRENAL MEDULLA

The adrenal glands are on top of the kidneys. The adrenal medulla, the central portion of the adrenal gland, actually does not conform to the definition of a gland in that it is not made up of glandular epithelium. It is composed, instead, of nervous tissue; in fact, it is, in essence, a large ganglion whose nerve endings secrete epinephrine and norepinephrine. These hormones raise blood pressure, stimulate respiration, dilate the respiratory passages, increase concentration of glucose in the bloodstream, and stimulate the general metabolic activity of cells. The adrenal medulla is stimulated by a nerve fiber of the sympathetic division and acts as an enforcer of sympathetic activity. Epinephrine and norepinephrine are destroyed by enzymes in the blood within minutes after their release, which is another mechanism for tight regulatory control. (The steroid and protein hormones are also broken down, although more slowly and principally by the liver.)

ADRENAL CORTEX

The outer layer of the adrenal gland, the adrenal cortex, is the source of a large number of steroids, of which cortisol and aldosterone are among the most active.

Cortisol and cortisol-like hormones promote the formation of glucose from carbohydrates and amino acids. At the same time, they inhibit the uptake of glucose by most cells, with the notable exception of brain cells, thus favoring mental activities at the expense of other body functions. Their release increases during periods of stress associated with events such as facing new situations, engaging in athletic competitions, and taking final exams. Thus they work in

Chemical structures of four hormones. (a) Cortisol is a steroid, as you can see by its characteristic four-ring structure. It is secreted by the adrenal cortex. (b) Epinephrine is a product of the adrenal medulla. The ring structure to the left is called a catechol, and epinephrine is characterized chemically

as a catecholamine, a nitrogen-containing catechol. (c) Thyroxine is the principal hormone produced by the thyroid gland. Note the four iodine atoms in its structure. Because we need iodine for thyroxine, it is an essential component of the human diet. (d)

Insulin consists of two polypeptide chains held together by disulfide bonds. It was the first protein whose primary structure was deciphered. Although it is small for a protein, insulin, as you can see, is much larger than the steroids or other hormones.

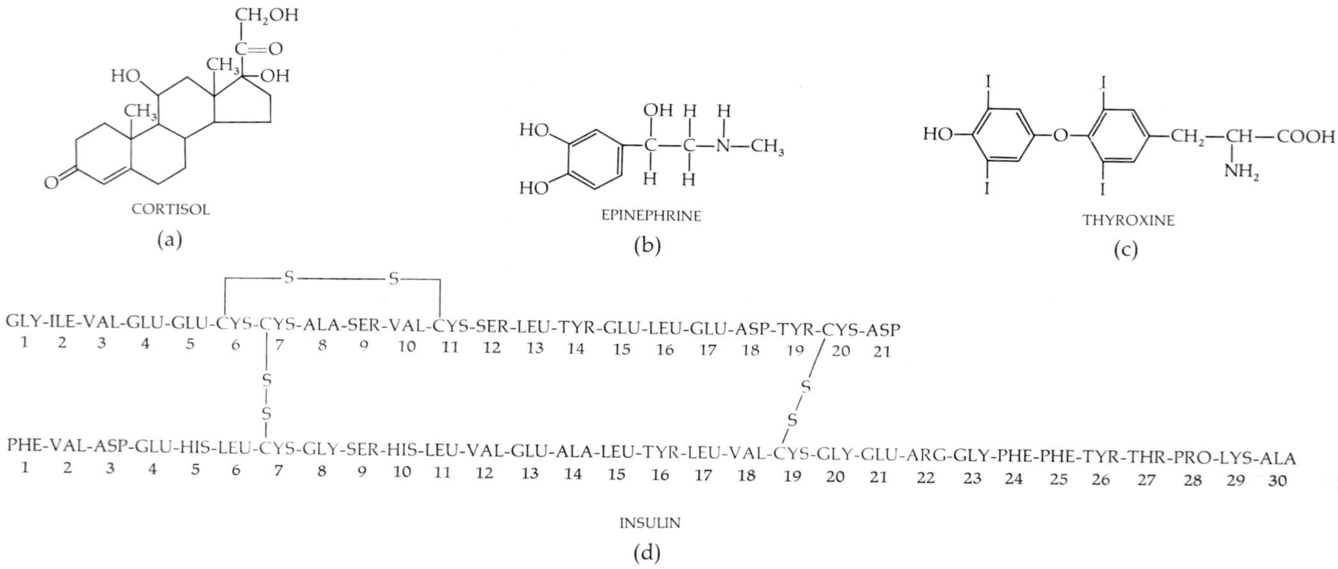

CORTISOL
(a)

EPINEPHRINE
(b)

THYROXINE
(c)

GLY-ILE-VAL-GLU-GLU-CYS-CYS-ALA-SER-VAL-CYS-SER-LEU-TYR-GLU-LEU-GLU-ASP-TYR-CYS-ASP
1   2   3   4   5   6   7   8   9   10 11 12 13 14 15 16 17 18 19 20 21

PHE-VAL-ASP-GLU-HIS-LEU-CYS-GLY-SER-HIS-LEU-VAL-GLU-ALA-LEU-TYR-LEU-VAL-CYS-GLY-GLU-ARG-GLY-PHE-PHE-TYR-THR-PRO-LYS-ALA
1   2   3   4   5   6   7   8   9   10 11 12 13 14 15 16 17 18 19 20 21 22 23 24 25 26 27 28 29 30

INSULIN
(d)

concert with the sympathetic nervous system. In large amounts, they inhibit inflammation and the production of antibodies.

Aldosterone and aldosteronelike hormones affect the concentration of ions, particularly sodium and potassium ions, in the blood.

The adrenal cortex is also the source of small amounts of male sex hormones in both men and women. An adrenal tumor may result in increased production of these hormones and, consequently, in women, the growth of facial hair and other masculine characteristics. Bearded ladies in circuses are often the victims of such tumors. Sex drive (libido) in women is associated with testosterone and other male hormones produced by the adrenal cortex rather than with estrogen.

THYROID GLAND

The thyroid, under the influence of the thyroid-stimulating hormone from the pituitary, produces thyroxine, which is an amino acid combined with atoms of iodine (Figure 25–16c). Thyroxine accelerates the rate of cellular respiration, so the rate at which oxygen is consumed at rest (basal metabolic rate) can be used as an index of how much thyroid hormone the tissues are receiving. Hyperthyroidism, the overproduction of thyroxine, produces excessive nervousness and excitability, increased rate of heartbeat and blood pressure, and weight loss. Hypothyroidism (too little thyroxine) in infancy affects development, particularly of the brain cells, and leads to permanent mental deficiency and dwarfism. In adults, hypothyroidism is associated with dry skin, intolerance to cold, and lack of energy. The thyroid also secretes calcitonin, which inhibits the release of calcium from bone.

PHEROMONES

Pheromones (from pher, meaning "to carry," and hormone) are chemical signals exchanged between organisms, usually members of the same species. They may have been the evolutionary forerunners of hormones, serving to coordinate the activities of single-celled organisms. They are secreted as liquids, usually by specialized cells or glands, and transmitted as liquids or gases.

Among modern species, communication by pheromones seems to be most prevalent among the insects, where these hormones play a variety of roles. Pheromones serve as sex attractants luring male insects to females. Ants lay down pheromones as trail markers, signposts to a food source. When you are stung by a honeybee, not only does the stinger remain in your skin but also a chemical substance (secreted by cells lining the sting pouch) that recruits other honeybees to the attack. Similarly, working ants of many species release pheromones as alarm substances when they are threatened by an invader; the pheromone spreads as a gas to alarm and recruit other workers. If these ants, too, encounter the invader, they will release the pheromone, so the signal will either die out or build up, depending on the magnitude of the threat. Pheromone action is also responsible for the striking physical differences among different castes of social insects, especially notable among termites. In these cases, the pheromones serve to regulate the internal hormone secretions of the insects.

Pheromones have also been found among mammals. The most familiar are the scent-marking substances excreted in the urine of male dogs and cats. Male mice, it has been found, release substances in their urine that alter the reproductive cycles of the females. The smell of urine from a strange male, for example,

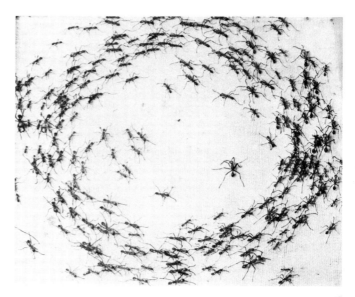

Ants following a pheromone trail.

can alter hormone balance and so interrupt the pregnancy of a female newly impregnated by another male, leaving her free to mate with the newcomer. Sex attractant pheromones have also been found among primates. None has been identified among humans, but the fact that there are striking differences in the capacity of men and women to detect certain odors may be a clue to their existence.

PARATHYROID GLANDS

The pea-sized parathyroid glands, the smallest of the known endocrine glands, are located behind or within the thyroid gland. They produce parathyroid hormone (parathormone), which maintains blood calcium levels by increasing its absorption from foods in the intestine and by reducing its excretion by the kidneys. It also stimulates the release into the bloodstream of calcium from bone, which contains 99 percent of the body's total calcium. Parathormone and calcitonin thus work as a fine-tuning mechanism, regulating blood calcium. Their production is regulated directly by the amount of calcium in the blood.

MECHANISM OF ACTION OF HORMONES

A chief difference between the nervous system and the endocrine system is that the nervous system consists of a physical network of (almost) touching cells. By contrast, endocrine glands are often at some distance from their target cells, and, in fact, all cells of the body are equally exposed to the hormones released into the bloodstream. Recent research has revealed the mechanisms underlying specificity of action of two major groups of hormones, the steroids and the protein hormones.

Autoradiograph (see page 30) showing the localization of progesterone within the nuclei of cells of the uterus of a hamster. The progesterone, which was labeled with tritium, was injected into a hamster. Fifteen minutes later, thin sections of the animal's uterine tissue were mounted on slides coated with a photographic emulsion. The slides were kept in darkness for about three months before the emulsion was developed. The black specks indicate concentrations of the labeled hormone.

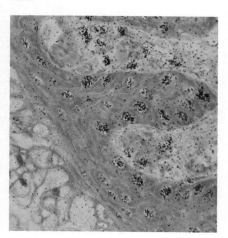

Steroid hormones are relatively small molecules. These molecules, it has been found, pass easily through cell membranes and so freely enter all the cells of the body. However, in their target cells, and only in their target cells, the hormones encounter a specific receptor molecule, a protein, which combines with them. The receptor and the steroid together act directly on the DNA of the cell to promote the synthesis of messenger RNA and so of specific enzymes and other proteins. (Note that this is similar to the operation of the operon, described on page 187. However, it differs in that hormone and receptor apparently serve to stimulate RNA synthesis in these cells, rather than to remove a repressor molecule, as occurs in the operon.) These findings explain how the very slight differences in configuration among steroid molecules can be correlated with such drastically different effects: The protein receptors, like enzymes, are highly specific in their combining properties.

By contrast, protein hormones, which are much larger molecules, do not enter cells but rather combine with molecules on the membrane surface, setting in motion a "second messenger" that is responsible for the sequence of events inside the cell. These findings may have important medical implications. For example, it has long been known that juvenile diabetes—diabetes in young persons—is caused by a deficiency of the hormone insulin. Insulin promotes the uptake of glucose by cells (a subject to be discussed further in Chapter 28). It was assumed by analogy that the diabetes found commonly in older persons and associated with obesity had the same cause. It has now been found, however, that this form of adult diabetes results from a decrease in the number of insulin binding sites on receptor cells. Such patients are treated most effectively by diet.

Another group of studies, for which Earl W. Sutherland was awarded the Nobel Prize in 1971, has shown that a chemical known as cyclic AMP (Figure 25-18) is the second messenger in a number of target cells. Hormones that trigger the action of cyclic AMP include ACTH, TSH, LH, ADH, epinephrine, and glucagon (a hormone secreted by the pancreas). The differences in their effects are due to the presence of different enzyme systems within the cells that respond to cyclic AMP.

At almost the same time that cyclic AMP was identified by these research workers in mammalian physiology, biologists studying a peculiar group of organisms known as the cellular slime molds isolated a chemical of great importance in this biological system. The cells of the cellular slime mold begin as individual amoebas and then come together to form a single organism. The chemical that calls them together was named acrasin, after Acrasia, the mythological siren who lured seamen in Homeric legend. Acrasin has now been identified as cyclic AMP, another example of the long thread of evolutionary history linking all organisms.

SUMMARY

The endocrine and nervous systems provide the precise and rapid communication necessary to coordinate the numerous internal functions which enable an animal to regulate its internal environment.

The unit of the nervous system is the neuron, or nerve cell, which consists of the dendrites and cell body, which receive impulses, and the axon, which relays the impulses to other cells. The nerve impulse travels along the fiber in the form of an action potential, which is a transient reversal in the polarity of the membrane of the axon or dendrite. The message carried along the fiber depends on the frequency with which action potentials are transmitted. Nerve cells

25-18

(a) *Cyclic AMP (adenosine monophosphate) acts as a "second messenger" within the cells of vertebrates. Following stimulation by various hormones—the "first messengers"—cyclic AMP is formed from ATP. "Cyclic" refers to the fact that the atoms of the phosphate group form a ring. (b) Cyclic AMP is also the chemical that attracts the amoebas of the cellular slime molds, causing them to aggregate into a sluglike body, which then behaves like a multicellular organism. The arrow shows the direction in which the cells are moving.*

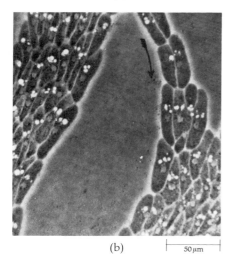

CYCLIC AMP

(a)

(b) |— 50 μm —|

transmit signals to other cells across a junction called the synapse. The signal crosses the synaptic gap by triggering the release of a chemical transmitter, which stimulates or inhibits the target cell. Nerves are bundles of motor and sensory fibers.

The central nervous system consists of the brain and the spinal cord, which are encased, in vertebrates, in the skull and vertebral column. The nervous system outside the central nervous system constitutes the peripheral nervous system. Spinal nerves emerge from the vertebral column in pairs, each pair consisting of motor fibers and sensory fibers. The motor fibers of each pair innervate the muscles of a different area of the body and the sensory fibers receive impulses from sensory receptors.

In vertebrates, the peripheral nervous system has two major divisions: (1) the somatic nervous system, which consists of motor and sensory neurons; and (2) the autonomic nervous system, which controls the muscles and glands involved in the digestive, circulatory, urinary, and reproductive functions. The autonomic system has its own network of motor neurons, separate from that of the somatic system, which are linked indirectly from the central nervous system to the target organs by secondary, or postganglionic, neurons. The autonomic system has two divisions: (1) the sympathetic system, which is largely responsible for reactions to the external environment; and (2) the parasympathetic system, which controls the restorative activities, such as digestion and rest.

The endocrine system consists of glands that secrete chemicals (hormones) into the bloodstream. These chemicals exert specific effects on certain organs and tissues. The effects of hormones on their target tissues depend on the presence in the target cells of specific receptors for that hormone. Table 25–4 lists the principal glands of vertebrates and their hormones. The nervous and endocrine systems influence one another's activities so profoundly that they may be considered a single regulatory agency.

Hormone production is characteristically regulated by a negative feedback system. In the case of thyroxine and the steroid hormones, the concentration of a particular hormone in the blood affects the release from the pituitary of the tropic hormone that stimulates that particular gland. In the case of other hormones, such as parathyroid hormone, which increases calcium concentration in the blood, the increased calcium acts directly on the gland to inhibit its secretion of the hormone. Such feedback systems are important mechanisms of homeostasis.

QUESTIONS

1. Distinguish between the following: neuron/nerve, afferent/efferent, somatic/autonomic, sympathetic/parasympathetic, axon/dendrite, resting potential/action potential, gray matter/white matter.

2. Draw a diagram of the negative feedback system regulating the production of thyroxine.

3. Both nervous and endocrine systems transmit information and thereby integrate and coordinate the activities of the body. Compare their operation with regard to speed of transmission, specificity of target, and the nature of the messages handled.

4. Women living together in college dormitories or other communal living quarters tend to synchronize their menstrual cycles. How this comes about is not known, but can you suggest an explanation?

CHAPTER 26

Circulation

As we discussed in the opening chapter of this book, every cell in the body of a complex organism builds its own membranes and organelles, makes its own ATP, and assembles its own enzymes and other proteins. To engage in these activities, cells need oxygen and nutrients and must dispose of carbon dioxide and other wastes.

In single-celled organisms and very small multicellular ones, the needs of each cell are supplied directly by the medium the organism lives in. In larger, more complex animals, these needs are supplied by a transport system. In man and other vertebrates, gases (oxygen and carbon dioxide), food molecules, hormones, and wastes are transported in the bloodstream. The blood circulates

26–1
(a) *Structure of blood vessels. Arteries have thick, tough, elastic walls that can withstand the pressure of the blood as it leaves the heart. Capillaries have walls only one cell thick. Exchange of gases, nutrients, and wastes between the blood and the cells of the body takes place through these thin capillary walls. Veins have larger lumens (passageways) and thinner, more readily extensible walls that minimize resistance to the flow of blood on its return to the heart. (b) Photomicrograph of a large vein and a small artery. The artery can be identified by its thick, muscular wall.*

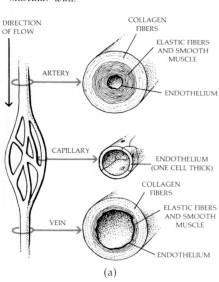

(a)

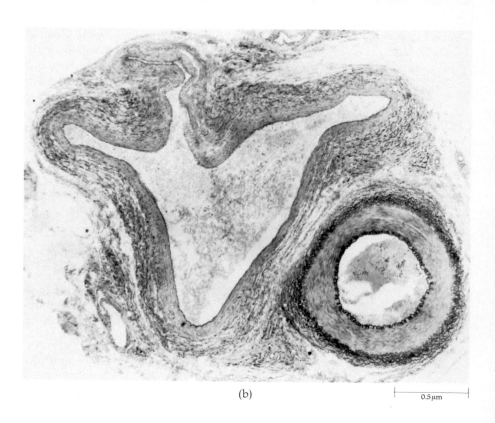

(b)

0.5 μm

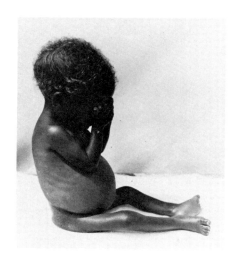

26-2

A child with kwashiorkor, a West African word that means "the sickness a child develops when another child is born." The swelling characteristic of this disease is caused by a deficit in blood proteins that results in osmotic movement of fluids out of the blood into the body tissues. Children with kwashiorkor are usually also weak and apathetic and have skin discoloration. The syndrome, which is the result of a deficiency in protein but not in total calories, usually develops after a child is weaned. Children with kwashiorkor can be found in the United States as well as in less developed countries.

through a closed circuit of continuous vessels propelled by the contractions of a specialized muscle, the heart. This system is known as the *cardiovascular system* (from *cardio*, meaning "heart," and *vascular*, meaning "vessel"). In this system, the heart pumps the blood into the large arteries, from which it travels to branching, smaller arteries (the *arterioles*) and then into very small vessels, the *capillaries*. From the capillaries, the blood passes back into small veins, the venules, then into larger veins, and, through them, back to the heart.

In man, the diameter of the opening of the largest artery, the *aorta*, is 2.5 centimeters, that of a capillary only 8 micrometers, and that of the largest vein, the *vena cava*, 3 centimeters. Arteries, veins, and capillaries differ not only in their diameters but also in the structure of their walls. Each is lined with endothelium, a type of epithelial cell. As we saw in Figure 1–11 on page 21, capillaries have walls only one cell in diameter. The walls of veins and arteries contain muscle and supporting tissues (Figure 26–1).

THE CAPILLARIES AND DIFFUSION

The thin-walled capillaries carry out the actual function of the cardiovascular system, which is to service the body tissues, supplying them with oxygen, food molecules, hormones, and other materials. To understand how this supply system works, it is necessary to review some of the principles of water transport described in Chapter 2.

Most of the molecules that cross the capillary walls pass in or out by diffusion. Some additional molecules cross by bulk flow; the pressure of the blood within the capillaries tends to force fluid out through the capillary walls. At the same time, because of the protein molecules present in the blood, fluid tends to reenter the bloodstream by osmosis. Thus, normally, there is a balance between inflow and outflow. When this balance is disturbed, and outflow is greater, excess fluid collects in the tissues, a condition known as edema (Figure 26–2).

There is a capillary within rapid diffusion distance of every cell in the body—the total length of the capillaries in a human adult, for example, is more than 80,000 kilometers (50,000 miles). Even the cells in the walls of the veins and arteries depend on this capillary system for their blood supply, as does every organ of the body, including the heart itself (Figure 26–3).

26-3

Electron micrograph of heart muscle showing a red blood cell within a capillary. The heart, like all the other tissues of the body, is dependent on the fine capillary network for its supply of oxygen and other blood-borne materials. The striated pattern of the heart muscle is clearly visible. Note also the large and numerous mitochondria and the extremely thin wall of the capillary.

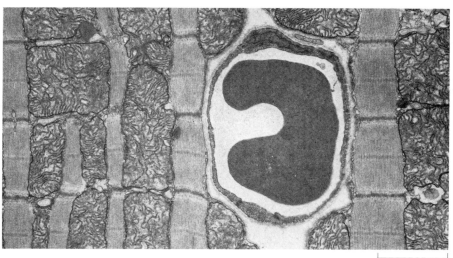

5 μm

In mammals, blood goes from the right heart to the lungs, from the lungs to the left heart, and, from the left heart, it enters the systemic circulation, moving from arteries to arterioles to capillaries to veins. Blood pressures vary in the different areas of the cardiovascular system. The fluctuations in blood pressure produced by three heartbeats are shown in each section of the diagram. Note the fall in pressure as blood traverses the arterioles of the systemic circulation.

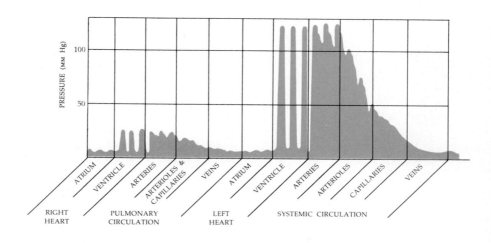

26-5

Valves in the veins open to permit movement of blood toward the heart but close to prevent backflow.

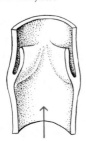
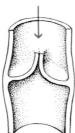

BLOOD PRESSURE

Blood pressure is the force with which blood pushes against the walls of the blood vessels. It is conventionally described in terms of how high it can push a column of mercury. For medical purposes, it is usually measured at the artery of the left upper arm. Normal blood pressure in a young man is 120 millimeters of mercury (120 mm Hg) when the ventricles are contracting (the systolic blood pressure) and 80 mm Hg when the ventricles relax (diastolic pressure). The blood pressure of a young woman is 8 to 10 mm Hg less.

The pressure is generated by the pumping action of the heart and changes with the rate at which it contracts. Blood flow is directly proportional to blood pressure; the greater the pressure, the greater the flow, and vice versa.

Blood pressure is not equal in the various parts of the cardiovascular system. As the blood gets far along the system of vessels, the pressure drops, reaching the lowest in the veins (Figure 26–4). Blood in the veins, which are soft-walled, is helped back toward the heart by the contractions of the body muscles. For example, when you walk, your leg muscles bulge and squeeze the veins between the contracting muscle. This raises the pressure within them and increases blood flow back to the heart. Valves in the veins prevent backflow (Figure 26–5).

REGULATING BLOOD FLOW

The blood flow to different parts of the body changes as the activities of the organism change. For example, blood flow through muscle increases during exercise, flow to the stomach and intestines increases during digestion, and flow through the skin increases in the heat and decreases in the cold.

The regulation of blood flow depends on a very simple physical principle: Fluid flow through a tube is proportional to the fourth power of the diameter of the tube (D^4). The diameter of the arterioles, which directly supply the capillaries, can be altered by rings of smooth muscle in the vessel walls. As the smooth muscle contracts, the opening in the vessel gets smaller, and blood flow into the vessel (and the capillary network it feeds) decreases (Figure 26–6). Conversely, when the smooth muscle relaxes, the vessel opens wider, and blood flow into the capillaries increases. These smooth muscles are controlled by

Diagram of a capillary network. The sphincter muscle at the end of the arteriole controls the blood flow through the capillaries. These muscles, which are innervated by the autonomic nervous system, make it possible to regulate the blood supply to different regions of the body depending on the physiological state of the organism.

autonomic nerves (chiefly sympathetic nerves), the hormones epinephrine and norepinephrine secreted by the adrenal medulla, and other chemicals that are produced locally in the tissues themselves. Activity of the nerves controlling smooth muscle in the blood vessels is coordinated with the activity of nerves regulating heart rate and strength of heart muscle contraction by the cardiovascular regulating center, which will be discussed later in this chapter.

EVOLUTION OF THE HEART

The heart, in its simplest form—such as the earthworm heart (page 235)—is a muscular contractile part of a circulatory vessel. In the course of vertebrate evolution, the heart has undergone some interesting adaptations, as shown in Figure 26–7.

In the fish, there is a single heart divided into the *atrium*, which is the receiving area for the blood, and the *ventricle*, which is the pumping area from which the blood is expelled into the vessels. The ventricle of the fish heart pumps blood directly into the capillaries of the gills, where it picks up oxygen and releases carbon dioxide. From the gills, oxygenated blood is carried to the tissues. By this time, however, most of the propulsive force of the heartbeat has been dissipated by the resistance of the capillaries in the gill, and the blood flow to and through the systemic circulation is relatively sluggish.

In amphibians, there are two atria; one receives oxygenated blood from the lungs and the other receives deoxygenated blood from the systemic circulation. Both atria empty into the single ventricle, which pumps the mixed blood simultaneously through the lungs and the systemic circulation. By this arrangement, the blood enters the systemic circulation under higher pressure.

26–7

Vertebrate circulatory systems. Oxygenated blood is indicated in color. (a) In the fish, the heart has only one atrium (A) and one ventricle (V). Blood oxygenated in the gill capillaries goes straight to the systemic capillaries without first returning to the heart. (b) In amphibians, the single primitive atrium has been divided into two separate chambers. Oxygenated blood from the lungs enters one atrium, and oxygen-poor blood from the tissues enters the other. The blood is mixed somewhat in the single ventricle and then pumped again through the lungs and body tissues. (c) In birds and mammals, the ventricle is divided into two separate chambers, so that there are, in effect, two hearts—one for pumping oxygen-poor blood through the lungs and one for pumping oxygen-rich blood through the body tissues.

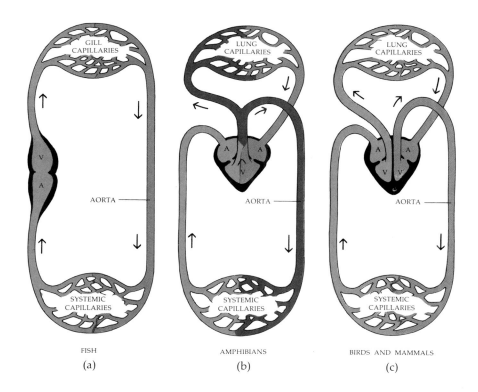

In the birds and mammals, the heart is functionally separated longitudinally into two organs. The right heart receives blood from the tissues and pumps it into the lungs, where it becomes oxygenated. From the lungs it returns to the left heart, from which it is pumped into the body tissues. This efficient, high-pressure circulation system is necessary to maintain the high metabolic rate of both birds and mammals, with their constant body temperature and their generally high level of physical and mental activity.

THE HEART AND REGULATION OF ITS BEAT

Figure 26–8 shows a diagram of the human heart. Its walls are made up of the kind of muscle known as cardiac muscle. Blood returning from the body tissues enters the right atrium through the superior and inferior venae cavae. Blood returning from the lungs enters the left atrium through the pulmonary veins. The atria, which are thin-walled compared with the ventricles, expand to receive the blood. Both atria then contract simultaneously, assisting the flow of blood through the open valves into the ventricles. Then the ventricles contract simultaneously, closing the valves (Figure 26–9). The right ventricle propels the blood into the lungs through the pulmonary arteries, and the left ventricle propels oxygenated blood into the aorta, the principal artery leading from the heart to the other body tissues. In a healthy man at rest, this rhythmic process takes place about 70 times a minute; under strenuous exercise, the rate more than doubles. The heart pumps 7,500 liters (2,000 gallons) a day.

What controls the heartbeat? In vertebrates, the heartbeat originates in the heart muscle itself. A vertebrate heart will continue to contract even after it is removed from the body if it is kept in a nutrient solution. In vertebrate embryos, the heart begins to beat very early in development, before the appearance of any nerve supply. In fact, embryonic heart cells isolated in the test tube will beat.

The beat of the cardiac muscle is synchronized by a special area of the heart known as the _sinoatrial node_, which is located in the right atrium (Figure 26–10).

26–8

The human heart. Blood returning from the systemic circulation through the superior and inferior venae cavae enters the right atrium and is pumped to the right ventricle, which propels it throughout the pulmonary arteries to the lungs, where it is oxygenated. Blood from the lungs enters the left atrium through the pulmonary veins and is pumped to the left ventricle and then through the aorta to the body tissues.

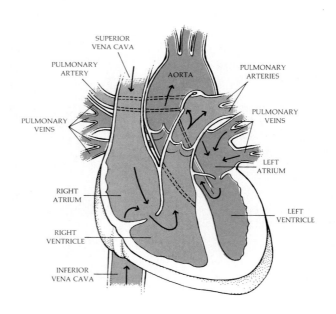

26–9

When blood enters the left atrium, the mitral valve opens, permitting blood to flow into the left ventricle. Ventricular contraction pushes the blood hard against the valve, closing it and thereby ensuring the movement of blood through the aortic semilunar valve into the aorta. A similar valve in the right heart, called the tricuspid, prevents the blood from reentering the atrium.

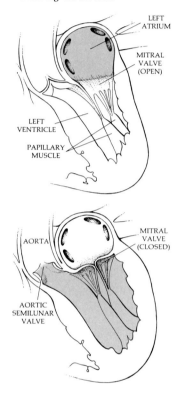

26–10

The beat of the mammalian heart is controlled by a region of specialized muscle tissue in the right atrium, the sinoatrial node, which functions as the heart's pacemaker. Some of the nerves regulating the heart have their endings in this region. Excitation spreads from the pacemaker through the atrial muscles, causing both atria to contract simultaneously. When the wave of excitation reaches the atrioventricular node, its conducting fibers pass the stimulation to the bundle of His, which triggers simultaneous contraction of the ventricles. Because the fibers of the atrioventricular node conduct relatively slowly, the ventricles do not contract until the atrial beat has been completed.

This node functions as the *pacemaker* for the heart, for it is the first part to undergo membrane depolarization and initiate an action potential. From the pacemaker this action potential spreads to the left atrium. As it passes along the surface of the individual heart muscle cells, it activates their contractile machinery, and they contract. The action potential travels very quickly, so many cells are activated almost simultaneously.

About 100 milliseconds after the pacemaker fires, impulses from both atria stimulate a second area of nodal tissue, the *atrioventricular node*. The atrioventricular node is the only electrical bridge between the atria and the ventricles. It consists of slow-conducting fibers. Thus the atrioventricular node imposes a delay between the atrial and ventricular contractions, so that the atrial beat is completed before the beat of the ventricles begins. From the atrioventricular node, stimulation passes to the *bundle of His* (named after its discoverer) and then to the walls of the right and left ventricles, which sets up a simultaneous contraction of the ventricles.

If you listen to a heartbeat, you hear "lubb-dup, lubb-dup." The deeper, first sound ("lubb") is the closing of the valves between the atria and the ventricles; the second sound ("dup") is the closing of the valves leading from the ventricles to the arteries. If any one of the four valves is damaged, as from rheumatic fever, blood may leak back through one of the valves, producing the noise characterized as a "heart murmur" (a "ph-f-f-t" sound).

Although the autonomic nervous system does not initiate the vertebrate heartbeat, it does control its rate. Fibers from the parasympathetic system travel through the vagus nerve, a large nerve that runs through the neck along the trachea. These nerve fibers secrete acetylcholine, which has a slowing effect on the pacemaker and thus decreases the rate of heartbeat. Conversely, fibers from the sympathetic nervous system secrete norepinephrine, which stimulates the pacemaker, increasing the rate of heartbeat. Epinephrine and norepinephrine from the adrenal medulla affect the heart in the same way that the sympathetic nerves do.

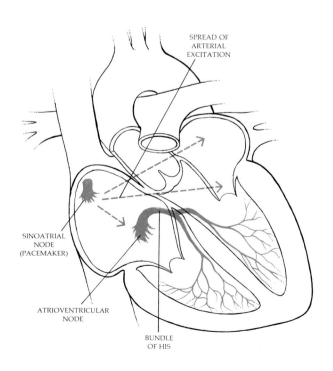

DISEASES OF THE HEART AND BLOOD VESSELS

Cardiovascular diseases cause more deaths in the United States than accidents and all other diseases combined. During 1975, there were 979,180 such deaths in the United States alone. According to recent estimates, more than 28 million people—13 percent of the total population—in this country have some form of cardiovascular disease.

About 65 percent of the cardiovascular deaths are caused by heart attacks. A heart attack is the result of an insufficient supply of blood to an area of heart muscle; with their oxygen supply cut off, the muscle cells die. It can be caused by a blood clot—a thrombus—that forms in the heart itself or by a clot that forms elsewhere in the body and travels to the heart and lodges there. (A wandering clot such as this is known as an embolus.) A heart attack can also result from a blocking of a blood vessel due to atherosclerosis (see below). Recovery from a heart attack depends on how much of the heart tissue is damaged and whether or not other blood vessels in the heart can enlarge their capacity and supply these tissues, which then may recover to some extent. Angina pectoris is a related condition in which the heart muscle receives an insufficient blood supply—often a result of a narrowing of the vessels. The symptoms are a pain in the center of the chest and, often, in the left arm and shoulder.

A stroke is caused by interference with the blood supply to the brain. This may be the result of a thrombus (cerebral thrombosis) or an embolus. It may also result from the bursting of a blood vessel in the brain. Its effects depend both on the severity of the damage and where it occurs in the brain.

Atherosclerosis contributes both to heart attacks and strokes. In this disease, the linings of the arteries thicken and their surface becomes roughened by deposits of cholesterol, fibrin (clotting material), and cellular debris. The arteries, becoming inelastic, no longer expand and contract, and blood moves with increasing difficulty through the narrowed vessels. (Hardening of the arteries—arteriosclerosis—may subsequently occur if calcium becomes deposited in the artery walls). Thrombi and emboli thus form more easily and are more likely to block the vessel. The causes are not clear-cut. Arteriosclerosis is associated with high levels of cholesterol (page 46) in the blood, and also with cigarette smoking, stress, and lack of exercise. Female hormones seem to protect against arteriosclerosis; the condition is much less

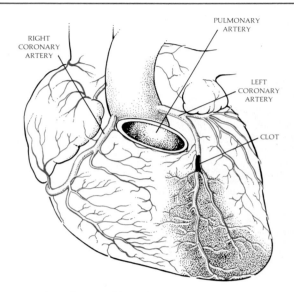

A heart attack can be caused by a clot in a blood vessel supplying the heart tissue. When such a clot occurs, the heart cells normally serviced by this vessel die from lack of oxygen.

common in premenopausal women than it is among men of the same age. This difference in susceptibility to arteriosclerosis is a major reason for the difference in longevity between men and women in the United States. (Women now live almost 10 years longer than men.)

High blood pressure—hypertension—also contributes significantly to cardiovascular disease. It places additional strain on the arterial walls and increases the chances of emboli. In most cases of hypertension, the cause is unknown, and about half the men and women with high blood pressure do not know they have this condition. Despite lack of knowledge about the causes, hypertension can be treated and blood pressure reduced to safer levels. In the United States, high blood pressure is twice as common among black males as among white males and almost twice as common among black females as among white females. Related to this disturbing statistic is the fact that the incidence of deaths among black men and women from cardiovascular diseases is even higher than in the white population.

CARDIOVASCULAR REGULATING CENTER

The sympathetic and parasympathetic nerves to the heart are controlled by the *cardiovascular regulating center*, which is located in the medulla, a small region of the brain close to the junction of brain and spinal cord. The cardiovascular regulating center is also the control center for the nerves controlling smooth muscle in the arterioles. Thus, if blood flow through a particular body area is increasing owing to dilation of blood vessels, the heart is simultaneously activated to develop greater pressure to support the greater flow.

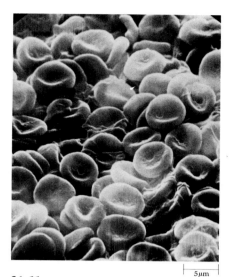

26–11

In vertebrates, oxygen is transported in red blood cells, shown here in a scanning electron micrograph. Red blood cells, which are only about 7 or 8 micrometers in diameter, are flexible and so are able to twist and turn in their passage through the capillaries.

The cardiovascular regulating center integrates the reflexes that control blood pressure. It receives information about existing blood pressure from specialized receptors in large arteries in the neck (the carotid arteries) and near the heart. The effector organs of the reflex are, as we have indicated, the heart and blood vessels.

The blood-pressure reflex is another example of a negative-feedback biological control system. When pressure falls, the activity of the heart is increased and the blood vessels are constricted, which raises the pressure again. Conversely, heart activity is decreased and the blood vessels dilated in response to high pressure. Diseases that interfere with this reflex are accompanied by sustained high or low blood pressure, called hypertension or hypotension, respectively.

COMPOSITION OF THE BLOOD

A normal adult male has about 5 liters of blood in his body. A straw-colored liquid called *plasma* makes up about 55 percent of the blood. Plasma, which is mostly water, contains thousands of different types of compounds in solution, such as fibrinogen, the protein from which clots are formed; nutrients, such as glucose, fats, and amino acids; various ions; antibodies, hormones, and enzymes; and waste materials such as urea and uric acid. A filtrate of plasma (lymph), containing no cells and essentially no protein, seeps out through the capillary walls, flows into the tissue spaces of the body, bathes the cells, and reenters the capillaries.

The remaining 45 percent of the blood is made up of cells: red blood cells, white blood cells, and platelets.

Red Blood Cells

Red blood cells, or *erythrocytes* (Figure 26–11), transport oxygen to all the tissues of the body. They are among the most highly specialized of all cells. As a mammalian red blood cell matures, it extrudes its nucleus, and its mitochondria and other cellular structures dissolve, so that almost the entire volume of the mature cell is filled with hemoglobin, the molecule that combines with oxygen to transport it from the lungs to the tissues. In man, each red blood cell has a life span of about 127 days; new ones are produced in the bone marrow. There are about 5 million (5×10^6) red blood cells in a cubic millimeter of blood—25 trillion (25×10^{12}) in the whole body.

White Blood Cells

For every 1,000 red blood cells in the human bloodstream, there is one white blood cell, or *leukocyte.* The chief function of the white blood cells is the defense of the body against invaders such as viruses, bacteria, and various foreign particles. This function is accomplished in two ways: (1) by taking part in the manufacture of antibodies, which are special protein molecules that combine with disease-causing agents and inactivate them; and (2) by phagocytosis, which is accomplished by white blood cells in much the same way as by amoebas (see page 59). The cell extends pseudopodia around the foreign substance, incorporates it within a food vacuole, and destroys it with the help of enzymes from the lysosomes (Figure 26–12).

White blood cells are often destroyed in the course of fighting infection; pus is composed largely of these dead cells. New white blood cells are formed

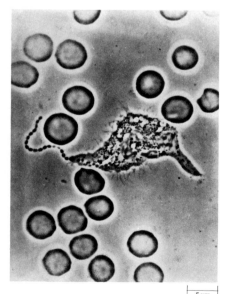

26–12

A white blood cell phagocytizing a chain of streptococci. Numerous red blood cells are also visible.

26–13

Human lymphatic system. Lymphatic fluid reenters the bloodstream through the thoracic duct.

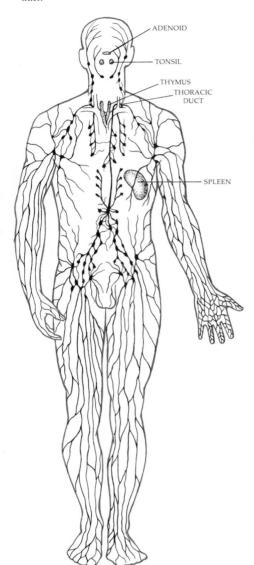

ADENOID

TONSIL

THYMUS
THORACIC
DUCT

SPLEEN

constantly in the spleen, in bone marrow, and in certain other tissues to take the place of those cells that are sacrificed.

White blood cells are unusual among somatic cells because of their active locomotion. They travel by amoeboid movement and are attracted to sites of infection, probably by the chemical substances diffused from such sites. They are quite capable of moving against the bloodstream, migrating through the walls of small blood vessels, and making their way into damaged or infected tissues.

Platelets

Platelets, found in the bloodstream of mammals, are colorless, round or biconcave disks less than half the size of red blood cells. They play an important role in plugging breaks in blood vessels and in the formation of clots.

LYMPHATIC SYSTEM

As was mentioned earlier, most of the tissue fluid passes back through the capillary walls to enter the general circulation. Other tissue fluid seeps through the thin walls of *lymph vessels*, which are present in most tissues, forming the lymph. In the lymph system (Figure 26–13), the lymph makes its way back into the bloodstream through progressively larger lymphatic ducts that empty into the veins. Frogs and many other vertebrates have lymph "hearts," which help to keep the fluid in circulation. Mammals rely chiefly on muscular movements and on valves to circulate the lymph, in a manner similar to that described earlier for venous circulation. Mammals also have lymph nodes, which are filtering areas for cellular debris, bacteria, and other infectious organisms picked up by the circulating lymph.

The spleen, tonsils, and adenoids are largely lymphoid tissue and are involved in the manufacture of lymphocytes, a type of white blood cell. The thymus, a spongy two-lobed organ that lies high in the chest, plays a critical role in the maturation of a particular type of lymphocyte known as the T-cell, an object of much current research. T-cells are apparently involved in several types of immune responses.

ANTIBODIES AND IMMUNITY

The invasion of the bloodstream by a foreign organism such as a bacterium or a virus induces the formation in the bloodstream of complex globular proteins known as *antibodies*. Antibodies are produced by plasma cells, which are specialized cells that arise from successive divisions of lymphocytes, a kind of white blood cell. Like enzymes, antibodies are highly specific, and their specificity, like that of enzymes, is based on their combining in a very precise way with another molecule. The molecule with which an antibody combines is known as an *antigen*; a single cell, such as a bacterial cell, may carry a number of antigens, all of which can combine with particular different antibodies.

Because it is programmed by evolution to act against foreign materials of all kinds, the immune system, particularly the T-cells, works vigorously against tissues—such as skin, kidney, or heart—transplanted from another individual (except an identical twin). Surgeons and medical research workers concerned wtih extending the use of tissue transplants are seeking ways to suppress or paralyze the immune response in such a way that these foreign but potentially lifesaving tissues can survive.

BLOOD TYPING

Blood typing ranks as one of the great advances of twentieth-century medicine. Before the discovery of blood types, transfusion was, in many cases, more hazardous than hemorrhage. Many types of injuries were more likely to be fatal than they are today, and most forms of major surgery could not be carried out.

Your particular blood type is determined by particular antigens on the surfaces of our red blood cells. There are two major types: A and B. Persons with type A blood are, simply, persons with antigen A on their red cells, type B means antigen B is present, and AB means both antigens are present. If you have neither antigen, you are type O. These blood types are inherited according to Mendelian laws. Persons with blood type A will form antibodies to red blood cells with antigen B, causing the blood cells to clump together (agglutinate). This reaction can be violent enough to cause death. Persons with type AB will not form antibodies to either A or B antigens and can therefore receive blood of any of the other major types. Persons with type O cannot receive blood from persons with Types A, B, or AB, but can give blood to persons with those types.

The Rh factor is also an antigen found on the red blood cells. If a woman is Rh-negative, meaning that she lacks the Rh factor, and her husband is homozygous for the Rh factor, any child that is conceived will be Rh-positive. At the time of birth, as the placenta is discarded, some Rh-positive cells may enter the mother's circulation and induce the formation of antibodies against the Rh antigen. If, in subsequent pregnancies, these antibodies cross the placental barrier and enter the circulation of the fetus, the fetus's Rh-positive cells may be destroyed. (Note that if the mother is Rh-positive, it does not matter whether the father is negative or positive, in terms of the Rh factor.) Methods for treating the infant at birth—or even still in the uterus—have been developed which can prevent this potentially fatal reaction.

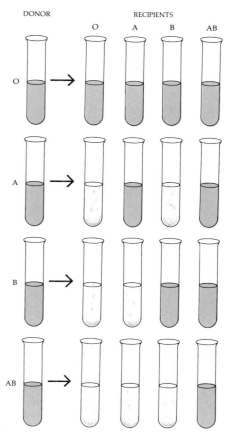

Transfusion reactions

Antibodies can act against invaders in one of three ways: (1) they may coat the foreign particle in such a way that it can be taken up by the scavenging white blood cells; (2) they may combine with it in such a way that they interfere with some vital activity—such as, for example, covering the site on the protein coat of a virus at which the virus attaches to the cell membrane; or (3) they may themselves, in combination with another blood component known as complement, actually lyse and destroy the foreign cell.

The first time a particular invader enters the bloodstream, the antibody-producing machinery, as we noted, works comparatively slowly, taking between four and five days to reach maximum antibody production. If the infection does not proceed too rapidly, antibody production will then catch up with the multiplication of the virus or bacteria and bring the infection under control. The second time this same invader is introduced, however, many lymphocytes, sensitized to that particular antibody, are present. Antibody production under these conditions proceeds so rapidly that the infectious agent rarely gains a foothold. It is for this reason that one infection with measles, mumps, or many other infectious diseases immunizes us against a second infection. Vaccines are agents that trigger the immune response of the body to a particular infectious agent without actually producing the disease; thus the invader, when it does appear, is met with a secondary response—the immediate and overwhelming production of antibodies—and so the disease is prevented.

The *immune response*, as the antibody-antigen system is called, is a powerful bulwark against disease, but it sometimes goes awry. Hayfever and other allergies are the result of interactions of pollen dust, or other substances which are weak antigens (antigens to which most people do not react), and a special kind of antibody which fixes onto the cells of the skin or mucous membranes where the antigen makes contact.

SUMMARY

Transport of oxygen, nutrients, hormones, and wastes is accomplished in vertebrates by closed circulation systems in which the blood is pumped by the muscular contractions of the heart into a vast circuit of arteries, arterioles, capillaries, venules, and veins. This network ultimately services every cell in the body. The essential function of the circulatory system is performed by the capillaries, through which substances diffuse into and out of the fluids surrounding the individual cells.

Heart structure in the vertebrates varies with the demands of their differing metabolic rates. The fish has a two-chambered heart; the atrium receives deoxygenated blood, and the ventricle expels the blood through the oxygenating gill capillaries into the systemic circulation. In amphibians, both oxygenated and deoxygenated blood are received and mingled in a two-chambered atrium and then are pumped simultaneously through the lungs and through the systemic circulation. Birds and mammals have a double circulation system, made possible by a four-chambered heart that functions as two separate pumping organs, one pumping deoxygenated blood to the lungs and the other pumping oxygenated blood from the lungs into the body tissues.

Synchronization of the heartbeat is controlled primarily by the sinoatrial node, an area of specialized heart muscle located in the right atrium, also called the pacemaker. The atrioventricular node delays the initiation of ventrical contraction until the atrial contraction is completed. The rate of heartbeat and strength of contraction are under neural control. Acetylcholine slows the pacemaker; epinephrine and norepinephrine stimulate it.

The blood is composed of plasma, red blood cells (erythrocytes), white blood cells (leukocytes), and platelets. The fluid part of the blood is plasma, which is chiefly water in which are dissolved or suspended nutrients, ions, antibodies, enzymes, waste substances, and compounds. Erythrocytes are the vehicles for oxygen-bearing hemoglobin. Leukocytes defend the body against invaders by the manufacture of antibodies and by phagocytosis. Platelets are involved in the formation of clots.

Fluid that seeps out of the capillaries can reenter the capillaries or be returned to the blood by the lymphatic system. The lymph also picks up bacteria, cellular debris, and other foreign particles that may be filtered out by the lymph nodes. The lymphatic system is involved in the production of lymphocytes, the special white blood cells involved in the immune reaction. The spleen, thymus, tonsils, and adenoids are also components of the lymphatic system.

The immune system is a defensive system of vertebrates which involves recognizing a substance entering the circulatory system as foreign and developing a specific defense against that substance. One important part of the system involves the production of antibodies by plasma cells. Antibodies are complex proteins that combine with specific antigens and inactivate them. The immune system is also involved in reactions against blood transfusions and tissue transplants.

QUESTIONS

1. What is the function of the cardiovascular system? What are its principal components?

2. Trace the course of a single red blood cell through the circulation in a mammal beginning and ending with the right atrium.

3. The smooth muscles of the arterioles have the capacity to alter the diameter of the arterioles by a factor of 5. What, then, is their total influence on blood flow?

4. Label the diagram at the left.

5. Fill in the blank spaces in the table below.

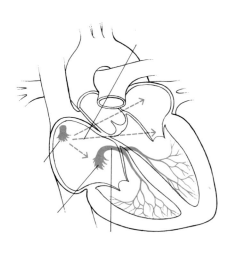

Inheritance of Major Blood Types

PHENOTYPES OF PARENTS		CHILDREN POSSIBLE	CHILDREN NOT POSSIBLE
A	A	A, O	AB, B
A	B		
		A, B, AB	O
A	O		
		B, O	A, AB
B	AB		
B	O		
AB	AB		
AB	O		
		O	

CHAPTER 27

Respiration

Respiration has two meanings in biology. At the biochemical level, "respiration" refers to the oxygen-requiring chemical reactions that take place in the mitochondria and are the chief source of energy for the cell. At the level of a whole organism, "respiration" refers to the process of taking in oxygen from the environment and returning carbon dioxide to it. The latter process—which is, of course, essential for the former—is the subject of this chapter.

The means by which organisms extract energy without oxygen are not very efficient, as we saw in Chapter 7. Before oxygen came to be present in the atmosphere, the only forms of life were one-celled organisms, and the only present-day forms that can carry on life processes without oxygen are a few types of bacteria and yeasts. The transition to an atmosphere containing free oxygen was responsible for one of the giant steps in evolution.

A second revolution in terms of energy came about with the transition to land. Air is a far better source of oxygen than water; one-fifth of the air of the modern atmosphere is made up of free oxygen, whereas at ordinary temperatures, even when water is saturated with air, only about 1 part in 250 (by weight) is free oxygen. Not only must more water be processed for oxygen, but water weighs a great deal more. A fish spends up to 20 percent of its energy in the muscular work associated with respiration, whereas an air breather expends only 1 or 2 percent of its energy in respiration. Also, oxygen diffuses much more rapidly through air than through water, about 300,000 times more rapidly, and so can be replenished much more quickly as it is used up by respiring organisms. All the higher vertebrates—the birds and the mammals—are air breathers, even those that live in the water.

Table 27-1 *Composition of Dry Air*

Oxygen	21%
Nitrogen	77%
Argon	1%
Carbon dioxide	0.03%
Other gases*	0.97%

* Includes hydrogen. neon, krypton, helium, ozone, and xenon.

AIR UNDER PRESSURE

At sea level, the air around us exerts a pressure on our skin of about 15 pounds per square inch. This pressure is enough to raise a column of water 32 feet high or a column of mercury 760 millimeters (29.91 inches) in the air. (Atmospheric pressure is generally measured in terms of mercury simply because mercury is relatively heavy—so the column will not be inconveniently tall—and because air does not dissolve in it.) Of this 760 millimeters, 159 results from the pressure of the oxygen in the air. It is as a result of the atmospheric pressure that oxygen enters living organisms.

27-1

Atmospheric pressure is usually measured by means of a mercury barometer. (a) To make a simple mercury barometer, fill a long glass tube, open at one end only, with mercury. Closing the tube with your finger, (b) invert it into a dish of mercury. Remove your finger. The mercury level will drop until the pressure of its weight inside the tube is equal to the atmospheric pressure outside. At sea level, the height of the column will be about 76 centimeters. (c) If the tube and dish are enclosed in an airtight covering and the air pressure is lowered by pumping air out, the mercury level will fall.

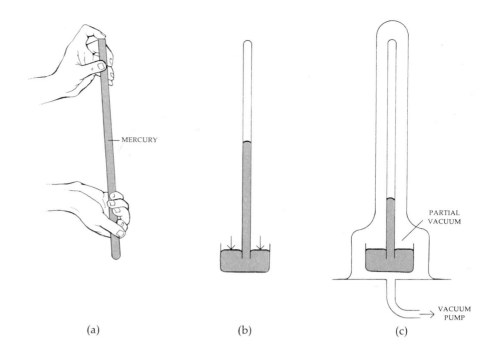

(a)　　　　　(b)　　　　　(c)

We are so accustomed to the pressure of the air around us that we are unaware of its presence or of its effects on us. However, if you visit a place—such as Mexico City—which is at a comparatively high altitude and therefore has a lower atmospheric pressure, you will feel lightheaded at first and will tire easily. Regular inhabitants of such places breathe more deeply, have enlarged hearts which circulate oxygen-carrying blood more rapidly, and have more red blood cells for oxygen transport.

The consequences of increased atmospheric pressures are seen in deep-sea divers. Early in the history of deep-sea diving, it was found that if divers come up from the bottom too quickly, they get the "bends," which is always painful and sometimes fatal. The bends develop as a result of breathing compressed air. Nitrogen is relatively insoluble in water, but high pressure forces nitrogen from the air in the lungs into solution in the blood and tissues. If the body is rapidly decompressed, the nitrogen bubbles out of the blood, like the bubbles that appear in a bottle of soda water when you first take off the top. These bubbles lodge in the capillaries, stopping blood flow, or invade nerves or other tissues.

EVOLUTION OF RESPIRATORY SYSTEMS

Oxygen enters and moves within cells by diffusion. Within the cell, it takes part in the oxidation of breakdown products of glucose and other carbon-containing compounds that serve as cellular energy sources. In this process, carbon dioxide is produced, which then diffuses out of the cell along the concentration gradient. This is true of all cells, whether an amoeba, a paramecium, a liver cell, or a brain cell. However, substances can move effectively by diffusion only for very short distances, less than 1 millimeter. These limits pose no problem for very small animals, in which each cell is quite close to the surface, or for animals in which much of the body mass is not metabolically active—like the "jelly" of jellyfish. Many eggs and embryos also respire in this simple way, particularly in their early stages of development.

27-2

(a) *Gaseous exchange across the entire surface of the body is found in a wide range of small animals from protozoans to earthworms. (b) Gaseous exchange across the surface of a flattened body. Flattening increases the surface-to-volume ratio and also decreases the distance over which diffusion has to occur within the body. Flatworms are examples. (c) External gills. These increase the* surface area, but note that they are unprotected and therefore easily damaged. Gaseous exchange usually takes place across the rest of the body surface as well as the gills. External gills are found in polychaete worms and some amphibians. (d) Highly vascularized internal gills. A ventilation mechanism draws water over the gill surfaces, as in fish. (e) Gaseous exchange at the terminal* ends of fine tracheal tubes that branch through the body and penetrate into all the tissues, found in insects and other arthropods. (f) Highly vascularized lungs consisting of sacs connected to a pharynx. Air is drawn into them by a ventilation mechanism. Lungs are found in all air-breathing vertebrates.*

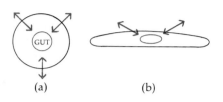

(a)   (b)

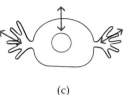

(c)

(d)

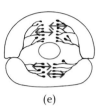

(e)

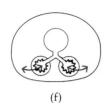

(f)

However, diffusion cannot possibly meet the needs of large organisms in which cells in the animal's interior may be many centimeters from the air or water serving as the oxygen source. As organisms increased in size in the course of evolution, there also evolved circulatory and respiratory systems that transport large numbers of gas molecules by bulk flow. (Remember that diffusion is the random movement of individual molecules down a concentration gradient, and bulk flow is the overall movement of molecules in response to pressure or gravity.)

An early stage in the evolution of gas-transport systems is exemplified by the earthworm (page 235). As is the case with most other kinds of worms, earthworms have a network of capillaries just one cell layer beneath the surface of their bodies. Oxygen and carbon dioxide diffuse directly through the body surface into and out of the blood as it travels through these capillaries. The blood picks up oxygen by diffusion as it travels near the surface of the animal and releases oxygen by diffusion as it travels past the oxygen-poor cells in the interior of the earthworm's body. Conversely, it picks up carbon dioxide from the cells and releases it by diffusion as it travels near the surface of the animal. Thus the gases move in and out of the animal by diffusion but are transported within the animal by bulk flow.

This system is particularly suitable for worms because their tube shape exposes a proportionally large surface area. Some worms can adjust their surface area in relation to oxygen supply. If you have an aquarium at home, you may be familiar with tubifex worms, which are frequently sold as fish food. If these worms are placed in water low in oxygen, such as a poorly aerated aquarium, they will stretch out as much as 10 times their normal length and thus increase the surface area through which oxygen diffusion occurs.

Evolution of the Gill

Gills and lungs are other ways of increasing the respiratory surface. Gills are outgrowths, whereas lungs are ingrowths, or cavities. The respiratory surface of the gill, like that of the earthworm, is a layer of cells, one cell thick, exposed to the environment on one side and to circulatory vessels on the other. The layers of gill tissue may be spread out flat, stacked in layers, or convoluted in various ways. The gill of a clam, for instance, is shaped like a steam-heat radiator, which is also designed to provide a maximum surface-area-to-volume ratio.

The vertebrate gill originated primarily as a feeding device. Primitive vertebrates breathed mostly through their skin. They filtered water into their mouths and out of what we now call their gill slits, extracting bits of organic matter from the water as it went through. (*Amphioxus*—page 243—which is believed to resemble closely the ancestral vertebrate, feeds in this way.)

In the course of time, numerous selection pressures, chiefly involved with predation, came into operation. As one consequence, there was a trend toward an increasingly thick skin, even one armored or covered with scales. Such a skin is not, of course, useful for respiratory purposes. At the same time, similar forces were operating to produce animals that were larger and swifter and so more efficient at capturing prey and escaping predators. Such animals also had larger energy requirements—and consequently larger oxygen requirements. These problems were solved by the "capture" of the gill for a new purpose: respiratory exchange. The surface area of the epithelium beneath the gill slowly increased in size and in blood supply over the millennia. The modern gill is the result of this evolutionary process.

In most fish, the water (in which oxygen is dissolved) is pumped in at the mouth by oscillations of the bony gill cover and flows out across the gills (Figure 27–3). The fish can regulate the rate of flow, and sometimes assist it, by opening and closing its mouth. Fast swimmers, such as mackerel, obtain enough oxygen only by keeping perpetually on the move. Such fish cannot be kept in an aquarium or any other space where their motion is limited because they will suffocate.

Evolution of the Lung

Lungs are internal cavities into which the oxygen-laden air is taken. They have certain disadvantages as compared with gills; it is more efficient from the point of view of diffusion to have a continuous flow across the respiratory surface. However, lungs have one overwhelming advantage in air: the respiratory surface can be kept moist without a large loss of water by evaporation. Lungs are largely a vertebrate "invention"; however, they are found in some lower animals. Land-dwelling snails, for example, have independently evolved lungs which are remarkably similar to the primitive lungs of some amphibians.

27–3

Rates of diffusion are proportional not only to the surface areas exposed but to differences in concentration of the diffusing molecules. The greater the difference in concentration of a molecule, the more rapid its diffusion. In the gill of the fish, the circulatory vessels are arranged so that the blood is pumped through them in a direction opposite to that of the oxygen-bearing water. This arrangement results in a far more efficient transfer of oxygen to the blood than if the blood flowed in the same direction as the water.

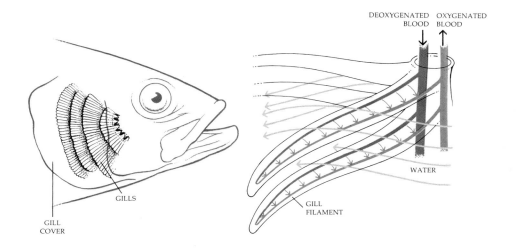

DEOXYGENATED BLOOD   OXYGENATED BLOOD

WATER

GILLS

GILL COVER

GILL FILAMENT

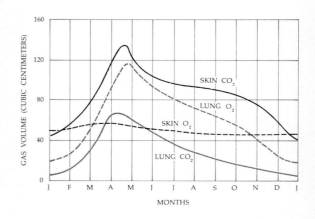

27-4

Frogs, such as this American bullfrog, have relatively small and simple lungs, and a major part of their external respiration takes place through the skin. The chart *shows the results of a study of a frog in which oxygen and carbon dioxide exchanges through the skin and lungs were measured simultaneously for one year. Can you ex-* *plain why the respiratory activity peaks in April and May and is lowest in December and January?*

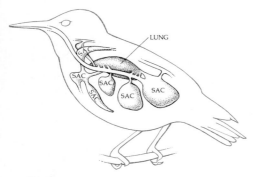

27-5

The lungs of birds are extraordinarily efficient. They are small and are expanded and compressed by movements of the body wall. Each lung has several air sacs attached to it which empty and fill like balloons at each breath. No gas exchange takes place in the sacs. They appear rather to act as bellows. Thus, the lung is almost completely flushed with fresh air at every breath; there is little residual "dead" air left in the lungs, as there is in mammals.

Some primitive fish had lungs as well as gills, although the lungs were not efficient enough to serve as more than accessory structures. These lungs were a special adaptation to fresh water, which, unlike ocean water, may stagnate and, because of decay or of algal bloom, become depleted of oxygen. A few species of lungfish still exist. These surface and gulp air into their lungs and so can live in water that does not have sufficient oxygen to support other fish life.

Amphibians and reptiles have relatively simple lungs, with small internal surface areas, although their lungs are far larger and more complex than those of the lungfish. The lungs of lungfishes developed directly from the pharynx, the posterior portion of the mouth cavity which leads to the digestive tract. In amphibians, reptiles, and other air-breathing vertebrates, we see the evolution of the windpipe, or *trachea,* guarded by a valve mechanism, the glottis, and nostrils, which make it possible for the animal to breathe with its mouth closed. Amphibians still rely largely on their skin for exchange of oxygen and carbon dioxide (Figure 27–4), but reptiles breathe almost entirely through their lungs.

An important feature of all vertebrate lungs is that the exchange of air with the atmosphere takes place as a result of changes in lung volume. Such lungs are known as ventilation lungs. Frogs gulp air and force it into their lungs in a swallowing motion; then they open the glottis and let it out again. In reptiles, birds, and mammals, air is sucked into the lungs as a consequence of changes in the size of the lung cavity, brought about by activity of the chest muscles.

Respiration in Large Animals: Some Principles

Let us stop a moment to sum up this discussion. In terms of the principles involved in large animals, both diffusion and bulk flow move oxygen molecules

ENERGY AND OXYGEN

Oxygen consumption is directly related to energy expenditure, and, in fact, energy requirements are usually calculated by measuring oxygen intake or the release of carbon dioxide. The energy expenditure at rest is known as basal metabolism. The basal metabolism of a warm-blooded (endothermic) animal is much higher than that of an exotherm, with 10 to 30 percent of the energy budget being expended on maintaining a constant temperature.

Metabolic rates increase sharply with exercise. A man exercising consumes 15 to 20 times the amount of oxygen as does a man sitting still. Oxygen consumption of a running animal increases linearly with its speed. If you are an aquatic animal, such as a fish or a porpoise, swimming is far less expensive,

in terms of energy requirements, than is running. However, a man, whose body is basically unfit for moving through the water, uses five times as much energy—as measured by his oxygen consumption—in swimming as in running.

In terms of energy consumption, flying is less expensive than walking and running. Many birds can fly more than 1,000 kilometers nonstop, whereas an animal of similar weight could not travel on the ground for that distance without stopping for food. (There are also rumors that migratory birds hitchhike on atmospheric currents.) However, flying, for a bird, is more expensive, in energy-consumption terms, than swimming is for a fish.

Oxygen is required for the energy-producing reactions that take place in the mitochondria. The oxygen consumption of a running animal—such as the cheetah—increases directly in proportion to its speed. The cheetah is the fastest land animal over short distances, attaining speeds of 110 kilometers per hour but only for distances of half a kilometer or less.

between the external environment and actively metabolizing tissues. This movement occurs in four stages:

1. Movement by bulk flow of the oxygen-containing external medium (gas or water) by means of lungs or gills to a thin membrane close to small blood vessels.
2. Diffusion of the oxygen across this membrane into the blood.
3. Movement of bulk flow of the oxygen with the circulating blood to the tissues where it will be used.
4. Diffusion of the oxygen from the blood into the tissue cells.

Carbon dioxide, which is produced in the tissue cells, follows the reverse path as it is eliminated from the body.

RESPIRATION IN MAN

The Human Lung

The human respiratory system is shown in Figure 27–6a. Air enters the nasal passages, where it is warmed and cleaned, after which it passes through the trachea. The trachea is strengthened by rings of cartilage that prevent it from collapsing during inspiration. The trachea leads into the _bronchi_ (singular: _bronchus_), which subdivide into smaller and smaller passageways, the bronchioles, and finally end in small air sacs known as _alveoli_, each only 1 or 2 millimeters in diameter (Figure 27–6b). The alveoli have very thin walls which contain an abundant supply of capillaries, and it is through these walls that the actual exchange (by diffusion) of gas molecules between blood and air occurs. The other parts of the respiratory system serve mainly to conduct air by bulk flow to and from the alveoli. A pair of human lungs has about 300 million alveoli, providing a respiratory surface area of some 70 square meters, approximately 40 times the surface area of the entire human body.

27–6

The human respiratory system. (a) Air enters through the nasopharynx and passes down the trachea, bronchi, and bronchioles to the alveoli (b) in the lungs. Within each alveolus, of which there are some 300 million in a pair of lungs, oxygen and carbon dioxide diffuse into and out of the bloodstream, through the capillary walls.

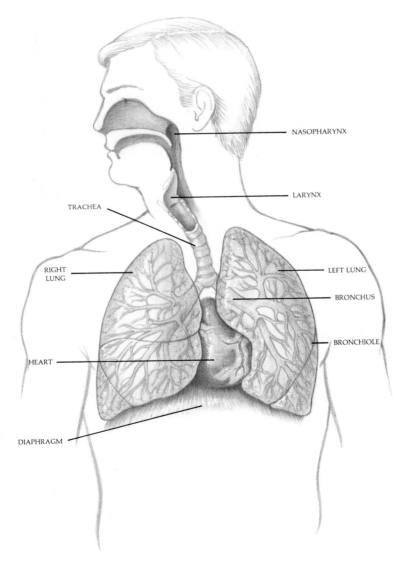

NASOPHARYNX

LARYNX

TRACHEA

RIGHT LUNG

LEFT LUNG

BRONCHUS

BRONCHIOLE

HEART

DIAPHRAGM

(a)

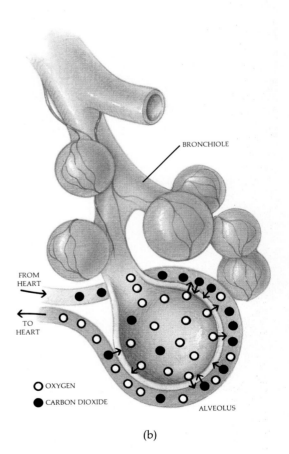

BRONCHIOLE

FROM HEART

TO HEART

○ OXYGEN
● CARBON DIOXIDE

ALVEOLUS

(b)

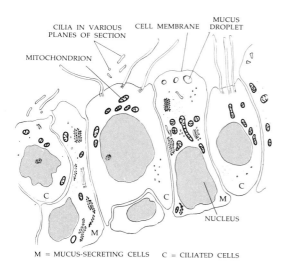

CILIA IN VARIOUS
PLANES OF SECTION
CELL MEMBRANE
MUCUS DROPLET
MITOCHONDRION

C
C
C
M
M
NUCLEUS

M = MUCUS-SECRETING CELLS   C = CILIATED CELLS

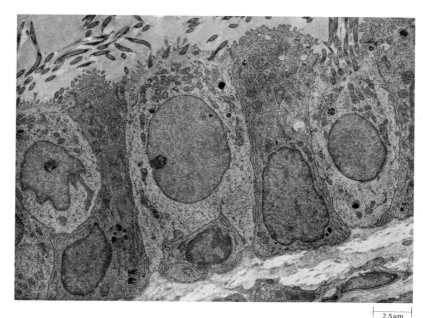

2.5 μm

27-7

Electron micrograph and diagram of tracheal epithelial cells. As you can see, the epithelium contains both ciliated cells (marked C in the diagram) and mucus-secreting cells (M), *sometimes called goblet cells. The beating of the cilia removes foreign particles from the trachea and distributes a protective* *coating of mucus. In the micrograph, the white circles in the cytoplasm of the middle goblet cell are secretion droplets of mucus.*

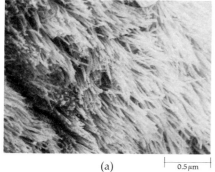

(a)   0.5 μm

(b)   0.3 μm

27-8

(a) *Ciliated surface of a normal bronchus.*
(b) *Surface of a bronchus with cancer. (See boxed essay, page 385.)*

The trachea and bronchi are lined with epithelial cells, which include mucus-secreting cells and ciliated cells. The mucus coats and protects the respiratory system. The cilia beat continuously, removing foreign particles from the surface of the respiratory tract and distributing the mucus (Figure 27-7). The cilia push the mucus and the foreign particles up toward the pharynx, from which they are generally swallowed. We are usually aware of the process only when the production of mucus is increased above normal as a result of irritation of the membrane.

Mechanics of Respiration

Air flows into or out of the lungs when the air pressure in the alveoli differs from the pressure of the external air (atmospheric pressure). When alveolar pressure is greater than atmospheric (and the air passages are open), air flows out of the lungs, and expiration occurs. When alveolar pressure is less than atmospheric, inspiration occurs. When the pressures are the same, as occurs momentarily at the end of inspiration and expiration, there is no air flow (Figure 27-9). Inspiration and expiration are, of course, forms of bulk flow.

The pressure in the lungs is varied by changing their volume; when the lungs are made bigger, as in inspiration, the pressure within them falls, and when they get smaller, the pressure rises. How do the lungs change size? They passively follow changes in the size of the thoracic cavity. We inhale by contracting the dome-shaped diaphragm and the muscles on the outside of our ribs. This flattens the diaphragm down against the abdominal organs and moves the ribs up and out. These movements enlarge the thoracic cavity; the pressure within it falls, and air moves into the lungs.

A model illustrating the way air is taken into and expelled from the lungs.

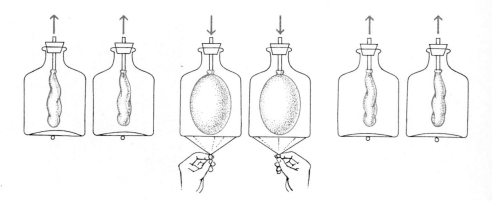

When inspiration ceases, the diaphragm and muscles between the ribs relax. This, in turn, allows the tissues of the thoracic cavity to return to their smaller, resting size. This decrease in volume is accompanied by an increase in pressure; the air pressure within the lungs rises above atmospheric pressure, and expiration occurs. Usually, only about 10 percent of the air in the lung cavity is exchanged at every breath, but as much as 80 percent can be exchanged by deliberate deep breathing.

Control of Respiration

The rate of respiration is controlled by a respiratory center in the medulla. This brain center receives and coordinates information about the carbon dioxide, hydrogen ion, and oxygen concentration of the blood. Action potentials are relayed from this center to nerves that originate in the spinal cord and that control the muscles of the chest and the diaphragm.

The system is so sensitive to even the smallest change in chemical composition of the blood, particularly to the concentration of carbon dioxide, that only the slightest variations are permitted to occur. If carbon dioxide concentration increases only slightly, breathing immediately becomes deeper and faster, permitting more carbon dioxide to leave the blood until the carbon dioxide level has returned to normal.

You can, of course, increase your breathing rate by voluntarily contracting and relaxing your chest muscles, but breathing is normally involuntary. It is impossible to commit suicide by deliberately holding your breath; as soon as you lose consciousness, the involuntary controls take over once more. Deaths due to barbiturates and other drugs with a depressant activity are usually the result of a damping of the activity of the respiratory center. The effect can often be reversed if artificial respiration is administered quickly and for a long enough period of time.

Transport of Oxygen in the Blood: Hemoglobin

Oxygen is not normally very soluble in water. The bloodstream's capacity to transport oxygen is greatly increased by the presence of a special carrier molecule, hemoglobin. Each hemoglobin molecule consists of four polypeptide chains with their associated heme groups (heme is a porphyrin ring containing iron). Each heme unit combines with one molecule of oxygen; thus each hemoglobin molecule can carry four molecules of oxygen. Hemoglobin enables our bloodstreams to carry about 60 times as much oxygen as could be transported by an equal volume of water or plasma (the fluid portion of the blood) alone.

CANCER OF THE LUNG

Lung cancer is the most rapidly increasing form of cancer in the United States and the most common cause of death from cancer in men. Almost 100,000 men and women will develop lung cancer in the United States in 1977, and more than 90 percent of them will die in less than three years. Most of these patients will be cigarette smokers.

The middle and lower lobes of a cancerous lung are shown in (a). The cancer is the solid grayish-white mass. Its rapid growth has replaced most of the normal lung tissue in the middle lobe. It has also probably begun to spread to other parts of the body. Notice the bronchial branch that leads into the cancer and is destroyed by it. The lung tissue remaining around the cancer is compressed, airless, and dark red. Away from the cancer, in the lower lobe, the lung is filled with air and is light pink.

There are several different kinds of lung cancer which can be distinguished microscopically. The three most common are epidermoid carcinoma (b), so called because the cells grow in sheets like the epidermis of the skin; adenocarcinoma (c) from adeno,

the Greek word for gland, because glandlike spaces are formed by the cancer cells; and oat-cell carcinoma (d) because the small, dark cancer cells resemble oats.

For many years the chest x-ray was the only means of detecting lung cancer early. Now we know that some cancers can be detected even before they show up on x-rays, because cancer cells are dislodged by coughing and can be identified microscopically in a specimen of sputum. A large epidermoid cancer cell and several white blood cells found in a sputum specimen are shown in (e). It is hoped that the combination of sputum cytology, as this examination is called, and chest x-rays will be more effective than either alone.

The lung is normally protected by ciliated cells lining the tracheobronchial tree. The cilia catch and sweep out particulates in the respired air. Cancer cells arising in the bronchus generally do not have cilia. The scanning electron micrographs on page 383 (Figure 27–8) show the normal ciliated surface of the bronchus and the surface of a bronchus with cancer.

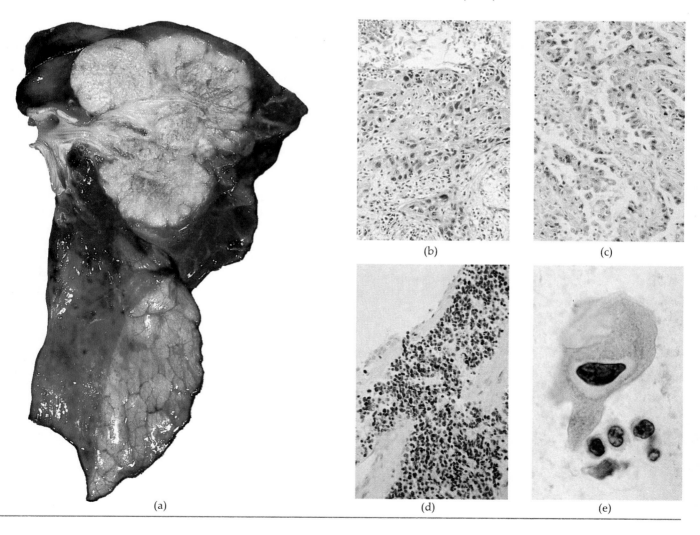

(a)

(b)

(c)

(d)

(e)

27–10

The hemoglobin molecule consists of four heme groups with their polypeptide chains intertwined in a quaternary structure (page 52). Each molecule can hold up to four oxygen molecules. (Adapted with permission from R. E. Dickerson and I. Geis, The Structure and Action of Proteins, *W. A. Benjamin, Inc., Menlo Park, Calif., 1969. Copyright 1969 by Dickerson and Geis.)*

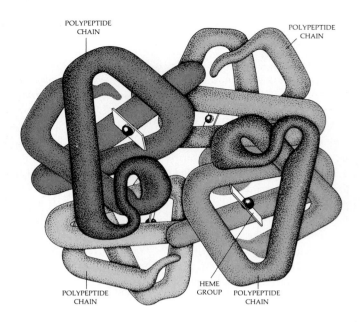

POLYPEPTIDE CHAIN

POLYPEPTIDE CHAIN

POLYPEPTIDE CHAIN

HEME GROUP

POLYPEPTIDE CHAIN

In invertebrates, hemoglobin and hemoglobinlike molecules often circulate free in the bloodstream. In vertebrates, hemoglobin is transported in highly specialized cells, the red blood cells, or erythrocytes. A mature red blood cell carries some 265 million molecules of hemoglobin. Hemoglobin is a red pigment that becomes redder upon oxygenation; for this reason blood flowing through arteries is redder than blood flowing through veins.

Whether oxygen combines with hemoglobin or is released from it depends on the oxygen concentration of the surrounding plasma. In the lung capillaries, where oxygen concentration is high, most of the hemoglobin is combined with oxygen. In the tissues, however, where the oxygen concentration is lower, fewer hemoglobin molecules are combined with oxygen, and the oxygen they carried is released into the plasma. This system compensates automatically for the oxygen requirements of the tissues. Thus, because a highly active tissue uses more oxygen, the oxygen concentration in the tissue is lower and more oxygen is released from the hemoglobin. Carbon monoxide is so extremely poisonous because it combines with the hemoglobin molecule more readily than does oxygen and fills the binding sites normally occupied by oxygen.

Carbon dioxide is more soluble than oxygen, and some of it is simply dissolved in the blood. Most is carried in combination with water as a bicarbonate ion (HCO_3^-):

$$CO_2 + H_2O \rightleftharpoons HCO_3^- + H^+$$

CARBON DIOXIDE ⠀⠀ WATER ⠀⠀ BICARBONATE ION ⠀⠀ HYDROGEN ION

The combination of carbon dioxide and water is catalyzed by the enzyme carbonic anhydrase, which is present in the red blood cell along with hemoglobin. Thus transport of carbon dioxide by the blood also depends on red blood cells.

As you can see by the arrows, this reaction can go in either direction. The direction it actually takes depends on the concentration of carbon dioxide. In the tissues, where carbon dioxide concentration is high, bicarbonate is formed. In the lungs, where carbon dioxide is low, bicarbonate dissociates to form carbon dioxide and water and, once released, the carbon dioxide diffuses from the plasma into the alveoli and flows out of the lung with the expired air.

SUMMARY

Organisms obtain most of their energy from oxidation of carbon-containing compounds in the mitochondria. This process requires oxygen and releases carbon dioxide. Respiration is the means by which an organism obtains the oxygen required by its cells and rids itself of carbon dioxide.

Oxygen is available in both water and air. It enters into cells and body tissues by diffusion. However, movement of oxygen by diffusion requires a relatively large surface area exposed to the source of oxygen and a short distance over which the oxygen has to diffuse. Selection pressures for increasingly efficient means of obtaining oxygen and getting rid of carbon dioxide led to the evolution in vertebrates of gills and lungs. Both gills and lungs present enormously increased surface areas for the exchange of gases and have a rich blood supply for transporting these gases to and from other parts of the animal's body. Respiration in large animals involves both diffusion and bulk flow. Bulk flow brings air or water to the lungs or gills and circulates oxygen and carbon dioxide in the bloodstream. Gases are exchanged between the blood and the atmosphere and between the blood and the tissues by diffusion.

In humans, air enters the lungs through the trachea, or windpipe, and goes from there into a network of increasingly smaller tubules, the bronchi and bronchioles, which terminate in small air sacs, the alveoli. Gas exchange actually takes place across the alveolar walls. The lungs empty and fill as a result of changes in the air pressure within the thoracic cavity which, in turn, result from changes in the size of the thoracic cavity. The muscles that bring about respiration are under the control of nerves that originate in the spine. These nerves are activated by a respiratory center in the medulla of the brain that responds to signals caused by very slight changes in the carbon dioxide or hydrogen ion concentration of the blood. They are responsive to larger changes in oxygen concentration.

Carbon dioxide is transported in the blood largely in the form of HCO_3^-. In vertebrates, oxygen is carried by hemoglobin molecules, each of which binds four molecules of oxygen. The hemoglobin is packed within red blood cells.

QUESTIONS

1. Define the following terms: bronchioles, ciliated epithelium, gill, alveoli.

2. Sketch the human respiratory system. When you have finished, compare your drawing to Figure 27-6.

3. Trace the path of an oxygen atom from the air into a body cell.

4. What are the advantages and disadvantages of obtaining oxygen from air rather than from water? You might be able to think of several of each besides those mentioned in the text.

5. Carbon monoxide (CO) attaches to hemoglobin; the resulting compound does not readily dissociate, can no longer combine with oxygen, and is a brighter red than normal hemoglobin. From these facts, suggest how you might recognize and give assistance to a victim of carbon monoxide inhalation.

CHAPTER 28

Digestion

Digestion is the process by which heterotrophs break down the tissues of other organisms to molecules that can be used by their own cells. Such molecules include monosaccharides (such as glucose), amino acids, fatty acids, and iron and other inorganic substances. Once within the cells, these molecules can be processed for entry into the Krebs cycle and the electron transport chain, where they are oxidized and their energy is packaged in the high-energy bonds of ATP. Or, after suitable alterations, these molecules can be incorporated into cellular structures or used to make enzymes, hormones, or other components of the cell's machinery. Digestion thus provides both the atoms and molecules needed by the individual cells and also the cellular fuel for their various activities.

DIGESTIVE TRACT IN VERTEBRATES

In higher animals, digestion takes place in a long tube, running from mouth to anus, which has different areas that are specialized for particular stages of the digestive process (Figure 28–2). The food is moved along this tube by successive waves of muscular contractions known as *peristalsis*.

Primary Processing: The Mouth

The mechanical breakdown of food begins in the mouth. Many vertebrates have teeth especially adapted for tearing or grinding. Some, such as birds and turtles, have horny beaks or bills. The tongue, also a vertebrate development, serves largely to move and manipulate food in mammals, although some vertebrate tongues, such as those of hagfish and lampreys, have horny "teeth," and the tongues of frogs and toads flip out (they are attached at the front, not the back) to catch insects. Mammalian tongues carry the taste buds (Figure 28–3), and in man the tongue has developed a secondary function of formulating sounds for communication, another example of preadaptation.

28–1
Although the requirements of animals are remarkably similar on a cellular level, the ways these requirements are met vary widely. An animal's diet determines not only its dentition and digestive enzymes but dictates, to a large degree, its way of life (and, of course, vice versa). Vegetarianism, however, does not ensure a peaceful disposition: the rhinoceros is among the most short-tempered of mammals.

The human digestive tract.

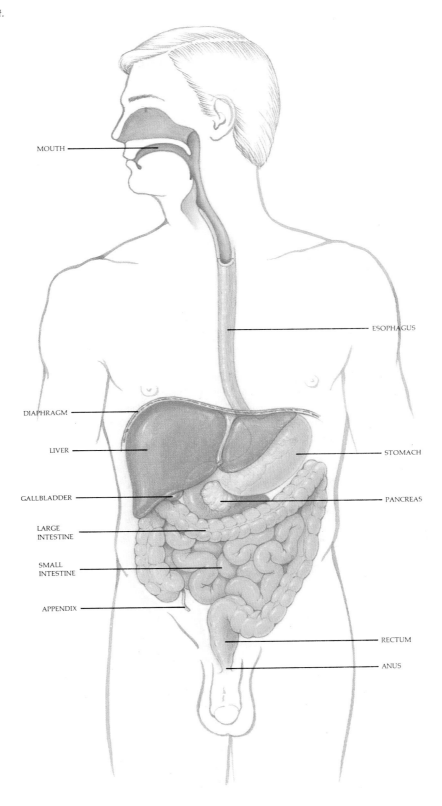

MOUTH

ESOPHAGUS

DIAPHRAGM

LIVER

STOMACH

GALLBLADDER

PANCREAS

LARGE
INTESTINE

SMALL
INTESTINE

APPENDIX

RECTUM

ANUS

SWALLOWING

Separation of the digestive and respiratory systems in mammals makes it possible for mammals to breathe while eating. The pharynx, the common cavity of the two systems, is at the back of the mouth and connects with the larynx and the esophagus. The pharynx also serves as a resonating chamber in sound production.

Swallowing begins as food is propelled from the mouth backward toward the pharynx. From the pharynx, the food mass moves to the esophagus. Notice how an inward bulge of the muscles of the posterior pharyngeal walls moves downward, pushing the food mass ahead of it.

In man, the upper part of the esophagus is striated muscle and the lower part is smooth, involuntary muscle. Dogs, cats, and other animals that gulp their food have striated muscle along the whole length of the esophagus.

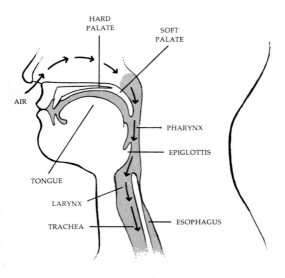

FACE AND NECK, SHOWING PARTS OF
RESPIRATORY AND DIGESTIVE SYSTEMS

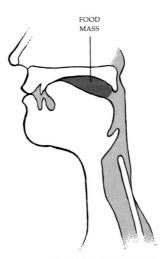

FOOD MASS IS IN MOUTH,
SOFT PALATE IS DRAWN UPWARD

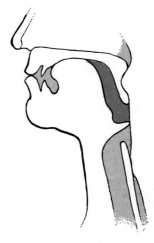

TONGUE PUSHES FOOD FURTHER BACK,
OPENING BETWEEN SOFT PALATE AND
PHARYNGEAL WALL CLOSES

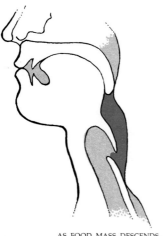

AS FOOD MASS DESCENDS,
EPIGLOTTIS TIPS DOWNWARD

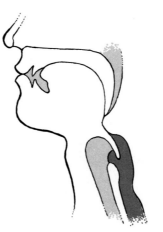

FOOD MASS PASSES INTO ESOPHAGUS

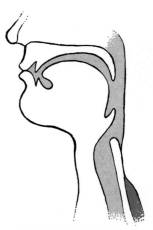

AS FOOD PASSES THE TRACHEA,
EPIGLOTTIS TURNS UP AGAIN AND
PHARYNX IS OPENED ONCE AGAIN

28-3

The outer pore of a taste bud. The receptor cells of the taste bud are just visible within the pore. This scanning electron micrograph shows the surface of the human tongue. Other taste buds are found on the roof of the mouth, the pharynx, and the larynx.

In mammals, enzymatic processes often start in the mouth. The salivary glands of man secrete amylases, which are enzymes that break starch down into sugars. In some human cultures, alcoholic beverages are made by chewing starchy vegetables, usually roots and tubers, long enough to begin the breakdown of the starch and then spitting out the pulp, leaving it for the glycolytic enzymes of airborne yeast cells to convert the sugars into alcohol.

The salivary glands, of which there are three pairs (Figure 28-4), also secrete water, mucus, and sodium bicarbonate, which makes the saliva slightly alkaline. The activity of the glands is under the control of the autonomic nervous system. The presence of food in the mouth, smelling or tasting food, or even anticipation of food makes the mouth "water."

Swallowing involves both skeletal and smooth muscles. It begins as a voluntary action, but once the food reaches the pharynx, swallowing is involuntary, controlled by reflexes. The swallowed material passes to the esophagus, a muscular tube about 25 centimeters long which leads to the stomach.

Secondary Processing: The Stomach

The stomach is an elastic, muscular bag that acts as a food reservoir. Distended, it can hold 2 to 4 liters of food. The walls of the stomach are lined by a layer of epithelial cells, which constitute the gastric mucosa (Figure 28-5). These gastric cells secrete mucus, pepsinogen, and hydrochloric acid (HCl), sometimes called the gastric juices. The HCl kills most bacteria and other living cells in the ingested food and activates the pepsinogen, converting it to pepsin, an enzyme that begins the breakdown of protein. The HCl also loosens the tough, fibrous components of ingested tissues and erodes the cementing substances between cells. Because of the HCl, the contents of the stomach are very acid, with a pH of between 1.5 and 2.5 in the normal person. (Pepsin does its enzymatic work best at a pH of 1.9.) If the sphincter muscle between the esophagus and the stomach should relax, as it sometimes does, gastric HCl will burn the sensitive lining of the esophagus, causing the sensation known as "heartburn." Few molecules are absorbed through the stomach walls, with the exception of alcohol, which is absorbed from the stomach as well as from the intestine.

The stomach is under both neural and hormonal control. Anticipation of food and the presence of food in the mouth stimulate gastric activity and the pro-

28-4

The bulk of the saliva is produced by three pairs of salivary glands. Additional amounts are supplied by minute glands in the mucous membrane lining the mouth. The parotid glands are the sites of infection of the mumps virus.

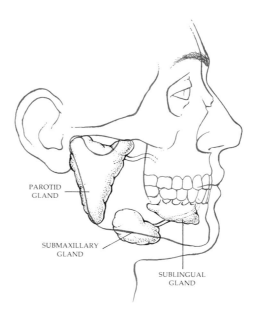

PAROTID
GLAND

SUBMAXILLARY
GLAND

SUBLINGUAL
GLAND

(a) *Surface of the stomach, as shown in a scanning electron micrograph. The numerous indentations are gastric pits. (b) A cross section of stomach mucosa. The parietal cells secrete hydrochloric acid and the chief cells produce pepsinogen. Mucus, secreted by cells of the surface epithelium, coats the surface of the stomach and lines the gastric pits, protecting the stomach surface.*

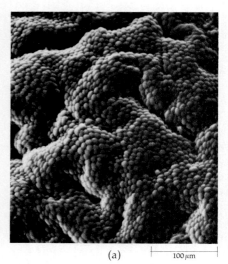

(a)   100 µm

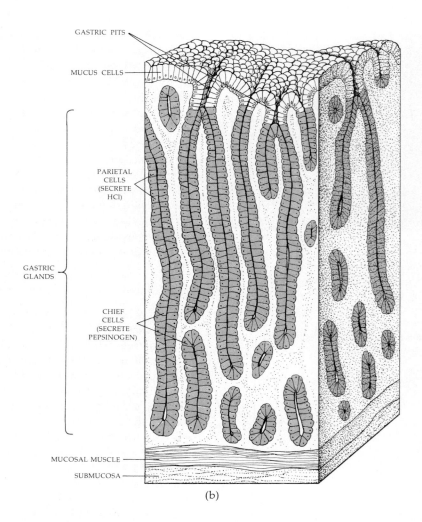

GASTRIC PITS

MUCUS CELLS

PARIETAL CELLS (SECRETE HCl)

GASTRIC GLANDS

CHIEF CELLS (SECRETE PEPSINOGEN)

MUCOSAL MUSCLE

SUBMUCOSA

(b)

duction of gastric juices by way of the autonomic nerves. When food reaches the stomach, particularly food rich in protein, its presence causes the release of a hormone, gastrin, from gastric cells into the bloodstream. This hormone acts on the epithelial cells of the stomach to increase their secretion of gastric juices. Food remains in the stomach for several hours, after which it is moved by peristalsis into the small intestine. The pyloric sphincter, a ring of muscle, surrounds the opening of the stomach into the small intestine.

Digestive Activity: The Duodenum, the Liver, and the Pancreas

Anatomically, the small intestine is characterized by numerous fingerlike projections, villi, on the mucosa, and tiny cytoplasmic projections, microvilli, on the surface of the individual epithelial cells (Figures 28–6 and 28–7). Both of these structural features increase the surface area of the small intestine. The epithelial cells of the mucosa secrete mucus. The small intestine is 3 to 4 meters long in the living adult; the total surface area of the human small intestine is almost 200 square meters, the area of a singles tennis court.

The upper 20 centimeters of the small intestine is known as the *duodenum*. Here, as a result of the action of a variety of enzymes, most of the digestion takes place. Most of these enzymes come from the pancreas and the liver through ducts opening into the duodenum; other enzymes are secreted by cells of the intestinal wall.

Longitudinal section of intestinal villi. Food molecules are absorbed through the walls of the villi and, with the exception of fat molecules, enter the bloodstream by means of the capillaries. Fats, hydrolyzed to fatty acids and glycerol, are taken up into the lymphatic system. The villi can move independently of one another; their motion increases after a meal.

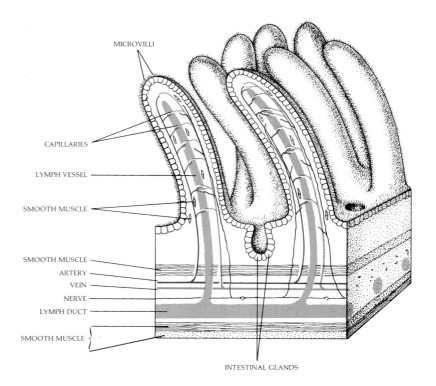

MICROVILLI

CAPILLARIES

LYMPH VESSEL

SMOOTH MUSCLE

SMOOTH MUSCLE
ARTERY
VEIN
NERVE
LYMPH DUCT

SMOOTH MUSCLE

INTESTINAL GLANDS

The pancreas secretes amylase, lipase, and enzymes that break down proteins. Amylases continue the breakdown of starch and glycogen begun in the mouth. Lipases hydrolyze fats into glycerol and fatty acids. Proteins are broken down by three types of enzymes. One group breaks apart the long protein chains. A second type of enzyme acts only on the carboxyl end of a chain to release free amino acids. Aided by a third type of enzyme, from the intestinal cells, the proteins in the food are broken into single amino acids, the form in which they are absorbed through the epithelial cells of the intestine and finally enter the bloodstream or, in the case of fats, the lymph.

In addition to these many different enzymes, the duodenum receives an alkaline fluid from the pancreas and liver that neutralizes the stomach acid, and bile, which is produced in the liver and stored in the gallbladder. Bile contains a mixture of salts that emulsify fats, breaking them apart into droplets so that they can be attacked by the lipases.

The digestive activities of the duodenum are almost entirely coordinated and regulated by hormones. In the presence of acid, the duodenum releases secretin, a hormone that stimulates the pancreas to secrete alkaline fluid and the liver to secrete alkaline bile. Fats and protein in the food in the duodenum stimulate production of another hormone, cholecystokinin, which triggers the release of enzymes from the pancreas and the emptying of the gallbladder. Fat inhibits gastric motility, slowing the rate at which food enters the intestine; this effect is produced in part by cholecystokinin but also seems to involve other as yet unidentified hormones. Thus a complex interplay of stimuli and checks and balances serve to activate and inactivate digestive enzymes and to adjust the chemical environment.

In the course of digestion, large amounts of water—some 2 or 3 liters—enter the stomach and small intestine from the body fluids. Most of this water is reabsorbed in the small intestine. This is an important function. When resorption of water is interfered with, as in vomiting or diarrhea, severe dehydration can result.

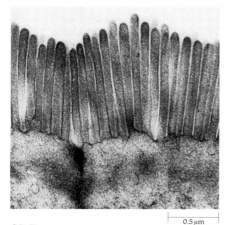

0.5 µm

Microvilli on two cells of intestinal epithelium. Near the outer surface of the cells, the membranes of the two cells are fused in what is known as a tight junction. Such tight junctions prevent extracellular materials from entering the space between the cells.

RUMINANTS

The ruminants—a group that includes cattle, sheep, goats, camels, and giraffes—are able to utilize cellulose as a source of energy. Ruminants do not themselves digest this tough polysaccharide; it is broken down by the action of resident microorganisms in their digestive tracts. In the ruminant digestive system is a series of four stomachs, the first two of which (the rumen) contain a rich concentration of bacteria and protozoans. These microorganisms secrete enzymes that break cellulose and fats down into simple fatty acids and gases (carbon dioxide and methane). The fatty acids pass through the walls of the rumen into the bloodstream of the animal and are utilized as energy sources in various parts of the body. The gases are belched forth by the animal. The last two stomachs of the ruminants secrete enzymes which break down the proteins of the bacteria and protozoans that continuously arrive from the first two stomachs. Thus, in effect, the ruminant obtains all its dietary essentials from this rich culture of bacteria and protozoans. The microorganisms, in turn, are provided with an environment rich in carbohydrates and maintained at a constant temperature favorable for growth.

Absorption and Excretion: The Lower Intestinal Tract

From the duodenum the food moves by peristalsis into the lower segment of the small intestine, the ileum, and from there into the large intestine. In the large intestine, most of the rest of the water is removed and some of the indigestible remains are broken down by the large population of bacteria. Many of these bacteria release materials useful to their host. In humans, for example, these bacteria are the chief source of vitamin K, which promotes blood clotting. The residue is stored in the rectum and finally excreted through the anus, as the feces.

Table 28–1 *Gastrointestinal Hormones*

HORMONE	SOURCE	STIMULUS FOR PRODUCTION	ACTION
Gastrin	Stomach	Protein-containing food in stomach, also parasympathetic nerves to stomach	Stimulates secretion of gastric juices
Secretin	Duodenum	HCl in duodenum	Stimulates secretion of alkaline bile and pancreatic fluids
Cholecystokinin	Duodenum	Fat and protein in duodenum	Stimulates release of pancreatic enzymes and stimulates release of bile from gall bladder

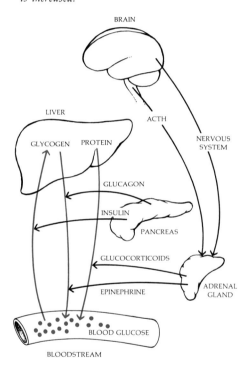

28–8

Hormonal regulation of blood glucose. When blood sugar concentrations are low, the pancreas releases glucagon, which stimulates the breakdown of glycogen and the subsequent release of glucose from the liver. When blood sugar concentrations are high, the pancreas releases insulin, which removes glucose from the bloodstream by stimulating its passage into cells and promoting its conversion to glycogen, the storage form. ACTH, produced by the pituitary, stimulates the adrenal cortex to produce glucocorticoids, which cause the liver to convert proteins to glucose. Under conditions of stress, the adrenal medulla releases epinephrine, which also raises blood sugar, and the production of ACTH is increased.

REGULATION OF BLOOD GLUCOSE

As we noted at the beginning of this chapter, a major function of digestion is to provide an energy source for each of the individual cells of which the body is composed. Although vertebrates rarely eat 24 hours a day, their blood sugar—the cellular energy supply—remains extraordinarily constant. The liver plays the central role in this critical process. Glucose and other monosaccharides are absorbed into the blood from the intestinal tract and are passed directly to the liver by way of the portal vein. The liver converts most of these monosaccharides to glycogen (the liver stores enough glycogen to satisfy the body's needs for about 4 hours) and to fat. Some of the fat is stored in the liver and some in fat cells. The liver similarly breaks down amino acids (excess amino acids are not stored) and converts them to glucose. The nitrogen from the amino acids is excreted in the form of urea, and the glucose is stored as glycogen.

Whether the liver takes up or releases glucose and the amount it takes up or releases are determined primarily by the concentration of glucose in the blood. The concentration of glucose is, in turn, regulated by a number of hormones and is influenced by the autonomic nervous system (Figure 28–8).

SOME NUTRITIONAL REQUIREMENTS

Because of the liver's activities in converting various types of food molecules into glucose, the energy requirements of the body can be met by carbohydrates, proteins, or fats—the three principal types of food molecules. Energy requirements are ordinarily met by a combination of the three. Carbohydrates and proteins supply about the same number of calories per gram, and fats about twice as much as either of them.

In addition to the need for calories, the cells of the body need the twenty amino acids required for assembling proteins. Of these twenty, vertebrates can synthesize twelve, either from a simple carbon skeleton or from another amino acid. The other eight, which must be obtained in the diet, are known as essential amino acids,* which, as you can see, is something of a misnomer. Plants are the ultimate source of these essential amino acids, but it is difficult (although by no means impossible) to obtain them by eating a completely vegetable diet, largely because plant proteins are relatively deficient in lysine and tryptophan.

Mammals also require but cannot synthesize certain polyunsaturated fats that provide fatty acids needed for fat synthesis. These can be obtained by eating plants or insects (or eating other animals that have eaten plants or insects).

Vitamins are an additional group of molecules required by living cells that cannot be synthesized by animal cells. They are characteristically required only in small amounts. Many of them function as coenzymes. Table 28–2 indicates some of the principal vitamins required in the human diet. There is no clear evidence that the ingestion of amounts of any particular vitamin in excess of the amounts available in well-balanced diets has any beneficial effects on a normally healthy individual.

The body also has a dietary requirement for a number of inorganic substances, or minerals. These include calcium and phosphorus for bone formation, iodine for thyroid hormone, iron for hemoglobin and cytochromes, and sodium and chloride and other ions essential for ionic balance. Most of these are present in the ordinary diet or in drinking water. Like the vitamins, however, they must be given in a supplementary form when the dietary intake is inadequate or when the individual is not able to assimilate them normally.

* The essential amino acids are isoleucine, leucine, lysine, methionine, phenylalanine, threonine, tryptophan, and valine.

28-9

Vitamin D is a steroid that is produced in the skin by the action of ultraviolet rays from the sun on a cholesterol-like compound (the absorbed energy opens the second ring of the molecule). According to one current hypothesis, the ancestors of modern Homo sapiens originated in the tropics and were all dark-skinned. In those populations that moved northward, however, the screening effects of the darker pigment, which inhibits the production of vitamin D, caused selection for lighter skin color, whereas no such selection pressures occurred in the sun-drenched areas nearer the equator.

VITAMIN D

Table 28-2 Vitamins

DESIGNATION LETTER AND NAME	MAJOR SOURCES	FUNCTION	DEFICIENCY SYMPTOMS
A, carotene	Egg yolk, green or yellow vegetables, fruits, liver, butter	Formation of visual pigments, maintenance of normal epithelial structure	Night blindness; dry, flaky skin
B-complex vitamins: B$_1$, thiamine	Brain, liver, kidney, heart, whole grains	Formation of enzyme involved in Krebs cycle	Beri-beri, neuritis, heart failure, mental disturbance
B$_2$, riboflavin	Milk, eggs, liver, whole grains	Part of electron carrier FAD	Photophobia, fissuring of skin
B$_6$, pyridoxine	Whole grains, liver, kidney, fish	Coenzyme for amino acid metabolism and fatty acid metabolism	Dermatitis, nervous disorder
B$_{12}$, cyanocobalamin	Liver, kidney, brain	Maturation of erythrocytes	Anemia, malformed erythrocytes
Biotin	Egg white, synthesis by intestinal bacteria	Concerned with protein synthesis, CO_2 fixation, and transamination	Scaly dermatitis, muscle pains, weakness
Folic acid	Liver, leafy vegetables	Nucleic acid synthesis, formation of erythrocytes	Failure of erythrocytes to mature, anemia
B$_3$, niacin (nicotinic acid)	Whole grains, liver and other meats	Part of electron carriers NAD, NADP, and of CoA	Pellagra, skin lesions, digestive disturbances
Pantothenic acid	Present in most foods	Forms part of CoA	Neuromotor disorders, cardiovascular disorders, GI distress
C, ascorbic acid	Citrus fruits, tomatoes, green leafy vegetables	Vital to collagen and ground (intercellular) substance	Scurvy, failure to form connective tissue fibers
D$_3$, calciferol	Fish oils, liver, milk and other dairy products, action of sunlight on lipids in the skin	Increases Ca^{2+} absorption from gut, important in bone and tooth formation	Rickets (defective bone formation)
E, tocopherol	Green leafy vegetables	Maintains resistance of red cells to hemolysis	Increased red blood cell fragility
K, naphthoquinone	Synthesis by intestinal bacteria, leafy vegetables	Enables synthesis of clotting factors by liver	Failure of blood coagulation

SUMMARY

Digestion is the process by which food is broken down into molecules that can be taken up by the bloodstream and so distributed to the individual cells of the body. It occurs in successive stages, regulated by an interplay of hormones and nervous stimuli. In mammals food is processed initially in the mouth, where the breakdown of starch begins. It moves through the esophagus to the stomach, where gastric juices destroy bacteria and begin to break down proteins.

Most of the digestion occurs in the duodenum, the upper portion of the small intestine; here, digestive activity, which is performed by enzymes, is almost completely under hormonal regulation. The breakdown of starch by amylases continues; fats are hydrolyzed by lipases; and proteins are reduced to single amino acids. Hormones (secretin and cholecystokinin) are secreted by duodenal cells, and these, in turn, stimulate the functions of the pancreas and the liver. The pancreas releases an alkalizing fluid containing digestive enzymes; the liver produces bile, which is also alkaline. Most absorption of food molecules occurs in the first half of the small intestine. Undigested remains are excreted from the large intestine.

The chief energy source for cells in the mammalian body is glucose circulating in the blood. The organ principally responsible for maintaining a steady supply of glucose is the liver, which stores glucose (in the form of glycogen or fat) when glucose levels in the blood are high and breaks down glycogen or fats, releasing glucose, when the levels drop. These activities of the liver are regulated by at least five different hormones. Requirements for good nutrition include calories (which can be obtained from carbohydrates, fats, or proteins), essential amino acids, certain essential fatty acids, vitamins, and minerals.

QUESTIONS

1. Define the following terms: digestion, sphincter, peristalsis, villi, glycogen, essential amino acids.

2. Diagram the human digestive tract, including all the major organs. Check your drawing against Figure 28-2.

3. Trace the chemical process of a hamburger on a bun through your digestive tract.

4. Many organisms are either herbivores or carnivores, and some are even specialists on a single organism as a food source. Humans are omnivores. Humans also have an unusually large number of substances specifically required in their diets. How might these two facts be related?

5. If you oxidize a pound of fat, whether butter or your own adipose tissue, about 3,500 kilocalories is released. Suppose that you go on a weight-reducing diet, limiting yourself to 1,000 kilocalories a day. You spend most of your time sitting in a well-heated library (studying, of course) and so you expend only about 3,000 kilocalories a day. How many pounds will you lose each week?

CHAPTER 29

Temperature Regulation, Water Balance, and Kidney Function

TEMPERATURE REGULATION

Some 200 years ago, Dr. Charles Blagden, then Secretary of the Royal Society of London, went into a room that had been heated to 126°C (260°F), taking with him a few friends, a small dog in a basket, and a raw steak. The entire group remained there for 45 minutes. Dr. Blagden and his friends emerged unaffected. So did the dog. (The basket had kept the dog's feet from being burned by the floor.) But the steak was cooked.

Temperature regulation is, of course, another example of homeostasis, a dominant theme of these chapters on animal physiology. The control of body heat is an important factor in regulation of the internal environment. For example, the rate of enzyme activity doubles, approximately, with every 10°C rise in temperature.

Temperature regulation, like water conservation, is a problem faced by animals as they become terrestrial. In general, large bodies of water, for the reasons discussed in Chapter 2, maintain a very stable temperature, and the

29–1

In laboratory studies, the internal temperature of reptiles was shown to be almost the same as the temperature of the surrounding air. It was not until observations were made of these animals in their own environment that it was found that they have behavioral means for temperature regulation that give them a surprising degree of temperature independence. By absorbing solar heat, as the desert iguana shown here is doing, reptiles can raise their temperature well above that of the air around them.

animals that live in them, with few exceptions, maintain a body temperature that is the same as the temperature of the water. Because the temperatures of large bodies of water are low (seldom exceeding about 5°C), the pace of life in the water, physically and intellectually, is generally a slow one. The few exceptions include the aquatic mammals and certain large fish, such as tuna, which retain inside their bodies heat generated by muscular activities.

Terrestrial reptiles, the snakes and lizards, are able to maintain remarkably stable body temperatures during their active hours by absorbing solar radiation. By careful selection of suitable sites, such as the slope of a hill facing the sun, and by orienting their bodies with a maximum surface exposed to sunlight, they can heat themselves rapidly. Such heating can occur as quickly as 1°C per minute, even on mornings when the air temperature is close to 0°C. By frequently changing position, these reptiles are able to keep their temperatures within quite a narrow range as long as the sun is shining. When it is not, they seek the safety of their shelters where they are not immobilized in an exposed position and thus vulnerable to attack. Such animals, which obtain their heat from the outside, are known as _exotherms_.

"WARM-BLOODED" ANIMALS

The so-called "warm-blooded" animals, the mammals and the birds, are _endotherms_. With a few exceptions, the endotherms are also _homeotherms_, maintaining a quite constant internal body temperature. The source of heat in endotherms is the oxidation of glucose and other energy-containing molecules within their body cells. Endotherms have a much higher metabolic activity than exotherms, with a relatively large proportion of their energy budget allocated to heat production. Also, small endotherms have a proportionally larger heat budget than large ones, because of surface/volume ratios (page 56). Endotherms are characterized by layers of insulating material—fur, feathers, fat—and by mechanisms for disposing of excess heat, such as panting in dogs.

29-2

Some adaptations to cold. (a) Prior to hibernating, mammals, such as the dormouse shown here, store up food reserves in body fat. (b) Note the heat-conserving reduction in surface-to-volume ratio in this hibernating chipmunk. (c) Bats have a number of special adaptations which enable them to conserve heat despite their small body size. Some hibernate. In others, the temperature drops when they are at rest (as it does in hummingbirds). Members of the species shown here also roll up their ears when they get cold.

(a)

(b)

(c)

As you know from the familiar experience of having your temperature taken, the human body maintains a remarkably constant temperature, deviating very little from 37°C. In fact, an elevation of even a degree or two—a fever—is considered a sign of illness. This constancy of temperature is maintained by an automatic system, a thermostat; the cells that perform this function are located in the hypothalamus. The hypothalamus also contains receptors that monitor the temperature of the blood flowing through the hypothalamus, and the skin as well contains hot and cold receptors, such as the one shown in Figure 25–3 on page 348. These relay information back to the integrating center in the hypothalamus.

Regulation at High Temperatures

As the external temperature rises, the blood vessels near the skin surface dilate and the supply of blood to the skin increases. If the air is cooler than the body surface, heat can be transferred directly to the air. Animals that live in hot climates characteristically have larger exposed surface areas than animals that live in the cold (Figure 29–3).

Heat can also be lost from the surface by the evaporation of perspiration or saliva. In man and most other large animals, when the external temperature rises above body temperature, perspiration begins. Evaporation requires heat energy (see page 28), which comes from the body surface, and the blood just below the skin surface cools as this heat energy is used. Dogs and some other animals pant, so that air passes rapidly over the large, moist tongue, evaporating saliva. Cats lick themselves all over as the temperature rises, and evaporation of this water cools their body surface. Horses and human beings sweat from all over their body surfaces. For animals that dissipate heat by water evaporation, temperature regulation at high temperatures necessarily involves water loss, which in turn stimulates thirst and, as we shall see, water conservation by the kidneys. Dr. Blagden and his friends were probably very thirsty.

(a)

29–3

The size of the extremities in a particular type of animal can often be correlated with the climate in which it lives. (a) The fennec of the North African desert has large ears

(b)

which help it to dissipate body heat. (b) The red fox of the eastern United States has ears of intermediate size, and (c) the Arctic fox has relatively small ears. Similar correla-

(c)

tions between size, weight, color, etc., and environment can also sometimes be made among animals of a single species living over an extended geographic range.

29–4

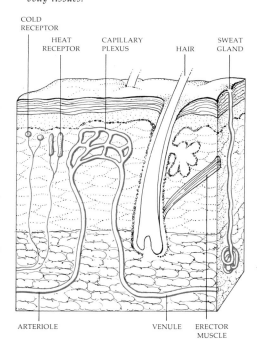

Cross section of human skin showing structures involved in temperature regulation. In cold, the arterioles constrict, reducing the flow of blood through the capillary network (plexus), and the hairs, each of which has a small erector muscle under nervous control, stand upright. In animals with body hair (Homo sapiens is one of the few terrestrial mammals that is virtually hairless), air trapped between the hairs insulates the skin surface, conserving heat. In heat, the arterioles dilate and the sweat glands secrete a salty liquid. Evaporation of this liquid cools the skin surface, dissipating heat (approximately 540 calories for every gram of H_2O). Beneath the skin is a layer of subcutaneous fat that serves as insulation, retaining the heat in the underlying body tissues.

Regulation at Low Temperatures

As external temperatures go down, cutaneous blood vessels constrict, limiting heat loss from the skin surface. Metabolic processes increase. Part of this increase is due to increased muscle activity (shivering, shifting from foot to foot), and part is due to hormones that stimulate metabolism. Muscular activity raises heat production by mobilizing glucose and ATP. Some energy is released directly as heat during the chemical reactions, and additional energy is released as heat during the muscle contractions.

Prime mobilizers of chemical energy are the autonomic nerves to fat, which increase its metabolic breakdown, and the hormone epinephrine, which is released from the adrenal medulla. In some animals, the thyroid gland increases its release of thyroxine in response to cold. Thyroxine increases the metabolic rate; it apparently exerts its effects directly on the mitochondria. Cortisol production may also increase under the stress of exposure to cold. Both of these effects, you will recall, are mediated by the hypothalamus acting on the pituitary, which, in turn, stimulates the target gland (Figure 29–5). So again we have an example of integration and control in which the nervous and endocrine systems function as a single unit.

29–5

Body temperature in mammals, regulated by a complex network of activities involving both the nervous and endocrine systems, is controlled largely by the hypothalamus. Many behaviorial responses are also involved.

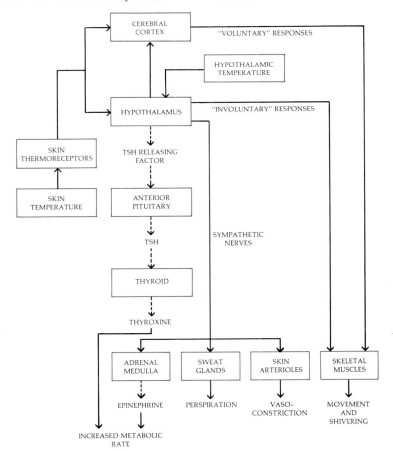

Fever

The elevation of temperature known as fever is due not to a malfunction of the thermostat but to a resetting. Thus, at the onset of fever, an individual typically feels cold and often has chills; although his body temperature is rising, it is still lower than his new thermostat setting. The substance primarily responsible for the resetting of the thermostat is a protein released by the white blood cells in response to a foreign organism. The adaptive value of fever, if there is one, is a subject of current research.

THE INTERNAL ENVIRONMENT

The human body is about 60 percent water. Two-thirds of this water is within the cells, and one-third, the extracellular fluid, surrounds, bathes, and nourishes the cells. Thus this extracellular fluid serves the same purpose for the body cells that the pre-Cambrian seas served for the earliest cells that arose in them. Indeed, for the modern prokaryotes and protists, seas, lakes, and other bodies of water—some as small as milliliters—still constitute the extracellular fluid. As they became multicellular in the course of evolution, animals began to produce their own extracellular fluid—close in composition to the salty fluid of the sea. As they did so, they also evolved mechanisms for regulating its composition.

In many invertebrates and all vertebrates, the internal chemical environment is to a large extent regulated by a special excretory system. Among terrestrial vertebrates, the most important components of this system are the kidneys. It is possible to relate major advances in vertebrate evolution—such as the transition to land and the development of homeothermy—to increasing efficiency of kidney function. In fact, one can trace vertebrate evolution in terms of the evolution of the kidney, as Homer Smith did in his classic book *From Fish to Philosopher*.

Regulation of the extracellular fluid involves the processing of three different classes of molecules. The first group consists of waste products released by cells into the bloodstream. Chief among these are carbon dioxide and the nitrogen compounds, such as ammonia, produced by the breakdown of amino acids. As we have seen, carbon dioxide diffuses out into water through the skin or gills or, in terrestrial vertebrates, is eliminated as a gas through lungs. In simple aquatic organisms, ammonia also diffuses out dissolved in water, but in larger, more complex organisms, it often must be converted to some other form since accumulated ammonia is highly toxic.

Birds and insects eliminate waste nitrogen in the form of *uric acid*, which can be excreted as crystals. In birds, the uric acid is mixed with the undigested wastes in the cloaca and the combination is dropped as a semisolid paste, familiar to frequenters of public parks and admirers of outdoor statuary. This nitrogen-laden substance forms a rich natural fertilizer. Guano, the excreta of seabirds, accumulates in such quantities on the small islands where these birds gather in great numbers that it is profitable to harvest it commercially. Mammals excrete nitrogen by-products largely in the form of *urea*, which must be dissolved in water for excretion (Figure 29–6).

The second class of compounds under regulatory control consists of ions. These include both positive ions, such as H^+, Na^+, K^+, Mg^{2+}, and Ca^{2+}, and the important negatively charged ions Cl^- and HCO_3^-. The precise regulation of their concentrations in the blood demands that the body fluids be analyzed and processed constantly.

29–6

Urea, the principal form in which nitrogen is excreted in mammals. It is formed in the liver by the combination of two molecules of ammonia with one of carbon dioxide through a complex series of energy-requiring reactions. What would be the other end product of this reaction?

$$HN_2-\underset{\underset{O}{\|}}{C}-NH_2$$

UREA

29-7

Some marine animals, such as the turtle, have special glands in their heads that can excrete sodium chloride at a concentration about twice that of seawater. Since ancient times, turtle watchers have reported that these great armored reptiles come ashore to lay their eggs with tears in their eyes, but it is only recently that biologists have learned that this is not caused by an excess of senti-ment—as is the case of Lewis Carroll's mock turtle—but is, rather, a useful solution to the problem of excess salt. Marine birds have similar salt-secreting glands.

The third class of compound is water. The concentration of a particular substance in the body depends not only on the absolute amount of the substance but on the amount of water in which it is dissolved. Thus, although the fundamental problem is always the same—the chemical regulation of internal environment—the solution to the problem varies widely, being strongly influenced by the habitat of the animal, that is, by the availability of water.

WATER BALANCE—AN EVOLUTIONARY PERSPECTIVE

The problem of water balance is such a universal one, biologically speaking, and is so important to the survival of the organism that we shall digress for a moment to discuss it in more detail before examining the anatomy and physiology of the kidney.

The earliest organisms probably had a salt and mineral composition much like that of the environment in which they lived. The early organism and its surroundings were probably also isotonic; that is, each had the same total effective concentration of dissolved substances, so water did not tend to move either in or out of the cell. When organisms moved to fresh water (a hypotonic—less concentrated—environment), they had to develop systems for "bailing themselves out," since fresh water tended to move into their bodies by osmosis; the contractile vacuole of *Paramecium* is an example of such a bailing device.

If the vertebrates evolved in fresh water, as is generally believed, the first function of the kidneys, phylogenetically speaking, was probably to pump water out and to keep salt and other desirable solutes, such as glucose, in. In freshwater fish, the kidney works primarily as a filter and reabsorber, and the urine is hypotonic—that is, it has a concentration of solutes lower than that of body fluids.

Saltwater fish have a different problem. Their body fluids are generally less concentrated than their environment and so they tend to lose water by osmosis. Their need is to conserve water and thereby keep their body fluids from becoming too concentrated. This problem has been solved in different ways by different groups of fish. In hagfish, for example, body fluids are about as salty as the salt waters of the surrounding ocean and so are isotonic with them.

Cartilaginous fish such as the shark are also isotonic to seawater, but they achieve their isotonicity in a different way. In the course of evolution, they developed an unusual tolerance for urea, so instead of constantly excreting it, as do all other fish, they retain a high concentration of it in the blood. This high concentration of urea makes their body fluids almost isotonic in relation to seawater; hence, they do not tend to lose water by osmosis.

The bony fish spread to the sea at a much later period than did the cartilaginous fish, and their body fluids are hypotonic in relation to the environment, having an osmotic pressure only about one-third that of seawater. Thus, like terrestrial animals, they have the problem of losing so much water to their environment that the solutes in the body fluids become too concentrated and the cells die. In bony marine fish, this problem has been solved by the evolution of special gland cells in the gills that excrete excess salt. Hence these fish can take in salt water and still remain hypotonic. (Freshwater fish, conversely, have salt-absorbing cells in their gills.)

Since terrestrial animals do not always have automatic access to either fresh or salt water, they must regulate water content in other ways, balancing off gains and expenditures (Figure 29–8).

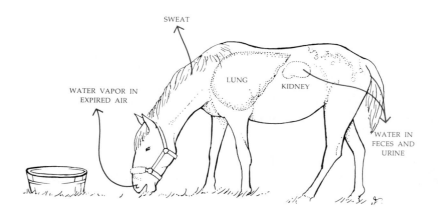

29–8
A mammal is in water balance when the total amount of water lost in expired air, evaporation from skin, and in urine and feces equals the total amount of water gained by the intake of food and drinking water and by the oxidation of food molecules.

SOURCES OF WATER GAIN AND LOSS

Animals gain water by drinking fluids, by eating water-containing foods, and as an end product of the oxidative processes that take place in the mitochondria, as we saw in Chapter 7. When 1 gram of glucose is oxidized, 0.6 gram of water is formed. When 1 gram of protein is oxidized, only about 0.3 gram of water is produced. Oxidation of 1 gram of fat, however, produces 1.1 grams of water because of the high hydrogen concentration in fat.

On the average, a human takes in about 2.3 liters of water a day in food and drink, and gains an additional 200 milliliters a day by oxidation of food molecules. Water is lost from the lungs in the form of moist exhaled air, is eliminated in the feces, and is lost by evaporation from the skin. However, as we shall see, the chief route of water loss is the urine.

In a normal adult, the rate of water excretion in urine averages 1.2 liters a day. Although the actual amount of urine produced may vary from a few hundred milliliters a day to several thousand, there will be a variation of less than 1 percent in the fluid content of the body. However a minimum output of 500 milliliters of water is necessary for health since this much water is needed to remove potentially toxic waste products, such as urea.

The control of urinary water loss is the major mechanism by which body water is regulated. First we shall describe the processes by which the blood is regulated chemically and then we shall examine the mechanisms for controlling water excretion.

CHEMICAL REGULATION OF THE BLOOD

The blood is chemically processed in the kidney. The structure in which the actual processing takes place is the *nephron,* which consists of a cluster of capillaries known as the *glomerulus,* a bulb called *Bowman's capsule,* and a long narrow tube, the *renal tubule* (Figure 29–9). Each of the two kidneys in man contains about a million nephrons with a total length of some 80 kilometers (50 miles) in an adult male.

The function of the nephron is intimately connected with the circulation of the blood. Blood enters the kidney through the renal artery, which divides into progressively smaller branches, the arterioles, which each lead, in turn, to a glomerulus. About one-fifth of the blood plasma that enters the kidney is forced out through the thin walls of the glomerular capillaries by the pressure of the circulating blood. Unlike other capillaries, a glomerular capillary lies

29-9

(a) *In longitudinal section, the human kidney is seen to be made up of an outer region, the cortex, which contains the fluid-filtering mechanisms, and an inner region, the medulla, through which collecting ducts carrying the urine merge and empty into the funnel-shaped renal pelvis, which enters into the ureter. (b) The nephron is the functional unit of the kidney. Blood enters the nephron through the afferent arteriole leading into the glomerulus. Fluid is forced out by the pressure of the blood through the thin capillary walls of the glomerulus into Bowman's capsule. The capsule connects with the long tubule that twists down into the medulla and up again. As the fluid travels through the tubule, almost all the water, ions, and other useful substances are reabsorbed into the bloodstream through the capillaries surrounding the tubule. Other substances are secreted from the capillaries into the tubules. Waste materials and some water pass along the entire length of the tubules into the collecting duct and are excreted from the body.*

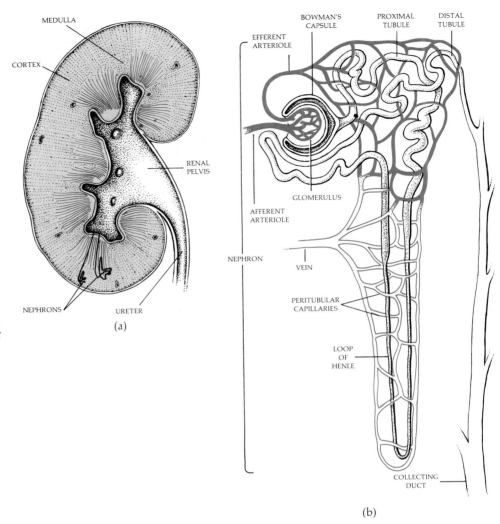

(a)

(b)

29-10

Urinary system in a human male. Urine formed in the kidneys travels through the ureters to the bladder, where it is stored until it is released through the urethra.

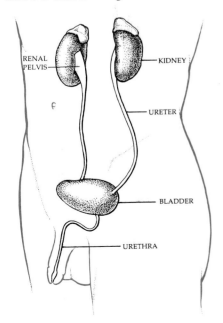

between two arterioles. The pressure in these capillaries is 50 to 70 mm Hg (about twice that in other capillaries). The fluid entering Bowman's capsule, which is lined only by a thin layer of cells, contains small molecules from the blood but not the large elements, such as the blood cells and large blood proteins. In the healthy kidney, blood cells and blood proteins, both of which are too large to pass through the capillary walls, are left behind. This stage of the process is known as *filtration* and the fluid that enters Bowman's capsule is called the *filtrate*.

The fluid then begins its long passage through the tubule, which is lined with a layer of epithelial cells especially adapted for active transport. The tubules of the nephrons terminate in collecting ducts, which lead to a common duct, the *ureter*. The ureter transmits the urine to the *bladder* for storage before excretion through the *urethra* (Figure 29-10). (The urinary system is also shown in Figure 23-12, on page 323.) Each day, some 180 liters of fluid is filtered through the nephrons (remember, the total blood supply is only about 2½ liters), where it is filtered, analyzed, and so carefully adjusted that it can be considered to be completely remade.

Most solutes and almost all the water contained in the filtrate that enters the tubule of the nephron are reabsorbed and returned to the blood through the

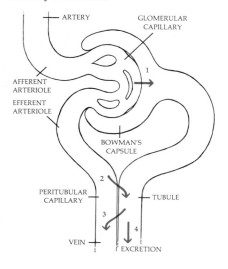

29–12

Aldosterone, a steroid hormone from the adrenal cortex, increases the reabsorption of sodium and chloride ions by the renal tubules, and thereby conserves sodium. Aldosterone secretion is regulated by the kidney itself. Reduction of sodium transport across the tubules of the kidney stimulates the kidney to produce renin. Renin, an enzyme, splits angiotensinogen, a large protein produced by the liver, to form angiotensin. Angiotensin stimulates the adrenal cortex to produce aldosterone.

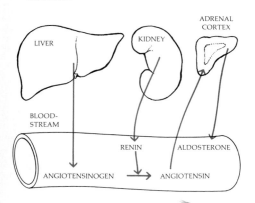

peritubular capillary system that surrounds the long, convoluted tubule. Some other substances from the blood may be secreted into the tubule from the capillary bed surrounding it. For example, glucose, most amino acids, and most vitamins are returned to the bloodstream by a healthy kidney. Products of amino acid breakdown, such as urea, are excreted. Most sodium and other ions are retained; some are excreted, depending on the physiological status of the body.

Aldosterone

As we noted in Chapter 25, the hormone aldosterone, which is produced in the adrenal cortex, controls the ionic composition of the blood. Aldosterone stimulates reabsorption of sodium from the distal tubule. When the adrenal glands are removed or when they function poorly (as in a disease called Addison's disease), sodium chloride and water are lost in the urine and the tissues of the body become depleted of them. Generalized weakness results, and if a patient with Addison's disease is untreated, the fluid loss can eventually become fatal. Treatment involves giving the patient enough sodium chloride and water to make up for that lost by excretion.

The kidney itself stimulates aldosterone production by releasing the enzyme renin when the sodium concentration within the renal tubule decreases. According to present evidence, renin acts on a blood protein, angiotensinogen, made by the liver, and converts it to angiotensin. Angiotensin, bypassing the hypothalamus-pituitary circuit, acts directly on the adrenal cortex to stimulate it to produce aldosterone (Figure 29–12). Angiotensin also acts on the brain to produce the sensation of thirst.

THE REGULATION OF WATER BALANCE

As we noted previously, the kidney also plays a main role in water conservation. Birds and mammals have developed the ability to excrete a hypertonic urine—that is, one that is more concentrated than their body fluids. This ability is associated with a hairpin-shaped section of the nephron known as the *loop of Henle*. By sampling the fluids in and around the nephron and analyzing the amount of sodium present in each sample, physiologists have been able to determine how such a deceptively simple looking structure makes this function possible. The function of the loop of Henle is described in Figure 29–13.

Antidiuretic Hormone

The amount of water excreted in the urine depends on the permeability of the walls of the collecting duct. This is under the control of a hormone known as ADH (antidiuretic hormone). ADH acts on the membranes of the collecting ducts of the nephrons and increases their permeability to water. When ADH concentrations are low, the collecting ducts are relatively impermeable to water and so little water passes back into the blood from the urine before it enters the ureters. In short, ADH allows our bodies to conserve water.

ADH, as we noted in Chapter 25, is a hormone secreted by the hypothalamus. The amount of ADH released depends on the osmotic concentration of the blood and also on the blood pressure. Osmotic receptors that monitor the solute content of the blood are located in the hypothalamus. Pressure receptors that indirectly detect changes in blood volume are found in the walls of the heart, in the aorta, and in the carotid arteries. Stimuli received by these recep-

The formation of hypertonic urine in the human nephron. As shown in this diagram, the fluid entering Bowman's capsule first enters the proximal convoluted tubule, descends the loop of Henle, ascends it, and then passes through the distal convoluted tubule into the collecting duct.

In the loop of Henle, water diffuses freely through the cells lining the walls of the descending branch while sodium is actively pumped from the ascending branch back to the fluid surrounding them. The sodium pumped out of the ascending branch passes back into the descending branch by simple diffusion since the fluids surrounding the descending branch are more concentrated than those within it. Thus the salt recirculates to the ascending branch, where it is pumped out again. Water cannot follow the salt out of

the ascending branch because its walls are impermeable to water. This recirculation of the sodium has two consequences: (1) As the urine passes through the loop of Henle, much salt but little water is removed; and (2) the lower part of the loop of Henle and also the lower part of the collecting duct are bathed in a fluid containing many times the level of salt normally found in tissue or blood.

The fluid entering the proximal convoluted tubule is isotonic with the blood plasma; that is, it has the same salt concentration as that of the blood plasma. As the urine descends the loop of Henle, it becomes increasingly concentrated. In the ascending branch, it becomes less and less concentrated, and as it enters the distal convoluted tubule, it is hypotonic. By the end of this distal tubule, it is once more isotonic. From this

point, the urine flows to the collecting duct, which traverses the zone of high salt concentration.

From this point onward, the urine concentration depends on the cells that form the walls of the collecting duct. If the duct walls are not permeable to water, the water entering the collecting duct from the distal tubule stays in the duct, and a less concentrated urine is excreted. If ADH is present, however, the cells of the collecting duct are permeable to water, and water diffuses out into the surrounding salty fluid, as shown in the diagram. In this case a highly concentrated (hypertonic) urine is passed down the duct to the renal pelvis, the ureter, the bladder, and finally out the urethra.

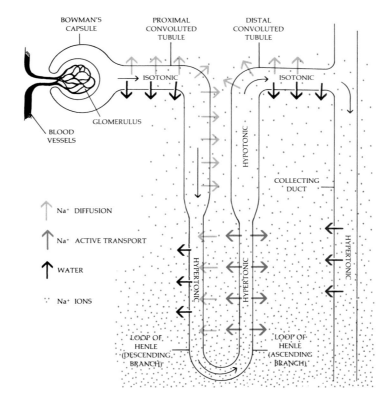

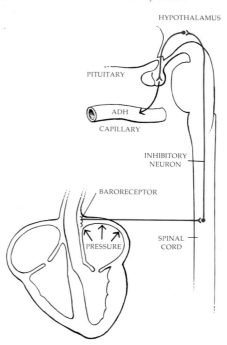

29–14

Regulation of ADH production. Increased blood pressure stimulates baroreceptors (pressure receptors) in the left atrium. These receptors then transmit impulses to the hypothalamus that inhibit ADH (antidiuretic hormone) production. ADH production is stimulated by increased concentration of solutes in the blood, detected by osmoreceptors located in the hypothalamus. As its name suggests, ADH inhibits the excretion of water.

HYPOTHALAMUS

PITUITARY

ADH

CAPILLARY

INHIBITORY NEURON

BARORECEPTOR

PRESSURE

SPINAL CORD

tors are transmitted to the hypothalamus (Figure 29–14). Factors that increase the concentration of solutes in the blood or decrease blood pressure, or both, stimulate the production of ADH and conservation of water in the body. Such factors include dehydration and hemorrhage. Factors that decrease blood concentration, such as the ingestion of large amounts of water, or that increase blood pressure—epinephrine, for instance—signal the hypothalamus to stop producing ADH, and so more water is excreted. Alcohol suppresses ADH secretion and thus increases urinary flow, a phenomenon familiar to imbibers of beer and other alcoholic beverages; pain and emotional stress trigger ADH secretion and thus decrease urinary flow.

ADAPTATIONS OF DESERT ANIMALS

We have seen in the course of this chapter, first, that water evaporation is a major mechanism for dissipating heat and, second, that water balance is essential to the smooth functioning of the internal environment. How do animals in the desert resolve these two conflicting demands? There are a variety of answers to this question.

Some animals can derive all their water from food and do not require drinking water. The kangaroo rat of the American desert, for example, can live its entire existence without drinking water if it eats the right type of food. It is not surprising that it prefers a diet of fatty seeds. If it is fed high-protein seeds, such as soybeans—the oxidation of which produces a large amount of nitrogen waste and a relatively small amount of water—the kangaroo rat will die of thirst unless some other source of water is available. Also, it is highly conservative in its water expenditures. The kangaroo rat has no sweat glands, and being nocturnal, it searches for food only when the external temperature is relatively cool. Its feces have a very small water content, and its urine is highly concentrated. Its major water loss is through respiration. Even this loss is reduced by the animal's long nose since some cooling of the expired air (with condensation of water from it) takes place in its nasal passages.

Many desert animals regulate their temperature by behavior, as do the reptiles mentioned earlier in this chapter. One of the most interesting examples is found in the architecture of the burrows built by the prairie dogs, which follow some basic principles of aerodynamics to air-condition their tunnels.

As you may know, airplane wings are shaped so that air will move along the top surface faster than along the bottom surface. This creates greater pressure beneath the wings so that the plane is, in effect, sucked upward. A prairie dog constructs a burrow about 18 meters long, with an opening at either end. Then, following the principle illustrated by the airplane wing, it constructs a high mound around one end and a low mound around the other, often gathering dirt from some distance around and working and reworking the structure. Air moves faster over the higher mound, and the pressure is less than at the other, lower opening of the burrow. As a consequence, air is pulled through the burrow to the lower end. Duke University scientists, who have studied the prairie dog's engineering accomplishments, calculate that when air is moving over the mounded openings at the rate of 1 mile an hour, a prairie dog burrow would get a complete change of air every 10 minutes.

As a final example, let us consider the camel, for so long man's desert companion. The camel has several advantages over man in the desert. For one thing, the camel excretes a much more concentrated urine; in other words, it does not need to use so much water to dissolve its waste products. (In fact, man

29-15
By facing the sun, a camel exposes as small an area of body surface as possible to the sun's radiation. Its body is insulated by fat on top, which minimizes heat gain by radiation. The under part of its body, which has much less insulation, radiates heat out to the ground cooled by the animal's shadow. Other adaptations of the camel to desert life include long eyelashes, which protect his eyes from the stinging sand, and flattened nostrils that retard water loss.

is very uneconomical with his water supply; even dogs and cats excrete a urine twice as concentrated as man's.)

Also, the camel can lose more water proportionally than man and still continue to function. If a man loses 10 percent of his body weight in water, he becomes delirious, deaf, and insensitive to pain. If he loses as much as 12 percent, he is unable to swallow and so cannot recover without assistance. Laboratory rats and many other common animals can tolerate dehydration up to 12 to 14 percent body weight. Camels can tolerate the loss of more than 25 percent of their body weight. They can go without drinking for one week in the summer months, three weeks in the winter.

Probably most important, the camel can tolerate a fluctuation in internal temperature of 5° to 6°C. This tolerance means that it can let its temperature rise during the daytime (which the human thermostat would never permit) and cool during the night. The camel begins the next day at below its normal temperature—storing up coolness, in effect. It is estimated that the camel saves as much as 5 liters of water a day as a result of these internal temperature fluctuations.

Finally, a camel orients itself to the sun in such a way as to expose as small an area as possible to the sun's radiation. Its fatty hump on top insulates the animal against the sun's rays striking its back. The relatively uninsulated underpart of its body radiates heat out to the ground cooled by the animal's shadow. As you could guess, when it goes to sleep on cold desert nights, it curls up.

SUMMARY

The control of body temperature is an essential feature of homeostasis. Exotherms, which include snakes and reptiles, are animals that adjust their body temperatures principally by regulating the amount of heat taken in from the external environment. For endotherms, such as man, the source of heat is the oxidation of glucose and other fuel molecules by the body cells and muscular activity.

Heat is carried by the bloodstream from the core of the body to the surface, where it is dissipated. In hot weather, the blood vessels near the surface of the skin dilate, increasing the flow of blood to the skin. As external temperatures rise above body temperatures, man and most other large animals begin to sweat. Evaporation of sweat from body surfaces requires heat and cools the surface. In the cold, blood vessels supplying the skin surface constrict. As temperatures drop, shivering begins; the increased cellular metabolism and muscular activity involved in shivering produces heat. Temperature is regulated by an automatic system, or thermostat, in the hypothalamus that measures the body temperature and sets in motion the response mechanisms.

The urinary system enables an animal to regulate its internal chemical environment. Regulation is accomplished by (1) the elimination of toxic by-products of metabolism, especially the nitrogen compounds produced in the breakdown of amino acids; (2) control of the ionic content of the body fluids; and (3) the maintenance of water balance.

The excretory unit of vertebrates is the nephron; each nephron consists of a long tubule attached to a closed bulb (Bowman's capsule) which encloses a twisted cluster of capillaries, the glomerulus. Fluids are forced from the glomerulus into Bowman's capsule (filtration). By a combination of filtration, reabsorption, and secretion, the circulating body fluids are processed during their passage along the tubules.

Problems of water balance are different for animals living in saltwater (hypertonic), freshwater (hypotonic), and terrestrial environments. Terrestrial animals generally need to conserve water. An important means of water conservation is the capacity for excreting a urine that is hypertonic in relation to the blood. The loop of Henle is the portion of the mammalian nephron that makes possible the production of a hypertonic urine.

Water and salt excretion are subject to hormonal regulation. Aldosterone, produced by the adrenal cortex, increases the reabsorption of sodium and chloride ions by the renal tubules. Aldosterone secretion is stimulated by angiotensin, a blood protein that is activated by renin, an enzyme produced by the kidney. ADH, a hormone produced by the pituitary gland, acts on the collecting ducts of the nephrons, increasing their permeability to water and thus decreasing water loss.

Desert animals have a number of solutions to the dual problem of temperature regulation and water conservation. These include dietary habits, the capacity to excrete a very concentrated urine, a comparatively wide tolerance for fluctuations in total water and temperature, and a number of behavioral adaptions.

QUESTIONS

1. Define the following terms: homeostasis, Bowman's capsule, nephron, urine, loop of Henle.

2. What is the difference between an endotherm and an exotherm? Between an endotherm and a homeotherm?

3. Describe what happens to the human organism as the temperature rises. As it falls.

4. Sketch a nephron, indicating the pathways of the blood and the glomerular filtrate.

5. Trace the path of urea through the human kidney.

6. Sketch the urinary tract of a human female, labeling the various components.

7. Explain the following terms in relation to kidney function: filtration, secretion, reabsorption, and excretion.

8. Distinguish between aldosterone and ADH in terms of their source, their effect, and their regulation.

9. Could a human being on a life raft survive by drinking seawater? By eating fresh fish? Explain your answers.

CHAPTER 30

The Brain and the Eye

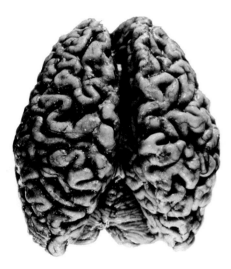

30-1

A brain, viewed from above. The corpus callosum is visible within the deep groove separating the two cerebral hemispheres, and the cerebellum can be seen below the hemispheres at the base of the brain. Note the many convolutions of the cerebral cortex. By these, you can immediately distinguish this brain as human.

The human brain, which weighs about 140 grams (3 pounds), has the consistency of semisoft cheese. Like the spinal cord, the brain is made up of both "white matter"—the fiber tracts—and gray matter. The gray matter consists of nerve cell bodies and glial ("glue") cells, which apparently support and nourish the neurons and may also play a role in ionic balance and thus in electric potentials. In some areas of the brain, neurons and glial cells are so densely packed that a single cubic inch of gray matter contains some 100 million cell bodies, with each neuron connected to as many as 60,000 others.

For the past 2,000 years, men have wondered about the relationship between the brain—this mass of semisoft substance—and the mind, the center of consciousness, thought, and emotion. Even René Descartes—"I think; therefore, I am"—who believed that the body is essentially a complex machine, conceived of mind and body as two separate entities, whose meeting place was a small gland, the pineal, located within the skull. The persistence of this dualistic concept is reflected in such phrases, still in use today, as "mind over matter." Many years after the cell doctrine was accepted for all other parts of the body, the brain was considered to be an exception. One reason was that it was not possible to visualize neurons as readily as other types of cells because special staining techniques are required to reveal them. Another and undoubtedly more important reason is the feeling that mental processes are somehow special, different from other physiological functions.

There is, of course, no longer any scientific doubt that mind and matter are one. Even today, however, many of us who accept without difficulty the fact that the tissues and cells of the digestive tract are responsible for digestion or that respiratory gases are exchanged across the surface of the lungs find it less easy to comprehend—to *really* believe—that the ideas, ideals, dreams, fantasies, thoughts, hypotheses, loves, hates, fears, and aspirations which make up the content of the mind somehow can be explained by the interactions of the cells within our heads. Nor, indeed, can scientists yet fully explain "mind" in terms of brain, despite the many intriguing discoveries that have been made in this field. This is one of the reasons why the brain is one of the most exciting of modern biological frontiers, and also why it is the remaining stronghold of the vitalists.

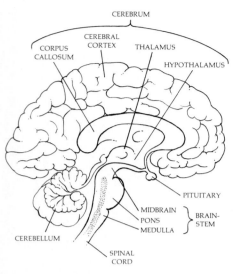

CEREBRUM

CORPUS
CALLOSUM

CEREBRAL
CORTEX

THALAMUS

HYPOTHALAMUS

PITUITARY

MIDBRAIN
PONS
MEDULLA
} BRAIN-
STEM

CEREBELLUM

SPINAL
CORD

30–2

Photograph of a longitudinal section of a human brain, with a diagram indicating some of the principal structures. The gray matter, characteristic of the cortex, consists of nerve cell bodies, glial cells, and unmyelinated fibers. The white matter is made up predominantly of myelinated fibers.

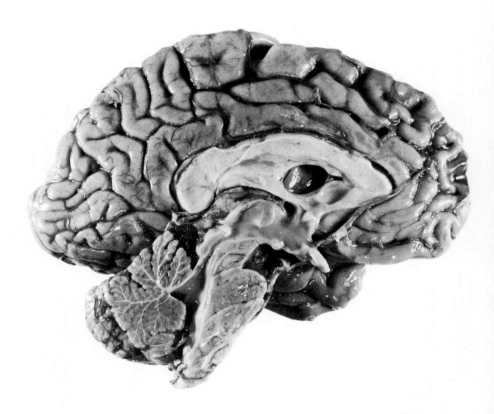

ANATOMY OF THE BRAIN

Anatomically, the human brain consists of a number of different structures (Figure 30–2). The three principal divisions are the *cerebrum*, the *cerebellum*, and the *brainstem*. The brainstem is the old brain, evolutionarily speaking, and is surprisingly similar from fish to man. The cerebrum is divided into two cerebral hemispheres, each of which, in man, is covered by the *cerebral cortex*. Both the brainstem and the cerebrum are made up of a number of anatomically distinct structures, which are listed in Table 30–1.

The brain, however, is not just an assembly of these parts, like a complicated hi-fi system with a number of components—amplifiers, tuners, etc.—that plug into one another. There is a vast, continuous series of interconnections among the nerve cells of all of the many parts of the brain, and the more that is learned about brain function, the more it becomes clear that it is the interconnections rather than the separate parts that determine how the brain works. Thus, the mapping studies we are about to describe represent only the tip of the proverbial iceberg.

For example, when someone speaks to you, the sound waves produced by his or her voice activate receptors in your ear that stimulate the auditory nerve. This information is relayed through a network of nerve cells that lead to an area in the cerebral cortex; this is called the auditory area of the cortex or, more briefly, the auditory cortex. From the auditory cortex, the information, which is processed in some way that is not understood, is transmitted to some other part of the brain; it is not until this happens that you have "heard" what was said. The auditory cortex can be quite precisely defined: If it is destroyed—which can occur as a result of brain injury or disease—hearing is lost. Yet it is only a small part of a larger and much more complex picture.

Table 30-1   *Structural Divisions of the Human Brain*

STRUCTURE	DESCRIPTION	FUNCTION
Cerebrum	Largest, most prominent part of human brain; divided by groove into right and left hemispheres; each hemisphere divided by furrows into four lobes: frontal, parietal, temporal, occipital	Receives and integrates incoming (afferent) information; makes associations between new data and information stored in cortex; coordinates responses
Cerebral cortex	Thin (2 millimeter) layer of gray matter, wrinkled and folded, overlying surface of cerebrum; reaches greatest development in higher primates, especially humans	Sensory, motor, and association functions (see text)
Inner cerebrum	Nerve tracts (white matter) connecting areas of cortex to each other and to rest of brain	Transmission and processing of information to and from cerebral cortex
Corpus callosum	Nerve fibers between hemispheres	Exchange of information between hemispheres
Thalamus	Two egg-shaped masses of gray matter below cerebral hemispheres	Main relay center between brainstem and cerebrum; nuclei sort and process information before transmission to cerebrum
Hypothalamus	Small (about 4 grams) area below thalamus, above pituitary	Source of ADH and oxytocin; controls pituitary, integration center for sex drive, anger, hunger, thirst, pleasure, temperature regulation
Cerebellum	Bulbous, convoluted mass of tissue that, in man, lies in back of and under the brain	Unconscious coordination of muscular activities; reaches largest size, proportionately, in birds
Brainstem	Stalk of brain; enlarged, knobby extension of spinal cord; consists of medulla, pons (bridge), and midbrain. Contains all afferent and efferent nerve fibers between spinal cord and higher brain centers. Fibers cross in brainstem, so left side of brain is associated with peripheral nerves in right side of body, and vice versa. Also contains nuclei involved with many autonomic reflexes.	Transmission of signals between spinal cord and higher brain centers; control of heartbeat, respiration, swallowing

30-3

The principal sulci (fissures) and lobes of the human cerebral cortex.

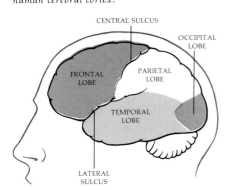

CENTRAL SULCUS
OCCIPITAL LOBE
FRONTAL LOBE
PARIETAL LOBE
TEMPORAL LOBE
LATERAL SULCUS

MAPPING THE BRAIN

The Cerebral Cortex

The cerebral cortex is the most recent evolutionary development of the vertebrate nervous system. Fish and amphibians have no cerebral cortex, and reptiles and birds have only a rudimentary indication of a cortex. More primitive mammals, such as rats, have a relatively smooth cortex. Among the primates, the cortex becomes increasingly complex, reaching a culmination in the human cerebral cortex, in which the highly convoluted surface greatly increases the total area. Of the approximately 12 billion nerve cells in the human brain, approximately 9 billion are in the cerebral cortex.

Because of its relative accessibility, just below the surface of the skull, the cerebral cortex is the most thoroughly studied area of the human brain, and some parts of it have been mapped in exquisite detail. In primates, each of the hemispheres is divided into lobes by two deep fissures, or grooves, in the surface. The principal fissures are the central sulcus and the lateral sulcus. There are four lobes: frontal, parietal, temporal, and occipital (Figure 30-3).

413   THE BRAIN AND THE EYE

The human cerebral cortex, showing the location of the motor and sensory areas, on either side of the central sulcus, and the auditory and visual zones.

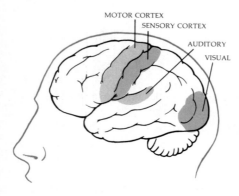

MOTOR CORTEX
SENSORY CORTEX
AUDITORY
VISUAL

30-5
The human cerebral cortex, showing the areas associated with speech. Damage to Broca's area—the more anterior area—affects motor control of speech. Damage to Wernicke's area affects conceptual aspects of language.

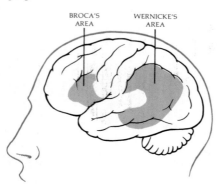

BROCA'S
AREA
WERNICKE'S
AREA

Motor and Sensory Cortex

The area just anterior to the central sulcus, in the frontal lobe, is the motor cortex, and the area posterior to it, in the parietal lobe, is the sensory cortex. These are involved in the integration of muscular actions and of tactile (touch) stimuli, respectively (Figure 30–4). Because of the crossing of fibers in the brainstem, the motor and sensory areas for the left side of the body are in the right hemisphere, and vice versa.

The temporal lobe of the brain lies below the frontal and parietal lobes, separated by the lateral sulcus. In the temporal lobe, partially buried within this fissure, is the auditory cortex. By measuring electrical discharges in this area of the brain in dogs and cats exposed to sounds of varying frequencies, investigators have been able to show that different regions of the auditory cortex respond to different frequencies of sound.

The area behind the central sulcus is divided into two main lobes, the most posterior of which, the occipital lobe, is concerned with vision. Stimulation of an animal's retina with light results in electrical discharges by cells in this area of the cortex.

In both experimental animals and humans, damage to specific areas of the posterior lobes seems to result in problems of discriminating specific stimuli. For example, if particular cortical areas of the posterior lobes are damaged, monkeys can no longer tell the difference between patterns, objects, or sounds that they were previously able to distinguish. In humans, damage to specific areas of the cortex on the left side of the brain (in most left-handed as well as all right-handed people) results in impairment of speech (Figure 30–5). The more anterior area (Broca's area) is adjacent to the region of the motor cortex that controls the movements of the muscles of the lips, tongue, jaw, and vocal cords. Damage to this area results in slow and labored speech but does not affect comprehension. When the more posterior area (Wernicke's area) is damaged, speech is fluent but often meaningless, and comprehension of both spoken and written words is impaired. The two speech areas are joined by a nerve bundle. The corresponding area in the right brain is associated with music and spatial relations. It is probable, although far from proved, that the functions of memory and learning and the organization of ideas take place in the "silent" areas of the posterior lobes.

Association Cortex

The unmapped area of the cortex is sometimes known as the association, or "silent," cortex. The term "association" was introduced when this part of the cortex was thought to function as a sort of giant switchboard, interconnecting the motor and sensory zones. More recent studies suggest, however, that organization of the brain is more vertical than horizontal; for example, severing the motor cortex from the sensory areas by deep vertical cuts appears to have no effect at all on an animal's behavior. Anatomical studies support the interpretation that communication between the sensory areas and motor areas of the cortex takes place primarily in lower brain centers, particularly the thalamus.

Nor are the association areas actually "silent." Although sensory stimuli do not evoke local responses in these areas, and direct electrical stimulation of these areas does not elicit muscle movements (as in the somatic sensory and motor areas, respectively), the association cortex shows steady background activity, apparent (but nonspecific) responses to stimuli, and increased activity following electrical stimulation of the other areas of the brain. Clearly, something is taking place in these "silent" areas. Since the proportion of association

A combined cross section of one hemisphere of the human cerebrum, indicating the functional areas of the motor and sensory cortices. The motor cortex is indicated in black; stimulation of these areas causes responses in corresponding parts of the body. The sensory cortex is in color; stimulation of various parts of the body produces electrical activity in corresponding parts of this cortex. Notice the relatively huge motor and sensory areas associated with the hand and the mouth. This map is based largely on studies done by neurosurgeon Wilder Penfield on patients undergoing surgical treatment for epilepsy. The motor and sensory cortices are located on either side of the central sulcus. Because nerve fibers cross in the brainstem, the motor and sensory cortices of the right hemisphere are associated with the left side of the body, and vice versa.

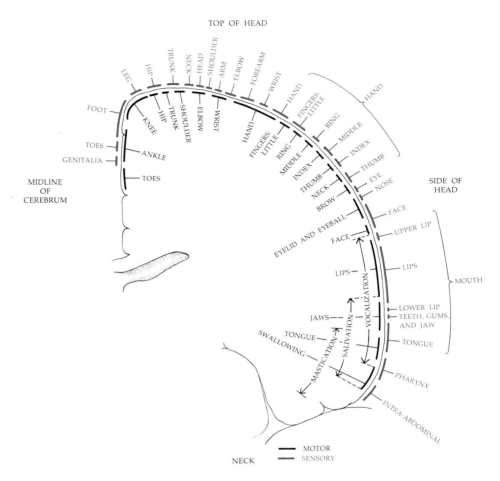

30-7

The reticular formation is a core of tissue running through the brainstem. Here, incoming stimuli are monitored and analyzed and relayed to other areas of the brain.

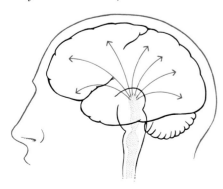

area to motor and sensory area is much larger in primates than in other mammals, even higher mammals such as cats, and is very large in man, we assume that these areas have something to do with what is special about the mind of man.

About half of the association area of the cortex is in the frontal lobes, the part of the brain developed most rapidly during the recent evolution of man. It is responsible for the high forehead of modern man, as compared with the beetle brow of our most immediate ancestors, and public appraisal of its function is reflected in the terms "high brow" and "low brow."

Reticular Formation

Another functional division of the brain is the *reticular formation*, which runs through the brainstem to the thalamus (Figure 30-7). The reticular formation is of particular interest to physiological psychologists because it is involved with arousal and attention—that hard-to-define state we know as consciousness. All the sensory systems have fibers that feed into the reticular formation, which apparently filters incoming stimuli and "decides" whether or not they may be important. Stimulation of this system, either artificially or by these incoming impulses, results in increased electrical activity in other areas of the brain. The existence of such a filtering system is well verified by ordinary experience. We are not generally conscious of the pressure of our clothing on our body, for

(a)

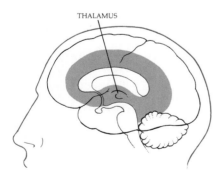

(b)

30–8

(a) *Rat with an electrode implanted in its hypothalamus. (b) By pressing a lever, the rat is able to stimulate its "pleasure center" with a weak electric current. Some animals press the lever as often as 8,000 times an hour.*

instance, or of background noise. A person may sleep through the familiar blare of a subway train or a loud radio or TV program but wake instantly at the cry of the baby or the stealthy turn of a doorknob. Similarly, we may be unaware of the contents of a dimly overheard conversation until something important—our own name, for instance—is mentioned, and then our degree of attention increases. The reticular formation also appears to be involved in sleep (page 418).

The Hypothalamus and the Limbic System

As we saw in Chapter 25, the hypothalamus regulates the activities of the pituitary gland and the several glands under pituitary feedback control. It is also the major central brain structure concerned with the functions of the autonomic nervous system, particularly with its sympathetic division. Thus it is, in effect, the chief coordinating center for the combined activities of the nervous and endocrine systems. The nuclei that make up the hypothalamus—which are paradoxically very small—are critically involved in appetite, thirst, sexual behavior, sleep, temperature regulation, and emotional behavior in general (Figure 30–8).

The hypothalamus interconnects with many regions of the brain, including regions of the cerebral cortex and the thalamus. These regions are viewed by many students of the brain as an integrated network of structures called the limbic system, which forms a loop around the center of the brain (Figure 30–9). The limbic system appears to be involved in complex behavior patterns associated with emotion, memory, motivation, and aggression.

CHEMICAL ACTIVITY IN THE BRAIN

As we noted in Chapter 25, transmission of the nerve impulses across synapses in the human nervous system involves the release of chemicals into the synaptic gap. In the peripheral nervous system, the chemical transmitters are acetylcholine and norepinephrine. Acetylcholine and norepinephrine are also transmitter substances in the brain. At least two additional transmitter substances have been found in the central nervous system: dopamine and serotonin. These are closely related chemically to norepinephrine, and together they constitute a group of chemicals known as the biogenic amines. They are formed from naturally occurring amino acids.

Acetylcholine is a transmitter substance of the reticular activating formation, and thus may play a role in arousal and paying attention. Norepinephrine is found in particularly large concentrations in the hypothalamus and limbic system. Dopamine is the transmitter substance for a relatively small group of neurons involved with muscular activity. It has recently been discovered that Parkinson's disease, which is characterized by involuntary muscular activity, is associated with a decrease in dopamine. Dopamine has now been used with great success in treating some persons with this fairly common condition. Dopamine also, for unknown reasons, appears to be of benefit in the treatment of some cases of depression.

Serotonin is found in many portions of the central nervous system, including, in particular, the hypothalamus and other parts of the limbic system. All of the cells producing serotonin, it has been found, however, have their cell bodies in a small area in the brainstem, which is part of the reticular formation. It thus appears that the serotonin system may be a very old portion of the brain dealing with basic functions such as sleep, consciousness, and emotional states.

Table 30–2 lists drugs that act on the brain to alter mood and behavior. As

THALAMUS

30–9

The limbic system, a group of interconnected structures within the cerebrum, is associated with emotional and motivational aspects of behavior.

Table 30-2 *Drugs That Affect Mood and Behavior*

DRUG	EFFECTS	MECHANISM OF ACTION
Caffeine (coffee, tea, cola drinks)	Stimulant; promotes alertness; increases motor activities, insomnia	Facilitates synaptic transmission
Nicotine (tobacco)	Stimulant	Mimics acetylcholine, facilitates synaptic transmission
Amphetamines (Benzedrine, Methedrine, Dexedrine)	Stimulant; lessens fatigue and depression; suppresses appetite; causes malnutrition, exhaustion, impairment of judgment, psychoses; strongly addictive	Increases activity of biogenic amines in brain
Alcohol	Sedative (after preliminary excitatory effect); loss of motor coordination and alertness; euphoria; cirrhosis of liver	Depresses CNS (exact mechanism unknown)
Chlorpromazine (Thorazine, Compazine, Stelazine)	Tranquilizer; depresses reticular-formation activity; suppresses hallucinations and delusions; antipsychotic	Blocks receptors of epinephrine and acetylcholine
Barbiturates (Seconal, Amytal, Nembutal, Tuinal)	Sedative; induces sleep, euphoria, irritability, hallucinations; highly addictive	Interferes with synthesis and release of norepinephrine and serotonin
Opium, morphine, heroin	Relieves pain, induces muscle relaxation, drowsiness, lethargy, euphoria, sleep; produces constipation, loss of appetite, addiction; kills by depressing respiratory center	CNS depressant
Cocaine	Local anesthesia, euphoria, alertness, addiction	Inhibits uptake of norepinephrine
MHO inhibiters	Antidepressant	Increases brain levels of norepinephrine
Psilocybin, mescaline	Increased sensory awareness, hallucinations, psychoses	Chemical similar to biogenic amines; may enhance their effects
LSD	Rich visual imagery, sensory awareness, visual hallucinations	Inhibits brain serotonin, perhaps by decreasing activity of serotonin-producing cells in reticular formation
Marijuana, hashish	Mild euphoria	Unknown

you can see, many of these agents appear to exert their effects by enhancing or inhibiting the effects of chemical transmitters.

Recently, scientists have isolated morphinelike chemicals from mammalian brains and also have found receptor sites for such chemicals on brain cells. These receptors are most concentrated in areas of the brain concerned with perception of pain, emotional reaction to pain, and other emotional centers. These chemicals, which have been called endorphins, are small peptides. The physiological role of endorphins is not known, but it is very possible that their function or malfunction may be involved in response to pain, in various mental disturbances, and in drug addiction.

ELECTRICAL ACTIVITY OF THE BRAIN

Another way in which the brain is studied is by the electroencephalogram (EEG). The electroencephalogram is a record of continuous electrical activity from the brain as measured by the difference in electric potential between electrodes placed on specific areas of the scalp and a "neutral" electrode placed elsewhere on the body or the difference between pairs of electrodes on the head. The voltages that arise are very weak—about 300 microvolts (a microvolt is a millionth of a volt) is the maximum in a normal adult—and extremely sensitive recording equipment is required.

30–10

A continuous discharge of alpha waves seems to be associated with tranquility and well-being. By concentrating on a musical tone that sounds when alpha waves are produced, it is possible to learn to turn on those waves at will. This technique is known as biofeedback training, or BFT.

Such recordings obviously reflect only gross electrical activity. Donald Kennedy of Stanford University, one of the leading investigators in the field of nerve cell function, compares the electroencephalogram to measurements that an observer on Mars might make of the roars emanating from the Houston Astrodome during a baseball game. They would certainly indicate that something was going on in Houston, but it would be very difficult to infer the rules or determine the progress or the nature of the game from these recordings.

Nevertheless, there are characteristic EEG patterns, and these patterns can be correlated with the level and type of brain activity.

Alpha and Beta Waves

One such pattern is the alpha wave, a slow, fairly irregular variation in electric potential, which has a frequency of about 8 to 12 cycles per second (Figure 30–11). Alpha waves are usually produced during periods of relaxation; in most persons they are conspicuous only when the eyes are shut.

About two-thirds of the general population have alpha waves that are disrupted by attention. Of the remaining third, about one-half have almost no alpha rhythm at all and about one-half have persistent alpha rhythms that are not easily disrupted by attention. According to a recent study, people who have almost no alpha rhythms under any conditions think almost exclusively by visual imagery, whereas people with a persistent alpha rhythm tend to be abstract thinkers.

Another EEG pattern is represented by the beta wave. Beta waves are of lesser amplitude than alpha waves, but their frequency is greater, being about 18 to 32 cycles per second. They occur in bursts in the anterior part of the brain and are associated with mental activity and excitement.

SLEEP AND DREAMS

Considering the prominence of sleep in our daily existence, it is remarkable how little we know about it. Why does it occupy one-third of our lives? What happens during sleep to restore our capacity to function? Why does sleep, or the lack of it, have such profound effects on our alertness, judgment, good temper, coordination, and emotional stability? Guinea pigs have survived, never sleeping, after destruction of areas in the midbrain. There are also two well-documented cases of persons, one in Italy and one in Australia, who never sleep, presumably because of some brain dysfunction, such as that inflicted on the guinea pigs. Thus, sleep is certainly not necessary for survival in the same way as oxygen, food, or water.

One of the oldest hypotheses concerning sleep is that it results from the buildup of some chemical in the brain that induces sleep and is then altered or

30–11

The frequencies (waves per second) and amplitudes of alpha and beta waves.

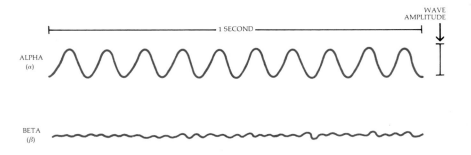

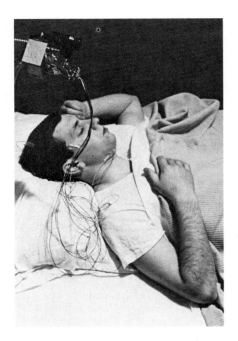

30–12
Volunteer wired to record brain waves, eye movements, breathing rate, and other physical functions.

destroyed during the sleep process. Certain chemicals, including serotonin and acetylcholine, have been discovered to induce sleep, but whether any of these is the natural sleep promoter—or if, indeed, there is a sleep-promoting chemical—is not known. In fact, some studies seem to refute this attractive hypothesis; for example, Siamese twins sharing a common circulation may sleep at different times.

During sleep, heart rate and arterial pressure decrease, breathing becomes more shallow, the carbon dioxide concentration of the blood increases, and rectal temperature drops as much as 2°C. There are also obvious changes in the electroencephalogram, which can be divided into five stages.

Stage 1. Drowsiness, in which alpha waves are slowed slightly. This stage is not classified as the true sleep but rather as a transition period between sleep and wakefulness. Sustained Stage 1 EEG patterns do not fulfill sleep requirements in experiments with human volunteers.
Stage 2. Light sleep, which is a mixture of slow (3 to 6 per second) low-amplitude waves interrupted by short bursts of faster (14 to 15 per second) waves known as sleep spindles.
Stage 3. Intermediate sleep, with high-amplitude waves at about 2 per second (delta) and with occasional sleep spindles.
Stage 4. Deep sleep, with very slow waves of comparatively high voltage and without spindles.
REM sleep. Every 80 to 120 minutes, there occurs a period of rapid, low-voltage activity similar to that seen in alert persons. This is also called paradoxical sleep and is associated with rapid eye movements (REMs), in which the eyeballs move jerkily as if the sleeper were watching some scene of intense activity.

As shown in Figure 30–13, sleepers pass through a regular sleep cycle, moving from one stage of sleep to another. The average young adult spends 50 percent of his sleeping time in Stage 2 and about 20 percent in REM sleep, divided into three to five episodes.

30–13
Stages of sleep as shown by an electroencephalogram. Five tracings are shown for each stage; the upper two are from electrodes attached beside the eye, and the lower three from electrodes attached on the scalp. The recorded length for each sleep stage represents 20 seconds.

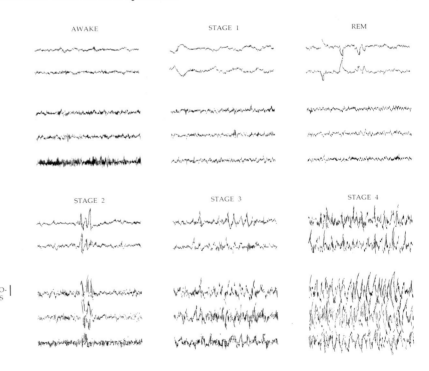

REM sleep differs from the other sleep stages. For example, although there is electroencephalographic evidence of alertness, the muscles are more relaxed and the sleeper is harder to awaken than in ordinary Stage 4 sleep, although, once awakened, he is more alert. Fluctuations in heart rate, blood pressure, and respiration are also seen during REM sleep. Many male subjects experience erections of the penis during REM sleep, and women have similar erections of the clitoris with secretions of vaginal fluid. A person awakened during a period of REM sleep nearly always reports that he or she has been dreaming. REM sleep was thus thought to be the period when all dreaming occurred and it was hypothesized that one of the reasons sleep is necessary is that dreaming is necessary and that REM sleep occurs in order to permit dreaming.

On the basis of more recent work, it is now generally believed, however, that dreaming can occur at any time during sleep but that conditions for recalling dreams are most favorable when subjects are awakened during REM sleep. Some alternative hypotheses have been suggested. One proposal is that REM sleep serves as an information-processing period during which data from the previous waking cycle are sorted, processed, and stored, and so this stage of sleep is essential to the memory process. Laboratory rats forget tasks they have learned if they are deprived of REM sleep. Similarly, evidence indicates that the student who stays up all night cramming for an exam will not have as good a recollection of the material as the student who studies and then sleeps.

According to other hypotheses, Stage 4 sleep of the total sleep cycle is necessary for the restorative functions of sleep, and the principal function of REM sleep is a partial arousal—somewhat similar to the occasional arousals seen in hibernating animals and presumably serving the same sort of sentinel function. These studies and the controversies they engender are continuing.

THE EYE AND THE BRAIN

Some of the most recent interesting studies on brain function have been concerned with interactions between brain and eye.

Structure of the Eye

The vertebrate type of eye is often called a camera eye. In fact, it has a number of features in common with an ordinary camera, as well as being equipped with several expensive accessories, such as a built-in cleaning and lubricating system, an exposure meter, and an automatic field finder. Light from the object being viewed passes through the transparent cornea and lens, which focus an inverted image of the object on the light-sensitive retina in the back of the eyeball. Near the center of the retina is the *fovea*. The lens focuses a perfect image on the fovea, which is the area of sharpest vision (Figure 30–14a).

The retina is made up of photoreceptor ("light-receiving") cells. In humans, these cells are of two types named, because of their shapes, rods and cones (Figure 30–15). The cones, of which the human eye contains about 5 million, provide greater resolution of vision, giving a "crisper" picture. They are generally concentrated in the center of the retina. Man has about 160,000 cones per square millimeter of eye.

Rods do not provide as great a degree of resolution as cones do, but they are more light-sensitive; in dim light, our vision depends entirely on rods. In man, rods are generally concentrated around the periphery of the retina. Some nocturnal animals, such as toads, mice, rats, and bats, have retinas made up entirely of rods, and some diurnal animals, such as some reptiles, have only cones.

(a) *A diagram of the human eye. An image, represented by the arrow, is focused (and inverted) on the retina, which contains the photoreceptor cells. The fovea, near the center of the retina, is the area of greatest visual acuity. The pupil is a hole in the center of the iris, which is the colored part of the eye. The cornea is the transparent outer layer of the eyeball. (b) The retina of the vertebrate eye. Light (shown here as entering from the right) must pass through a layer of cells to reach the photoreceptors (the rods and cones) at the back of the eye. Signals from the photoreceptor cells are then transmitted through the bipolar cells to the ganglion cells, whose axons converge to become the optic nerve. The transmission paths are not direct but involve elaborate interconnections.*

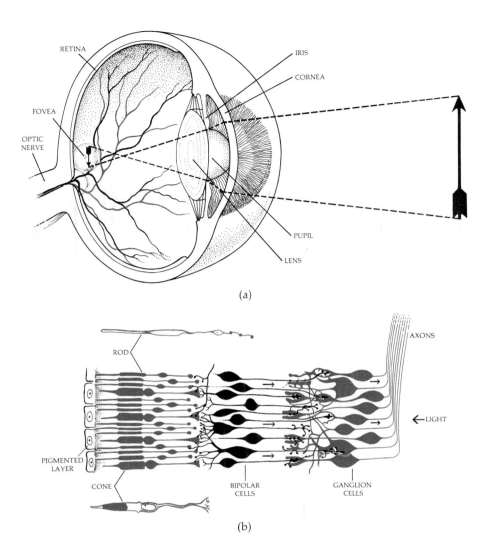

(a)

(b)

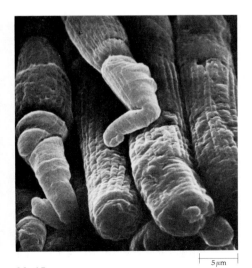

30-15
Rods and cones as shown by the scanning electron microscope.

Cones are responsible for color vision, which is why the world becomes gray to us at twilight, when the light is too dim to stimulate the cones and we must depend on the rods.

When the photoreceptor cells are stimulated by light, they transmit a signal to neurons known as bipolar cells, which pass it on to the ganglion cells whose axons form the optic nerve. As you can see in Figure 30-14b, these pathways of transmission are not direct; varied and elaborate interconnections are made among the cells even before nerve impulses leave the retina, indicating that some preliminary analysis of visual information takes place in the retina itself. The human retina contains about 125 million photoreceptors and the optic nerve about 1 million fibers.

The fibers of the optic nerve can be traced to a visual relay center in the thalamus, where the fibers from the two eyes come together on each side of the brain and synapse with fibers leading to the visual cortex. Various experiments have shown that the spatial arrangement of neurons on the cortex corresponds topographically with the spatial arrangement of the image as it is received on the retina, except for "over-representation" of the fovea. (The fovea, which occupies only about 1 percent of the area of the retina, projects to nearly 50 percent of the visual cortex.)

Vision and Behavior

Vision is a direct stimulus to action in many of the vertebrates. Predatory fish are vision-stimulated, and many such fish literally "cannot help" striking at a moving object of appropriate size. The frog can learn not to strike at a small moving object, but only with difficulty. If it comes to associate an unpleasant experience with a particular shape, it can restrain itself, although apparently with effort, from striking at that shape. On the other hand, it can never learn to take an object that is not moving. Even a frog that has eaten thousands of live flies will starve to death surrounded by motionless ones.

Recently, it has been found that certain ganglion cell axons leaving the retina of the frog produce impulses only when the frog's eye sees a small moving object, rounded on the top. In the laboratory, the visual stimulus for these ganglion cells can be simulated by a variety of objects—such as paper clips and pencil erasers—provided they are in motion. In a frog's natural environment, however, the objects likely to elicit this response most frequently are flying insects. A large, moving object or a small, still one provokes no response. These ganglion cell axons terminate in a special layer of cells in the frog's midbrain, and although no further connections have been traced, it is tempting to speculate that signals are relayed from this part of the midbrain to motor nerves that control the muscles involved in striking.

The Perception of Form

How do we perceive form? The logical answer is that we perceive form simply because of the way the visual image is arranged on the visual cortex. However, recent studies on cats by David H. Hubel and Torsten N. Wiesel of Harvard University reveal that quite a different mechanism is involved (Figure 30–16). They implanted microelectrodes into single cells of the visual cortex, recorded the responses of individual cells to visual stimuli, and came up with some very surprising results. One cortical neuron, for example, was found to respond only to a horizontal bar; as the bar was tipped away from the horizontal, discharges

30–16

The experimental method of Hubel and Wiesel. A cat with a small electrode implanted in a single cell in its visual cortex observes a screen on which a moving shape is projected. (a) and (b) represent tracings from two different neurons. The first neuron (a) responds to a horizontal bar; as the bar is turned toward the vertical, the action potential ceases. The second neuron (b) responds to a right angle. Note the reduction in response as the angle increases.

from the cell slowed and ceased. Another responded to a vertical bar, ceasing to fire as the bar was oriented toward the horizontal. A third neuron fired only when the bar was moved from left to right; another only to a right-to-left movement. One cell turned out to be a right-angle detector; it responded only to an angle moving across the visual field and was most excited when the angle was a right angle. Thus an important component of the visual information-processing system is a set of very specifically tuned neurons, each responsive to one aspect of the shape of the object perceived. In other words, we see forms the way we do not simply because they are there; we see them because of the organization of our visual systems. The information-processing system synthesizes the final perception by integrating the information received from its individual elements.

The neurons in the frog's retina also apparently carried out much this same kind of data processing. These very fundamental discoveries would seem to be leading to a new and far deeper level of understanding of the ancient dilemma about the relationship between the mind of man and the outside world.

SPLIT-BRAIN STUDIES

Additional interesting implications concerning processing of information in the cortex have come from split-brain studies carried out by Roger Sperry and coworkers at the California Institute of Technology. Sperry severed the connections between the two cerebral hemispheres in cats and monkeys by cutting through the corpus callosum and the optic chiasma (see Figure 30–17). The animals' everyday behavior was normal, but by using specially contrived testing techniques, the investigators were able to show some remarkable and unexpected effects. The animals responded to test procedures as if they had two brains. When the right eye of an animal was covered, for instance, it could learn tasks and discriminations using only the left eye. But then if the left eye were covered and the same tests were presented to the right eye, the animal had to learn all over again. In fact, it was possible, without creating any confusion at all, to have both sides of the brain learn opposite solutions to the same problem.

More recently, similar observations have been made in patients in whom the corpus callosum has been severed for the treatment of epilepsy. The patients are able to carry out their normal activities, but specially devised tests indicate the presence of what Sperry calls "two realms of consciousness." If such a patient is asked to identify by touch objects that he cannot see, he can name those he can feel with his right hand but not those he can feel only with his left hand; apparently, because the speech centers of the brain are located in the left hemisphere, the right brain is mute. If, for instance, the patient's left hand is given a plastic object shaped like the number 2, which the patient cannot see, he is unable to identify the object verbally but he can readily tell the experimenter what it is by extending two fingers.

In a person whose optic chiasma is intact and whose head is held in a fixed position, the right brain sees only the left side of a picture briefly presented and the left brain sees only the right side. If a picture is flashed before the eyes of such a person, and he is asked what he saw, he reports only what he saw on the right side of the picture; however, the left side of the picture can have an emotional effect on him. The patient laughs, for instance, but cannot explain why.

One conclusion that can be drawn from these experiments is that there is a great deal of redundancy in the human cortex. This redundancy has long been suspected by surgeons who have observed patients functioning normally after

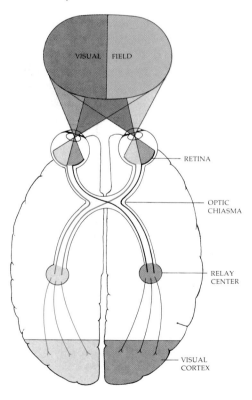

30-17

The optic chiasma viewed from below. The chiasma is the structure formed by the crossing-over of fibers traveling from the retina to the visual cortex. Because of this crossing-over, the right and left halves of the center each "see" only the opposite half of the visual field.

VISUAL FIELD

RETINA

OPTIC CHIASMA

RELAY CENTER

VISUAL CORTEX

30-18

A split face (a) used in tests of patients whose brains have been surgically divided is made up from two of the faces shown in the selection (b). The patient, wearing a headgear that restrains eye movements, sees the picture projected briefly on a screen. The left side of the brain recognizes the child; the right side sees the woman with glasses.

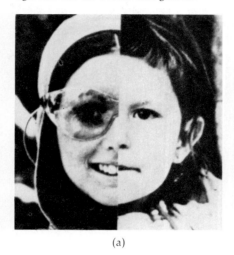

(a)

(b)

removal of or damage to extensive areas of the right temporal lobes. In a number of cases, children have, after an interval, learned to speak again after left temporal lobe damage, as a result of the right lobe taking over the speech functions. (Around puberty, apparently some of this plasticity is lost and the right speech centers cannot take over.)

Sperry's experiments also demonstrate quite clearly that the site of complex learned behaviors in humans is the cerebral cortex, rather than the brainstem, since this structure was still intact in all the subjects tested.

There is some indication that there are differences in mental capacity between the two brains, with the left brain excelling in verbal skills, obviously, and the right one in discriminations involving shape, form, tune, and texture. The implication of this finding for the normal person is not clear, however.

LEARNING AND MEMORY

For scientists interested in brain research, the biggest present challenge is to understand the mechanisms of memory and learning. If we define learning as a change in behavior based on experience, and if the functions of the brain—the mind—are to be explained in terms of the atoms and molecules and the structures composed of them, then learning must involve changes in these atoms or molecules or structures. But what is this change, and where and how does it take place?

Almost 50 years ago, the late Karl Lashley of Harvard set out to locate this physical change, the "scar" of memory, which he called the engram. He taught rats and other animals to solve particular problems and then performed operations to see whether he could remove the portion of the brain containing that particular engram. But he never found the engram. As long as he left enough brain tissue to enable the animal to respond to the test procedures at all, he left

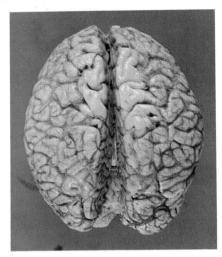

30-19
The human cerebral cortex viewed from above.

memory as well. And the amount of memory that remained was generally proportional to the amount of remaining brain tissue. Lashley concluded that memory is "nowhere and everywhere present."

Although much more sophisticated techniques are now available for brain research, modern neurologists have had little more success than Lashley in finding memory's hiding place. Moreover, the problem has become, if anything, more complex. It is now clear that there are two types of memory, short-term and long-term. A simple example of the former is looking up an unfamiliar number in a phone book; you can usually remember it just long enough to dial it. A less common but also well-known phenomenon is a loss of memory for the events immediately preceding a blow to the head; only rarely (except in soap operas) is there any loss of the records of earlier experiences.

The existence of long-term and short-term memory has also been convincingly demonstrated in memory tests in human volunteers who are called upon to memorize, for instance, lists of unrelated words. Not surprisingly, it was found that the memorizing of new lists hastens the forgetting of the previous ones, suggesting that the short span of short-term memory may serve the useful function of clearing the mind for new information. In experimental animals, it has been shown that a physical insult to the brain, such as electric shock, given soon after the learning of a simple task, can apparently erase the memory just acquired without interfering with earlier ones.

The working hypothesis held by most neurobiologists concerning memory and learning is that these processes are the result of changes in the connections among neurons—in the synapses. These changes may involve the production of new proteins; brain cells have an unusually high content of RNA, and the RNA content appeared to increase during learning.

The human brain is the most complicated structure ever known. The average human brain contains about 12 billion neurons, and it has been calculated that the possible number of interconnections among them in a single brain is greater than the total number of atomic particles making up the universe. The human brain, as we have noted, is capable of many remarkable activities. It remains to be seen if it can meet this most difficult challenge of all—understanding itself.

SUMMARY

The function of the brain is to process information received from the environment and to initiate, by means of motor neuron impulses, appropriate actions. Mapping studies of the human cerebral cortex have succeeded in identifying a motor cortex, in which stimulation of particular areas results in motor responses in particular parts of the body; a sensory cortex, which appears to be a terminal for incoming somatic sensory impulses; and auditory and visual cortices, where impulses from the ear and eye, respectively, are received and analyzed.

On the left cerebral hemisphere are also the so-called speech areas; damage to these areas interferes with the complex processes involved in the formulation and organization of speech. The rest of the cortex—which constitutes about three-fourths of the total area—is known as the association cortex. The relative amount of association cortex is much larger in man than it is in other mammals, and these areas are believed to be involved in planning, learning, memory, and other abstract mental processes.

Other functional divisions of the brain are the reticular formation, concerned with arousal, and the limbic system, which involves the emotions.

Chemical transmitters in the brain include acetylcholine, norephinephrine, dopamine, and serotonin. Many drugs appear to act on the brain by increasing or decreasing the concentration of these chemicals.

Brain function can be studied by recording wavelike variations in electric potential from the brain surface (electroencephalography). Three principal types of brain waves can be recorded in normal subjects: beta waves, of relatively high frequency, associated with alertness and attention; alpha waves, of lower frequency, associated with relaxation; and delta waves, of still lower frequency, which normally appear only during sleep.

Anatomically, the brain is made up of white matter (myelinated nerve fiber tracts) and gray matter (unmyelinated fibers, nerve cell bodies, and glial cells). The brain consists of the brainstem, which includes the medulla, pons, and midbrain; the cerebellum; and the cerebrum, which includes the thalamus, hypothalamus, and corpus callosum. The cerebrum is divided into two cerebral hemispheres and, in the higher mammals, has a greatly convoluted outer layer of cells, the cerebral cortex.

Sleep involves metabolic changes, such as decrease in heart rate, in ventilation, and in body temperature and also changes in the electroencephalogram. The slow, rhythmic waves characteristic of deep sleep are interrupted by periods of rapid, high-frequency waves (REM, or paradoxical, sleep) typically accompanied by rapid eye movements. The function of sleep is not understood.

Studies on the eye and the brain reveal that visual form is coded by single neurons in the brain, each of which responds to one or two specific features of the visual image.

Split-brain studies have been carried out on experimental animals and human patients in which connections have been severed between the cerebral hemispheres. Results of such studies indicate that the two cerebral hemispheres of the same individual can function independently and that they differ somewhat in their capacities.

Studies on the mechanisms of memory and learning involve the search for changes in the physicochemical structure of brain cells or the connections among them. No convincing evidence regarding the nature of these changes in the organizational relationship among brain cells is as yet available.

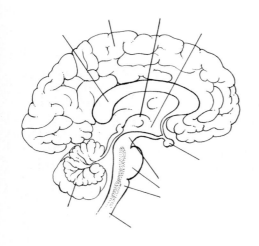

QUESTIONS

1. Define the following terms: gray matter, white matter, association cortex, reticular system.

2. On the drawing at the left, label or note the approximate position of the following structures: brainstem, midbrain, medulla, cerebellum, thalamus, hypothalamus, pituitary, cerebral cortex.

3. Draw a picture of the eye and label its parts.

SUGGESTIONS FOR FURTHER READING

BECK, WILLIAM S.: *Human Design*, Harcourt Brace Jovanovich, Inc., New York, 1971.

A comprehensive, modern treatise, designed for the undergraduate student, covering "molecular, cellular, and systematic physiology" of man.

BODEMER, CHARLES W.: *Modern Embryology,* Holt, Rinehart and Winston, Inc., New York, 1968.

Designed for a college course in embryology, this book strikes a good comprehensive balance between classical embryology and the emerging field of molecular embryology.

BRECHER, EDWARD M., et al.: *Licit and Illicit Drugs: The Consumers Union Report on Narcotics, Stimulants, Depressants, Inhalants, Hallucinogens, and Marijuana—Including Caffeine, Nicotine, and Alcohol,* Little, Brown and Company, Boston, 1972.

A well-balanced, straightforward, and nonsensational review.

CALDER, NIGEL: *The Mind of Man,* The Viking Press, Inc., New York, 1970.

A well-written, fast-moving, journalistic report on the "drama of brain research."

FREEDMAN, RUSSELL, and JAMES E. MORRIS: *The Brains of Animals and Man,* Holiday House, Inc., New York, 1972.

A simple, well-illustrated account of the brain and modern research on its function.

GREGORY, R. L.: *Eye and Brain: The Psychology of Seeing,* 2nd ed., McGraw-Hill Book Company, New York, 1973.*

A vivid introduction to the science of vision.

HANDLER, PHILIP (ed.): *Biology and the Future of Man,* Oxford University Press, New York, 1970.*

A widely acclaimed survey of the state of biology today, written by a committee of experts in the various fields. Its purpose is to inform the general public both of present knowledge in major fields of biology and of current problems and probable areas of future investigation. It deals with a wide variety of subjects, including most of the ones covered in this book. There are outstanding sections on animal physiology, development, and behavior.

OATLEY, KEITH: *Brain Mechanisms and Mind,* E. P. Dutton, New York, 1972.

A readable, up-to-date, well-illustrated survey written by an experimental psychologist for the general reader.

ROMER, ALFRED: *The Vertebrate Story,* 4th ed., The University of Chicago Press, Chicago, 1959.

The history of vertebrate evolution, written by an expert but as readable as a novel.

RUGH, ROBERTS, and LANDRUM B. SHETTLES: *From Conception to Birth: The Drama of Life's Beginnings,* Harper & Row, Publishers, Inc., New York, 1971.

This is an account of the history of life before birth. The book describes in detail the development of the unborn child from the moment of fertilization and also the changes in the mother during pregnancy. It discusses such related topics as birth control, congenital malformation, labor, and delivery of the baby. There are a large number of illustrations, including a group of magnificent color photographs of the developing fetus.

SCHMIDT-NIELSEN, KNUT: *Desert Animals,* Oxford University Press, New York, 1964.

Although considered the definitive work on the physiological problems relating to heat and water, this readable book also contains numerous anecdotes—such as that about Dr. Blagden—and many fascinating personal observations.

SCHMIDT-NIELSEN, KNUT: *Animal Physiology: Adaptations and Environment,* Cambridge University Press, New York, 1975.*

Schmidt-Nielsen is concerned with underlying principles of animal physiology—the problems animals have to solve in order to survive. The emphasis is on comparative physiology, and the lucid exposition is illuminated by many interesting examples.

* Available in paperback.

SMITH, HOMER W.: *From Fish to Philosopher*, Doubleday & Company, Inc., Garden City, N.Y., 1959.*

Smith is an eminent specialist in the physiology of the kidney. Writing for the general public, he explains the role of this remarkable organ in the story of how, in the course of evolution, organisms have increasingly freed themselves from their environments.

THOMPSON, R. F.: *Introduction to Physiological Psychology*, Harper & Row, Publishers, Inc., New York, 1975.

Intended for the undergraduate student, this text presents an up-to-date survey of the biological foundations of psychology.

VANDER, A. J., J. H. SHERMAN, and DOROTHY S. LUCIANO: *Human Physiology*, 2d ed., McGraw-Hill Book Company, New York, 1975.

Most highly recommended. The text is a model of clarity, and the diagrams, many of which we have borrowed, are splendid.

* Available in paperback.

PART III   Ecology and Evolution

SECTION 7   Evolution and Population Biology

31-1

Galapagos tortoises, near Alcedo Volcano on Isabela Island, one of the Galapagos. Thirteen different races of tortoise inhabit the archipelago; this particular race is found only near the Alcedo Volcano. The Galapagos archipelago consists of thirteen volcanic islands that pushed up from the sea more than a million years ago. Craters still covered with black basaltic lava rise to 1,000 meters or more. The major vegetation is a dreary grayish-brown thornbush, making up vast areas of dense, leafless thicket, and a few tall tree cactuses— "what we might imagine the cultivated parts of the Infernal regions to be," young Charles Darwin wrote in his diary.

CHAPTER 31

Darwin and the Theory of Evolution

31–2
"Afterwards, on becoming very intimate with Fitz Roy [the captain of the Beagle*], I heard that I had run a very narrow risk of being rejected on account of the shape of my nose! He . . . was convinced that he could judge of a man's character by the outline of his features; and he doubted whether anyone with my nose could possess sufficient energy and determination for the voyage. But I think he was afterwards well satisfied that my nose had spoken falsely." (Charles Darwin,* The Voyage of the Beagle.*)*

In 1831, a young Englishman, Charles Darwin, sailed from Devonport on what was to prove the most consequential voyage in the history of biology. Not yet 23, Darwin had already abandoned a proposed career in medicine—he describes himself as fleeing a surgical theater in which an operation was being performed on an unanesthetized child—and was a reluctant candidate for the clergy, a profession deemed more suitable for the younger son of an English gentleman. An indifferent student, Darwin was an ardent hunter and fisherman, a collector of beetles, mollusks, and shells, and an amateur botanist and geologist. When the captain of the surveying ship H.M.S. *Beagle,* himself only a little older than Darwin, offered passage to any young man who would volunteer to go without pay as a naturalist, Darwin eagerly seized the opportunity to escape from Cambridge. This voyage, which lasted five years, shaped the course of Darwin's future work. He returned to an inherited fortune, an estate in the English countryside, and a lifetime of work and study.

THE ROAD TO EVOLUTIONARY THEORY

That Darwin was the founder of the modern theory of evolution is well known. In order to understand the meaning of his theory, however, it is useful to look briefly at the intellectual climate in which it was formulated. (To some degree, this is true, as you may be beginning to see, of all theories in biology.)

The Ladder of Life

Aristotle, the first great biologist, believed that all living things could be arranged in a hierarchy. This hierarchy became known as the *Scala Naturae,* or ladder of nature, in which the simplest creatures had a humble position on the bottommost rung, man occupied the top, and all other organisms had their proper places between. Up until the end of the last century, many European biologists believed in such a natural hierarchy. But whereas to Aristotle living organisms had always existed, the later Europeans, in harmony with the teachings of the Scriptures, believed that all living things were the products of a divine creation. They believed, moreover, that most were created for the service or pleasure of mankind. Indeed, it was pointed out, even the lengths of day and night were planned to coincide with man's need for sleep.

That each type of living thing came into existence in its present form—specially and specifically created—was a compelling idea. How else could one explain the astonishing extent to which every living thing was adapted to its environment and to its role in nature? It was not only the authority of the church but the evidence before one's own eyes that gave such strength to the concept of special creation.

One of the more famous biologists who believed in divine creation was Carolus Linnaeus, the great eighteenth-century Swedish systematist. All the time that Linnaeus was at work on his encyclopedic *Systema Naturae* and *Species Plantarum*, explorers of the New World were continuing to return to Europe with new species of plants and animals and even, apparently, new kinds of human beings. Linnaeus revised edition after edition to accommodate these findings, but he did not change his opinion that all modern species were created on the sixth day of God's labor and have remained fixed ever since. During Linnaeus's time, however, it became clear that the pattern of creation was far more complex than had been originally envisioned.

Preevolutionists

The French scientist Georges-Louis Leclerc de Buffon (1707–1788) was among the first to suggest that species might undergo some changes in the course of time. Buffon believed that these changes took place by a process of degeneration. He suggested that, in addition to the numerous creatures that were produced by divine creation at the beginning of the world, "there are lesser families conceived by Nature and produced by Time." In fact, as he summed it up, ". . . improvement and degeneration are the same thing, for both imply an alteration of the original constitution." As you can see, Buffon's hypothesis explained the bewildering variety of creatures in the modern world.

Another early doubter of the fixity of species was Erasmus Darwin (1731–1802), Charles Darwin's grandfather. Erasmus Darwin was a physician, a gentleman naturalist, and a prolific writer, often in verse, on both botany and zoology. Erasmus Darwin suggested, largely in asides and footnotes, that species have historical connections with one another, that competition plays a role in the development of different species, that animals may change in response to their environment, and that their offspring may inherit these changes. He maintained, for instance, that a polar bear is an "ordinary" bear that, by living in the Arctic, becomes modified and passes these modifications along to its cubs. These ideas were never clearly formulated but are interesting because of their possible effects on Charles Darwin, although the latter, born after his grandfather died, did not hold his grandfather's views in high esteem.

The Age of the Earth

It was the geologists far more than the biologists who paved the way for evolutionary theory. One of the most influential of these was James Hutton (1726–1797). Hutton proposed that the earth had been molded not by sudden, violent events but by slow and gradual processes—wind, weather, and the flow of water—the same processes that can be seen at work in the world today. This theory of Hutton's, which was known as uniformitarianism, was important for three reasons. First, it implied that the earth has a living history, and a long one. This was a new idea. Christian theologians, by counting the successive generations since Adam (as recorded in the Bible), had calculated the maximum life span of the world at about 6,000 years. No one had ever thought in terms of a

31-3

A fossil is a remnant or trace of a once-living organism. (a) Among the most common vertebrate fossils are bones that have become impregnated with a hard mineral such as lime or silica, as in this hominid skull. (b) Similarly, minerals may fill the hollows left by the decay of soft tissues, as, for example, in these fossilized leaves. (c) Among the most common early fossils are the discarded exoskeletons of trilobites, a type of arthropod that flourished during the Cambrian period, which began 500 to 600 million years ago. (d) A more recent fossil is an insect caught in amber, the hardened resin of a long-dead tree. Famous but less common are the frozen remains of wooly mammoths and a giant rhinoceros. One of the most thoroughly studied wooly mammoths was discovered in Siberia in 1900.

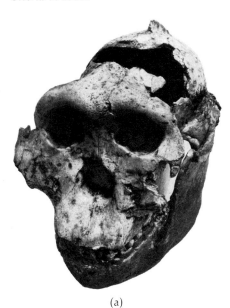

(a)

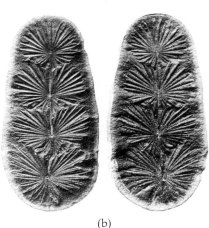

(b)

(c)

(d)

longer period. And 6,000 years is not enough time for such major evolutionary changes as the formation of new species to have taken place. Second, the theory of uniformitarianism stated that change is the *normal* course of events, as opposed to the concept of a normally static system interrupted by an occasional unusual event, such as a flood. Third, although this was never explicit, uniformitarianism suggested that there might be interpretations of the Bible other than the literal one.

The Fossil Record

During the late eighteenth century, there was a revival of interest in fossils (Figure 31-3). In previous centuries, fossils had been collected as curiosities, but they had generally been regarded either as accidents of nature—stones that somehow looked like shells—or as evidence of great natural catastrophes, such as Noah's Flood. The English surveyor William Smith (1769–1839) was the first to make a systematic study of fossils. Whenever his work took him down into a mine or along canals or cross-country, he carefully noted the order of the different layers of rock, which are called strata, and collected the fossils from each layer. He eventually established that each stratum, no matter where he came across it in England, contained a characteristic group of specimens and that these fossils were actually the best way to identify a particular stratum. (The use of fossils to identify strata is still widely practiced, for instance, by geologists looking for oil.) Smith did not interpret his findings, but the implication that the present surface of the earth had been formed layer by layer over the course of time was an unavoidable one.

Like Hutton's world, the world seen and reported by William Smith was clearly a very ancient one. A revolution in geology was beginning; earth science was becoming a study of time and change rather than a mere cataloging of types of rocks. As a consequence, the history of the earth became inseparable from the history of living organisms, as revealed in the fossil record.

Catastrophism

Although the way was being prepared by the revolution in geology, the time was not yet ripe for a parallel revolution in biology. The dominating force in European science in the early nineteenth century was Georges Cuvier (1769–

THE RECORD IN THE ROCKS

The earth's long history is recorded in the rocks that lie at or near its surface, layer piled upon layer, like the chapters in a book. These layers, or strata, are formed as rocks in upland areas are broken down to pebbles, sand, and clay and are carried to the lowlands and the seas. Once deposited, they slowly become compacted and cemented into a solid form as a new material is deposited above them. As continents and ocean basins change shape, some strata sink below the surface of an ocean or a lake, others are forced upward into mountain ranges, and some are worn away in turn by water, wind, or ice or are deformed by heat or pressure.

Individual strata may be paper-thin or many meters thick. They can be distinguished from one another by the types of parent material from which they were laid down, the way the material was transported, and the environmental conditions under which the strata were formed, all of which leave their traces in the rock. They can be distinguished, moreover, by the types of fossils they contain. Small marine fossils, in particular, can be associated with specific periods in the earth's history. The fossil record is never complete in any one place, but because of the identifying characteristics of the strata, it is possible to piece together the evidence from many different sources. It is somewhat like having many copies of the same book, all with chapters

missing—but different chapters, so it is possible to reconstruct the whole.

The geologic eras—Precambrian, Paleozoic, Mesozoic, and Cenozoic—which are the major subdivisions of the geologic "book," were established and named in the early nineteenth century. These eras were subdivided into periods, each named, quite simply, for the areas in which the particular strata were first studied or studied most completely: the Devonian for Devonshire in southern England, the Permian for the province of Perm in Russia, the Jurassic for the Jura Mountains between France and Switzerland, and so on.

Early attempts to date the various eras and periods were based simply on their relative ages compared to the age of the earth; obviously, a stratum occurring regularly above another was younger than the one below it. The first scientific estimate was made in the mid-1800s by the famous British physicist Lord Kelvin. On the basis of his calculations of the time necessary for the earth to have cooled from its original molten state, Kelvin maintained that the planet could not have been more than 25 million years old. This estimate greatly troubled Charles Darwin, who knew that it was too little time for evolution to have taken place. (Kelvin was not aware of the existence under the earth's surface of radioactive materials that heat the planet from within.) In the last 20 years, however, new methods for determining the ages of strata have been developed involving measurements of the decay of radioactive isotopes. As a result, the estimated age of the earth has increased, in little more than a century, from 25 million years to more than 4.5 billion.

Darwin's concept of the time necessary for major evolutionary changes to take place has been more than borne out by subsequent studies. For instance, in the evolution of the horse (page 460), a major index of change is relation of tooth width to tooth height as horses gradually changed from browsers (leaf eaters) to grazers (grass eaters). The largest actual change—and this is a fast evolutionary line—was a difference of 1.61 millimeters in the height of a particular tooth in 1 million years. In the entire sequence of dawn horse to Equus, there were more than eight successive genera over 60 million years, or about 7.5 million years per genus.

Geologic strata (and the fossils within them) are now dated whenever possible by analysis of the radioactive isotopes they contain. Because the rate of decay of an isotope is constant (page 30), the proportion in a particular sample of ^{238}U to ^{206}Pb, for example, can be used to calculate the number of years since a particular stratum was formed. However, because the daughter atoms (^{206}Pb, for instance) are fewer and so more likely than parent atoms (^{238}U, in this case) to go undetected, dates set by radioactive isotopes are more likely to be too recent. Hence, as methods become more precise and more samples are analyzed, geologic time continues to stretch farther and farther backward.

Strata visible in the Grand Canyon.

31-4

Drawing by Georges Cuvier of a mastodon. Although Cuvier was one of the world's experts in reconstructing extinct animals from their fossil remains, he was a powerful opponent of evolutionary theories.

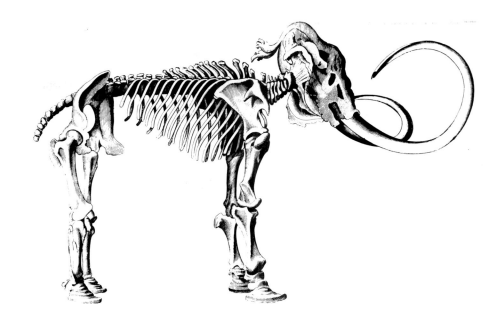

1832). Cuvier was the founder of vertebrate paleontology, the scientific study of the fossil record. An expert in anatomy and zoology, he applied his special knowledge of the way in which animals are constructed to the study of fossil animals, and he was able to make brilliant deductions about the form of an entire animal from a few fragments of bone. We think of paleontology and evolution as so closely connected that it is surprising to learn that Cuvier was a staunch and powerful opponent of evolutionary theories. He recognized the fact that many species that had once existed no longer did. (In fact, according to modern estimates, considerably less than 1 percent of all species that have ever lived are represented on the earth today.) Cuvier explained the extinction of species by postulating a series of catastrophes. After each catastrophe, the most recent of which was the Flood, new species filled the vacancies.

Cuvier hedged somewhat on the source of the new animals and plants that appeared following the extinction of older forms; he was inclined to believe they moved in from parts unknown. A contemporary and supporter of his, the first professor of paleontology at the Paris museum, was more straightforward: He taught that all species were wiped out at each catastrophe and wholly new ones were specially created to take their places. (One symptom of the terminal illness of a scientific theory is the accumulation of new hypotheses required to bolster its shaking foundations.)

THE THEORIES OF LAMARCK

The first scientist to work out a systematic theory of evolution was Jean Baptiste Lamarck (1744–1829). "This justly celebrated naturalist," as Darwin himself referred to him, boldly proposed in 1801 that all species, including man, are descended from other species. Lamarck, unlike most of the other zoologists of his time, was particularly interested in one-celled organisms and invertebrates. Undoubtedly it was his long study of these forms of life that led him to think of living things in terms of constantly increasing complexity, each form derived from an earlier, less complex form.

31-5

According to Lamarck's hypothesis, the necks of giraffes became longer when they stretched to reach high branches, and this acquired characteristic was transmitted to their offspring.

Like Cuvier and others, Lamarck noted that older rocks generally contained fossils of simpler forms of life. Unlike Cuvier, however, Lamarck interpreted this as meaning that the higher forms had risen from the simpler forms by a kind of progression. According to his hypothesis, this progression, or "evolution," to use the modern term, is dependent on two main forces. The first is the inheritance of acquired characteristics. Organs in animals become stronger or weaker, more or less important, through use or disuse, and these changes, according to Lamarck's theory, are transmitted from the parents to the progeny. His most famous example, and the one that Cuvier used most often to ridicule him, was that of the giraffe, which stretched its neck longer and longer to reach leaves on higher branches and transmitted this longer neck to its offspring, which again stretched its neck, and so on.

The second important factor in Lamarck's theory of evolution was a universal creative principle, an unconscious striving upward on the *Scala Naturae* that moved every living creature toward greater complexity. Every amoeba was on its way to man. Some might get waylaid—the orangutan, for instance, by being caught in an unfavorable environment had been diverted off its course—but the will was always present. Life in its simplest forms was constantly emerging by spontaneous generation to fill the void left at the bottom of the ladder. In Lamarck's formulation, the ladder of nature of the ancients had been transformed into a steadily ascending escalator powered by a universal will.

Lamarck's concept of the inheritance of acquired characteristics is an attractive one. Darwin borrowed from it consciously and unconsciously. In fact, successive editions of *The Origin of Species* show that Darwin was maneuvered into a Lamarckian point of view by some of his critics and by his own inability (owing to the primitive state of the science of genetics at the time) to explain some of the ways in which animals changed. Belief in the inheritance of acquired characteristics persisted into the twentieth century in the work of the Russian biologist Trofim Lysenko.

Lamarck's contemporaries did not object to his ideas about the inheritance of acquired characteristics, as we would today with our more advanced knowledge of genetics, nor did they criticize his belief in a metaphysical force, which was actually a common element in many of the theories of the time. But these vague, untestable postulates provided a very shaky foundation for his radical proposal that higher forms evolved from simpler forms, and Lamarck personally was no match for the brilliant and witty Cuvier. As a result, Lamarck's career was ruined, and both scientists and the public became even less prepared for an evolutionary doctrine.

DARWIN'S THEORY

The Earth Has a History

The person who most influenced Darwin, it is generally agreed, was Charles Lyell (1797–1875), a geologist who was Darwin's senior by 12 years. One of the few books that Darwin took with him on his voyage was the first volume of Lyell's newly published *Principles of Geology*, and the second volume was sent to him while he was on the *Beagle*. On the basis of his own observations and those of his predecessors, Lyell opposed the theory of catastrophes. Instead, he produced new evidence in support of Hutton's earlier theory of uniformitarianism. According to Lyell, the slow, steady, and cumulative effect of natural forces had produced continuous change in the course of the earth's history. Since this process is demonstrably slow, its results being barely visible in a single lifetime, it must have been going on for a very long time. In his earlier works, Lyell did

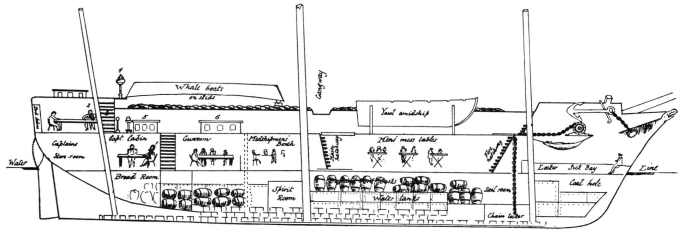

H.M.S. Beagle 1832

1 Mr. Darwin's seat in Capt. Cabin
2 "   "   "   Poop
3 "   "   "   drawers "
4 Azimuth Compass
·5 Captain's skylight
6 Gunroom "

31-6

Cutaway view of the Beagle. *Only 28 meters in length, this "good little vessel" set sail on its five-year voyage with 74 people aboard. Darwin shared the poop cabin with a midshipman and 22 chronometers belonging to Captain Fitz Roy, who had a passion for exactness. His sleeping space was so confined that he had to remove a drawer from a locker to make room for his feet.*

not discuss the biological implications of his theory, but apparently they were clear to Darwin. If the earth had a long continuous history and if no forces other than outside physical agencies were needed to explain the events as they were recorded in the geologic record, might not living organisms have had a similar history? What Darwin's theory needed, as he knew very well, was time, and it was time that Lyell gave him.

The Voyage of the Beagle

This, then, was the intellectual equipment with which Charles Darwin set sail from Devonport. As the *Beagle* moved down the Atlantic coast of South America, through the Straits of Magellan, and up the Pacific coast, Darwin traveled the interior, fished, hunted, and rode horseback. He explored the rich fossil beds of South America (with the theories of Lyell fresh in his mind) and collected specimens of the many new kinds of plant and animal life he encountered. He was impressed most strongly during his long, slow trip down the coast and up again by the constantly changing varieties of organisms he encountered. The birds and other animals on the west coast, for example, were very different from those on the east coast, and even as he moved slowly up the western coast, one species would give way to another one.

Most interesting to Darwin were the animals and plants that inhabited a small barren group of islands, the Galapagos, that lie some 950 kilometers off the coast of Ecuador. The Galapagos were named after the islands' most striking inhabitants, the tortoises (*galápagos* in Spanish), some of which weigh 100 kilograms or more (Figure 31–1). Each island has its own type of tortoise; sailors who took these tortoises on board and kept them as convenient sources of fresh meat on their sea voyages could readily tell which island any particular tortoise had come from. Then there was a group of finchlike birds, 13 species in all, which differed from one another in body size, and particularly in the type of food they ate. In fact, although still clearly finches, they had taken on many characteristics seen only in completely different types of birds on the mainland.

31-7

The Beagle's *voyage around South America.*

31-8

*The woodpecker finch of the Galapagos.
Was it the product of special creation or the
modification of an existing species? Darwin's
answer to this question was an important
step toward the formulation of his theory of
evolution. This finch is also of interest as
one of a very few species of tool-using
animals.*

One finch, for example, feeds by routing insects out of the bark of trees. It is not fully equipped for this, however, lacking the long tongue with which the woodpecker flicks out insects from under the bark. Instead, the woodpecker finch carries with it a small stick to pry the insects loose (Figure 31–8).

From this knowledge of geology, Darwin knew that these islands, clearly of volcanic origin, were much younger than the mainland. Yet the plants and animals of the islands were different from those of the mainland, and in fact the inhabitants of different islands in the archipelago differed from one another. Were the living things on each island the product of a separate special creation? "One might really fancy," Darwin mused at a later date, "that from an original paucity of birds in this archipelago one species had been taken and modified for different ends." For years after his return, this problem continued, in his own word, to "haunt" him.

Development of the Theory

Not long after Darwin's return, he came across a book by the Reverend Thomas Malthus that had first appeared in 1798. In this book, Malthus warned, as economists have warned frequently ever since, that the human population was increasing so rapidly that it not only would soon outstrip the food supply but would leave "standing room only" on earth. Darwin saw that Malthus's conclusion—that food supply and other factors hold populations in check—is true for all species, not just the human one. For example, a single breeding pair of elephants, which are the slowest breeders of all animals, could produce 19 million elephants in 750 years, yet the average number of elephants generally remains the same over the years. Where there might have been 19 million elephants in theory, there are, in fact, only two. The process by which the two survivors are "chosen" was termed by Darwin *natural selection*. He saw it as a process analogous to the type of selection exercised by breeders of cattle, horses, or dogs—with which, as a country squire, he was very familiar. In the case of artificial selection, man chooses variants for breeding on the basis of characteristics that seem to him to be desirable. In the case of natural selection, environmental conditions are the principal forces that operate on the variations continually produced in all species of living organisms. These forces "favor," or select for, some varieties and "discourage" or eliminate others.

Where do the variations come from? According to Darwin's theory, variations occur absolutely at random. They are not produced by the environment, by a "creative force," or by the unconscious striving of the organism. In themselves, they have no direction; *direction is imposed entirely by natural selection.* A variation that gives an animal even a slight advantage makes that animal more likely to leave surviving offspring. Thus, to return to Lamarck's giraffe, an animal with a slightly longer neck has an advantage in feeding and so is apt to leave more offspring than one with a shorter neck. If the longer neck is an inherited trait, some of these offspring will also have long necks, and, if the long-necked animals in this generation have an advantage, members of the next generation will include more long-necked individuals. Finally, the population of short-necked giraffes will give way to a population of long-necked ones.

As you can see, the essential difference between Darwin's formulation and that of any of his predecessors is the central role he gave to variation. Others had thought of variations as mere disturbances in the overall design, whereas Darwin saw that variations among individuals are the real fabric of the evolutionary process. Species arise, he saw, when differences among individuals within a group are gradually converted into differences between groups as the groups become separated in space and time.

The Origin of Species, which Darwin pondered for more than 20 years before its publication in 1859, is, in his own words, "one long argument." No experiments are performed. No new form of evidence is revealed. Fact after fact, observation after observation, culled from the most remote Pacific island to a neighbor's pasture, is recorded, analyzed, and commented upon. Every objection is anticipated and countered. Because the process of evolution is so slow, Darwin did not believe that direct proof of his theory was possible. However, as we shall see, the twentieth century has produced clear evidence of evolution in progress. No scientist now doubts that species have originated in the past and are still originating, that species have become extinct in the past and are still becoming extinct, and that all living things today have an ancestral species in the past.

The real difficulty in accepting Darwin's theory has always been that it seems to diminish man's significance. The new astronomy had made it clear that the earth is not the center of the universe or even of our own solar system. Now the new biology was asking man to accept the fact that, like all other organisms, he is the product of a random process and that, as far as science can show him, he is not created for any special purpose or as a part of any universal design. We are still dealing with this problem today.

SUMMARY

Until the eighteenth century, it was generally accepted that species were the products of divine creation and remained unchanged. This concept, however, came into question as a result of several developments, including: the discovery of many new species, as a consequence of New World exploration; studies in geology that indicated a gradual, constant change in the surface of the earth (uniformitarianism); the accompanying recognition that the earth has a long history (without which evolution would have been impossible); and the discovery of fossils and the recognition of their origin.

Darwin was not the first to propose a theory of evolution. His most notable predecessor was Lamarck, whose theory of evolution was based on the inheritance of characteristics acquired by an organism during its lifetime. Darwin's theory differed from others in that it envisioned evolution as a two-part process, depending upon (1) the existence in nature of inheritable variations among organisms and (2) the process of natural selection by which some organisms, by virtue of their inheritable variations, leave more offspring than others. Darwin's theory is rightfully regarded as the greatest unifying principle in biology.

QUESTIONS

1. What is the essential difference between Darwin's theory of evolution and that of Lamarck?

2. What is the essential difference between the theories of Darwin and Lamarck on the one hand and the account of the creation of living things set forth in the Book of Genesis? Do you think the theory of special creation set forth in the Bible should be presented in a biology textbook? Are they incompatible?

3. The phrase "chance and necessity" has been applied to the process by which organisms evolve. Relate this to the fact that snails living on grass do not have green shells but that there are, for example, green frogs and green insects.

4. What tool-using animals other than the woodpecker finch can you think of?

CHAPTER 32

Some Principles of Population Genetics

Although it is now more than a hundred years since the first publication of *The Origin of Species*, Darwin's original concept of how evolution occurs still provides the basic framework for our understanding of the process. His concept rests on four premises:

1. Like begets like—in other words, there is stability in the process of reproduction.
2. In any given population, there are variations among individual organisms, and some of these variations are inheritable.
3. In every species, the number of individuals that survive to reproductive age is much less than the number produced (twentieth-century man is an exception).
4. Which individuals will survive and reproduce and which will not is determined in large part by the nature of these variations. Those with favorable variations are more likely to pass their characteristics on to the next and future generations in greater numbers.

Modern evolutionary theory conceives of evolution as Darwin conceived of it, with the important addition of an understanding of the mechanisms of inheritance. Twentieth-century genetics answers two questions that Darwin was never able to resolve: (1) why genetic traits are not "blended out" but can disappear and reappear (like whiteness in pea flowers) and (2) how the variations appear on which natural selection acts. This combination of evolutionary theory and genetics is known as the synthetic theory of evolution. (Here "synthetic" does not mean artificial, which is the connotation it has for us in these days of man-made fabrics and artificial colors and flavors, but has its original meaning of the putting together of two or more different elements.)

The branch of genetics that emerged from this synthesis of Darwinian evolution and Mendelian principles is known as *population genetics*. A population, for the geneticist, is an interbreeding group of organisms. For instance, all the fish of one particular species in a pond are a population, and so are all the fruit flies in one bottle. The population is defined and united by its gene pool, which is simply the sum of all the alleles of all the genes of all the individuals in a population.

In every population, the relative proportion of alleles of certain genes in the gene pool will change from one generation to the next. In population genetics, evolution is, by definition, this change in the composition of the gene pool.

32–1

A pair of cichlids, surrounded by some of their offspring. Despite the large numbers of young produced by each mating, there is no remarkable increase in the number of cichlids from generation to generation.

THE HARDY-WEINBERG FORMULA

In population genetics, it is customary to express the varying proportions, or frequencies, of alleles of a particular gene in decimals rather than in fractions, as we did in our earlier discussion of Mendel's laws (pages 130 to 140), but the principles are exactly the same.

In the case of a gene for which there are two alleles in the gene pool, the frequency (represented by the symbol p) of one allele plus the frequency (q) of the other equal the frequency of the whole, which is always 1 ($p + q = 1$). Or, to put it another way, $1 - p = q$ and $1 - q = p$. This is equivalent to saying that if there are only two alleles, A and a, for instance, of a particular gene, and if half (0.50) of the alleles in the gene pool are A, the other half (0.50) have to be a. Similarly, if 99 out of 100 (0.99) are A, 0.01 are a. Notice that in this example, we are talking not of phenotypes or genotypes but strictly of genes and their alleles.

Now let us take an imaginary example. Suppose we have a gene pool in which p (the frequency of allele A) equals 0.8 and q (the frequency of a) equals 0.2, and suppose that mating takes place at random. We can calculate the genotypes and phenotypes by drawing a Punnett square (Figure 32-2).

As you can see from the square, the population produced by random mating would consist of 64 percent AAs, 32 percent AAs, and 4 percent aas.

We can do the same thing algebraically. Because we know that $p + q = 1$, it follows that:

$$(p + q)(p + q) = 1 \times 1 = 1$$

or, as you probably remember from high school algebra:

$$p^2 + 2pq + q^2 = 1$$

In this example, p^2 would be AAs, $2pq$ would represent the heterozygotes, and q^2 would designate the recessives.

The algebraic expression of gene frequency is known as the *Hardy-Weinberg formula.*

Although normally any one diploid individual has no more than two alleles of the same gene, there may, of course, be many alleles in the gene pool. For instance, in the case of the gene for red eyes in *Drosophila* (see Chapter 12), there are a variety of intermediate forms, known by names such as apricot, buff, eosin, and honey, as well as the mutant that results in white eyes. More than 15 alleles have been recognized, and probably more could be detected if there were ways of distinguishing finer gradations of color.

In the case of three alleles, r is used to represent the frequency of the third allele, $p + q + r$ still equals 1, and the results of random mating are expressed by

$$(p + q + r)(p + q + r) = 1$$

which is a little more difficult mathematically but not different in principle.

THE HARDY-WEINBERG PRINCIPLE

The Hardy-Weinberg formula holds true only under certain conditions. They are (1) that no mutations occur, (2) that there is no movement in or out of the population under study, (3) that the population is large enough, and (4) that natural selection does not occur. This set of circumstances is never found; Hardy-Weinberg equilibrium never exists. Why, then, is it of central impor-

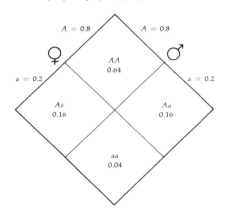

32-2

Results of random breeding in a population in which the frequency of A (p) *equals 0.8 and the frequency of* a (q) *equals 0.2.*

tance? Its importance lies not in its mathematics, which are very simple, but in the underlying principle it demonstrates, which, though even simpler, is not immediately obvious.

When a pea plant homozygous for red flowers is crossed with a white-flowered pea plant, three-fourths of the seeds, on the average, will produce plants with red flowers. When a person with brachydactylism (Figure 32–3) marries a person with normal fingers, three-fourths of their children, on the average, will have brachydactylism. Does this mean that eventually all pea plants will have red flowers and all persons will have short, stubby fingers? This question was raised by students of genetics in the early 1900s and was answered in 1908 by G. H. Hardy, an English mathematician, and G. Weinberg, a German physician. The answer is emphatically no; under the theoretical circumstances just defined, alleles are returned to the gene pool in the same proportions generation after generation. Test this statement yourself. To make the calculation simple, begin by crossing homozygotes for the dominant allele (*AA*) with homozygotes for the recessive allele (*aa*), just as Mendel did in his experiments. Then cross the individuals resulting from this mating through enough generations so you can satisfy yourself that the frequencies of the alleles in the gene pool do not change as a result of this process. If you run into difficulties, consult Appendix D at the back of the book.

To recapitulate, population genetics is concerned with the changes that take place over time in the relative proportions of alleles in the gene pool. The Hardy-Weinberg principle states that the repeated recombination and segregation of these alleles does not in itself affect their relative proportions in the gene pool. In every population, however, frequencies of alleles do change, which is another way of saying that evolution has occurred.

FORCES OF EVOLUTION

Let us now look at the factors that disturb the Hardy-Weinberg equilibrium and produce changes in the gene pool: mutations, gene flow, genetic drift, and natural selection. We shall consider each in turn.

Mutations

Mutations are defined as inheritable changes in the genotype. A mutation may involve a large part of a chromosome or only a single nucleotide in a DNA molecule. As we saw in Chapter 15, the change in a single base pair is responsible for the abnormal hemoglobin associated with sickle cell anemia.

Mutations, as we noted previously, can be produced by exposure to x-rays, radioactive compounds, ultraviolet rays, and other agents. Most occur "spontaneously"—meaning simply that we do not know the reasons for them. The rate of spontaneous mutation is low; for humans, the chance of a mutation occurring in any given gene has been estimated at less than 1 in 10,000. However, given the large number of genes in every gamete, there is an excellent possibility that any given gamete will carry a mutation in at least one gene.

Different genes have different rates of mutation. The reason for these differences is not known but probably has to do with the chemical nature of the gene itself or with its position in the chromosome. Mutations can go in both directions; that is, a gene that mutates "forward" to the less common allele can also mutate "backward" to the more common, "normal" allele. From the point of view of the individual, such mutations may be all-important, but in the evolutionary perspective, the forward mutations and backward mutations may cancel each other out.

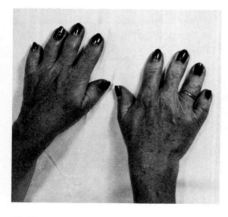

32–3

A dominant gene is responsible for the trait known as brachydactylism (short fingers). In the brachydactylous hands shown here, the first bones of the fingers are of normal length but the second and third bones are abnormally short. If brachydactylism is caused by a dominant gene, why is it such a rare trait?

An example of the sometimes dramatic effects of mutation. The ewe in the middle is an Ancon, an unusually short-legged strain of sheep. The first Ancon on record was born in the late nineteenth century into the flock of a New England farmer. By inbreeding (the trait was transmitted as a recessive), it was possible to produce a strain of animals with legs too short to jump the low stone walls that traditionally enclosed the New England sheep pastures. A similar strain was produced in northern Europe as a result of an independent mutation. At one time, it was thought that evolution took place in sudden, large jumps such as this—a concept sometimes referred to as the "hopeful monster" theory. One reason this concept was abandoned is that nearly all mutations producing dramatic changes in the phenotype are harmful, as this one would be in a wild population.

At one time, many students of evolution viewed mutations as an important element in determining the *direction* of evolutionary change (see Figure 32–4). Today, most population biologists believe that mutations are important only as the original source of the small cumulative variations upon which selection acts.

Gene Flow

Changes in the Hardy-Weinberg equilibrium of a population can, of course, be produced by the movement of new breeding individuals into the population or, as in the case of plants, the introduction of gametes (in the form of pollen) from other populations. In the human population, the effects of gene flow are particularly evident on the Hawaiian Islands, where the breaching of a once-formidable geographic barrier ended the long genetic isolation of the Polynesians who were the islands' original inhabitants. This phenomenon is often referred to as *gene flow*. As we shall see, gene flow often counteracts the effects of natural selection.

Genetic Drift

As we stated previously, the Hardy-Weinberg law holds true only if the population is large. This qualification is necessary because the Hardy-Weinberg equilibrium depends on the laws of probability. These laws—the laws of chance—apply equally as well to flipping coins, rolling dice, betting at roulette, and taking a poll to see who is going to win an election. In flipping coins, it is possible for heads to show up five times in a row, but on the average, heads will show up half the time and tails half the time; the more times the coin is flipped, the more closely the expected frequencies of half (0.50) and half (0.50) are approached. Similarly, in a presidential poll, it is possible that the first five people interviewed will be Republicans, and it is not until a sufficiently large sample of the voting population is queried that accurate predictions become possible. In a small population, as in a small sample, chance plays a large role.

Consider, for example, an allele, say *a*, that is present in 2 percent of the individuals. In a population of 1 million, 20,000 *a* alleles would be present in the gene pool. But in a population of 50, only one individual would carry this allele. If this individual failed to mate or were destroyed by chance before leaving offspring, allele *a* would be completely lost. (If a gene has two alleles, *A* and *a*, and if *a* is completely lost so that the gene pool consists only of *A*, *A* is said to be fixed.) Similarly, if 10 of the 49 without allele *a* were lost, the frequency would jump from 1 in 50 to 1 in 40. This phenomenon, a change in the gene pool that takes place as a result of chance, is known as *genetic drift*.

The Founder Principle

A small population that branches off from a larger one may or may not be genetically representative of the larger population from which it was derived. Some rare alleles may be overrepresented (like the Republicans in the poll) or may be lost completely. As a consequence, even when and if the small population increases in size, it will have a different genetic composition—a different gene pool—from that of the parent group. This phenomenon is known as the *founder principle.* In a small population, there is likely to be less variation in the population. Thus, even though matings may be at random, the mates will be more closely related to each other—more similar genetically—with a consequent increase in homozygosity.

An example of the founder principle is found in the Old Order Amish of Lancaster, Pennsylvania (Figure 32–5). Among these people, there is an unprecedented frequency of a gene which, in the homozygous state, causes a combination of dwarfism and polydactylism (extra fingers). Since the group was founded in the early 1770s, some 61 cases of this rare congenital deformity have been reported, about as many as in all the rest of the world's population. Approximately 13 percent of the persons in the group, which numbers some 17,000, are estimated to carry this rare mutant gene.

The entire colony, which is virtually isolated from the rest of the world, is descended from three couples. By chance, one of the six must have been a carrier of this gene. Inbreeding in itself does not cause disease or deformity. However, when recessive alleles are present, they are more likely to be expressed in an inbreeding population. We shall discuss the importance of the founder principle in more detail in Chapter 34.

Genetic Drift and Non-Darwinian Evolution

Changes in the gene pool that occur as a result of random mutations and that increase or become fixed as a result of genetic drift are sometimes referred to as non-Darwinian evolution. Now that scientists are beginning to analyze the genotype at the molecular level, more and more variations are being found that might be "neutral"—that is, that might have no effect on the phenotype. Such variations will be described in more detail in the following chapter. Whether or not any variation can, in fact, be neutral and the extent to which changing frequencies of variation can be caused merely by random sampling (genetic drift) are matters currently under debate by population biologists.

Natural Selection

Natural selection is the fourth and by far the most important agent of evolution. Darwin coined the term "natural selection" as an analogy to what he called "artificial selection," the process by which breeders of domestic animals or crops

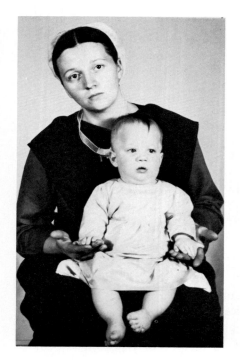

32–5
Among the Old Order Amish, a group founded by only a few couples some 200 years ago, there is an unusually high frequency of a rare allele. In its homozygous state, the allele results in extra fingers and dwarfism. This Amish child is a six-fingered dwarf.

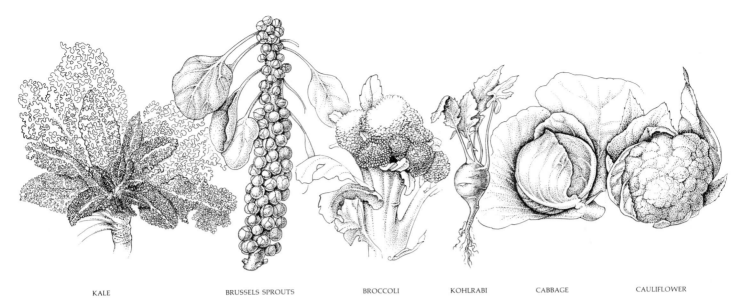

KALE   BRUSSELS SPROUTS   BROCCOLI   KOHLRABI   CABBAGE   CAULIFLOWER

32-6

Six vegetables produced from a single species of plant (Brassica oleracea, a member of the mustard family). They are the result of selection for leaves (kale), lateral buds (brussels sprouts), flowers and stem (broccoli), stem (kohlrabi), enlarged terminal buds (cabbage), and flower clusters (cauliflower). Kale most resembles the wild plant. This array of different varieties reveals the variation latent within a single species.

32-7

A white-barked pine growing near the timberline on a mountain in California. Genotypically a tall straight tree, the constant strong winds in which it has grown have produced this phenotype.

select certain individuals and discard others as they seek to develop desirable characteristics in their stock. In Darwin's time, pigeon fanciers had produced a number of different and often very exotic breeds by means of artificial selection. In our own time, a familiar example is found among "purebred" dogs. Another example is shown in Figure 32–6.

It is important to emphasize that selection acts upon the phenotype, the physical expression of the genotype. The phenotype, you will recall, is the result of the interactions of the genes of an individual with one another and with the external environment in the course of the individual's life (Figure 32–7). Natural selection is the process of interaction between an organism and its environment that results in the differential rate of survival and reproduction of different phenotypes in a population. It may result, in time, in changes in relative frequencies of alleles in the gene pool—that is, in evolution.

Types of Selection

Three general types of selection operate within populations: stabilizing, disruptive, and directional (Figure 32–8).

Stabilizing selection, a process that goes on at all times in all populations, is the continual elimination of extreme individuals. Most mutant forms are probably immediately weeded out in this way, many of them in the zygote or the embryo. Clutch size in birds is an example of stabilizing selection. Clutch size (the number of eggs a bird lays) is determined genetically. As you can see in Table 32–1, it is a disadvantage, in terms of surviving young, for the Swiss starling to have a clutch size of less than 4 or more than 5. The female whose genotype

Table 32–1 *Survival in Relation to Number of Young in Swiss Starling*

Number of young in brood	1	2	3	4	5	6	7	8
Number of young marked	65	328	1,278	3,956	6,175	3,156	651	120
Number of individuals recaptured after 3 months, per 100 birds ringed	—	1.8	2.0	2.1	2.1	1.7	1.5	0.8

The three different types of natural selection. The dots represent individuals that failed to reproduce or that left less than the average number of offspring. Stabilizing selection involves the elimination of extremes. In disruptive selection, intermediate forms are eliminated, producing two divergent populations. Directional selection, which is the gradual elimination of one phenotype in favor of another, produces adaptive change. In these graphs, the vertical axes denote the proportion of individuals in a population with a particular characteristic, and the horizontal axes, the varying dimensions of whatever characteristic is being considered.

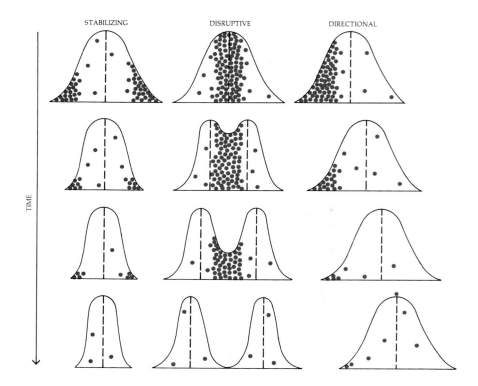

32-9

Male and female black widow spiders are conspicuously different in size, a result of disruptive selection. Here the tiny male is courting the female. He will deposit a drop of semen on the web, pick it up with a special receptacle on one of his legs, and insert it into the female. After the ceremony, he will probably be eaten.

dictates a clutch size of 4 or 5 will have more surviving young, on the average, than a member of her species that lays more or fewer eggs.

A somewhat similar example is seen in a study that correlated birth weight and survival of 13,730 babies born in London over a ten-year period (1935-1945). As you can see in Figure 32-10, the optimum weight for this population of babies was about 3.8 kilograms. At this birth weight, 98.2 percent of the babies survived. The overall survival rate was 95.5 percent. To the extent that birth weight of offspring is determined by heredity, stabilizing selection was at work in this population to maintain this birth weight.

The second type of selection, *disruptive selection*, occurs when two extreme types in a population increase at the expense of intermediate forms. Disruptive selection is probably a major factor in sexual dimorphism—the differences between males and females of the same species. Another example of disruptive selection is given on page 535, in the discussion of mimics and models.

The third type, which is the one we shall be most concerned with, is *directional selection*. Directional selection acts either for or against an extreme phenotypic characteristic and so is likely to result in the gradual replacement of one allele by another in the gene pool.

The phrase "survival of the fittest" is often used in describing the Darwinian theory. In the early twentieth century, the doctrine of survival of the fittest in natural populations was used by some businessmen to defend ruthless competitive tactics in industry on the grounds that they were merely following a "law" of nature. This business philosophy was referred to by some as social Darwinism. However, in actual fact, very little in the process of change fits the concept of "nature red in tooth and claw." One fuchsia flowering a little brighter than its neighbors and better able to catch the attention of a passing hummingbird is a more pertinent model of the struggle for survival. There is only one criterion of fitness as it is measured by population geneticists: the relative number of descendents of an individual in a future population.

32–10

Relationship between weight at birth and survival in a particular group of human babies. The death rate is higher for babies weighing either more or less than about 3.8 kilograms.

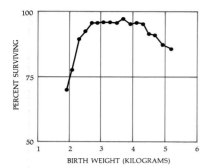

SUMMARY

Popular genetics is a synthesis of the Darwinian theory of evolution with the Mendelian principles of genetics. A population, for the population geneticists, is an interbreeding group of organisms, defined and united by its gene pool (the sum of all the alleles of all the genes of all the individuals in the population). Evolution is, by definition, a change in the composition of the gene pool.

The Hardy-Weinberg law is the application of Mendelian genetics to population biology. It states that the proportion of different alleles in a population will remain the same if (1) no mutations occur, (2) there is no gene flow, (3) the population is large, and (4) there is no natural selection. For a gene with two alleles, the mathematical expression of the Hardy-Weinberg equilibrium is $(p + q)^2 = 1$, which represents the proportion of genotypes in the population, where p equals the frequency of one allele and q equals the frequency of the other allele. Thus, in any given population, the homozygotes for one allele are represented by p^2 or q^2, and the heterozygotes by $2pq$.

The Hardy-Weinberg equilibrium applies only if the population is large enough. In a small population, chance events may cause the frequency of certain alleles to increase or to decrease and perhaps even disappear; this phenomenon is known as genetic drift. A small population that has branched off from a larger population may not be a representative sample of the larger population; this phenomenon is known as the founder principle.

Natural selection is the cause of changes in the Hardy-Weinberg equilibrium resulting from differences in the number of offspring left by different organisms because of variations in their phenotypes. It may take the form of stabilizing selection (elimination of extreme individuals), disruptive selection (elimination of intermediate forms), or directional selection (elimination of an extreme phenotype by another). Of these three, the first is the most frequent and the third is primarily responsible for the changes we usually associate with evolution.

QUESTIONS

1. Define the following terms: allele, gene frequency, natural selection, evolution.

2. In terms of molecular genetics, what is a possible explanation for the many different eye colors in *Drosophila*?

3. What is the difference between gene flow and genetic drift? How do they each affect the gene pool of a population?

4. How does stabilizing selection affect the gene pool? Directional selection?

5. Directional selection can occur without any change in the gene pool. Define the circumstances under which this might take place.

6. What characteristics in the human population are probably being maintained by stabilizing selection? Which might be subject to directional selection?

CHAPTER 33

Variability: Its Extent, Preservation, and Promotion

As we emphasized previously, Darwin was the first to recognize the importance of widespread, inheritable variations in the process of evolution. Much of the research in modern population genetics is concerned with the extent of such variability (far greater than Darwin could have realized) and with the way variations are preserved and fostered in the gene pool.

THE EXTENT OF VARIABILITY

Darwin's awareness that variations exist among individuals in a population was based largely on his observations as a naturalist, both in England and during his voyage on the *Beagle*. He also recognized that the traits that emerged in the course of selective breeding of domestic animals and plants reflected variability somehow latent in the gene pool. This important hypothesis of Darwin has now been confirmed by a variety of laboratory investigations.

33–1

Even in populations in which the individuals appear almost identical, such as the members of this group of gannets, we know that variations exist because individuals have no difficulty in identifying their own mates or offspring. To gannets, all people probably look alike.

The results of an experiment with Drosophila melanogaster, demonstrating the extent of latent variability in a natural population. From a single parental stock, one group was selected for an increase in the number of bristles on the ventral surface (high selection line) and one for a decrease in the bristle number (low selection line). As you can see, the high selection line rapidly reached a peak of 56, but then the stock began to become sterile. Selection was abandoned at generation 21 and begun again at generation 24. This time, the previous high bristle number was regained and there was no apparent loss in reproductive capacity. Note that after generation 24 the stock interbreeding without selection was also continued, as indicated by the line of color. After 60 generations the freely breeding group from the high selection line had 45 bristles. The low selection line died out owing to sterility.

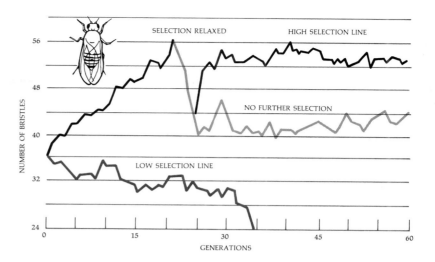

Bristle Number in Drosophila

In one group of studies, for example, the extent of latent variability in a natural population was demonstrated in the laboratory by experiments with the fruit fly *Drosophila melanogaster*. An easily observable hereditary trait, the number of bristles on the ventral surface of the fourth and fifth abdominal segments, was chosen for study. In the starting stock, the average number of bristles was 36. Two selection groups were run, one for increase of bristles and one for decrease. In every generation, individuals with the fewest bristles were selected and crossbred, and so were individuals with the highest number of bristles. Selection for low bristle number resulted in a drop after 30 generations from 36 to an average of 30 bristles. In the high-bristle-number line, progress was at first rapid and steady. In 21 generations, bristle number rose steadily from 36 to an average of 56 (Figure 33–2). No new genetic material had been introduced. It was apparent that within the single population, a wide range of variety existed. Subsequent experiments in *Drosophila* and other organisms have shown that the choice of bristle number was not merely a fortunate accident. No matter what single characteristic is selected for in breeding experiments, all reveal a comparable range of natural variability.

There is a second part to the bristle-number story. The low-bristle-number line soon died out, owing to sterility. Presumably, changes in factors affecting fertility had also taken place during selection. When sterility became severe in the high-bristle line, a mass culture was started; members of the high-bristle line were permitted to interbreed without selection. The average number of bristles fell sharply, and in five generations went from 56 to 40. Thereafter, as this line continued to breed without selection, the bristle number fluctuated up and down, usually between 40 and 45, which still was higher than the original 36. At generation 24, selection for high bristle number was begun again for a portion of this line. The previous high bristle number of 56 was regained, and this time there was no loss in reproductive capacity. Apparently, the genotype had rearranged and reintegrated itself so that the genes controlling bristle number were present in more favorable combinations.

Mapping studies have shown that bristle number is controlled by a large number of genes, at least one on every chromosome and sometimes several at different sites on the same chromosome. Therefore, although we do not know how important bristle number is to the survival of the animal, selection for this

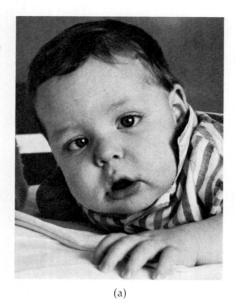

(a)

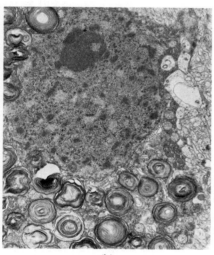

(b)

33-3

Certain diseases are the result of the coming together of two recessive alleles. Even though this condition is fatal in the homozygote, the allele will continue to persist in the population because it is protected in the heterozygous state by diploidy. (a) In this one-year-old infant with Tay-Sachs disease, deterioration of the brain, already begun, will progress rapidly. The child will probably die before he is four years old. The disease is caused by absence of an enzyme involved with lipid metabolism. (b) In the absence of the enzyme, harmful fatty deposits accumulate on the brain. Among American Jews, there is 1 chance in 27 of carrying an allele for this disease.

trait in some way disrupted the entire genotype. Livestock breeders are well aware of this consequence of artificial selection. Loss of fertility is a major problem in virtually all circumstances in which animals have been purposely inbred for particular traits. This result emphasizes the fact that in natural selection it is the entire phenotype that is selected rather than certain isolated traits, as is often the case in artificial selection.

VARIABILITY IN GENE PRODUCTS

Molecular biology provides another method for studying latent variability. Genes, as we saw in Section 3, code for and regulate the productions of proteins. Proteins therefore provide clues about the genes coding for them. J. L. Hubby and R. C. Lewontin ground up a population of fruit flies and extracted proteins from them. From these proteins they were able to isolate 18 different enzyme groups, which they detected on the basis of each group's enzymatic activity. Then they analyzed each enzyme group separately to see if it was composed of a single protein or of proteins that were slightly different structurally.

Of the 18 enzyme groups studied in this way, nine were found to be composed of structurally identical proteins; in other words, the genes that had produced these proteins were the same throughout the entire population of fruit flies studied. However, nine of the groups contained enzymes that differed structurally within each group. Therefore, without any direct analysis of the genes themselves, the investigators were able to conclude that among the fruit flies studied, there were two or more alleles of the gene responsible for each of the nine enzyme groups. In one group of enzymes, there were as many as six slightly different structural forms; that is, six alleles for the gene coding for that enzyme group were shown to exist in the species as a whole.

Each fruit-fly population tested was heterozygous for almost a third of the genes tested. Each individual, it was estimated, was probably heterozygous for about 12 percent of its genes. Similar studies in human populations, using an accessible tissue such as blood or placenta, indicate that, in any given group, at least 25 percent of the genes are represented by two or more alleles and individuals are heterozygous for at least 7 percent of their genes, on the average.

Explaining the Extent of Variation

The experiments of Hubby and Lewontin and those patterned after them disclosed a far greater extent of variability than had been previously imagined and, like most important scientific discoveries, raised major new questions. Most geneticists had previously thought that the individuals of a population should be close to genetic uniformity, as a result of a long history of selection for "optimal" genes. Yet, as these studies revealed, there is a great deal of genetic variation present in natural populations. One school of geneticists, the "selectionists," claim that this variation is maintained by a balance of forces of natural selection which favor some genotypes at some times and in some areas and others at other times or in other localities. An opposing school, the "neutralists," claim that the observed variations in the protein molecules are so slight that they do not make any difference in the function of the organism and so are not affected by natural selection. This latter group believes that the variations result from the accumulation over a long period of time of random mutations, many of which are lost from the population by genetic drift, but some of which persist.

PRESERVATION OF VARIABILITY

Diploidy

The most important factor in the preservation of variability in eukaryotes is diploidy. In a haploid organism, any genetic variations are immediately exposed to the selection process, whereas in the diploid organism, such variations may be stored as recessives, as with the allele for white flowers in Mendel's pea plants. The extent to which a rare allele is protected is revealed by the following table:

FREQUENCY OF ALLELE a IN GENE POOL	GENOTYPE FREQUENCIES			PERCENTAGE OF a IN HETEROZYGOTES
	AA	Aa	aa	
0.9	0.01	0.18	0.81	10
0.1	0.81	0.18	0.01	90
0.01	0.9801	0.0198	0.0001	99

As you can see, the lower the frequency of allele a, the smaller the proportion of it exposed in the homozygotes becomes, and the removal of the allele by natural selection slows down accordingly (see also Figure 33–4). This result is of special interest to students of eugenics, the improvement of the human gene pool through controlled breeding. For instance, consider a human genetic disorder expressed only in the homozygous recessive (like PKU) with a frequency of about 0.01 in the human population; individuals with the aa genotype make up 0.0001 of the population (1 child for every 10,000 born). You can calculate that it would take 100 generations, roughly 2,500 years, of a program of sterilization of defective homozygote individuals to halve the gene frequency and reduce the number to 1 in 40,000.

Suppose, however, that the recessive allele in the homozygous state conferred an advantage under certain environmental circumstances or in conjunction with another particular set of genes. Because it would be protected against selection by appearing mostly in the heterozygous state, it would continue to recur and so by chance might appear under favorable selective circumstances.

33–4

The relationship between the frequency of allele a in the population and the frequency of the genotypes AA, Aa, and aa. Naturally, the more AA genotypes there are, the lower the frequency of a alleles. Because of the interrelationship of AA, Aa, and aa, a change in the frequency of one allele, say a, results in a corresponding and symmetrical change in the frequencies of the other allele, say A, and of the three genotypes.

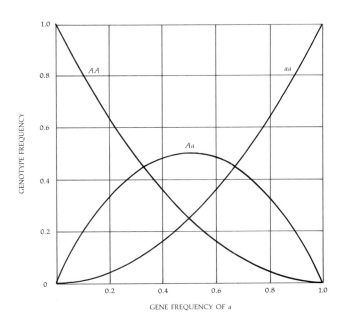

The Evolution of Diploidy

Once sexual reproduction was established among the unicellular eukaryotes (page 122), the stage was set for the evolution of diploidy. By "accident"—an accident that apparently took place in a number of separate evolutionary lines—the zygote, originally the only diploid stage of the organism's life cycle, divided mitotically instead of meiotically, producing an organism with two complete sets of chromosomes. The fact that all organisms we consider most highly evolved are diploid is an indication of the important role of diploidy in the evolutionary process.

Heterozygote Superiority

Recessive alleles, even though they may be harmful in the homozygous state, may not only be sheltered in the heterozygous state, but may actually be selected for. This phenomenon, known as *heterozygote superiority*, is another way that variability is preserved. A dramatic example of heterozygote superiority is found in association with sickle cell anemia. Individuals homozygous for sickling almost never live to maturity. Therefore, almost every time one sickling gene encounters another, two sickling genes are removed from the population. At one time, it was thought that the sickling gene was maintained in the population by a steady influx of new mutations. Yet in some African tribes, as much as 45 percent of the population is heterozygous for sickling, and to replace the loss of sickle genes by mutations alone would require a rate of about 1 mutant per 100 genes. This is about 5,000 times greater than any other known mutation rate in man.

In the search for an alternative explanation, it was discovered that the sickling allele is maintained at high frequencies because the heterozygote has a selective advantage. In many African tribes, malaria is one of the leading causes of illness and death, especially among young children. Studies of the incidence of malaria among young children showed that susceptibility to malaria is significantly lower in individuals heterozygous for sickling. Moreover, for reasons that are not known, women who carry the sickling gene are more fertile.

Among blacks in the United States, only about 9 percent are heterozygous for the sickling allele. Since no more than half of this loss of the sickling allele can be explained by the black-white admixture in America, the conclusion is that once selection pressure for the heterozygote is relaxed, the mutant will tend to be eliminated slowly from the population.

Other genes that are harmful in the homozygous state appear to be maintained in the population by heterozygote superiority. For example, again for reasons that are not known, women who are carriers of hemophilia (heterozygotes) have a fertility rate about 20 percent greater than that of the rest of the human female population.

PROMOTING VARIATION

Sexual Reproduction

By far the most important method by which organisms promote variation in the phenotype of their offspring is by sexual reproduction. As we saw in Chapter 9, sexual reproduction combines parental genes in new ways so that wholly new genotypes appear in the next generation. Assuming that we are correct in our assertion that efficient utilization of energy and other resources is under selective pressure, consider the expenditure of energy and resources involved in sex. In the words of Edward O. Wilson:

33–5
Plants such as the violet shown here reproduce both sexually and asexually. The larger flowers are cross-pollinated (sexual reproduction) by insects, and the windborne seeds bearing new alleles in new combinations are carried some distance from the parent plant.

The smaller flowers, closer to the ground, are self-pollinated and never open. Seeds from these flowers will carry only the alleles of the parent plant, although they may be present in new combinations. These asexually produced seeds drop close to the parent

plant, most likely in a very similar environment. Underground stems can also produce a new series of genetically identical plants close to the parent by vegetative (asexual) reproduction.

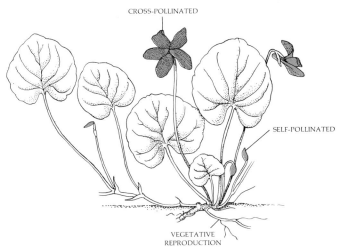

CROSS-POLLINATED

SELF-POLLINATED

VEGETATIVE
REPRODUCTION

33–6
Like many solitary slow-moving animals, slugs are hermaphrodites. However, they seldom self-fertilize. The photo shows two slugs mating, suspended from a branch by a cord of mucus.

Sexual reproduction is in every way a consuming biological activity. Reproductive organs tend to be elaborate in structure, courtship activities lengthy and energetically expensive, and genetic mechanisms of sex determination finely tuned and easily wrecked. But more important, a female that elects to reproduce by sex cuts her genetic contribution to each gamete by one-half, without, in the vast majority of species, receiving any material aid from the male. If an egg develops parthenogenetically without meiosis—the simplest way to proceed—all of the genes in the resulting offspring will be identical to those of the mother. In sexual reproduction, which entails reduction division during meiosis, only half are identical. The female, in other words, has thrown away half of her investment.*

Note that the only selective advantage conferred by sexual reproduction is the production of totally new phenotypes that are available to be tested in the next generation.

Some eukaryotic species, such as amoebas and *Euglena*, reproduce only asexually. Most students of the evolution of genetic systems agree, however, that all eukaryotes did reproduce sexually at one time but that some species eventually lost this capacity in the course of their evolutionary history. In effect, in balancing the costs, they opted for the stability conferred by asexual reproduction. Many modern organisms, particularly members of the plant kingdom, hedge their bets, reproducing both sexually and asexually (Figure 33–5).

Mechanisms That Promote Outbreeding

Many ways have evolved by which new genetic combinations are promoted in sexually reproducing populations. Among animals, even in those invertebrates that are hermaphrodites (each organism produces both egg and sperm cells), such as earthworms, snails, and slugs (Figure 33–6), an individual seldom fertilizes its own eggs. Among plants, there is a variety of mechanisms that ensure

* E. O. Wilson, *Science,* **188**:1139, 1975.

33–7

Flowers have various morphological adaptations that prevent self-pollination. Orchid pollen is bound together in a mass, called a pollinium, which contains a sticky extension that attaches the pollinium to the pollinating insect. In a bee-pollinated orchid, a bee with
a pollinium on her back lands on the lip of the flower and crawls inside to gather the nectar. As she backs out, the flap separating the stigma from the pollen mass first scrapes the transported pollinium off her back onto the sticky surface of the stigma (from which
the pollen sperm enter the style and fertilize the egg cell in the ovule). Next, as she pushes the anther cap outward, the pollinium from the flower she is leaving becomes deposited on her back and is thus carried to another flower.

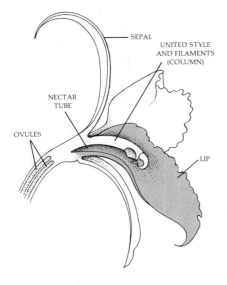

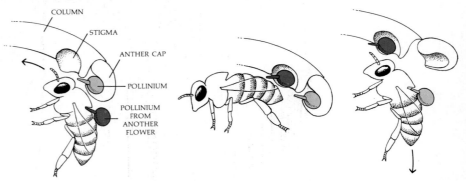

that the sperm-bearing pollen is from a different individual than the stigma it lights upon. Many plants, such as the holly and the ginkgo, have male flowers on one tree and female on the other. In others, such as the avocado, the pollen of a particular plant matures at a time when its own stigma is not receptive. In some species, anatomical arrangements inhibit self-pollination (Figures 33–7 and 33–8).

Some plants have genes for self-sterility. Typically, such a gene has multiple alleles—s^1, s^2, s^3, and so on. A plant carrying the allele s^1 cannot pollinate a plant with an s^1 allele, one with an s^1/s^2 genotype cannot pollinate any plant with either of those alleles, and so forth. In one population of about 500 evening primrose plants, 37 different self-sterility alleles were found, and it has been estimated that there are more than 200 alleles for self-sterility in red clover. As a consequence, a plant with a rare self-sterility allele (which is nearly always accompanied by a rare genotype) is more likely to be able to pollinate another plant than a plant with a common self-sterility allele. Such a system strongly encourages variability in a population: selection for the rare allele makes it more common, whereas more common alleles become rarer.

The existence of such a wide variety of mechanisms, some of them very elaborate, reaffirms the selective value of genetic variability.

33–8

Diagrams of a pin type and a thrum type of a primrose. Notice that the pollen-bearing anthers of the pin flower and the pollen-receiving stigma of the thrum flower are both situated about halfway up the length of the flower and that the pin stigma is level with the thrum anthers. An insect foraging for nectar in these plants would collect pollen in different areas of its body, so that thrum pollen would be deposited on pin stigmas, and vice versa.

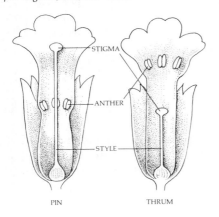

TYPES OF VARIABILITY

Variations within populations may be continuous or polymorphic. Continuous variation, as we noted in Section 3, is characteristic of traits, such as height and weight, that are governed by several or more genes (polygenic inheritance). *Polymorphism* is the coexistence within a population of two or more phenotypically different forms with no intermediate forms connecting them. There are many examples in nature, of which we shall cite only three.

Polymorphism

Color and Banding in Snails

One of the best-studied cases of polymorphism is found among land snails of the genus *Cepaea*. In one species (*Cepaea nemoralis*), for instance, the shell of the

(a)

(b)

33–9

(a) *Polymorphism in land snails: a banded and an unbanded snail.* (b) *An "anvil," where song thrushes break land snails open in order to obtain the soft, edible parts. From the evidence left by the empty shells, investigators have been able to show that in areas where the background is fairly uniform, unbanded snails have a survival advantage over the banded type. Conversely, in colonies of snails living on dark, mottled backgrounds (such as woodland floors), banded snails are preyed upon less frequently.*

snail may be yellow, brown, or any shade from pale fawn through pink and orange to red. The lip of the shell may be black or dark brown (normally) or pink or white (rarely), and up to five black or dark-brown longitudinal bands may decorate it (Figure 33–9). Fossil evidence shows that these different types of shells have coexisted more than 10,000 years.

Studies among English colonies of *Cepaea nemoralis* have revealed some of the selective forces at work in some of the colonies. An important predator of the snail is the song thrush. Song thrushes select snails from the colonies and take them to nearby rocks, where they break them open, eating the soft parts and leaving the shells. By comparing the proportions of types of shells around the thrush "anvils" with the proportions in the nearby colony, investigators have been able to show correlations between the types of snails seized by the thrushes and the habitats of the snails. For instance, of 560 individuals taken from a small bog near Oxford, 296 (52.8 percent) were unbanded; whereas of 863 broken shells collected from around the rocks, only 377 (43.7 percent) were unbanded.* In other words, in bogs, where the background is fairly uniform, unbanded snails are less likely to be preyed upon than banded ones.

Studies of a wide variety of colonies have confirmed these correlations. In uniform environments, a higher proportion of snails is unbanded, whereas in rough, tangled habitats, such as woodland floors, far more tend to be banded. Similarly, the greenest habitats have the highest proportion of yellow shells, but among snails living on dark backgrounds, the yellow shells are much more visible and are clearly disadvantageous, judging from the evidence conveniently assembled by thrushes. In other words, polymorphism in this situation is produced by a balance of selection pressures. The agent of selection in this case is the predaceous song thrush.

Blue and White Snow Geese

At one time the lesser white snow goose (*Chen hyperborea*, "the goose from beyond the north wind") and the blue goose (*Chen caerulescens*) were believed to represent distinct species. More recently it has been found that they actually represent one species, polymorphic for color. The white is recessive; heterozygotes and homozygotes for blue are both the same dark blue color. However, birds mate preferentially with animals of their own color, probably because of imprinting (page 543). As a result, there are more homozygotes than if mating were random.

Blue and white geese differ significantly in their physiological adaptations for breeding. One major difference is that the white geese, which breed earlier, are less visible on snow, whereas the later-breeding blue geese are better camouflaged on the muddy ground that follows. Apparently, the chances for the survival of young are approximately equal, so both the blue and white varieties are maintained.

However, as the snow geese make their long fall migration from the Arctic to the Gulf of Mexico, they are hunted intensively by man. The white-plumaged birds, which are more conspicuous and also more highly prized as trophies, are shot in much larger numbers than those with blue plumage, thus introducing a new selection pressure.

* This difference (about 10 percent) may seem too small to be significant, but actually the number of shells counted was sufficiently large that there is only 1 chance in 1,000 that this difference could have occurred by chance.

Blood Groups in Man

The human blood types A, B, AB, and O are familiar examples of polymorphism. Apparently, the three alleles associated with these blood types are a part of our ancestral legacy, since the same blood types are also found in other primates. Some population biologists regard the blood types as probably neutral in terms of their selective value. Others maintain that polymorphism in human blood groups is a result of selection factors. For example, among white men, the life expectancy is greatest for those with group O and least with group B; exactly the opposite is true for white women with these blood types. Persons with type A blood run a relatively higher risk of cancer of the stomach and of pernicious anemia. Men with type B blood are about 5 centimeters taller, on the average, than those with other blood types. Persons with type O blood have a higher risk of duodenal ulcers. Thus there are selective pressures at work on each of the groups that apparently serve to keep them in balance. The geographic distributions of the A, B, AB, and O groups are irregular and may reflect some differences in the selective forces favoring particular blood groups under particular conditions.

Geographic Variation: Clines

Sometimes variations within the same species follow a geographic distribution and are correlated with temperature, humidity, or some other environmental condition. Such a graded variation in a trait or a complex of traits is known as a *cline*.

Many species exhibit north-south clines of various characteristics. Small mammals, for example, tend to have a smaller body size in the warmer parts of the range of the species and a larger body size in the cooler parts; in cooler regions, a larger body size is advantageous for heat conservation. Conversely, tails, ears, bills, and other extremities of animals are relatively longer in the warmer areas of a species' range; such adaptations allow for heat radiation from the animal. Plants growing in the south often have slightly different requirements for flowering or for ending dormancy than the same plants growing in the north, although they all may belong to the same species. Most geographic variations, however, are too irregular to be classified into clines.

A species that occupies many different habitats may appear to be slightly different in each one. Each group of distinct phenotypes is known as an *ecotype*. Are the differences among ecotypes determined entirely by the environment or are there genetic differences, too? Figure 33–10 illustrates a series of experiments

33–10

Ecotypes of Potentilla glandulosa, *a relative of the strawberry. Notice the correlation between the height of the plant and the altitude at which it grows; there are other phenotypic differences between the plants as well. When plants of the four geographical races are grown under identical conditions, many of these phenotypic differences persist and are passed on to the next generation, indicating that these plants are genotypically, as well as phenotypically, different.*

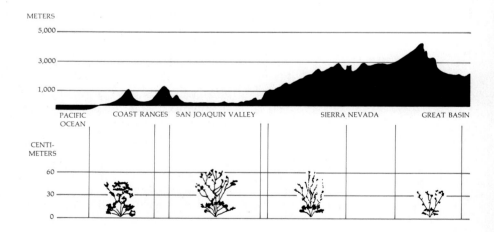

carried out in California under the auspices of the Carnegie Institution of Washington with the perennial plant *Potentilla glandulosa,* a relative of the strawberry. Experimental gardens were established at various altitudes, and wild plants collected from near each of the experimental sites were grown in all the gardens. (*Potentilla glandulosa,* like the strawberry, reproduces asexually by runners, making it possible to study genetically uniform groups.) Under these conditions it was possible to demonstrate that many of the phenotypic differences among the ecotypes of *Potentilla glandulosa* were due to genetic variations. It is not surprising that in their very different environments, different traits were selected for and so, over time, genetic differences arose.

Variations and the Origin of Species

Darwin believed, as do modern students of evolution, that the differences between individuals within a population could, with time, be translated into the differences between species. We shall explore this concept further in the next chapter.

SUMMARY

Natural populations can be shown, by a variety of methods, to harbor a wide spectrum of latent variations. Variability is preserved in populations by diploidy (in which an organism has two sets of chromosomes) and by heterozygote superiority (in which a recessive allele detrimental in its homozygous state confers an advantage in its heterozygous state). Variability is promoted by sexual reproduction and by mechanisms, such as self-sterility alleles and anatomic adaptations in plants, that discourage self-fertilizing and promote interbreeding.

Variability may be continuous, as in traits that are governed by several or more genes (polygenic inheritance), or polymorphic. Polymorphism is the coexistence within a population of two or more distinct phenotypes; it is the result of a balance of selection pressures. Variations that follow a geographic distribution are known as clines. Ecotypes are distinct groups of phenotypes of a species occupying different habitats.

QUESTIONS

1. Define the following terms: heterozygote superiority, sexual reproduction, self-sterility, polymorphism, cline.

2. In what sense does a female reproducing sexually throw away half of her investment? If you are not sure of the answer, review the diagram of meiosis, page 128. Does the male also throw away part of his "investment"?

3. In the experiment on bristle number in *Drosophila,* the average number of flies at the end of the experiment had 56 bristles. No fly at the beginning of the experiment had as many as 56, however. How do you explain this fact?

4. In bacteria, the exchange of genetic information and reproduction take place as separate events. In organisms other than prokaryotes, genetic recombination is generally linked with reproduction. Why do you think this might be so?

5. Describe in your own words, the steps in the Hubby-Lewontin experiment and the significance of the results.

CHAPTER 34

On the Origin of Species

Darwin believed evolution to be such a slow process that it could never be observed directly. However, the effects of human civilization have produced such extremely strong selection pressures on some organisms that it has been possible to observe not only the results but the actual process of selection.

EVOLUTION IN ACTION

The Peppered Moth

One of the most striking and best-studied examples is that of *Biston betularia*, the peppered moth. These moths were well known to British naturalists of the nineteenth century, who remarked that they were usually found on lichen-covered trees and rocks. Against this background, the light coloring of these moths made them practically invisible, concealing them from predatory birds. Until 1845, all reported specimens of *Biston betularia* had been light-colored, but in that year one black moth of this species was captured at the growing industrial center of Manchester.

With the increasing industrialization of England, smoke particles began to cover the foliage in the vicinity of industrial towns, poisoning the lichens and leaving the tree trunks bare. In heavily polluted districts, the trunks and even the rocks and the ground became black. During this period, more and more black *Biston betularia* were found. This mutant black form spread through the population until black moths made up 99 percent of the Manchester population.

Where did the black moths come from? Eventually, it was demonstrated that the black color is the result of a recurring mutation. Why did their numbers increase? Figure 34–1 indicates the answer. When England was unpolluted, the mutant black forms were more easily spotted by predators; in polluted areas, however, the light variety is at a disadvantage.

Until recently, only a few of the light-colored populations could be found, and these were far from industrial centers. Because of the prevailing westerly wind in England, the moths to the east of industrial towns tended to be of the black variety right up to the east coast of England, and the few light-colored populations were concentrated in the west, where lichens still grew. A similar tendency for dark-colored forms to replace light-colored forms has been found among some 70 other moth species in England and some 100 species of moths in the Pittsburgh area of Pennsylvania.

The two varieties of Biston betularia, *the peppered moth, resting on (a) a lichen-covered tree trunk near Birmingham, England, and (b) a dark tree trunk, near Manchester. A striking example of an evolutionary change resulting from a drastic environmental change, the black form of the moth began to appear in the latter part of the nineteenth century as the English countryside became increasingly polluted by industrial smoke. Note that there are two moths in each photograph.*

(a)

(b)

Recent pollution controls instituted in Great Britain have considerably reduced the particulate content of smoke. The light-colored moths are already increasing in proportion to the black forms, but it is not yet known whether a complete reversal either in pollution or in adaptive coloration will come about.

Insecticide Resistance

Chemicals poisonous to insects, such as DDT, were originally hailed as major saviors of the health and property of mankind. They have fallen into disfavor not only because of their tendency to accumulate in the environment, but because of the extraordinary increase in resistant strains of insects. In fact, one insect is known to be able to remove a chlorine atom from a DDT molecule and use the remainder as food.

A particularly striking example of insecticide resistance has been found in the scale insects that attack citrus trees in California. In the early 1900s, a concentration of hydrocyanic gas sufficient to kill nearly 100 percent of the insects was applied to orange groves at regular intervals with great success. By 1914, orange growers near Corona began to notice that the standard dose of the fumigant was no longer sufficient to destroy one type of scale insect, the red scale. A concentration of the gas that had left less than 1 in 100 survivors in the nonresistant strain left 22 survivors out of 100 in the resistant strain. By crossing resistant and nonresistant strains, it was possible to show that the two differed in a single gene. The mechanism for this resistance is not known, but one group of experiments has shown that the resistant individual can keep its spiracles closed for 30 minutes under unfavorable conditions whereas the nonresistant insect can do so for only 60 seconds.

MACROEVOLUTION

Can minor changes, such as those observed in populations of peppered moths, add up over the eons to the major changes associated with the emergence of entirely new forms? The evidence that this is indeed the case is found in the fossil record.

34–2

Phyletic evolution: the modern horse and some of its ancestors. Over the past 60 million years, small several-toed browsers, such as Eohippus, were replaced in gradual stages by members of the genus Equus, characterized by, among other features, a larger size, broad molars adapted to grinding coarse grass blades, a single toe surrounded by a tough, protective keratin hoof, and a leg in which the bones of the lower leg had fused, with joints becoming more pulleylike and motion restricted to a single plane. Only a few of the evolutionary lines are shown here.

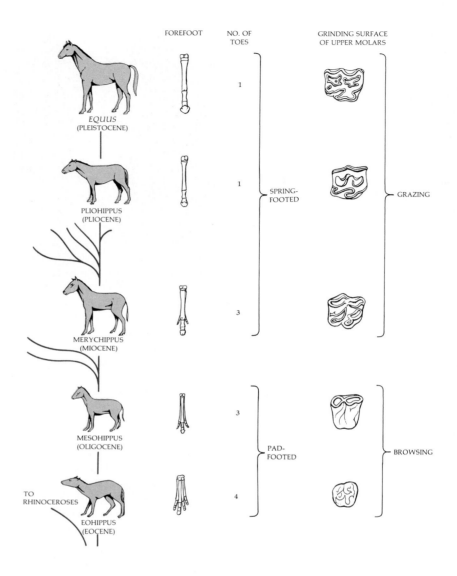

Paleontologists—students of fossils—recognize three principal patterns of organic evolution: phyletic evolution, splitting evolution, and adaptive radiation. As we shall see, these three patterns are not really so distinct as they first appear to be.

Phyletic Evolution

Phyletic evolution describes the changes taking place in a single line of organisms over a long period of time. One of the best-known examples is the evolution of the horse (Figure 34–2), whose history is abundantly represented in the fossil record. At the beginning of the Tertiary period, some 65 million years ago, the horse lineage was represented by Eohippus ("dawn horse"). It was a small herbivore (25 to 50 centimeters high at the shoulders) with three toes on its hind feet, four toes on its front feet, and doglike footpads on which its weight was carried. Its eyes were halfway between the top of its head and the tip of its nose, and its teeth had small grinding surfaces and low crowns, which probably could not have stood much wear. Its teeth indicate that it did not eat grass; in fact, there was probably not much grass to eat. It probably lived on succulent leaves. Eohippus evolved by stages into a group of larger, three-toed herbivores. Some of these lines survived, and some did not. The molar teeth of some of the successful lines had large crowns that continued to grow as they were worn away (as do those of modern horses).

During the Pliocene, horses, of which there were still several distinct species, became still larger. The middle or third digit of each foot expanded and became the weight-bearing part of the foot. Some of the species remained three-toed, while in others the other digits became greatly reduced. The three-toed species did not survive, but those in whom the digits were greatly reduced—to one—did, and were the direct ancestors of the modern horse.

By hindsight, it is possible to correlate these changes with changes in the environment and so to interpret them as adaptive. In the time of little Eohippus, the land was marshy and the chief vegetation was leaves; the teeth of Eohippus were adapted for browsing. By the Miocene, the grasslands began to spread; groups of horses whose teeth became adapted to grinding grasses (which have much coarser blades than leaves) survived, whereas those who remained browsers did not. The placement of the eye higher in the head may have facilitated watching for predators while grazing. The climate became drier and the ground became harder; reduction of the number of toes, with the development of the spring-footed gait characteristic of the modern horse, was an adaptation to harder ground and to larger size. An animal twice as high as another tends to weigh about eight times as much. Little Eohippus was probably as fast as the modern horse, but a larger, heavier horse with the foot and leg structure of Eohippus would have been too slow to escape from predators. (During this same period predators were developing adaptations that rendered them better able to catch large herbivores, including horses.)

Thus the evolution of *Equus,* viewed from a long retrospective of the geological record, represents a fairly straightforward accumulation of adaptive changes related to pressure for increased size and for grazing.

Splitting Evolution

A second mode of evolution involves the splitting of phylogenetic lineages. For example, *Ursus arctos*—the brown bear, a species that also includes the Kodiak and grizzly bears—is distributed throughout the northern hemisphere, ranging through the deciduous forests up through the taiga and into the tundra, as it was some 1½ million years ago. As is characteristic of such widespread species, there are many local ecotypes or races. During one of the massive glaciations of the Pleistocene, a population of *Ursus arctos* was split off from the main group and, according to the fossil evidence, this group, under extreme pressure from the harsh environment, evolved into the polar bear, *Ursus maritimus* (Figure 34–3). Brown bears, although they are members of the order of carnivores and closely related to dogs, are mostly vegetarians, supplementing their diet only occasionally with fish and game. The polar bear, however, is almost entirely carnivorous, with seals being its staple diet. It spends most of its time in the water and, unlike the brown bear, does not hibernate. The polar bear differs physically from the other bears in a number of ways, including, besides its white color, its carnivore-type teeth, its streamlined head and shoulders, and the stiff bristles that cover the soles of its feet, providing insulation and traction on the slippery ice.

34–3
The white coloring of the polar bear is probably related to its need for camouflage while hunting rather than a need to hide from predators since it is one of the world's largest carnivores. (Its greatest enemy of all time is modern man, who, prizing the skins for trophies, has taken to hunting the bears by helicopter.) Unlike their southern cousins, the brown bears, polar bears do not hibernate (except for females with young) but roam the ice and icy waters all winter long, in search of fish, seals, young walruses, and other game.

The cave bear *(Ursus spelaeus),* which was twice as large as any existing bear, is pictured in cave drawings of early man but is now extinct. It also appears from fossil remains to have originated by splitting evolution from *Ursus arctos.*

Adaptive Radiation

Adaptive radiation is believed by many experts to be the major pattern of evolution. It involves the sudden (in geologic time) diversification of a group of organisms that share a common ancestor, often itself newly evolved. It also involves the opening up of a new biological frontier which may be as vast as the land or the air, or, as in the case of the Galapagos finches (page 468), as small as an archipelago.

The fossil record contains many examples of adaptive radiation. For example, some 425 million years ago, the first terrestrial plants and animals suddenly emerged. A large number of different populations apparently began the same move at about the same time. Most undoubtedly perished, but a few survived and then differentiated into the many different types of terrestrial plants and animals existing today. A similar, even more rapid burst of evolution gave rise to the birds. Apparently, according to the fossil record, a number of populations of dinosaurs must have taken to climbing and gliding at about the same time.

The shift was made very rapidly from the land to the air, however, and these intermediate populations, the gliding reptiles, disappeared quickly and permanently, as the birds evolved. The mammals similarly burst forth on the evolutionary scene, with many different kinds appearing simultaneously in the fossil record.

It has recently been suggested that the sudden bursts of evolution associated with adaptive radiation may involve changes in regulatory genes rather than in structural genes (Chapter 16). Such changes in the off-on mechanisms could produce major phenotypic changes more rapidly than mutations involving structural genes.

Common Features in Patterns of Evolution

Actually, as we noted earlier, these three major patterns of evolution have a great deal in common. Although we can trace an overall pattern of evolution in the *Equus* and its progenitors, it did not proceed in a straight line and it involved many branches. Conversely, although a branching off was the crucial event in the creation of *Ursus maritimus,* the modern polar bear is itself a product of phyletic evolution, as is every modern organism. Adaptive radiation is a combination of both phyletic and splitting evolution.

ORIGIN OF SPECIES

The crucial event in splitting evolution and in adaptive radiation is the branching off of a single population into two or more populations that do not interbreed. (Some population biologists contend that such splitting off also plays a major role in phyletic evolution, holding that major changes take place primarily in small, separate populations.) The polar bear could never have developed its distinctive characteristics had gene flow continued between it and *Ursus arctos.* Despite the title of Darwin's major work, he never dealt directly with the problem of speciation. It has been, however, a chief concern of twentieth-century evolutionists.

What is a species? In the words of Terrell H. Hamilton, "A species may be envisioned as an isolated pool of genes flowing through space and time, constantly adapting to changes in its environment as well as to the new environments encountered by its extension into other geographic regions."*

The word that we shall examine more closely in this definition is "isolated." How does one pool of genes split off from another and begin its solitary journey? How do two species, often very similar to one another, inhabit the same place and the same time and yet maintain the genetic isolation essential to their identity? Almost all population biologists agree that new species can arise only when populations are genetically isolated from one another. If, under such circumstances of genetic isolation, the populations are subject to different selective pressures, they will begin to diverge genetically. As a consequence of such genetic diversion, the populations will eventually be unable to interbreed. At this point, speciation is said to have occurred.

* Terrell H. Hamilton, *Process and Pattern in Evolution,* The Macmillan Company, New York, 1967.

THE BREAKUP OF PANGAEA

In the past 10 years, the theory of continental drift—viewed at one time with as much suspicion as reports of flying saucers and extrasensory perception—has become firmly established.

According to this theory, the outermost layer of the earth is divided into a number of segments, or plates. These plates, on which the continents rest, slide across the surface of the earth, moving in relationship to one another. The geologic expression of this relative motion occurs mainly at the boundaries between the adjacent plates. Where plates collide, volcanic islands such as the Aleutians may be formed or mountain belts such as the Andes or Himalayas may be uplifted. At the boundaries where plates are separating, volcanic material wells up to fill the void. It is here that ocean basins are created. Plates may also move parallel to the boundary that joins them but in opposite directions. Such is the case of the San Andreas fault.

About 200 million years ago, all the major continents were locked together in a supercontinent, Pangaea. Several reconstructions of Pangaea have been proposed, one of which is shown here. It is generally agreed that Pangaea began to break up about 190 million years ago, about the time the dinosaurs approached their zenith and the first mammals began to appear. First, the northern group of continents (Laurasia) split apart from the southern group (Gondwana). Subsequently, Gondwana broke into three parts, Africa-South America, Australia-Antarctica, and India. India drifted northward and collided with Asia about 50 million years ago. This collision initiated the uplift of the Himalayas, which continue to rise today as India still pushes northward into Asia.

By the end of the Cretaceous period, according to current reconstructions, about 65 million years ago, South America and Africa had separated sufficiently to have formed half the South Atlantic, and Europe, North America, and Greenland had begun to drift apart; however, final separation between Europe and North America–Greenland did not occur until the Eocene (43 million years ago). During the Cenozoic era, Australia finally split from Antarctica and moved northward to its present position, and the two Americas were joined by the Isthmus of Panama, which was created by volcanic action.

Geology now finds itself in somewhat the same position as astronomy did at the time of Galileo and Copernicus, and geology texts are being hastily rewritten in accordance with the new doctrines. From the outset, of course, the theory of continental drift has been closely interwoven with that of evolution. One of the earliest and most impressive pieces of evidence in favor of the new theory was the discovery of fossil remains of a small, snaggle-toothed reptile, Mesosaurus, found in coastal regions of Brazil and South Africa but nowhere else. Students of evolution have not yet felt the full impact of the new geology; however, it is already beginning to call forth some important new interpretations of fossil history.

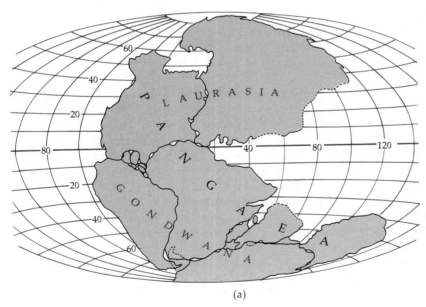

(a)

(b)

(a) Pangaea. (b) The San Andreas fault is a boundary between two giant plates that are moving past one another. The fault, running through San Francisco and continuing southeast of Los Angeles, is responsible for California's notorious earthquakes.

(a)

(b)

(c)

34-4
While the placental mammals were under-going adaptive radiation on the other continents, the marsupials, geographically isolated in Australia, went through a parallel pattern of speciation. (a) The ecological role of large herbivore, usually filled by hooved mammals, is occupied in Australia by kangaroos. There are some 45 herbivorous species of kangaroo; adults of the larger species weigh up to 90 kilograms and can jump 9 meters in a single leap. (b) Though the koala is called a bear, it has no true placental equivalent. It is arboreal and lives exclusively on the oily aromatic leaves of certain Eucalyptus trees. As for the Eucalyptus, more than 600 different species are known in Australia; none are native to other continents. (c) Wombats, of which there are four species, are the marsupial equivalent of large rodents. Nocturnal animals, they spend the day in tunnels and underground burrows dug with their shovel-like paws.

Speciation with Geographic Isolation

Geographic Races

Every widespread species that has been carefully studied has been found to contain geographically representative populations that differ from each other to a greater or lesser extent, as with the ecotypes of *Potentilla glandulosa* (page 456). A species composed of geographic races of this sort is obviously particularly susceptible to speciation if geographic barriers arise, as with the polar bear.

Because of the work of John Moore, the leopard frog *(Rana pipiens)* has become one of the best-known examples of geographic races and speciation. Leopard frogs are found in North America as far north as Quebec and as far south as Mexico. Moore has studied and compared a large number of characteristics in 29 separate populations of this species. He has found that there are marked variations, as might be expected, from population to population. Crossbreeding of individuals from different populations produces normal offspring when parents are drawn from populations that are geographically adjacent, such as central and southern Florida, or that lie at roughly the same latitude, such as Texas and central Florida. However, the greater the north-south gap separating the home populations of the parents, the greater also is the proportion of defective, often inviable, offspring. In Texas-Vermont hybrids, for example, mortality among the developing embryos may reach as high as 100 percent. Thus what was clearly once a single species is becoming fragmented in time and space into separate subspecies and, as we shall see, species.

Geographic barriers are of many different types. Islands are frequent sites for the development of unusual species. The breakup of Pangaea (see the essay on the facing page) profoundly altered the course of evolution, setting Australia adrift, for example, as a veritable Noah's ark of marsupials. However, populations of many organisms can become cut off from one another by far less dramatic barriers (Figure 34-5).

Depending on the species, islands, genetically speaking, may be (a) true islands, (b) mountaintops, (c) ponds, lakes, or even oceans, or (d) isolated clumps of vegetation.

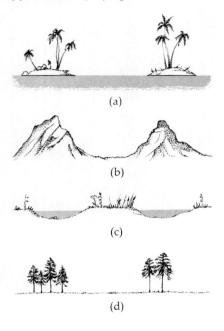

(a)

(b)

(c)

(d)

Speciation without Geographic Isolation

The only proven means of producing new species without geographic isolation occurs by hybridization or polyploidy. Hybrids, the offspring of parents of different species, occur in both animals and plants. In nature, they are far more frequent in plants, and also because plants often reproduce vegetatively (page 453), hybrid populations are not uncommon. Some hybrids spread by runners and eventually cover large areas. Kentucky bluegrass, for instance, occurs widely throughout the United States, is a promiscuous hybridizer, and has formed crosses with a large number of related species. Occasional offspring of these matings have been better adapted than either of the parents and have become successful in their own right.

Hybrids in both plants and animals are nearly always sterile because the chromosomes cannot pair at meiosis (since they have no homologues), a necessary step for producing viable gametes (Figure 34–6a). In plants, however, fertile forms may arise from these infertile hybrids by the process of polyploidy, in which the cells acquire more than two sets of chromosomes. Polyploid cells arise at a low frequency as the result of a "mistake" in mitosis, so that the chromosomes divide but the cell does not. If such cells divide by further mitosis so that they give rise to a new individual, either sexually or asexually, that individual will have twice the number of chromosomes of its parent. Polyploid individuals can be produced deliberately in the laboratory by the use of a drug called colchicine, which prevents separation of chromosomes during mitosis.

If polyploidy occurs in a sterile hybrid, each chromosome from both parents will be present in duplicate. It then can pair with its duplicate chromosome, meiosis will be normal, and fertility restored (Figure 34–6b).

Approximately half of the 235,000 kinds of flowering plants have had a polyploid origin, and many important agricultural species, including wheat, are hybrid polyploids.

34–6

(a) A hybrid organism (such as a mule) produced from two 1n gametes can grow normally because mitosis is normal, but it cannot reproduce because the chromosomes cannot pair at meiosis. (b) However, a hybrid organism produced from two 2n gametes can both divide normally mitotically and undergo meiosis. Since each chromosome will have a partner, the chromosomes can pair at meiosis.

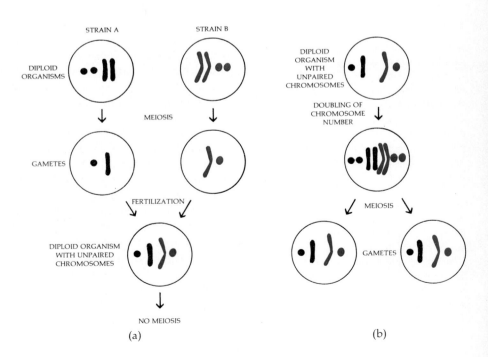

Sticklebacks, small freshwater fish, have elaborate mating behavior. The male at breeding time, in response to increasing periods of sunlight, changes from dull brown to the radiant colors shown here. He builds a nest and begins to court females, zigging toward them and zagging away from them. A female ready to lay eggs responds by displaying her swollen belly. The male then leads her into the tunnel-like nest, prodding her through. In response to his caresses, she lays her eggs and swims off; he follows her through, fertilizes the eggs, and stays to tend the brood. If either partner fails in any step of this quite elaborate ritual, no young are produced.

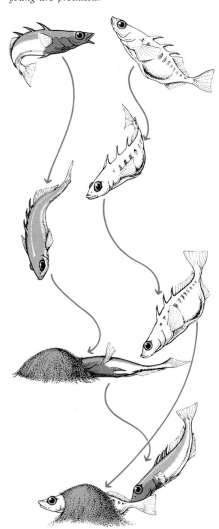

Isolating Mechanisms

Some species of organisms are so similar phenotypically that only an expert with a microscope can tell them apart—*Drosophila* offers several examples. Yet such species may live together in the same area without interbreeding. What factors operate to maintain genetic isolation of closely related species? Isolating mechanisms may be conveniently divided into two categories: those distinguishable on an anatomical or physiological basis and those that can be identified in terms of behavior.

Physiological Isolating Mechanisms

These barriers arise because of physiological incompatibilities between members of the two species. There are a number of such incompatibilities. For instance:

1. Differences in the shape of the genitalia may prevent insemination.
2. The sperm may not be able to survive in the reproductive tract of the female.
3. The pollen tube (in the case of plants) may not be able to grow.
4. The sperm cell may not fuse with the ovum.
5. The ovum, once fertilized, may not be able to develop.
6. The young, or some of the young, may survive but may not become reproductively mature.
7. The offspring may be hardy—the mule, for instance—but sterile. Sterility in such cases is apparently often caused by the fact that the dissimilar chromosomes cannot pair at meiosis.

Behavioral Isolating Mechanisms

One of the most significant things about the physiological (postmating) isolating mechanisms is that, in nature, they are rarely tested. In animals, physiologically incompatible genotypes are usually prevented from mating by behavioral isolating mechanisms. These premating, behavioral mechanisms can take many forms.

Visual recognition is important, particularly among birds, which are visually oriented in all their behavior. According to observers, Galapagos finches (page 468) recognize each other by their beaks. A male finch may mistakenly pursue a finch of another species—the finches often look very much alike from the rear—only to lose interest as soon as he sees the beak. He will not court a female of another species. The beak is a conspicuous feature in courtship, during which food is passed from the beak of the male to that of the female.

Among many invertebrates, pheromones are the chief isolating mechanisms (page 361). They may serve as signals, as in the case of the cecropia moth, to attract the male or, with oysters, for instance, to trigger the release of gametes by the female.

Bird songs, frog calls, the strident love notes of cicadas and crickets serve to identify members of a species to one another. For example, field studies of leopard frogs have recently revealed that some of the populations that still can crossbreed under laboratory conditions do not. They are effectively isolated from one another by differences in their mating calls. Thus, new species have evolved.

The flashings of fireflies (which are actually beetles) are sexual signals. Each species of firefly has its own particular flashing pattern, which is different from

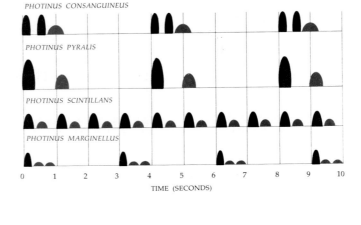

34-8

Flashing patterns of various species of fireflies found in Delaware. Each division of the horizontal lines represents one second, and the height of each curve represents the brilliance of the flash. The black curves are male flashes, and the colored ones are the female responses. The male flashes first and the female answers, returning the species-specific signal. Firefly flashes differ from species to species, not only in duration, intensity, and timing, but also in color.

that of other species both in duration of flashes and in intervals between them (Figure 34–8). For example, one common flash pattern consists of two short pulses of light separated by about 2 seconds, with the phrase repeated every 4 to 7 seconds. The lights also vary among the species in intensity and color. The male flashes first, and the female answers, returning the species-specific signal. By mimicking signals with a flashlight, it is possible to attract the males of a given species.

Temporal mechanisms also play an important role in sexual isolation. Species differences in flowering times are important isolating mechanisms in plants. Most mammals—man is a notable exception—have seasons for mating, often controlled by temperature or by day length. Figure 34–9 shows the mating calendar of species of frogs near Ithaca, New York. The spawning of freshwater fishes is often regulated by water temperatures. Members of the species *Drosophila pseudoobscura* and *Drosophila persimilis*, which often are found in the same areas, congregate and mate almost every day but differ in their mating hour.

The fact that hybrid progeny are often less viable serves to reinforce behavioral isolating mechanisms in the species. A female cricket that answers to the wrong song or a frog whose individual calendar is not synchronized with that of the rest of the species will contribute less to the gene pool. As a consequence, there will be a steady selection for behavioral isolating mechanisms.

An Example: Darwin's Finches

Admirers of Charles Darwin find it particularly appropriate that one of the best examples of speciation is provided by the finches observed by Darwin on his voyage to the Galapagos Islands. All the Galapagos finches are believed to have arisen from one common ancestral group—perhaps a single pregnant female—transported from the South American mainland, some 950 kilometers away. How she or they got there is, of course, not known, but it is probable that they were the survivors of some particularly severe storm. (Every year, for instance, some American birds and insects appear on the coasts of Ireland and England after having been blown across the North Atlantic.) It is very likely that finches were the first birds to colonize the islands.

From this ancestral group, 13 different species arose (plus one to the northeast on the Cocos Island, a few hundred kilometers away). Apparently the various islands were near enough to one another so that, over the years, small

34-9

Mating timetable for various frogs and toads that live near Ithaca, New York. In the two cases where two different species have mating seasons that coincide, the breeding sites differ. Peepers prefer woodland ponds and shallow water; leopard frogs breed in swamps; pickerel frogs mate in upland streams and ponds; and common toads use any ditch or puddle.

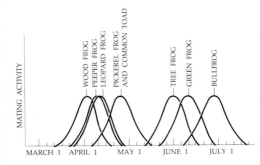

468   Evolution and Population Biology

The Galapagos Islands, some 950 kilometers west of the coast of Ecuador, have been called "a living laboratory of evolution." Species and subspecies of plants and animals that have been found nowhere else in the world inhabit these islands. "One is astonished," wrote Charles Darwin in 1837, "at the amount of creative force . . . displayed on these small, barren, and rocky islands. . . ."

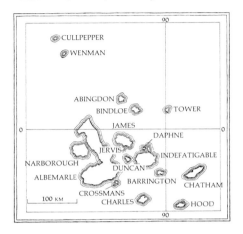

34-11

Six of the 13 different species of Darwin's finches. There are six species of ground finch, six species of tree finch, and one warbler-like species—all derived from a single ancestral species. Except for the warbler finch, which resembles a warbler more than a finch, the species look very much alike; the birds are all small and dusky-brown or blackish, with stubby tails. The differences between them lie mainly in their bills, which vary from small, thin beaks to huge, thick ones. (a) The small ground finch (Geospiza fuliginosa) and (b) the medium ground finch (Geospiza fortis) are both seed eaters. Geospiza fortis, with a somewhat larger beak than G. fuliginosa, is able to crack larger seeds. The cactus ground finches Geospiza scandens (c) and Geospiza conirostris (d) live on cactus blooms and fruit. Notice that their beaks are much larger and more pointed than those of the other two ground finch species. The tree finches Camarhynchus parvalus (e) and Camarhynchus pauper (f), both insectivorous, take prey of different sizes.

founding groups could land, yet far enough apart so that once a group was established, there would be no gene flow between it and the parent group for a period of time long enough for genetic barriers to develop. (Finches are not very good fliers; if they had been, this natural experiment in evolution would have been a failure.) Development of a new species requires, it is estimated, at least 10,000 years of genetic isolation, but since there are many islands, several species could have been evolving at the same time.

The ancestral type was a finch, a smallish bird with a short, stout conical bill especially adapted for seed-crushing; canaries and sparrows are finches and so are cardinals. The ancestor is believed to have been a ground-feeding finch, and six of the Galapagos finches are ground finches. Four species of ground finches live together on most of the islands. Three of them eat seeds and differ from one another mainly in the size of their beaks, which in turn, of course, influences the size of seeds they eat. The fourth lives largely on the prickly pear and has a much longer and more pointed beak. The two other species of ground finch are usually found only on outlying islands, where some supplement their diet with cactus.

In addition to the ground finches, there are six species of tree finches, also differing from one another mainly in beak size and shape. One has a parrotlike beak, suited to its diet of buds and fruit. Four of these tree finches have insect-

(a)

(b)

(c)

(d)

(e)

(f)

eating beaks, each adapted to a different size range of insects. The sixth, and most remarkable of the insect eaters, is the woodpecker finch, which has a beak like a woodpecker's that it uses as a chisel. Lacking the woodpecker's long, prying tongue, it carries about a twig or cactus spine to dislodge insects from crevices in the bark (Figure 31–8).

Classical taxonomists, using all ordinary standards of external appearance and behavior, would classify the thirteenth species of Galapagos finch as a warbler, but its internal anatomy and other characteristics clearly place it among the finches, and there is general agreement that it, too, is a descendant of the common ancestor or ancestors.

Notice also that the evolution of the Galapagos finches—the diversification of a single type to fill a vacant ecological space—provides, as we noted previously, a good example of adaptive radiation.

EXTINCTION

Some students of the evolutionary process believe that many or even most new species originated as a small, isolated population—perhaps of just a few individuals. As we noted in Chapter 32, such populations would not necessarily be genetically representative of the species as a whole, and they would, of course, become intensively inbred. If, in addition, they were placed under strong selective pressures—as they might be in a new environment—and if they survived, evolution would be very rapid.

Only a small fraction of all the species that have ever lived are presently in existence—certainly less than $\frac{1}{10}$ of 1 percent, perhaps less than $\frac{1}{1000}$ of 1 percent. Extinction is very much a part of the evolutionary process, with the elimination of species making room for new ones. Certain times in the earth's history have seen the elimination of major life forms: the trilobites that dominated the early Paleozoic seas; the dinosaurs that disappeared at the end of the Mesozoic era; and the great land mammals that became extinct during the Pleistocene, victims of its harsh climatic changes or, perhaps, of the predations of a new life form, man.

34–12

Among the thousands of plants and animals now threatened with extinction are (a) the monkey-eating eagle, (b) the Bengal tiger, and (c) Franklinia alatamaha, a handsome tree of the camellia family. Man causes other organisms to become extinct as a direct result of his predation, as a consequence of toxins (such as DDT) released into the biosphere, and, most commonly, by alteration of natural habitats, such as, for example, the widespread destruction now taking place in the tropical rain forest.

(a)

(b)

(c)

MOLECULAR CLOCKS

Evolutionary relationships may be established by comparisons of living species and by examination of fossil forms. Most recently a new method has become available: the study of macromolecules. Amino acids from similar proteins of different species are different, and the degree of difference between two species appears to correspond well with the degree of evolutionary divergence between them. For example, the protein chain of cytochrome c is exactly the same in men and chimpanzees but differs from the cytochrome c of Neurospora by 44 out of 100 amino acids. The composition of cytochrome c from 40 species has now been determined, and it appears that about 20 million years is required to produce a change of 1 percent in this particular protein.

The altered protein is able to function because the changes involve only certain amino acids; most of the amino acid sequence is the same for all species. Moreover, when a substitution is made, it almost always involves an amino acid of similar chemical properties, so that the conformation of the chain is altered little if at all. Presumably mutations occur that involve other amino acids, but these are weeded out as disadvantageous.

Other proteins change at different rates. Hemoglobin, for example, appears to change at a rate of 1 percent in 6 million years, or approximately one substitution in 3.5 million years, presumably because it is more tolerant of change.

Although the macromolecules evolve at different rates, they are internally consistent; that is, if you compare two proteins, you will get the same answer as to the divergence of the species. Moreover, they are generally, but not always, in accord with accepted interpretations of the fossil evidence. Perhaps most important, these ancient proteins may hold the key to important and unsolved evolutionary problems, such as the relationships of the invertebrate phyla to one another and to vertebrate origins.

ADAPTATION

Adaptation is the result of natural selection; it generally produces individuals more suited to their environment. (We say generally, because, as R. C. Lewontin points out, ". . . genetic changes take place to increase the fitness of the population in *today's* environment but the result of that change is a population in the next generation living in *tomorrow's* environment.")

All organisms are, of course, adapted to their environment; otherwise they would not be in existence. Look just a little more closely than usual at some common plants and animals. For example, watch a squirrel, sometimes unkindly characterized as a tree-dwelling rat. Regard the way in which its tail serves as counterbalance as the squirrel leaps and turns; in addition, the same marvelous structure serves as a parasol, a blanket, an aerial rudder, and, in case of mishap, a parachute. Pluck a burr from your pants leg and consider the ingenuity with which it clings there. Consider the love and devotion characteristic of the domesticated dog; these are adaptations related to procurement of food and shelter as stringently selected for as the beak of a hawk. Thread a needle; your capacity to do so represents the cumulative effect of millions of years of selective pressures for digital dexterity and eye-hand coordination. (The needle itself made its appearance a mere 10,000 years ago.)

We shall mention only a few special types of adaptation.

Preadaptation

Early critics of Darwin were fond of arguing that it was impossible by chance alone for so intricate and marvelous an organ as, for example, the vertebrate eye to be assembled. They, of course, were missing an important point: evolution is very conservative. Based on studies of comparative anatomy, the eye began as a patch of light-sensitive epithelium and gradually accumulated, by mutation and recombination, modifications that increased its survival value to the organism.

34–13

Special adaptations of a woodpecker include two toes pointing forward and one toe directed to the side with which it clings to the tree bark, a strong beak that can chisel holes in the bark, strong neck muscles that make the beak work as a hammer, and a very long tongue that can reach grubs under the bark. The bird shown here is a pileated woodpecker.

(a)

(b)

(c)

(d)

34–14

Some unusual adaptations. (a) Among a number of different species of ants that live in arid habitats, certain individuals, known as repletes, are used as storage containers for honey. (b) A cave-dwelling crayfish, like other animals adapted for life in the lightless interior of caves, is sightless and unpigmented. Its antennae are longer than those of an ordinary crayfish, and, moving slowly and silently, it is able to detect even slight disturbances in the water. (c) A red-eyed tree frog of Central America. These animals, like many others that dwell in the canopy of the tropical rain forest, never descend to the forest floor. The frogs' eggs are laid on leaves that overhang water. After they hatch, the tadpoles drop into the water where they complete their development. The bulges at the ends of the toes are suction pads. (d) Anteaters include marsupials and placentals, Old World and New World forms, closely related to one another only by certain special adaptations. They all have long noses, very long, wormlike tongues, and stout digging claws. This is the giant anteater of tropical America and, like most anteaters, it eats termites.

The other side of this argument, however, is that organic changes accumulate and are conserved because they are of immediate—not just future—survival value. To take a simple example, the lungs and walking fins of certain Devonian fishes did not evolve in preparation for the transition to land; they improved the survival chances of the fishes in the receding waters of that time. Similarly, the increased visual acuity and upright stance associated with life in the treetops fortuitously prepared a group of large primates for life on the open tropical grasslands, the savanna. This phenomenon is often seen on the biochemical level as well, where a chemical such as AMP is captured for another use in the course of evolution. Possession of specific traits such as these on which wholly new adaptations can be constructed is, as we have noted previously, called preadaptation. In one sense, of course, all organisms possess preadaptations. However, as with most aspects of evolution, these can be accurately identified only by hindsight.

Adaptations Related to Sexual Selection

One of the most powerful forces of natural selection is that exercised by individuals in choosing a mate.

Among most insects and vertebrates, males do the courting and females do the choosing. This division of labor makes sense. Sperm are less costly, energetically speaking, than eggs and are always more abundant. All other factors being equal, therefore, a male optimizes its contribution to the gene pool by mating as frequently as possible. On the other hand, a female seeks quality rather than quantity.

As a consequence, the process of sexual selection is more apt to produce striking adaptations among males. Females are usually quick to reject suitors who deviate in any way from the "ideal." For instance, female fruit flies will not mate with a white-eyed male if a red-eyed male is present. Clearly, sexual selection is often stabilizing, but this is not necessarily the case. Sexual selection pressure led, for example, to the evolution of the elaborate plumage and huge tails of certain birds, the giant antlers of some deer, and the long, intricate mating rituals of many different types of organisms.

34–15
A male frigate bird displays his crimson pouch. Throughout the courtship period, the pouch remains bright and inflated, even when the bird is flying or sleeping. The pouch's size and bright color are the result of sexual selection.

Convergent Evolution

Organisms that occupy similar environments often come to resemble one another even though they may be only very distantly related phylogenetically. The whales, a group that includes the dolphin and porpoises, are very similar in many exterior features to sharks and other large fish (Figure 34–16); the fins of whales conceal the remnants of a vertebrate hand, they are warm-blooded, like their land-dwelling ancestors, and they have lungs rather than gills. (Gills are clearly more useful than lungs in aquatic respiration—another example of the fact that evolutionary processes work only with what is already available.) Similarly, two families of plants invaded the desert, giving rise to the cacti and the euphorbs. Both evolved large fleshy stems with water-storage tissues and protective spines and appear superficially similar. However, their quite different flowers reveal their widely separate evolutionary origins.

Coevolution

When two or more populations interact so closely that each is a strong selective force on the other, simultaneous adjustments occur that result in coadaptation, or, as it is more commonly called, _coevolution._

We have previously mentioned several examples of coevolution. One of the most important, in terms of sheer numbers of species and individuals involved, was the coevolution of flowers and their pollinators, described in Chapter 21. Others will be described in the next section. Here we shall give just one example.

Ants and Acacias

Trees and shrubs of the genus _Acacia_ grow throughout the tropical and subtropical regions of the world. In Africa and tropical America, acacia species are protected by thorns. (The acacias of Australia, which has few large grazing animals except recently introduced species, have no thorns.)

(a)

(b)

34–16

Although porpoises (a) _are descendants of primitive mammals, their shapes closely resemble those of large ocean-dwelling fish, as_ _exemplified by the blue shark_ (b). _This independent development of similar characteris-_ _tics by animals living in similar environments is known as convergent evolution._

(a)

(d)

(b)

(c)

34–17

Ants and acacias. (a) The beginning. A queen ant cuts an entrance into a thorn on a seedling bull's-horn acacia. She will hollow out the thorn and raise her first brood inside it. (b) The tip of an acacia leaf showing a worker on patrol. The orange structures at the tips of the leaves are Beltian bodies. They are a source of food for the ants. (c) A worker ant drinking from the nectary. (d) Warriors in a battle for possession of an acacia. Such battles occur when the branches of acacias grow to touch each other. The largest colony usually wins.

On one of the African species of *Acacia*, ants of the genus *Crematogaster* gnaw entrance holes in the walls of the thorns and live permanently in them. Each colony of ants inhabits the thorns on one or more trees. The ants obtain food from nectar-secreting glands on the leaves of the acacias and eat caterpillars and other herbivores that they find on the trees. Both the ants and the acacias appear to benefit from this symbiotic association.

In the lowlands of Mexico and Central America, the ant-acacia relationship has been extended to even greater lengths. The bull's-horn acacia, a common plant of that area, is found particularly frequently in cutover or disturbed areas, where the competition for light is intense. It grows extremely rapidly. This species of acacia has a pair of greatly swollen thorns several centimeters in length at the base of most leaves. The petioles bear nectaries, and at the very tip of each leaflet is a small structure rich in oils and proteins known as a Beltian body. Thomas Belt, the naturalist who first described these bodies, noted that their only apparent function was to nourish the ants. Ants live in the thorns, obtain sugars from the nectaries, eat the Beltian bodies, and feed them to their larvae.

Worker ants, which swarm over the surface of the plant, are very aggressive toward other insects, and indeed toward animals of all sizes. They become alert at the mere smell of a man or a cow, and when their tree is brushed by an animal, they swarm out and attack at once, inflicting painful burning stings. The effect has been described as similar to walking into a large nettle. Moreover, and even more surprisingly, alien plants sprouting within as much as a meter of occupied acacias are chewed and mauled, and twigs and branches of other trees that touched an occupied acacia are similarly destroyed. Not so surprisingly, acacias inhabited by these ants grow very rapidly, soon overtopping other vegetation.

Daniel Janzen of the University of Michigan, who first analyzed the ant-acacia relationship in detail, removed ants from acacias artificially, by insecticides or by removing thorns or entire occupied branches. Acacias without their ants grew slowly and usually suffered severe damage from insect herbivores. Their stunted bodies were soon overshadowed by competing species of plants and vines. As for the ants, according to Janzen, these particular species can live only on acacias.

SUMMARY

Examples of the effects of natural selection on contemporary populations can be seen in the peppered moth *(Biston betularia)* and in the phenomenon of insecticide resistance.

The effects of natural selection on past populations can be seen in the fossil record. Three principal patterns of evolution are observed: phyletic, or "straight line," evolution; splitting, or "branching," evolution; and adaptive radiation. These are not distinct entities; phyletic evolution always involves branching, splitting evolution always involves phyletic evolution, and adaptive radiation involves a combination of both forms.

A species is defined in population biology as an isolated gene pool moving through time and space. Genetic isolation is required for a new species to originate. Among animals, genetic isolation appears always to require geographical isolation. Among plants, it may also be achieved by hybridization or polyploidy. Polyploidy is often a mechanism whereby hybrids become fertile. Genetic barriers among species are maintained by a variety of physiological and behavioral isolating mechanisms.

Adaptation, which is a property of all organisms, is the result of interactions between the organism and its environment. Coevolution is produced by the close interaction of two species upon one another.

QUESTIONS

1. Define the following terms: polyploidy, adaptive radiation, ecological race, behavioral isolation.

2. Describe the separate steps involved in the formation of the distinct species of Galapagos finches.

3. Explain why genetic isolation is important for speciation.

4. In your opinion, why is there at present only one species in the genus *Homo?* Is a second *Homo* species likely to arise?

SUGGESTIONS FOR FURTHER READING

CALDER, NIGEL: *The Restless Earth,* The Viking Press, Inc., New York, 1972.

   A handsomely illustrated report on the new geology, written for the general reader.

CARLQUIST, SHERWIN: *Island Life: A Natural History of the Islands of the World,* The Natural History Press, Garden City, N.Y., 1965.

   An exploration of the nature of island life and of the intricate and unexpected evolutionary patterns found in island plants and animals.

DARWIN, CHARLES: *On the Origin of Species by Means of Natural Selection, or The Preservation of Favored Races in the Struggle for Life*, Doubleday & Company, Inc., Garden City, N.Y., 1960.*

Darwin's "long argument." Every student of biology should, at the very least, browse through this book to catch its special flavor and to begin to understand its extraordinary force.

DARWIN, CHARLES: *The Voyage of the Beagle*, Natural History Library, Doubleday & Company, Inc., Garden City, N.Y., 1962.*

Darwin's own chronicle of the expedition on which he made the discoveries and observations that eventually led him to his theory of evolution. The sensitive, eager young Darwin that emerges from these pages is very unlike the image many of us have formed of him from his later portraits.

DOBZHANSKY, THEODOSIUS: *Genetics and the Origin of Species*, 3d ed., Columbia University Press, New York, 1951.*

A presentation of the basic themes of population genetics by one of the foremost leaders in studies using Drosophila.

EHRLICH, PAUL R., RICHARD W. HOLM, and DENNIS R. PARNELL: *The Process of Evolution*, McGraw-Hill Book Company, New York, 1974.

A good short text for an undergraduate course in evolution theory, appropriate for students who have completed a year of general biology.

EICHER, DON L.: *Geologic Time*, Prentice-Hall, Inc., Englewood Cliffs, N.J., 1968.*

A short supplementary text that describes the fossil strata and their significance.

HAMILTON, TERRELL H.: *Process and Pattern in Evolution*, The Macmillan Company, New York, 1967.*

This short book, designed as a supplementary text, is outstanding for its clarity of definition and presentation of evolutionary concepts.

LACK, DAVID: *Darwin's Finches*, Harper & Row, Publishers, Inc., New York, 1961.*

This short, readable book, first published in 1947, gives a marvelous account of the Galapagos, its finches and other inhabitants, and of the general process of evolution.

LEWONTIN, R. C.: *The Genetic Basis of Evolutionary Change*, Columbia University Press, New York, 1974.

This book is intended for the advanced student or another worker in this field, but it is also highly recommended for any reader interested in the problems and perspectives of population genetics.

MAYR, E.: *Animal Species and Evolution*, Harvard University Press, Cambridge, Mass., 1963.

A masterly, authoritative, and illuminating statement of contemporary thinking about species—how they arise and their role as units of evolution.

MOOREHEAD, ALAN: *Darwin and the Beagle*, Harper & Row, Publishers, Inc., New York, 1969.

A delightful narrative of Darwin's journey, beautifully illustrated with contemporary or near-contemporary drawings, paintings, and lithographs.

SIMPSON, GEORGE GAYLORD: *Horses*, Oxford University Press, New York, 1951.

The story of the horse family in the modern world and through 60 million years of history, as recorded by one of the nation's leading paleontologists.

* Available in paperback.

SMITH, JOHN MAYNARD: *The Theory of Evolution,* 3d ed., Penguin Books, Inc., Baltimore, Md., 1975.*

Written for the general public, this book is notable for its many concrete examples of evolution, past and present.

STEBBINS, G. LEDYARD: *Processes of Organic Evolution,* Prentice-Hall, Inc., Englewood Cliffs, N.J., 1966.*

A brief review of the entire field by one of its outstanding practitioners.

SULLIVAN, WALTER: *Continents in Motion, the New Earth Debate,* McGraw-Hill Book Company, New York, 1974.

An account, by the nation's leading science reporter, of the discoveries that led to the current revolution in geology.

TOULMIN, STEPHEN, and JUNE GOODFIELD: *The Discovery of Time,* Harper & Row, Publishers, Inc., New York, 1965.*

The historical development of our concepts of time as they relate to nature, human nature, and human society.

WILSON, E. O., and WILLIAM H. BOSSERT: *A Primer of Population Biology,* Sinauer Associates, Inc., Stamford, Conn., 1971.*

An introduction to the mathematics of population biology.

* Available in paperback.

SECTION 8   Ecology

35-1
The energy on which life depends enters
the biosphere in the form of light. It is
converted to chemical energy by photo-
synthetic organisms, such as these sunlit
corn plants. Such photosynthetic organisms
are, in turn, the energy source for all
heterotrophs, including man.

CHAPTER 35

The Biosphere

Ecology is the study of the interactions of organisms with their physical environment and with each other. As a science, it seeks to discover how an organism affects, and is affected by, its environment and to define how these interactions determine the kinds and numbers of organisms found in a particular place at a particular time.

We shall begin as if we were examining our planet from outer space. There is a thin film at the surface of the earth. It extends about 8 or 10 kilometers up into the atmosphere and down into the soil, as far as roots penetrate and microorganisms are found. It includes all of the surface waters and some of the ocean depths. This thin film—the part of the earth where life exists—is known as the *biosphere*.

DISTRIBUTION OF LIFE

Approaching the biosphere more closely, we can see that it is not uniform but varies greatly in depth and density from place to place. In fact, in some areas of the biosphere, such as portions of the deserts or the ice sheets of the Antarctic and Greenland, the only evidence of life to be found is in the form of dormant spores.

The Earth and the Sun

A number of factors are involved in producing variations in the biosphere. The major one is the radiation the planet receives from the sun. The earth is about 150 million kilometers from this star, which is the source of its heat and light energy and, indirectly, of its chemical energy as well. At the equator, the sun's rays are almost perpendicular to the earth's surface, and this sector receives more energy from the sun than the areas to the north and south, with the polar regions receiving the least (Figure 35–2a). Moreover, because the earth, which is tilted on its axis, rotates once every 24 hours and completes an orbit around the sun once about every 365 days, the amount of energy reaching different parts of the surface varies hour by hour and season by season (Figure 35–2b).

35-2

(a) *A beam of solar energy striking the earth near one of the poles is spread over a wider area of the earth's surface than is a similar beam striking the earth near the equator.*
(b) *In the northern and southern hemispheres, temperatures change in an annual* cycle because the earth is slightly tilted on its axis in relation to its pathway around the sun. In winter, the northern hemisphere tilts away from the sun, which decreases the angle at which the sun's rays strike the surface and also decreases the duration of daylight, both of which result in lower temperatures. In the summer, the northern hemisphere tilts toward the sun. Note that the polar region of the northern hemisphere is continuously dark during the winter and continuously light during the summer.

(a)

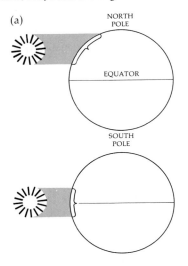

(b)

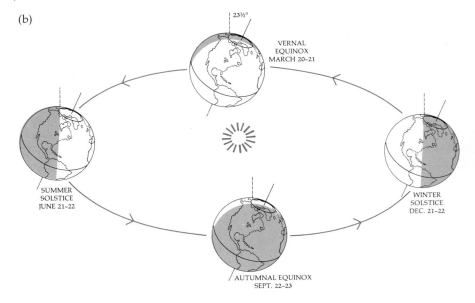

35-3

The earth's surface is covered by many belts of air currents, which determine the major patterns of rainfall and wind. Air rising at the equator loses moisture in the form of rain, and falling air at latitudes of 30° north and south is responsible for the great deserts found at these latitudes.

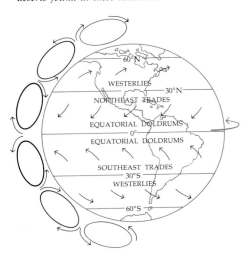

Wind and Water

Temperature variations over the surface of the earth and the earth's rotation establish major patterns of air circulation and rainfall. These patterns depend, to a large extent, on the fact that cold air is denser than warm air. As a consequence, hot air rises and cold air falls. As air rises, it encounters lower pressure and consequently expands, and as a gas expands, it cools. Cooler air holds less moisture, so rising, cooling air tends to lose its moisture in the form of rain or snow.

The air is hottest along the equator, the region heated most by the sun. This air rises, creating a low-pressure area (the doldrums), moves away from the equator, cools, loses its moisture, and falls again at latitudes of about 30° north and south, the regions where most of the great deserts of the world are found. This air warms, picks up moisture, and rises again at about 60° latitude (north and south); this is the polar front, another low-pressure area. A third, weaker belt rising at the polar front descends again at the poles, producing a region in which again, as in other areas of descending air, there is virtually no rainfall. The earth's spinning movement twists the winds caused by the heat transfer from equator to poles, creating the major wind patterns (Figure 35-3).

The worldwide patterns are modified locally by a variety of factors. For example, along our own West Coast, where the winds are prevailing westerlies, the western slopes of the Sierra Nevadas have abundant rainfall, while the eastern slopes are dry and desertlike (Figure 35-4). As the air from the ocean hits the western slope, it rises, is cooled, and releases its water. Then, after passing the crest of the mountain range, the air descends again, becomes warmer, and its water-holding capacity increases, resulting in a so-called "rain shadow" on the eastern slope.

The mean annual rainfall (vertical columns) in relation to altitude at a series of stations from Palo Alto on the Pacific Coast across the Coastal Range and the Sierra Nevada mountains. The prevailing winds are from the west and there are rain shadows on the eastern slopes of the two mountain ranges.

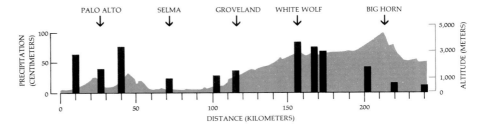

The Earth's Atmosphere

Another major influence on the biosphere is the mixture of gas molecules and, usually, dust and other solid particles held in a blanket over the earth's surface by gravity. Of the radiant energy directed toward the earth from the sun, about 29 percent (averaged over the surface of the globe) is reflected back into space. About 20 percent is absorbed by the atmosphere. The remaining 51 percent travels through the atmosphere and is absorbed at the earth's surface. Much of the light energy absorbed by the earth is reradiated as heat energy. The atmosphere impedes the loss of this heat energy from the earth's surface, just as a blanket impedes the loss of heat from a person's body. The major contributors to this blanketing effect are carbon dioxide and water vapor.

The Earth's Surface

Much of the iron and other heavier materials of which the earth is composed are collected in a dense core in the center. Surrounding the core is a lighter layer of solid and molten rock. The outermost layer is made up of interlocking, mobile plates on which the continents rest.

The continents themselves are composed of granite, which is an igneous rock (from the Latin word *ignis*, for "fire"); igneous rock is rock formed directly from molten material. The surfaces of the continents change constantly. They are crumpled by contractions and collisions as the continents rise, sink, and collide because of the motion of the plates beneath them. As a consequence, the earth's surface is not at all uniform but varies widely from place to place in its composition and in its height above sea level. Both the mineral content of the earth's surface and the altitude affect the growth of plants and other living organisms, and as we saw in the example of the Sierra Nevadas, the mountain ranges of the continents do much to determine the patterns of rainfall.

The interplay of all these factors makes various parts of the earth's surface very different from one another, and these differences—in the availability of water, temperature, light, altitude, and soil composition—are the strongest factors (with the possible exception, as we shall see, of man) in determining the pattern of growth of living things.

LIFE ON LAND: THE BIOMES

The major different life formations on land are called *biomes* (Figure 35-5). These biomes, the big landscapes, are generally characterized and identified by their vegetation.

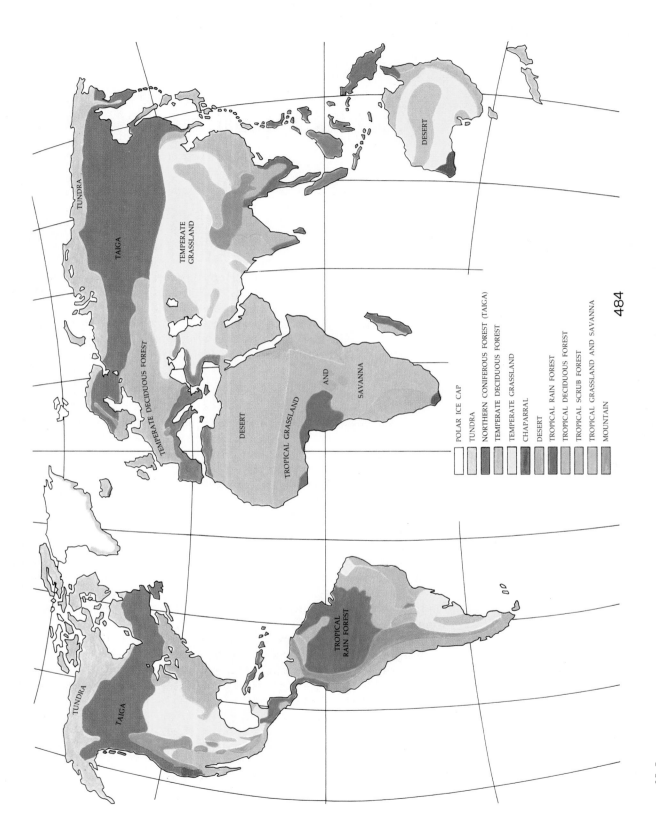

POLAR ICE CAP

TUNDRA

NORTHERN CONIFEROUS FOREST (TAIGA)

TEMPERATE DECIDUOUS FOREST

TEMPERATE GRASSLAND

CHAPARRAL

DESERT

TROPICAL RAIN FOREST

TROPICAL DECIDUOUS FOREST

TROPICAL SCRUB FOREST

TROPICAL GRASSLAND AND SAVANNA

MOUNTAIN

35-5
Biomes of the world.

484

35–6
A tropical rain forest, Rancho Grande National Park, in Venezuela. Notice the height of the trees and the huge, buttressed trunk of the tree in the center. The vine growing up it is a philodendron.

A biome is thus an abstraction. When we speak of the tropical forest biome we are not speaking of a particular geographical region, but rather of all the tropical forests on the planet. As with most abstractions, important details are omitted. For example, the boundaries are not so sharp as can be shown on a map nor are all areas of the world easy to categorize. However, the biome concept emphasizes one important truth: Where the climate is the same, the organisms are also very similar, even though these organisms are not related genetically and are far apart in their evolutionary history. This phenomenon, as we noted in the previous chapter, is called convergent evolution.

Tropical Forests

In the equatorial zone, where most of the tropical forest formations are found, the mean daily temperature is the same throughout the year and the length of day varies by less than one hour. Rainfall is seasonal, however, with maxima at the time of the equinoxes (Figure 35–2). Variations in total rainfall from one area to another within these zones are caused principally by mountains and their rain shadows.

Tropical Rain Forest

In the tropical rain forest, rainfall is abundant all year around; total rainfall is between 200 and 400 centimeters per year, and a month with less than 100 millimeters of rain is considered relatively dry. More species of plants and animals live in the tropical rain forest than in all the rest of the biomes of the world combined. As many as 100 species of trees can be counted on 1 hectare (2.47 acres). Conversely, in contrast to the deciduous forests of the northeastern United States, where, typically, only a few species are represented in any particular area, in the tropical rain forest a species may be represented only once per hectare.

The competition among plants of the tropical forest is for light. About 70 percent of all species of plants are trees. The upper tree story consists of solitary giants 50 to 60 meters tall. A lower story of trees characteristically forms a solid canopy. They are remarkably similar in appearance. The trunks of the trees are usually slender and branch only near the crown. The crowns are high up and relatively small as a result of crowding. Because the soil is perpetually wet, the roots do not reach deep into the soil, and trunks often end in thick buttresses that provide firm, broad anchorage. Their leaves are large, leathery, and dark green; their bark is thin and smooth; and their flowers are generally inconspicuous and greenish or whitish in color.

Woody vines, or lianas, are abundant, especially where an opening has appeared in the forest, as a result, for example, of a tree's falling; vines as long as 240 meters have been measured. There are also many *epiphytes,* which are plants that grow on the branches of other plants high above the forest floor (Figure 35–7). These epiphytes germinate in the branches of trees and obtain water and minerals from the humid air of the canopy. Unlike the plants that have contact with the moist floor, epiphytes need to conserve water between rainfalls. Some epiphytes resemble desert succulents, having fleshy, water-storing leaves and stems. Others have spongy roots or cup-shaped leaves that capture moisture and organic debris; many of these epiphytes can take up nutrients from decaying organisms in these storage tanks. A variety of plants, including ferns, orchids, mosses, and lichens, have exploited this life style.

35-7

Epiphytes, such as this bromeliad, grow in the canopy of the tropical rain forest, obtaining water and minerals from the moist air. Bromeliads are members of the pineapple family that are especially adapted to the tropical forest biome.

An extraordinary variety of insects, birds, and other animals, including mammals, have moved into the treetops along with the vines and epiphytes to make it the area of the tropical rain forest that is most abundantly and diversely populated.

Little light reaches the forest floors (from 0.1 to 1 percent of the total), and the few plants that are found there are adapted to growing at low light intensities. Many of these, such as the African violet, are familiar to us as house plants.

There is almost no accumulation of leaf litter on the forest floor such as we find in our northern forests; decomposition is too rapid. Everything that touches the ground disappears almost immediately—carried off, consumed, or rapidly decomposed. In many places the ground is bare.

The soils of tropical rain forests are relatively infertile. Many are chiefly composed of a red clay; these red soils are known as laterites, from the Latin word *later,* or "brick." When laterite soils are cleared, in many cases they either erode rapidly or form thick, impenetrable crusts that cannot be cultivated after a season or two. Tropical soils are generally deficient in minerals; most of the nitrogen, phosphorus, calcium, and other nutrients are found in the plants rather than the soil. Also, those minerals that are present in the soil are leached out by the heavy rainfall. As a consequence, soils where tropical forests have stood are very poor agriculturally and will support crops for only a few years. The rapidly expanding human population in the tropics has made the traditional tropical agricultural practices of clearing and short-term cultivation immensely destructive because they are now carried out on such a wide scale. Although the tropical rain forest now forms about half of the forested area of the earth, some ecologists predict that, at the present rate of destruction, it may almost all have disappeared by the year 2000, and with it thousands of species of plants and animals found nowhere else in the world.

Other Tropical Forests

Where the rainfall is more seasonal, tropical rain forests give way to tropical deciduous forests, which contain trees that lose their leaves during the dry seasons. Many of the species flower before they put out new leaves (transpiration, or water loss, is extremely low in leafless, flowering trees because petals have few stomata). Tropical scrub forests, which have trees with small, water-conserving leaves, are found where rainfall is limited all year around.

35-8

Very few kinds of plants and animals are found in both the Old and New World jungles. Many species, however, closely resemble each other, such as (a) the toucans of South America and (b) the hornbills of Africa and Asia, both of which have bills adapted for eating fruit and crushing nuts. This is an example of convergent evolution.

(a)   (b)

A water hole on a savanna in Kenya, with zebras and hartebeests. The trees in the background are acacias.

Savanna

Savannas are tropical grasslands with scattered clumps of trees (Figure 35–9). The transition from open forest with grassy undergrowth to savanna is gradual and is determined by the duration and severity of the dry season and, often, by fire and grazing and browsing animals.

In the savanna, the competition is for water, and grasses are the victors. Grasses are well suited to a fine, sandy soil with seasonal rain because their roots form a dense root system capable of extracting the maximum amount of water during the rainy period. During dry seasons, the above-ground portions of the plants die, but the roots are able to survive even many months of desiccation. The balance between woody plants and grasses is a delicate one. If rainfall decreases, the trees die entirely. If rainfall increases, the trees increase in number until they shade the grasses, which in turn die. If the grasses are overgrazed (which often happens when man begins to use the savanna for agricultural purposes and introduces livestock), enough water is left in the soil so that the woody plants can increase in number and the grassland is eventually destroyed.

The best-known savannas are those of Africa, which are inhabited by the most abundant and diverse group of large herbivores in the world, including the gazelle, the impala, the eland, the buffalo, the giraffe, the zebra, and the wildebeest.

The Desert

The great deserts of the world are located at latitudes of about 30°, both north and south, and extend poleward in the interiors of the continents. These are areas of falling, warming air and, consequently, little rainfall. The Sahara Desert, which stretches all the way from the Atlantic coast of Africa to Arabia, is the largest in the world, almost equal to the size of the United States. Only about 5 percent of North America is desert.

35–10

The North American deserts. The Sonoran stretches from southern California to western Arizona and down into Mexico. A dominant plant, the giant saguaro cactus (a), is often as much as 15 meters high, with a widespreading network of shallow roots. Water is stored in a thickened stem which expands, accordionlike, after a rainfall.

To the southeast is the Chihuahuan desert, one of whose principal plants is the agave (b), or century plant, a monocot.

North of the Sonoran is the Mojave, whose characteristic plant is the grotesque Joshua tree (c). This plant was named by early Mormon colonists who thought that its strange, awkward form resembled a bearded patriarch gesticulating in prayer. The Mojave contains Death Valley, the lowest point on the continent (90 meters below sea level), only 130 kilometers from Mt. Whitney, whose elevation is more than 4,000 meters.

The Mojave blends into the Great Basin, a cold desert bounded by the Sierra Nevada to the west and the Rockies to the east. It is the largest and bleakest of the American deserts. The dominant plant form is sagebrush (d), shown here in the background. The large green plant in the foreground is shadscale, and the yellow-flowered plant to the left is rabbitbrush.

(a)

GREAT BASIN DESERT

MOJAVE DESERT

SONORAN DESERT

CHIHUAHUAN DESERT

(b)

(c)

(d)

35–11

Inhabitants of the North American deserts: (a) Gamble quail, (b) giant hairy scorpion, (c) golden carpet, found only (and rarely) in Death Valley, (d) horned lizard with a grasshopper, (e) roadrunner with a whiptail lizard, (f) bighorn sheep of the southwestern desert mountains, and (g) gila monster.

(a)

(b)

(c)

(d)

(e)

(f)

(g)

Desert regions are characterized by less than 25 centimeters of rain a year. Because there is little water vapor in the air to moderate the temperature (see page 25), the nights are often extremely cold. The temperature may drop as much as 30° C at night, in comparison with the humid tropics, where day and night temperatures vary by only a few degrees.

Many of the plants in the desert are annuals that race from seed to flower to seed during periods when water is available, and during the brief growing seasons, the desert may be carpeted with flowers. The perennials are succulents, such as cacti, which store water, are drought-deciduous (dropping their leaves in dry seasons), or have small, leathery, water-conserving leaves.

The animals that live in the desert have special adaptations to this extreme climate. Reptiles and insects, for example, have waterproof outer coverings and dry, and therefore water-conserving, excretions. The few mammals of the desert are small and nocturnal and get what little water they require from the plants they eat.

Chaparral

Regions with mild, rainy winters and long, hot, dry summers, such as the southern coast of California, are dominated by small trees or often by spiny shrubs with broad, hard, thick evergreen leaves. In the United States, such areas are known as chaparral. Similar communities are found in areas of the Mediterranean (where they are called the *maquis*, which became the name of the French underground in World War II), in Chile (where they are the *matorral*), in southern Africa, and along portions of the coast of Australia. Although the plants of these various areas are essentially unrelated, they closely resemble one another in their growth patterns and characteristic appearance.

35–12
The bushy vegetation that characterizes the chaparral grows as dense as a mat on the foothills of southern California. It is the result of long, dry summers, during which much of the plant life is semidormant, followed by a brief, cool rainy season. The name comes from chapparo, *the Indian word for the scrub oak that is one of the prominent components of the chaparral. Chaps, the leather leggings worn by the cowboys making their way through this dense, dry growth, has the same derivation.*

Mule deer live in the North American chaparral during the spring growing season, moving out to cooler regions during the summer. The resident vertebrates—brush rabbits, lizards, wren-tits, and brown towhees—are generally small and dull-colored, matching the dull-colored vegetation.

Grasslands

Grasslands, which are transitional areas between the deserts and the temperate forests, are found in the interior areas of continents. They are characterized also by periodic droughts, rolling to flat terrain, and hot-cold seasons (rather than the wet-dry seasons of the savanna). The great grasslands of the world include the plains and prairies of North America, the steppes of Russia, the veld of South Africa, and the pampas of Argentina.

The vegetation is largely bunch or sod-forming grasses, often mixed with legumes (clover and alfalfa) and various annuals. In North America, there is a transition from the more desertlike, western, short-grass prairie (the Great Plains), through the moister, richer, tall-grass prairie (the Corn Belt), to the eastern temperate deciduous woodland. Grasslands become drier and drier at increasing distances from the Atlantic Ocean and the Gulf of Mexico, which are the major sources of moisture-bearing winds in the eastern half of the continent.

The grasslands of the world support small, seed-eating rodents and also large herbivores, such as the bison of early America, the gazelles and zebras of the African veld, the wild horses, wild sheep, and ibex of the Asiatic steppes, and now the domestic herbivores raised by man. These large, grass-eating mammals, in turn, support the carnivores, such as lions, tigers, and wolves, as well as omnivorous man. The herds of grazing animals serve, as do ground fires, to maintain the nature of the grassland landscape, destroying tree seedlings and preventing their encroachment.

(a)

35–14

(a) *A prairie dog, a characteristic inhabitant of short-grass prairies. Large animals graze in herds for protection on the open plains, while smaller animals, such as the prairie dog shown here, often seek shelter and safety below ground. Prairie dogs (actually burrowing squirrels) live in vast underground cities, often inhabited by millions of individuals. (b) A June day on a tall-grass prairie in North Dakota. The cottonwood grove by the prairie creek is characteristic of this biome. A thunderstorm is gathering on the horizon.*

(b)

Temperate Deciduous Forest

Deciduous forests occupy areas where there is a warm, mild growing season with moderate precipitation, followed by a colder period less suited to plant growth. Leaf-shedding in the deciduous, as in the tropical, forest is a protection against water loss.

The dominant trees of the deciduous forests vary from region to region, depending largely on the local rainfall. In the northern and upland regions, oak, birch, beech, and maple are the most prominent trees. Before the chestnut blight struck North America, an oak-chestnut forest ran from Cape Ann, Massachusetts, and the Mohawk River valley of New York to the southern end of the Appalachian highland. Maple and basswood predominate in Wisconsin and Minnesota, and maple and beech in southern Michigan, becoming mixed with hemlock and white pine as the forest moves northward. The southern and lowland regions have forests of oak and hickory. Along the southeastern coast of the United States, the wet, warm climate supports an evergreen forest of live oak and magnolia.

In deciduous woodlands there are up to four layers of plant growth.

1. The tree layer, in which the crowns form a continuous canopy. The canopy is usually between 8 and 30 meters high.
2. The shrub layer, which grows to a height of about 5 meters. Shrubs and bushes resemble trees in that they are woody and deciduous, but they branch at or close to the ground.
3. The field layer, made up of grasses and other herbaceous (nonwoody) plants, including the wild flowers, which typically bloom in the spring before the trees regain their leaves. Bracken and other ferns, whose large leaf areas make them efficient interceptors of light, are also often conspicuous members of the field layer.
4. The ground layer, which consists of mosses and liverworts. It is also often covered with a layer of leaf litter.

Beneath the ground layer is the soil of the forest, often a rich, gray-brown topsoil. Such soil is composed largely of organic material—decomposing leaves and other plant parts and decaying insects and other animals—and the bacteria, protozoans, fungi, worms, and arthropods that live on this organic matter. The roots of plants penetrate the soil to depths measurable in meters and add organic matter to the soil when they die. Carnivorous arthropods carry fragments of their prey to considerable depths in the soil. The myriad passageways left by dead roots and fungi and by the earthworms and other small animals that inhabit the forest make the soil into a sponge that holds the water and minerals. The land where deciduous forests have stood makes good farmland.

The deciduous forest supports an abundant animal life. Smaller mammals, such as chipmunks, voles, squirrels, raccoons, opossums, and white-footed mice, live mainly on the nuts and fruits of the forest. The wolves, bobcats, gray foxes, and mountain lions, in the areas where they have not been driven out by the encroachments of civilization, feed on these smaller mammals. Deer live mainly on the forest borders, where shrubs and seedlings grow.

Temperate deciduous forests once covered most of eastern North America, as well as most of Europe, part of Japan and Australia, and the tip of South America. In the United States, only scattered patches of the original forest still remain.

35–16   **492**

Among the common harbingers of spring in the deciduous forest is the white-flowered bloodroot, a member of the poppy family. It derives its name from the orange-colored juice that flows from a broken stem.

35–17

The deciduous forest and some of its inhabitants. (a) A birch forest, showing layers of plant growth. The shrub layer is made up of thimbleberry and mountain maple, a favorite browse of deer. (b) Bracket fungi on dead elm. (c) Black-throated blue warbler. (d) Raccoon prowling through snow. (e) Eastern chipmunk.

(a)

(b)

(c)

(d)

(e)

Coniferous Forests

With the exception of a few groups, conifers are evergreens with small compact leaves protected by a thick cuticle that guards against water loss. The dropping of leaves is a more efficient adaptation to a hot-cold season, and conifers generally cannot compete with deciduous trees in temperate zones with adequate summer rainfall. However, deciduous trees require a warm summer period of at least four months to permit their leaves to regrow. The northern boundary between the deciduous forest and the coniferous forest occurs where the summers become too short and the winters too long for deciduous trees to grow well.

Taiga

The northern coniferous forest is characterized by long, severe winters and a constant cover of winter snow. It is composed chiefly of evergreen needle-leaved trees such as pine, fir, spruce, and hemlock. A thick layer of needles and dead twigs, matted together by fungus mycelium, covers the ground. Along the stream banks grow deciduous trees, such as tamarack, willow, birch, alder, and poplar. There are virtually no annual plants.

The principal large animals of the North American coniferous forest are elk, moose, mule deer, black bears, and grizzlies. Among the smaller animals are porcupines, red-backed mice, snowshoe hares, shrews, wolverines, lynxes, warblers, and grouse. The small animals use the dense growths of the evergreens for breeding and for shelter. Wolves feed upon these other mammals, particularly the larger ones. The black bear and grizzly bear eat everything—leaves, buds, fruits, nuts, berries, fish, the supplies of campers, and occasionally the flesh of other mammals. Porcupines are bark eaters, and many seriously damage trees by girdling them. Moose and mule deer are largely browsers. The ground layer of the coniferous forest is less richly populated by invertebrates than that of the deciduous forest because the accumulated litter is slower to decompose.

35–19

The great horned owl ranges the American forest from the North American timberline to the Straits of Magellan. The eyes of owls are adapted to night vision, their hearing is acute, and their flight is completely silent (owing in part to their soft feathers), all of which are advantageous in capturing prey.

35–20

An example of a coniferous forest. Red pines. Fires, which have left their marks on the tree trunks, have destroyed all undergrowth. The forest floor is carpeted in pine needles.

35–21

A coniferous forest of balsam spruce and white pine. Both of these photographs were taken near the Canadian border.

(a)

(b)

35–22

(a) *Redwoods (Sequoia sempervirens)*
grow only along the coasts of California
and southwestern Oregon. Their distri-
bution is determined by the moisture-laden
coastal fog. The tallest trees in the world,
some reach heights of up to 115 meters

and diameters in excess of 10 meters.
(b) *Sequoiadendron giganteum, the giant*
sequoia, is found only in widely scattered
groves on the western slopes of the Sierra
Nevadas. Giant sequoias are not as tall
as the coastal redwoods (the tallest

known sequoia is not quite 90 meters), but
they are heavier. In fact, the giant sequoia
is the largest living organism. These two
are among the last surviving species of a
redwood empire that, 25 million years
ago, stretched across North America.

The Pacific Coast

An unusual and magnificent coniferous forest grows along the mountain range
of the Pacific Coast of North America. In Alaska, it is composed largely of
hemlock and Sitka spruce. Between British Columbia and Oregon, the Douglas
fir is the dominant tree. In southern Oregon and northern California are found
the coastal redwoods. *Sequoia sempervirens,* the "always green" redwood, has a
life span of more than 2,000 years. Nearly all these redwoods grow within 32
kilometers of the ocean.

The trees of the Pacific forests are all very tall, taller than those of the
deciduous forests. The tallest living specimen of redwood measures 110 meters.
Douglas firs range up to 80 meters tall and 15 meters in circumference. The
giant sequoia, *Sequoiadendron giganteum,* occurs only on the western slopes of the
Sierra Nevadas. Some of these trees are known to be more than 2,000 years old.
Recent studies have shown that these forests may owe their existence to fires
that cleared the ground of the seedlings of other trees, while leaving the thick-
barked sequoias virtually unharmed.

(a)

(b)

(c)

35–23

(a) The ptarmigan and (b) the varying hare, also known as the snowshoe hare because of its large feet, which enable it to walk on snow. Both are year-round residents of the tundra that change their coats with the changing seasons. The ptarmigan is unusual in having three changes of costume: brown for summer, gray for autumn, and white for winter. (The changes are regulated by photoperiodism.) (c) The arctic tern represents one of a number of bird species that breed in the tundra, taking advantage of the long summer days to gather food for their nestlings. The terns winter in the Antarctic, following migration routes of 13,000 to 18,000 kilometers. Within three months after hatching, the young must be ready to migrate.

35–24

Tundra of North America on a long Arctic day. A glacial range is visible in the background.

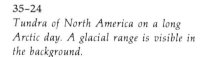

The Tundra

Where the climate is too cold and the winters too long even for the conifers, the taiga grades into the tundra. The tundra is a form of grassland that occupies one-tenth of the earth's land surface, forming a continuous belt across northern North America, Europe, and Asia. Its most characteristic feature is permafrost, a layer of permanently frozen subsoil. During the summer the ground thaws to a depth of a few centimeters and becomes wet and soggy; in winter it freezes again. This freeze-thaw process, which tears and crushes the roots, keeps the plants small and stunted. Drying winter winds and abrasive driven snow further reduce the growth of the arctic tundra.

The virtually treeless vegetation of the tundra is dominated by herbaceous plants, such as grasses, sedges, rushes, and heather, beneath which is a well-developed ground layer of mosses and lichens, particularly the lichen known as reindeer moss. All the flowering plants are perennials.

The largest animals of the arctic tundra are the musk oxen and caribou of North America and the reindeer of the Old World. Lemmings (small, furry rodents with short tails) and ptarmigans (pigeon-sized grouse) feed off the tundra. The white fox and the snowy owl of the Arctic live largely on the lemmings.

During the brief arctic summer, insects emerge in great numbers, and migratory birds visit, taking advantage of the insect hordes and the long periods of daylight. The growing season in many areas of the tundra is less than two months.

35–25
Most of the major lakes of the northern hemisphere were scraped into the continents by glaciations. This photograph shows Found Lake in Glacier Peak Wilderness, Washington. Composed of glacial meltwater, the lake contains white pulverized rock from the glacier and therefore reflects a chalky green color.

FRESHWATER BIOMES

In general, limnologists—scientists who study natural fresh waters in all their aspects—classify freshwater habitats into two groups: standing water (lakes and ponds) and running water (rivers and streams).

Lakes and Ponds

Lakes vary in size from large seas, covering thousands of square kilometers, to small ponds. Lakes contain three distinct zones: littoral, limnetic, and profundal.

The *littoral zone*, at the edge of the lake, is the most richly inhabited. Here the most conspicuous plants are angiosperms rooted to the bottom, such as cattails and rushes. Water lilies grow farther out from shore. There is often a green blanket of duckweed, a small free-floating angiosperm. Other pond weeds grow entirely beneath the water, where plants lack a waxy outer cuticle and so can absorb minerals through their epidermis as well as through their roots. Their submerged plant surfaces harbor large numbers of small organisms. Snails, small arthropods, and mosquito larvae feed upon the plants. Other insects that live among the submerged plants, such as the larvae of the dragon-fly and damselfly, and the water scorpion, are carnivorous. Clams, worms, snails, and still other insect larvae burrow in the mud. Frogs, salamanders, water turtles, and water snakes are found almost exclusively in these littoral zones, where they feed primarily upon the insects. Fish, too, are found in greater numbers along the lake margins. Ducks, geese, and herons feed on the plants, insects, mollusks, fish, and amphibians abundant in this zone. The shallow margins of some lakes and ponds are marshy. Among the inhabitants of marshes are such invertebrates and vertebrates as snails, frogs, ducks, herons, bitterns, muskrats, otters, and beavers.

In the *limnetic zone*, the zone of open water, small, floating algae—phyto-plankton—are usually the only photosynthetic organisms found. This zone, which extends down to the limits of light penetration, is the habitat, for example, of smallmouth bass, bluegills, and, in colder waters, of trout.

The deepwater *profundal zone* extending down from the limnetic zone has no plant life. Its principal occupants are bacteria and fungi that decompose the organic debris filtering down from the overlying water.

35–26
(a) A great blue heron at the edge of a lake. The water is covered with duckweed. (b) Pond with water lilies.

(a)

(b)

Seasonal cycle of temperature changes in a temperate-zone lake. Water, like air, increases in density as it cools, reaching a maximum density at 4°C. In the summer, the top layer of water, called the epilimnion, becomes warmer than the lower layers and therefore remains on the surface. Only the water in this warm, oxygen-rich layer circulates. In the middle layer, which is called the thermocline, there is an abrupt drop in temperature. Since the thermocline water does not mix, it cuts off oxygen from the third layer, the hypolimnion, producing summer stagnation (a). In the fall, the temperature of the epilimnion drops until it is the same as that of the hypolimnion. The warmer water of the thermocline then rises to the surface, producing the fall overturn (b). Aided by the fall winds, all the water of the lake begins to circulate (c), and oxygen is returned to the depths. As the surface water cools below 4°C, it expands, becoming lighter, and remains on the surface; in many areas, it freezes. The result is winter stratification (d). In spring, as ice melts and the water on the surface warms to 4°C, it sinks to the bottom, producing the spring overturn (e), after which the water again circulates freely (f).

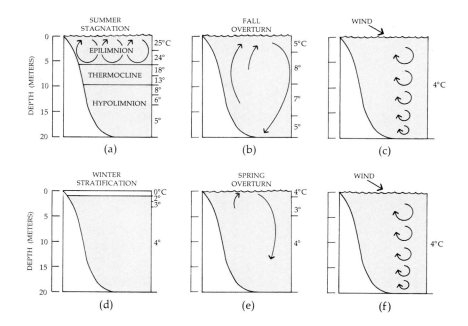

Ice

Liquid water, like other liquids, contracts as it gets colder. However, when water reaches 4°C, it expands, and because it expands, it becomes less dense and therefore lighter. Hence, ice cubes float at the top of a water glass, icebergs cruise the surface of the oceans, and ponds and lakes freeze from the top down.

But what if water behaved like an "ordinary" liquid and became denser and denser as it froze? Not only would ice cubes and icebergs plummet to the bottom, but ponds and lakes and perhaps even oceans would begin to freeze from the bottom up. Spring and summer might stop the freezing process, but laboratory experiments have shown that if ice is held to the bottom of even a relatively shallow tank, water can be boiled on the top without melting the ice. Thus if water did not expand when it froze, it would continue to freeze from the bottom up, year after year, and never melt again, leaving half of our planet locked in ice.

Because water increases in density as its temperature drops (until it reaches 4°C), a lake during the summer will consist of an upper layer of water that has been heated by the sun and a lower layer of denser, cold water. In winter, the coldest, least dense water (below 4°C) will remain on top, and the slightly warmer water will sink to the bottom. This stratification effectively seals off the depths from any contact with the air. Also, nutrients derived from the organic debris that collects on the lake bottom cannot rise to the surface. Consequently, during the summer and winter, the supply of oxygen in the profundal zone and of nutrients in the limnetic zone may run short, imposing special stresses on the inhabitants of these zones. In the spring and fall, when the entire body of water approaches the same temperature, mixing occurs, redistributing oxygen and nutrients and often causing "blooms" of phytoplankton to appear.

Rivers and Streams

Rivers and streams are characterized by continuously moving water. They may begin as outlets of ponds or lakes, as runoffs from melting ice or snow, or they may arise from springs (flows of groundwater emerging from bedrock).

*A fast-moving stream in Glacier Park,
Montana.*

The character of life in a stream is determined to a large extent by the swiftness of the current, which characteristically changes as a stream moves downward from its source and, fed by tributaries, increases in volume and decreases in speed. In swift streams, most organisms live in the riffles, or shallows, where small photosynthetic algae and mosses cling to rock surfaces. Many insects, both adult and immature forms, live on the underside of rocks and gravel in the riffles. For those small organisms that can survive the swiftly moving current, there is an abundance of oxygen and of nutrients swept along by the flowing waters.

As the stream travels along its course, the riffles are often interrupted by quieter pools, where organic material may collect and be decomposed. Few plants can gain footholds on the shifting bottoms of stream pools, but some invertebrates, such as dragonflies and water striders, are typically found in or about the pools. Some organisms, notably trout, move back and forth between the riffles and the pools.

As the streams broaden and become slower, they begin to take on the characteristics of lakes and ponds.

The Oceans

The oceans cover almost three-fourths of the surface of the earth. Life extends to its deepest portions, but photosynthesizing organisms are restricted to the upper, lighted zones. The sea has an average depth of more than 3 kilometers and, except for a very small surface fraction, is dark and cold. Most of it, therefore, is inhabited by bacteria, fungi, and animals, rather than plants.

The sea absorbs light readily. Even in clear water, less than 40 percent of the sunlight reaches a depth of 1 meter, and less than 1 percent of the sunlight that

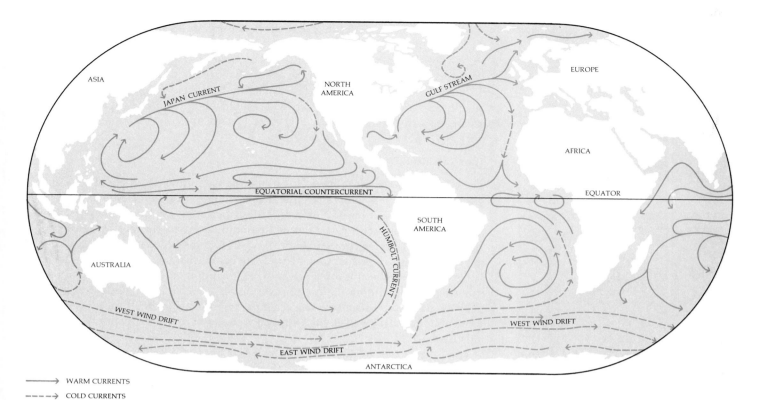

Map labels: ASIA, JAPAN CURRENT, NORTH AMERICA, GULF STREAM, EUROPE, AFRICA, EQUATOR, EQUATORIAL COUNTERCURRENT, SOUTH AMERICA, HUMBOLT CURRENT, AUSTRALIA, WEST WIND DRIFT, WEST WIND DRIFT, EAST WIND DRIFT, ANTARCTICA

⟶ WARM CURRENTS

---→ COLD CURRENTS

35–29

*The major currents of the ocean have pro-
found effects on climate. Because of the
warming effects of the Gulf Stream, Europe
is milder in temperature than is North
America at similar latitudes. The eastern
coast of South America is warmed by water
from the equator, and the Humboldt Current
brings cooler weather to the western coast of
South America.*

35–30

*A sample of living phytoplankton showing
several species of diatoms. Plankton is the
main component of the pelagic division of the
sea.*

reaches the surface penetrates below 50 meters. Red, orange, and yellow wavelengths of the light are absorbed first, so that only the shorter wavelengths,
specifically blue and green, penetrate deeply. Thus, below depths of a few
meters, only those photosynthesizing organisms capable of utilizing the short
wavelengths of light can grow.

There are two main divisions of life in the open ocean: *pelagic* (free-floating)
and <u>benthic</u> (bottom-dwelling). A major component of the pelagic division is
plankton. It is composed of photosynthetic algal cells (phytoplankton), intermingled with small shrimp and other crustaceans and the eggs and larval forms
of many fish and invertebrates (zooplankton). These planktonic forms provide
food for the fish and other animals that are also part of the pelagic division. The
benthic division contains the bottom dwellers, the sessile animals, such as
sponges, sea anemones, and clams, and many mobile animals, such as worms,
starfish, snails, crustaceans, and deepwater fish. A variety of fungi and bacteria
also inhabit the benthic zone, subsisting on the sparse but steady accumulation
of debris drifting from the more populated levels of the ocean.

The major ocean currents, which are produced by a combination of winds
and the earth's rotation, profoundly affect life in the oceans and alter the climate
along the ocean coasts. These patterns of water circulation—clockwise in the
northern hemisphere and counterclockwise in the southern hemisphere—move
currents of warm water north and south from the equator. One such current,
the Gulf Stream, warms a portion of the eastern coast of North America and the
western shores of Europe, and another warms the eastern coast of South
America. The same patterns of circulation bring cold waters to the western
coasts of North and South America. Where the winds move the water continuously away from the shores, as off the coasts of Portugal and Peru, cold water
rich in nutrients is brought to the surface in upwellings. Such areas traditionally
have highly profitable fishing industries.

35-31

An inhabitant of the benthic division, before and after eating. The prey was attracted by wiggling the lure at the end of the appendage above the mouth. This angler fish is a female. The males are much smaller and are parasitic upon the females, an arrangement which solves the problem of finding partners in the dark.

Despite the fact that oceans cover three times more surface area of the planet than does the land, the total productivity of the open ocean—as measured by the amount of carbon converted to organic compounds by photosynthesis—is only about one-third as great. In fact, the open ocean is only slightly more productive per square meter than the desert, presumably because of the low concentration of minerals in the areas of ocean where light penetrates and photosynthetic organisms can survive.

The Coral Reef

The coral reef is the most diverse of all marine communities. The reef structure itself is formed by colonial coelenterates. Each polyp in the colony secretes its own calcium-containing skeleton that then becomes a part of the reef. The photosynthetic activity of the reef is carried out almost entirely by symbiotic algae living within the corals. Carbon, oxygen, and dissolved minerals flow over the reef as a result of the movement of the ocean currents. The reef furnishes both food and shelter for other sea animals, including numerous species of reef fishes and a tremendous variety of invertebrates, such as sponges, sea urchins, polychaetes, and crustaceans. The coelenterates and algae that form the reef can grow only in warm, well-lighted surface water, where the temperature seldom falls below 21°C.

The longest reef in the world is the Great Barrier Reef of Australia, which extends some 2,000 kilometers. Other reefs are found throughout tropical waters and as far north as Bermuda, which is warmed by the Gulf Stream.

The Seashore

The edges of the continents extend 10 to 20 kilometers out into the sea. Along these edges, known as the continental shelves, nutrients are washed out from the land, and life is much denser than in the open seas. Sessile animals, such as

35-32

A reef in the New Hebrides, islands in the Pacific, east of Australia. Staghorn coral is in the foreground.

(a)

(b)

35-33

(a) *Sea otter in a kelp bed off the coast of California. (b) Tidal pools at low tide on Alaska's rocky southeast coast.*

sponges and corals, are found all over the ocean bottom, but they are abundant only in areas that are close to the shores. They are especially numerous in warmer waters. Predators, such as mollusks, echinoderms, crustaceans, and many kinds of fish, roam over the bottoms of the continental shelves. Eel grasses, turtle grasses, and seaweeds provide shelter for many animals and increase the supply of oxygen. Snails, slugs, and worms crawl over the plant surfaces, eating off the encrusted growth, and fish nibble the smaller animals that cling to the plants.

Seashores—where the sea and the land join—are of three general types along most of the shores of the temperate zones: rocky, sandy, and muddy. The organisms that live on rocky coasts, like those that live in the riffles of fast-moving streams, have special adaptations for clinging to rocks. The algae have strong holdfasts. The starfish of the rocky coasts lies spread-eagled on the rocks, clinging with its suction cups. The abalone holds tight with its well-developed muscular foot. Mussels secrete coarse, ropelike strands that anchor them to rocky surfaces.

The organisms of the rocky coast face the additional problem of the rising and falling tides that threaten them alternately with too much and too little water. Largely as a consequence of the effects of the tidal ebb and flow and the adaptations of organisms to it, life along the rocky coasts is highly stratified. The supralittoral zone, which is flooded only occasionally by high tides, is a zone of dark algal and lichen growth. The littoral zone, submerged and exposed daily, is characterized by rockweed (a brown alga), often intermixed with Irish moss (a red alga) and other seaweeds. Animal life includes barnacles, oysters, blue mussels, limpets, and periwinkles. The sublittoral zone, exposed only occasionally, contains forests of the large brown alga *Laminaria*, sea squirts, starfish, and various other invertebrates.

Sandy beaches have fewer bottom dwellers because of the constantly shifting sands. Clams, numerous ghost crabs, sand fleas, lugworms, and other small invertebrates live below the surface of the sand, sometimes emerging at low tide to feed on the debris washed in and out by the tide. Along the sandy beaches, beach grasses, which spread by means of underground stems, are important for stabilizing the shifting dunes.

35-34

A ghost crab. These crabs, also known as racing crabs, run sideways at high speeds across the sandy tropical beaches that are their habitat.

(a)

(b)

35–35

(a) *A salt marsh. The grasses are* Spartina.
(b) *Whooping cranes probing for marine worms.*

The mud flat, while not so rich or diverse in species as the rocky coast, is more suitable than either rocky coasts or sandy beaches for animal growth, with many animals living not only on but beneath its surface. A mud flat can support tens of thousands of individuals per cubic meter.

Mud flats, salt marshes, and estuaries (areas where the fresh water of streams and rivers drains into the sea) are the receiving grounds for a constant flow of nutrients drained off from the land and so are extremely rich in animal life. They serve an important role as the spawning places and nurseries for many forms of marine life.

In the tropics and subtropics (including parts of Florida, Puerto Rico, and Hawaii), mangrove forests are important tideland communities, spawning and nourishing marine organisms and exporting minerals and nutrients.

Because they are often located in prime recreational and commercial areas and because they cannot be directly exploited by man for agriculture or lumbering, thousands of kilometers of mud flats, salt marshes, and mangrove forests are destroyed each year as these wetlands are filled and paved and rendered sterile for man's occupation. Their protection is of special importance because of their role in nurturing the life of the oceans.

SUMMARY

The biosphere is the part of the earth that contains living organisms. It is a thin film on the surface of the planet. The biosphere is affected by the position and movements of the earth in relation to the sun and the movements of air and water over the earth's surface. These conditions can cause wide differences in temperature and rainfall from place to place and season to season on the earth. There are also differences in the surfaces of the continents, both in composition and in altitude.

The major life formations on land are called biomes. Organisms occupying the same biome tend to be similar, even though they are genetically unrelated.

The richest terrestrial biome is the tropical rain forest, where neither water shortage nor extreme temperatures limit plant growth. Here the trees are broad-leaved evergreens, which are characteristically covered with vines and epiphytes. There is almost no collection of decomposing material for humus on the floor of the tropical rain forest. Tropical soils are often red clays (called laterites), which erode or solidify when the forest is cleared.

The savanna biome comprises tropical grasslands on which trees are present. Rain is limited and seasonal.

Desert regions have very little rainfall and high daytime temperatures. The chief vegetation is made up of annual plants with extremely short growing seasons, as well as cacti and other perennial desert plants highly adapted for the conservation of water.

Chaparral, a shrubland biome found on the southern California coast and in the Mediterranean area, is characterized by dry summers and mild, rainy winters. The chief vegetation consists of spiny evergreen shrubs.

The grasslands lie between the deciduous forests and the deserts and have a rainfall intermediate between the two. The growth of trees is prevented not only by a shortage of rain but also by grazing animals and by recurring prairie fires.

The temperate deciduous forest is an important biome of eastern North America and Eurasia, where temperatures are warm in the summer and cold in the winter and where the warm growing season is at least four months long. In the deciduous forest, the trees lose their leaves in the fall, which reduces water loss by the tree during the winter months, when the water in the soil is locked in ice.

North of the deciduous forest is the taiga, the worldwide, subarctic coniferous forest. The trees of this forest, the conifers, have a number of special adaptations that conserve water, protect against extreme cold, and take advantage of the relatively short growing season. Other needle-leaved evergreen forests are found along the Pacific Coast.

Between the taiga and the northern polar region is the tundra, on which there are few trees and the dominant vegetation consists of low-growing perennials. The tundra is characterized by permafrost, a layer of permanently frozen subsoil.

Freshwater communities include lakes and ponds, which are standing water, and rivers and streams, which are running water.

Lakes are conventionally divided into three ecological zones: the littoral zone, along the shore, characterized by rooted vegetation; the limnetic zone of open water extending as far down as the light penetrates; and the deepwater, profundal zone. The littoral zone contains a rich variety of plant and animal life, including snails, many insects, and other small invertebrates. The photosynthetic organisms of the limnetic zone are chiefly phytoplankton; the larger fish are found principally in this zone. There are no plants in the profundal zone; the principal organisms here are decomposers, including bacteria and fungi. Lakes in temperate zones characteristically are thermally stratified and turn over in spring and fall, redistributing oxygen and nutrients.

The character of rivers and streams is determined in large part by the swiftness of the current. In fast-moving fresh water, living organisms are found principally in the riffles, clinging to rocks and wedged into crevices.

The open ocean has two distinct life divisions: the pelagic (planktonic and swimming) division and the benthic (bottom-dwelling) division. The photosynthetic vegetation of the open ocean is almost entirely plankton—floating single-celled algae. The plant plankton (phytoplankton), mixed with animal plankton (zooplankton), provides the basic food supply for fish and deepwater mammals and also for the sparse but diverse animal, fungal, and bacterial life of the ocean bottoms.

Patterns of air circulation combined with the earth's rotation produce the major currents of the oceans. These modify the temperatures of the coastal waters and the continents.

The richest zone of ocean life is the coral reef, which is composed of the bodies and skeletons of coelenterates and occurs primarily in shallow tropical waters.

The seashores include rocky coasts, sandy beaches, and mud flats, each with its characteristic variety of life. Mud flats, salt marshes, and estuaries shelter and nourish many immature forms of marine life.

QUESTIONS

1. Define the following terms: biome, littoral, limnetic, pelagic, taiga, epiphytes, convergent evolution.

2. What are the eight principal biomes? How are they distinguished climatically?

3. Name a plant and an animal associated with each biome and describe the special adaptations of each.

4. Describe the conditions resulting in spring and fall overturn in a lake.

CHAPTER 36   Ecosystems

36-1
At the top of the food chain, a Kodiak bear feeds on salmon.

In the preceding chapter, we described the major biomes of the world, some of the forces that determine their characteristics, and some of the organisms that inhabit each of them. If we look more closely at any one part of the biosphere—at a patch of woodland, a pasture, a pond, or a coral reef—we begin to see that none of the organisms that lives in a particular area exists in isolation; rather, each is involved in a number of relationships both with the other organisms and with factors in the nonliving environment. The details of these relationships vary from place to place. In all cases, however, the interactions of the various organisms in a natural setting have two consequences: (1) a flow of energy through photosynthetic autotrophs (green plants or algae) to heterotrophs, which eat either the autotrophs or other heterotrophs; and (2) a cycling of inorganic materials, which move from the abiotic (nonliving) environment through the bodies of living organisms and back to the abiotic environment.

Such a combination of living and nonliving—biotic and abiotic—elements through which energy flows and minerals recycle is known as an *ecosystem*. Taking a large, astronautical view, the entire biosphere can be seen as a single ecosystem. This view is useful when studying materials that are circulated on a worldwide basis, such as carbon dioxide, oxygen, and water. A suitably stocked aquarium or terrarium is also an ecosystem, and such man-made models may be useful in studying certain ecological problems, such as the details of transfer of a particular mineral element. Most studies of ecosystems have been made, however, on more or less self-contained natural units—on a pond, for example, or a swamp, or a meadow.

The portion of an ecosystem in which a particular organism lives is known as the organism's *habitat*. Thus, within a pond ecosystem, an amoeba's habitat would be a square millimeter of bottom ooze, the surface of the pond would be the habitat of a water strider, and the open water the habitat of a largemouth bass.

The role of each species of organism within an ecosystem is known as an *ecological niche*. The ecological niche, which is an abstract concept, involves such factors as where an organism lives, what kind and size of prey it eats, and the season in which it appears. We shall have more to say about this concept in Chapter 37.

Calculation of the productivity of a field in Michigan in which the vegetation was mostly perennial grasses and herbs. Measurements are in terms of calories per square meter per year. In this field, the net production—the amount of chemical energy stored in plant material—was 4,950,000 calories per square meter per year. Thus, slightly more than 1 percent of the 471,000,000 calories per square meter per year of sunlight reaching the field was converted to chemical energy and stored in the plant bodies.

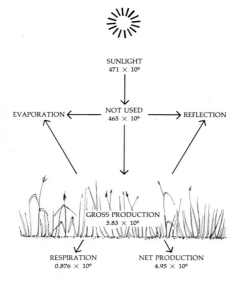

THE SOURCE OF ENERGY

The ultimate source of energy for all natural ecosystems is the sun. The earth receives from the sun an average of 2 calories of radiant energy per minute for every square centimeter of earth's surface—a total of 13×10^{23} (13 followed by 23 zeros) every year. About half of this energy, however, is reflected from the clouds and dust of the atmosphere and never reaches the earth's surface. (Because of these reflections, earth, from outer space, is a shining planet, as bright as Venus.)

Of the solar energy that reaches the ground, some is dissipated in the evaporation of water. Most of it is absorbed into the ground and radiated back in the form of heat. Only a small fraction of the total available energy from the sun enters the food chain of living organisms. Even when the light falls where vegetation is abundant, as in a forest, a cornfield, or a marsh, only 1 or 2 percent of that light (calculated on an annual basis) is used in photosynthesis. Yet this fraction, as small as it is, may result in the production—from carbon, oxygen, water, and a few minerals—of several thousand grams (dry weight) of organic matter per year in a single square meter of field or forest, a total of about 90 billion metric tons of such organic matter per year on a worldwide basis.

TROPHIC LEVELS

The passage of energy from one organism to another takes place along a particular food chain, which is made up of *trophic levels,* or feeding levels. The circumstances of this passage of energy and each organism's relative position on the chain constitute one of the principal distinguishing features of an ecological niche. In most communities, food chains form complex food webs involving many different types of organisms, especially on the lower trophic levels.

Diagram of a food web. The arrows point from consumers to their energy source. This food web is much simplified; in reality, many more animals and plants would be involved.

The Producers

The first step in the food chain is always a primary producer, which on land is usually a green plant and in aquatic ecosystems is usually a photosynthetic alga. These photosynthetic organisms use light energy to make carbohydrates and other compounds, which then become sources of chemical energy. Producers far outweigh consumers; 99 percent of all the organic matter in the biosphere is made up of plants and algae. All heterotrophs combined account for only 1 percent.

Ecologists speak of the *productivity* of ecosystems; productivity is measured by the amount of energy (measured in calories) stored in chemical compounds or by the increase in biomass (measured in grams or metric tons) in a particular length of time. (*Biomass* is a convenient shorthand term meaning the weight of all the living organisms in a given area.) *Net productivity* represents the amount of light energy converted to organic matter less the amount of glucose and other compounds used in respiration; in agricultural communities, the standing crop at the end of the season represents the net primary production.

The Consumers

Energy enters the animal world through the activities of the herbivores, animals that eat plants and algae. A herbivore may be a blue whale, a caterpillar, an elephant, a sea urchin, a snail, or a field mouse; each type of ecosystem has its characteristic complement of herbivores. Of the organic material consumed by herbivores, much is excreted undigested. Some of the chemical energy is transformed to other types of energy—heat or motion—or used in the digestive process itself. A fraction of the material is converted to animal biomass.

The next level in the food chain, the second consumer level, involves a carnivore, a meat-eating animal that devours herbivores. The carnivore may be a lion, a minnow, a starfish, a robin, or a spider, but in each of these cases, only a small part of the organic substance present in the body of the herbivore be-

36–4
Wild barley, a producer.

36–5
Two consumers. (a) Harvest mouse on ripe wheat. The flowers in the background are corn marigolds. (b) Leopard with dead prey in tree.

(a)

(b)

36–6

Pyramids of numbers for (a) a grassland eco-system, in which the number of primary producers (grass plants) is large, and (b) a temperate forest, in which a single primary producer, a tree, can support a large number of herbivores.

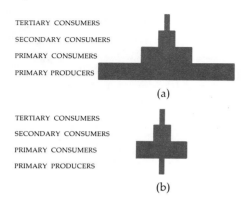

(a)

(b)

36–7

Pyramids for biomass for (a) a field in Georgia and (b) the English Channel. Such pyramids reflect the mass present at any one time, hence the seemingly paradoxical relationship between phytoplankton and zooplankton.

comes incorporated into the body of the consumer. Some chains have third and fourth consumer levels, but five links are usually the absolute limit, largely because of the waste involved in the transfer of energy from one trophic level to another. Other important consumers include the decomposers, whose role will be described in more detail later in this chapter.

Ecological Pyramids

The flow of energy through a food chain is often represented by a graph of quantitative relationships among the various trophic levels. Because large amounts of energy and biomass are dissipated at every trophic level, so that each level retains a much smaller amount than the preceding level, these diagrams nearly always take the form of pyramids. An *ecological pyramid,* as such a diagram is called, may be (1) a pyramid of numbers, showing the numbers of individual organisms at each level; (2) a pyramid of biomass, based either on the total dry weight of the organisms at each level or on the number of calories at each level; or (3) a pyramid of energy flow, showing the productivity of the different trophic levels.

The shape of any particular pyramid tells a great deal about the ecosystem it represents. Figure 36–6a, for example, shows a pyramid of numbers for a grassland ecosystem. In this type of food chain, the primary producers (the individual grass plants) are small, and so a large quantity of them is required to support the primary consumers (the herbivores). In a food chain in which the primary producers are large (for instance, trees), one primary producer may support many herbivores, as indicated in Figure 36–6b.

A pyramid of biomass for a grassland ecosystem, like the pyramid of numbers for that system, takes the form of an upright pyramid, as shown in Figure 36–7a. Pyramids of biomass are inverted only when the producers have very high reproduction rates. For example, in the ocean, the standing crop of phytoplankton (plant plankton) may be smaller than the biomass of the zooplankton (animal plankton) that feeds upon it (Figure 36–7b). Because the growth rate of the phytoplankton is much more rapid than that of the zooplankton, a small biomass of phytoplankton can supply food for a larger biomass of zooplankton. Like pyramids of numbers, pyramids of biomass indicate only the quantity of organic material present at any one time; they do not give the total amount of material produced or, as do pyramids of energy, the rate at which it is produced.

Pyramids of energy, in keeping with the second law of thermodynamics, have the same basic shape for every ecosystem (Figure 36–8). In general, of the total amount of energy in the biomass that passes from one trophic level to another in a food chain, only about 10 percent is stored in body tissue. About 90 percent of the total calories is either unassimilated or "burned" in respiration by the eating animals at that level. Hence the tertiary consumers, the carnivores that

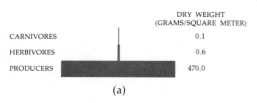

(a)

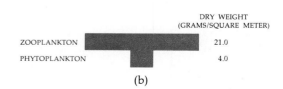

(b)

ENERGY COSTS OF FOOD GATHERING

How much does a calorie cost, in terms of calories? For the expenditure of 1 calorie, an organism in most natural populations obtains from 2 to 20 calories in food energy. This is true for organisms whose expenditures are very high—such as the hummingbird, which spends up to 330 calories a minute—as well as for organisms whose expenditures are very low—such as the damselfly, which uses less than a calorie a day.

In simple human societies, in which individuals obtain their food without fossil-fuel energy subsidies, the ratio of food calories gained to calories invested is similar to that which prevails for the rest of the animal kingdom. Hunter-gatherers average 5 to 10 calories for each calorie spent; shifting agriculture (which requires no fertilizer) yields about 20 calories per calorie spent.

As is true in most societies (the social insects and civilized man are among the exceptions), almost the entire adult population has a share in the business of getting food. In the United States, about 20 percent of the population is involved in the food supply system. (Only about 2 percent are actually farmers; the rest are involved in food processing, transportation, and marketing.) Thus 80 percent are, for better or worse, free for other pursuits.

On the surface, it would seem that we expend less energy than most animals on the mundane work of food supply. Not so. At the turn of the century, for each calorie expended, including human labor, fuel for farm machinery and food transport, and the energy cost of fertilizer, we received about a calorie in return. Today, in the United States, as in other "advanced" technological societies, for every calorie invested we get a return of 0.1 calorie. This cost figure does not include the energy used for heating or lighting or running private automobiles (even those that bring the food home from the market) or electric can openers.

Obviously, other animals cannot live so profligately; their energy income must exceed their expenditures. Like these other organisms, we also are dependent almost exclusively on solar energy. There is an important difference, however; because of our technology, we have been able to draw upon energy stored millions of years ago. It is only in the last decade that we have come to realize that not only are these resources finite but also they may soon be expended.

From Robert M. May, "Energy Costs of Food Gathering," Nature, 255:669, June 26, 1975.

36-8

Pyramid of energy flow for a river system in Florida. This type of pyramid, regardless of the ecosystem under analysis, is never inverted.

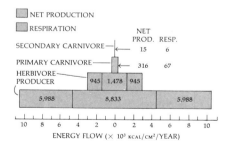

eat the carnivores that eat the herbivores, are reduced to approximately $\frac{1}{10} \times \frac{1}{10} \times \frac{1}{10} = \frac{1}{1000}$ the energy stored in the plants that are eaten. Supercarnivores, which eat these tertiary carnivores, are reduced to one-tenth of this, or $\frac{1}{10,000}$ the energy in the plant material. Individuals at the top of the ecological pyramid, such as man, may have to be quite large, as individuals, to capture and eat other animals, but even if they are larger as individuals, they are usually smaller in total number and total biomass. And they are always lower in total captured energy than the organisms they eat.

Thus, when we eat beef, only 10 percent of the energy present in the grasses or grain that fed the steer is converted to animal biomass, and even less to edible meat. Consequently, in terms of energy, those who eat meat are consuming 10 times as much of the world's energy resources as those who eat grains and other vegetables, a matter not without relevance in the face of worldwide starvation.

CYCLING OF MINERALS

Energy takes a one-way course through an ecosystem, but many inorganic substances cycle through the system. Such substances include water, nitrogen, carbon, phosphorus, potassium, sulfur, magnesium, calcium, sodium, chlorine, and also a number of minerals, such as iron and cobalt, which are required by some living systems in only very small amounts.

36–9
The phosphorous cycle.

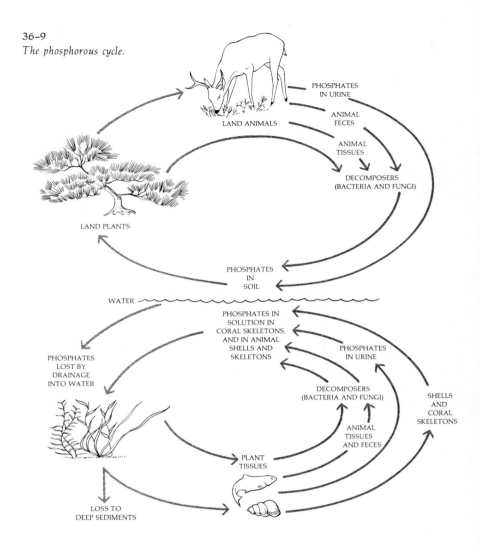

Movements of inorganic substances are referred to as *biogeochemical cycles* because they involve geological as well as biological components of the ecosystem. The geological components are the atmosphere, which is made up largely of gases, including water vapor; the lithosphere, the solid crust of the earth; and the hydrosphere, comprising the oceans, lakes, and rivers, which cover three-fourths of the earth's surface.

The biological components of biogeochemical cycles include the producers, the consumers, and the decomposers. The decomposers, which are primarily bacteria and fungi, break down dead and discarded organic matter, completing the oxidation of the energy-rich compounds formed by photosynthesis. As a result of the metabolic work of the decomposers, waste products—dead leaves and branches, the roots of annual plants, feces, carcasses, even the discarded exoskeletons of insects—are broken down to inorganic substances that are returned to the soil or water. From the soil or water, they once more enter the bodies of plants and begin their cycle again.

The cycling of one important mineral, phosphorus, through the ecosystem is shown in Figure 36–9. The chief reservoir of phosphorus, like that of all other minerals except nitrogen, is in the soil. All the minerals in the ecosystem undergo similar cycles.

Nitrogen Cycle

The chief reservoir of nitrogen is the atmosphere; in fact, nitrogen makes up 78 percent of the gases in the atmosphere. Since most living things, however, cannot use elemental atmospheric nitrogen to make amino acids and other nitrogen-containing compounds, they must depend on nitrogen present in soil minerals. So, despite the abundance of nitrogen in the biosphere, a shortage of nitrogen in the soil is often the major limiting factor in plant growth. The process by which this limited amount of nitrogen is circulated and recirculated throughout the world of living organisms is known as the *nitrogen cycle*. The three principal stages of this cycle are (1) ammonification, (2) nitrification, and (3) assimilation.

Much of the nitrogen found in the soil, a result of the decomposition of organic materials, is in the form of complex organic compounds, such as proteins, amino acids, nucleic acids, and nucleotides. However, these nitrogenous compounds are usually rapidly decomposed into simple compounds by soil-dwelling organisms. Certain soil bacteria and fungi are mainly responsible for the decomposition of dead organic materials. These microorganisms use the proteins and amino acids as a source of their own needed proteins and release the excess nitrogen in the form of ammonia (NH_3) or ammonium (NH_4^+). This process is known as *ammonification*.

Several species of bacteria common in soils are able to oxidize ammonia (NH_3) or ammonium (NH_4^+). The oxidation of ammonia or ammonium, known

36–10

The nitrogen cycle. Nitrogen is returned to the soil in the form of dead tissues, feces, and urine.

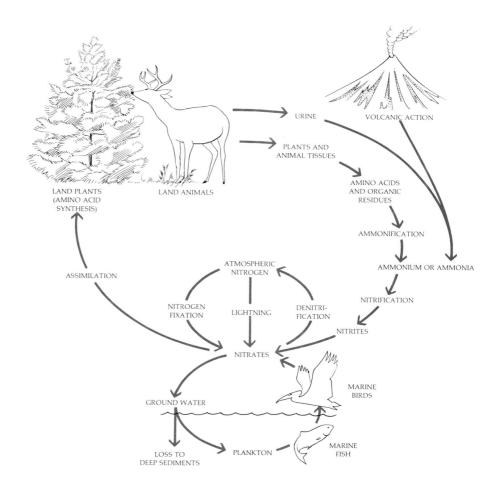

as *nitrification,* is an energy-yielding process, and the energy released in the process is used by these bacteria as their primary energy source. One group of bacteria oxidizes ammonia (or ammonium) to nitrite (NO_2^-):

$$2NH_3 + 3O_2 \longrightarrow 2NO_2^- + 2H^+ + 2H_2O$$

Nitrite is toxic to higher plants, but it rarely accumulates. Members of another genus of bacteria oxidize the nitrite to nitrate, again with a release of energy:

$$2NO_2^- + O_2 \longrightarrow 2NO_3^-$$

Although plants can utilize ammonium directly, nitrate is the form in which most nitrogen moves from the soil into the roots.

Once the nitrate is within the plant cell, it is reduced back to ammonium. In contrast to nitrification, this assimilation process requires energy. The ammonium ions thus formed are transferred to carbon-containing compounds to produce amino acids and other nitrogen-containing organic compounds needed by the plant.

Nitrogen Fixation

The nitrogen-containing compounds of green plants are returned to the soil with the death of the plants (or of the animals that have eaten the plants) and are reprocessed by soil organisms and microorganisms, taken up by the plant roots in the form of nitrate dissolved in the soil water, and reconverted to organic compounds. In the course of this cycle, a certain amount of nitrogen is always "lost," in the sense that it becomes unavailable to plants.

The main source of nitrogen loss is the removal of plants from the soil. Soils under cultivation often show a steady decline of nitrogen content. Nitrogen may also be lost when topsoil is carried off by soil erosion or when ground cover is destroyed by fire. Nitrogen is also leached away by water percolating down through the soil to the groundwater. In addition, numerous types of bacteria are present in the soil that, when oxygen is not present, can break down nitrates, releasing nitrogen into the air and using the oxygen for the oxidation of carbon compounds (respiration). This process, known as *denitrification,* takes place in poorly drained (hence, poorly aerated) soils.

As you can see, if the nitrogen lost from the soil were not steadily replaced, virtually all life on this planet would finally flicker out. The "lost" nitrogen is returned to the soil by *nitrogen fixation,* the process by which gaseous nitrogen from the air is incorporated into organic nitrogen-containing compounds and thereby brought into the nitrogen cycle. It is carried out to a small extent by abiotic processes, such as the production of nitrogen oxides by interactions between nitrogen and other atmospheric gases; lightning is a possible energy source.

Nitrogen fixation is also carried out commercially by combining 1 mole of nitrogen gas (N_2) with 3 moles of hydrogen gas (H_2) to produce ammonia (NH_3). Commercial nitrogen fixation, however, requires a large input of energy, which is supplied by fossil fuels. As the costs of these fuels increase, so do the prices of the nitrogen-containing fertilizers produced. As a result, almost one-third of the energy expended for corn production in the United States is due to the fertilizer used.

Most nitrogen fixation, however, is carried out by a few types of free-living microorganisms, including blue-green algae, some free-living bacteria, and some species of bacteria that live in symbiosis with plants (Figure 36–11). Just as all organisms are ultimately dependent on photosynthesis for energy, they all depend on nitrogen fixation for their nitrogen.

36-11

(a) Nitrogen-fixing nodules on the roots of a bird's-foot trefoil, a legume. These nodules are the result of a symbiotic relationship between a soil bacterium (Rhizobium) and root cells. (b) Cross section of an infected nodule, showing bacteria. The plant supplies the bacteria with an energy source; the bacteria supply the plant with fixed nitrogen.

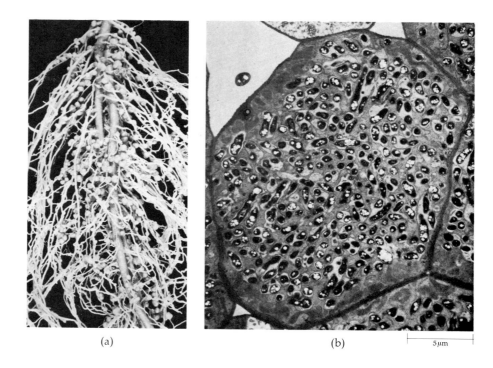

(a)   (b)   5μm

36-12

Diagrams of soil layers of three major soil types. (a) The litter of the northern coniferous forest is acid and slow to decay, and the soil has little accumulation of humus, is very acid, and is leached of minerals. (b) In the cool, temperate deciduous forest, decay is somewhat more rapid, leaching less extensive, and the soil more fertile. Such soils have been used extensively for agriculture, but they need to be prepared by lime (to reduce acidity) and fertilizer. (c) In the grasslands, almost all of the plant material above the ground dies each year as do many of the roots, and so much organic matter is constantly returned to the soil. In addition, the finely divided roots penetrate the soil extensively. The result is highly fertile soil, often black in color, with a topsoil sometimes more than a meter in depth.

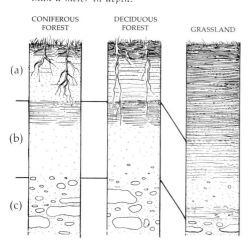

Fifty million metric tons of nitrogen are added to the soil each year, of which 45 million metric tons are biological in origin. (The other 10 percent is largely in the form of chemical fertilizers.) Of the various classes of nitrogen-fixing organisms, the symbiotic bacteria are among the most important in terms of total amounts of nitrogen fixed. The most common of the nitrogen-fixing symbiotic bacteria is *Rhizobium*, a type of bacterium that invades the roots of leguminous plants, such as clover, peas, beans, vetches, and alfalfa.

Soils and Mineral Cycles

Soil, the uppermost layer of the earth's crust, is composed of weathered rock associated with organic material, both living and in various stages of decomposition. It typically has three layers: the A horizon, the B horizon, and the C horizon. The A horizon, or topsoil, is the zone of maximum organic accumulation (humus). The B horizon, or subsoil, consists of inorganic particles in combination with mineral nutrients that have leached down from the A horizon. The C horizon is made up of loose rock that extends down to the bedrock beneath it. Figure 36–12 shows profiles of three common soil types.

The mineral content of the soil depends in part on the parent rock from which the soil is formed. These differences in mineral content can be extremely localized, with sharp lines of demarcation. Geologists sometimes use types of vegetation or modifications in color or growth patterns of plants as indicators of mineral deposits.

In most soils, however, the mineral content is more dependent on the biotic component of the ecosystem. In an undisturbed environment, as we noted earlier, most of the mineral nutrients stay within the ecosystem. If, however, the vegetation is repeatedly removed, as when crops are harvested or grasslands are overgrazed, or the top, humus-rich layer of the A horizon is eroded, the soil rapidly becomes depleted and can be used for agricultural purposes only if it is heavily fertilized.

Table 36-1 *Soil Classification*

	DIAMETER OF FRAGMENTS (MICROMETERS)
Coarse sand	200–2,000 (0.2–2 millimeters)
Sand	20–200
Silt	2–20
Clay	Less than 2

Another factor influencing the mineral content of soils is the soil composition. The smaller fragments of rock are classified as sand, silt, or clay, according to size (see Table 36–1). Water and minerals drain rapidly through soil composed of large particles (sandy soil). Soil composed of small particles (clay) holds the water against gravity. However, a pure clay soil is not suitable for plant growth because it is usually too tightly packed to let in enough oxygen for the respiration of plant roots, soil animals, and most soil microorganisms. Clay soils that contain enough large particles to keep the soil from packing are known as loams, and these are generally the best soils for plant growth.

Soils and plant life interact. Plants constantly add to the humus, thereby changing not only the content of the soil but also its texture and its capacity to hold minerals and water. In turn, the plants are dependent upon the mineral content of the soil and its holding capacity. As these improve, plants increase in biomass and also often change in kind, thereby producing further changes in the soil. Thus, under natural conditions, the soil is constantly changing its composition.

36-13
Prairie soil. Note how the roots of the grasses bind the topsoil.

ECOLOGICAL SUCCESSION

If land is bare as the result of a landslide, erosion, excavation, the eruption of a volcano, the retreat of a glacier, the rising of a new island from the sea, or some other such phenomenon, it will, if the environment is not too harsh, slowly become covered with vegetation and its accompanying animal life. The vegetation that initially colonizes the bare land is usually replaced in the course of time by a second type, which gradually crowds out the first and which itself may eventually be replaced. Moreover, it has been observed that this replacement process takes place in a predictable and orderly sequence. This sequence is known as *ecological succession.*

The process of ecological succession is carried out by the living organisms themselves. Each temporary community changes the local conditions of temperature, light, humidity, soil content, and other abiotic factors, and so sets up favorable conditions for the next temporary community. When the site has been modified as much as possible (here the limits are set by the environment), succession ceases, or at least slows down considerably. The final, mature community is known as the *climax community.*

The physical characteristics of the environment determine the nature of the mature community. In regions where conditions are particularly unfavorable, such as the tundra, the process of succession involves relatively few stages and the climax community is correspondingly simple. Where physical conditions are less limiting, the mature community is rich and diverse. Major changes in the external environment—as occurred during the Ice Age—would, of course alter the nature of the climax community.

Primary Succession

The occupation by plants of an area not previously covered by vegetation is known as *primary succession.* Rocks and cliffs are common sites of primary succession. The first stage in such areas is the formation of soil. The solid rock is broken down by weathering processes, such as freezing and thawing or heating and cooling, which cause substances in the rocks to expand and contract, thus splitting the rocks apart. Water and wind exert a scouring action that breaks the fragmented rock into smaller particles, often carrying the fragments great dis-

(a)

(b)

36-14

(a) *Yapoah Crater, a volcanic cinder cone east of the Cascade Mountains in central Oregon. Succession leading to the establishment of climax forest on such a cone may* *take centuries and may often be interrupted by further volcanic activity long before it is complete. (b) An early stage of succession on* *a rocky slope. Lichens have begun to accumulate soil, and a bladder fern has sprung up in a small crevice.*

tances. Water enters between the particles, and soluble materials such as rock salt dissolve in the water. Water in combination with carbon dioxide from the air forms a mild acid which dissolves substances that will not dissolve in water alone. Chemical reactions that contribute to the disintegration of the rock begin to take place.

Soon, if other conditions, such as light and temperature, permit, bacteria, fungi, and then small plants begin to gain a foothold. Growing roots split rock particles, and the disintegrating bodies of the plants and those of the animals associated with them add to the accumulating material. Finally, the larger plants move in, anchoring the soil in place with their root systems, and a new community has begun. Primary succession may also take place on sand dunes or with the filling in of a pond or lake.

Secondary Succession

When succession occurs on a site previously occupied by vegetation, it is known as *secondary succession*. Secondary succession commonly occurs on areas laid bare by man, such as abandoned farmland and strip mines, roadsides, and landfills. Such an open area is bombarded by the seeds of numerous plants and is captured by those that can germinate most quickly.

(a)

(b)

(c)

36–15

Secondary succession. (a) A young stand of loblolly pine is taking over an abandoned field in Wake County, North Carolina. The plow furrows are still clearly visible. The field was abandoned perhaps 20 years ago. (b) Seedling trees of balsam fir growing up under and replacing quaking aspen in northern Minnesota—a stage in forest succession leading to a climax community of white spruce and balsam fir. (c) The seedling of a red maple rising above needles of white pine. Mature white pines filter the light so that their own seedlings cannot survive and only those tolerant of shade, such as maple and oak seedlings, can gain a foothold.

In an open field, the plants that take hold are those that can survive the sunlight and drying winds, for example, weeds, grasses, and such trees as cedars, white pines, poplars, and birch. For awhile, these plants are dominant, but eventually they eliminate themselves because their seedlings cannot compete in the shade cast by the parent trees. Some seedlings, however, can thrive in partial shade—for instance, oaks, red maples, white ash, and tulip trees. As the forest matures, these trees grow tall, and finally they shut out so much light that even *their* seedlings cannot grow. Eventually, the only young trees that can grow in the forest are those that can survive in the dimmest light, such as hemlock, beech, and sugar maple, and these ultimately take over the forest. Nothing else can compete with them in the conditions that have been established. This is the climax forest.

Fire

In some areas, fire plays an important part in determining the final stage in forest succession. Young seedlings of deciduous trees are very susceptible to fire, while pines are resistant to it. Jack pines, in fact, open their cones and release their seeds only after they have been heated, so they tend to spread most readily after a fire. In the southern pine forests and the northern lake area of the United States, recurrent ground fires maintain the pinewoods, keeping the forest perpetually young. Similarly, the forests of sugar pines and giant sequoias on the western coast of the United States are maintained by fires, which destroy competitive trees that are faster-growing and less fire-resistant than the giant sequoias and lay bare the ground, making it possible for the small seeds to germinate.

These light ground fires, characteristic of an ecosystem regulated by fire, differ greatly from the uncontrolled crown fires of northern forests, which spread through the treetops, destroying entire communities of plants and animals and leaving the ground barren.

When fire sweeps through a forest, second-ary succession—with regeneration from nearby unburned stands of vegetation—is initiated. Some plants produce sprouts from the stumps; others seed abundantly on the burn. In one group of pines, the closed-cone pines, the cones do not open to release their seeds until they have been heated by fire.

36–17
The primary producers of a man-made ecosystem, such as this wheat field, must be constantly defended against disease, predation, and competition. Also, because organic matter is removed from the system, minerals must be constantly replaced, in the form of fertilizer.

Immature and Mature Ecosystems

Although the changes that take place as the ecosystem matures differ in detail as to the exact species involved and the rates of change, they have certain results in common. First, there is an increase in total biomass. Compare, for example, a recently abandoned field, which is an immature ecosystem, with a deciduous forest, which is a more mature one. Second, there is an initial increase in net productivity followed by a decrease in productivity in relation to biomass and to respiration. Third, there is a general increase in the size of organisms, the lengths of their lives, and the complexity of their life cycles. Fourth, the number of species increases, and with this increase there is a greater complexity in the relationships between species, including more symbiotic associations and more intricate food webs. Fifth, the ecosystem becomes more stable as it matures. Mature ecosystems show a greater capacity for homeostasis and are less likely to be drastically affected by changes in the environment.

Agriculture and Ecological Succession

An area under intensive agriculture is an immature ecosystem. It has high productivity, relatively little biomass, and few species. It is also a very unstable system compared to a mature, complex ecosystem, which, with its complicated food webs, has many built-in checks and balances. Individual members of the plant and animal community may be sick or dying, but the natural, mature ecosystem itself is healthy, and species tend to endure in relatively stable numbers. In areas under cultivation, plants do not grow in complex communities, as they do in a forest, but in pure stands. A cornfield, for example, has little inherent stability. If not constantly guarded by man, it will be immediately overrun with insects and weeds. It is for this reason that insecticides and herbicides play such a large and, indeed, indispensable role in our modern life.

The susceptibility of modern crops to predators and parasites was tragically illustrated by the great potato famine of Ireland, which was caused by a fungus

COMPOST

Composting, a practice as old as agriculture itself, has recently been attracting increased interest as a means of disposing of organic wastes by converting them to fertilizer. The starting product is any collection of organic matter—leaves, garbage, manure, straw, lawn clippings, sewage sludge, sawdust—and the population of bacteria and other microorganisms normally present. The only other requirements are oxygen and moisture. Grinding of the organic matter is not essential, but it provides greater surface area for microbial attack and so speeds the process.

In a compost heap, microbial growth accelerates rapidly, generating heat, much of which is conserved in the pile because the outer layers of organic matter act as an insulator. In a large pile (2 meters $\times$ 2 meters $\times$ 1.5 meters, for instance), the interior temperature rises to 70°C; in small piles, it usually reaches 40°C. As the temperature rises, the population of decomposers changes, with heat-tolerant forms replacing the organisms originally present. As the previous forms die, their organic matter also becomes part of the product. A useful side effect of the temperature increase is that most of the common pathogenic bacteria present in sewer sludge are destroyed, as are cysts, eggs, and other immature forms of plant and animal parasites.

Changes in pH also occur. The initial pH is usually slightly acid (about pH 6), as is the liquid portion of most plant material. During the early stages of decomposition, the production of organic acids causes a further acidification, to about pH 4.5 to 5.0. However, as the temperature rises, the pH also increases, leveling off at slightly alkaline values (pH 7.5 to 8.5).

An important factor in composting (as in any biological growth process) is the ratio of carbon to nitrogen. About 30 to 1 (by weight) is optimal. If the carbon ratio is higher, microbial growth slows. If the nitrogen ratio is higher, some escapes as ammonia. Adding limestone (calcium carbonate) to the pile, although a common practice, increases the nitrogen loss.

Studies with municipal compost piles at Berkeley, California, demonstrated that if large piles were kept moist and aerated, composting could be completed in as little as 2 weeks. Three months or more during the winter is a more usual schedule. If compost is added to the soil before the composting process is complete, it may actually rob the soil of nitrogen.

Because it greatly reduces the bulk of plant wastes, composting can be a very useful means of waste disposal. In Scarsdale, New York, for example, leaves composted in a municipal site were reduced to one-fifth their original volume. At the same time, they formed a useful soil conditioner, improving aeration and water-holding capacities. Chemical analyses indicate, however, that the value of compost as fertilizer is limited. A rich compost commonly contains, in dry weight, about 1.5 to 3.5 percent nitrogen, 0.5 to 1.0 percent phosphorus, and 1.0 to 2.0 percent potassium, far less than a commercial fertilizer. It is also, in terms of manpower, far less efficient to prepare. However, as commercial fertilizers are becoming more expensive and less available and our waters are becoming increasingly polluted with fertilizer runoff and organic wastes, composting, at least under certain circumstances, is coming to seem an increasingly attractive alternative.

---

infection. The famine of 1845–1847 was responsible for more than a million deaths from starvation and initiated large-scale emigration from Ireland to the United States; within a decade, the population of Ireland dropped from 8 million to 4 million. Virtually the entire Irish potato crop was wiped out in a single week in the summer of 1846. A number of plant geneticists are warning that the new strains of wheat and rice, which promise major contributions toward feeding the growing populations, will be particularly susceptible, because of their genetic uniformity and widespread distribution, to such disasters.

SUMMARY

An ecosystem, or ecological system, is a unit of biological organization made up of all the organisms in a given area and the environment in which they live. It is characterized by interactions between the living (biotic) and nonliving (abiotic) components that result in (1) a flow of energy from the sun through autotrophs to heterotrophs and (2) a cycling of minerals and other inorganic materials.

Within an ecosystem, there are trophic (feeding) levels. All ecosystems have at least two such levels: autotrophs, which are plants or photosynthetic algae, and herbivores, which are usually animals. The autotrophs, the primary producers in the ecosystem, convert a small proportion (about 1 percent) of the sun's energy into chemical energy. The herbivores, which eat the autotrophs, are the primary consumers. A carnivore that eats the herbivore is a secondary consumer, and so on. About 10 percent of the energy is transferred at each trophic level. There are seldom more than five links in a food chain.

The movements of water, carbon, nitrogen, and minerals through ecosystems are known as biogeochemical cycles. In such cycles, inorganic materials from the air, water, or soil are taken up by primary producers, passed on to consumers, and eventually transferred to decomposers, chiefly bacteria and fungi. The decomposers break down dead and discarded organic material and return it to the soil or water in a form that can be used again by the primary producers. Characteristics of soil affect the presence, retention, and recycling of minerals.

The cycling of nitrogen from the soil, through the bodies of plants and animals, and back to the soil again is known as the nitrogen cycle. It involves several stages. Nitrogen reaches the soil in the form of organic material of plant and animal origin. This material is decomposed by soil organisms. Ammonification, the breakdown of nitrogen-containing molecules to ammonia (NH_3) or ammonium (NH_4^+), is carried out by certain soil bacteria and fungi. Nitrification is the oxidation of ammonia or ammonium to form nitrites and nitrates; these steps are carried out by two different types of bacteria. Nitrogen enters plants almost entirely in the form of nitrates. Nitrogen-containing organic compounds are eventually returned to the soil, principally through death and decay, completing the nitrogen cycle.

Nitrogen is lost from the soil by harvest, erosion, fire, leaching, and denitrification. Nitrogen is increased in the soil by nitrogen fixation, which is the incorporation of elemental nitrogen into organic components. Biological nitrogen fixation is carried out entirely by microorganisms, including blue-green algae, or by a symbiotic association of bacteria *(Rhizobium)* and legumes.

Ecological succession is an orderly sequence of changes in the type of vegetation and other organisms on a particular site. Primary succession occurs in areas previously unoccupied by living organisms, whereas secondary succession occurs in disturbed areas where vegetation has been removed. Succession culminates in the establishment of a climax community, which is the characteristic community for that area.

Maturation in ecosystems is accompanied by an increase in biomass, a decrease in productivity, an increase in the number of species, and an increase in the complexity of the relationships among species. The mature system is characterized by greater stability than is present in the immature system.

QUESTIONS

1. Define the following terms: trophic level, ecological pyramid, biomass, B horizon, nitrification, ecological succession.

2. Describe what happens to the energy in light striking a temperate forest ecosystem. What happens when it strikes a cornfield? A pond? A field on which cattle are grazing?

3. Describe what happens to a nutrient mineral in each of those areas.

4. A leading ecologist has stated: "The plough is the most deadly agent of extinction ever devised; not even thermonuclear weapons pose such a threat to the beauty and diversity of life on earth." Explain.

CHAPTER 37

Communities and Populations

At the beginning of this section, we defined ecology as the scientific study of the interactions that determine the distribution and abundance of organisms. In Chapter 35, we examined broad factors, such as light, temperature, and precipitation, that dictate, in very general terms, the kinds and numbers of plants and other living things found in various parts of the biosphere. In Chapter 36, we subdivided these broad areas, the biomes, into functioning units, ecosystems, and saw how the flow of energy and recycling of minerals that define an ecosystem influence the kinds and numbers of organisms. For example, an ecosystem will probably contain herbivores and first-level carnivores, and the biomass of the former will usually be at least 10 times that of the latter.

In this chapter we are going to move in still closer and look at communities and populations. A _community_ consists simply of all the plants, animals, and other organisms that live in a particular area. A _population_ is a group of organisms of the same species occupying a particular area at the same time. Thus communities are made up of populations, and, as we shall see, interactions among populations within a community are major factors in determining the numbers and kinds of organisms found there.

37-1

Some members of a seashore community: starfish, barnacles, whelks, rockweed, and mussels.

The principle of limiting factors. Every species has a characteristic limiting factor curve for each factor in its environment. The three critical points on the curve are the lower limit of tolerance, the point of greatest abundance, and the upper limit of tolerance.

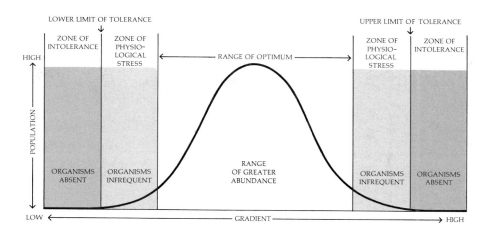

LIMITING FACTORS

Whether or not a particular type of organism can survive in a given area depends in part on the organism's range of tolerance for various environmental factors, such as light, temperature, available water, salinity, nesting space, and shortages (or excesses) of required nutrients. If any one of these factors does not fall within the particular organism's range of tolerance, even though all the other requirements are met, survival of the organism is not possible. Whatever it is that determines whether or not an organism can survive in a given environment is known as a *limiting factor*. For example, as we noted previously, the length of the growing season is a limiting factor for deciduous trees. It would be wrong to suppose, however, that one would necessarily find a particular type of organism in a given area just because its minimum requirements are met. Conifers and prairie grasses are well able to grow in the areas now occupied by deciduous trees but do not because they are unable to compete. (We shall discuss competition later in the chapter.)

The way in which limiting factors affect growth of organisms is illustrated by a common pollution problem. When phosphorus is the limiting factor in the growth of freshwater algae in a lake or slow-moving stream, phosphate-containing detergents added to the water from sewage systems will produce a spectacular bloom of algae. This period of rapid growth continues until the available supply of another essential element—perhaps calcium—is consumed. Then the algae begin to die, the decomposing bacteria take over, and the respiration of the bacteria begins to use up the oxygen in the water. Eventually the amount of oxygen present drops below the tolerance of fish and other organisms, and these, too, begin to die. If the process is not interrupted, the lake becomes completely stagnant, and only bacteria and other microorganisms can survive in it.

THE GROWTH OF POPULATIONS

Birth Rates

Birth rates, as we saw in the example cited in Table 32-1 (page 445), are a compromise between quantity and quality. In terms of an organism's energy budget, an individual can produce many ova or seeds with little nutrient supply or proportionately fewer with greater food reserves. An animal can opt, evolutionarily speaking, for many untended offspring—as in the case of invertebrates, which may produce millions of eggs a season—or few offspring with long periods of intensive care—as with many higher mammals and birds.

As we have seen, the number of young produced is under constant selection pressure. A large mammal can produce and care for only one infant every two or three years. It has been calculated, for instance, that a single female housefly producing an average of 120 eggs per laying, with half developing into females and with seven generations per year per female, could produce 6,182,442,727,320 houseflies in the space of one year. At the other end of the reproduction scale is a pair of elephants, which have a gestation period of 20 to 22 months and usually give birth to only one offspring at a time. If all offspring and their offspring, in turn, survived and reproduced, this one pair of elephants would have 19 million descendants at the end of 750 years, as noted by Darwin.

These hypothetical figures of the housefly and the pair of elephants represent examples of exponential growth, which operates on the same principle as compound interest. The more money in a compound savings account, the more dollars are added to the account per unit time. In short, the interest earns interest. Similarly, in a population increasing exponentially, the more individuals that are added to the population, the faster it increases.

Exponential growth curves, such as that seen in Figure 37–3a, are characteristic of seasonal blooms of algae, which flourish until all the nutrients are used up and then rapidly die off. Such growth curves are also typical of insect invasions, such as the gypsy moth infestations that recurrently plague the northeastern United States. The populations grow either until they are overcome by an epidemic of disease organisms or until they reduce their food supply so that they are not able to complete their metamorphoses and reach the reproductive stages. (Spraying of the caterpillars with insecticides often reduces the size of the population just sufficiently to delay the population crash.)

The more common natural growth curve is the sigmoid (S-shaped) curve shown in Figure 37–3b. Increase in the number of individuals is slow at first, because there are relatively few individuals reproducing. Then, as the number increases, the curve comes to resemble the exponential curve. Typically, it then slows down and finally levels off, reaching a point at which there is no net change in the population. In well-established communities, most of the populations have reached the leveling-off point of the sigmoid curve. Thereafter, the numbers continue to oscillate. An immature ecosystem, with its relatively few species, will have more marked oscillation. In a more mature ecosystem, the greater complexity of the relationships among organisms tends to diminish the oscillations, but they occur nonetheless.

37–3

(a) *Exponential growth curve. After an initial establishment phase, the population increases compound-interest fashion until an environmental limit causes a population crash. (b) A more usual growth curve is sigmoid, or S-shaped. As with exponential growth, there is an establishment phase (1), and a phase of rapid acceleration (2). Then, as the population approaches environmental limits, the growth rate slows down (3 and 4), and finally stabilizes (5).*

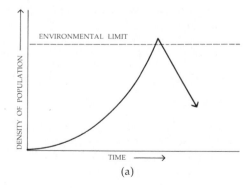
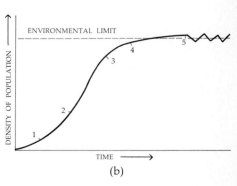

INTERSPECIFIC COMPETITION: THE ECOLOGICAL NICHE

Competition affects both the kinds and numbers of organisms found in a community. Closely linked to the phenomenon of competition is the concept of the ecological niche, which, as we noted earlier, is defined as an organism's position in the ecosystem. According to the principle of *competitive exclusion*, formulated by the Russian biologist G. F. Gause—in any given community, only one species (that is, one population) can occupy any given ecological niche for an extended period of time.

Gause bolstered his hypothesis by a number of laboratory experiments. His simplest, now classic, experiment involved laboratory cultures of two species of paramecia, *Paramecium aurelia* and *Paramecium caudatum*. When the two species were grown under identical conditions in separate containers, *P. aurelia* grew much more rapidly than *P. caudatum*, indicating that the former used the available food supply more efficiently than the latter. When the two were grown together, the former rapidly outmultiplied the latter, which soon died out (Figure 37–4).

A similar situation occurs with two species of duckweed, *Lemna gibba* and *Lemna polyrrhiza*. *Lemna gibba* grows more slowly in pure culture than *L. polyrrhiza*; *L. gibba*, however, always replaces *L. polyrrhiza* when they are grown together. Again, evolution has provided one with an advantge. The plant bodies of *L. gibba* have air-filled sacs that serve as little pontoons, so that these plants form a mass over the other species, cutting off the light. As a consequence, the shaded *L. polyrrhiza* dies out (Figure 37–5).

37–4

Results of Gause's experiment with two species of paramecia demonstrate Gause's principle that if two species are in competition for the same resource—in this case, food—one eliminates the other. Paramecium caudatum and Paramecium aurelia were first grown separately under controlled conditions and with a constant food supply. As you can see, Paramecium aurelia grew much more rapidly than Paramecium caudatum, indicating that the latter uses available food supplies more efficiently. When the two protozoans were grown together, the more rapidly growing species outmultiplied and eliminated the slower-growing species.

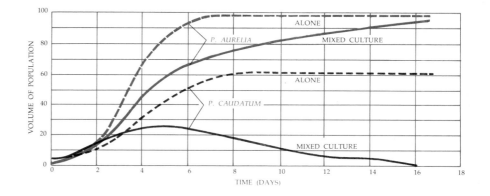

37–5

An experiment with two species of floating duckweed, tiny angiosperms found in ponds and lakes. One species, Lemna polyrrhiza, grows more rapidly in pure culture than the other species, Lemna gibba. But the Lemna gibba has tiny air-filled sacs which, like pontoons, float it on the surface, and so it shades the other species, making it the victor in the competition for light.

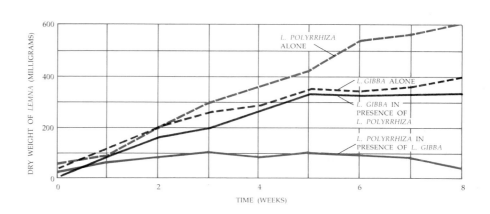

A demonstration of Gause's principle: the feeding zones in a spruce tree of five species of North American warblers. The colored areas in the tree indicate where each species spends at least half its feeding time. In this way, all five species feed in the same trees with diminished competition.

CAPE MAY WARBLER

BAY BREASTED WARBLER

BLACKBURNIAN WARBLER

BLACK-THROATED GREEN WARBLER

MYRTLE WARBLER

It is possible to devise different culture conditions under which the outcomes of both the *Paramecium* and *Lemna* experiments can be reversed. However, regardless of conditions, one species always wins.

The Niche in Nature

Gause's principle leads one to predict that in a situation in which two very similar species coexist in nature, it will be possible to demonstrate differences in the niches they inhabit. The prediction has been fulfilled by a number of observations. For example, some New England forests are inhabited by five closely related species of warbler, all about the same size and all insect eaters. Why do these birds not eliminate one another by competitive exclusions? An analysis by Robert MacArthur showed that these warblers have different feeding zones in the canopy (Figure 37-6), and because of these different zones they are able to coexist without direct competition.

J. H. Connell studied competition between two genera of barnacles in Scotland. Barnacles are crustaceans. Before they change from their immature, larval forms, in which they are free-swimming, into their adult, feeding forms, they cement themselves to rocks and secrete shells. One species of barnacle, *Chthamalus stellatus*, occurs in the high part of the intertidal seashore, and another, *Balanus balanoides*, occurs lower down. Although *Chthamalus* larvae, after their short period of drifting in the plankton, often attach to rocks in the lower, *Balanus*-occupied zone, no adults are ever found there.

The history of a population of barnacles can be recorded very accurately by holding a pane of glass over a patch of barnacles and noting with glass-marking ink each spot where a barnacle is. Once attached, barnacles remain fixed, so that by returning later, one can check exactly which barnacles have died and which new ones have arrived.

By doing this, Connell was able to show that in the lower zone, *Balanus*, which grows faster, ousts *Chthamalus* by crowding it off the rocks or growing over it. When *Chthamalus* was isolated from contact with *Balanus*, it lived with no difficulty in the lower zone, showing that competition with *Balanus* restricted *Chthamalus* to the higher, less favorable zone. However, when *Chthamalus* was removed, *Balanus* was unable to live in its range. *Chthamalus* survives in the intertidal community by special physiological adaptations that enable it to inhabit a marginal area.

A final example is afforded by Darwin's finches. As we noted in Chapter 34, the large, medium, and small ground finches are very similar except for differences in overall body size and in the sizes of their beaks. These differences in beak size are correlated with the fact that they eat seeds of different sizes. Because of these different eating habits, competition for food is reduced. On islands such as Abingdon and Bindloe, where all three species of ground finch exist together, there are clear-cut differences in beak size. On Charles and Chatham Islands (see Figure 34–11), the large species is not found, and the beak size of the medium ground finches found on these islands overlaps the beak size of the large finches found on Abingdon and Bindloe. Daphne and Crossmans, which are very small islands, each have only one species; Daphne has the medium-sized finch and Crossmans the small finch. These two populations have similar beak sizes, which are intermediate between those of the medium-sized and small finches on the larger islands (Figure 37–7).

Thus it would appear that although competing species may eliminate one another in the laboratory, in nature the selection pressure of competition may result in simply subdividing the niche.

Beak sizes in three species of ground finch found on the Galapagos Islands. Beak measurements are plotted horizontally, and the percentage of specimens of each species is shown vertically. Daphne and Crossmans, which are very small islands, each have only one species of ground finch. These species have beak sizes halfway between those of the medium-sized and small finches on the larger islands.

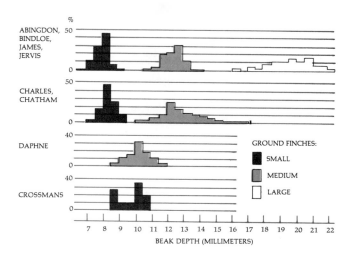

Definition of the Niche

At one time, the niche was regarded simply as the organism's role in the ecosystem and defined in relation to its energy source; according to this concept, habitat was an organism's address and the niche its occupation. Most ecologists now view the niche of an organism as the range of each environmental variable, such as temperature, humidity, and energy resources within which a species can exist and reproduce.

SYMBIOSIS

Symbiosis ("living together") is a close and permanent association between organisms of different species. There are three types of symbiotic relationships. If the relationship is beneficial to both, it is called *mutualism*. If one species benefits from the association while one is neither harmed nor benefited, it is called *commensalism*. If one species benefits and the other is harmed, the relationship is known as *parasitism*. However, since all the details of symbiotic relationships between species are often not fully understood, these distinctions are not always useful.

The examples of symbioses in nature are many and varied. One, which we have mentioned previously, is the association between a ruminant animal, such as a cow, and the bacteria that inhabit its stomach and break down cellulose. A classic example of symbiosis is provided by the lichens, which are part alga, part fungus. The body of the lichen is composed largely of fungal mycelium; held within the mycelium are numerous photosynthetic algal cells. The two organisms together form a closely integrated unit that can grow under conditions where neither the fungus nor the alga alone could survive. Lichens occur from arid desert regions to the Arctic, growing on bare soil, tree trunks, sun-baked rocks, windswept alpine peaks, and rocky intertidal shores all over the world. They are often the first colonists of bare rocky areas.

One of the most important symbioses from the ecological point of view is the one that exists between certain plants, particularly legumes, and the bacteria that infect their roots. As we saw in Chapter 36, this relationship is crucial to the essential process of nitrogen fixation by the plant. Many plants have symbiotic relationships with fungi that form a feltlike coating (mycorrhizae) on their roots. These plants grow less well if they are cultivated in soil that does not contain their characteristic root fungi, although it may have all of the same nutrients. Some other examples are shown in Figure 37–8.

Parasites and Hosts

In parasitism an individual of one species lives at the expense of an individual of another species. Most plants and animals in a natural community together support hundreds of parasites of many species—in fact, perhaps millions, if one were to count viruses.

Parasites do not usually kill their host, and they almost never destroy entire populations. We know this from our own experience. Bacterial disease is fatal only when the bacteria find themselves a particularly favorable place to multiply, such as the blind pouch of an infected appendix, an open wound, or a host with lowered resistance. Most bacteria are usually harmless, and some are helpful, such as those in the stomachs of herbivores and the vitamin-supplying bacteria in our own intestinal tracts. Modern antibiotics often cause nausea and diarrhea because they destroy the useful and protective bacteria of the intestinal tract. Many virus infections cause no symptoms at all; some, such as polio, are usually very mild and only in rare cases cause permanent disability or death. Only one known viral disease of man, rabies, is regularly fatal if untreated.

In animal and plant communities, diseases are most likely to wipe out the very young, the very old, and the disabled—either directly or, often, indirectly, by making them more susceptible to other predators or to the effects of climate. It is logical that a parasite-caused disease should not be too virulent or too efficient. If a parasite were to kill all the hosts for which it is adapted, it, too, would perish. This principle is particularly well illustrated by a series of misadventures on the continent of Australia.

There were no rabbits in Australia until 1859, when an English gentleman imported a dozen from Europe to grace his estate. Six years later he had killed a total of 20,000 on his own property and estimated he had 10,000 remaining. In 1887, in New South Wales alone, Australians killed 20 million rabbits. By 1950, Australia was being stripped of its vegetation by giant rabbit hordes. In that year, rabbits infected with myxoma virus were released on the continent of Australia. (Myxoma virus causes only a mild disease in the South American rabbits, its normal host, but is usually fatal to the European rabbit.) At first, the effects were spectacular and the rabbit population steadily declined, yielding a share of pastureland once more to the sheep herds, on which much of the economy of the country depends. But then occasional rabbits began to survive, and their litters also showed resistance to the myxoma virus.

A double process of selection had taken place. The virus, as originally introduced, was so rapidly fatal that often a rabbit died before it could be bitten by a mosquito and thereby infect another rabbit; the virus strain then died with the rabbit. Strains less drastic in their effects, on the other hand, had a better chance of survival since they had a greater opportunity to spread to a new host. (After the initial infection, the rabbit is immune to the virus, just as human beings usually become immune to mumps or measles after one infection.) So, first, selection began to work in favor of a less virulent strain of myxoma virus. Almost simultaneously, rabbits that were resistant to the original virus began to appear. Now, as a result of coevolution, the two are reaching a peaceful commensal equilibrium, like most hosts and parasites.

37–8

Symbioses. (a) *Sea anemones on the back of a hermit crab. The anemone protects and camouflages the crab and, in turn, gains mobility—and so a wider feeding range—from its association with the crab. Hermit crabs, which periodically move into new, larger shells, will coax their anemones to move with them.* (b) *Cleaner fish are permitted to approach larger fish with impunity because they feed off the algae, fungi, and other microorganisms on the fish's body. The fish recognize the cleaners by their distinctive markings. Other species of fish, by closely resembling the cleaners, are able to get close enough to the large fish to remove large bites of flesh. What would probably happen if cleaner mimics began to outnumber cleaners?* (c) *Oxpeckers live on ticks which they remove from their hosts. An oxpecker forms an association with one particular animal, such as the young impala shown here, conducting most of its activities, including courtship and mating, on its back.* (d) *Aphids suck phloem, removing certain amino acids, sugars, and other nutrients from it and excreting most of it as "honeydew," or "sugar-lerp," as it is called in Australia, where it is harvested as food by the aborigines. Some species of aphids have been domesticated by some species of ants. These aphids do not excrete their honeydew at random, but only in response to caressing movements of the ant's antennae and forelimbs. The aphids involved in this symbiotic association have lost all their own natural defenses, including even their hard outer skeletons, relying upon their hosts for protection.*

(a)

(b)

(c)

(d)

529 COMMUNITIES AND POPULATIONS

PREDATOR-PREY RELATIONSHIPS

As we saw in the previous chapter, the various trophic levels of an ecosystem are linked by predator-prey relationships. Herbivores prey on plants, and other animals prey on the herbivores. Predator-prey relationships have an important effect on the distribution and abundance of organisms. In complex ecosystems, most predators prey on a wide variety of other organisms. When one species begins to decrease in number, its members are preyed upon less frequently, and the predators turn their attention to one increasing in number. In very simple ecosystems, such as the desert, the tundra, or areas of human agriculture, predator and prey numbers may fluctuate widely. When prey is abundant, the population of the prey is then reduced drastically in number, and many of the predators, lacking other prey, starve to death. The prey than increases in number, and the cycle begins again.

The introduction of a new species in a community where there are no natural predators for that species may sometimes have disastrous results, as we have noted previously in the case of the rabbits in Australia. Similarly, when prickly pear cactus was brought to Australia from South America, it too escaped from the garden of the gentleman who imported it and spread into fields and pastureland until more than 12 million hectares were so densely covered with prickly pears that they could support almost no other vegetation (Figure 37-9). The cactus then began to take over the rest of Australia at the rate of about 400,000 hectares a year. It was not brought under control until a natural predator was imported. This was a South American moth, whose caterpillars live only on the cactus. Now only an occasional cactus and a few moths can be found.

Predation is not necessarily detrimental to the population as a whole, particularly when there is competition within the population for a limited resource such as food. Wolves, for instance, have great difficulty overtaking healthy adult caribou or moose or even healthy calves. A study of Isle Royale, an island on Lake Superior, showed that in some seasons more than 50 percent of the moose the wolves killed had lung disease, although the incidence of such individuals in the population was less than 2 percent. (Human hunters, however, with their superior weapons and their desire for a "prize" specimen, are more likely to injure or destroy strong, well-adapted animals.)

37-9

Prickly pear cactus on a homestead in Australia. Such rapid and environmentally destructive spread is often seen among alien organisms introduced into a region where they have no natural enemies.

37-10

Moose under attack by a pack of wolves. The moose stood its ground for 5 minutes, after which the wolves gave up and left.

37-11

Shrubs of purple sage produce chemicals that inhibit the growth of other plants in their vicinity, including seedlings of their own species. Secretions from leaves of the bushes collect in the soil. Here, near Santa Barbara, California, bushes are surrounded first by completely bare ground and then by a zone of grassland inhabited by stunted annual herbs that have evolved some tolerance for the chemicals.

NATURAL DEFENSES

Just as natural selection favors the most efficient predator, it also favors the prey best able to avoid predation, so predator and prey coevolve. Under strong evolutionary pressures from predators, plants and animals have developed some interesting and, to the human eye, extremely ingenious defenses.

Natural Defenses in Plants

Some natural defenses in plants are structural, such as the sharp-toothed edges of the holly leaf, the thorn of the rosebush, the spine of the cactus, and the sting of the nettle. Plants also produce a number of chemical substances for which the plant itself has no physiological use and which appear to act against predators and also against competing plants. Eating foxgloves (*Digitalis purpurea*) can cause convulsive heart attacks in vertebrates. Other plants containing digitalis-like toxins include oleander *(Nerium oleander),* of which a single leaf may be fatal to man, and the members of a large family of plants known as the milkweeds. A number of gymnosperms have been found to contain chemicals with molting-hormone (ecdysone) activity that fatally accelerate insect metamorphosis. The balsam fir has been found to produce a chemical that resembles juvenile hormone and arrests the development of certain insect species. The leaves of several species of shrubs in the chaparral produce toxic substances that are carried to the ground by rainfall. There they prevent the germination of seeds or the encroachment of roots of other plants. The characteristic aromatic odor of some chaparral communities is the result of the release into the air of volatile chemicals, which settle to the ground and attach to dry soil particles. These chemicals also are toxic to plant seedlings (Figure 37–11).

Plant defense chemicals include many substances presently useful to man, among them digitalis, quinine, castor oil, and peppercorns. Plant defense chemicals of more questionable value to man include nicotine, caffeine, morphine, and others, such as the active principles in marijuana, mescaline, and peyote, whose desirability for the animal world is a matter of opinion. Plants appear to have been waging such chemical warfare long before the coming of man, and it is possible that some small leafhopper was the first of all animals to have its mind expanded in a psychedelic experience.

Natural Defenses in Animals

Natural defenses in animals include structural adaptations and the capacity to produce noxious chemicals, both of which are seen in plants, and behavioral strategies as well. Some animals are formidably armored—for example, the armadillo, the porcupine, and the sea urchin. Some have ingenious behavioral devices. The armadillo and the pillbug roll themselves up in tight armor-plated balls. Squids and octopuses vanish, jet-propelled, leaving behind only an ink cloud. Many lizards have brightly colored tails that break off when they are attacked and, conspicuous and wriggling, divert the predator from the prey—which is escaping, tailless but with all its vital organs intact. Hermit crabs often carry sea anemones as passengers on their shells. The anemone protects the crab from predators while enjoying the benefits of mobility (and perhaps some stray morsels of food). When the crab changes shells, it often coaxes its sea anemone from its old home to the new one. (See Figure 37–8a.)

The periodical cicadas avoid predation in quite a different way. By emerging all together only after a long interval underground as larvae (17 years in one species), they greatly reduce the likelihood that a natural predator dependent

(a)

(b)

37-12   (c)

(a) When attacked, a blue-tailed skink leaves its tail to distract predators while it escapes. (b) When threatened, porcupines release barbed quills. (c) The arrow poison frog produces a deadly nerve toxin.

on cicadas will be present; most predators simply cannot wait that long. (An exception is a species of fungus parasitic on the cicada that lies dormant between the emergences of the cicadas.)

Some birds emit alarm cries that warn the flock of an approaching predator. This seeming altruism has been interpreted by some scientists as running counter to theories of natural selection, since the bird that calls attention to itself in this way is more likely to be lost—and its genes with it—to the predator than the bird that is not so altruistic. Other scientists point out, however, that since the bird is nearly always closely related genetically to the other members of its own flock, the genotype it saves, while perhaps not its own, is very similar. Therefore, genotypes dictating warning behavior would be more likely to be preserved than those that do not. A somewhat similar situation is seen in the worker bee that dies on stinging an intruder; in so doing, she may save her hive of sisters, and, most important, her queen.

Concealment and Camouflage

Hiding is one of the chief means of escape from predators. The young of many birds and of some other vertebrates respond to the warning cries of their parents by "freezing" or by running to cover. Small mammals often have nests or burows or makeshift residences in hollow logs or beneath tree roots in which they conceal themselves from their enemies.

Protective coloration is common. Mice, lizards, and arthropods that live on the sand are often light-colored, and such light-colored species, if removed from their home territories, will immediately try to return to them. Snails that live on mottled backgrounds are often banded. Grass snakes are grass-colored, as are many of the insects that live among the grasses.

Some animals are countershaded for camouflage. For example, fish are nearly always darker on the doral (top) surface than on the ventral (bottom) surface. Countershading reduces the contrast between the shaded and unshaded areas of the body when the sun is shining on the organism from overhead. A fish was once found—the Nile catfish—that was reverse-countershaded; that is, its dorsal surface was light and its ventral surface dark. The selective theory of camouflage was momentarily threatened, but scientific order was restored when it was discovered that the Nile catfish characteristically swims upside down.

Other organisms hide by looking like something else. One type of insect, the treehopper, looks like a thorn; another, the walkingstick, like a twig. Young larvae of some swallowtail butterflies look like bird droppings. In order for such disguises to work, animals must behave appropriately. *Biston betularia*, the peppered moth, lies very flat and motionless on its tree trunk, so that its colors blend with those of the tree. These and other moths that sit exposed on the bark of trees even habitually orient themselves so that the dark markings on their wings lie parallel to the dark cracks in the bark. Some desert succulent plants look like smooth stones, revealing their vegetable nature only once a year when they flower.

Some insects manage through warning coloration to frighten off their would-be predators. Large spots that look like eyes are commonly found on the backs of butterflies or the bodies of caterpillars, where they will suddenly appear when the insect spreads its wings or arches its body. Small birds reported to flee at the sight are probably birds that themselves are likely to be the prey of larger, large-eyed birds, such as owls or hawks. Smaller eyespots, while probably not frightening, seem to have the effect of deflecting the point of attack away from the head. Examination of wounded butterflies has shown that if a part of the wings bears beak marks or is missing, it is most often the part that

(a)

(b)

(c)

(d)

(e)

37–13

Concealment and camouflage. (a) Butterfly larva, disguised as a bird dropping, (b) leaf-like insect (Anaea), (c) bark katydid, (d) horned lizard in desert, (e) American woodcock on her nest, and (f) butterfly larva, a "measuring worm," that holds itself rigid and sticklike, resembling a twig, when disturbed. The physical adaptation of these animals is dependent, in large part, on a crucial behavioral adaptation. In times of danger, all of these animals remain absolutely still.

(f)

37-14

Monarch butterflies lay their eggs on milkweeds, such as the swamp milkweed shown here. The bitter sap of plants of the milkweed family contains a potential heart poison, to which larvae of the monarch butterfly are immune. By feeding on milkweeds, monarch caterpillars incorporate the poison into their bodies, and they and the butterflies are protected from predators by the poison's bitter taste and toxic effects.

contains the eyespots. One investigator has tested and confirmed this conclusion by painting eyespots on the wings of living insects, releasing them, and later recapturing them for examination.

On Being Obnoxious

Some animals defend themselves by having a disagreeable taste, odor, or spray. In insects, these may be derived from distasteful chemicals in the plants they eat. Monarchs and other butterflies, for example, feed in their larval stages on milkweeds (which they, in turn, as butterflies, pollinate), and the digitalislike compounds of the plants are concentrated in their tissues. The caterpillars are immune to these poisons, but the birds that eat the caterpillars become violently ill.

Obviously, tasting bad, while useful, may not be an ideal defense from the point of view of the individual, since making this fact known may demand a certain amount of personal sacrifice. (Actually, birds drop the monarch butterfly after the first bite, but they often inflict fatal injury in the process.) Obnoxious sprays and odors have the advantage of warding off predators before they harm the prey. To man, the most familiar of such animals is the skunk, which advertises its malodorous threat by its distinctive coloration. Many insects and other arthropods have developed defense secretions. In some millipedes, for example, the secretion oozes out of a gland onto the surface of the animal's body. Other arthropods are able to spray their secretion over a distance, in some instances even aiming it precisely at the attacker. The caterpillar *Schizura concinna*, whose single spray gland opens ventrally just behind its head, directs the spray simply by aiming its front end. In the beetle *Eleodes longicollis*, the spray glands are in the rear, and the beetle, when disturbed, does a quick headstand and spreads a secretion from its abdominal tip. The soldiers of certain termite species possess a pointed cephalic nozzle from which their defensive spray is ejected. This spray not only can incapacitate a small predator

(a)

(b)

(c)

37-15

(a) *The beetle* Eleodes longicollis *has glands in its abdomen which secrete a foul-smelling liquid. When disturbed, it stands on its head and sprays the liquid at the potential predator.* (b) Eleodes longicollis *is on the right. On the left is* Megasida obliterata,

which, as you can see, resembles Eleodes *and emphasizes this resemblance by also standing on its head. However,* Megasida *has no similar glands, no noxious secretion, and no spray.* (c) *A grasshopper mouse which has*

solved the problem of how to eat Eleodes. *The mouse drives the posterior of the beetle into the ground and eats it head first. It would probably eat* Megasida *the same way.*

but also acts as an attractant to summon more troops. In the whip scorpion, two glands open at the tip of a short knob that moves like a gun turret. Many arthropods possess a number of glands but discharge only from those closest to the point of attack, thus saving ammunition and gaining efficiency. The secretions usually act as topical irritants, especially to the mouth, nose, and eyes of the predator. Because of their relatively permeable skin, frogs and toads, common predators of arthropods, are sensitive to these irritants over their entire bodies.

Müllerian Mimicry: Advertising

For animals that have a highly effective protective device, such as a sting, a revolting smell, or a poisonous or bad-tasting secretion, it is advantageous to advertise. The more inconspicuous or rare such an animal is, the larger the proportion of individuals that must be sacrificed before the bird or other predator learns to avoid it. Müllerian mimics, named after F. Müller, who first described the phenomenon, are groups of organisms that, although not closely related phylogenetically, all have effective obnoxious defenses and all resemble one another. Bees, wasps, and hornets probably offer the most familiar example; even if we cannot tell which is which, we recognize them immediately as stinging insects and keep a respectful distance. Similarly, large numbers of bad-tasting butterflies are look-alikes. Müllerian mimicry is advantageous for the individuals of all species involved because each prospers from a predator's experience with another.

Batesian Mimicry: Deception

Batesian mimicry, first described by the British naturalist H. W. Bates in 1862, is deceptive mimicry. In Batesian mimicry, the innocuous mimic fools its predator by resembling a stinging or bad-tasting "model" which the predator has learned to avoid. Some species of harmless flies resemble bees or hornets, and many species of butterflies resemble monarchs or other unpalatable butterflies or moths.

Laboratory experiments have clearly demonstrated Batesian mimicry in operation. Jane Brower, working at Oxford, made artificial models by dipping

37–16
(a) *A monarch butterfly (below) and one of its mimics, a viceroy butterfly. A Batesian mimic has been compared to an unscrupulous retailer who copies the advertisement of a successful firm. Müllerian mimics, by contrast, are reputable tradesmen who share a common advertisement and divide its costs. Viceroys are Batesian. (b) A brightly colored, distasteful beetle, and (c) its Batesian mimic, a moth.*

(a)

(b)

(c)

mealworms in a solution of quinine, to give them a bitter taste, and then marking each one with a band of green cellulose paint. Other mealworms, which had first been dipped in distilled water, were painted green like the models, so as to produce mimics, and still others were painted orange to indicate another species. These colors were chosen deliberately: Orange is a warning color, since it is clearly distinguishable, and green is usually found in species that are not repellent and for whom, therefore, there would be no survival value in advertising.

The painted mealworms were fed to caged starlings, which ordinarily eat mealworms voraciously. Each of the nine birds tested received models and mimics in varying proportions. After initial tasting and violent rejection, the models were generally recognized by their appearance and avoided. In consequence, their mimics were protected also. Even when mimics made up as much as 60 percent of the green-banded worms, 80 percent of the mimics escaped the predators.

Batesian mimicry obviously works to the advantage only of the mimic. The model, on the other hand, suffers from attacks not only from inexperienced predators but from predators who have had their first experience with mimic rather than with model. The mimetic pattern will be at its greatest advantage if the mimic is rare, that is, less likely to be encountered than the model, and also if the mimic makes its seasonal appearance after the model, thus reducing its chances of being encountered first. However, if the model is sufficiently distasteful—as in the case of the quinine-soaked mealworm—it may protect mimics even if the latter are very common.

Among some butterfly species, the females may mimic more than one model, with the proportions of mimetic forms varying from population to population according to the prevalence in the area of one type of model or the other. Such mimicry is a product of disruptive selection (page 446), with the selection process presumably always working against the intermediate form. (Among such species, the males are always the same. If they were not, it would disrupt sexual selection, which, as you will recall, is a female prerogative in most societies.)

Complex patterns of mimicry (and also of symbioses) are most frequent in the tropical rain forest biome, where there are the largest number of species and where the evolutionary processes that moderate the interactions between them have been at work, undisturbed, for the longest period of time.

SUMMARY

A population is a group of organisms of the same species occupying a particular area at the same time. A community consists of all of the populations in a particular area. Whether or not an organism is found in a particular area depends in part on its range of tolerances for many environmental variables. Any single variable—such as temperature, water supply, or the availability of a particular required nutrient—that prevents survival of a particular organism is known as a limiting factor. The position, or role, of a population within a community is called its ecological niche. The hypothesis of competitive exclusion predicts that only one species can occupy the same ecological niche at the same time and that when two species compete for the same niche, one will be eliminated.

The theoretical growth rate of a population—its reproductive potential—is exponential (that is, 2, 4, 8, 16, 32); the more individuals in the population, the faster the population grows. The actual growth rate of an expanding population can usually be graphed by a sigmoid curve, beginning slowly, increasing exponentially for a time, and then leveling off. Even after the leveling-off phase has occurred, most natural populations exhibit periodic oscillations in numbers of members, with more striking oscillations in simpler ecosystems.

The kinds and numbers of organisms in a community are shaped not only by abiotic factors, such as those described in the previous chapter, but also by biotic factors, the individual interactions among the various populations.

Among the types of interactions are competition, which may result in the local elimination of one species (as with the duckweed species) or the restriction of one or both species to a noncompetitive position (barnacles and Galapagos finches). Symbiosis is the close association between organisms of different species. The association may be beneficial to both (mutualism), beneficial to one and harmless to the other (commensalism), or beneficial to one and harmful to the other (parasitism).

The majority of diseases in organisms are caused by parasites. Most parasites do not kill their host, and they almost never wipe out entire populations. Parasites tend to become so completely adapted to their hosts that they are entirely dependent on them.

Predator-prey associations affect the distribution of species, exert a regulating force on the size of populations, and have profound evolutionary effects on the various species involved. Plants and animals have developed a variety of defenses against predation. These include "armor" and other forms of physical protection, as seen in cacti, armadillos, turtles, and numerous other organisms, and chemical weapons, such as the plant poisons and noxious secretions of insects. Many organisms are camouflaged.

Some insects have come to resemble organisms of other species either to advertise an effective protective device that they possess in common with the other species (Müllerian mimicry) or to "pretend" they possess such a device when they actually do not (Batesian mimicry).

All these associations are the results of evolutionary processes and all help to determine the character of the community and of the organisms within it.

QUESTIONS

1. Define the following terms: population, exponential growth, sigmoid curve, ecological niche, mimicry (Müllerian and Batesian).

2. Introducing a new species into a community can have a number of possible effects. Name some of these possible consequences both to the community and to the species. What types of studies should be made before the importing of an "alien" organism? Some states and many countries have laws restricting such importations. Has your own state adopted any such laws? Are they, in your opinion, ecologically sound?

CHAPTER 38

Societies and Social Behavior

A _society_ is a group of individuals belonging to the same species and organized in a cooperative manner. It is thus something more than an aggregation of individuals. Often, similar organisms—bacteria, for instance, or paramecia, or mealworms—are found gathered in the same place because of environmental conditions, such as humidity, shelter, temperature, food supply, and so on. These aggregations do not represent societies in the sense that the word is used by students of animal behavior. In a society, stimuli exchanged among members of the group serve to hold the group together. These exchanges of stimuli, which are essentially forms of communication, result in what is defined as _social behavior._

We shall begin with a description of insect societies, since they are by far the most ancient of all societies, and also, with the single exception of those of modern man, by far the most complex. We shall follow with a description of some of the foundations of vertebrate social behavior and close with a brief analysis of one vertebrate society, that of the hamadryas baboon.

38–1

Societies of animals are held together by various forms of communication. Here a homecoming worker (at center left) shares nectar with other members of the hive. The hivemates extend their tongues to receive droplets of regurgitated fluid.

(a)

(b)

38–2
Solitary insects do not tend their young but they often provide for them. In subsocial species, the mother feeds the larvae after they have hatched. (a) Wasps of the Apanteles, *a solitary and parasitic genus, inject their eggs under the skins of caterpillars. The resultant larvae eat the internal tissues of the caterpillar, chew their way to the surface, and spin the cocoons shown here. Adults emerge from these cocoons. (b) Sand wasps, which are subsocial species, feed their larvae. Here a female is shown pushing paralyzed sandflies into the larvae's burrow. A larva is shown at left.*

38–3
Bumblebees are large, hairy bees, primarily adapted to colder climates. They are social bees, but their societies are smaller and simpler than those of the highest social bees, the honeybees. The life cycle is annual; only the fertilized queen survives the winter.

INSECT SOCIETIES

Eusocial, or "truly social," insects include ants, termites, wasps, and bees; all ants and termites are eusocial, as are some species of wasps and bees.

Of all animal organizations, the insect societies are probably the best understood, in terms both of their evolution and of the interplay of forces that keep them together. As with other animals, the social insects evolved from forms that were originally solitary. In fact, eusociality has evolved on at least eight separate occasions in bees and four times among wasps. The type and degree of care provided to the eggs and larvae by the wasp or bee are coming to be used as a taxonomic characteristic for determining relationships among the many species, just as slight differences in the shape of the body or the color of the wing might be used.

Stages of Socialization

Most species of bees and wasps are solitary. Among the solitary species, the female builds a small nest, lays her eggs in it, stocks it with a food supply, seals it off, and leaves it forever. She usually dies before the larvae mature.

Among subsocial or presocial species, the mother returns to feed the larvae for some period of time, and the emerging young may subsequently lay their eggs in the same nest or comb. However, the community is not permanent (usually being destroyed over the winter), there is no division of labor, and all females are fertile.

Eusocial insects are characterized by cooperation in caring for the young and a division of labor, with sterile individuals working on behalf of reproductive ones. The honeybees are the most familiar example.

Honeybees

The Workers

A honeybee society usually has a population of 30,000 to 40,000 workers and one adult queen. The life span of a worker is usually only about six weeks. Each worker, always a female, begins life as a fertilized egg deposited by the queen in a separate wax cell. (Drones, or male bees, develop from unfertilized eggs.) The fertilized egg hatches to produce a white, grublike larva that is fed almost continuously by the nurse workers; each larval bee eats about 1,300 meals a day. After the larva has grown until it fills the cell, a matter of about six days, the nurses cover the cell with a wax lid, sealing it in. It pupates for about 12 days, after which an adult emerges.

(a)

(b)

38-4

Honeybee workers. (a) The first segment of each of the three pairs of legs (tarsi) has a patch of bristles on its inner surface. Those of the first and second pairs are pollen brushes which gather the pollen that sticks to the bee's hairy body. On the third pair of legs, the bristles form a pollen comb that collects pollen from the brushes and the abdomen. From the comb, the pollen is forced up into the pollen basket, a concave surface fringed with hairs on the upper segment of the third pair of legs. Transfer of pollen to the pollen basket occurs in midflight. The sting is at the tip of the abdomen. (b) The mouth parts are fused into a sucking tube containing a tongue with which the bee obtains nectar. The antennae, attached to the head by a ball-and-socket joint, contain both touch receptors and olfactory organs. The large compound eyes cannot see red (which is black, or colorless, to them) but can see ultraviolet, which is colorless to human eyes.

The newly emerged adult worker rests for a day or two and then begins successive phases of employment. She is first a nurse, bringing honey and pollen from storage cells to the queen, drones, and larvae. This occupation usually lasts about a week, but it may be extended or shortened, depending on the conditions of the community. Then she begins to produce wax, which is exuded from the abdomen, passed forward by the hind legs to the front legs, chewed thoroughly, and then used to enlarge the comb. During this stage of employment as a houseworking bee, she may also remove sick or dead comrades from the hive, clean emptied cells for reuse, or serve as a guard at the hive entrance. During this period, she begins to make brief trips outside, seemingly to become familiar with the immediate neighborhood of the hive. It is only in the third and final phase of her existence, that the worker honeybee forages for honey and nectar.

The Queen

The queen begins life as an egg genetically identical to that of the worker. The differences between the two depend on the substance fed the queen-to-be in the larval stage and on the pheromonal influences she, in turn, exerts upon her subjects. For the first two days of life, all bee larvae are fed "brood food," a white paste produced in the glands of young workers and secreted from the mouth. Worker and drone larvae are then fed honey and pollen, while the larva being made ready for queenhood is fed only the glandular secretions (hence known as royal jelly) all during its larval stage. Attempts have been made to identify the substance in royal jelly that confers queenhood, but so far they have not been successful.

Queens are raised in special cells larger than the ordinary cells and shaped somewhat like a peanut shell. If a hive loses its queen, workers will notice her absence very quickly and will become quite agitated. Very shortly, they begin enlarging worker cells to form emergency queen cells. The larvae in the enlarged cells are then fed exclusively on royal jelly until a queen develops. Any diploid larva so treated will become a queen.

The Queen's Pheromones

The queen exerts influences on her subjects by means of pheromones, of which there appear to be several. As shown by the British entomologist C. G. Butler and his co-workers, the influence of one of the pheromones, known as queen substance, inhibits ovarian development in the worker bees and prevents them from becoming queens or producing rival queens. This pheromone passes through the workers of the hive orally. As the workers meet, they often exchange the contents of their stomachs. Studies in which queen substance has been tagged with a radioactive label have shown that, as a result of this activity, the pheromone travels through the hive with remarkable rapidity. Within only half an hour after removal of the queen, the shortage of the queen substance is already noticed, and the hive begins to grow restless. It is difficult to understand how a single queen can produce enough pheromone to influence the entire hive of as many as 30,000 to 40,000 workers, as well as tend to her stupendous egg-laying chore (commonly more than 1,000 per day). It has been suggested that, after the pheromone is passed among the workers, it is fed back to the queen in a reduced form and she need simply oxidize it to reactivate it.

Note that the words "queen" and "royal" imply, by analogy, that the queen bee's life is more desirable than that of her sisters. However, she can also be viewed, from a slightly different perspective, as an egg-laying machine held captive and operated by the workers.

(a)

(b)

(c)

38–5

The life of the hive. (a) Workers tending honey and pollen storage cells. The honey is made from nectar processed by special enzymes in the workers' bodies and is thickened by evaporation during repeated regurgitation and swallowing. (b) Queen inspecting a cell before laying an egg in it. (c) Longitudinal section of pupa showing the larva within it.

Winter Organization

The honeybee colony differs from that of subsocial bees in that it survives the winter. This means that the bees must stay warm despite the cold. Honeybees cannot fly if the temperature falls below 10°C and cannot walk if the temperature is below 7°C. Within the wintering hive, bees maintain their temperature by clustering together in a dense ball; the lower the temperature, the denser the cluster. The clustered bees produce heat by constant muscular movements of their wings, legs, and abdomens. In very cold weather, the bees on the outside of the cluster keep moving toward the center, while those in the center move to the colder outside periphery. The entire cluster moves slowly about on the combs, eating the stored honey from the combs as it moves.

New Colonies

In the spring, when the nectar supplies are at their peak, so many new broods are raised that the group separates into two colonies. The new colony is always founded by the old queen, who leaves the hive taking about half of the workers with her. This helps to ensure survival of the new colony since this queen is of proven fertility. The group stays together in a swarm for a few days, gathered around the queen, after which the swarm either will settle in some suitable hollow tree or other shelter found by its scouts. The swarm can also be picked up by a beekeeper and transferred to an empty hive, which it will then furnish with wax.

In the meantime, in the old hive, new queens have begun to develop, often even before the old queen has left. Ovarian development has begun in some of the workers, a few of which lay eggs. The unfertilized eggs develop into drones. After the old queen leaves the hive, a new young queen emerges, and any other developing queens are destroyed. The young queen then goes on her nuptial flight, exuding a pheromone (apparently also the queen substance) that entices the drones of the colony to follow. She mates only on this one occasion, although she may mate with more than one male, and then returns to the hive to settle down to a life devoted to egg production.

During her nuptial flight, the queen receives enough sperm to last her entire life, which may be some five to seven years. These are stored in her spermatheca and are released, one at a time, to fertilize each egg as it is being laid. The queen usually lays unfertilized eggs only in the spring, at the time males are required to inseminate the new queens, so the release of the sperm cells from the spermatheca is apparently under the control of her internal or external environment.

The drones' only service to the hive is their participation in the nuptial flight. Since they are unable to feed themselves, they become an increasing liability to the hive. As nectar supplies decrease in the fall, they are stung to death or driven out to starve by their sisters.

Because of the reproductive pattern of the honeybee, members of the hive are very similar genetically and, thus, their apparent altruism on behalf of their hivemates is actually just another way of ensuring that their own genes have maximum representation in the next generation.

Organizational Principles of Insect Societies

Insect societies, although they resemble human societies in many different ways, differ from them in at least two important respects. The first major point of difference is that reproduction in insect societies is carried out by only a very few individuals. For this reason, as we noted previously, all of the insects in one society are likely to be very closely related genetically. In fact, in some respects, an insect society is more like a superorganism with different specialized parts than a collection of individuals.

Second, recognition among members of the insect society is based entirely on chemical or other signals; there is no personal recognition of individual members as there is in primate or other mammalian societies. This stereotyping of behavior is, of course, in keeping with (and demanded by) the relatively short life span of insects. A worker honeybee, for example, with her short life span, has little time for complex learning. This impersonality is also in keeping with the large, sometimes enormous, size of the societies themselves.

VERTEBRATE SOCIETIES

Vertebrate societies range from small, often transient groups of closely related individuals, in which the social focus is on the raising of the young, to larger, permanent, quite stable groups, such as those found among some of the primates. Some of the societies—such as flocks of birds, schools of fish, and large migratory herds of herbivores—may number in the thousands, but they are not usually as large as the societies of most eusocial insects.

Man is among the most social of the social animals, and so, from the point of view of the human observer, social living would appear to be the norm. However, in fact, living in a society has decided disadvantages for the individual organism. The most important of these is the sharing of food resources. Another is the much greater vulnerability to disease. Finally, an individual in a society may have to compete for a mate, thereby diminishing his or her chances of leaving viable offspring. The advantage of social living seems to be related almost entirely to avoidance of predation. In addition, in some societies, such as those of wolves, social behavior permits the hunting of large animals, such as moose or caribou, that could not be caught or killed by a solitary individual. As we shall see, hunting large game was undoubtedly also a major factor in the socialization of early man.

Thus, whether or not animals are social at all, the degree of social organization, and the ways in which individuals are held together and organized within a society, are, like all other characteristics of an organism, the products of often-conflicting forces.

38-6

In the summer, when the vegetation of the tundra is abundant, the caribou form large herds, such as these, in which they feed and migrate. In the winter, when food is scarce, they disperse and forage in isolated groups.

How are vertebrate groups held together? There is no simple answer to this question, partly because the different kinds of groups are undoubtedly bound together in different ways and partly because most explanations depend heavily on concepts such as "sexual drive," "maternal instinct," or "affectional systems," which in themselves are almost impossible to define. As an indication of the complexities involved, we shall touch on two types of behavior that serve to keep animals in groups: imprinting and rituals.

Imprinting

Certain birds, such as swans, chickens, and turkeys, which are physiologically mature enough to leave the nest soon after they are hatched, follow the first moving object that they see. For ducklings, the effective objects can range from a matchbox on a string to a walking man. Once one of these newly hatched birds has become attached to a particular object, it will follow only that object. Under natural circumstances, of course, this object will be a parent, and it is this following response that keeps the young birds close behind and well within the protective range of the mother until the end of their juvenile period, when the response is lost.

38-7

Many species of precocial birds (birds that are able to walk as soon as they're born) will follow the first moving object they see after hatching and will continue to show this following response as they mature. The phenomenon is known as imprinting. The birds shown here are goslings, and the object of their affection is animal behaviorist Konrad Lorenz.

The learning pattern that involves this act of recognition is called *imprinting*, and it differs from other types of learning in that it can occur only within a limited period. Studies of the newly hatched mallard duckling, using a mechanically operated decoy, showed that imprinting was most effective between 13 and 16 hours after hatching. After 24 hours, attempts to imprint were not very successful, and after 30 hours, unimprinted ducklings not only failed to follow but actually avoided moving objects. Similarly, chicks will not become imprinted to a moving object when they are only a few hours old or when they are several days old but only during the intervening period.

Imprinting also seems to influence mate selection in the adult birds. Lorenz reports many examples of ducks and geese becoming sexually fixated on objects or on members of other species, including Lorenz himself.

(a)

(b)

(c)

38–8

Rituals and ceremonies are common among birds. (a) Greeting ceremonies between storks at the nest. Pair bonding between storks characteristically endures for a lifetime, based, in part, on the repeated performance of this ritual. (b) Courtship display between two albatrosses of the Galapagos. (c) Female tern begging for a fish from a male tern. If he is ready to mate, he will give her the fish (although she may have to ask several times). If she is ready, she will eat it; otherwise she returns it. Such mating ceremonies serve to promote genetic isolation (see page 467).

The term "imprinting" is usually applied only to phenomena associated with the following response in birds, but similar effects are seen in other animals. For example, if a baby lamb is taken from its mother right after birth and bottle-raised, it will not follow other sheep when it is mature, as do lambs raised normally.

Rituals

Although few vertebrates are cannibals, many of them, fish and birds in particular, have a distinct preference for maintaining discreet distances between themselves and other animals, whether of their own or another species. Among birds lined up on a telephone wire, for instance, each will be found almost equidistant from the next, as though the spaces were measured off by an invisible yardstick. Even those animals that fly in flocks or swim in schools tend not to touch one another.

The mating ceremonies of these animals are interpreted as a means for allaying those highly adaptive feelings of hostility and suspicion so that mating can take place. As a consequence, many of these ceremonies incorporate both aggression and appeasing behavior. In the mating behavior of the stickleback, for example, described in Figure 34–7, the male's "zig" toward the female is identical to a motion of attack, and when the male "zags," the motion entices her toward the nest. Most females flee the attack motion, but the one whose need to lay her eggs is sufficiently great stands still, turns sideways (an appeasing gesture on her part since an antagonist would zig back), and then follows the male's zag.

In mating ceremonies between birds, feelings of fear and aggression brought into play by the closeness of another individual may be handled by being redirected at either a real or an imaginary antagonist. Konrad Lorenz describes the so-called "triumph ceremony" in the graylag goose, one of his favorite research subjects and companions. As a form of greeting to his partner, the gander proceeds to attack an "enemy." This attack is performed, as is a real attack, with the head and neck pointing obliquely forward and upward and is accompanied by a raucous trumpeting. After the "enemy" is routed or defeated, the gander returns to his partner. On his return, the gander holds his head lowered and pointed forward, but instead of pointing directly at the goose, as he would at an enemy, he points obliquely past her. She comes forward to meet him, her head inverted submissively, and he cackles to her triumphantly.

This ceremony, which probably had a purely sexual origin, now serves to hold entire flocks together, and even small goslings participate in elements of the triumph ceremony. When a young male performs the ceremony with a strange young female, it usually marks the beginning of a mating bond that may last for the entire lifetime of the individuals. The ceremony is typically performed between young geese the year before mating and breeding begin, and it will continue to be performed by the partners throughout their entire lives whenever they encounter one another even after a short separation. By the intensity with which the ceremony is performed, an experienced observer can judge the length and strength of the bond between the partners.

Although few animals—except perhaps humans of ambassadorial rank—have such elaborate social rituals, higher animals of many species exhibit obligate social formalities upon meeting. Think, for example, of the greeting ceremonies between domestic dogs, and how they vary with sex, rank, and familiarity. Rituals serve the function both of allaying the anxieties of the individuals involved and of identifying them to one another as members of the "in" group.

38-9

Wolves have dominance hierarchies of both males and females. Here a subordinate female is licking the muzzle of the dominant female. This same muzzle-nuzzle gesture is used by pups begging for food.

38-10

A subordinate baboon turns his buttocks toward a superior. This gesture, known as presenting and used by females to indicate their readiness to mate, is also used between males and between females to signify submission or conciliation or to beg for special favors. The superior is reassuring the subordinate with a pat on the back.

Social Dominance

Animal societies are often arranged in hierarchies. One type of social dominance that has been studied in some detail is what is called the *pecking order* in chickens. A pecking order is established whenever a flock of hens is kept together over any period of time. In any one flock, one hen usually dominates all the others; she can peck any other hen without being pecked in return. A second hen can peck all hens but the first one; a third, all hens but the first two; and so on through the flock, down to the unfortunate pullet that is pecked by all and can peck none in return.

Hens that rank high in pecking order have privileges such as first chance at the food trough, the roost, and the nest boxes. As a consequence, they can usually be recognized at sight by their sleek appearance and confident demeanor. Low-ranking hens tend to look dowdy and unpreened and to hover timidly on the fringes of the group.

During the period when a pecking order is being established, frequent and sometimes bloody battles may ensue, but once rank is fixed in the group, a mere raising or lowering of the head is sufficient to acknowledge the dominance or submission of one hen in relation to another. Life then proceeds in harmony. If a number of new members are added to a flock, the entire pecking order must be reestablished, and the subsequent disorganization results in more fighting, less eating, and less tending to the essential business, from the poultry dealer's point of view, of growth and egg laying.

Pecking orders reduce the breeding population. Cocks and hens low in the pecking order copulate much less frequently than socially superior chickens. Thus the final outcome is the same as if the social structure did not exist: The stronger and otherwise superior animals eat better, sleep better, and leave the most offspring. However, because of the social hierarchy, this comes about at minimum expense to dominant and subordinate individuals alike.

Territoriality in Birds

Territoriality was first recognized by an English amateur naturalist and bird watcher, Eliot Howard, who observed that the spring songs of male birds served not only to court the females but also to warn other males of the same species away from the terrain that the prospective father had selected for his own. In general, a territory is established by a male. Courtship of the female, nest building, raising of the young, and often feeding are carried out within this territory. Frequently the female also participates in territory defense.

By virtue of territoriality, a mating pair is assured of a monopoly of food and nesting materials in the area and of a safe place to carry on all the activities associated with reproduction and care of the young. Some pairs carry out all their domestic activities within the territory. Others perform the mating and nesting activities in the territories, which are defended vigorously by the males, but do their food gathering on a nearby communal feeding ground, where the birds congregate amicably together. A third type of territory functions only for courtship and mating, as in the bower of the bowerbird or the arena of the prairie chicken. In these territories, the male prances, struts, and postures—but very rarely fights—while the females look on and eventually indicate their choice of a mate by entering his territory. Males that have not been able to secure a territory for themselves are not able to reproduce; in fact, there is evidence from studies of some territorial species, such as the Australian magpie, that adults that do not secure territories do not mature sexually.

Examples of several territorial animals are given in Figure 38-11.

38-11

Territories come in many shapes and sizes. (a) The male Uganda kob displays on his stamping ground, which is about 15 meters in diameter and is surrounded by similar stamping grounds on which other males display. A female signifies her choice by entering one of the stamping grounds and grazing there. Only a small proportion of males possess stamping grounds, and those that do are the only ones that breed. For a fiddler crab (b), it is a burrow, from which he signals with his large claw, beckoning females and warning off other males. (c) Howler monkeys shift their territories as they move through the jungle canopy, but maintain spacing between groups by chorusing. (d) Territoriality is common among reef fish. For many species, a territory is a crevice in the coral, but for others, such as the skunk clown fish shown here, the territory is a sea anemone. The fish is covered by thick slime that partially protects it from the poison of the tentacles, but its acceptance by the anemone is chiefly a consequence of behavioral adaptation of the fish, which even mates and raises its brood among the tentacles.

(a)

(b)

(c)

(d)

Territorial Defense

Even though territorial boundaries may be invisible, they are clearly defined and recognized by the territory owner. With birds, for example, it is not the mere proximity of another bird of the same species that elicits aggression, but his presence within a part of a particular area. The territory owner patrols his territory by flying from tree to tree. He will ignore a nearby rival outside his territory, but he will fly off to attack a more distant one that has crossed the border. Animals of other species are generally ignored unless they are prey or predators.

Once an animal has taken possession of a territory, he is virtually undefeatable on it. Among territory owners, prancing, posturing, scent marking, and singing and other types of calls usually suffice to dispel intruders, which are at a great psychological disadvantage. For example, a male cichlid, a tropical freshwater fish, will dart toward a rival male within his territory but as he chases the rival back into his own territory, he begins to swim more slowly, his caudal fin seemingly working harder and harder, just as if he were making his way against a current that increases in strength the farther he pushes into the other male's home ground.

Similarly, the expulsion from communal territories is typically accomplished by ritual rather than by force. For example, among the red grouse of Scotland, the males crow and threaten only very early in the morning, and then only when the weather is good. This ceremony may become so threatening that weaker members of the group leave the moor. Those that leave often starve or are killed by predators. Once the early-morning contest is over, the remaining birds flock together and feed side by side for the rest of the day.

Territories and Population Regulation

As we mentioned previously, animals that do not gain possession of territories or that are excluded from the "home range" or breeding area do not produce young. As older animals die, younger ones will contend for their places, keep-

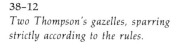

38–12
Two Thompson's gazelles, sparring strictly according to the rules.

SOCIAL DOMINANCE AND SEX REVERSAL

An unusual form of social dominance is seen among the wrasses, a species of small iridescent blue fish that inhabit the waters around Australia's Great Barrier Reef. Each group of wrasses consists of a male and a number of females; the group is territorial, with territoriality exhibited largely among males. The single male and his females are organized in a hierarchy; the largest, oldest individual is the male, who dominates all the females of the group. Next in rank is the largest, oldest female, who occupies the center of the territory with the male and is dominant over the other smaller and younger females. When the male dies, either neighboring males invade the territory, or, as occurs in the majority of cases, the ranking female becomes dominant. Within 1½ to 2 hours after the male's death, she begins to show male aggressive behavior and within a few hours she is exerting dominance over the other females and patroling the borders of the territory. Within a few days, the new dominant has begun male courtship and spawning behavior, and small regions of dormant testicular tissue within her

Wrasses at a reef off New Providence Island, Bahamas.

ovaries have begun to develop. By the end of about two weeks, the ovaries have become testes and "she" is fully male, able to produce sperm and fertilize the eggs of the subordinate female family members.

ing the breeding population stabilized. If conditions are particularly favorable, or if the range is enlarged, more animals can gain access to the breeding community.

Limitation of the reproductive rate, as by social dominance or territoriality, is sometimes referred to as "self-regulation of animal populations." Some ecologists object to this term, pointing out that in these populations, as in all others, external environmental factors, such as food supply, are limiting and not the population itself. However, within the limits imposed by the environment, dominance and territoriality, while allowing the survival and reproduction of dominant animals (which would occur anyway), minimize physical combat.

Some Notes on Aggression

The examples cited in the discussion of social hierarchies and territoriality serve to remind us that actual combat is rare among members of the same species, whether or not they are members of complex societies.

When fighting does occur among animals of the same species, it is often ritualized in such a way that both animals remain unharmed. For instance, male iguanas of the Galapagos Islands fight by pushing their heads against one another; the one that drops to its belly in submission is no longer attacked. Male cichlid fish of one species first display, presenting themselves head on and then side on, with their dorsal fins erected, and then beat water at each other with their tails. If this does not bring about a decision, each grasps the other by its thick strong lips. They then pull and push with great force until one lets go and, unharmed but defeated, swims away. Rattlesnakes, which could kill each other with a single bite, never bite when they fight but instead glide along side by side, each pushing its head against the head of the other, trying to push it to the ground in a form of Indian wrestling. Many antlered animals, such as stags of the fallow deer, which have long and vicious horns, follow an equally careful ceremony and attack only when they are facing each other, so that their antlers are used only for dueling and not for goring.

38–13

(a) *A male baboon. Males are much larger than the females. Other conspicuous differences are their bright hindquarters (pale pink in the nonestrous female) and, in this species, the long mane, or mantle. (b) Female and infant. The young are born black and their coats become lighter as they mature.*

These prohibitions against killing members of one's own species (intraspecific killing) make sense biologically. It is to the advantage of each individual, in terms of its own survival, not to waste its strength in needless bloodshed, and therefore genetic traits that channel aggressions and reduce conflict will be selected under evolutionary pressures. Except for certain fish that regularly cannibalize their young, and rats operating under severe population pressures, man is one of the few vertebrates that regularly kills large numbers of its own kind.

Social Organization among Baboons

Baboons are large, quadrupedal African monkeys with protruding, doglike muzzles. One genus *(Papio)* and four species are commonly recognized: a savanna species *(Papio anubis),* of which there are at least two and probably more distinct races; two forest species (the drill and the mandrill), both short-tailed baboons living in West Africa; and a desert species found in North Africa and Arabia, *Papio hamadryas,* the hamadryas baboon.

The baboons are of particular ecological interest for two reasons. First, unlike most monkeys, they live on the ground and have done so for several million years. For this reason, differences between baboons and other monkeys can be related to some degree to their environments. Second, there is a marked difference in social organization between the savanna and forest species, on the one hand, and the desert species, on the other. Since the groups are closely related genetically, obviously sharing a common ancestor in the not-too-distant past, it is very likely that at least some of these differences can be explained ecologically rather than genetically. This question is of particular current interest because of the recent (and recurrent) discussion concerning the relative effects of nature and nurture in the human species.

The Genus Papio

Baboons are large, with adult males weighing about 55 kilograms. Their size is clearly related to a terrestrial existence; among the primates, only large animals, such as baboons, some of the great apes, and man, are terrestrial.

Among all baboons, adult males are very different in appearance from adult females (Figure 38–13). This phenomenon, known as sexual dimorphism, is much less pronounced among other primates, and particularly other monkeys. The males have mantles and prominent canines and, most obviously, they are about twice the size of the females. Sexual dimorphism, which is much less pronounced among tree-dwelling primates, appears to be related to the increased hazards of life on the ground and to the male role of defender.

Like most primates, baboons are nearly exclusively vegetarian, living on leaves, fruits, flowers, and young stems, all of which they pick and consume on the spot. They also dig for roots, bulbs, and tubers. This vegetable diet is supplemented by insects and occasionally by meat, such as snakes, lizards, fledgling birds, or young gazelles, when they are encountered accidentally. There is no organized hunting for prey animals and no sharing of prey. Baboons sleep in tall trees or on cliffs, returning to a nesting site every night. Hence they must often travel to find food, sometimes for many miles across open country. Baboons have a harder time finding food than do tree-dwelling primates, which are more likely to live in the lush tropical rain forests and which sleep where they eat. This hardship is probably related to the difference in size between the two sexes of baboons. The male is large enough to act as protector, whereas the female's smaller size reduces the food requirements of the family unit.

38–14

Baboons sleep in groups, either in tall trees, as shown here, or on ledges or steep cliffs. The size of the group and the choice of a sleeping site appear to depend more on the terrain than on the species.

The Structure of the Band

Baboons are organized into multimale bands; those observed have ranged in size from eight to more than 185 individuals. Adult females outnumber adult males by about 2 to 1. Much of this difference can be accounted for by the much slower maturation rate of the males. Although an adult male can probably produce viable sperm by the time he is about four years old, he does not reach full size nor are his canines fully erupted until he is eight or more, and therefore is not counted as an adult male either in the social organization of the band or by the field observer peering through his binoculars. Females, on the other hand, are adults by the age of two or three.

The bands are territorial, with fairly distinct home ranges, and bands rarely meet. When they do, the one farthest from the center of its customary range generally moves away, sometimes after an exchange of vocalizations, canine displays, and gestures, but more often with no visible reaction at all, except perhaps a slight display of nervousness. Bands that know each other may share large sleeping groves. Overt fighting has been reported once; it occurred between two groups trying to sleep in the same clump of trees.

When the bands move in open country, females and infants characteristically are in the middle of the group, close to adult males. Mass counterattacks by adult males, who placed themselves between the group and the attackers, have been observed to rout a leopard and disperse a large dog pack, killing or wounding several of its members.

Another function of the band is the pooling of information. Hans Kummer, one of the most experiencd and imaginative observers of baboons, describes a band at the beginning of a day's march as resembling a giant amoeba, stretching out and then withdrawing one pseudopod after another, as small segments begin to move off in one direction, only to rejoin the band if the others do not follow. Finally, a movement is initiated that meets with general approval, and the day's march begins.

The size of the group seems to be somewhat related to its habitat. Very small groups are correlated with small areas of vegetation, quite widely separated from one another, whereas larger groups may be found where a typical feeding site might be a clump of trees. Logically speaking, troop size must be a compromise between the availability of food resources and the need of the troop to defend itself.

Social Dominance

In all baboon species except the hamadryas, the band is organized around a dominance hierarchy of adult males. As in other dominance hierarchies, the band is maintained mostly by behavioral conventions, with only occasional fights, which rarely result in injury. Dominant males threaten subordinates by staring, yawning (with the ears laid back), raising the eyebrows, and, at close range, by grinding their teeth. Subordinate males demonstrate submissiveness by looking away, grinning, turning their backs, and presenting their hindquarters. The dominant male may mount the presenting male briefly. The gesture of presentation, also used by females in estrus as an invitation to copulate, is the most reliable index of hierarchical position. The top-ranking male presents to no other individual, the second-ranking male only to the top one, and so on down the social ladder. (Presentation as a gesture of submission or appeasement is seen among many primate species.) Adult females are subordinate to males but also have a separate dominance hierarchy, although not as clearly structured as that of the males. Adult females present to superior females.

38–15

A dominant male, threatening. The huge canines are found only in the males, indicating that they are related less to diet (male and female eating habits are the same) than to dominance and aggression. However, altercations among baboons seldom result in serious injury.

Socially superior males mate more often than inferior males. In some groups, it was observed that a number of males copulated with females when they were not in full estrus, when the likelihood of fertilization is considerably less. But when a superior female is in full estrus, as revealed by maximum swelling and coloring of the genital area, only the highest-ranking male mates with her. Thus the highest-ranking male is the one most likely to father offspring. Dominant males also have first choice of nesting sites. However, because there is no shared food supply, dominants are not allocated a larger share of food than subordinates, as they are in some other pecking orders.

Dominant males play a leading role in territorial disputes, thus maintaining the spacing among bands. They are also the most aggressive in defending the band against intraspecific aggression. If another member of the group barks a warning or screams in fear, the dominant male steps forth and inspects the danger.

Social Bonds

As among almost all higher primates, grooming is a prominent form of social behavior among baboons. It involves a brisk parting of the fur of another individual with the fingers of both hands and the picking off of small particles, including insects; these items are often swallowed. Dominant males, females in estrus, and females with infants are groomed most frequently, but all members of the group receive some grooming attention. The function of grooming is clearly not only hygienic but also a continual reinforcement of social bonds.

Infants are another clear social bond. The infant, which is all black and therefore very conspicuous for its first few weeks of life (perhaps so it can be guarded more zealously), usually clings to its mother, who may support it with her arm. With the mother's permission, it may be touched or even held briefly by other members of the troop. The infant or infant-mother pair are clearly attractive to other members of the group, who tend to cluster around, especially when the infant is newborn. Mothers with infants are groomed frequently, particularly by other females, and dominant males attend them closely as the

38–16

(a) *Mother and nursing infant.* (b) *A female baboon presenting, in social deference, to a mother and newborn infant because she wants to approach and touch the infant.*

38–17

As among all social primates (except man), grooming is one of the primary bonds among group members. The amount of grooming an individual receives is closely related to his (or her) position in the social hierarchy. Here an adult female grooms a dominant male. Being groomed is not only (obviously) pleasurable but also removes insects and other parasites from the skin.

troop moves. An infant that loses its mother will be adopted, sometimes by a childless female, but often by a young male.

The Hamadryas Baboon

The hamadryas baboon is unlike the other members of the genus *Papio* in that there are three distinct levels of social organization. The principal unit is a male with one or more females and their young, totaling up to seven or eight members. These one-male units group together into bands, similar in size and organization to the bands of the other baboon species. Finally, the bands come together in troops, some of which have been counted as containing 750 members.

Students of primate behavior hypothesize a direct relationship between these social organizations and the environmental resources of the hamadryas. The habitat, which is on the edges of deserts, is typically arid grassland, interspersed with thorny acacias and other small trees and bushes. There are no tall trees, and the baboons of this area sleep on ledges on the vertical slopes of steep cliffs. (This is learned behavior. The hamadryas will sleep in tall trees if they are available, and other species will sleep on cliffs if there are no tall trees.) The baboons move across open country, often for long distances, between their sleeping cliffs and their feeding sites. They travel in bands, break up into one-male units for feeding—typically one unit to a tree—regroup into bands, return to the cliffs, where they often sleep in troops, and then recongregate each morning in the band. Thus each social unit serves a clear and important function. The one-male unit is an optimal foraging unit, provided with a protector; the band provides for mutual defense when traveling; and the troop makes possible maximum utilization of safe sleeping sites within the habitat.

The strongest bonds are those that hold the one-male unit together. The females mate exclusively with the unit leader, and almost all social interactions, such as grooming, are carried out within the unit. Juveniles sometimes leave to play with those of other units within the band.

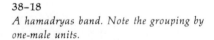

38–18
A hamadryas band. Note the grouping by one-male units.

A young male hamadryas with a juvenile female that he has stolen from another one-male unit. New family groups originate in this way.

The structure of the hamadryas band is generally similar to that of other baboon species. There is a male dominance hierarchy, and males act as defenders. About 20 percent of the adult males are not part of any family unit, but, as with other baboon species, these males are part of the band. The band moves as a whole, apparently acting on information from knowledgeable members of the group.

The ties that hold the band together are less strong than those that hold the one-male unit of the hamadryas together. If a member, because of age, illness, or injury, drops out of a nonhamadryas band, he or she is left behind. If a disabled female drops out of a hamadryas band, her male leaves the band to stay with her. The band, however, like the one-male unit, is a stable social group, always composed of the same members. The troop, on the other hand, is made up of varying groups of bands from the same territory that apparently recognize one another and are mutually tolerant.

Because the male-dominated band is the primary social structure of the other *Papio* species (and also of the macaques, arboreal monkeys who are their closest relatives), the hamadryas band appears to be a legacy from their common ancestors. Similarly, the troop seems easy to understand as a direct outgrowth of environmental pressures; baboons of other species share water holes and large sleeping groves when they have to.

What, however, are the bonds that hold the one-male unit together? The cohesion of the unit turns out to depend primarily on two behavioral patterns, both those of the male. The first is a herding instinct toward the females. The females are taught to follow the male. He constantly watches them over his shoulder and, if one drops back or attempts to slip away, he stares, threatens, and sometimes nips her on the back of the neck, after which she immediately follows. If a male hamadryas is presented with a female olive baboon *(Papio anubis)*, he accepts her quite readily and immediately trains her to follow him.

The second behavioral pattern is a strong inhibition among baboon males against taking another male's females. This, too, was tested experimentally. Two adult males, who knew each other, and a strange adult female were trapped. One male was enclosed with the female while the other was permitted to watch from a cage 10 meters away. The male immediately made grooming, mounting, and herding advances toward the female, as he would toward a

38-20
Societies of animals are held together by various forms of communication. In insect societies, which are often very large, these communications are entirely impersonal, depending primarily on exchange of chemical signals (pheromones). Responses to these stimuli are genetically programed. In nonhuman primate societies, by contrast, relationships are personal and are based on recognition of other members of the group as individuals. The identity of other members of the group and one's behavior toward them depends on learning as well as instinct. The photograph shows two young baboons engaged in this learning process.

member of his harem. Fifteen minutes later, the second male was put in the same enclosure with the pair. He not only refrained from fighting over the female, but he avoided even looking at them. The animals were separated, and two days later the experiment was repeated again, this time switching the males. Once again, the male that had watched the activities between the male-female pair was strongly inhibited in his behavior.

Another adaptive feature of this inhibition is that it applies only to adult males and females. Young males are attracted to juvenile females, which they "kidnap" from their family groups before they are sexually mature and train to follow them. Thus the younger male has a detour around the inhibition barrier, and new family units can come into existence.

Lest you be tempted to extrapolate to human behavior from these accounts of baboon behavior, it is of some interest that, among the baboonlike geladas, one-male groups are formed and held together by activities of both sexes. Dominant males pair with dominant females, wife number one is dominant over wife number two, and so on. The geladas are large, baboonlike monkeys that live in the mountainous grasslands of Ethiopia; though they are referred to as baboons in common usage, they are different enough to be recognized as a separate genus (*Theropithecus gelada*).

The baboons remind us that social behavior in animals is ecologically adaptive—that is, that it promotes the survival of members of the social organization (and particularly of their young) within a given habitat or range of environmental conditions. Social behavior is undoubtedly acquired in a number of different ways—ranging from the rigid genetic programing seen among the social insects to the apparently much more flexible behavior patterns seen among baboons and in human cultures.

SUMMARY

A society is a group of individuals of the same species organized in a cooperative manner and communicating with one another in some way.

Among the most complex societies are those of insects. These societies are matriarchies, centering around the care of the queen and the raising of the brood. The behavior of members of the society is determined by chemical substances—pheromones—exchanged among members of the society. In eusocial insects, relatively few members of the society reproduce; the others care for the reproducing individuals and the young.

Vertebrate societies are varied both in their size and their organization and also in the behavioral factors that hold them together. Examples of two such behavioral influences are imprinting and rituals. Imprinting involves a very specific sort of learning that takes place in a very narrow developmental time limit and results, in nature, in the recognition of members of one's own species. Rituals are stylized behavioral patterns, shared by members of a group, that allay anxieties and promote recognition of group members by one another.

Some animal societies are organized in social hierarchies. Socially inferior animals—those low in the pecking order—reproduce less frequently than their social superiors and are often psychological castrates. They are also the first to starve if food is limited or to be driven out if shelter is limited.

Territories are areas defended by an individual or a society against others of the same species. Territories may be "real"—that is, they may be actual areas of land containing food and nesting material to support a mating pair and young—or they may be symbolic, such as an arena. In either case, the only animals that breed are those with territories. Animals without territories provide replacements for territory owners and a reserve for population expansion if the range or food supply of the population is increased.

Acts of aggression are restricted in animal societies by social dominance, territoriality, and other forms of social behavior. Intraspecific killing is found only rarely among animals other than man.

Baboons are large, quadrupedal, terrestrial monkeys, with pronounced sexual dimorphism. They are organized into multimale bands. The males have a pronounced social hierarchy; superior males mate more often, have first choice of nesting sites, and play a more important role in defending the band than their social inferiors. The hamadryas baboon is unusual in that the band is subdivided into one-male family units. These units group into bands, and the bands come together in large troops, which utilize the sleeping cliffs. Thus social organization is hypothesized to be related to the habitat. The one-male family is an optimal foraging unit, the band is the traveling unit, and the troop provides for the most efficient use of the sleeping resources.

QUESTIONS

1. Define the following terms: eusocial insects, pheromone, drone, imprinting, territoriality, social dominance.

2. Give a specific example of each of the terms in Question 1.

3. In what ways are human societies different from insect societies? How are they similar?

4. What behavioral conventions limit intraspecific aggression in man? Do any foster it?

SUGGESTIONS FOR FURTHER READING

"The Biosphere," *Scientific American*, September 1970.*

    A reprint of the September 1970 issue of Scientific American. *Its 11 chapters by different authors are devoted entirely to energy flow and biogeochemical cycles and their relationship to current human problems of food consumption and pollution.*

CARSON, RACHEL: *The Sea around Us*, New American Library, Inc., New York, 1954.*

    Miss Carson was a rare combination of scientist and poet. This deservedly popular book traces the history of the formation of the oceans, describes their role in the origin of life and in its evolution, and discusses their present-day importance to human life.

COLINVAUX, PAUL: *Introduction to Ecology*, John Wiley & Sons, Inc., New York, 1973.

    A good, general text by a teacher who clearly enjoys his subject, designed primarily for second-year students and science majors. It is notable for its stress on and criticism of the concepts of ecology.

EMMEL, THOMAS C.: *An Introduction to Ecology and Populations*, W. W. Norton & Company, Inc., New York, 1973.*

    A short, sound text for the beginner, logically presented and sensibly organized.

FARB, PETER: *Face of North America*, Harper & Row, Publishers, Inc., New York, 1963.

    An interesting and comprehensive account for a general audience of the natural history of the continent—the interplay of geological change, climate, and plant and animal life.

* Available in paperback.

HOGUE, CHARLES L.: *The Armies of the Ant*, World Publishing Company, New York, 1972.

An excellent description of the life of a man studying insects in the field.

KREBS, CHARLES J.: *Ecology: The Experimental Analysis of Distribution and Abundance*, Harper & Row, Publishers, Inc., New York, 1972.

This excellent, modern text focuses on populations and on interactions among them. Requires some knowledge both of biology and of mathematics.

KRUTCH, J. W.: *The Desert Year*, The Viking Press, Inc., New York, 1960.*

A description by one of the best contemporary American nature writers of the animal and plant life of the American desert.

KUMMER, HANS: *Primate Societies: Group Techniques of Ecological Adaptation*, Aldine, Atherton, Inc., Chicago, 1971.

Most of this book is concerned with behavior in baboons. This book is greatly enriched both by the author's personal experience in observing baboons and other primates and by his interpretation of primate behavior patterns in terms of evolutionary and ecological principles. Some wonderful and unusual photographs, also by the author.

MORAN, JOSEPH M., MICHAEL D. MORGAN, and JAMES H. WIERSMA: *An Introduction to Environmental Sciences*, Little, Brown and Company, Boston, 1973.

A clear, matter-of-fact presentation of the basic features of our planet and, in particular, of the biosphere. It provides a good background for the study of ecology in general and problems of environmental pollution and technological change in particular.

SMITH, ROBERT L.: *Ecology and Field Biology*, 2d ed., Harper & Row, Publishers, Inc., New York, 1974.

The outstanding sections of this text are those that deal with descriptions of biomes and communities and the plants and animals within them. Although meant to accompany a course in field biology, this book would be an asset and a pleasure to the amateur naturalist.

STORER, JOHN H.: *The Web of Life*, New American Library, Inc., New York, 1966.*

One of the first books ever written on ecology for the layman. In its simple presentation of the interdependence of living things, it remains a classic.

TINBERGEN, NIKO: *Curious Naturalists*, Natural History Library, Doubleday & Company, Inc., Garden City, N.Y., 1968.*

Some charming descriptions of the activities and discoveries of scientists studying the behavior of animals in their natural environment.

WALTER, HEINRICH: *Vegetation of the Earth*, Springer-Verlag, New York, 1973.

Probably the best description of the major biomes and the forces that determine them.

WICKLER, WOLFGANG: *Mimicry in Plants and Animals*, World University Library, London, 1968.*

Many examples and illustrations of a delightful subject.

WILSON, E. O.: *The Insect Societies*, Harvard University Press, Cambridge, Mass., 1971.

A comprehensive and fascinating account of the social insects. An unusual nominee for the National Book Award.

WILSON, E. O.: *Sociobiology: The New Synthesis*, Harvard University Press, Cambridge, Mass., 1975.

In this extremely interesting, beautifully written, and beautifully illustrated book, Wilson undertakes to set forth the biological principles that govern social behavior in all kinds of animals. The last chapter, which concerns the sociobiology of man, has become a focus of the current controversy about the inheritance of behavioral traits in Homo sapiens.

* Available in paperback.

Evolution and Ecology of Man

CHAPTER 39

Antecedents of the Human Species

According to the fossil record, the first mammals—the class to which man belongs—arose from a primitive reptilian stock about 200 million years ago, at about the time of the first dinosaurs. From the fossil evidence, we believe that the first mammals were about the size of a cat. They had sharp teeth, indicating that they were basically carnivorous, but since they were too small to attack most other vertebrates, they are assumed to have lived on insects and worms, supplementing their diet with tender buds, fruit, and perhaps eggs. These first mammals were probably nocturnal, judging by the large size of their eye sockets, and they were almost certainly warm-blooded. If such an animal were alive today, it would be classified as an insectivore, something like a ground shrew.

For about 130 million years, these small mammals led furtive existences in a land dominated by reptiles. Then suddenly, as geologic time is measured, the giant reptiles, the dinosaurs, disappeared. Their disappearance occurred at a time when, geologists believe, there was a drop in the average temperature and, perhaps more important, a marked increase in seasonal temperature fluctuations. Perhaps it is significant that the warm, essentially unchanging climate under which the dinosaurs evolved and thrived lasted 130 million years. Perhaps these ideal conditions—from a reptilian point of view—led to such a high degree of stabilizing selection that the dinosaur populations no longer contained enough genetic variability to undergo a major adaptive change. In any case, by the end of the Cretaceous period all of the dinosaurs had disappeared forever, and about 65 million years ago an explosive radiation of the mammals began.

The early mammals immediately diverged into the two dozen or so different lines that included (1) the monotremes, or egg-laying mammals, of which the duckbilled platypus is one of the few remaining examples; (2) the marsupials, such as the kangaroos, opossums, koala bears, and others whose young are born in embryonic form and continue their development in pouches; and (3) the placentals, by far the largest group. Among the placentals are carnivores, ranging in size from the saber-toothed tiger down to small, weasel-like creatures; herbivores, which include not only the many wild grazing animals but also most of our domesticated farm animals; the omnipresent rodents; and such odd groups as the whales and dolphins, the bats, the modern insectivores, and the primates. Man is a placental mammal and a member of the primate order, as are tarsiers, lemurs, monkeys, and apes, among others.

39–1
Cave painting from Lascaux, France. This wild ox is the auroch, Bos primigenius, *which became extinct in Europe during the seventeenth century. They were much larger than modern cattle, with the bulls often as much as 2 meters high at the shoulders. The hunting of large game such as this bull was probably an important factor in the evolution of modern man.*

TRENDS IN PRIMATE EVOLUTION

Primate evolution began when a group of the small, shrewlike mammals took to the trees; fossils of early primates indicate that they closely resembled modern-day tree shrews (Figure 39-2). Most trends in primate evolution seem to be related to various adaptations to arboreal life.

The Primate Hand and Arm

With a few exceptions, primates have five digits and a divergent thumb. The divergent thumb, which can be brought into opposition to the forefinger, greatly increases gripping powers and dexterity. There is an evolutionary trend among the primates toward finer manipulative ability that reaches its culmination in man (Figure 39-3).

Compared to other mammals, however, primates are relatively unspecialized. Their extremities resemble those of the primitive mammals—indeed, of the reptiles—more closely than do the extremities of mammals of most of the other major orders (see page 247). The first four-legged mammals all had five separate digits on each hand and foot, and each digit except the thumb and the first toe had three separate segments that made it flexible and capable of independent movement. In the course of evolution, most mammals developed hooves and paws more suited for running, seizing prey, and digging; other mammals developed flippers for swimming. The primates retained and elaborated on the primitive five-digited pattern.

In the basic quadrupedal structure of the early mammals and reptiles, the forelimb is supported by two bones (the radius and the ulna), a pattern that provides for flexibility. Among mammals, it is the primates, in particular, that have retained the ability to twist the radius, the bone on the thumb side, over the ulna so that the hand can be rotated through a full semicircle without moving the elbow or the upper arm.

Many other mammals have similarly lost the ability to move the upper arm freely in the shoulder socket. A dog or horse, for instance, usually moves its legs in only one plane, forward and backward; some South American monkeys, apes, and man are among the few higher mammals that can rotate the arm widely in the socket.

39-2

A modern tree shrew, which the earliest primates probably resembled. If you look closely you will see five-digited paws. Although clawed, they can be spread out and used for grasping. In some classification systems, tree shrews are grouped with the primates, and in others, with the insectivores, which indicates the closeness of the two evolutionary lines.

39-3

Some primate hands. The hand of the tarsier has enlarged skin pads for grasping branches. In the orangutan, the fingers are lengthened and the thumb reduced, which

provide for efficient brachiating. The gorilla's hand, which is used in walking as well as handling, has shortened fingers. Man's thumb is larger proportionately than

that of any other primate, and opposition of thumb and fingers, on which the handling ability depends, is greatest in man.

TARSIER

ORANGUTAN

GORILLA

MAN

39-4
Eye-hand coordination is another characteristic of higher primates, such as the female olive baboon shown here with her child.

Primates also have nails rather than claws. Nails leave the tactile surface of the digit free and so greatly increase the sensitivity of the digits for exploration and manipulation.

Visual Acuity

Another result of the move to the trees is the high premium placed on visual acuity, with a decreasing emphasis on the role of olfaction, the most important of the senses among many of the other mammalian orders. (Flying produced similar evolutionary pressures among the birds, which were also evolving rapidly during this same period.) This shift from dependence on smell to dependence on sight has anatomical consequences. Tree shrews, like many other animals, have eyes that are directed laterally, but among the other primates there can be traced a steady evolutionary trend toward frontally directed eyes and stereoscopic vision.

Almost all primate retinas have cones as well as rods; cones, as we discussed on page 420, are concerned with color vision and with fine visual discrimination. All primate retinas have foveas, areas of closely packed cones that produce sharp visual images.

Care of the Young

Another principal trend in primate evolution is toward increased care of the young. Because mammals, by definition, nurse their young, they tend to have longer, stronger mother-child relationships than other vertebrates (with the exception, in some cases, of birds). In the larger primates, the young mature slowly and have long periods of dependency and learning.

Uprightness

Another adaptation to arboreal life is an upright posture. Even quadrupedal primates, such as monkeys, sit upright. One consequence of this posture is a

39-5
Life in the treetops made maternal care a major factor in infant survival. Also the necessity for carrying the young for long periods resulted in strong selection pressures for reduced numbers of offspring. (a) Anthropoids, such as this vervet monkey, usually have single births. (b) A mother chimpanzee with her infant. Field studies suggest that bonds between mother and offspring and perhaps also among siblings last well into adulthood, perhaps for a lifetime.

(a)

(b)

(a)

(b)

(c)

39-6

*Some prosimians. (a) A lesser bush baby.
Bush babies move over level ground by
hopping like kangaroos, and sometimes
leap three or four meters in the air from
one tree to another. The fingers work
together, not independently like human
fingers. (b) A ring-tailed lemur, combing his*

*tail. The second digit of each foot is a
special grooming claw. (c) A native of
Indonesia, the little tarsier (about the size
of a kitten) has existed relatively unchanged
for some 50 million years. Its upper lip, like
ours, is free from the gum below it, giving
it the ability (as you can see) to make faces.*

*It has stereoscopic vision and a larger brain
than the lemur's. Living entirely in trees, it
has hands and feet with enlarged skin pads
for grasping branches. As you may have
guessed from the owl-like eyes, tarsiers
are nocturnal.*

change in the orientation of the head, allowing the animal to look straight ahead while in a vertical position; it is this characteristic, above all others, that makes primates look so "human" to us. Vertical posture was an important preadaptation for the upright stance of modern man.

MAJOR LINES OF PRIMATE EVOLUTION

Primates are divided into two major groups: the *prosimians* (lorises, bush babies, tarsiers, and lemurs) and the higher primates, the *anthropoids* (monkeys, apes, and man). During the Paleocene and the Eocene (about 65 to 38 million years ago), a great abundance and variety of prosimians inhabited the tropical and subtropical forests that spread much farther north and south of the equator than they do today. Modern prosimians (Figure 39-6) are small (typically about the size of a cat), arboreal, furry, eat fruit and insects, and are almost all nocturnal. The prosimians are generally considered an unnatural group (as are the protists, for example) in the sense that they are not closely related.

39-7

*Only New World monkeys, such as the spi-
der monkey, can hang by their tails. Note
the infant on her back. The platyrrhine monkeys are the only primates native to the
Americas.*

The anthropoids arose from one prosimian line, apparently during the Eocene (53 to 38 million years ago). There are two major groups of anthropoids: the platyrrhines (flat-nosed), which includes the New World Monkeys, and the catarrhines (downward-nosed), which includes the Old World monkeys, apes, and man (Figure 39-8). These two groups diverged early in their history, with the platyrrhines evolving in South America and the catarrhines in Africa, both apparently during the Oligocene, some 38 to 26 million years ago.

CULTURAL EVOLUTION AMONG MACAQUES

The little Japanese island of Koshima, a high, wooded mountain surrounded by a beach, is inhabited by macaques. The monkeys lived and fed in the forest until a decade or so ago, when a group of Japanese researchers began throwing sweet potatoes on the beach for them. The group quickly got used to venturing onto the beach, brushing sand off the potatoes, and eating them. One year after the feeding started, a two-year-old female the scientists had named Imo was observed carrying a sweet potato to the water, dipping it in with one hand, and brushing off the sand with the other. Soon, other macaques began to wash their potatoes, too. Only macaques that had close associations with a potato washer took up the practice themselves. Thus it spread among close companions, siblings, and their mothers, but adult males, which were rarely part of these intimate groups, did not acquire the habit. However, when the young females that learned potato washing matured and had offspring of their own, all of them learned potato washing from their mothers. Today all of the macaques of Koshima dip their potatoes in the salt water to rinse them off, and many of them, having acquired a taste for salt, dip them between bites.

This was only the beginning. Later, the scientists began scattering wheat kernels on the beach. Like the others, Imo, now four years old, had been picking the grains one by one out of the sand. One day she began carrying handfuls of sand and wheat to the shore and throwing them into the water. The sand sank, the wheat kernels floated to the top, and Imo collected the wheat and ate it. The researchers were particularly intrigued by this new behavior since it involved throwing away food once collected, much less a part of the macaques' normal behavioral repertory than holding on to food and cleaning it off. Washing wheat spread through the group in much the same way that washing potatoes had. Now the macaques, which had never even been seen on the beaches before the feeding program began, have taken up swimming. The youngsters splash in the water on hot days. Some of them dive and bring up seaweed, and at least one has left Koshima and swum to a neighboring island, perhaps as a cultural missionary.

39–8
Early in their evolution, the anthropoids split into two main lines, the platyrrhine, or flat-nosed (left), and the catarrhine, or downward-nosed (right). New World monkeys are platyrrhines: Old World monkeys and the hominoids are catarrhines. There are other characteristic anatomical differences between the two groups.

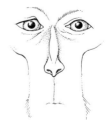

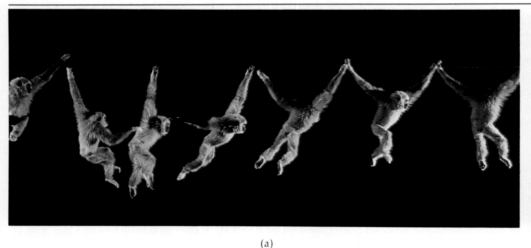

(a)

(b)

ANTHROPOID APES

The anthropoid apes include gibbons, orangutans, gorillas, and man's closest living relatives, chimpanzees. Gibbons are the smallest of the apes. They live in pairs or in small family groups. Gibbons and some human beings are the only monogamous anthropoids. They can stand and walk upright, but usually they move through the trees, swinging arm over arm (brachiating) as shown above (a). The rarest and least studied of the large primates, the modern orangutan (b), is found only in the tropical rain forests of Borneo and Sumatra. Males, which can weigh as much as 220 pounds (100 kilograms), are usually found alone.

A male gorilla is about as tall as a tall man but weighs more than three times as much (some 180 to 275 kilograms). (c) Gorillas have a beetling brow, a strong heavy jaw, and, on top of the skull of adult males, a bony crest to which the strong jaw muscles are attached. Its legs are shorter in proportion to its body than are man's, but its arms are much longer—with a span of about 3 meters—and extremely powerful.

Gorillas usually walk on all fours but they can stand erect, as they do when challenging an enemy, and they also walk erect for short distances. They have almost completely abandoned trees, with the exception of some of the smaller animals that sleep in nests in the lower branches. The larger gorillas sleep on the ground, and all feed on ground plants. In (d), a mature male and family are feeding on a banana plant.

Gorillas live in groups ranging from 8 to 24 individuals, with about twice as many females as males, and a number of juveniles and infants. Each gorilla troop has a large, mature (silver-backed) male as a leader. Although reportedly placid by nature, when threatened, males will give fearsome exhibitions of power, thrashing with broken branches, standing erect, beating their chests, barking and roaring, even hitting themselves under the chin to make their teeth rattle. Given the massive size of an adult male, such as the silver-backed male lunging forward to charge (e), the display is impressive.

(c)

(d)

(e)

Chimpanzees are somewhat smaller than man; an adult male weighs about 45 kilograms. They feed and sleep in the trees, but spend a large part of their waking hours on the ground. They walk on all fours, on the flats of their feet and the knuckles of their hands. Although their fingers are shorter than man's, they have a fine precision grip. They use simple tools, such as a stick to pry out termites, a leaf as a blotter, or a stick as a weapon. Whereas the gorilla is shy by nature, chimps are gregarious, curious, boisterous, and extroverted. (f) A male pant-hoots; (g) a young male relaxes in a tree; (h) a chimpanzee group sits and grooms; (i) a juvenile and infant wrestle in play; and (j) a mother shares food with her child.

(f)

(g)

(h)

(i)

(j)

The Hominoids

The _hominoids_, the group comprising the apes and man, are larger than the monkeys and have proportionately larger brain cases. They are tailless. They are adapted for brachiation, moving through the trees arm over arm instead of on all fours (or fives) like the monkeys. Among modern forms, however, only the gibbons move primarily by brachiating.

The hominoid line diverged, forming several major branches. One led to the modern gibbon, which is by far the smallest of the apes. Then the line branched again. During the Miocene, there emerged a group of apes _(Dryopithecus)_, comprising several species, which are considered to be ancestral to the modern great apes, the orangutan, the chimpanzee, and the gorilla. A contemporary of _Dryopithecus, Ramapithecus_, is believed by many anthropologists to be ancestral to man.

Ramapithecus

Ramapithecus is known only from fossil fragments of upper and lower jaws from India and Africa (Figure 39-9). The fossils are dated about 12 to 14 million years ago. Protein clocks (page 471), however, suggest a much more recent evolutionary divergence between man and the apes. What do these jaw fragments indicate? First, when the bits of jawbone are pieced together, it is clear that _Ramapithecus_ had a smaller and broader dental arch, as compared to other large contemporary primates or, indeed, to modern apes. The jaw fragments also indicate that _Ramapithecus_ was comparatively small, about the size of a chimpanzee. Adjacent molars on the jaw fragments of _Ramapithecus_ show marked differences in wear, in comparison to _Dryopithecus_ specimens, in which no sharp variations in wear are evident. This suggests that the molar eruption sequence of _Ramapithecus_ was much slower than that of _Dryopithecus_; as a consequence, there is reason to believe that _Ramapithecus_ matured more slowly than its hominoid contemporaries—as does modern man.

Finally, the teeth and their condition indicate that tasks such as biting off and tearing up vegetation, for which apes use their front teeth, were not carried out to the same extent by the teeth of _Ramapithecus_. Physical anthropologists hypothesize that he (or it, as you prefer) used his forelimbs for these purposes, and on the basis of this hypothesis, they further hypothesize a trend toward bipedalism. Additional support for bipedalism is found in the fact that the back teeth are specialized for more powerful chewing, which may indicate that _Ramapithecus_ had abandoned the fruits and other moist vegetation of the forest for

39-9

A comparison of the upper jaws of man, Ramapithecus, and a chimpanzee. The jaw of Ramapithecus _is more rounded and thus more manlike than that of the ape._

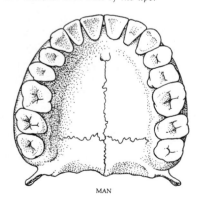

MAN

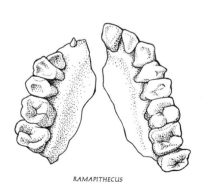

RAMAPITHECUS

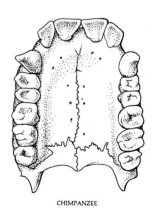

CHIMPANZEE

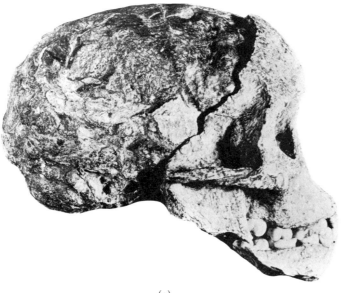

(a)

(b)

39-10

The skull of a child, found in a limestone quarry in Taung, South Africa, in 1924, was the first specimen discovered of the earliest known hominid, an australopithecine. The oldest australopithecine fossils, according to radioisotope dating methods, are specimens found near Lake Omo in eastern

Africa, representing remains of creatures that lived about 3½ million years ago. Shown here are (a) the skull and (b) a reconstruction drawing of the head of the child, which was about five or six years old when it died. The australopithecines were erect-

walking creatures with brains somewhat larger proportionately than the gorilla's and with teeth that are more similar to man's than to the ape's. Australopithecus lived on the ground, and some of them probably ate meat and used tools of their own making.

the harder, drier food (such as tough rhizomes and dried seeds) of the bush and savanna.

And then the curtain falls for another 10 to 11 million years, leaving many questions unanswered.

Australopithecines

The earliest hominoid universally accepted as a _hominid_—a member of the family of man, Hominidae—is the form generally known as *Australopithecus* ("southern ape"). The first *Australopithecus* specimen (Figure 39-10) was described in 1925 by anatomist Raymond Dart. Dart's evidence for its hominid status included the rounded appearance of the skull, the size of the fossilized brain within it, and the roundness of the jaw. Also, the point of attachment of the vertebral column to the skull indicated that the young animal was a biped.

Subsequent fossil finds built up a picture of a small, lightly built hominid, not more than 1½ meters tall, with cranial capacities ranging from 450 to 700 cubic centimeters, as compared to 340 to 750 for male gorillas (a much larger animal) and 1,000 to 2,000 for modern man. Their hands had a broad, flat thumb and the beginnings of a human-style grip. The teeth were very much like our own, although the molars were larger. Australopithecines walked upright, probably covering the ground with quick, short steps in a sort of jog trot, rather than by our far more efficient heel-and-toe striding movement.

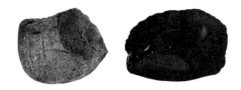

39-11

Pebble tools such as these have been found at Olduvai and other sites in eastern Africa in fossil strata of 2 million or more years ago. Australopithecines (presumably) took pebbles of lava and quartz and either struck off flakes in two directions at one end, making a somewhat pointed implement, or in a row on one side, making a chopper. They measure up to 10 centimeters in length and were probably used to prepare plant food and to butcher game. The flakes were also used apparently as scrapers.

They used simple tools (Figure 39-11) and ate meat; collections of bones of small animals—frogs, lizards, rats, and mice—and of the young of larger animals, such as antelopes, have been found at various australopithecine sites.

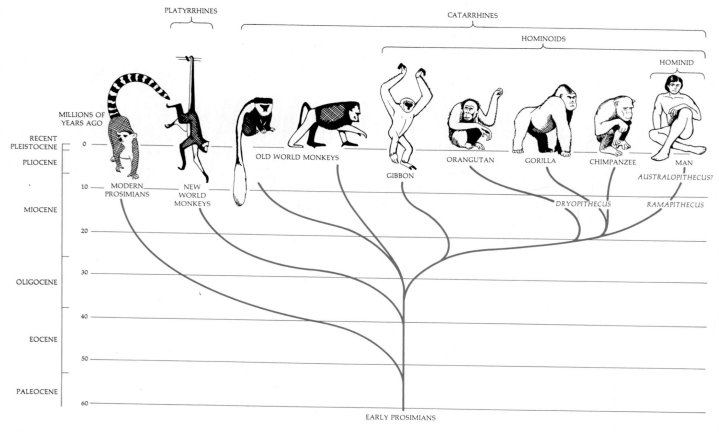

PLATYRRHINES

CATARRHINES

HOMINOIDS

HOMINID

MILLIONS OF
YEARS AGO

RECENT
PLEISTOCENE
PLIOCENE

0

OLD WORLD MONKEYS

MAN
AUSTRALOPITHECUS?

MODERN
PROSIMIANS

NEW
WORLD
MONKEYS

10

GIBBON

ORANGUTAN

GORILLA

CHIMPANZEE

MIOCENE

DRYOPITHECUS

RAMAPITHECUS

20

30

OLIGOCENE

40

EOCENE

50

PALEOCENE

60

EARLY PROSIMIANS

39–12

*A tentative phylogenetic tree of the primates,
based on fossil evidence. The monkeys, apes,
and man are anthropoids.*

The evidence indicates that they lived in small groups, like most anthropoids.

In the past decade, discoveries of australopithecine fossils have increased in number and in antiquity. According to radioisotope dating, the oldest remains were deposited about 3½ million years ago. The picture has also become more complicated in the last few years. Some authorities now contend that two (or more) species of hominids lived in this area at the same time. One, known as *Australopithecus africanus*, is the slender form; the other, *Australopithecus robustus*, is larger than the first, with more massive molars and premolars and comparatively small incisors and canines. Some anthropologists find these two groups of hominids so different that they place them in different genera: *Australopithecus* and *Paranthropus*. Others contend that these forms are basically similar, merely representing the varied members of a single species. Perhaps the differences are simply sexual, female and male: There is evidence of considerable sexual dimorphism, such as that seen among baboons.

The situation is further complicated by evidence (still controversial) put forth by the Leakeys (Figure 39–13) that still another manlike species, which they call *Homo habilis*, existed during this same period. The specimens they identify as *Homo habilis* have larger brains and, apparently, more manlike extremities. The Leakeys contend that *Homo habilis* was the true antecedent of man and that the other australopithecines coexisted and became extinct.

This mystery of man's origins, a subject of lively transatlantic controversy, can be resolved only by the discovery and analysis of more fossil evidence.

39–13

Louis and Mary Leakey, the British anthropologists, with one of their sons at Olduvai Gorge in the Serengeti plain of Tanzania. Olduvai, from which many important australopithecine fossils have been recovered, represents a geologic sequence that is almost continuous from the present to nearly 2 million years ago. Like our Grand Canyon, it was cut by river action. Because it was by the side of a lake, it was visited by many animals and was apparently a popular hominid camping ground. Dr. Leakey died in 1975, but Mary Leakey and their son Richard are still leading figures in the search for early man.

39–14

The fossil jaw of a child, discovered by Mary Leakey in 1975 and dated at about 3½ million years ago. A canine tooth is pushing up through the milk teeth, indicating that the young hominid had a long period of maturation (and hence of learning). The small size of the teeth is interpreted to mean use of tools and perhaps weapons and a diet of meat.

How Did It Happen?

What prompted one group of primitive apes to make the first fateful step to bipedalism? We know from pollen specimens that grasslands were spreading and forests contracting during the Miocene, the period during which *Ramapithecus* made its first appearance. The grasslands may have represented a new adaptive zone. The horse and other grazing animals evolved rapidly during this time, filling the new niches available to herbivores. Baboons also seem to have become terrestrial in this period. We do not know why one group of apes and not another moved into the grasslands. One possible explanation is that this group was driven out of the woodland by competition with other primates. The fact that *Ramapithecus* was smaller than other contemporary apes is in keeping with this speculation. Perhaps they possessed some important preadaptation that we do not know about.

Standing upright on the ground would have greatly increased the range of vision of a ground-dwelling species with highly developed eyesight; other mammals, including bears, squirrels, and prairie dogs, often raise themselves on their two hind legs to look around. Standing upright also, of course, frees the hands, which then can be employed for tool making and using, for carrying food, and for brandishing weapons. Use of weapons for defense correlates well with the reduction in the size of the canines among manlike fossil primates, as compared with baboons, for example, or gorillas, among which the males have large canines that are used in aggressive display. The advantages of bipedalism must have been very great for natural selection to have operated in its favor, because the first bipedal primates must have moved slowly and inefficiently

39-15

Bipedalism frees the upper extremities for a variety of activities. An adult female chimpanzee "fishes" for termites with a twig.

compared with the swift movement of other large animals, either on the ground or in the trees. As we shall discuss in the next chapter, among the consequences of bipedalism is man's enormous brain, the most characteristic feature of the genus *Homo*.

SUMMARY

The first mammals arose from primitive reptilian stock about 200 million years ago and coexisted with the dinosaurs for 130 million years. The extinction of the dinosaurs was followed by a rapid adaptive radiation of the mammals. The primates are an order of mammals that became adapted to arboreal life. Primates are characterized by five-digited extremities adapted for grasping with nails rather than claws and by freely movable limbs. They are dependent more upon vision than upon smell, and the higher primates all have stereoscopic vision with foveas for fine focus and cones for color vision.

The two principal groups of living primates are the prosimians and the anthropoids. Fossil prosimians were widespread and abundant during the Paleocene and Eocene epochs, some 65 to 38 million years ago. Modern prosimians include lorises, bush babies, lemurs, and tarsiers.

The anthropoids include the New World monkeys, the Old World monkeys, and the hominoids (apes and man). Early in hominoid evolution, in the Miocene epoch, the hominoids diverged. *Ramapithecus,* known only from some jaw fragments dated 12 to 14 million years ago, is believed to be ancestral to man. *Ramapithecus* was smaller than its contemporaries, with a broader dental arch and teeth adapted for powerful chewing. Evidence suggests that *Ramapithecus* was bipedal.

The earliest hominids belong to the genus *Australopithecus. Australopithecus* is known from several hundred bone fragments, ranging in age from $3\frac{1}{2}$ million years or older to 1 million years. The australopithecines were small, bipedal, probably used simple tools, and ate meat. It is not known at this time whether or not they were direct antecedents of man.

QUESTIONS

1. Define the following terms: primate, prosimian, hominid, hominoid, anthropoid.

2. Give some distinguishing features of each of these genera: *Ramapithecus, Tarsius, Australopithecus, Homo.*

3. Name five evolutionary trends among primates and name the probable selective value for each.

4. Can you offer a possible relation between these two facts: (1) the only modern lemurs are a widely diversified group found on the island of Madagascar; (2) the only other primate on Madagascar is man, and he is a late arrival.

40-1

A rock painting from Spain. Such paintings date from postglacial times, about 8000 to 3000 B.C. In contrast to the earlier cave paintings, paintings such as this were executed on sheltered rock surfaces, frequently show human figures, and are small in scale.

These paintings typically show forest animals rather than the great plains animals that stalked the caves of Lascaux, France, and Altamira, Spain. The deer are apparently being driven toward the archers by

beaters. The hunting of small, nonmigratory animals such as these is believed to mark the transition between the nomadic, hunting-gathering existence of early man and the agricultural revolution.

THE ICE AGES

During most of our planet's history, its climate appears to have been warmer than it is at the present time. However, these long periods of milder temperatures have been interrupted periodically by Ice Ages, so called because they are characterized by glaciations, persistent accumulations of ice and snow. Such glaciations occur whenever the summers are not hot enough and long enough to melt ice that has accumulated during the winter. In many parts of the world, an alteration of only a few degrees in temperature is enough to begin or end a glaciation.

An early Ice Age appears to have occurred at the beginning of the Paleozoic era, some 600 million years ago. Another, marked by extensive glaciations in the Southern Hemisphere, closed the Paleozoic, some 250 million years ago. The conifers evolved during this period and possibly the angiosperms, as older forest types disappeared. A more recent, less severe cold, dry period occurred at the end of the Mesozoic, about 65 million years ago, and was perhaps a principal cause of the extinction of the dinosaurs.

The most recent Ice Age began during the Pleistocene epoch, about 1½ million years ago. The Pleistocene has been marked by four extensive glaciations that have covered large areas of North America, England, and northern Europe. Between the glaciations there have been intervals, called interglacials, during which the climate has become warmer. In each of these four Pleistocene glaciations, sheets of ice, thicker than 3 kilometers in some regions, spread out locally from the poles, scraped their way over much of the continents—reaching as far south as southern Illinois in North America and covering Scandinavia, most of Great Britain, northern Germany, and northern Russia—and then receded again. We are living at the end of the fourth glaciation, which began its retreat only some 10,000 years ago.

During these periods of violent climatic changes, the fossil record shows that the populations of these regions were under extraordinary evolutionary pressures. Plant and animal populations moved, changed, or became extinct. In the interglacial periods, during which the average temperatures were at times warmer than those of today, the tropical forests and their inhabitants spread up through today's temperate zones. During the periods of glaciations, only animals of the northern tundra could survive in these same locations. Rhinoceroses, great herds of horses, large bears, and lions roamed Europe in the interglacial periods, and in North America, as the fossil record shows, there were camels and horses, saber-toothed cats, and great ground sloths, one species as large as an elephant. In the colder periods, reindeer ranged as far south as southern France, while during the warmer periods, the hippopotamus reached England. It was during this most recent period of glaciations and interglacials that hominids of the genus Homo evolved.

The reason for these large changes in temperature is one of the most controversial issues in modern science. They have been variously ascribed to changes in the earth's orbit, variations in the earth's angle of inclination toward the sun, migration of the magnetic poles, fluctuations in the solar energy, migration of the continents, higher elevations of the continental masses, continental drift, and combinations of these and other causes. Also under debate is the question of whether this period of glaciations is over—perhaps for another 200 million years—or whether we are merely enjoying a brief interglacial before the ice sheets begin to creep toward the equator once more.

Glaciers are accumulations of snow and ice that flow across a land surface as a result of their own weight. This glacier is flowing into the sea. As the glacier moves forward, its edges melt; the rocks carried within it have become concentrated in the dark band at the right. This is part of the Greenland ice sheet, which has an area of about 1,726,000 square kilometers and at some places is more than 3 kilometers thick. Glaciers covered much of North America and Europe during the period of the evolution of the genus Homo.

Home and Hearth

All higher primates except modern man are continuously on the move, and though they range back and forth over the same territory, they sleep in a different place every night. Any member of the group that cannot keep up is left behind. Early men were also nomads, as are most modern hunter-gatherers, but there is evidence that they maintained base camps. Some of these camps appear to have been fairly permanent, in the sense that they were returned to year after year.

The australopithecines and the earliest men probably lived mostly in the open, although one australopithecine site is in the natural shelter of overhanging rocks. The campsites found in southeastern Africa are near streams or lakes, which would not only have provided water for the camper but would also have attracted game.

The Choukoutien caves, near Peking, China, inhabited 350,000 years ago, are among the earliest caves known to have been used for human habitation. It is probably no coincidence that within these same caves are also the first traces of the human use of fire. Before men had fires, caves were inhabited by bears and other large carnivores and were too dangerous for humans to occupy. Fire, however, made them available for human habitation, and they have been used by man up to modern times. The caves near Shanidar, for example, in the remote mountains of Iraq have been occupied continuously—to the present day—for some 100,000 years (Figure 40-5).

Increase in Intelligence

In mammals, the proportion of brain size to body size can be correlated with the intelligence of the species under consideration. It is therefore reasonable to assume that the enormous increase in cranial capacity that took place in the course of human evolution signifies strong selection pressures for intelligence.

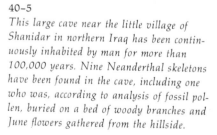

40-5

This large cave near the little village of Shanidar in northern Iraq has been continuously inhabited by man for more than 100,000 years. Nine Neanderthal skeletons have been found in the cave, including one who was, according to analysis of fossil pollen, buried on a bed of woody branches and June flowers gathered from the hillside.

40-6

Increase in brain volume in the course of hominid evolution. The first figure for Australopithecus is for the type generally known as Australopithecus africanus; the second is for the type sometimes called Australopithecus habilis or Homo habilis. Some of the increase in brain volume can be correlated with the increase in body size that was also taking place during this period, but most of it is believed to represent the results of strong selection pressures for intelligence.

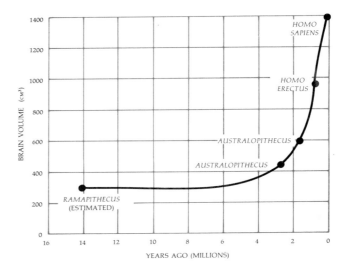

In fact, insofar as one can compare the evolution of the human brain with the evolution, for instance, of the foreleg of the horse (which is considered very rapid), the human brain evolved 100 times more quickly. The increase in cranial capacity (Figure 40-6) took place well after bipedalism and the early use of tools and well before the development of sophisticated tools or weapons, which did not take place until a mere 30,000 to 40,000 years ago, as we shall see.

HOMO SAPIENS

The earliest fossils that are classified as *Homo sapiens* come from Swanscombe in England and Steinheim in Germany. They consist only of some skull fragments, but these clearly indicate that the brains of these men were larger than those of *Homo erectus* and that their skulls were less massive. These skull fragments are dated at about 250,000 to 150,000 years ago, during an interglacial period when Europe was warm, probably warmer than it is today.

Neanderthal Man

Then we have another of those large and frustrating gaps in the fossil record. The next specimens that are known to us date from around 80,000 to about 40,000 years ago, the period of the last glaciation. This period abounds with specimens of what we have come to call Neanderthal man. They have been found largely in Europe but also in the Near East and Central Asia. Neanderthal man stood erect, had a brain capacity somewhat larger than modern man, a thick skull, a prognathous (protruding) muzzle, a low forehead, and heavy brow ridges. He is now classified as a variety of *Homo sapiens.*

Neanderthals also used simple, hand-held stone tools (Figure 40-7). Some of the stone tools appear to have been used for scraping hides, suggesting that Neanderthals wore clothing made of animal skins, which would certainly have been in keeping with the climate in which they lived.

40-7

Fossils of Neanderthal man are associated with a decrease in the use of core tools, such as the hand ax, and a predominance of flake tools. Multiple flakes were struck off from a central core and then retouched. They are known as points and scrapers, which also presumably describes their functions. They are usually 5 to 8 centimeters long. Some may have been used as points for spears or javelins, but, in general, it appears that they were hand-held. Spears, which were in use at that time, were of fire-hardened wood.

FLAKE STRUCK FROM CORE

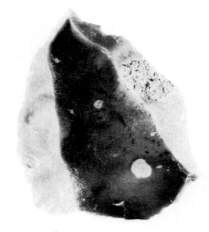

40-8

Cro-Magnon culture was characterized by a great increase in the types of specialized tools made from long and relatively thin flakes with parallel sides, called blades. These beautifully worked "laurel leaf" blades are often more than 30 centimeters long and only ½ centimeter thick.

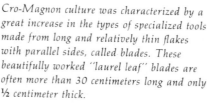

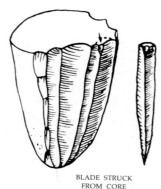

BLADE STRUCK
FROM CORE

Neanderthals buried their dead, often with food and weapons and, in at least one instance, with spring flowers. Formal burials such as these suggest a belief in life after death. Whether or not Neanderthals were able to speak is a matter of current debate, but it is difficult to understand how a society could hold such abstract concepts without some means of exchanging ideas among its members.

Cro-Magnon Man

About 30,000 to 40,000 years ago, specimens of Neanderthal man disappear abruptly from the fossil record and are replaced by what is known as Cro-Magnon man, or sometimes as Upper Pleistocene man, who is physically indistinguishable from modern man. We do not know what became of the Neanderthals. Perhaps they were exterminated in warfare, although there is no evidence of this. Perhaps they were simply unable to compete for food and living space with the better-equipped Cro-Magnon type. Perhaps they interbred, although there are no clear traces of intermediate forms. Some have suggested that modern men brought with them some disease to which they themselves were resistant and the Neanderthals were not. In any case, soon after the appearance of Cro-Magnon man, there were no other hominids in Europe, and within the course of 10,000 to 20,000 years, this new variety of primate had spread over the face of the planet.

Tools and Art

Cro-Magnon man, when he first appeared in Europe, came bearing a new, quite different, and far better tool kit (Figure 40-8). His stone tools were essentially flakes—which, of course, had been in use for more than 2½ million years—but they were struck from a carefully prepared core with the aid of a punch (a tool made to make another tool). These flakes, usually referred to as blades, were smaller, flatter, and narrower, and most important, they could be and were shaped in a large variety of ways. They included, from the beginning, various scraping and piercing tools, flat-backed knives, awls, chisels, and a number of different engraving tools. Using these tools to work other materials, especially bone and ivory, Cro-Magnon man made a variety of projectile points, barbed points for spears and harpoons, fishing hooks, and needles. Thus, although

In almost every cave, a small number of the animals show wounds, such as these seen in this horse from the Lascaux Caves in France. Although the animals are shown in realistic detail, the weapons and the hunters, if shown at all, are abstractions, perhaps to keep their identity a secret. The illusion of movement is greatly enhanced by the patterns of light and shadow in the dark, narrow recesses of the cave.

Cro-Magnon man lived much the same sort of existence as that of his forebears, he apparently lived it with more possessions, more comfort, and more style.

Perhaps our closest emotional links to this most immediate ancestor are the cave paintings of southern France and western Spain (Figure 40-9). The examples that remain to us, many surprisingly untouched by time, clearly form a part of a rich artistic tradition that has endured for at least 10,000 years. The cave drawings are almost entirely of animals, nearly all game animals, and they are deep within the caves, so they must have been viewed (as they must have been painted) by the light of crude lamps or torches.

The meaning of these drawings and paintings has long been a matter of debate. Some of the animals are marked with darts or wounds (although very few appear to be seriously injured or dying). Such markings have led to the suggestion that the figures are examples of sympathetic magic, in which there is the notion that one can do harm to one's enemy by sticking needles in his image. The fact that many of the animals appear to be pregnant suggests that they may symbolize fertility. Many appear also to be in movement. Perhaps these animals, so vital to the hunters' welfare, were migratory in these areas, and they may have seemed to vanish at certain times in the year, mysteriously returning, heavy with young, in the springtime. This return of the animals might have been an event to be solicited or celebrated in much the same spirit as the rites of spring of more recent peoples or Easter are celebrated.

Cave art came to an end perhaps 8,000 or 10,000 years ago. Not only were the tools and pigments laid aside, but the sacred places—for such they seem to have been—were no longer visited. New forces were at work to mold the course of men's existences, and the end of this first great era in human art is our clearest line of demarcation between the old life of man the hunter and the new life that was to come.

THE AGRICULTURAL REVOLUTION

The most important event in the cultural evolution of man was the change to an agricultural way of life. The reasons for such a transition are not clear, but one

THE GREEN REVOLUTION

The greatest challenge of botanical research today is a very practical one: to feed the hungry. Of the more than 4 billion people of the world today, it is estimated that almost half are undernourished or malnourished. The effort to increase agriculture by the development of new crop plants—especially grains—is called the Green Revolution.

Enormous progress is being made. From 1950 to 1970, the production of wheat in Mexico increased from 270,000 metric tons per year to 2.35 million; the corn harvest increased a more modest 250 percent, with yields per hectare almost doubling. Between 1950 and the present, India has increased its production of food grains about 2.8 percent a year. (Its population during this period has increased about 2.1 percent a year, which is significantly less.) Since 1971, China, the most populous nation in the world, has become agriculturally self-sustaining. Most of this success has come about as a result of improved varieties of crop plants, combined with better techniques of irrigation and fertilization. Moreover, and perhaps most important, the full potential of the Green Revolution has not yet been realized.

Despite its acknowledged success, this massive effort has come under criticism in recent years. One reason is the increasing cost of fertilizer; these new grains require intensive cultivation.

Because the large landowners are able to afford the investment in fertilizer and farming equipment that the small-scale farmers cannot, these new agricultural developments are seen as accelerating the consolidation of farm lands into a few large holdings by the very wealthy. Bad weather—both droughts and floods—diminished yields in 1972, 1973, and 1974. Most serious of all, although food production is still outstripping population growth, the Green Revolution, at its most productive, will not be able to keep pace indefinitely with the rapid growth of the world's population.

Finally, there is a more fundamental though more elusive reason for the dissatisfaction with the Green Revolution. When it was first introduced, it appeared to many to be an almost magical solution to problems so enormous and distressing that they had seemed insoluble. It is now clear, however, that poverty and famine and the unrest and violence they may bring will not be solved by a "technological fix." The Green Revolution must of course go forward. At the same time, we must recognize that the broader solutions are social, political, and ethical, involving not only the growth of crops but their distribution, not only the limiting of populations but the raising of living standards of these populations to tolerable levels.

(a)

(b)

(a) Norman Borlaug, who was awarded the Nobel Peace Prize in 1970. Borlaug is the leader of a research project sponsored by The Rockefeller Foundation under which new strains of wheat have been developed in Mexico. Widely planted, these new strains have changed the status of Mexico from that of a wheat importer when the program began in 1944 to that of an exporter by 1964.
(b) Field workers weeding a rice plot at the International Rice Research Institute in the Philippines. The new strains of rice are dwarf varieties with short, stiff stalks that respond well to heavy fertilization. Older, nondwarf strains become too tall when heavily fertilized and tend to fall over, spilling the grain.

factor seems to have been a change in the climate. The most recent of the glaciations began to retreat about 18,000 years ago, withdrawing slowly for about 10,000 years. As the glaciers retreated, the plains of northern Europe and of North America, once cold grasslands, or steppes, gave way to forest. Many of the great herbivores that roamed these steppes retreated northward and eventually vanished; the woolly mammoth was last seen in Siberia more than 10,000 years ago. While some animals became extinct, humans adapted, as they had during the entire period of violent climatic changes that have marked their evolutionary history. With the migratory animals gone, man shifted his attention to smaller game. The hunting and gathering of small animals—instead of large migratory herbivores—undoubtedly resulted in a less nomadic existence for the hunters and formed a prelude to the agricultural revolution. (However, it should be noted that similar fluctuations had occurred previously in the history of man without leading to a revolution in man's way of life.)

The earliest traces of agriculture, dating back about 11,000 years, are found in areas of the Near East, which are now parts of Iran, Iraq, and Turkey. Here were the raw materials required by an agricultural economy: cereals, which are grasses with seeds capable of being stored for long periods without serious deterioration, and herbivorous herd animals, which can be readily domesticated. The grasses in these areas were wild wheats and barley, which still grow wild in the foothills. The animals were wild sheep and goats.

By 8,100 years ago, agricultural communities were established in eastern Europe. By 7,000 years ago (about 5000 B.C.), agriculture had spread to the western Mediterranean and up the Danube River into central Europe, and by 4000 B.C., to Britain. During this same period, agriculture originated separately in Central and South America, and perhaps slightly later in the Far East.

THE POPULATION EXPLOSION

One immediate and direct consequence of the agricultural revolution was an increase in population. It is estimated that about 25,000 years ago there may have been as many as 3 million people. By the close of the Pleistocene epoch, some 10,000 years ago, the human population probably numbered a little more than 5 million, spread over the entire world. By 4000 B.C., about 6,000 years ago, the population had increased enormously, to about 90 million, and by the time of Christ, it is estimated, there were 130 million people. In other words, the population increased more than 25 times between 10,000 and 2,000 years ago.

By 1650, the world population had reached 500 million, many people lived in urban centers, and the development of science, technology, and industrialization had begun, bringing about further profound changes in the life of man and his relationship to nature.

By 1977, there were more than 4 billion people on our planet (Figure 40-10) and the number is increasing at an unprecedentedly rapid rate. Although the birth rate in the United States has decreased drastically since 1972, the world's population is growing at about 2.2 percent per year. This means that about 175 people are added to the world population every minute, about 250,000 each day, and 90 million every year. If this rate of increase is sustained, there will be 7 billion people on earth by the year 2000.

Life versus Death

The size of a population is, of course, a factor both of its birth rate and its death rate (see page 524), and the enormous recent increase in the world's population

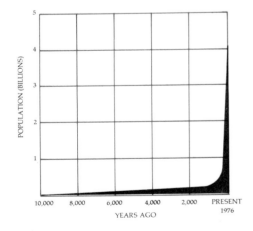

40-10

Growth of the human population. The total biomass of human beings is now estimated at 135 million metric tons, far greater than that of any other animal, with 1 million metric tons added every year.

In many tropical countries, the death rate has fallen rapidly since 1940, resulting in a rapid growth of the population. The drop in death rate is the result of increased medical services, control of malaria by DDT, and the availability of new antibacterial drugs, especially the antibiotics. Note that the birth rate has also begun to fall. The area on the graph marked in color indicates population growth. The data shown here are from Sri Lanka (formerly Ceylon).

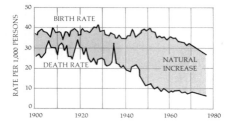

is actually more a product of the decline in the death rates, especially among the young, than an increase in births. For example, in Pakistan, the population was 20 million in 1911, 40 million in 1951, 71 million in 1976, and will be well over 80 million by 1980. During this period, birth rates seem to have remained the same—about 45 per 1,000—but the death rate is one-third of what it was only 20 years ago. Figure 40-11 charts similar data for Sri Lanka (formerly Ceylon), an island off the coast of India.

Population growth also depends on the age of the population. In Pakistan, again, almost half of the population is under 15, so if the ideal of a two-child family were to be obtained in the near future, Pakistan's population could still double in the next half century. (The effects of population structure on population growth are diagramed in Figure 40-12.)

Birth Rates, Death Rates, and Social Security

Despite the fact that the most recent phase of the population increase is related to a reduction in death rates, particularly from infectious diseases, some experts believe that further reductions in death rates and a general increase in the standard of living will reduce the rate of population growth. There seems to be a correlation between economic deprivation and high birth rates. It is in the less developed countries that the greatest rates of population growth are found. In India, for example, where large numbers of people are now starving, the annual rate of population increase is $2\frac{1}{2}$ percent. Yet, most Indian women do not seek help in birth control until they have three or four children. The desire for large families is deeply rooted in the Indian culture. Moreover, in Indian tradition, children provide security for their parents in old age. Two recent studies of life expectancy in India show that, with the high death rate among children, it is necessary for a mother to bear five children if the couple is to be 95 percent certain that one son will survive the father's sixty-fifth birthday.

Charting populations by age and sex permits predictions about future growth rates. In India, for example, nearly 45 percent of the population is under 15 years of age. Even if these young men and women limit their fam-

ily size enough to produce only enough children to replace themselves (which means cutting the current birth rate in half), population growth will not level off until

about the year 2040—and at a level of well over a billion. The population of Sweden, by contrast, will remain the same unless the birth rate is dramatically increased.

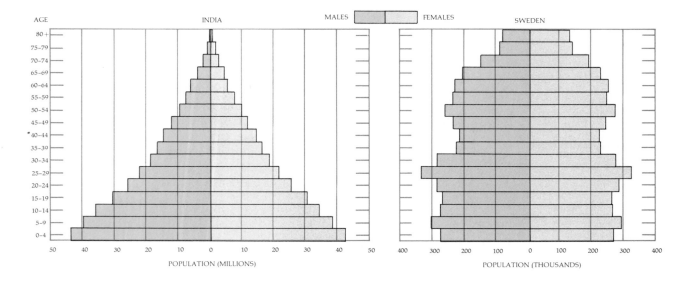

Barry Commoner of Washington University in St. Louis, who has long been in the front ranks of our environmental protectors, is of the opinion that population pressures can be significantly reduced by reducing death rates, particularly among infants. Wherever income rises and death rates drop below a critical point, birth rates begin to fall too, he points out. This was true historically in industrialized nations, and it is becoming true in some countries where living standards are just beginning to rise.

China offers an example of a country with a falling birth rate. The birth rate in rural Chinese communities apparently has now been reduced to 16 per 1,000 (about what it is in the United States). This reduction has been achieved, in part, by providing free contraceptives, abortions, and sterilization as part of the general health program. More important would appear to be social factors such as late marriages (combined with premarital chastity), the active employment of women outside the home, and a guarantee, by the state, of security for the aged, which children used to provide.

However, as Paul Ehrlich and others point out, there is no assurance that such a transition will take place in other countries (and, in fact, some view the statistics from China with skepticism), even if they are able to improve the living conditions of their inhabitants. Indeed, for many countries, a significant rise in the standard of living seems an almost unattainable goal and, in some cases, a rapidly receding one. It is clear that a reduction in birth rate is not directly correlated with the availability of birth-control measures. United States' birth rates are very low now, but they were equally low in the 1930s during the Depression. The Caribbean island of Jamaica has a Family Planning Center in almost every small town dispensing pills and contraceptive devices at very low cost, yet the current birth rate is 31.3 per 1,000 population.

OUR HUNGRY PLANET

Earlier in this book, we have mentioned some of the consequences of the increase in the human population, including the problems of pollution, depletion of fossil fuels, destruction of natural resources, extinction of other species. By far the most difficult problem and also the most urgent is hunger and starvation. Of the world's 4 billion people, at least 2 billion are inadequately nourished. It is estimated that about one-third of the deaths that now occur worldwide are due directly or indirectly to malnutrition. Perhaps even more important, in terms of the world's future, are the effects, both physical and psychological, of the prolonged chronic hunger of so large a proportion of our population.

In the age of the dinosaurs, the earliest primates survived, it would appear, largely by their wits; now, if man is to survive the monsters of his own creating, he will have to do it, again, by the contents of his own skull. For within the mind of man—that complex collection of neurons and synapses—resides the uniquely human capacity to accumulate knowledge, to plan with foresight, and so to act with enlightened self-interest and even, on occasion, with compassion.

SUMMARY

Two species of the genus *Homo* are now widely recognized: *Homo erectus*, now extinct, and *Homo sapiens*. Fossil remains of *Homo erectus*, dating from about 1 million to 300,000 years ago, have been found in Africa, Asia, and Europe.

Individuals of the species *Homo erectus* were successful hunters of very large animals, indicating that they lived in fairly large groups, shared food, and worked together cooperatively. The hand ax is associated with *Homo erectus,* and its widespread use indicates cultural exchanges between groups. Some groups lived in caves and had fire, two developments that are probably related.

Neanderthal man is considered a variety of *Homo sapiens.* Neanderthal fossils date from about 80,000 to 30,000 years ago, the period of the last glacial advance. The majority of specimens have been found in Europe. Neanderthals had fire, inhabited caves, hunted large animals, and probably wore clothing. They used stone tools of a characteristic type. They buried their dead, sometimes with food and weapons. Neanderthals disappeared 30,000 to 40,000 years ago.

Cro-Magnon man, or Upper Pleistocene man, who replaced the Neanderthal, were nomads and hunters, like their predecessors. They had tools (blade tools) distinctly different, smaller, and far more varied than had been seen previously. Their art, which covers a period of some 10,000 years, is evidence of a culture rich not only in beauty but in mystery, magic, and tradition.

The most important event in the cultural evolution of man was the advent of agriculture, about 10,000 years ago.

About 8000 B.C., when agriculture had its beginning, there were probably about 5 million people in the world. By 4000 B.C., the population had increased to about 90 million; by the time of Christ, there were probably around 130 million people in the world; and by 1650, there were about 500 million. In 1977, there were more than 4 billion, and the population continues to increase, especially in the less developed countries.

The most serious consequence of this rapid increase in world population is widespread malnutrition and starvation. A more equable allocation of the resources we now have, the development of new resources, and the halting of the growth of the human population are the most serious issues now confronting the species *Homo sapiens.*

QUESTIONS

1. Suppose the birth rate were to drop immediately to the so-called "replacement level"—two children per couple. Assuming that the great majority of children are born to women between 15 and 30 years of age, what would happen to the population of a country with a triangular population profile (Figure 40–12)? One with a barrel-shaped profile? One whose profile is an inverted pyramid?

2. In an editorial entitled "Food, Overpopulation, and Irresponsibility" that appeared recently in a scientific journal, the authors concluded: "Because it creates a vicious cycle that compounds human suffering at a high rate, the provision of food to the malnourished nations of the world that cannot, or will not, take very substantial measures to control their own reproductive rates is inhuman, immoral and irresponsible." What is your opinion?

3. Buckminster Fuller, the architect, has said, "Pollution is resources we are not harvesting." Give some examples to support this statement. Do you agree with his definition?

SUGGESTIONS FOR FURTHER READING

BAKER, HERBERT G.: *Plants and Civilization*, Wadsworth Publishing Company, Inc., Belmont, Calif., 1965.*

An introduction to the study of plants in relation to man, this book illustrates the profound influence of plants on man's economic, cultural, and political history.

CAMPBELL, BERNARD: *Human Evolution*, Heinemann Educational Books, Ltd., London, 1967.

Campbell emphasizes the way that the special adaptations characterizing the hominids can be related to their environment, ecology, and culture.

CARSON, RACHEL: *Silent Spring*, Houghton Mifflin Company, Boston, 1962.*

This is the book that awakened the nation to the dangers of pesticides. Once seen as highly controversial, it is now regarded as a classic of modern ecology.

COMMONER, BARRY: *The Closing Circle*, Alfred A. Knopf, Inc., New York, 1971.

The author traces the relationship between the present ecological crisis and technological and social factors in our society. His title, The Closing Circle, refers to his conviction that if we are to survive, "we must learn how to restore to nature the wealth we borrow from it." Commoner's book is criticized by Ehrlich and some other ecologists and demographers as presenting an oversimplified and too optimistic solution to environmental problems.

EHRLICH, PAUL, and ANNE EHRLICH: *Population/Resources/Environment*, 2d ed., W. H. Freeman and Company, San Francisco, 1972.

Required, although not cheerful, reading for all concerned with the ecological crisis. The Ehrlichs present a wealth of useful data, even though you may not agree with all their conclusions.

HOWELLS, W. W.: *Mankind in the Making*, rev. ed., Doubleday & Company, Inc., Garden City, N.Y., 1967.

A clear and lively treatment of the story of human evolution, written for the general reader. The line drawings are charming.

HOWELLS, WILLIAM: *Evolution of the Genus* Homo, Addison-Wesley Publishing Company, Inc., Reading, Mass., 1973.

A brief, up-to-date survey of human evolution.

JOLLY, ALISON: *The Evolution of Primate Behavior*, The Macmillan Company, New York, 1972.*

Few areas of modern biology are expanding as rapidly as studies of behavior of living primates. This is an excellent survey, covering the broad fields of ecology, society, and intelligence, from pottos to people.

LAPPÉ, FRANCES MOORE: *Diet for a Small Planet*, Ballantine Books, Inc., New York, 1971.*

Lappé clearly establishes the feasibility of eating "low on the food chain," thereby greatly increasing the availability of proteins to other human populations.

LE GROS CLARK, W. E.: *Antecedents of Man: An Introduction to the Evolution of the Primates*, Harper & Row, Publishers, Inc., New York, 1963.*

Le Gros Clark's emphasis is on the anatomy and physiology of the entire primate order.

PFEIFFER, JOHN E.: *The Emergence of Man*, 2d ed., Harper & Row, Publishers, Inc., New York, 1972.

Pfeiffer is one of the outstanding interpreters of complex scientific matters for the general public. This book is comprehensive, accurate, and thoroughly readable.

* Available in paperback.

PILBEAM, DAVID: *The Ascent of Man: An Introduction to Human Evolution,* The Macmillan Company, New York, 1972.

A thorough survey of primate evolution, including evolution of the hominids.

SANDERS, N. K.: *Prehistoric Art in Europe,* Penguin Books, Ltd., Harmondsworth, England, 1968.

An illustrated history of art's first 30,000 years.

UCKO, PETER J., and ANDRÉE ROSENFELD: *Paleolithic Cave Art,* McGraw-Hill Book Company, New York, 1967.*

This book not only presents the cave art, in photos and drawings, but also examines its contents and context and discusses the various interpretations of it.

VAN LAWICK–GOODALL, JANE: *In the Shadow of Man,* Houghton Mifflin Company, Boston, 1971.

An absorbing personal account of eleven years spent observing the complex social organization of a single chimpanzee community in Tanzania.

* Available in paperback.

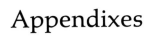

Appendixes

METRIC TABLE

	QUANTITY	NUMERICAL VALUE	ENGLISH EQUIVALENT	CONVERTING ENGLISH TO METRIC
Length	kilometer (km)	1,000 (10^3) meters	1 km = 0.62 mile	1 mile = 1.609 km
	meter (m)	100 centimeters	1 m = 3.28 feet	1 yard = 0.914 m
			= 1.09 yards	1 foot = 0.305 m
	centimeter (cm)	0.01 (10^{-2}) meter	1 cm = 0.394 inch	= 30.5 cm
	millimeter (mm)	0.001 (10^{-3}) meter	1 mm = 0.039 inch	1 inch = 2.54 cm
	micrometer (μm)	0.000001 (10^{-6}) meter		
	nanometer (nm)	0.000000001 (10^{-9}) meter		
	angstrom (Å)	0.0000000001 (10^{-10}) meter		
Area	square kilometer (km²)	100 hectares	1 km² = 0.3861 square mile	1 square inch = 6.4516 cm²
	hectare (ha)	10,000 square meters	1 ha = 2.471 acres	1 square foot = 0.0929 m²
	square meter (m²)	10,000 square centimeters	1 m² = 1.1960 square yards	1 square yard = 0.8361 m²
			= 10.764 square feet	1 acre = 0.4047 ha
	square centimeter (cm²)	100 square millimeters	1 cm² = 0.155 square inch	1 square mile = 2.590 km²
Mass	metric ton (t)	1,000 kilograms = 1,000,000 grams	1 t = 0.9842 ton	1 ton = 1.0160 t
	kilogram (kg)	1,000 grams	1 kg = 2.205 pounds	1 pound = 0.4536 kg
	gram (g)	1,000 milligrams	1 g = 0.0353 ounce	1 ounce = 28.35 g
	milligram (mg)	0.001 gram		
	microgram (μg)	0.000001 gram		
Time	second (sec)	1,000 milliseconds		
	millisecond	0.001 second		
	microsecond	0.000001 second		
Volume (solids)	1 cubic meter (m³)	1,000,000 cubic centimeters	1 m³ = 35.315 cubic feet	1 cubic yard = 0.7646 m³
			= 1.3080 cubic yards	1 cubic foot = 0.0283 m³
	1 cubic centimeter (cm³)	1,000 cubic millimeters	1 cm³ = 0.0610 cubic inch	1 cubic inch = 16.387 cm³
Volume (liquids)	kiloliter (kl)	1,000 liters	1 kl = 264.17 gallons	1 gal = 3.785 l
	liter (l)	1,000 milliliters	1 l = 1.06 quarts	1 qt = 0.94 l
	milliliter (ml)	0.001 liter	1 ml = 0.034 fluid ounce	1 pt = 0.47 l
	microliter (μl)	0.000001 liter		1 fluid ounce = 29.57 ml

TEMPERATURE CONVERSION SCALE

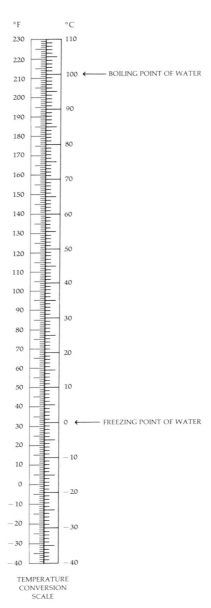

TEMPERATURE
CONVERSION
SCALE

FOR CONVERSION OF FAHRENHEIT TO CENTIGRADE,
THE FOLLOWING FORMULA CAN BE USED:

$$°C = \frac{5}{9}(°F - 32)$$

FOR CONVERSION OF CENTIGRADE TO FAHRENHEIT,
THE FOLLOWING FORMULA CAN BE USED:

$$°F = \frac{9}{5}°C + 32$$

CLASSIFICATION OF ORGANISMS

There are several ways to classify organisms. The one presented here is based closely upon that of Whittaker (in *Science*, **163**:150–160, 1969). Organisms are divided into five major groups, or kingdoms: Monera, Protista, Fungi, Plantae, and Animalia.

The chief taxonomic divisions are kingdom, phylum, class, order, family, genus, species. The following classification includes all of the major phyla. Certain classes and orders, particularly those mentioned in this book, are also included, but the listing is far from complete. The number of species given for each group is the estimated number of living species described and named.

KINGDOM MONERA

Prokaryotes, i.e., cells which lack a nuclear envelope, chloroplasts and other plastids, mitochondria, and 9 + 2 flagella. Monera are unicellular but sometimes aggregate into filaments or other superficially multicellular bodies. Their predominant mode of nutrition is absorption, but some groups are photosynthetic or chemosynthetic. Reproduction is primarily asexual, by fission or budding, but conjugation occurs in some species.

PHYLUM

PHYLUM SCHIZOPHYTA: the bacteria. Unicellular prokaryotes; reproduction usually asexual by cell division; nutrition usually heterotrophic. About 1,600 species.

PHYLUM CYANOPHYTA: blue-green algae. Unicellular or colonial; prokaryotic; chlorophyll, but no chloroplasts; nutrition usually autotrophic; reproduction by fission. Common on damp soil and rocks and in fresh and salt water. About 200 distinct, nonsymbiotic species.

KINGDOM PROTISTA

Eukaryotic organisms, including unicellular heterotrophs (protozoans) and unicellular or multicellular algae. Their modes of nutrition include ingestion, photosynthesis, and sometimes absorption. Reproduction both sexual (in some forms) and asexual. They move by 9 + 2 flagella or pseudopods or are nonmotile.

PHYLUM

PHYLUM PROTOZOA: microscopic, unicellular or simple colonial heterotrophic organisms; reproduction usually asexual by mitotic division; classified by type of locomotion. About 30,000 species.

CLASS

Class Mastigophora: protozoans with flagella, including a number of symbiotic forms such as *Trichonympha* and *Trypanosoma*, the cause of sleeping sickness.

Class Sarcodina: protozoans with pseudopods, such as amoebas. No stiffening pellicle; some secrete shells.

Class Ciliophora: protozoans with cilia, including *Paramecium* and *Stentor*.

Class Sporozoa: parasitic protozoans; usually without locomotive organs during a major part of their life cycle. Includes *Plasmodium*, several species of which cause malaria.

PHYLUM

PHYLUM EUGLENOPHYTA: euglenoids. Unicellular photosynthetic (or sometimes secondarily heterotrophic) organisms with chlorophylls *a* and *b*. They store food as paramylon, an unusual carbohydrate. Euglenoids usually have a single apical flagellum and a contractile vacuole. Sexual reproduction is unknown. Euglenoids occur mostly in fresh water. There are some 450 species.

PHYLUM CHRYSOPHYTA: golden algae and diatoms. Unicellular photosynthetic organisms with chlorophylls *a* and *c* and the accessory pigment fucoxanthin. Food stored as the carbohydrate leucosin or as large oil droplets. Cell walls consisting mainly of pectic compounds, sometimes heavily impregnated with siliceous materials. Some 6,000 to 10,000 living species.

CLASS

Class Bacillariophyceae: diatoms. Chrysophyta with double siliceous shells, the two halves of which fit together like a pillbox. They are sometimes motile by the secretion of mucilage fibrils along a specialized groove, the raphe. There are many extinct and 5,000 to 9,000 living species.

Class Chrysophyceae: golden algae. A diverse group of organisms including flagellated, amoeboid, and nonmotile forms, some naked and others with a cell wall that may be ornamented with siliceous scales. At least 1,000 species.

PHYLUM

PHYLUM PYRROPHYTA: "fire" algae, sometimes called golden-brown algae. Unicellular photosynthetic organisms with chlorophylls *a* and *c*. Food is stored as starch. Cell walls contain cellulose. The phylum contains some 1,100 species, mostly biflagellated organisms, of which the great majority belong to the following class:

CLASS

Class Dinophyceae: dinoflagellates. Pyrrophyta with lateral flagella, one of which beats in a groove that encircles the organism. They probably have no form of sexual reproduction and their mitosis is unlike that in any other organism. More than 1,000 species.

PHYLUM

PHYLUM XANTHOPHYTA: Autotrophic unicellular, amoeboid, and colonial organisms with chlorophyll *a* and carotenoids. Food stored as leucosin. Cell walls consisting primarily of cellulose and pectin. About 450 species.

PHYLUM CHLOROPHYTA: green algae. Unicellular or multicellular plants characterized by chlorophylls *a* and *b* and various carotenoids. The carbohydrate food reserve is starch. Motile cells have two whiplash flagella at the apical end. True multicellular genera do not exhibit complex patterns of differentiation. Multicellularity has arisen at least three times, and quite possibly more often. There are about 7,000 known species and possibly many more.

PHYLUM PHAEOPHYTA: brown algae. Multicellular marine plants characterized by the presence of chlorophylls *a* and *c* and the pigment fucoxanthin. Their food reserve is a carbohydrate called laminarin. Motile cells are biflagellate, with one forward flagellum of the tinsel type and one trailing one of the whiplash type. A considerable amount of differentiation is found in some of the kelps, with specialized conducting cells for transporting photosynthate to dimly lighted regions of the plant present in some genera. There is however no differentiation into leaves, roots, and stem, as in the land plants. About 1,100 species.

PHYLUM RHODOPHYTA: red algae. Primarily marine plants characterized by the presence of chlorophyll *a* and pigments known as phycobilins. Their carbohydrate reserve is a special type of starch (floridian). No motile cells are present at any stage in the complex life cycle. The plant body is built up of closely packed filaments in a gelatinous matrix and is not differentiated into leaves, roots, and stem. It lacks specialized conducting cells. There are some 4,000 species.

PHYLUM GYMNOMYCOTA: the slime molds. Heterotrophic amoeboid organisms that mostly lack a cell wall but form sporangia at some stage in their life cycle. Predominant mode of nutrition is by ingestion. There are three classes:

CLASS

Class Myxomycetes: plasmodial slime molds. Slime molds with multinucleate plasmodium which creeps along as a mass and eventually differentiates into sporangia, each of which is multinucleate and eventually gives rise to many spores. About 450 species.

Class Acrasiomycetes: cellular slime molds. Slime molds in which there are separate amoebas which eventually swarm together to form a mass but retain their identity within this mass, which eventually differentiates into a compound sporangium. Seven genera and about 26 species.

Class Protostelidomycetes: In this recently discovered group, the amoebas may remain separate or mass, but each one eventually differentiates into a simple stalked sporangium with one or two spores at its apex. Five genera and more than a dozen species.

KINGDOM FUNGI

Eukaryotic unicellular or multinucleate organisms in which the nuclei occur in a basically continuous mycelium; this mycelium becomes septate (partitioned off) in certain groups and at certain stages of the life cycle. They are heterotrophic, with nutrition by absorption. Reproductive cycles often include both sexual and asexual phases. Some 100,000 species of fungi have been named.

Class Oomycetes: Mostly aquatic fungi with motile cells characteristic of certain stages of the life cycle, their cell walls are composed of glucose polymers including cellulose. There are several hundred species.

Class Zygomycetes: terrestrial fungi, such as black bread mold, with the hyphae septate only during the formation of reproductive bodies; chitin predominant in the cell walls. The class includes several hundred species.

Class Ascomycetes: terrestrial and aquatic fungi, including *Neurospora*, powdery mildews, morels, and truffles. The hyphae are septate but the septa perforated; complete septa cut off the reproductive bodies, such as spores or gametangia. Chitin is predominant in the cell walls. Sexual reproduction involves the formation of a characteristic cell, the ascus, in which meiosis takes place and within which spores are formed. The hyphae in many ascomycetes are packed together into complex "fruiting bodies." Yeasts are unicellular ascomycetes that reproduce asexually by budding. About 30,000 species.

Class Basidiomycetes: terrestrial fungi, including the mushrooms and toadstools, with the hyphae septate but the septa perforated; complete septa cut off reproductive bodies, such as spores or gametangia. Chitin is predominant in the cell walls. Sexual reproduction involves formation of basidia, in which meiosis takes place and on which the spores are borne. There are some 25,000 species.

Fungi Imperfecti: Mainly fungi with the characteristics of Ascomycetes but in which the sexual cycle has not been observed; a few probably belong to other classes. The Fungi Imperfecti are classified by their asexual spore-bearing organs. There are some 25,000 species, including a *Penicillium*, the original source of penicillin, fungi which cause athlete's foot and other skin diseases, and many of the molds which give cheese, such as Roquefort and Camembert, their special flavor.

KINGDOM PLANTAE

Multicellular photosynthetic terrestrial eukaryotes and closely related forms. The photosynthetic pigment is chlorophyll *a*, with chlorophyll *b* and a number of carotenoids serving as accessory pigments. The cell walls contain cellulose. There is considerable differentiation of organs and tissues. Their reproduction is primarily sexual with alternating gametophytic and sporophytic phases; the gametophyte phase has been progressively reduced in the course of evolution.

PHYLUM BRYOPHYTA: mosses, hornworts, and liverworts. Multicellular plants with the photosynthetic pigments and food reserves similar to those of the green algae. They have gametangia with a multicellular sterile jacket one cell layer thick. The sperm are biflagellate and motile. Gametophytes and sporophytes both exhibit complex multicellular patterns of development, but the conducting tissues are usually completely absent and not well differentiated when present. Most of the photosynthesis in these primarily terrestrial plants is carried out by the gametophyte, upon which the sporophyte is initially dependent. More than 23,500 species.

Class Hepaticae: liverworts. The gametophytes are either thallose (not differentiated into roots, leaves, and stem) or leafy, and the sporophytes relatively simple in construction. About 9,000 species.

Class Antherocerotae: hornworts. The gametophytes are thallose. The sporophyte grows from a basal meristem for as long as conditions are favorable. Stomata are present on the sporophyte. About 100 species.

Class Musci: mosses. The gametophytes are leafy. Sporophytes have complex patterns of spore discharge. Stomata are present on the sporophyte. About 14,500 species.

PHYLUM TRACHEOPHYTA: vascular plants. Plants with complex differentiation of organs into leaves, roots, and stem. The only motile cells are the male gametes of some species, which are propelled by many cilia. The vascular plants have well-developed strands of conducting tissue for the transport of water and organic materials. The main trends of evolution in the vascular plants involve a progressive reduction in the gametophyte, which is green and free-living in ferns but heterotrophic and more or less enclosed by sporophytic tissue in the others; the loss of multicellular gametangia and motile sperm; and the evolution of the seed. The phylum includes the following subphyla with living representatives:

SUBPHYLUM	SUBPHYLUM LYCOPHYTINA: lycophytes. Homosporous and heterosporous vascular plants with microphylls; extremely diverse in appearance. All lycophytes have motile sperm. There are five genera and about 1,000 species.

SUBPHYLUM SPHENOPHYTINA: horsetails. Homosporous vascular plants with jointed stems marked by conspicuous nodes and elevated siliceous ribs and sporangia borne in a strobilus at the apex of the stem. Leaves are scalelike. Sperm are motile. Although now thought to have evolved from a megaphyll, the leaves of the horsetails are structurally indistinguishable from microphylls. There is one genus, *Equisetum*, with about two dozen living species.

SUBPHYLUM PTEROPHYTINA: ferns, gymnosperms, and flowering plants. Although diverse, these groups possess in common the megaphyll, which in certain genera has become much reduced. About 260,000 species. |
| **CLASS** | *Class Filicineae:* the ferns. They are mostly homosporous although some are heterosporous. The gametophyte is more or less free-living and usually photosynthetic. Multicellular gametangia and free-swimming sperm are present. About 11,000 species.

Class Coniferinae: the conifers. Seed plants with active cambial growth and simple leaves, in which the ovules are not enclosed and the sperm are not flagellated. There are some 50 genera and about 550 species, the most familiar group of gymnosperms.

Class Cycadinae: cycads. Seed plants with sluggish cambial growth and pinnately compound, palmlike or fernlike leaves. The ovules are not enclosed. The sperm are flagellated and motile, but are carried to the vicinity of the ovule in a pollen tube. Cycads are gymnosperms. There are nine genera and about 100 species.

Class Ginkgoinae: ginkgo. Seed plants with active cambial growth and fan-shaped leaves with open dichotomous venation. The ovules are not enclosed and are fleshy at maturity. Sperm are carried to the vicinity of the ovule in a pollen tube, but are flagellated and motile. They are gymnosperms. There is one species only.

Class Angiospermae: flowering plants. Seed plants in which the ovules are enclosed in a carpel (in all but a very few genera), and the seeds at maturity are borne within fruits. They are extremely diverse vegetatively but characterized by the flower, which is basically insect-pollinated. Other modes of pollination, such as wind pollination, have been derived in a number of different lines. The gametophytes are much reduced, with the female gametophyte often consisting of only eight cells or nuclei at maturity. Double fertilization involving two of the three nuclei from the mature microgametophyte gives rise to the zygote and to the primary endosperm nucleus; the former becomes the embryo and the latter a special nutritive tissue, the endosperm. About 250,000 species. |
| **SUBCLASS** | *Subclass Dicotyledonae:* dicots. Flower parts are usually in fours or fives; leaf venation is usually netlike, pinnate, or palmate; there is true secondary growth with vascular cambium commonly present; there are two cotyledons; and the vascular bundles in the stem are in a ring. About 190,000 species.

Subclass Monocotyledonae: monocots. Flower parts are usually in threes, leaf venation is usually parallel, true secondary growth is not present, there is one cotyledon, and vascular bundles in the stem are scattered. About 60,000 species. |

KINGDOM ANIMALIA

Eukaryotic multicellular organisms. Their principal mode of nutrition is by ingestion. Many animals are motile, and they generally lack the rigid cell walls characteristic of plants. Considerable cellular migration and reorganization of tissues often occurring during the course of embryology. Their reproduction is primarily sexual, with male and female diploid organisms producing haploid gametes which fuse to form the zygote. More than a million species have been described and the actual number may be close to 10 million.

PHYLUM		
		PHYLUM PORIFERA: sponges. Simple multicellular animals, largely marine, with stiff skeletons, and bodies perforated by many pores that admit water containing food particles. All have choanocytes, "collar cells." About 4,200 species.
		PHYLUM COELENTERATA: coelenterates. Animals with radially symmetrical, "two-layered" bodies of a jellylike consistency. Reproduction is asexual or sexual. They are the only organisms with cnidoblasts, special stinging cells. All are aquatic and most are marine. About 11,000 species.

	CLASS	
		Class Hydrozoa: Hydra, Obelia, and other *Hydra*-like animals. They are often colonial, and often have a regular alternation of asexual and sexual generations. The polyp form is dominant.
		Class Scyphozoa: marine jellyfishes or "cup animals," including *Aurelia.* The medusa form is dominant. They have true muscle cells.
		Class Anthozoa: sea anemones ("flower animals") and colonial corals. They have no medusa stage.

PHYLUM		
		PHYLUM CTENOPHORA: comb jellies and sea walnuts. They are free-swimming, often almost spherical animals. They are translucent, gelatinous, delicately colored, and often bioluminescent. They possess eight bands of cilia, for locomotion. About 80 species.
		PHYLUM PLATYHELMINTHES: flatworms. Bilaterally symmetrical with three tissue layers. The gut has only one opening. They have no coelom or circulatory system. They have complex hermaphroditic reproductive systems and excrete by means of special (flame) cells. About 15,000 species.

	CLASS	
		Class Turbellaria: planaria and other nonparasitic flatworms. They are ciliated, carnivorous, and have ocelli ("eyespots").
		Class Trematoda: flukes. They are parasitic flatworms with digestive tracts.
		Class Cestoidea: tapeworms. They are parasitic flatworms with no digestive tracts; they absorb nourishment through body surface.

PHYLUM		
		PHYLUM RHYNOCHOCOELA: proboscis, nemertine, or ribbon worms. They are nonparasitic, usually marine, and have a tubelike gut with mouth and anus, a protrusible proboscis armed with a hook for capturing prey, and simple circulatory and reproductive systems. About 600 species.
		PHYLUM NEMATODA: roundworms. The phylum includes minute free-living forms, such as vinegar eels, and plant and animal parasites, such as hookworms. they are characterized by elongated, cylindrical, bilaterally symmetrical bodies. About 80,000 species.
		PHYLUM ACANTHOCEPHALA: spiny-headed worms. They are parasitic worms with no digestive tract and a head armed with many recurved spines. About 300 species.
		PHYLUM CHAETOGNATHA: arrow worms. Free-swimming planktonic marine worms, they have a coelom, a complete digestive tract, and a mouth with strong sickle-shaped hooks on each side. About 50 species.

		PHYLUM NEMATOMORPHA: horsehair worms. They are extremely slender, brown or black worms up to 3 feet long. Adults are free-living, but the larvae are parasitic in insects. About 250 species.
		PHYLUM ROTIFERA: microscopic, wormlike or spherical animals, "wheel animalcules." They have a complete digestive tract, flame cells, and a circle of cilia on the head, the beating of which suggests a wheel; males are minute and either degenerate or unknown in many species. About 1,500 species.
		PHYLUM GASTROTRICHA: These are microscopic, wormlike animals which move by longitudinal bands of cilia. About 140 species.
		PHYLUM BRYOZOA: "moss" animals. Microscopic aquatic organisms, they are characterized by a U-shaped row of ciliated tentacles, with which they feed; they usually form fixed and branching colonies; they superficially resemble hydroid coelenterates but are much more complex; having anus and coelum, they retain larvae in special brood pouch. About 4,000 species.

594   Appendix C

PHYLUM BRACHIOPODA: lamp shells. Marine animals with two hard shells (one dorsal and one ventral), they superficially resemble clams. Fixed by a stalk or one shell in adult life, they obtain food by means of ciliated tentacles. About 260 living species; 3,000 extinct.

PHYLUM PHORONIDEA: sedentary, elongated, wormlike animals that secrete and live in a leathery tube. They have a U-shaped digestive tract and a ring of ciliated tentacles with which they feed. Marine. About 15 species.

PHYLUM ANNELIDA: ringed or segmented worms. they usually have a well-developed coelom, a one-way digestive tract, head, and circulatory system, nephridia, and well-defined nervous system. About 8,800 species.

CLASS

Class Archiannelida: small, simple, probably primitive, marine worms. About 35 species.

Class Polychaeta: mainly marine worms, such as *Nereis.* They have a distinct head with palps and tentacles and many-bristled lobose appendages. Parapodia are often brightly colored. About 4,000 species.

Class Oligochaeta: soil, freshwater, and marine annelids, including the earthworm *(Lumbricus).* They have scanty bristles and usually a poorly differentiated head. About 2,500 species.

Class Hirudinea: leeches. They have a posterior sucker and usually an anteior sucker surrounding the mouth. They are freshwater, marine, and terrestrial; either free-living or parasitic. About 300 species.

PHYLUM

PHYLUM MOLLUSCA: unsegmented animals, with a head, a mantle and a muscular foot, variously modified. They are mostly aquatic; soft-bodied, often with one or more hard shells, and a three-chambered heart. All mollusks, except bivalves, have a radula (rasplike organ used for scraping or marine drilling). About 110,000 species.

CLASS

Class Amphineura: chitons. The simplest type of mollusks, they have an elongated body covered with a mantle in which are embedded eight dorsal shell plates. About 700 species.

Class Pelecypoda: two-shelled mollusks, including clams, oysters, mussels, scallops. They usually have a hatchet-shaped foot and no distinct head. Generally sessile. About 15,000 species.

Class Scaphopoda: tooth or tusk shells. They are marine mollusks with a conical tubular shell. About 350 species.

Class Gastropoda: asymmetrical mollusks including snails, whelks, slugs. They usually have a spiral shell and a head with one or two pairs of tentacles. About 80,000 species.

Class Cephalopoda: octopus, squid, *Nautilus.* They are characterized by a "head-foot" with eight or ten arms or many tentacles, mouth with two horny jaws, and well-developed eyes and nervous system. The shell is external *(Nautilus),* internal (squid), or absent (octopus). All except *Nautilus* have ink glands. About 400 species.

PHYLUM

PHYLUM ARTHROPODA: The largest phylum in the animal kingdom, arthropods are segmented animals with paired jointed appendages, a hard jointed exoskeleton, a complete digestive tract, reduced coelom, no nephridia, a dorsal brain, and a ventral nerve cord with paired ganglia in each segment. About 765,257 species.

CLASS

Class Merostomata: horseshoe crabs. They are aquatic, with book gills. About 5 species.

Class Crustacea: lobsters, crabs, crayfish, shrimps. Crustaceans are mostly aquatic, with two pairs of antennae, one pair of mandibles, and typically two pairs of maxillae. The thoracic segments have appendages, and the abdominal segments are with or without appendages. About 30,000 species.

Class Arachnida: spiders, mites, scorpions. Most members are terrestrial, air-breathing; usually have 4 or 5 pairs of legs; first pair of appendages used for grasping; have chelicerae (pincers or fangs) in place of jaws or antennae. About 35,000 species.

Class Onychophora: simple, terrestrial arthropods. All belong to one genus, *Peripatus.* They have many short unjointed pairs of legs. About 73 species.

Class Insecta: insects, including bees, ants, beetles, butterflies, fleas, lice, flies, etc. Most insects are terrestrial, and most breathe by means of trachea. They have one pair of antennae, three pairs of legs, and three distinct parts of the body (head, thorax, and abdomen). Most have 2 pairs of wings. About 700,000 species.

Class Chilopoda: centipedes. They have 15 to 173 trunk segments, each with one pair of jointed appendages. About 2,000 species.

Class Diplopoda: millipedes. They have an abdomen with 20 to 100 segments, each with two pairs of appendages. About 7,000 species.

PHYLUM

PHYLUM ECHINODERMATA: starfish and sea urchins. Echinoderms are radially symmetrical in adult stage, with a well-developed coelom, an endoskeleton of calcareous ossicles and spines, and a unique water vascular system. They have tube feet and are marine. About 6,000 species.

CLASS

Class Crinoidea: sea lilies and feather stars. Sessile animals, they often have a jointed stalk for attachment, and they have ten arms bearing many slender lateral branches. Most species are fossils.

Class Asteroidae: starfish. They have five to fifty arms, an oral surface directed downward, and two to four tube feet.

Class Ophiuroidea: brittle stars, serpent stars. They are greatly elongated, with highly flexible slender arms and rapid horizontal locomotion.

Class Echinoidea: sea urchins and sand dollars. Skeletal plates form rigid test which bears many movable spines.

Class Holothuroidea: sea cucumbers. They have a sausage-shaped or wormlike elongated body.

PHYLUM

PHYLUM HEMICHORDATA: small group of wormlike marine animals, including the acorn worms. They have a notochordlike structure in the head end, gill slits, and a solid nerve cord. About 91 species.

PHYLUM CHORDATA: animals having at some stage a notochord, pharyngeal gill slits, and a hollow nerve cord on the dorsal side. About 44,794 species.

SUBPHYLUM

SUBPHYLUM TUNICATA: tunicates or ascidians. Adults are saclike, usually sessile, often forming branching colonies. They feed by ciliary currents, have gill slits, a reduced nervous system, and no notochord. Larvae are active, with well-developed nervous system and notochord. They are marine. About 1,600 species.

SUBPHYLUM CEPHALOCHORDATA: lancelets. This small subphylum contains only *Amphioxus* and related forms. They are somewhat fishlike marine animals with a permanent notochord the whole length of the body, a nerve cord, a pharynx with gill slits, and no cartilage or bone. About 13 species.

SUBPHYLUM VERTEBRATA: the vertebrates. The most important subphylum of Chordata. In the vertebrates the notochord is replaced by cartilage or bone, forming the segmented vertebral column or backbone. A skull surrounds a well-developed brain. They usually have a tail. About 43,090 species.

CLASS

Class Agnatha: lampreys and hagfish. These are eel-like aquatic vertebrates without limbs, with a jawless sucking mouth, and no bones, scales, or fins.*

Class Chondrichthyes: sharks, rays, skates, and other cartilaginous fish. They have complicated copulatory organs, scales, and no air bladders. They are almost exclusively marine.*

Class Osteichthyes: the bony fish, including nearly all modern freshwater fish, such as sturgeon, trout, perch, anglerfish, lungfish, and some almost extinct groups. They usually have an air bladder or (rarely) a lung.*

Class Amphibia: salamanders, frogs, and toads. They usually breathe by gills in the larval stage and by lungs in the adult stage. They have incomplete double circulation and a usually naked skin. The limbs are legs. They were the first vertebrates to inhabit the land and are ancestors of the reptiles. Their eggs are unprotected by a shell and embryonic membranes. About 2,000 species.

Class Reptilia: turtles, lizards, snakes, crocodiles; includes extinct species such as the dinosaurs. Reptiles breathe by lungs and have incomplete double circulation. Their skin is usually covered with scales. The four limbs are legs (absent in snakes). They are cold-blooded. Most live and reproduce on land though some are aquatic. Their embryo is enclosed in an egg shell and has protective membranes. About 5,000 species.

Class Aves: birds. Birds are warm-blooded animals with complete double circulation and a skin covered with feathers. The forelimbs are wings. Their embryo is enclosed in an egg shell with protective membranes. Includes the extinct *Archaeopteryx.* About 8,590 species.

Class Mammalia: mammals. Mammals are warm-blooded animals with complete double circulation. Their skin is usually covered with hair. The young are nourished with milk secreted by the mother. They have four limbs, usually legs (forelimbs sometimes arms, wings, or fins), a diaphragm used in respiration, a lower jaw made up of a single pair of bones, 3 bones in each middle ear connecting eardrum and inner ear, and 7 vertebrae in the neck. About 4,500 species.

* The total number of species of fish is estimated to be about 23,000.

SUBCLASS

Subclass Prototheria: monotremes. These are the oviparous (egg-laying) mammals with imperfect temperature regulation. There are only two living species: the duckbill platypus and spiny anteater of Australia and New Guinea.

Subclass Metatheria: marsupials, including kangaroos, opossums, and others. Marsupials are viviparous mammals without a placenta (or with a poorly developed one); the young are born in an undeveloped state and are carried in an external pouch of the mother for some time after birth. They are found chiefly in Australia.

Subclass Eutheria: mammals with a well-developed placenta. This subclass comprises the great majority of living mammals. There are 12 principal orders of Eutheria:

Order

Insectivora: shrews, moles, hedgehogs, etc.
Edentata: toothless mammals—anteaters, sloths, armadillos, etc.
Rodentia: the rodents—rats, mice, squirrels, etc.
Artiodactyla: even-toed ungulates (hoofed mammals)—cattle, deer, camels, hippopotamuses, etc.
Perissodactyla: odd-toed ungulates—horses, zebras, rhinoceroses, etc.
Proboscidea: elephants
Lagomorpha: rabbits and hares
Sirenia: the manatee, dugong, and sea cows. Large aquatic mammals with the forelimbs finlike, the hind limbs absent.
Carnivora: carnivorous animals—cats, dogs, bears, weasels, seals, etc.
Cetacea: the whales, dolphins, and porpoises. Aquatic mammals with the forelimbs fins, the hind limbs absent.
Chiroptera: the bats. Aerial mammals with the forelimbs wings.
Primates: the lemurs, monkeys, apes, and man.

THE HARDY-WEINBERG EQUILIBRIUM

The Hardy-Weinberg formula for equilibrium under random mating ($p^2 + 2pq + q^2 = 1$) is based, as are Mendel's laws, on the mathematics of probability. Suppose, for instance, that in a large group of female mice, one-fourth of the females are brown. The chance of picking a brown female, at random, is therefore 1 in 4, or $\frac{1}{4}$. In the group of male mice, one-fourth are also brown, and so the chance of picking a male is 1 in 4, or $\frac{1}{4}$. But what are your chances of picking a brown female and a brown male, at random, at the same time? There are 4×4, or 16, possible combinations of male and female mice. (You can easily work this out in a diagram.) Therefore, your chances of picking a brown female and brown male together are 1 in 16. According to the laws of chance, multiplying the probability of one event (picking a brown female) by the probability of another event (picking a brown male) will give you the probability of the two events occurring simultaneously (picking a brown female and a brown male): $\frac{1}{4} \times \frac{1}{4} = \frac{1}{16}$.

Hardy-Weinberg in Mendelian Terms

To demonstrate the operation of the Hardy-Weinberg principle, let us begin by crossbreeding two populations, one of which is homozygous AA and one of which is homozygous aa. (These are the sort of populations with which Mendel began his studies.) All the first filial generation will be heterozygous (Aa). When members of this generation interbreed ($Aa \times Aa$), the typical Mendelian ratio results; one-fourth of the population will be AA, one-fourth Aa, one-fourth aA, and one-fourth aa:

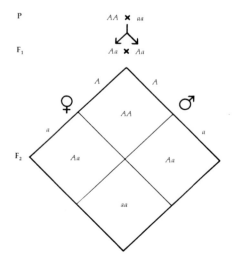

Remember that in these Punnett squares, like those in Section 3, each square represents a group of individuals rather than a single individual.

What will happen to these genotype proportions in the next generation? Let us consider first all the different kinds of crosses that will produce AA in the next generation. The first combination, female AA × male AA, will produce all AAs, and the chances of this occurring are $\frac{1}{4}$ × $\frac{1}{4}$, or $\frac{1}{16}$:

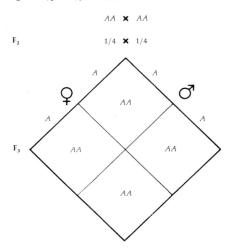

Female AA × male Aa (or aA) will produce some AAs. The chances of these matings occurring are $\frac{1}{4}$ × $\frac{1}{2}$, or $\frac{1}{8}$, but only one-half of the offspring will be AAs, so the chances of getting AA from these crossings are $\frac{1}{4}$ × $\frac{1}{2}$ × $\frac{1}{2}$, or $\frac{1}{16}$:

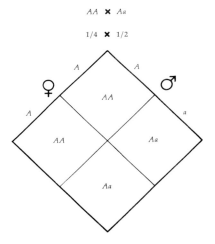

The third combination that will produce some AAs is female Aa × male AA. Here again, the chance of the offspring being AA is $\frac{1}{16}$:

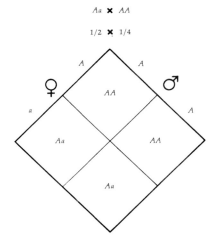

The fourth combination that will produce *AA*s is *Aa* × *Aa*. There is one chance in four that such a mating will occur and one chance in four that it will produce an *AA*, again ¹⁄₁₆.

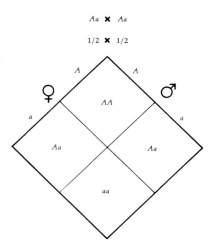

Now add these up. The chances of producing an *AA* in the third generation are ¹⁄₁₆ plus ¹⁄₁₆ plus ¹⁄₁₆ plus ¹⁄₁₆, or ⁴⁄₁₆ (= ¼), the same as in the F_2 generation. In other words, the probability is that the proportion of dominants in the third generation will be just the same as the proportion in the second generation. And, in fact, the proportions of *Aa*, *aA*, and *aa* will remain the same also, as you can readily demonstrate in this same way. In short, in a large population in which there is random mating, in the absence of forces that change the proportion of genes (conditions 3 to 5), the original frequencies of the two alleles will be retained from generation to generation.

Hardy-Weinberg in Algebraic Terms

For studies in population genetics, the Hardy-Weinberg principle is usually stated in algebraic terms, with the fractions that we used as examples before expressed as decimals. In the case of a gene for which there are two alleles in the gene pool, the frequency (represented by the symbol p) of one gene and the frequency (q) of the other must together equal the frequency of the whole, or 1; $p + q = 1$. Or, to put it another way, $1 - p = q$ and $1 - q = p$. This is equivalent to saying that if there are only two alleles, *A* and *a*, for instance, of a particular gene and if half (0.50) of the alleles in the gene pool are *A*, the other half (0.50) has to be *a*. Similarly, if 99 out of 100 (0.99) are *A*, 0.01 are *a*.

Then how do we find the relative proportions of individuals who are *AA*, *Aa*, and *aa*? As we established previously, these proportions can be calculated by multiplying the frequency of *A* (male) by that of *A* (female), *A* (male) times *a* (female), *a* (male) times *A* (female), and *a* (male) times *a* (female). We can express these multiplications in algebraic terms as $p^2 + 2pq + q^2$—which equals, as you probably know, $(p + q)^2$. Using this expression, we can see that if half (0.50) of the gene pool is *A* and half *a*, the proportion of *AA* will be 0.25, the proportion of *Aa* will be 0.50, and the proportion of *aa* will be 0.25, which is exactly what Mendel said—although in slightly different terms—in the first place. [$A^2 + 2Aa + a^2 = (0.50)^2 + 2(0.50)(0.50) + (0.50)^2 = (0.25) + 2(0.25) + (0.25) = 0.25 + 0.50 + 0.25 = 1.$]

Now let us look at a gene pool in which p (the frequency of *A*) equals 0.8 and q (the frequency of *a*) equals 0.2. The genotypic frequencies are *AA* equals 0.64, *Aa* equals 0.32, and *aa* equals 0.04, providing that mating takes place at random. The resulting population will thus be made up of 64 percent *AA*s, 32 percent *Aa*s, and 4 percent *aa*s. And, as we have just proved, the percentages will tend to remain in this equilibrium.

GLOSSARY

This list does not include units of measure or names of taxonomic groups, which can be found in Appendixes B and C, or terms that are used only once in the text and defined there.

ABDOMEN: In vertebrates, that portion of the trunk containing visceral organs except for heart and lungs; in arthropods, the posterior portion of the body, composed of a group of similar segments and containing the reproductive organs and posterior portion of the digestive tract.

ABSCISSION: In plants, the dropping of leaves, flowers, fruits, or stems at the end of a growing season, as the result of formation of an abscission layer, a layer of specialized cells, and the action of a hormone (abscisic acid).

ABSORPTION: The passage of water and dissolved substances into a cell or organism.

ABSORPTION SPECTRUM: The spectrum of light waves absorbed by a particular pigment.

ACETYLCHOLINE (a-**sea**-tell-**co**-leen): One of the known chemical transmitters of nerve impulses across synaptic junctions.

ACID [L. *acidus*, sour]: A substance that, on dissociation, releases hydrogen ions (H^+) but not hydroxyl ions (OH^-); having a pH of less than 7; the opposite of a base.

ACTIN: One of the two major proteins of muscle; makes up the thin filaments.

ACTION POTENTIAL: A rapid change in electrical potential across a membrane, resulting in transmission of a nerve impulse.

ACTION SPECTRUM: The spectrum of light waves that elicit a particular reaction.

ACTIVE SITE: That part of the surface of an enzyme molecule into which the substrate fits during a catalytic reaction.

ACTIVE TRANSPORT: The energy-expending process by which a cell moves a substance across the cell membrane from a point of lower concentration to a point of higher concentration, against the diffusion gradient.

ADAPTATION [L. *adaptare*, to fit]: (1) The acquiring by a group of organisms, through evolutionary processes, of characteristics that make them better suited to live and reproduce in their environment. (2) A peculiarity of structure, physiology, or behavior of an organism that aids the organism in its particular environment.

ADAPTIVE RADIATION: The evolution from a relatively primitive and unspecialized type of organism to several divergent forms specialized to fit numerous distinct and diverse ways of life.

ADENOSINE TRIPHOSPHATE (ATP): The major source of usable energy in cell metabolism; composed of adenine, ribose, and three phosphate groups. On hydrolysis, ATP loses one phosphate and one hydrogen to become adenosine diphosphate (ADP), releasing energy in the process.

ADHESION [L. *adhaerere*, to stick to]: A sticking together of substances.

ADRENAL GLAND: A vertebrate endocrine gland. The cortex (outer surface) is the source of cortisol, aldosterone, and other steroid hormones; the medulla (inner core) secretes epinephrine.

ADRENALINE: *See* Epinephrine.

AEROBIC [Gk. *aēr, aeros,* air + *bios,* life]: Requiring free oxygen for respiration.

AFFERENT [L. *ferre,* to bear]: Bringing inward to a central part, applied to nerves and blood vessels.

ALDOSTERONE: A hormone produced by the adrenal cortex that stimulates the reabsorption of sodium from the kidney.

ALGA (**al**-gah): A photosynthetic organism lacking multicellular sex organs.

ALKALINE: Pertaining to substances that release hydroxyl (OH^-) ions in water; having a pH greater than 7; basic; opposite of acidic.

ALLELE: (al-**eel**) (ALLELOMORPH) [Gk. *allelon,* of one another + -morphe, form]: One of the alternative forms of the same functional gene. Alleles occupy the same position (locus) on homologous chromosomes and so are separated from each other at meiosis.

ALLOPOLYPLOID: A polyploid in which the different sets of chromosomes come from different species or widely different strains.

ALTERNATION OF GENERATIONS: A reproductive cycle in which a haploid ($1n$) phase, the gametophyte, gives rise to gametes, which, after fusion to form a zygote, germinate to produce a diploid ($2n$) phase, the sporophyte. Spores produced by meiotic division from the sporophyte give rise to new gametophytes, completing the cycle.

AMINO ACIDS (am-**ee**-no) [Gk. *Ammon,* referring to the Egyptian sun god, near whose temple ammonium salts were first prepared from camel dung]: Organic molecules containing nitrogen in the form of NH_2; the "building blocks" of protein molecules.

AMNION (**am**-neon) [Gk. dim. of *amnos,* lamb]: Inner, fluid-filled sac composed of a thin double membrane that surrounds the embryo in reptiles, birds, and mammals.

AMNIOTE EGG: An egg that is isolated from the environment by a more or less impervious shell during the period of its development and that is completely self-sufficient, requiring only oxygen from the outside.

AMPHI— [Gk., on both sides]: Prefix, meaning "on both sides," "both," or "of both kinds."

AMPHIBIAN [Gk. *amphibios,* living a double life]: A class of vertebrates intermediate in many characteristics between fish and reptiles, which live part of the time in water and part on land.

ANAEROBIC: Applied to cells (largely bacterial) that can live without free oxygen; obligate anaerobes cannot live in the presence of oxygen; facultative anaerobes can live with or without oxygen.

ANALOGOUS [Gk. *analogos,* proportionate]: Applied to structures similar in function but different in evolutionary origin, such as the wing of a bird and the wing of an insect.

ANAPHASE (**anna**-phase) [Gk. *ana,* up + *phasis,* form]: A stage in mitosis or meiosis in which the chromatids of each chromosome separate and move to opposite poles.

ANDRO— [Gk. *aner, andros*, man]: Prefix, meaning "man" or "male."

ANDROGEN: Male sex hormone.

ANGIOSPERM (**an**-jee-o-sperm) [Gk. *angion*, vessel + *sperma*, seed]: Literally, a seed borne in a vessel; thus, one of a group of plants whose seeds are borne within a matured ovary (fruit).

ANNUAL: A plant that completes its life cycle from seed germination to seed production, followed by death, in a single growing season.

ANTENNA: Long, paired sensory appendage on the head of many arthropods.

ANTERIOR [L. *ante*, before, toward, in front of]: The front end of an organism; in human anatomy, the ventral surface.

ANTHER [Gk. *anthos*, flower]: In plants, the pollen-bearing portion of a stamen.

ANTHROPO— [Gk. *anthropos*, man, human]: Prefix, meaning "man" or "human."

ANTHROPOID: A higher primate; includes monkeys, apes, and man.

ANTHROPOMORPHISM: Assignment of human characteristics, abilities, or feelings to nonhuman organisms.

ANTIBIOTIC: An organic compound formed and secreted by an organism that is toxic to organisms of other species.

ANTIBODY: A globular protein that is produced in the body fluids, in response to a foreign substance (antigen), with which it reacts specifically.

ANTIDIURETIC HORMONE (ADH): A hormone secreted by the hypothalamus that inhibits urine excretion by inducing the reabsorption of water from the nephrons of the kidneys, also called vasopressin.

ANTIGEN: A foreign substance, usually a protein or polysaccharide, that stimulates the formation of specific antibodies.

AORTA (a-**ore**-ta) [Gk. *airein*, to lift, heave]: The major artery in blood-circulating systems.

APICAL MERISTEM: The growing point, composed of meristematic tissue, at the tip of the root or stem in vascular plants.

ARBOREAL: Tree-dwelling.

ARCH—, ARCHEO— [Gk. *arche, archos*, beginning]: Prefix, meaning "first," "main," or "earliest."

ARCHEGONIUM, *pl.* ARCHEGONIA [Gk. *archegeonos*, first of a race]: In plants, multicellular egg-producing organ.

ARTERY: A vessel carrying blood from the heart to the tissues; it is usually thick-walled, elastic, and muscular.

ARTHROPOD [Gk. *arthron*, joint + *pous, podos*, foot]: An invertebrate animal with jointed appendages; a member of the phylum Arthropoda.

ASCUS, *pl.* ASCI (**as**-kus, **as**-i): A specialized cell, characteristic of the Ascomycetes, in which two haploid nuclei fuse to produce a zygote which immediately divides by meiosis; at maturity, an ascus contains ascospores.

ASEXUAL REPRODUCTION: Any reproductive process, such as fission or budding, that does not involve the union of gametes.

ATMOSPHERIC PRESSURE: The weight of the earth's atmosphere over a unit area of the earth's surface.

ATOM [Gk. *atomos*, indivisible]: The smallest unit into which a chemical element can be divided and still retain its characteristic properties.

ATOMIC NUCLEUS: The central core of an atom, containing protons and neutrons, around which electrons orbit.

ATOMIC NUMBER: The number of protons in the nucleus of an atom; equal to the number of electrons in the neutral atom.

ATOMIC WEIGHT: The average weight of an atom of an element relative to the weight of an atom of carbon (^{12}C), which is assigned the integral value of 12.

ATP: Abbreviation of adenosine triphosphate.

ATRIUM, *pl.* ATRIA (a-tree-um) [L., yard, court, hall]: A chamber of the heart that receives blood and passes it on to a ventricle.

AUTO— [Gk. *autos*, same, self]: Prefix, meaning "same" or "same self."

AUTONOMIC [Gk. *nomos*, law]: Self-controlling, independent of outside influences.

AUTONOMIC NERVOUS SYSTEM: A special system of motor nerves and ganglia in vertebrates that is not under voluntary control and that innervates the heart, glands, visceral organs, and smooth muscle. It is subdivided into the sympathetic and the parasympathetic nervous systems.

AUTOPOLYPLOID: A polyploid in which the chromosomes all come from the same source.

AUTOSOME: Any chromosome other than the sex chromosomes. Man has 22 pairs of autosomes and 1 pair of sex chromosomes.

AUTOTROPH [Gk. *trophos*, feeder]: An organism that is able to synthesize organic molecules from inorganic substances, in contrast to heterotroph. Plants, algae, and some bacteria are autotrophs.

AUXIN [Gk. *auxein*, to increase]: One of a group of plant hormones with a variety of growth-regulating effects, including promotion of cell elongation.

AXIS: An imaginary line passing through a body or organ around which parts are symmetrically aligned.

AXON: The part of a neuron that carries impulses away from the cell body.

BACKCROSS: A cross between an individual that is heterozygous for a pair of alleles and a parent that is homozygous for the recessive alleles involved in the experiment; a test cross involving a homozygous recessive parent.

BACTERIOPHAGE [L. *bacterium* + Gk. *phagein*, to eat]: A virus that parasitizes a bacterial cell.

BACTERIUM [Gk. dim. of *baktron*, staff]: A small unicellular prokaryotic organism.

BARK: All plant tissues outside the cambium in a woody stem.

BASAL BODY: A cytoplasmic organelle that organizes cilia or flagella, identical in structure to a centriole.

BASAL METABOLISM: The metabolism of an organism when it is using just enough energy to maintain vital processes; measured as the amount of heat (in terms of quantity of oxygen consumed or carbon dioxide given off) produced by an animal at rest (but not asleep), determined at least 14 hours after eating and expressed as Kilocalories per square meter of body surface (or per unit weight) per hour.

BASE: A substance that, on dissociation, releases hydroxyl (OH^-) ions but not hydrogen (H^+) ions; having a pH of more than 7; the opposite of an acid.

BASE-PAIRING RULE: The requirement that adenine must always pair with thymine (or uracil) and guanine with cytosine, as in a nucleic acid double helix.

BASIDIUM, *pl.* BASIDIA (ba-**sid**-i-um): A specialized reproductive cell of the Basidiomycetes, often club-shaped, in which nuclear fusion and meiosis occur.

BEHAVIORAL ISOLATING MECHANISMS: Modes of behavior, such as display rituals of courtship patterns, that serve to prevent mating between species.

BI— [L. *bis*, twice, double, two]: Prefix, meaning "two," "twice," or "having two points."

BIENNIAL [L. *annus*, year]: Occurring once in two years; a plant that requires two years to complete its reproductive cycle, with vegetative growth occurring in the first year and flowering, seed production, and death in the second.

BILATERAL SYMMETRY: An anatomical arrangement in which the right and left halves of an organism are approximate mirror images of each other.

BILE: A yellow secretion of the vertebrate liver, temporarily stored in the gallbladder and composed of organic salts that emulsify fats in the small intestine.

BINOMIAL SYSTEM: A system in which the name of an organism consists of two parts, the first designating the genus and the second designating the species.

BIO— [Gk. *bios*, life]: Prefix, meaning "pertaining to life."

BIOLOGICAL CLOCK: An unidentified internal factor (or factors) in plants and animals that governs the innate biological rhythms (growth and activity patterns) of the organism.

BIOMASS: Total weight of all organisms in a particular habitat or area.

BIOMES: A worldwide complex of communities, characterized by distinctive vegetation and climate; for example, the grassland areas collectively form the grassland biome, the tropical rain forests form the tropical rain forest biome, etc.

BIOSPHERE: The whole zone of air, land, and water at the surface of the earth occupied by living things.

BIOSYNTHESIS: Formation of organic compounds from elements or simple compounds by living organisms.

BLADE: The broad, expanded part of a leaf or leaflike organ.

BLASTULA [Gk. *blastos*, sprout]: An animal embryo after cleavage and before gastrulation; usually consists of a hollow sphere the walls of which are composed of a single layer of cells.

BOND ENERGY: The energy required to break a bond.

BOWMAN'S CAPSULE: The bulbous unit of the nephron, including the glomerulus. It is the site of filtration of the renal fluid from the blood, the initial process in urine formation.

BRAINSTEM: The most posterior portion of the brain; includes medulla, pons, and midbrain.

BRONCHUS, *pl.* BRONCHI (**bronk**-us, **bronk**-eye) [Gk. *bronchos*, windpipe]: One of a pair of respiratory tubes branching into either lung at the lower end of the trachea; it subdivides into progressively finer passageways, the bronchioles, culminating in the alveoli.

BUD: (1) In plants, an embryonic shoot, including rudimentary leaves, often protected by specialized bud scales. (2) In animals, an asexually produced protuberance that develops into a new individual.

BUFFER: A substance that prevents appreciable changes of pH in solutions to which small amounts of acids or bases are added.

BULB: A modified bud with thickened leaves adapted for underground food storage.

BULK FLOW: The overall movement of a liquid induced by gravity, pressure, or an interplay of both.

CALORIE [L. *calor*, heat]: The amount of energy in the form of heat required to raise the temperature of 1 gram of water 1°C; in making metabolic measurements the kilocalorie (Calorie) is generally used. A Calorie is the amount of heat required to raise the temperature of 1 kilogram of water 1°C.

CALVIN CYCLE: The process by which carbon dioxide is reduced to carbohydrates during photosynthesis.

CAMBIUM [L. *cambiare*, to exchange]: *See* Cork cambium, Vascular cambium.

CAPILLARY [L. *capillaris*, relating to hair]: A small, thin-walled blood vessel through which diffusion and filtration into the tissues occurs; connects arteries with veins.

CAPILLARY ACTION: The movement of water along a surface, against the action of gravity, resulting from the combined effect of water molecules cohering to each other and adhering to the molecules of the surface material.

CAPSULE (**kap**-sul): (1) A slimy layer around the cells of certain bacteria. (2) The sporangium of Bryophyta. (3) A dehiscent, dry fruit that develops from two or more carpels.

CARBOHYDRATE: An organic compound consisting of a chain of carbon atoms to which hydrogen and oxygen are attached in a 2:1 ratio; includes sugars, starch, glycogen, cellulose, etc.

CARNIVORE: An organism that obtains its food energy by eating animals.

CAROTENE [L. *carota*, carrot]: A yellow or orange pigment found in plants and some algae; converted into vitamin A in the vertebrate liver.

CAROTENOIDS: A class of pigments that includes the carotenes (yellows and oranges) and the xanthophylls (yellow); accessory pigments in photosynthesis.

CARPEL: A leaflike floral structure enclosing the ovule or ovules of the angiosperm, typically divided into ovary, style, and stigma; a flower may have one or more carpels, either single or fused.

CARTILAGE: The skeletal connective tissue of vertebrates; forms much of the skeleton of adult lower vertebrates and immature higher vertebrates.

CASPARIAN STRIP [After Robert Caspary, German botanist]: A thickened waxy strip that extends around and seals the walls of endodermal cells in plants, thus restricting the diffusion of solutes across the endodermis into the vascular tissues of the root.

CATALYST [Gk. *katalysis*, dissolution]: A substance that controls the rate of a chemical reaction but is not used up in the reaction; enzymes are catalysts.

CELL: The structural unit of protoplasm, composed of cytoplasm and one or more nuclei and surrounded by a membrane. In most plants, fungi, and bacteria there is a cell wall outside the membrane.

CELL MEMBRANE: The outermost membrane of the cell. Also called the plasma membrane.

CELL PLATE: A flattened structure that forms at the equator of the spindle in the dividing cells of plants and a few green algae during early telophase; the predecessor of the middle lamella.

CELL WALL: A relatively rigid structure, produced by the cell and located outside the cell membrane in most plants, fungi, and bacteria; in plant cells, it consists mostly of cellulose.

CELLULOSE: The chief constituent of the cell wall in all green plants; an insoluble complex carbohydrate formed of microfibrils of glucose molecules.

CENTRAL NERVOUS SYSTEM: In vertebrates, the brain and spinal cord; in invertebrates it usually consists of one or more cords of nervous tissue plus their associated ganglia.

CENTRIOLE (**sen**-tree-ole) [Gk. *kentron*, center]: A cytoplasmic organelle generally found in animal cells and in flagellated cells in other groups, usually outside the nuclear membrane, identical in structure to a basal body.

CENTROMERE (**sen**-tro-mere) [Gk. *kentron*, center + *meros*, a part]: *See* Kinetochore.

CEPHALO— [Gk. *kephale*, head]: Prefix, meaning "head."

CEREBELLUM [L. dim. of *cerebrum*, brain]: An enlarged part of the dorsal side of the vertebrate brain; chief muscle-coordinating center.

CEREBRAL CORTEX: A layer of neurons (gray matter) forming the upper surface of the cerebrum, well developed only in mammals; the seat of conscious sensations and voluntary muscular activity.

CEREBRUM [L., brain]: The principal portion of the vertebrate brain, occupying the upper part of the cranium, consisting of two cerebral hemispheres united by the corpus callosum.

CHELICERA: First pair of appendages in arachnids; used for seizing and crushing prey.

CHEMICAL REACTION: A change of one or more substances into different substances by recombination of their constituent atoms into different kinds of molecules.

CHEMORECEPTORS: A cell or organ that detects substances according to their chemical structure; includes smell and taste receptors.

CHEMOTROPISM: The behavioral response of an organism to chemical stimulation; the turning to or away from a chemical stimulus.

CHIASMA, *pl.* CHIASMATA (**kye**-az-ma) [Gk., a cross]: The X-shaped figure formed by the meeting of two nonsister chromatids of homologous chromosomes; the site of crossing over.

CHITIN (**kite**-n): A tough, resistant, nitrogen-containing polysaccharide present in the exoskeleton of arthropods, the epidermal cuticle or other surface structures of many other invertebrates, and the cell walls of certain fungi.

CHLORO— [Gk. *chlōros*, green]: Prefix, meaning "green."

CHLOROPHYLL: The green pigments of plant cells, necessary for photosynthesis.

CHLOROPLAST: A membrane-bound, chlorophyll-containing or-

ganelle in green plant cells; site of photosynthesis in eukaryotes.

—CHORD, CHORDA— [L. *chorda*, cord, string]: Suffix or prefix, meaning "cord."

CHORDATE: Member of the animal phylum (Chordata) in which all members possess a notochord, dorsal nerve cord, and pharyngeal gill slits, at least at some stage of the life cycle.

CHORION (**core**-ee-on): The outermost embryonic membrane of reptiles, birds, and mammals; in placental mammals it contributes to the structure of the placenta.

CHROM— [Gk. *chrōma*, color]: Prefix, meaning "color."

CHROMATID (**crow**-ma-tid): One of the two daughter strands of a duplicated chromosome which are joined by a single kinetochore.

CHROMATIN (**crow**-ma-tin): The deeply staining nucleoprotein complex of the chromosomes.

CHROMOSOME [Gk. *soma*, body]: One of the bodies in the cell nucleus containing genes in a linear order; visualized as threads or rods of chromatin which appear in a contracted form during mitosis and meiosis.

CHROMOSOME MAP: A plan showing the relative position of the genes on the chromosome, determined chiefly by analysis of the relative frequency of crossing over between any two genes.

CILIUM, *pl.* CILIA (**silly**-um) [L., eyelash]: A short hairlike structure present on the surface of some cells, usually in large numbers and arranged in rows. Each cilium has a highly characteristic internal structure of two inner fibrils surrounded by nine pairs of outer fibrils.

CIRCADIAN RHYTHMS [L. *circa*, about + *dies*, day]: Regular rhythms of growth and activity that occur approximately on a 24-hour basis.

CLEAVAGE: The successive cell divisions of the fertilized egg to form the multicellular blastula.

CLIMAX COMMUNITY: Final or relatively stable community in a successional series.

CLINE [Gk. *klinein*, to lean]: A correlation between a series of gradual differences within a species with differences in climate or other geographical factors.

CLITORIS (**klit**-o-ris) [Gk. *kleitoris*, small hill]: A small erectile body at the anterior part of the vulva, homologous to the penis.

CLONE [Gk. *klon*, twig]: A line of cells all of which have arisen from the same single cell by mitotic division; a population of individuals descended by asexual reproduction from a single ancestor.

CODON (**code**-on): Three adjacent nucleotides on a molecule of mRNA that form the code for a single amino acid.

—COEL, COELA—, COELO— [Gk., from *koilos*, hollow]: Suffix or prefix, meaning "cavity."

COELENTERON (see-**len**-t-ron): A digestive cavity with only one opening, characteristic of the phylum Coelenterata (jellyfish, hydra, corals, etc.).

COELOM (**seal**-um) [Gk. *koilos*, a hollow]: A body cavity formed between layers of mesoderm.

COENZYME: An organic molecule that plays an accessory role in enzyme-catalyzed processes, often by acting as a donor or acceptor of a substance involved in the reaction; NAD, NADP, and FAD are common coenzymes.

COHESION [L. *cohaerere*, to stick together]: The union or holding together of like molecules or like substances.

COLEOPTILE (coal-ee-**op**-tile) [Gk. *koleon*, sheath + *ptilon*, feather]: A sheathlike structure covering the shoot of grass seedlings.

COLLAGEN [Gk. *kolla*, glue]: A fibrous protein material in bones, tendons, and other connective tissues.

COLON: The large intestine of vertebrates leading to the rectum or cloaca.

COLONY: A group of unicellular or multicellular organisms living together in close association.

COMMENSALISM [L. *com*, together + *mensa*, table]: *See* Symbiosis.

COMMUNITY: The organisms inhabiting a common environment and interacting with one another.

COMPETITION: Interaction between members of the same population or of two or more populations resulting from a greater demand for their supply of a mutually required resource.

COMPOUND [L. *componere*, to put together]: A combination of atoms in definite ratios, held together by chemical bonds; a substance containing only one kind of molecule, each molecule composed of two or more kinds of atoms.

COMPOUND EYE: In arthropods, a complex eye composed of a number of separate elements (ommatidia), each with light-sensitive cells and a refractory system that can form an image.

CONE: (1) In plants, the reproductive structure of a conifer. (2) In animals, a type of light-sensitive neuron in the vertebrate retina, concerned with the perception of color and with the most acute discrimination of detail.

CONJUGATION: The process in unicellular organisms by which genetic material is passed from one cell to another.

CONNECTIVE TISSUE: A type of tissue which lies between groups of nerves, glands, and muscle cells and beneath epithelial cells, in which the cells are irregularly distributed through a relatively large amount of intercellular material; includes bone, cartilage, blood, and lymph.

CONVERGENT EVOLUTION [L. *convergere*, to turn together]: The independent development of similar structures in forms of life that are unrelated or only distantly related; often found in organisms living in similar environments, such as porpoises and sharks.

CORK: A secondary tissue produced by a cork cambium; made up of polygonal cells, nonliving at maturity, with walls infiltrated with suberin, a waxy or fatty material resistant to the passage of gases and water vapor.

CORK CAMBIUM: A lateral meristem producing cork in woody and some herbaceous plants; also called phellogen.

COROLLA (ko-**role**-a) [L. dim. of *corona*, wreath, crown]: Petals, collectively; usually the conspicuously colored flower parts.

CORTEX [L., bark]: (1) The outer layer. (2) In a stem or root, the primary tissue bounded externally by the epidermis and internally by the central cylinder of vascular tissue and consisting of parenchyma cells.

COTYLEDON (cottle-ee-don) [Gk. *kotyledon*, a cup-shaped hollow]: A leaflike structure of the embryo of a seed plant, concerned with digestion and storage of food.

COVALENT BOND: A chemical bond formed between atoms as a result of the sharing of a pair of electrons.

CROSS-FERTILIZATION: The mutual exchange of sperm between two hermaphroditic individuals and subsequent union of eggs and sperm, as in snails and earthworms.

CROSSING OVER: The exchange of corresponding segments of genetic material between chromatids of homologous chromosomes at meiosis.

CUTICLE (**ku**-tik-l) [L. *cuticula*, dim. of *cutis*, the skin]: Layer of waxy substance (cutin) on outer surface of plant cell walls.

CYCLOSIS (si-**klo**-sis) [Gk. *kyklosis*, circulation]: The circulation of protoplasm within a cell.

—CYTE, CYTO— [Gk. *kytos*, vessel, container]: Suffix or prefix, meaning "pertaining to cell."

CYTOCHROME: Heme-containing protein serving as an electron carrier in electron transport chains, involved in cellular respiration and photosynthesis.

CYTOKINESIS [Gk. *kinesis*, motion]: Division of the cytoplasm of a cell.

CYTOKININ [Gk. *kytos*, vessel + *kinesis*, motion]: One of a group of chemically related plant hormones that promote cell division, among other effects.

CYTOPLASM (**sight**-o-plazm): The living matter of a cell excluding the nucleus.

DECIDUOUS [L. *decidere*, to fall off]: Refers to plants that shed their leaves at a certain season.

DECOMPOSERS: Organisms (bacteria, fungi) in an ecosystem that convert dead organic material into plant nutrients.

DENDRITE [Gk. *dendron*, tree]: Nerve fiber, typically branched, that conducts impulses toward a nerve body.

DEOXYRIBONUCLEIC ACID (DNA) (dee-ox-y-rye-bo-new-**klee**-ick): The carrier of genetic information in cells, composed of two chains of phosphate, sugar molecules (deoxyribose), and purines and pyrimidines wound in a double helix; capable of self-replication as well as of determining RNA synthesis.

DIAPHRAGM [Gk., from *diaphrassein*, to barricade]: A sheetlike muscle forming the partition between the abdominal and thoracic cavities.

DICOTYLEDON (**dye**-cottle-ee-don) [Gk. *kotyledon*, a cup-shaped hollow]: A subclass of angiosperms having two seed leaves or cotyledons, among other distinguishing features; often abbreviated as dicot.

DIFFERENTIATION: The developmental process by which a relatively unspecialized cell or tissue undergoes a progressive change to become a more specialized cell or tissue.

DIFFUSION [L. *diffundere*, to pour out]: The movement of suspended or dissolved particles from a more concentrated to a less concentrated region as a result of the random movement of individual particles; the process tends to distribute them uniformly throughout a medium.

DIGESTION: The conversion of complex, usually insoluble foods into simple, usually soluble forms by means of enzymatic action.

DIOECIOUS (dye-ee-shuss) [Gk. *di*, two + *oikos*, house]: In plants, having the male and female (or staminate and ovulate) elements on different individuals of the same species.

DIPLOID: The chromosome state in which each type of chromosome except for the sex chromosomes is represented twice (2n), in contrast to haploid (1n).

DOMINANT ALLELE: An allele that exerts its full phenotypic effect regardless of its allelic partner.

DORMANCY: A period during which growth ceases and is resumed only if certain requirements, as of temperature or day length, have been fulfilled.

DORSAL [L. *dorsum*, back]: Pertaining to or situated near the back; opposite of ventral.

DOUBLE FERTILIZATION: The fusion of the egg and sperm (resulting in a 2n fertilized egg, the zygote) and the simultaneous fusion of a second male gamete with the polar nuclei (resulting in a 3n primary endosperm nucleus); a unique characteristic of angiosperms.

DUODENUM (duo-**dee**-num) [L. *duodeni*, twelve each—from its length, about 12 fingers' breadth]: The upper portion of the small intestine in vertebrates, where food particles are broken down into molecules that can be absorbed by cells.

ECO— [Gk. *oikos*, house, home]: Prefix, meaning "house" or "home."

ECOLOGICAL DOMINANT: A species that, by virtue of size, number, or behavior, exerts a controlling influence on its environment and, as a result, determines what other kinds of organisms exist in that ecosystem.

ECOLOGICAL NICHE: The place occupied by a species in the community structure of which it is a part; the way in which an organism utilizes the resources of its ecosystem.

ECOLOGICAL SUCCESSION: The process by which a community goes through a number of temporary developmental stages, each characterized by a different species composition, before reaching a more stable condition, known as a climax community.

ECOLOGY [Gk. *logos*, a discourse]: The study of the interactions of organisms with their physical environment and with each other and of the results of such interactions.

ECOSYSTEM: All organisms in a community plus the associated environmental factors with which they interact.

ECTO— [Gk. *ektos*, outside]: Prefix, meaning "outside" or "outer."

ECTODERM [Gk. *derma*, skin]: (1) The outermost layer of body tissue. (2) One of the three cell layers of the gastrula that give rise to all the tissues of the organism; the ectoderm gives rise to the epithelial tissue of the skin and sense organs, to the nerve cells, etc.

EFFECTOR: Cell, tissue, or organ (such as muscle or gland) capable of producing a response to stimuli.

EFFERENT [L. *ex*, out + *ferre*, to bear]: Carrying away from a center, applied to nerves and blood vessels.

EGG: A female gamete, or germ cell, which usually contains abundant cytoplasm and yolk; usually immotile, often larger than a male gamete.

ELECTRON: A subatomic particle with a negative electric charge equal in magnitude to the positive charge of the proton but with a mass 1/1,837 that of the proton; normally orbits the atom's positively charged nucleus.

ELECTRON ACCEPTOR: Substance acting to receive electrons in an oxidation-reduction reaction.

ELECTRON CARRIER: A specialized protein, such as a cytochrome, that can gain and lose electrons reversibly and that functions to transfer electrons from organic nutrients to oxygen.

ELECTRON DONOR: Substance acting to donate electrons in an oxidation-reduction reaction.

ELEMENT: A substance composed of only one kind of atom; one of about 100 distinct natural or man-made types of matter which, singly or in combination, compose all materials of the universe.

EMBRYO [Gk. *en*, in + *bryein*, to swell]: The early developmental stage of an organism produced from a fertilized egg; a young organism before it emerges from the seed, egg, or the body of its mother.

ENDERGONIC: Energy-requiring, as in a chemical reaction; applied to an "uphill" process.

ENDO— [Gk. *endon*, within]: Prefix, meaning "within."

ENDOCRINE GLAND [Gk. *krinein*, to separate]: Ductless gland whose secretions (hormones) are released into the circulatory system; in vertebrates, includes pituitary, sex glands, adrenal, thyroid, and others.

ENDODERM [Gk. *derma*, skin]: In animals, (1) the innermost layer of body tissue; (2) one of the three cell layers of the gastrula that give rise to all the tissues of the organism; the endoderm gives rise to the epithelium that lines certain internal structures, such as most of the digestive tract and its outgrowths, most of the respiratory tract, and the urinary bladder.

ENDODERMIS: In plants, a one-celled layer of specialized cells that lies between the cortex and the vascular tissues in young roots. The Casparian strip of the endodermis prevents diffusion of materials across the root.

ENDOMETRIUM: The glandular lining of the uterus in mammals; thickens in response to progesterone secretion during ovulation and is sloughed off in menstruation.

ENDOPLASMIC RETICULUM [L. *reticulum*, network]: An extensive system of double membranes present in most cells, dividing the cytoplasm into compartments and channels, often coated with ribosomes.

ENDOSPERM [Gk. *sperma*, seed]: In plants, a 3n tissue containing stored food that develops from the union of a male nucleus and the polar bodies of the egg; found only in angiosperms.

ENTROPY: The randomness or disorder of a system.

ENZYME: A protein molecule that regulates the rate of (catalyzes) a chemical reaction.

EPIDERMIS [Gk. *derma*, skin]: In plants and animals, the outermost layers of cells.

EPINEPHRINE: A hormone produced by the medulla of the adrenal gland which increases the concentration of sugar in the blood, raises blood pressure and heartbeat rate, and increases muscular power and resistance to fatigue; also a chemical transmitter across synaptic junctions. Also called adrenaline.

EPITHELIAL TISSUE [Gk. *thele*, nipple]: In animals, a type of tissue that covers a body or structure or lines a cavity; epithelial cells form one or more regular layers with little intercellular material.

EQUILIBRIUM: The state of a system in which no further net change is occurring.

ERYTHROCYTE (eh-**rith**-ro-site) [Gk. *erythros*, red + *kytos*, vessel]: Red blood cell, the carrier of hemoglobin.

ESTRUS [Gk. *oistros*, frenzy]: The mating period in female mammals, characterized by intensified sexual urge.

EUKARYOTE (you-**car**-ry-oat) [Gk. *eu*, good + *karyon*, nut, kernel]: A cell having a membrane-bound nucleus, membrane-surrounded organelles, and chromosomes in which the DNA is combined with special proteins; an organism composed of such cells.

EUSOCIAL: Applied to insect societies in which sterile individuals work on behalf of individuals involved in reproduction.

EVOLUTION [L. *e-*, out + *volvere*, to roll]: Any change in the gene pool from one generation to the next; Darwinian evolution is the result of natural selection operating on random genetic variations.

EXERGONIC: Energy-producing, as in a chemical reaction; applied to a "downhill" process.

EXOSKELETON: A skeleton covering the outside of the body; common in arthropods.

EXTRAEMBRYONIC MEMBRANES: Membranes formed of embryonic tissues that lie outside the embryo and are concerned with its protection and metabolism; include amnion, chorion, allantois, and yolk sac.

F_1 (FIRST FILIAL GENERATION): The offspring resulting from the crossing of plants or animals of the parental generation.

F_2 (SECOND FILIAL GENERATION): The offspring resulting from crossing members of the F_1 generation among themselves.

FALLOPIAN TUBES: *See* Oviduct.

FERTILIZATION: The fusion of two gametes, and so of their nuclei, to form a diploid or polyploid zygote.

FETUS [L., pregnant]: An unborn or unhatched vertebrate that has passed through the earliest developmental stages; a developing human from about the third month after conception until birth.

FIBRIL: Any minute, threadlike structure within a cell.

FIBROUS PROTEIN: Insoluble structural protein, in which the polypeptide chain is extended or coiled along one dimension.

FILAMENT: (1) A chain of functionally separate cells. (2) In plants, the stalk of a stamen.

FITNESS: The relative ability to leave offspring.

FLAGELLUM, *pl.* FLAGELLA (fla-**jell**-um) [L. *flagellum*, whip]: A fine, long, threadlike structure, which protrudes from a cell body; it is longer than a cilium but has the same internal structure of nine pairs of microtubules encircling two central microtubules. Flagella are used in locomotion and feeding.

FLOWER: The reproductive structure of angiosperms; a complete flower includes calyx, petals, stamens (male sex organs), and carpels (female sex organs), but some of these are absent in flowers of many species.

FOOD CHAIN, FOOD WEB: A sequence of organisms, including producers, herbivores, and carnivores, through which energy and materials move within an ecosystem.

FOSSIL [L. *fossilis*, dug up]: The remains of an organism, or direct evidence of its presence (such as tracks). May be an unaltered hard part (tooth or bone), a mold in a rock, petrifaction (wood or bone), unaltered or partially altered soft parts (a frozen mammoth).

FOSSIL FUELS: The remains of once-living organisms that are burned to release energy. Examples: coal, oil, and natural gas.

FREE ENERGY: That part of the total energy of a system that can do work under conditions of constant temperature and pressure.

FRUIT [L. *fructus*, fruit]: In plants, a matured, ripened ovary or group of ovaries and associated structures; formed from the ovule case of an angiosperm; contains the seeds.

FUNCTION: Characteristic role or action of any structure or process in the maintenance of normal metabolism or behavior of an animal.

GAMETE (**gam**-meet) [Gk., wife]: The mature functional haploid reproductive cell whose nucleus fuses with that of another gamete of an opposite sex (fertilization), with the resulting cell (zygote) developing into a new individual.

GAMETOPHYTE (gam-**meet**-o-fight): In plants having alternation of generations, the haploid ($1n$) gamete-producing generation.

GANGLION, *pl.* GANGLIA (**gang**-lee-on): Aggregate of nerve cell bodies.

GASTRULA [Gk. *gaster*, stomach]: An embryo in the process of gastrulation, the stage of development during which the blastula with its single layer of cells turns into a three-layered embryo, made up of ectoderm, mesoderm, and endoderm, often enclosing an archenteron.

GAUSE'S PRINCIPLE [after G. F. Gause, German geneticist]: The hypothesis that any two species with the same ecological requirements cannot coexist in the same locality and that occupation of that space will go to the species that is more efficient in utilizing the available resources.

GENE: A unit of heredity which is transmitted in the chromosome and which by interaction with internal and external environment controls the development of a trait; capable of self-replication. The sequence of nucleotides in a DNA molecule that dictates the nucleotide sequence of an RNA molecule.

GENE FREQUENCY: The incidence (relative occurrence) of a particular allele in a population.

GENE POOL: All the alleles of all the genes in a population.

GENETIC CODE: The three-symboled system of base-pair sequences in DNA; referred to as a code because it determines the amino acid sequence in the enzymes and other protein components synthesized by the organism.

GENETIC ISOLATION: The inhibition of gene exchange between different species by morphological, behavioral, or physiological mechanisms.

GENOTYPE (**jean**-o-type): The genetic constitution, latent or expressed, of an organism, as contrasted with the phenotype; the sum total of all the genes present in an individual.

GENUS (**jean**-us): Inclusive group of related species.

GERM CELLS: Gametes or the direct antecedent cells of gametes.

GERMINATION: The resumption of growth.

GIBBERELLINS (jibb-e-**rell**-ins) [Fr. *gibberella*, genus of fungi]: A group of chemically related plant growth hormones, whose most characteristic effect is stem elongation in dwarf plants and bolting.

GILL: The respiratory organ of aquatic animals, usually a thin-walled projection from some part of the external body surface or, in vertebrates, from some part of the digestive tract.

GLAND: A cell or organ producing one or more secretions which are discharged to the outside of the gland.

GLOBULAR PROTEIN: Protein in which the polypeptide chain is folded in three dimensions to form a globular shape.

GLOMERULUS (glom-**mare**-u-lus) [L. *glomus*, ball]: Cluster of capillaries enclosed by the Bowman's capsule; blood plasma minus large molecules filters through the walls of the glomerular capillaries into the renal tubules.

GLUCAGON: Hormone produced in pancreas that acts to raise concentration of blood sugar.

GLUCOSE: A six-carbon sugar ($C_6H_{12}O_6$); the most common monosaccharide in animals.

GLYCOGEN: A complex carbohydrate (polysaccharide); one of the main stored food substances of most animals and fungi; it is converted into glucose by hydrolysis.

GLYCOLYSIS (gly-**coll**-y-sis): The process by which the glucose molecule is changed anaerobically to two molecules of pyruvic acid with the liberation of a small amount of useful energy; reaction takes place in the cytoplasm.

GOLGI BODY (**goal**-jee): An organelle present in eukaryotic cells consisting of flat, disk-shaped sacs, tubules, and vesicles. It functions as a collecting and packaging center for substances the cell manufactures.

GONAD [Gk. *gone*, seed]: Gamete-producing organ of multicellular animals; ovary or testes.

GRANUM, *pl.* GRANA: A structure within the choloroplast, seen as a green granule with the light microscope and as a series of stacked thylakoids with the electron microscope. The grana contain chlorophylls and carotenoids and are the site of the light reactions of photosynthesis.

GROUNDWATER: Water in the zone of saturation where all openings in rocks and soil are filled, the upper surface of which forms the water table.

HABITAT: The place normally occupied by a particular type of organism.

HALF-LIFE: The time required for disappearance or decay of one-half of a given substance.

HAPLOID [Gk. *haploos*, single]: Having only one of each type of chromosome ($1n$), in contrast to diploid ($2n$); characteristic of gametes and of the gametophyte generation in plants and of some microorganisms.

HARDY-WEINBERG LAW: The mathematical expression of the relationship between relative frequencies of two or more alleles in a population; it demonstrates that the frequencies of alleles in a gene pool are not changed by the process of sexual recombination.

HEME: The iron-porphyrin group of heme proteins.

HEMO—, HEMATO— [Gk. *haima*, blood]: Prefix, meaning "blood."

HEMOGLOBIN: The iron-containing protein in the blood that carries oxygen.

HEMOPHILIA: A hereditary disease in man characterized by failure of the blood to clot and excessive bleeding from even minor wounds.

HERBACEOUS (her-**bay**-shus): In plants, nonwoody.

HERBIVORE: An organism that eats plants or other photosynthetic organisms to obtain its food energy.

HEREDITY: The transfer of characteristics from parent to offspring by the transmission of genes from ancestor to descendent through the germ cells.

HERMAPHRODITE [Gk. *Hermes* and *Aphrodite*]: An organism possessing both male and female reproductive organs.

HETERO— [Gk. *heteros*, other, different]: Prefix, meaning "other" or "different."

HETEROTROPH [Gk. *trophos*, feeder]: An organism that cannot manufacture organic compounds and so must feed on complex organic food materials that have originated in other plants and animals; in contrast to autotroph.

HETEROZYGOTE: A diploid organism that has two different alleles of the same gene.

HETEROZYGOTE SUPERIORITY: The greater fitness of a heterozygote as compared with the two homozygotes.

HIBERNATION: A period of dormancy and inactivity, varying in length depending on the organism and occurring in dry or cold seasons. During hibernation, metabolic processes are greatly slowed and, in mammals, body temperature may drop to just above freezing.

HOMEO—, HOMO—, HOMOLO— [Gk. *homos*, same, similar]: Prefix, meaning "similar" or "same."

HOMEOSTASIS (home-e-o-**stay**-sis) [Gk. *stasis*, standing]: The maintaining of a relatively stable internal physiological environment or equilibrium in an organism, population, or ecosystem.

HOMEOTHERM: Organism capable of maintaining a uniform body temperature independent of the environment; warm-blooded. Birds and mammals are homeotherms.

HOMINID: A member of the family of man; includes modern man and fossil man but not the apes.

HOMINOID: A hominid or one of the great apes.

HOMOLOGUES: Chromosomes that associate in pairs in the first stage of meiosis; each member of the pair is derived from a different parent.

HOMOLOGY [Gk. *homologia*, agreement]: Similarity in structure resulting from a common ancestry, regardless of function, such as the wing of a bird and the foreleg of a mammal.

HOMOZYGOTE: A diploid organism that has identical alleles of a particular gene.

HORMONE [Gk. *hormaein*, to excite]: A chemical substance secreted, usually in minute amounts, in one part of an organism and transported to another part of that organism where it has a specific effect.

HOST: An organism on or in which a parasite lives.

HYBRID: Offspring of two parents that differ in one or more heritable characters; offspring of two different varieties or of two different species.

HYDROGEN BOND: A weak molecular bond linking a hydrogen atom that is covalently bonded to another atom—usually oxygen, nitrogen, or fluorine—to the oxygen, nitrogen, or fluorine atom of another molecule.

HYDROLYSIS [Gk. *hydro*, water + *lysis*, loosening]: Splitting of one molecule into two by addition of H^+ and OH^- ions of water.

HYPER— [Gk., above, over]: Prefix, meaning "above" or "over."

HYPERTONIC [Gk. *hyper*, above + *tonos*, tension]: Having a concentration of solutes high enough to gain water across a selectively permeable membrane from another solution.

HYPHA [Gk. *hyphe*, web]: A single tubular filament of a fungus; the hyphae together make up the mycelium.

HYPO— [Gk., less than]: Prefix, meaning "under" or "less."

HYPOTHALAMUS [Gk. *thalamos*, inner room]: The floor and sides of the vertebrate brain just below the cerebral hemispheres; controls the autonomic nervous system and the pituitary gland and contains centers that regulate body temperature and appetite.

HYPOTHESIS [Gk. *hypo-*, under + *tithenai*, to put]: A temporary working explanation or supposition based on accumulated facts and suggesting some general principle or relation of cause and effect; a postulated solution to a scientific problem that must be tested by experimentation and, if not validated, discarded.

HYPOTONIC: Having a concentration of solutes low enough to lose water across a membrane to another solution.

IMPRINTING: A rapid and extremely narrow form of learning which occurs during a very short period in the early life of an organism, such as the following response in certain birds.

INBREEDING: The mating of individuals closely related genetically.

INDEPENDENT ASSORTMENT: *See* Mendel's second law.

INSTINCT: A genetically determined pattern of behavior or response not based on the previous experience of the individual.

INSULIN: Hormone produced by the pancreas that lowers the concentration of sugar in the blood.

INTER— [L., between]: Prefix, meaning "between"; for example, intercellular, "between cells."

INTERPHASE: The stage between two mitotic or meiotic cycles.

INTRA— [L., within]: Prefix, meaning "within"; for example, intracellular, "within cells."

INVAGINATION [L. *in*, in + *vagina*, sheath]: The local infolding of a layer of tissue, especially in animal embryos, so as to form a depression or pocket opening to the outside.

ION (**eye**-on): An atom or molecule that has lost or gained one or more electrons. By this process, known as ionization, the atom becomes electrically charged.

ISO— [Gk. *isos*, equal]: Prefix, meaning "equal."

ISOGAMY: A type of sexual reproduction in algae and fungi in which the gametes are alike in size.

ISOLATING MECHANISMS: Mechanisms that prevent gene exchange between different species; may be morphological, behavioral, or physiological.

ISOTONIC: Having the same solute concentration as that of the substance on the other side of a selectively permeable membrane.

ISOTOPE [Gk. *topos*, place]: One of several possible forms of a chemical element, differing from other forms in the number of neutrons in the atomic nucleus and so in mass; some are unstable and emit radioactivity.

KARYOTYPE: The general appearance of the chromosomes with regard to number, size, and shape.

KERATIN [Gk. *keras*, horn]: One of a group of tough, fibrous proteins; a horny tissue formed by certain epidermal tissues, especially abundant in skin, claws, hair, feathers, and hooves.

KIDNEY: In vertebrates, the organ that regulates the balance of water and solutes and the excretion of nitrogen wastes in the form of urine.

KINETOCHORE: Region of constriction of chromosomes that holds sister chromatids together and to which the spindle fibers attach; also called the centromere.

LAMELLA (lah-**mell**-ah) [L. dim. of *lamina*, plate or leaf]: Layer, thin sheet.

LARVA [L., ghost]: An immature animal such as a caterpillar or tadpole that is morphologically very different from the adult.

LATERITE: A leached tropical soil with a high iron and aluminum concentration.

LEACHING: The removal of minerals and other elements from soil by the downward movement of water.

LEARNING: The process that produces adaptive change in individual behavior as the result of experience.

LEUKOCYTE [Gk. *leukos*, white + *kytos*, vessel]: White blood cell. Two of the principal types are neutrophiles, active in phagocytosis, and lymphocytes, involved in immune reactions.

LEUKOPLAST [Gk. *plastes*, molder]: In plants, a colorless cell organelle that serves as a starch repository; usually found in cells not exposed to light, such as roots and internal stem tissue.

LICHEN: Organism composed of a symbiotic alga and fungus.

LIFE CYCLE: The entire sequence of phases in the growth and development of any organism from time of zygote formation until gamete formation.

LIMITING FACTOR: Any environmental factor that determines whether or not an organism will grow in a particular environment.

LINKAGE: The tendency for certain genes to be inherited together owing to the fact that they are located on the same chromosome.

LIPID [Gk. *lipos*, fat]: One of a large variety of organic fat or fatlike compounds; includes fats, waxes, steroids, phospholipids, and carotenes.

LOCUS, *pl.* LOCI [L., place]: In genetics, the position of a gene on a chromosome.

LOOP OF HENLE [after F. G. J. Henle, German pathologist]: A hairpin-shaped portion of the kidney tubule system found in birds and mammals in which a hypertonic urine is formed by processes of osmosis and active transport.

LYMPH: Colorless fluid occurring in special lymph ducts, derived from blood by filtration through capillary walls.

—LYSIS, —LYTIC, —LYTE [Gk. *lysis*, a loosening]: Suffix, meaning "pertaining to dissolving."

LYSIS: Disintegration of cell.

LYSOGENIC BACTERIA (lye-so-**jenn**-ick): Bacteria carrying viruses (bacteriophages) which eventually break loose from the bacterial chromosome and set up an active cycle of infection, producing lysis in their bacterial hosts.

LYSOSOME: A membrane-surrounded organelle in which hydrolytic enzymes are segregated.

MACRO— [Gk. *makros*, large, long]: Prefix, meaning "large" or "long"; opposite of "micro-."

MACROMOLECULE: A molecule of very high molecular weight; refers specifically to proteins, nucleic acids, polysaccharides, and complexes of these.

MANTLE: (1) In mollusks, the outermost layer of the body wall or a soft extension of it; usually secretes a shell. (2) In geology, the region of earth below the crust extending down to the molten outer core.

MARINE [L. marini(us), from mare, the sea]: Living in salt water.

MARSUPIAL [Gk. marsypos, pouch, little bag]: A nonplacental mammal in which the female has a ventral pouch or folds surrounding the nipples; the premature young leave the uterus and crawl into the pouch, where each one attaches itself by the mouth to a nipple until development is completed.

MEDULLA (med-dull-a) [L., the innermost part]: (1) The inner as opposed to the outer part of an organ, as in the adrenal gland. (2) The most posterior region of the vertebrate brain; connects with the spinal cord.

MEDUSA: The free-swimming bell- or umbrella-shaped stage in the life cycle of many coelenterates; a jellyfish.

MEGA— [Gk. megas, great, large]: Prefix, meaning "large."

MEGASPORE: In plants, a haploid (1n) spore that develops into a female gametophyte.

MEIOSIS (my-o-sis) [Gk. meioun, to make smaller]: The two successive nuclear divisions in which the chromosome number is reduced from diploid (2n) to haploid (1n) and segregation and reassortment of the genes occur; gametes or spores may be produced as a result of meiosis.

MENDEL'S FIRST LAW: The factors for a pair of alternative characters are separate and only one may be carried in a particular gamete (genetic segregation).

MENDEL'S SECOND LAW: The inheritance of a pair of factors for one trait is independent of the simultaneous inheritance of factors for other traits, such factors "assorting independently" as though there were no other factors present (later modified by the discovery of linkage).

MENSTRUAL CYCLE [L. mensis, mouth]: In certain primates, the periodic discharge of blood and disintegrated uterine lining through the vagina.

MERI—, MERO—, —MER: Prefix or suffix, meaning "part" or "portion."

MERISTEM: The undifferentiated plant tissue from which new cells arise.

MESODERM [Gk. derma, skin]: One of the three primary cell layers in the early embryo that give rise to the tissues of the organism; the mesoderm gives rise to muscle, connective tissue, the circulatory system, and most of the excretory and reproductive systems.

MESSENGER RNA (mRNA): The RNA that carries genetic information from the gene to the ribosome, where it determines the order of the amino acids in the formation of a polypeptide.

METABOLISM [Gk. metabole, change]: The sum of all chemical reactions occurring within a living unit.

METAMORPHOSIS [Gk. metamorphoun, to transform]: Abrupt transition from larval to adult form, such as the transition from tadpole to adult frog.

METAPHASE [Gk. meta, middle + phasis, form]: The stage of mitosis or meiosis during which the chromosomes lie in the central plane of the spindle.

MICRO— [Gk. mikros, small]: Prefix, meaning "small."

MICRONUTRIENT: A mineral required in only minute amounts for plant growth, such as iron, chlorine, copper, manganese, zinc, molybdenum, and boron.

MICROSPORE: A spore which develops into a male gametophyte; in seed plants, it becomes a pollen grain.

MIDDLE LAMELLA: Distinct layer between adjacent cell walls, rich in pectic compounds, derived from the cell plate.

MIMICRY [Gk. mimos, mime]: The superficial resemblance in form, color, or behavior of certain organisms (mimics) to other more powerful or more protected ones (models), resulting in protection, concealment, or some other advantage for the mimic.

MINERAL: A naturally occurring chemical element or inorganic compound.

MITOCHONDRION, pl. MITOCHONDRIA: An organelle bound by a double membrane in which energy is captured in the form of ATP in the course of cellular respiration.

MITOSIS [Gk. mitos, thread]: Nuclear division characterized by exact chromosome duplication and the formation of two identical daughter cells; the means by which somatic cells divide.

MOLECULAR WEIGHT: The sum of the atomic weights of the constituent atoms in a molecule.

MOLECULE [L. moles, mass]: Smallest possible unit of a compound substance, consisting of two or more atoms.

MOLTING: Shedding of all or part of outer covering; in arthropods, periodic shedding of the exoskeleton to permit an increase in size.

MONO— [Gk. monos, single]: Prefix, meaning "one'" or "single."

MONOCOTYLEDON [Gk. kotyledon, a cup-shaped hollow]: A subclass of angiosperms, characterized by a variety of features, among which is the presence of a single seed leaf (cotyledon); abbreviated as monocot.

MONOECIOUS (mo-nee-shuss) [Gk. monos, single + oikos, house]: In plants, having the anthers and carpels on the same individual but on different flowers.

MONOMER [Gk. meros, part]: A simple molecule of relatively low molecular weight that can be linked to others to form a polymer.

MONOSACCHARIDE [Gk. sakcharon, sugar]: A simple sugar, such as the five- and six-carbon sugars.

—MORPH, MORPHO— [Gk. morphe, form]: Suffix or prefix, meaning "form."

MORPHOGENESIS: The development of size, form, and other structural features of organisms.

MORPHOLOGY: The study of form and structure, at any level of organization.

MOTOR NEURON: Neuron that transmits nerve impulses from the central nervous system to skeletal muscle; efferent neuron.

MUSCLE FIBER: Muscle cell; a long, cylindrical, multinucleated cell containing numerous myofibrils.

MUTAGEN: An agent that increases the mutation rate.

MUTANT: A mutated gene or an organism carrying a gene that has undergone a mutation.

MUTATION [L. from *mutare,* change]: An inheritable change in the chromosomes; usually a change of a gene from one allelic form to another.

MUTUALISM: *See* Symbiosis.

MYCELIUM [Gk. *mykes,* fungus]: The mass of hyphae forming the body of a fungus.

MYELIN SHEATH: A fatty material surrounding the axons of some nerve cells in vertebrates; made up of the membranes of Schwann cells.

MYO— [Gk. *mys,* muscle]: Prefix, meaning "muscle."

MYOFIBRIL: Contractile element of a muscle fiber, made up of thick and thin filaments arranged in sarcomeres.

MYOSIN: One of the principal proteins in muscle; makes up the thick filaments.

MYX,— MYXO— [Gk. *myxa,* slime]: Prefix, meaning "slime."

NAD: Abbreviation of nicotinamide adenine dinucleotide, a coenzyme that functions as a hydrogen acceptor.

NATURAL SELECTION: The nonrandom reproduction of genotypes, resulting from interactions among a variety of phenotypes and the environment.

NECTAR [Gk. *nektar,* the drink of the gods]: A sugary fluid that attracts insects to plants.

NEPHRIDIUM: A type of excretory organ found in many invertebrates.

NEPHRON [Gk. *nephros,* kidney]: A unit of the kidney structure in reptiles, birds, and mammals; a human kidney contains about 1 million nephrons.

NERVE: A group or bundle of nerve fibers with accompanying connective tissue.

NERVE FIBER: A filamentous process of a neuron; either dendrite or axon.

NERVE IMPULSE: A rapid, transient change in electrical potential propagated along a nerve fiber from one part of an animal to another.

NERVOUS SYSTEM: All the nerve cells of an animal; the receptor-conductor-effector system; in man, the nervous system consists of brain, spinal cord, and all nerves.

NEUROHORMONE: A hormone secreted by a nerve ending; neurohormones include adrenaline, noradrenaline, and acetylcholine.

NEURON: Nerve cell, including cell body, dendrites, and axon.

NEUROSECRETORY CELL: A neuron that produces one or more hormones.

NEUTRON (**new**-tron): An uncharged particle with a mass slightly greater than that of a proton. Found in the atomic nucleus of all elements except hydrogen, in which the nucleus consists of a single proton.

NICHE: *See* Ecological niche.

NITRIFY: To convert organic nitrogen compounds into ammonium compounds, nitrates, and nitrites, as by nitrifying bacteria and fungi.

NITROGEN BASE: A nitrogen-containing molecule having basic properties (tendency to acquire an H atom); a purine or pyrimidine.

NITROGEN CYCLE: Worldwide circulation and reutilization of nitrogen atoms, chiefly due to metabolic processes of living organisms; plants take up inorganic nitrogen and convert it into organic compounds (chiefly proteins) which are assimilated into the bodies of one or more animals; excretion and bacterial and fungal action on dead organisms return nitrogen atoms to the inorganic state.

NITROGEN FIXATION: Incorporation of atmospheric nitrogen into inorganic nitrogen compounds available to plants, a process that can be carried out only by certain microorganisms, or by certain plants in symbiotic association with microorganisms.

NODE [L. *nodus,* knot]: In plants, a joint of a stem; the place where branches and leaves are joined to the stem.

NONDISJUNCTION: The failure of homologous chromosomes to separate during meiosis, resulting in one or more extra chromosomes in the cells of some offspring.

NOTOCHORD: A longitudinal, solid, elastic, rodlike structure serving as the internal skeleton in the embryos of all chordates; in most adult chordates the notochord is replaced by a vertebral column which forms around (but not from) the notochord.

NUCLEAR ENVELOPE: The double membrane surrounding the nucleus within a cell.

NUCLEIC ACID: An organic acid consisting of nucleotides; the principal types are deoxyribonucleic acid (DNA) and ribonucleic acid (RNA).

NUCLEOLUS (new-**klee**-o-lus) [L. *nucleolus,* a small kernel]: A spherical body, containing DNA, RNA, and protein, present in nucleus of eukaryotic cells; site of production of ribosomal RNA.

NUCLEOTIDE: A single unit of nucleic acid composed of phosphate, a five-carbon sugar (either ribose or deoxyribose), and a purine or a pyrimidine.

NUCLEUS [L. *nucleus,* a kernel]: (1) The cellular structure that contains the genetic information in the form of DNA. In the eukaryotic cell, a specialized body bound by a double membrane, containing the chromosomes. (2) The central part of an atom. (3) A group of nerve cell bodies in the central nervous system.

OCELLUS, *pl.* OCELLI [L. dim. of *oculus,* eye]: A simple light receptor common among invertebrates.

OLFACTORY [L. *olfacere,* to smell]: Pertaining to smell.

OMMATIDIUM, *pl.* OMMATIDIA [Gk. *ommos,* eye]: The single visual unit in the compound eye of arthropods; contains light-sensitive cells and a refractory system able to form an image.

OMNI— [L. *omnis,* all]: Prefix meaning "all."

OMNIVOROUS: Eating "everything," for example, using both plants and animals as food.

ONTOGENY [Gk. *on,* being + *genesis,* origin]: The developmental history of an individual organism from zygote to maturity.

OO— [Gk. *oion,* egg]: Prefix, meaning "egg."

OOCYTE (**o**-uh-sight) [Gk. *oion,* egg + *kytos,* vessel]: A cell that gives rise by meiosis to an ovum.

OPERATOR: A segment of DNA to which a repressor protein is attached, thus controlling the function of an operon.

OPERON: A group of adjacent structural genes whose functions are related to a particular biochemical pathway and which are controlled by a single repressor protein.

OPPOSABLE THUMB: Thumb that rotates at the joint so that the tip of the thumb can be placed opposite the tip of any one of the four fingers.

ORGAN [Gk. *organon*, tool]: A body part composed of several tissues grouped together in a structural and functional unit.

ORGANELLE: A formed body in the cytoplasm of a cell.

ORGANIC: Pertaining to (1) organisms or living things generally, or (2) compounds formed by living organisms, or (3) the chemistry of compounds containing carbon.

ORGANISM: Any individual living creature, either unicellular or multicellular.

OSMOSIS [Gk. *osmos*, impulse, thrust]: The movement of water between two solutions separated by a membrane that permits the free passage of water and that prevents or retards the passage of the solute; the water tends to move from the side containing a lesser concentration of solute to the side containing a greater concentration.

OSMOTIC PRESSURE: Pressure generated by osmotic flow of water.

OV—, OVI— [L. *ovum*, egg]: Prefix, meaning "egg."

OVARY: (1) In animals, the egg-producing organ. (2) In flowering plants, the enlarged basal portion of a carpel or a fused carpel, containing the ovule or ovules; the ovary matures to become the fruit.

OVIDUCT [L. *ductus*, duct]: The tube serving to transport the eggs to the uterus or to the outside; Fallopian tubes (in humans).

OVULATION: In animals, release of an egg or eggs from the ovary.

OVULE: In seed plants, the megasporangium, a structure composed of a protective outer coat, a tissue specialized for food storage, and a female gametophyte with egg cell; becomes a seed after fertilization.

OVUM, *pl.* OVA: The egg cell; female gamete.

OXIDATION: Loss of an electron by an atom. Oxidation and reduction (gain of an electron) take place simultaneously since an electron that is lost by one atom is accepted by another. Oxidation-reduction reactions are an important means of energy transfer within living systems.

PACEMAKER: Area of heart that initiates the heartbeat; located where the superior vena cava enters the right atrium; sinoatrial node.

PALEO— [Gk. *palaios*, old]: Prefix, meaning "old."

PALEONTOLOGY: The study of the life of past geologic times, principally by means of fossils.

PALISADE CELLS: In plants, columnar chloroplast-containing cells [part of the mesophyll ("middle leaf")].

PANCREAS (**pang**-kree-us) [Gk. *pan*, all + *kreas*, meat, flesh]: A small complex gland in vertebrates, located between the stomach and the duodenum, which produces digestive fluids and the hormones insulin and glucagon.

PARASITE [Gk. *para*, beside + *sitos*, food]: An organism that lives on or in an organism of a different species and derives nutrients from it. *See* Symbiosis.

PARASYMPATHETIC NERVOUS SYSTEM: A subdivision of the autonomic nervous system of vertebrates with centers located in the brain and in the most anterior part and most posterior parts of the spinal cord; stimulates digestion; generally inhibits other functions. Chemical transmitter: acetylcholine.

PARENCHYMA (pah-**renk**-ee-ma) [Gk. *en*, in + *chein*, to pour]: A plant tissue composed of living, thin-walled, randomly arranged cells with large vacuoles; usually photosynthetic or storage tissue.

PECKING ORDER: The dominance hierarchy in poultry flocks in which each individual according to its rank can peck a number of subordinate fowl with impunity and must submit to pecking by a number of superiors without retaliation.

—PED, —PEDIA, PEDI— [L. *pes, pedis*, foot]: Suffix or prefix, meaning "foot"; like the Greek suffix "-pod"; for example, bipedal, "two-footed."

PELLICLE [L. dim. of *pellis*, skin]: A thin translucent envelope located outside the cell membrane in many protozoans.

PENT—, PENTA— [Gk. *pente*, five]: Prefix, meaning "five"; for example, pentose, "five-carbon sugar."

PEPTIDE: Two or more amino acids linked together; molecules made up of a relatively small number of amino acids (2 to about 100) are called peptides, while those formed of a larger number of amino acids are called polypeptides or proteins.

PEPTIDE BOND: The type of bond formed when two amino acid units are joined end to end; the acidic group ($-COOH$) of one amino acid is attached to the basic group ($-NH_2$) of the next, and a molecule of water (H_2O) is removed.

PERENNIAL [L. *per*, through + *annus*, year]: A plant that persists in whole or in part from year to year and flowers in more than one year.

PERI— [Gk., around]: Prefix, meaning "around"; for example, peristomal, "around the mouth"; peristalsis, "wavelike compression around a tubular organ," such as the gut.

PERITONEUM [Gk. *peritonos*, stretched over]: A mesodermal epithelial membrane lining of the body cavity and forming the external covering of the visceral organs.

PERMEABLE [L. *permeare*, to pass through]: Penetrable, usually applied to membranes which let given substances pass through.

PETIOLE (**pet**-ee-ole): The stalk of a leaf.

pH: A symbol denoting the relative concentration of hydrogen ions in a solution; pH values range from 0 to 14; the lower the value, the more acid a solution, that is, the more hydrogen ions it contains; pH 7 is neutral, less than 7 is acid, more than 7 is alkaline.

PHAGO—, —PHAGE [Gk. *phagein*, to eat]: Prefix or suffix, meaning "eating."

PHAGOCYTE (**fag**-o-sight): Any cell that engulfs foreign particles.

PHAGOCYTOSIS: Cell "eating"; the intake of solid particles by a cell, by flowing over and engulfing them; characteristic of amoebas, digestive cells of some invertebrates, and vertebrate white blood cells.

PHENOTYPE [Gk. *phainein*, to show]: The observable properties of an organism resulting from interaction between its genetic constitution (genotype) and the environment.

PHEROMONE (**fair**-o-moan): Substance secreted by an animal that influences the behavior or morphological development or both of other animals of the same species, such as sex attractants of moths, odor trail of ants.

PHLOEM (**flow**-em) [Gk. *phloos*, bark]: Vascular tissue that conducts sugars and other organic molecules from the leaves to other parts of the plant; composed of sieve cells (in gymnosperms) or sieve tubes and companion cells (in angiosperms), parenchyma, and fibers.

PHORO—, —PHORE [Gk. -*phoros*, bearing, carrying]: Prefix or suffix, meaning "bearing."

PHOTO— [Gk. *photos*, light]: Prefix, meaning "light."

PHOTON: Unit of light energy.

PHOTOPERIODISM: The response to relative day and night length, a mechanism by which organisms measure seasonal change.

PHOTOPHOSPHORYLATION: Formation of ATP molecules in the chloroplast using radiant energy.

PHOTORECEPTOR: A cell or organ capable of detecting light.

PHOTOSYNTHESIS [Gk. *syn*, together + *tithenai*, to place]: The conversion of light energy to chemical energy; the production of carbohydrate from carbon dioxide and water in the presence of chlorophyll, using light energy.

PHOTOTROPISM [Gk. *trope*, turning]: Movement in which the direction of the light is the determining factor, such as the growth of a plant toward a light source; turning or bending response to light.

PHYLOGENY [Gk. *phylon*, race, tribe]: Evolutionary relationships among organisms; developmental history of a group of organisms.

PHYSIOLOGY [Gk. *physis*, nature + *logos*, a discourse]: The study of function in cells, organs, or entire organisms.

PHYTO—, —PHYTE [Gk. *phyton*, plant]: Prefix or suffix, meaning "plant."

PHYTOCHROME: A pigment found in plants. It is a photoreceptor for red or far-red light and is involved with a number of developmental processes, such as flowering, dormancy, leaf formation, and seed germination.

PHYTOPLANKTON [Gk. *planktos*, wandering]: Aquatic free-floating microscopic photosynthetic organisms.

PIGMENT: A substance that absorbs light.

PINOCYTOSIS [Gk. *pinein*, to drink]: Cell "drinking"; the intake of fluid droplets by a cell.

PITUITARY [L. *pituita*, phlegm]: Endocrine gland in vertebrates; the anterior lobe is the source of tropic hormones, growth hormone, and prolactin and is stimulated by neurosecretory cells in the hypothalamus; the posterior lobe stores and releases oxcytocin and ADH produced by the hypothalamus.

PLACENTA [Gk. *plax*, a flat object]: A structure formed in part from the inner lining of the uterus and in part from the extraembryonic membranes; develops in most species of mammals and serves as the connection between the mother and the embryo during pregnancy through which exchanges occur between the blood of mother and of embryo.

PLANKTON [Gk. *planktos*, wandering]: Free-floating, mostly mi-croscopic, aquatic organisms, both photosynthetic (phytoplankton) and heterotrophic (zooplankton).

PLANULA [L. dim. of *planus*, flat]: The ciliated, free-swimming larval form occurring in many coelenterates.

PLASMA: The clear, colorless fluid component of blood, containing dissolved salts and proteins; blood minus the blood cells.

PLASMODESMATA [Gk. *plassein*, to mold + *desmos*, band, bond]: In plants, minute cytoplasmic threads that extend through the pores in cell walls and connect the protoplasts of adjacent cells.

PLASTID: A cytoplasmic, often pigmented organelle in plant cells (three types are leucoplasts, chromoplasts, and chloroplasts).

PLATELET (**plate**-let): In mammals, a minute, granular body suspended in the blood and involved in the formation of blood clots.

PLEIOTROPY (**plee**-o-trope-ee): The capacity of a gene to affect a number of different characteristics.

POLAR: Having parts or areas with opposed or contrasting properties, such as positive and negative charges, head and tail.

POLAR BODY: Minute nonfunctioning cell produced during meiotic divisions in egg cells; contains a nucleus but very little cytoplasm.

POLLEN [L., fine dust]: The male gametophytes of seed plants at the stage in which they are shed.

POLLINATION: The transfer of pollen from where it was formed (the anther) to a receptive surface (the stigma, in angiosperms) associated with an ovule of a flower.

POLY— [Gk. *polys*, many]: Prefix, meaning "many."

POLYGENIC INHERITANCE: The determination of a given characteristic, such as weight or height, by the complex interaction of many genes.

POLYMER: A large molecule composed of many molecular subunits.

POLYMORPHISM [Gk. *morphe*, form]: Occurrence together of two or more morphologically distinct forms of a population.

POLYP [Gk. *poly-*, many + *pous*, foot]: The sessile stage in the life cycle of coelenterates.

POLYPEPTIDE: A molecule consisting of numerous amino acids linked together by peptide bonds.

POLYPLOIDY [Gk. *ploos*, fold or times]: The possession of more than two complete sets of chromosomes per cell.

POLYSACCHARIDE: A carbohydrate composed of many joined monosaccharide units in a long chain; for example, glycogen, starch, cellulose.

POPULATION: Any group of individuals of one species; in genetic terms, an interbreeding group of organisms.

POST—, POSTERO— [L., behind, after]: Prefix, meaning "at," "near," or "toward the hind part"; opposite of "pre-" and "antero-."

POSTERIOR: Of or pertaining to the rear end; in man, the back of the body is said to be posterior.

PRE— [L., before, in front of]: Prefix, meaning "before" or "in front of"; opposite of "post-."

PREADAPTATION: The possession by an organism of a structure or function that develops new uses in the course of evolution.

PRIMARY GROWTH: In plants, growth originating in the apical meristem of the shoots and roots, as contrasted with secondary growth; results in increase in length.

PRIMARY STRUCTURE OF PROTEINS: The amino acid sequence of a protein.

PRIMATE: A member of the order of mammals that includes anthropoids and prosimians.

PRIMITIVE [L., from *primus*, first]: Not specialized; at an early stage of evolution or development.

PROCAMBIUM: In plants, a primary meristematic tissue; gives rise to vascular tissues of primary plant body and to the vascular cambium.

PROGESTERONE: In mammals, steroid hormone produced by corpus luteum that helps prepare the uterus for reception of the ovum.

PROKARYOTE [L. *pro*, before + Gk. *karyon*, with nut, kernel]: A cell lacking a membrane-bound nucleus or membrane-bound organelles; a bacterium or a blue-green alga.

PROPHASE [Gk. *pro*, before + *phasis*, form]: An early stage in nuclear division, characterized by the condensing of the chromosomes and their movement to the equator of the spindle.

PROSIMIAN: A lower primate; includes lemurs, lorises, tarsiers, and tree shrews, as well as many fossil forms.

PROSTAGLANDIN: One of a group of fatty acid hormones discovered in semen, now found to be present in many other tissues. Believed to play a role in fertilization.

PROSTATE GLAND: A mass of muscle and glandular tissue surrounding the base of the urethra in male mammals, through which the seminal vesicle passes and which secretes an alkaline fluid that has a stimulating effect on the sperm as they are released.

PROTEIN [Gk. *proteios*, primary]: A complex organic compound composed of one or more polypeptide chains, each made up of many (about 100 or more) amino acids joined by peptide bonds.

PROTO— [Gk. *protos*, first]: Prefix, meaning "first"; for example, Protozoa, "first animals."

PROMOTER: Segment of DNA to which RNA polymerase attaches to initiate transcription of mRNA from an operon.

PROTON: A subatomic particle with a single positive charge equal in magnitude to the charge of an electron and with a mass of 1; a component of every atomic nucleus.

PROTOPLASM [Gk. *plasma*, anything molded]: The living substance of all cells.

PROTOPLAST: A plant cell body, not including the cell wall.

PSEUDO— [Gk. *pseudes*, false]: Prefix, meaning "false."

PSEUDOPOD [Gk. *pous*, pod-, foot]: A temporary cytoplasmic protrusion from an amoeboid cell which functions in locomotion or in feeding by phagocytosis.

PULMONARY ARTERY [L. *pulmonis*, lung]: In vertebrates, the artery carrying blood to the lungs.

PULMONARY VEIN: In vertebrates, a vein carrying oxygenated blood from the lungs to the left atrium, from which blood is pumped into the left ventricle and from there to the body tissues.

PUNNETT SQUARE: The checkerboard diagram used for analysis of gene segregation.

PUPA [L., girl, doll]: A developmental stage, nonfeeding, immotile, and sometimes encapsulated or in a cocoon, between the larval and adult phases in insects.

PURINE: A nitrogenous base such as adenine or guanine; one of the components of nucleic acids.

PYLORIC SPHINCTER: A ring of muscle at the junction of the stomach and duodenum which controls the entrance of the stomach contents into the duodenum.

PYRAMID OF ENERGY: Energy relationships between various feeding levels involved in a particular food chain; plants (at the base of the pyramid) represent the greatest amount of energy utilization, herbivores next, then primary carnivores, secondary carnivores, etc.

PYRIMIDINE: A nitrogenous base such as cytosine, thymine, or uracil; one of the components of nucleic acids.

QUEEN: The fertile, or fully developed, female of social bees, ants, and termites whose function is to lay eggs.

RADIAL SYMMETRY: The regular arrangement of parts around one longitudinal axis; any line drawn through this oral-aboral axis will divide similar halves; seen in coelenterates and adult echinoderms.

RADIATION [L., a spoke of a wheel, hence, a ray]: The movement of energy from one place to another.

RADIOACTIVE ISOTOPE: An isotope with an unstable nucleus that stablizes itself by emitting radiation.

RECESSIVE ALLELE [L. *recedere*, to recede]: An allele whose phenotypic expression is masked by a dominant allele and so is manifest only in the homozygous condition. Heterozygotes involving recessives may be phenotypically indistinguishable from dominant homozygotes.

RECOMBINATION: In genetics, the appearance of gene combinations in progeny which differ, as a result of the sexual process, from the combinations present in the parents.

REDUCTION [L. *reducere*, to lead back]: Gain of an electron by a compound; takes place simultaneously with oxidation (loss of an electron by an atom) since an electron that is lost by one atom is accepted by another.

REFLEX [L. *reflectere*, to bend back]: Unit of action of the nervous system involving a sensory neuron, often an interneuron or -neurons, and one or more motor neurons.

REGULATOR GENE: A gene that influences whether or not a protein is produced by another structural gene.

RENAL [L. *renes*, kidneys]: Pertaining to the kidney.

REPRESSOR: The substance produced by a regulator gene that represses protein formation.

RESOLVING POWER: Ability of a lens to distinguish two lines as separate.

RESPIRATION [L. *respirare*, to breathe]: (1) In organisms, the intake of oxygen and the liberation of carbon dioxide. (2) In cells, the oxygen-requiring stage in the breakdown and release of energy from fuel molecules.

RETICULAR FORMATION: A core of tissue which runs centrally through the entire brainstem; a weblike network of fibers and neurons; involved with consciousness.

RETICULUM [L., network]: Any network of weblike structure, as in endoplasmic reticulum.

RETINA: The innermost nerve-tissue layer of the eyeball; contains several layers of neurons and light-receptor cells (rods and cones); receives the image formed by the lens and is connected to the brain by the optic nerve.

RHIZOID [Gk. *rhiza*, root]: Rootlike anchoring structure in nonvascular plants.

RHIZOME [Gk. *rhizoma*, mass of roots]: In plants, a horizontal underground stem, often enlarged for storage.

RIBONUCLEIC ACID (RNA) (rye-bo-new-**klee**-ick): A nucleic acid formed on chromosomal DNA and involved in protein synthesis; similar in composition to DNA, except that the pyrimidine uracil replaces thymine. RNA is the genetic material of many viruses.

RIBOSOME: A small organelle composed of protein and ribonucleic acid; the site of protein synthesis.

ROD: Light-sensitive nerve cell found in the vertebrate retina; sensitive to very dim light, responsible for "night vision."

ROOT: The descending axis of a plant, normally below ground and serving both to anchor the plant and to take up and conduct water and minerals.

SARCOMERE: Functional and structural unit of contraction in striated muscle.

SECONDARY GROWTH: In plants, growth derived from secondary meristem, that is, the vascular cambium and the cork cambium; results in an increase in diameter and in the production of woody tissue.

SECONDARY SEX CHARACTERISTICS: External characteristics that distinguish between the two sexes but which have no direct role in reproduction, such as the rooster's comb.

SECRETION: Product of any cell, gland, or tissue which is released through the cell membrane.

SEED: A complex organ formed by the maturation of the ovule of seed plants following fertilization. In conifers, it consists of seed coat, embryo, and female gametophyte storage tissue. In angiosperms, it consists of seed coat and embryo; some angiosperm seeds also contain endosperm, a storage tissue.

SEGREGATION: The separation of two alleles into different gametes during meiosis; *see* Mendel's first law.

SELF-FERTILIZATION: The union of egg and sperm produced by a single hermaphroditic organism.

SELF-POLLINATION: The transfer of pollen from another to stigma in the same flower or to another flower of the same plant, leading to self-fertilization.

SEMEN [L., seed]: Product of male reproductive system; includes sperm and the sperm-carrying fluid.

SENSILLUM, *pl.* SENSILLA: Sensory-receptor unit on body surface of arthropods; responsive to touch, smell, taste, and vibration (sound).

SENSORY NEURON: A neuron that carries impulses from a receptor to the central nervous system or central ganglion.

SENSORY RECEPTOR: A cell, tissue, or organ that detects internal or external stimuli.

SESSILE [L., from *sedere*, to sit]: Attached, not free to move about.

SEX CHROMOSOMES: Special sex-determining chromosomes, not occurring in identical numbers or shapes in both sexes.

SEX-LINKED CHARACTERISTIC: A genetic characteristic, such as color blindness, determined by a gene located either on the X or Y chromosome.

SEXUAL REPRODUCTION: Reproduction involving meiosis and syngamy (the union of gametes).

SHOOT: The above-ground portions, such as the stem and leaves, of a vascular plant.

SIEVE CELL: A long and slender sugar-conducting cell with relatively unspecialized sieve areas and with tapering end walls that lack sieve plates; found in the phloem of gymnosperms.

SIEVE TUBE: A vertical series of sugar-conducting cells (sieve-tube elements) of the phloem of angiosperms.

SMOOTH MUSCLE: Nonstriated muscle; lines the walls of internal organs and arteries and is under involuntary control.

SOCIAL DOMINANCE: A hierarchical pattern of physical domination of some members of a group by other members in a relatively orderly and long-lasting pattern.

SOLUTION: A mixture (usually liquid) in which one or more substances (the solute) are dispersed in the form of separate molecules or ions throughout the entire substance.

SOMATIC CELLS: The differentiated, usually diploid ($2n$) cells composing body tissues of multicellular plants and animals; all body cells except the germ cells.

SOMATIC NERVOUS SYSTEM: In vertebrates, the motor, sensory, and interneurons; the "voluntary" system, as contrasted with the "involuntary," or autonomic, nervous system.

SOMATO— [Gk. *soma*, body]: Prefix, meaning "body" or "of the body."

SOMITE: One of the segments into which the body of many animals is divided, especially an incompletely developed embryonic segment.

SPECIALIZED: (1) Of organisms, having special adaptations to a particular habitat or mode of life. (2) Of cells, having particular functions in a multicellular organism.

SPECIES, *pl.* SPECIES [L., kind, sort]: A group of organisms that actually (or potentially) interbreed and are reproductively isolated from all other such groups.

SPECIES-SPECIFIC: Characteristic of (and limited to) a particular species.

SPECIFIC HEAT: The amount of heat (in calories) required to raise the temperature of 1 gram of a substance 1°C; the specific heat of water is 1 calorie.

SPECIFICITY: Uniqueness, as in proteins in a given organism and of enzymes in given reactions.

—SPERM, SPERME—, SPERMA—, SPERMATO— [Gk. *sperma*, seed]: Suffix or prefix, meaning "seed."

SPERM [Gk. *sperma*, seed]: A mature male sex cell or gamete, usually motile and smaller than the female gamete.

SPERMATID: Each of four haploid cells resulting from the meiotic divisions of a spermatocyte, which becomes differentiated into a sperm cell.

SPERMATOCYTES: The diploid cells formed by the enlargement of the spermatogonia; give rise by meiotic division to the spermatids.

SPERMATOGENESIS [Gk. *genesis,* origin]: The process by which spermatogonia develop into sperm.

SPERMATOGONIA: The unspecialized diploid (2*n*) germ cell on the walls of the testes which by meiotic division become spermatocytes, then spermatids, then spermatozoa or sperm cells.

SPERMATOZOON, *pl.* SPERMATOZOA: A sperm cell.

SPINAL CORD: The part of the vertebrate central nervous system that consists of a thick longitudinal bundle of nerve fibers extending from the brain posteriorly along the dorsal side.

SPIRACLE [L. *spirare,* to breathe]: One of the external openings of the respiratory system in terrestrial arthropods.

SPORA—, SPORA— [Gk. *spora,* seed]: Prefix, meaning "seed."

SPORANGIOPHORE (spo-**ran**-ji-o-for) [Gk. *phore,* from *phorein,* go bear]: A branch bearing one or more sporangia.

SPORANGIUM, *pl.* SPORANGIA: A hollow unicellular or multicellular structure in which spores are produced.

SPORE: An asexual reproductive cell capable of developing into an adult without fusion with another cell; in contrast to a gamete.

SPOROPHYLL: Spore-bearing leaf; the carpels and stamens of flowers are modified sporophylls.

SPOROPHYTE: The spore-producing diploid (2*n*) phase in the life cycle of a plant having alternation of generations.

STAMEN [L., a thread]: The male organ of a flower which produces microspores or pollen; usually consists of a stalk, the filament, bearing an anther at its apex.

STARCH [M.E. *sterchen,* to stiffen]: A complex insoluble carbohydrate, the chief food-storage substance of plants, which is composed of several hundred glucose units ($C_6H_{12}O_6$); it is readily broken down enzymatically into these glucose units.

STEM: The part of the axis of vascular plants that is above ground, as well as anatomically similar portions below ground (such as rhizomes).

STEREOSCOPIC VISION [Gk. *stereos,* solid + *optikos,* pertaining to the eye]: Three-dimensional viewing of an object.

STIGMA: In plants, the region of a carpel serving as a receptive surface for pollen grains on which they germinate.

STIMULUS [L., goad, incentive]: Any internal or external change which influences the activity of an organism or of part of an organism.

STOMA, *pl.* STOMATA (**sto**-ma) [Gk. *stoma,* mouth]: A minute opening bordered by guard cells in the epidermis of leaves and stems through which gases pass. Also used to refer to the entire stomatal apparatus, the guard cells plus their included pore.

STRIATED MUSCLE [L., from *striare,* to groove]: Skeletal voluntary muscle and cardiac muscle.

STROMA: The ground substance that makes up the interior of the chloroplast in which the grana are dispersed.

STRUCTURAL GENE: One of the genes in an operon which produces a protein and whose function is controlled by the operator and the regulator gene.

SUB—, SUS— [L., under, below]: Prefix, meaning "under" or "below"; for example, subepidermal, "underneath the epidermis."

SUBSTRATE [L. *substratus,* strewn under]: (1) The foundation to which an organism is attached. (2) A substance acted on by an enzyme.

SUCCESSION: In ecology, the slow, orderly progression of changes in species composition of a community during development of vegetation in any area.

SUCROSE: A common sugar; a disaccharide found in many plants.

SUGAR: Any monosaccharide or disaccharide.

SYM—, SYN— [Gk. *syn,* together, with]: Prefix, meaning "together."

SYMBIOSIS [Gk. *syn,* to live]: An intimate, protracted, and dependent relationship between two or more organisms of different species. Includes mutualism, in which the association is beneficial to both; commensalism, in which one benefits and the other is neither harmed nor benefitted; and parasitism, in which one benefits and the other is harmed.

SYMPATHETIC NERVOUS SYSTEM: A subdivision of the autonomic nervous system, with centers in the midportion of the spinal cord; slows digestion; generally excites other functions. Chemical transmitters: epinephrine or norepinephrine.

SYNAPSE [Gk. *synapsis,* a union]: The region of nerve impulse transfer between two neurons.

SYNGAMY (**sin**-gamy): The union of gametes in sexual reproduction; fertilization.

SYNTHESIS: The formation of a more complex substance from simpler ones.

TAXONOMY [Gk. *taxis,* arrange, put in order + *nomos,* law]: The science of the classification of organisms.

TELOPHASE [Gk. *telos,* end + *phasis,* form]: The last stage in mitosis and meiosis, during which the chromosomes become reorganized into two new nuclei.

TEMPLATE: A pattern or mold guiding the formation of a negative or complement. DNA replication is explained in terms of a template hypothesis.

TENTACLES [L. *tentare,* to touch]: Long, flexible protrusions located about the mouth in many invertebrates; usually prehensile or tactile.

TERRITORY: An area or space occupied by an individual or a group, trespassers into which are attacked (and usually defeated); may be the site of breeding, nesting, food gathering, or any combination thereof.

TEST CROSS: A mating between a homozygous recessive individual and a dominant phenotype to determine the genetic constitution of the latter, that is, whether it is homozygous or heterozygous for a particular gene.

TESTIS, *pl.* TESTES [L., witness]: The male gamete-producing organ; also the source of male sex hormone.

TESTOSTERONE: A hormone secreted by the testes in higher vertebrates and stimulating the development and maintenance of male sex characteristics and the production of sperm; an androgen.

TETRAD [Gk. *tetras,* four]: In plants, a group of four spores formed by meiosis within a mother cell.

THALAMUS [Gk. *thalamos,* chamber]: A part of the vertebrate forebrain just posterior to the cerebrum; an important intermediary between all other parts of the nervous system and the cerebrum.

THEORY [Gk. *theorein,* to look at]: A generalization based on observation and experiments that were conducted to test the validity of a hypothesis and found to support the hypothesis.

THERMODYNAMICS [Gk. *therme,* heat + *dynamis,* power]: The study of energy, using heat as the most convenient form of measurement of energy. The first law of thermodynamics states that in all processes, the total energy of the universe remains constant. The second law states that the entropy, or degree of randomness, tends to increase.

THORAX: (1) In vertebrates, that portion of the trunk containing the heart and lungs. (2) In insects, the three leg-bearing segments between head and abdomen.

THYLAKOID: A saclike membranous structure in the chloroplast; stacks of thylakoids form the grana.

THYROID: An endocrine gland of vertebrates, located in the neck; source of an iodine-containing hormone (thyroxine) that increases the rate of oxidative processes (metabolic rate).

TISSUE [L. *texere,* to weave]: A group of similar cells organized into a structural and functional unit.

TRACHEA (**trake**-ee-a), *pl.* TRACHEAE [Gk. *tracheia,* rough]: An air-conducting tube, such as the windpipe of mammals and the breathing systems of insects.

TRACHEID (**tray**-key-idd) [Gk. *tracheia,* rough)]: An elongated, thick-walled conducting and supporting cell of xylem, characterized by tapering ends and pitted walls without true perforations. It is found in nearly all vascular plants and is dead at maturity.

TRANSCRIPTION: The enzymatic process, involving base pairing, by which the genetic information contained in DNA is used to specify a complementary sequence of bases in an RNA molecule.

TRANSDUCTION: The transfer of genetic material (DNA) from one bacterium to another by a lysogenic bacteriophage.

TRANSFER RNA (tRNA): The type of RNA that becomes attached to an amino acid and guides it to the correct position on the ribosome-mRNA complex for protein synthesis. There is at least one tRNA molecule for each amino acid. Each tRNA molecule is only about 80 nucleotides in length.

TRANSFORMATION: A genetic change produced by the incorporation into a cell of DNA from another cell.

TRANSLATION: The process by which the genetic information contained in a messenger RNA molecule dictates the sequence of amino acids in a protein molecule.

TRANSLOCATION: (1) In plants, the transport of the products of photosynthesis. (2) In genetics, the breaking off of a piece of a chromosome and its attachment to a nonhomologous chromosome.

TRANSPIRATION [L. *spirare,* to breathe]: In plants, the loss of water vapor from the stomata.

—TROPH, TROPHO— [Gk. *trophos,* feeder]: Suffix or prefix, meaning "feeder" or "feeding"; for example, autotrophic, "self-nourishing."

TROPIC [Gk. *trope,* a turning]: Pertaining to behavior or action brought about by specific stimuli, for example, phototropic

("light-oriented") motion, gonadotropic ("stimulating the gonads") hormone.

TUBER [L. *tuber,* bump, swelling]: A much-enlarged, short, fleshy underground stem, such as that of the potato.

TURGOR [L. *turgere,* to swell]: The rigid distention of a plant cell by its fluid contents.

UREA [Gk. *ouron,* urine]: An organic compound formed in the vertebrate liver from ammonia and carbon dioxide and excreted by the kidneys; the principal form of disposal of nitrogenous wastes in mammals.

URETER [Gk. from *ourein,* to urinate]: The tube carrying urine from the kidney to the cloaca (in reptiles and birds) or to the bladder (in amphibians and mammals).

URETHRA: The tube carrying urine from the bladder to the exterior in mammals.

URIC ACID: An insoluble nitrogenous waste product that is the principal excretory product of birds, reptiles, and insects.

URINE: The liquid waste filtered from the blood by the kidney and excreted by the bladder.

UTERUS [L., womb]: The muscular, expanded portion of the female reproductive tract modified for the storage of eggs or for housing and nourishing the developing embryo.

VACUOLE [L. *vacuus,* empty]: A space within a cell, bound by a membrane and filled with water and solid materials in solution.

VAGINA [L. *vagina,* sheath]: The part of the female reproductive duct in mammals that receives the male penis during copulation.

VAGUS NERVE [L. *vagus,* wandering]: A nerve arising from the medulla of the vertebrate brain that innervates the visceral organs; carries parasympathetic fibers.

VALENCE [L. *valere,* to have power]: A measure of the bonding capacity of an atom, based on the number of electrons in its outer shell; equal to the charge which ions of that atom display in chemical reactions.

VAS DEFERENS (vass **deaf**-er-ens): In mammals, the tube carrying sperm from the testes to the urethra.

VASCULAR: Containing or concerning vessels that conduct fluid.

VASCULAR BUNDLE: In plants, a group of longitudinal supporting and conducting tissues (xylem and phloem).

VASCULAR CAMBIUM: A cylindrical sheath of meristematic cells which divide, producing secondary phloem and secondary xylem, but always with a cambial cell remaining.

VEIN: In plants, a vascular bundle forming a part of the framework of the conducting and supporting tissue of a leaf or other expanded organ. In animals, a blood vessel carrying blood from the tissues to the heart.

VENA CAVA (**vee**-na **cah**-va): A large vein which brings blood from the tissues to the right atrium of the heart. The superior vena cava collects blood from the forelimbs, head, and anterior or upper trunk; the inferior vena cava collects blood from the kidneys, the liver or lower gonads, and the general posterior body region.

VENTRAL [L. *venter,* belly]: Pertaining to the undersurface of an animal that moves on all fours; to the front surface of an animal that holds its body erect.

VENTRICLE [L. *ventriculus,* the stomach]: A chamber of the heart that receives blood from an atrium and pumps blood out of the heart.

VERTEBRAL COLUMN: The backbone; in nearly all vertebrates, it forms the supporting axis of the body and a protection for the spinal cord.

VESICLE [L. *vesicula,* a little bladder]: A small, intracellular membrane-bound sac.

VESSEL: A tubelike structure of the xylem of angiosperms composed of dead cells (vessel elements) placed end to end. Its function is to conduct water and minerals from the soil.

VILLUS [L. *villus,* a tuft of hair]: In vertebrates, one of the minute fingerlike projections lining the small intestine that serve to increase the absorptive surface area of the intestine.

VIRUS [L., slimy, liquid, poison]: A submicroscopic noncellular particle, composed of a nucleic acid core and a protein shell; parasitic; reproduces only within a host cell.

VISCERA: The collective term for the internal organs of an animal.

VITAMIN [L. *vita,* life]: Any of a number of unrelated organic substances that cannot be synthesized by a particular organism and are essential in minute quantities for normal growth and function.

VULVA: External genitalia of the human female; includes the clitoris and the labia.

WATER TABLE: The upper limit of permanently saturated soil; the top layer of the groundwater.

WILD TYPE: In genetics, the phenotype that is characteristic of the vast majority of individuals of a species in a natural environment.

WORKER: A member of the nonreproductive laboring caste in social insects.

XYLEM [Gk. *xylon,* wood]: A complex vascular tissue through which most of the water and minerals are conducted from the roots to other parts of the plant; consists of tracheids or vessel elements, parenchyma cells, and fibers; constitutes the wood of trees and shrubs.

YOLK: The stored food material in egg cells that nourishes the embryo.

ZOOPLANKTON: A collective term for the nonphotosynthetic organisms present in plankton.

ZYGOTE (**zi**-got) [Gk. *zygon,* yolk, pair]: The diploid ($2n$) cell resulting from the fusion of male and female gametes.

ILLUSTRATION ACKNOWLEDGMENTS

1-1 © California Institute of Technology and Carnegie Institution of Washington

1-2 NASA

1-3 NASA

1-4 Icelandic Photo and Press Service

Page 14 Grant Heilman; J. C. Thompson, National Audubon Society Collection/PR; Stouffer Productions, Ltd., Bruce Coleman; Douglas P. Wilson

Page 15 Craig MacFarland; Roberts Rugh and Landrum B. Shettles, M. D., *From Conception to Birth: The Drama of Life's Beginnings,* Harper & Row, Publishers, Inc., New York, 1971; E. S. Ross

1-5 Photograph by C. Ponnamperuma

1-6 A. Ryter

1-7 Photograph by Norma J. Lang, *Journal of Phycology,* **1:**127–134, 1965

1-9 Photograph by Ursula Goodenough

1-10 Michael A. Walsh

1-11 From William Bloom and Don W. Fawcett, *A Textbook of Histology,* 9th ed., W. B. Saunders Company, Philadelphia, 1968

2-1 Oskar Kreisel, *Photography Annual,* 1956

2-2 Grant Heilman

2-3 Larry Pringle, Photo Researchers

2-4 Jack Dermid

2-5 © After James Sutcliffe, *Plants and Water,* St. Martin's Press, Inc., New York, 1968

2-9 Calgon Corporation

Page 41 From James D. Watson, *The Dou-ble Helix,* Atheneum Publishers, New York, 1968

3-2 Max and Kit Hunn, Photo Researchers

3-3 (c) Myron C. Ledbetter

3-4 Keith R. Porter

3-5 (a) R. D. Preston

3-6 Larry West

3-9 B. E. Juniper

Page 48 Photograph by U. N. Food and Agriculture Organization

3-16 (a) Keith R. Porter; (b) Emil Bernstein; (c) Jack Dermid

4-1 Gregory Antipa

4-3 Gregory Antipa

4-4 (a) Mercedes E. Edwards; (b) J. D. Robertson

Page 58 J. Pickett-Heaps

4-8 Gregory Antipa

4-9 Birgit Satir

4-10 Don W. Fawcett

4-11 Ursula Goodenough

4-12 After Stephen Wolfe, *Biology of the Cell.* © 1972 by Wadsworth Publishing Company, Inc., Belmont, Calif. By permission of the publisher.

4-13 After Edward O. Wilson et al., *Life on Earth,* Sinauer Associates, Inc., Stamford, Conn., 1973

4-14 Photograph by Myron C. Ledbetter

4-16 J. F. M. Hoeniger, *Journal of General Microbiology,* **40:**29, 1965

4-17 Gregory Antipa

4-18 (a) After E. J. DuPraw, *Cell and Molecular Biology,* Academic Press, Inc., New York, 1968; (b) A. V. Grimstone

Page 68 Eric V. Gravé; Eugene B. Small and Donald S. Marszalek, *Science,* **163:**1064–1065, 1969. Copyright 1969 by the American Association for the Advancement of Science; Eric V. Gravé

Page 69 Rugh and Shettles, op. cit.; John H. Troughton; Keith R. Porter; L. Heimer

4-20 D. Branton

4-21 Gregory Antipa

5-1 R. D. Estes

6-1 Stephen B. Gough and William J. Woelkering

6-3 (a) Howard Towner; (b) Myron C. Ledbetter; (c), (d) L. K. Shumway

7-1 Photo Researchers

7-2 Keith R. Porter

7-11 (a) Don W. Fawcett; (b) Jack Dermid; (c) Grant Heilman

8-1 Kurt Hirschhorn, M. D., from *Birth Defects: Original Article Series,* vol. 4, no. 4, *Guide to Human Chromosome Defects* by Audrey Redding and Kurt Hirschhorn, M. D., The National Foundation

8-2 Ursula Goodenough

8-6 (a) Étienne de Harven, "The Nucleus," 1968, Academic Press Inc., New York; (b) M. Friedlander

8-7 S. Inoué, "Polarization Optical Studies of the Mitotic Spindle. I. The Demonstration of Spindle Fibers in Living Cells," *Chromosome Bd.,* **5:**487–500, Springer, 1953

8-8 William Tai

8-10 General Biological Supply Company

8–11 James Cronshaw

8–13 Hugh Spencer, National Audubon Society Collection/PR

9–1 Carnegie Institution of Washington

9–8 B. John

9–9 Mary E. Clutter

9–12 Adapted from DuPraw, op. cit.

10–1 The Bettmann Archive, Inc.

10–3 After Karl von Frisch, *Biology*, Harper & Row, Publishers, Inc., New York, 1964, translated by Jane Oppenheimer

10–8 Mary E. Clutter

Page 138 (left) Gary Laurish; (center and right) Patricia Farnsworth

11–1 C. G. G. J. van Steenis

Page 143 Photograph by M. E. Browning, National Audubon Society Collection/PR

11–3 Photograph by F. B. Hutt, *Journal of Genetics*, **22**:126, 1930

11–4 After E. D. Merrell, 1964

Page 150 Photograph by Shirley Baty

12–4 Photograph by B. John

12–7 B. P. Kaufman

13–3 Culver Pictures

13–4 Photograph by The Bettmann Archive, Inc.; geneology after I. Michael Lerner, *Heredity, Evolution and Society*, W. H. Freeman and Company, San Francisco, 1968

14–1 A. K. Kleinschmidt, D. Land, D. Jacherts, and R. K. Zahn, *Biochemica Biophysica Acta*, **61**:857–864, 1962, Fig. 1

14–5 Robert Austrian, *Journal of Experimental Medicine*, **92**:21, 1953

14–6 After Derry D. Koob and William E. Boggs, *The Nature of Life*, Addison-Wesley Publishing Company, Inc., Reading, Mass., 1972

14–8 After James D. Watson, *Molecular Biology of The Gene*, 2d ed., W. A. Benjamin, Inc., Menlo Park, Calif., 1970

14–9 Lee D. Simon

14–10 From Watson, op. cit., 1968

Page 172 From Watson, op. cit., 1968

15–4 Hans Ris

15–6 O. L. Miller, Jr., and Barbara R. Beatty, Oak Ridge National Laboratory

16–1 T. F. Anderson, from E. L. Wollman, F. Jacob, and W. Hayes, *Cold Spring Harbor Symposium on Quantitative Biology*, **21**:141, 1950

16–6 Jack Griffith

16–9 J. Gall

16–12 Donald E. Olins and Ada L. Olins, University of Tennessee, Oak Ridge Graduate School of Biomedical Sciences and The Oak Ridge National Laboratory

17–4 Tsuyoshi Kakefuda

17–5 Stanley N. Cohen

17–6 George Ancona

18–1 Carl W. Rettenmeyer

18–2 (a) G. D. Dodge and D. R. Thompson, Bruce Coleman; (b) Dale P. Hansen, National Audubon Society Collection/PR; (c) Hans Reinhard, Bruce Coleman

18–3 (a) H. Forest; (c) Michael P. Godomski, National Audubon Society Collection/PR; (d) Les Blacklock; (e) Alvin E. Staffan

18–4 (a) Martin Dworkin; (b) Center for Disease Control; (c) Parke, Davis and Company

18–5 I. Polunin, Bruce Coleman

18–7 (a), (b) J. D. Almeida and A. F. Howatson, *Journal of Cell Biology*, **16**:616, 1963; (c) Frederick A. Murphy; (e) Lee D. Simon

18–9 (a)–(e) Eric V. Gravé; (f), (g) Eric V. Gravé, Photo Researchers

18–11 (a), (c) D. P. Wilson

18–12 D. P. Wilson

18–13 D. P. Wilson

18–14 (a) Valerie Taylor, Ron Taylor Film Productions; (b), (c) Oxford Scientific Films, Bruce Coleman

18–16 Larry West

18–17 (a) Agenzia Fotografica, Luisa Ricciarini, Milan; (b) Alvin E. Staffan; (c) S. J. Kraseman, Photo Researchers; (d) Jane Burton, Bruce Coleman

18–19 (a) Jack Dermid, Photo Researchers; (b) Larry West

18–20 Alvin E. Staffan

18–21 (a) Bruce Coleman; (b) Robert Carr; (c) A–Z Collection, 1976/PR

18–22 (a) William Harlow; (b) Larry West

18–25 After T. Elliot Weir, C. Ralph Stocking, and Michael G. Barbour, *Botany*, 4th ed., John Wiley & Sons, Inc., New York, 1970

18–26 (a) Karl Maslowski, Photo Researchers; (b) Charles W. Mann, National Audubon Society Collection/PR

18–27 Nat Fain

18–29 Douglas Faulkner

18–32 Al Giddings, Bruce Coleman

18–35 (a) James McConnell; (b) Maria Wimmer; (c) David Pramer

18–37 Maria Wimmer

18–41 (a) R. Marischal, Bruce Coleman; (b) C. W. Schwartz, Photo Researchers; (c) D. P. Wilson; (d) Jane Burton, Bruce Coleman

18–42 D. P. Wilson

18–43 E. S. Ross

18–46 (a), (e) E. S. Ross; (b) Alexander Klots; (c), (d), (f) Larry West

18–47 (a), (b) Larry West; (c), (d) E. R. Degginger, Bruce Coleman

18–48 Thomas Eisner

18–50 William M. Stephens, Photo Researchers

18–53 Donald H. Fritts

18–54 Tom McHugh, Photo Researchers

18–55 Grant Heilman

18–56 Lynwood M. Chace, National Audubon Society Collection/PR

18–57 (a) The American Museum of Natural History; (b) Tomas W. Friedman, Photo Researchers

18–58 (a) The New York Zoological Society; (b) Gordon S. Smith, National Audubon Society Collection/PR

18–59 (a) L. L. Rue, III, Photo Researchers; (b) Russ Kinne, Photo Researchers; (c) Mark N. Boulton, National Audubon Society Collection/PR; (d) Eric Hosking; (e) T. R. Pierson, Rapho/PR; (f) Ylla, Rapho/PR

19–1 Jack Dermid

19–4 Grant Heilman

19–5 (a) B. E. Juniper; (b) John H. Troughton

19–7 John H. Gerard

19–8 Ray F. Evert

20–1 Ray F. Evert

20–2 Jack Dermid

20–3 After Peter Ray, *The Living Plant*, 2d ed., Holt, Rinehart and Winston, Inc., New York, 1971

20–4 Photograph by Ray F. Evert

20–5 Photograph by Ray F. Evert

20–6 (a) Lynwood M. Chace, National Audubon Society Collection/PR; (b) Jack Dermid; (c) Grant Heilman

20-7 From Myron C. Ledbetter and Keith R. Porter, *Introduction to the Fine Structures of Plant Cells*, Springer-Verlag, New York, 1970

20-8 Ray F. Evert

20-9 (a) Jack Dermid; (b) Mary M. Thacher, Photo Researchers

20-11 Ray F. Evert

20-12 Larry West

20-13 After A. C. Leopold, *Plant Growth and Development*, McGraw-Hill Book Company, New York, 1964 (after Kramer, 1937)

20-14 After M. Richardson, *Translocation in Plants*, St. Martin's Press, Inc., New York, 1968 (after Stout and Hoagland, 1939)

Page 277 F. W. Went

20-15 Jeanne White, National Audubon Society Collection/PR

20-16 Ledbetter and Porter, op. cit.

20-17 (a) Martin H. Zimmerman; (b) George A. Schaefers

21-1 Jack Dermid

21-3 E. S. Ross

21-4 (c) Gene Ahrens, Bruce Coleman; (d) E. S. Ross

Page 286 Jack Dermid

21-5 (a) Larry West; (b)–(f) E. S. Ross; (d) D. J. Howell; (g) William Calder

21-6 Patrick Echlin

Page 290 Ray F. Evert

21-10 After Ray, op. cit.

21-11 Peter Ray

21-14 From Richard Ketchum, *The Secret Life of the Forest*, American Heritage Press, New York, 1970

22-3 Fredric Winkowski

22-5 J. P. Nitsch, *American Journal of Botany*, **37:3**, 1950

22-6 S. H. Wittwer

22-7 J. E. Varner

22-8 H. R. Chen

22-11 After Aubrey W. Naylor, "The Control of Flowering," *Scientific American*, May 1952

22-12 Hugh Spencer, National Audubon Society Collection/PR

22-16 Jack Dermid

22-17 Jack Dermid

22-18 William Harlow

23-1 Ken Heyman

23-4 Keith R. Porter

23-5 M. H. Ross and Edward J. Reith

23-7 Photograph by Keith R. Porter

23-9 Photograph by Keith R. Porter

23-10 Raymond C. Truex

23-11 After David Kirk, *Biology Today*, 2d ed., CRM/Random House, Inc., New York, 1975

Page 325 Hugh E. Huxley

24-2 © After M. B. V. Roberts, *Biology; A Functional Approach*, The Ronald Press Company, New York, 1971

24-5 Bruce Coleman; Mario Fantin, Photo Researchers; L. L. Rue, III, Photo Researchers; T. P. Dickinson, Photo Researchers

24-13 Rugh and Shettles, op. cit.

24-14 Carnegie Institution of Washington

24-16 Carnegie Institution of Washington

24-17 Rugh and Shettles, op. cit.

24-25 Paoli Koch, Photo Researchers

25-2 (a) Keith R. Porter

25-4 After Keith R. Porter and Mary A. Bonneville, *Fine Structure of Cells and Tissues*, Lee and Febiger, Philadelphia, 1968

25-8 After Wilson et al., op. cit.

25-11 David H. Hubel

25-12 Photograph by Keith R. Porter

Page 361 The American Museum of Natural History

25-17 Walter E. Stumpf and Madhabananda Sar

25-18 (b) J. T. Bonner

26-1 (a) After Roberts, op. cit.

26-2 U.N. Food and Agriculture Organization

26-3 Keith R. Porter

26-6 After Roberts, op. cit.

26-11 Thomas L. Hayes and L. McDonald, *Experimental and Molecular Pathology*, vol. 10, Academic Press, Inc., New York, 1969

26-12 Pfizer, Inc.

27-4 Photograph by S. C. Bisserôt, Bruce Coleman

Page 381 John Dominis, *Life* © 1971

27-7 Keith R. Porter

27-8 Myron Melamed

Page 385 Myron Melamed

28-1 Wildlife Photographers, Bruce Coleman

28-3 I. Kaufman Arenberg

28-5 (a) Dr. Jeanne M. Riddle

28-7 Bloom and Fawcett, op. cit.

Page 394 U.S. Department of Agriculture

28-8 After Wilson et al., op. cit.

29-1 Jack Dermid

29-2 (a) Nicholas Mrosovsky; (b) L. L. Rue, III; (c) S. C. Bisserôt

29-3 (a) Ron Garrison, Zoological Society of San Diego; (b) L. L. Rue, III, National Audubon Society Collection/PR

29-5 After A. J. Vander, J. H. Sherman, and Dorothy S. Luciano, *Human Physiology*, 2d ed., McGraw-Hill Book Company, New York, 1975

29-7 From Lewis Carroll, *Alice's Adventures in Wonderland*, illustrated by John Tenniel

29-12 After Vander, Sherman, and Luciano, op. cit.

29-14 After Vander, Sherman, and Luciano, op. cit.

29-15 Ylla, Rapho/PR

30-1 Arthur Jacques

30-2 Photograph by Arthur Jacques

30-8 James Olds

30-10 Aquarius Electronics, Mendocino, California; copyright 1971

30-12 Jerry Hecht, National Institute of Mental Health

30-13 Laverne C. Johnson, "Are Stages of Sleep Related to Waking Behavior?," *American Scientist*, **61:3**, 1973

30-15 E. R. Lewis

30-18 Jerre Levy, Colwyn Trevarthen, and Roger W. Sperry, *Brain: The Journal of Neurology*, 1972

30-19 Arthur Jacques

31-1 T. A. DeRoy, Bruce Coleman

31-2 Radio Times Hulton Picture Library

31-3 (a) George Gerster, Rapho/PR; (b), (c) Tom McHugh, Photo Researchers; (d) E. S. Ross

Page 434 Fritz Henle, Photo Researchers

31-4 The American Museum of Natural History

31-5 E. S. Ross

31-6 Medical Illustration Unit, Royal College of Surgeons of England

31-8 I. Eibl-Eibesfeldt

32–1 George Barlow

32–3 Victor McKusick

32–4 Edmund B. Gerard

32–5 Victor McKusick

32–6 Jack R. Harlow, "The Plants and Animals that Nourish Man," *Scientific American,* September 1976; courtesy The Green Thumb, Water Mill, N.Y.

32–7 Philip Hyde

32–8 Charles W. Brown, Santa Rosa Junior College

32–9 James H. Robinson, Photo Researchers

33–1 Eric Hosking

33–2 Mather and Harrison, *Heredity,* vol. 3, 1949

33–3 (a) Herbert A. Fischler, Isaac Albert Research Institute of the Kingsbrook Jewish Medical Center; (b) John S. O'Brien

33–6 Lynwood M. Chace, National Audubon Society Collection/PR

33–7 After Grover C. Stephens and Barbara Best North, *Biology,* John Wiley & Sons, Inc., New York, 1974

33–8 After G. Ledyard Stebbins, *Processes of Organic Evolution,* Prentice-Hall, Inc., Englewood Cliffs, N.J., 1966

33–9 P. M. Sheppard

33–10 After J. Clausen and W. M. Hiesey, Carnegie Institution of Washington, publication 615, 1958

34–1 H. B. D. Kettlewell

34–2 After Stephens and North, op. cit.

34–3 Department of Indian Affairs and Northern Development

Page 464 Photograph by John S. Shelton

34–4 (a), (b) John Wallis, Bruce Coleman; (c) K. W. Fink, Bruce Coleman

34–5 After Edward O. Wilson and William H. Bossert, *A Primer of Population Biology,* Sinauer Associates, Inc., Stamford, Conn., 1971

34–7 After N. Tinbergen, *Social Behavior in Animals,* John Wiley & Sons, Inc., New York, 1953

34–8 After Bruce Wallace and Adrian Srb, *Adaptation,* Prentice-Hall, Inc., Englewood Cliffs, N.J., 1964

34–9 After Wallace and Srb, op. cit.

34–11 M. P. Harris

34–12 (a) Joe Van Wormer, Bruce Coleman; (b) Bruce Coleman; (c) Edward S. Ayensu, National Museum of Natural History

34–13 R. Austing, Bruce Coleman

34–14 (a) M. W. Larson, Bruce Coleman; (b) Robert Mitchell; (c) Alan Blank, Bruce Coleman; (d) Jeanne White, National Audubon Society Collection/PR

34–15 Jen and Des Bartlett, Bruce Coleman

34–16 (a) Barry Hennings, Photo Researchers; (b) Tom McHugh, Photo Researchers

34–17 Daniel Janzen

35–1 Elihu Blotnick, BBM Associates

35–6 Karl Weidman, National Audubon Society Collection/PR

35–7 Russ Kinne, Photo Researchers

35–8 (a) G. Harrison, Bruce Coleman; (b) Bruce Coleman

35–9 John Flannery, Bruce Coleman

35–10 (a) Allan D. Cruickshank, National Audubon Society Collection/PR; (b) Max Thompson, National Audubon Society Collection/PR; (c) N. Myers, Bruce Coleman; (d) Verna R. Johnston, Photo Researchers

35–11 (a) Jen and Des Bartlett, Bruce Coleman; (b) Robert H. Wright, National Audubon Society Collection/PR; (c) Bill Ratcliffe; (d) Anthony Mercieca, National Audubon Society Collection/PR; (e) Grace A. Thompson, National Audubon Society Collection/PR; (f) Willis Peterson; (g) Alan Blank, Bruce Coleman

35–12 Dennis Brokaw

35–14 (a) Les Blacklock; (b) Patricia Caulfield

35–16 Jack Dermid

35–17 (a) Les Blacklock; (b), (c) Larry West; (d) Norman R. Lightfoot, Photo Researchers; (e) Larry West

35–19 R. Austing, Bruce Coleman

35–20 Les Blacklock

35–21 Les Blacklock

35–22 (a) Gene Ahrens, Bruce Coleman; (b) R. and J. Spurr, Bruce Coleman

35–23 (a) Les Blacklock; (b) Marty Stouffer, Photo Researchers; (c) Alvin E. Staffan

35–24 Charlie Ott, National Audubon Society Collection/PR

35–25 K. Gunnar, Bruce Coleman

35–26 Les Blacklock

35–28 Les Blacklock

35–30 D. P. Wilson

35–31 Bruce H. Robison

35–32 Allan Power, Bruce Coleman

35–33 (a) Jeff Foote, Bruce Coleman; (b) Charlie Ott, National Audubon Society Collection/PR

35–34 Jack Dermid

35–35 Jack Dermid

36–1 Jeff Foote, Bruce Coleman

36–4 Jack Dermid

36–5 (a) Jane Burton, Bruce Coleman; (b) N. Myers, Bruce Coleman

36–11 (a) The Nitragin Company; (b) R. R. Hebert, R. D. Holsten, and R. W. F. Hardy, E. I. DuPont de Nemours and Co., Inc.

36–13 R. H. Wright, National Audubon Society Collection/PR

36–14 (a) Larry West; (b) U.S. Forest Service

36–15 (a), (c) Jack Dermid; (b) U.S. Forest Service

36–16 Jack Dermid

36–17 Grant Heilman

37–1 Jeanne White, National Audubon Society Collection/PR

37–2 After S. Charles Kendeigh, *Animal Ecology,* Prentice-Hall, Inc., Englewood Cliffs, N.J., 1961 (after Shelford, 1911)

37–5 After John L. Harper, *Symposia Society for Experimental Biology,* **15:**25, 1961

37–8 (a) Douglas Faulkner; (b) R. Marischal, Bruce Coleman; (c) C. Haagner, Bruce Coleman; (d) N. Smythe, National Audubon Society Collection/PR

37–9 Biological Section Laboratory, Australian Department of Lands

37–10 L. David Mech

37–11 C. H. Muller, Bulletin of the Torrey Botanical Club, **93:**334, 1966

37–12 (a) E. S. Ross; (b) Doug Fulton, Photo Researchers; (c) S. C. Bisserôt, Bruce Coleman

37–13 (a)–(c) E. S. Ross; (d) Alan Blank, Bruce Coleman; (e) G. R. Austing, Bruce Coleman; (f) E. S. Ross

37–14 Photo Researchers

37–15 Thomas Eisner

37–16 E. S. Ross

38–1 E. S. Ross

38–2 E. S. Ross

38–3 John R. Clawson, National Audubon Society Collection/PR

38–4 (a) E. S. Ross; (b) Treat Davidson,

National Audubon Society Collection/PR

38-5 (a), (b) E. S. Ross; (c) Stephen Dalton, Photo Researchers

38-6 Charlie Ott, National Audubon Society Collection/PR

38-8 (a) M. P. Kahl, Photo Researchers; (b) George Holton, Photo Researchers; (c) Jeff Foote, Bruce Coleman

38-9 Patricia Caulfield

38-10 T. W. Ransom

38-11 (a) R. R. Pawlaski, Bruce Coleman; (b) S. Gaulin, Anthro/Photos; (c) L. L. Rue, III, Bruce Coleman; (d) Douglas Faulkner

38-12 Tanaka Kojo, Animals Animals

Page 548 Douglas Faulkner

38-13 Bruce Coleman

38-14 Irven deVore, Anthro/Photos

38-15 Irven deVore, Anthro/Photos

38-16 (a) Tom McHugh, Photo Researchers; (b) Irven deVore, Anthro/Photos

38-17 Irven deVore, Anthro/Photos

38-18 J. Popp, Anthro/Photos

38-19 J. Popp, Anthro/Photos

38-20 Masud Quraishy, Photo Researchers

39-1 Archives Photographique, Ministère des Affaires Culturelles Caisse Nationale des Monuments Historiques et des Sites

39-2 Zoological Society of San Diego

39-4 L. L. Rue, III, National Audubon Society Collection/PR

39-5 (a) E. S. Ross; (b) H. Albrecht, Bruce Coleman

39-6 (a) George Holton, Photo Researchers; (b) Tom McHugh, Photo Researchers; (c) M. P. L. Fogden, Bruce Coleman

39-7 F. Erize, Bruce Coleman

Page 563 C. R. Carpenter

39-8 After W. W. Howells, *Mankind in the Making* (rev. ed.), Doubleday & Company, Inc., Garden City, N.Y., 1967

Page 564 (a) Ralph Morse, Time-Life Pictures; (b) Irven deVore, Anthro/Photos; (c) George Holton, Photo Researchers; (d), (e) Lee Lyon, Bruce Coleman

Page 565 (f), (j) Teleki/Baldwin; (g)–(i) S. Wrangham, Anthro/Photos

39-10 The American Museum of Natural History

39-11 The American Museum of Natural History

39-13 Robert F. Sisson, © National Geographic Society

39-14 © National Geographic Society

39-15 Geza Teleki

40-1 The American Museum of Natural History

40-2 Tom McHugh, Photo Researchers

40-3 Geza Teleki

40-4 Lee Boltin

Page 575 George Herben, Photo Researchers

40-5 Ralph S. Solecki

40-6 From David Pilbeam, *The Ascent of Man: An Introduction to Human Evolution,* The Macmillan Company, New York, 1972

40-7 Drawings from Jacques Bordaz, *Tools of the Old and New Stone Age,* The Natural History Press, Garden City, N.Y., 1970; photographs by Lee Boltin

40-8 Drawings from J. Bordaz, op. cit.; photographs by Peabody Museum, Harvard University, Cambridge

40-9 J. J. Languepin, Rapho/PR

Page 580 Wide World Photos; International Rice Research Institute

Blindness, color, **155–156**
Blood (*see also entries under* Blood . . .)
  calcium levels in, *357*, 361
  chemical regulation of, 384, **404–406**
  circulation of, 364–365
  composition of, **371–372**, 406
  as connective tissue, 318
  in hemophilia, 156
Blood cells
  in kidney, *69*, 405
  red, *69*, *319*, **371**, *373*, 377
    androgen and, 328
    carbon dioxide carried by, 386
    hemoglobin carried by, 164, 188–189, 386
    and malaria, *214*
    in sickle cell anemia, 164
  white, **371–372**, *373*
    and fever, 402
    phagocytosis by, 60, *62*, *165*
    in preparation of karyotypes, *160*
Blood clots, 372, 394
  and cardiovascular disease, *370*
  in sickle cell anemia, *138*
Blood flow, regulation of, **366–367**, *370*, 371
Blood glucose, regulation of, **395**
Blood plasma, 274
Blood pressure
  defined, 366
  effect of ADH on, 358, 406, 408
  of giraffe, *15*
  human, *303*, **366**, *370*, 371
    during rage, 324
    during sleep, 419
  of mammals, *366*
Blood proteins, 405, 406
Bloodstream
  foreign organisms in, 373, 374
  oxygen transport in, 324, 364, 365, 369, *370*, 377, 378, 381, 382, **384**, 386 (*see also* Circulatory system)
Blood sugar, regulation of, **395**
Blood transfusion, 373
Blood types, human, *138*, *373*, **456**
Blood vessels [*see also* Arteries; Blood pressure; Capillaries; Veins (of circulatory system)]
  constriction of, *219*
  diseases of, 46, *370*, 371
  of erectile tissues, 327, *328*
  in human embryo, 338, *341*
  and nervous system, 322, 353
  smooth muscle in, 366–367, 370
  in stomach, 323
  structure of, *364*
  and temperature regulation, 400, 401
Blue-green algae, 18, *19*, 207, 208, **209**, 590
  nitrogen fixation by, 514
Body hair, 328, 330, 401
Body plans, of animal phyla, 235 (*see also specific organisms*)
Body temperature (*see also* Endotherms; Exotherms)
  of birds, 247, 368
  human, *303*, 400, 402
  of mammals, 248, 368, *401*
  regulation of, **398–402**, 408, 409
Bolting, 302
Bond(s) (chemical)
  between atoms, 27
  covalent, 27, 39
  double, *40*
  high-energy
    of ATP molecules, 105
    defined, 85
  hydrogen [*see* Hydrogen bond(s)]
  ionic, defined, 29
  in oxidation, 42
Bonds (social)
  among baboons, **551–552**, 553–554
  of birds, 247
Bone(s)
  growth of, *357*, 358
  human, 318, 319
    calcium in, *357*, 360
    osteocyte of, *318*
  of vertebrate skeleton, 243–244
Bone cell, *318*
Bone marrow, 371, 372

Bonner, James, 308, 309
Bony fish, 403
Book gills, 595
Book lungs, 239, *241*
Borlaug, Norman, *580*
Boron, 278
Bowman's capsule, 404, 405, 406
Brachiation, *560*, 564, 566
Brachiopoda, 595
Brachydactylism, 442
Brain
  of earthworm, 352
  evolution of, 232, 413, 414
  eye and, **420–423**
  of *Homo erectus*, 573
  human, 317, 411, 570
    anatomy of, *411*, **412**, *413*
    blood supply to, *370*
    chemical activity in, **416–417**
    development of, 338, 339, 340, 342
    electrical activity in, **417–418**
    evolution of, **576–577**
    mapping of, **413–416**
    in nervous system, *69*, 347, 348, 349, *351*
    and sleep, **418–420**
    specialization of, 250
    split-, 423, 424
  and learning, **424–425**
  of mammals, 413, 576
  and memory, **424–425**
  of octopus, 237
  of vertebrates, 243, 323
Brain cells
  effect of hypothyroidism on, 360
  of human, *69*, 411
  of human fetus, 342
  RNA in, 425
  uptake of glucose by, 359
Brain damage, 412, 413, 423, 424
Brainstem, *351*, 412, 413, 414, *415*, 417, 424
Brain waves, 342, **418**
Branch roots, 268, *269*, 270
Bread
  fungi in production of, 218
  mold, 220 (*see also Neurospora*)
Breasts, human, 330, 331, *333*
Breathing (*see also* Respiration)
  rate of, **384**
  during sleep, 419
Breeding (*see also* Mating)
  cross- (*see* Crossbreeding)
  hierarchies and, 545
  selective, 448 (*see also* Artificial selection)
  and territoriality, 545, 546, 547–548
Bristle number, in *Drosophila*, **449–450**
Broca's area, 414
Bromeliads, 486
Bronchi, 382, 383
Bronchioles, 382
Brower, Jane, 535–536
Brown algae, *212*, 215, *216*, 503, 591
Brown bear, 461, 462
Bryophytes (Bryophyta), **222**, 592
Bryozoa, 594
Büchner, Eduard, 3
Buds, growth of, 300, 304
Buffon, Georges-Louis Leclerc de, 432
Bulbs, *263*, *286*
Bulk flow, **31–32**
  capillaries and, 365
  gas transport by, 378, 380, 381, 383
  in plants, 281
Bullfrog, lungs of, *380*
Bumblebees, *539*
Bundle of His, 369
Burial, among Neanderthals, 576, 578
Burrows, of prairie dogs, 408
Butterfly
  defense mechanisms of, 532, 534, 535, 536
  development of, *240*
  mimicry by, *535*
  as pollinator, 287, 288

Cabbage, 121, *263*, 266, 302, 445
Cactus
  as example of convergent evolution, 474

prickly pear, 530
  water conservation by, 262, 263, 270, *488*, 490
Caffeine, *417*, 531
Calcitonin, *357*, 360, 361
Calcium
  on artery walls, *370*
  in blood, *357*, 361
  in bone, *357*, 360
  excretion of, *303*
  in human diet, 395
  ions, 304
  needed by plants, *278*
Calico cats, *150*
Callose, *279*
Callus, plant, *286*, 304
Calories
  from chemical reactions, 42, 43
  defined, 25
  energy costs of, *511*
  need for, 395
Calvin, Melvin, 94
Calvin cycle, **94–95**, *96*
Calyx, 283
Cambium
  cork, 295, 296
  vascular, 295, *297*, 298
Cambrian period, 229, 238, 433
Camels, 229, 394, *408–409*
Camouflage, *462*, **532**, *533*, **534** (*see also* Coloration)
CAM photosynthesis, 262
Cancer
  and blood type, 456
  lung, 383, *385*
  from radiation exposure, 142
Cancer cells, *119*, 385
Canis, classification of, 205
Cannibalism, 549
Capillaries, *319*, *364*, 365, 366, 371
  cells of, *21*
  and diffusion, **365**
  of earthworms, 378
  food molecules through, *393*
  of gills, 367
  of kidneys, *69*, 404, 405, 406
  in respiratory system, 382, 386
  in sickle cell anemia, *138*
Capillary action, **24**, *27*
Carbohydrates [*see also* Glucose; Sugar(s)]
  calories from, 395
  energy from, 45, 84, 86
  oxidation of, 80, 84, 86, 222
  from photosynthesis, 42, 83, 84, 92, 222, 261
  sugars as, 42
Carbon (*see also* Carbohydrates; Hydrocarbons)
  atomic structure of, **38–39**
  compounds (*see also specific compounds*)
    animal use of 279
    oxidation of, 98, 99, 100, 101, 103, 104, 105
    plant use of, 278
  fixation of, 226
  and origin of life, 16
  in photosynthesis, 39, 92, 93, 94, 95
  radioactive, 94
Carbon dioxide (*see also* Carbon; Photosynthesis; Respiration, cellular)
  atmospheric, *105*, 376, 483
  in blood, 384, 386, 419
  in breakdown of rocks, 517
  from cellular respiration, 62, 84, 103, 104
  diffusion of, 59, 377, 378, 381, 382
  in human respiratory system, *382*
  in photosynthesis, 42, 80, 83, 84, 92–96, 221, 222, 262
  radioactive, 30, 94, 280
  and regulation of internal environment, 402
  in ruminant digestion, 394
Carboniferous period, 222, 224, *229*
Carbon monoxide, 386
Carboxyl group, 46, 48, 49
Cardiac muscle, *319*, 365, 366, *370*
  beat of, 367–368
  and nervous system, 349, *350*
Cardiovascular diseases, *370*, 371
Cardiovascular regulating center, 367, **370–371**
Cardiovascular system, 365 (*see also specific parts*)
  of mammals, *366*

resistance to, *185*, 196–197, 198
Dryopithecus, 565
Dubos, René, *143*
Duckbilled platypus, 248, *337*, 559, 597
Ducklings, imprinting in, 543
"Ductless glands" (*see* Endocrine glands)
Duodenal ulcers, and blood type, 456
Duodenum
  defined, 392
  of vertebrates, 392, *394*
DuPraw, E. J., 190
Dwarfism, 302, 358, 360, 444
Dysentery, *185*

Ear (*see also* Hearing; Sound, frequencies of)
  of human embryo, 340
  of katydid, 238
Early man (*see also Australopithecus*; Cro-Magnon man;
    Man, evolution of; Neanderthal man)
  socialization of, 542
Earth (*see also* Biosphere)
  age of, **432–433**, *434*
  atmosphere of, *483*
  formation of, **11–12**
  life on
    distribution of, **481–483**
    reasons for, **12**
  rotation of, 501
  structure of, *464*
  surface of, **483**
Earth science (*see* Geology)
Earthworms, 234, 235, 595 (*see also* Segmented worms)
  nervous system of, *352*
  reproduction in, 453
  respiration in, 378
Ecdysone, *189*, 531
Echinoderms (Echinodermata), **242**, 503
  classification of, 596
Echinoidea, 596
E. coli (*see Escherichia coli*)
Ecological niche, 507, 508, 525–526, 527, 536
Ecological pyramids, **510–511** (*see also* Food chain)
Ecological succession, **516–518**, 521 [*see also*
    Community(ies)]
  agriculture and, **519–520**
Ecology [*see also* Biosphere; Community(ies);
    Ecological niche; Ecological succession; Ecosystem;
    Environment(s); Population(s)]
  defined, 481, 522
  of modern man, **581–583**
Eco RI, *191*, 196, *197*
Ecosystem [*see also* Biosphere; Biome(s); Ecological
    niche; Environment(s); Habitat(s); Population(s)]
  cycling of minerals through, 507, **511–516**
  defined, 507, 520
  and ecological succession, **516–520**
  energy source for, **508**
  immature and mature, 519, 524
  producers in manmade, *519*
  productivity of, 508, 509
  trophic levels of, 530
Ecotypes, 456–457, 461
Ectoderm, *230*, 232
Edema, 365
Edentata, 597
EEG, 417, 418, 419
Effectors
  in human nervous system, 347, *349*, 350
  in *Hydra*, *352*
Efferent neurons (*see* Motor neurons)
Eggs
  amniote, *337*
  of amphibians, 245
  of birds
    chicken, *337*
    number of, 445–446
  of fish, 248
  of insects, 241
    bees, 539, 540, *541*
    butterflies, *240*
    houseflies, 524
    wasps, 539
  of primitive mammals, 248
  of reptiles, 245–246, 248
  respiration in, 377
Egg cell(s)

age of, and chromosomal abnormalities, 158
  of angiosperms, *131*, 284, 288, 289, *454*
  of *Aurelia*, 231
  of bryophytes, 222
  as carriers of heredity, 130, 132, 133, *139*
  of ferns, *123*
  of frogs, 188, *189*, 198
  of green algae, 216
  of gymnosperms, 223, *224*, *225*
  human (*see* Ovum)
  lampbrush chromosomes of, *189*
  of marine worm, *116*
  mitotic divisions in, 188
  of sea urchin, 111–112
  size of, 57
Ehrlich, Paul, 583
Einstein, Albert, 81–82
Ejaculation, in human male, 127, 327, 328, *335*
Elastic fibers, in connective tissue, 318
Electrical activity, of human brain, **417–418**
Electrical energy, 16 [*see also* Energy, flow of; Energy
    exchange(s); Thermodynamics, laws of]
  light energy converted to, 81
  in photosynthesis [*see also* Electron(s), in
    photosynthesis]
Electric current, 81
Electric impulses, in plants, 311
Electric potential (*see also* Action potential, of nerve
    fiber; Resting potential)
  of cell membrane, 353, 354
  of human brain, 417, 418
Electroencephalogram (*see* EEG)
Electromagnetic spectrum, 81, *82*
Electron(s) [*see also* Ion(s); *specific elements*]
  defined, 26
  as energy source, **79–80**, 83
  in molecular models, *41*
  in oxidation-reduction reactions, 80
  and photoelectric effect, 81, *82*
  in photosynthesis, 91, 92–93, 94
  in water molecule, 26–27
Electron carrier molecules, 91–92, 93, 99, *103*, 104 [*see
    also* NAD]
Electron microscope, 58
Electron shells, 26, 79, 80
Electron transport chain, 92, 93, *103*, **104–105**
Electrophoresis, 165
Elements
  in living systems, 38 (*see also* Organic compounds;
    specific elements)
  mineral, essential to organisms, *278*
  radioactive, 30
  trace, 278
Elephants, 249, 524, 597
Embolus, 370
Embryo
  in amniote egg, *337*
  of angiosperms, *131*, 265, 270, **289**, 290, 291
  of club moss, 223
  human, 69 (*see also* Fetus, human)
    development of, 333, 336, **337–339**, 340
    nutrients for, 127
    traces of gill slits in, 243
  of pine, *224*, *225*
  of reptiles, 245, 246
  respiration in, 377
  vertebrate, 243–244, 368
  of wheat, *290*
Embryophyta, 206
Embryo sac, 223–224
Emotions, and human brain, 416, 417
Endangered species, 470
Endergonic reaction, 86
Endocrine glands (*see also specific glands, specific
    hormones*)
  of arthropods, 239
  defined, 356
  disorders involving, 358, 360
  human, **346–362**, 416
  and nervous system, 358, 361, 401
  pancreas as, *356*
  of vertebrates, *357*
Endocrinology, defined, 356
Endoderm
  of coelenterates, *230*, 234
  of flatworms, 232

of roundworms, 235
Endodermis, root, 268, *269*, 292
Endogenous rhythms, *303*
Endometrium, 331, *332*, 333, 336, 337
Endoplasmic reticulum, 60–61, *65*, *67*, *99*, 210
Endorphins, 417
Endoskeleton, 243–244, *319*
Endosperm
  of angiosperms, 288, 289, 290, *291*, 302, 304, 593
Endothelium, of blood vessels, 365
Endotherms, 247, 248, *381*, **399–402**
Energy (*see also* Energy exchange(s); Energy levels, of
    electrons; *specific types of energy*)
  from ATP, 84, 85–86, *106*
  for cells (*see* Cell(s), energy for)
  in chemical reactions, 42, 43
  consumption of, in U.S., *583*
  electrons as source of, **79–80**, 83
  expended on obtaining food, *511*
  flow of, *76*, **77–86**, 105
    through a food chain, 507, **508–511**
    pyramid of, 510–511
  in glucose oxidation, 98, 99, 100, 101, *103*, 104, 105
  a human body's requirements for, 395
  and laws of thermodynamics, 3–5
  and living systems, 3–5, 77, 79, 105
  of motion (*see* Kinetic energy)
  for muscle contraction, 320
  new sources of, 222
  from nitrification, 514
  obtained by bacteria, 208–209
  for organisms, 376 (*see also* Glycolysis; Respiration)
  and origin of life, 16
  in oxidation-reduction reactions, 80
  oxygen consumption and, *381*
  of photons, 82
  in photosynthesis, 83–84, 94, 95 (*see also specific
    types of energy*)
  for plant cells, 279 (*see also* Photosynthesis)
  during rage, 324
  reactions producing, *381*
  in reduction of nitrate, 514
  for reproduction, 452–453, 523
  resources, consumed by meateaters, 511
  stored in lipids, 45
  sugar as, *42*, 43
  sun as source of, *480*, 481, *482*, 483, 508
Energy-acceptor molecules (*see* Electron carrier
    molecules)
Energy exchange(s)
  involving ATP, 84, 85–86, *106*
  laws of thermodynamics applied to, 77–79
  in living systems, 14, 77, 79
  organelles for, **19–20**
  in photosynthesis, 88, 91, 92, 93, 94
Energy levels, of electrons, 104
  in photosynthesis, 91, 92, 93
Engineering, genetic, **194–198**
Engram, 424–425
Entropy, 79 (*see also* Second law of thermodynamics)
Envelope, nuclear (*see* Nuclear envelope)
Environment(s) [*see also* Biosphere; Ecological niche;
    Ecological succession; Ecological pyramids;
    Ecology; Ecosystem; Habitat(s)]
  adaptation to [*see* Adaptation(s)]
  of baboons, 549, 550, 552, 553, 554
  and differences among ecotypes, 456–457
  ecological succession within, **516–520**
  external response to, 324
  "fitness of the," 83
  vs. heredity, **199–201**, 549
  as influence on height, 144–145
  interaction of genes with, *143*
  internal, regulation of, 324, 398, **402–403**
  and leaf abscission, 265
  limiting factors within, **523**
  and phenotypic differences, 456, 457
  plant's relationship with, 289 (*see also*
    Photoperiodism)
  and testosterone production, 330
Enzyme(s), 3, 165 (*see also specific enzymes*)
  action of, **52–53**
    examples of, 94, *103*, 239, 302, 386, *541*
  biosynthesis of, **185–188**, 195, 196, 210, 362
  defined, 52
  digestive, 230, 239, 318, *356*, 388, 391, 392–393, *394*

Enzyme(s) continued
  fungal, 218
  plant, *279*, 311
  in experiments with *Neurospora*, 163–164
  in four-carbon plants, 95, *96*
  gene control of, 163–164, 450
  glycolytic, 100, 391
  and hormones, 359, 362, 406
  hydrolytic, 61–62
  inducible, *185*, 186
  in leaf abscission, 265
  locations of, 59, 61, 66, 212, 331, 336
  manufacture of, in cells, 60, *61*
  and nerve impulse transmission, 356
  in phagocytosis, 371
  in protein biosynthesis, 176, 178, *179*, 180
  protein digestion by, 190
  repressible, *186*, 188
  in respiration, 99
  restriction, 191, 196, 197
  for splitting of ATP, 320
  and Tay-Sachs disease, *450*
  and temperature, 398
Eocene epoch, *229*, 562
Eohippus, 460, 461
Epidermis
  of arthropods, 239
  of coelenterates, 231
  plant, 295, 296, *297*
    leaf, 261, 262, *263*
    root, **267**, 268, *292*, 293
    shoot, 293
    stem, 270
Epididymis, 327
Epilimnion, of lake, *499*
Epinephrine
  as adrenal hormone, *357*, 359
  chemical structure of, *360*
  effects of, 324, 359, 362, 367, 369, *395*, 401, 408
Epiphytes, 485, *486*
Epithelial cells
  of human respiratory tract, 383
  of *Hydra*, *352*
  mucus-secreting, 318
  of renal tubule, 405
  of small intestine, 392, *393*
  of sponges, 228
  of stomach walls, 391, 392
Epithelial tissues (*see also* Epithelium; Glands)
  human, **317–318**
Epithelium
  beneath gill, 379
  glandular, 318, 323
  of human lung, *385*
  of stomach, 323
  of uterine lining, 336
Equilibrium, dynamic, 33
Equus (*see* Horse)
Eras, geologic, *229*, 434
Erection
  of human clitoris, 420
  of human penis, *326*, 327, *328*, 420
Ergot, 219
Erosion, soil, 514, 515
Erythrocytes (*see* Blood cells, red)
Escherichia coli, 17
  attacked by bacteriophages, 167–168, *169*
  chemotactic responses of, *209*
  conjugation in, 183–184
  drug resistance of, *185*, 196–197, *198*
  Eco RI from, 191
  enzyme production in, 185–186
  genetic experiments with, 196–197, *198*
  nitrogen bases in DNA of, 169
  proteins in, 47–48, 176, 180
Esophagus, *390*, 391
Essential amino acids, *48*, 279, *290*, 395
Estrogen, 330, 331, *332*, *333*, 334, 338
Estrus, in female baboons, 550, *551*
Estuaries, 504
Ethnic groups, IQ of, 200–201
Ethology, defined, 200
Ethyl alcohol, *102*
Ethylene, 305
Etiolation, *309*, 310
Eugenics, defined, 451

Euglena, *212*, 453
Euglenoids, 590
Euglenophyta, *212*, 590
Eukaryotes, **19–20**
  cell division in, **114–120**
  chromosomes of, 183, 188–190
    DNA of, 189 , **190–191**, 198
    proteins of, *192*
  cilia of, **65–66**
  diploidy in, 451, 452
  evolution of, **210, 212**
  flagella of, **65–66**
  nucleus of [*see* Nucleus (of cell)]
  vs. prokaryotes, *210*
  reproduction in, 453
  viruses and, 196
Euphorbs, 474
Eusociality, of insects, 539
Eutheria, 597 (*see also* Mammals, placental)
Evaporation
  defined, 25
  heat loss by, 28, 400, *401*
  from plants, 221, 275, 276, *297*
Evening primrose, 141, 454
Evergreens [*see* Coniferous forests; Conifers (Coniferinae)]
Evolution
  of animals, 210, 227, *238*
  of arthropods, 238, 239
  of bears, 461, 462, 463
  of birds, *229*, 462, 463, 561
    Galapagos finches, 470
  of brain, 232, 413, 414
  and breakup of Pangaea, 464, 465
  co-, **474–476**, 530
  of coelenterates, 231
  convergent, *474*, 485, *486*
  cultural, among macaques, *563*
  defined in population genetics, 440, 447
  of diploidy, **452**
  of echinoderms, 242
  extinction as part of, 470
  of fish, *229*, 244
  forces of, **442–444**
  of genetic systems, 453
  and homeostasis, 324
  of homologous structures, 246
  of horses, 434, **460–461**, 463
  of invertebrates, 233, *238*
  macro-, **459–463**
  macromolecules in study of, *471*
  of mammals, *229*, 248, *250*, 463, 559, 560
  of man, *229*, 559, **566–568**, *569*, **573–579**
  of mollusks, 236, 237
  and mutations, **141–142**
  non-Darwinian, 444
  phyletic, **460–461**, 463
  of plants, 216, **221**, 222, 223, 224, *229*, 259, 261, 263, 277, **284–288**, 462
  of primates, *229*, 559–560
  recent examples of, **458–459**
  of respiratory systems, **377–381**
  splitting, **461–462**, 463
  theory of, 1
    concepts preceding, **431–436**
    Darwin's, 2, 141, 431, 434, **436–439**, 440, 446, 448, 457, 458
    "hopeful monster," *443*
    "leaping," 141
    synthetic, 440
  and use of energy resources, 5
  of vertebrates, 242, *243*, 402
    eye, 471
    heart, **367–368**
  of worms, 231, 234, 237
Exclusion, principle of competitive, **525–526**
Excretion
  in arthropods, 239
  in flatworms, 232
  human, *326*, 327, 404, 405, *406*
  by kidneys, 358
  in mammals, *323*, 402, 406
  in mollusks, 237
  and regulation of internal environment, 402
  in segmented worms, 234

  of urea, 402
  of uric acid, 402
  in vertebrates, 394, 395
Exercise, and metabolic rates, *381*
Exobiology, defined, 13
Exocrine glands, 356
Exocytosis, 60
Exoskeleton, arthropod, 44, 45, 46, 218, 238–239, *433*
Exotherms, 247, *381*, 398, 399, 409
Expiration, 383, 384 (*see also* Breathing; Respiration)
Exponential growth curves, 524
Extinction, **470**
  of dinosaurs, 470, 559, *575*
Extraembryonic membranes, 336, *337*
Eye(s)
  compound, of arthropods, *239*, 241, *540*
  evolution of, 471
  human, *421*
    development of, 338, 340
    resolving power of, *58*
  of octopus, 237
  of owls, *494*
  of primates, *561*
  of scallop, 237
  vertebrate, **420–422**, 471
Eye color, of *Drosophila*, 148–149, 441
Eye-hand coordination, of primates, *561*
Eyespots, 532, 534

Fahrenheit temperature scale, 589
Fallopian tube (*see* Oviduct)
Families, in classification system, 205, *206*, 590
Famine, potato, 519–520
Farming (*see also* Agricultural revolution; Agriculture; Crops; Cultivation)
  of oceans, 215
Far-red light, 309, 310
Fat(s), 38, 45, **46** (*see also* Lipoproteins)
  biosynthesis of, 395
  calories from, 395
  in cellular respiration, 103
  digestion of, 393, *394*
  oxidation of, 103, 404
  saturated vs. unsaturated, 46
  storage of food as, 227
  in temperature regulation, 401
Fatty acids, 46, 47, 393, *394*, 395
Fermentation, anaerobic, **102**, 103
Ferns, 222, 229, 485, 517, 593 (*see also* Vascular plants)
  of deciduous forests, 492, *493*
  life cycle of, 122, *123*
  reproduction in, 122, *123*, *126*, 223, 224
Fertility
  effect of inbreeding on, 449, 450
  and heterozygote superiority, 452
Fertility factor in bacteria, 183–184
Fertilization (*see also* Sexual reproduction)
  in amphibians, 245
  in angiosperms, 226, 288, 289, 454
    pea plants, 131–132
  in *Aurelia*, 231
  and chromosome number, 121
  defined, 121
  double, 289
  human, 330, 331, *332*, 336
  in life cycles, 122, *123*
  in pines, 224, 225
  in sea urchin, 112
  self-, 453–454
  sex determination at, *148*
Fertilizer, 402, 514, 515, 519, 520, 580
Fetus
  hemoglobin of, 188–189
    and RH factor, *373*
  human, 337
    androgen production in male, 328
    detection of extra chromosomes in, 159
    development of, 330, **339–341** (*see also* Embryo, human development of)
  mammalian, *37*
Fever, 400, **402**
  rheumatic, 369
  typhoid, *185*
  undulant, *185*
Fever blisters, 196

of gymnosperms, *224, 225*, 284
of pea plants, *131*
in secondary succession, 517, *519*
Seed coat, *25*, 224, 289, 290, 291, 292, 297
Seed dispersal, 225, 226, 227, *259*
Seed dormancy, 289, **290**, 291
Seedlings
    effect of auxin on, 299–300, *301*
    phytochrome and, 309–310
    in secondary succession, 518
Seed plants, **224–226**
Segmentation
    of arthropods, 238, 239
    of muscles, 339
    of worms, 234, *235*, 238
Segmented worms, **234,** *235,* 237, 238, 239, 242, 595
Segregation, principle of, **132–134**
Selection
    artificial, 438, 444–445, 450 (*see also* Selective
        breeding)
    natural (*see* Natural selection)
    sexual, **473**, 536
Selective breeding, 448, 449 (*see also* Artificial
    selection)
Selective pressure(s) [*see also* Adaptation(s); Natural
    selection]
    of competition, 526
    for conflict reduction, 549
    for defense against predators, 531
    on energy utilization, 452
    in evolution of horse, 461
    on gills, 379
    for heterozygote, 452
    from human civilization, **458–459**
    for intelligence, 576, *577*
    on isolated populations, 470
    on number of young, 524, *561*
    polymorphism produced by, 455, 456
    sexual, 473
    for skin color, *396*
    and speciation, 463
    and use of oxygen, *105*
Self-pollination, *131,* 132, 133, 134, 142, 453
Self-sterility, genes for, 454
Semen, *323, 326,* 328
Seminal vesicle, *326, 327,* 328
Seminiferous tubules, *326,* 327, 328, 329
Sensory cortex, of brain, 414
Sensory neurons, *322, 323,* 347, *349,* 353
Sensory receptors
    human, 347, 348, 349, 353
    of *Hydra,* 352
Sensory response (*see also* Nervous system; Sensory
    neurons)
    in arthropods, 241
    of association cortex, 414
    in bacteria, **209**
    in coelenterates, *230*
    in flatworms, 232
    in plants, **310–311**
    and reticular formation, 415
    in segmented worms, 234
Sepals, 283
Sequoias, 496, 518
Serotonin, 416, 419
Sertoli cells, *327*
Sessile animals, 501, 502–503, *596,* 597
Setae, *234*
Sex cell(s) [*see also* Egg cell(s); Gametes; Sperm cell(s)]
    aging of female, 158
    as carriers of heredity, 130, 132
    chromosomes on, 121, 147–148
    effects of radiation on, 142
    of fungi, 218
    of plants, 221
Sex chromosomes, **148–149**
    human, *111,* 190, 335
        abnormalities in, **156, 158–161,** 200
        in karyotype, *155, 158, 160*
        and sex-linked characteristics, 155–156
Sex determination, 147–148
Sex drive, in women, 360
Sex-linked characteristics, **155–156,** *157*
Sex reversal, social dominance and, *548*
Sexual behavior (*see also* Courtship; Mating)
    and imprinting, 543

of insects, *106*
Sexual dimorphism, 446, 549, 568
Sexual reproduction [*see also* Fertilization; Mating;
    Sexual behavior]
    adaptations related to, *473,* 536
    advantage of, 128
    in animals, 227 (*see also specific animals*)
    defined, 121
    energy expenditure in, 452–453
    fertilization as stage of, 121
    in fungi, 220, 592
    in green algae, 216, *217*
    meiosis in, **121–127**
    in plants, 453, 592
        in angiosperms, 259
        in bryophytes, 222
    for promoting genetic variation, **452–454**
    in protists, 210
Sharks, 229, 244, 403, 596
Sheep, *169, 394,* 443, 489
Shells
    of brachiopods, 595
    of mollusks, 236, 237
Shigella, 185
Shoot, plant, *260,* 289, 291, **293,** 297, 299, 300, *301*
Shores, 215, 216
Short-day plants, 307, 308–309
Shotgunning, 197, 198
Siamese cats, *143*
Siamese twins, 419
Sickle cell anemia, *138,* 164–165, 180–181, 197, 452
Sieve plates, 279, 297
Sieve tube, 279, 280, 281, 297
Sieve-tube elements, 279, 280, 297
Sight (*see* Vision)
Sigmoid growth curve, 524
Silicon, *223,* 279
Silt, 516
Silurian period, *229*
Singer, S. J., *59*
Single-celled organisms, **17–20** [*see also* Protozoans
    (Protozoa); *specific organisms*]
    asexual reproduction in, 119, 198
    cell cycles in, 119
    flagella of, 65
    life cycle of, *122*
    lysosomes of, 62
    nuclei of, 57
    and osmosis, 34
Sinoatrial node, 368–369
Sinsheimer, Robert, 197
Sirenia, 597
Skates, 244, 596
Skeleton (*see also* Endoskeleton; Exoskeleton,
    arthropod)
    of birds, 247
    of corals, 231, 502
    human
        embryonic, *339, 340*
        fetal, 341
        muscles of, **319–320,** *321*
        size of male, 328
    hydraulic, 234
    muscles of, **319–320,** *321,* 349, 350, 391
    of sponges, 228
Skin
    of amphibians, 245, 380
    cell division in, 119
    human, 318, 347, 348, 401
        color of, 145, *396*
    receptors in, 400
    selection pressures involving, 379
    of sharks and skates, 244
Skull, 348
    of *Homo erectus,* 573
    human, 317
Sleep, 416, **418–420**
Sleeping sickness, 213, 590
Slime molds, 212, **218,** 362, 591
Slugs, 236, 453
Small intestine, 391, **392–393**
Smell, in primates, 561
Smith, Homer, 402
Smith, William, 433
Smoking
    and arteriosclerosis, *370*

and cancer, *385*
Smooth endoplasmic reticulum, 61, *67*
Smooth muscle, 319, **322,** 390
    and autonomic nervous system, 349, 350
    in blood vessel walls, 366–367, 370
    in stomach, 323
    in swallowing, 391
    in uterus, 358
Snails, 236, 379, 498, 501, 503
    color and banding in, **454–455,** 532
    reproduction in, 453
Snakes, *14, 106,* 245, 246, 247, 498, 532, 548, 597
    body temperature of, 399
Snow geese, polymorphism for color in, **455**
Social behavior [*see also* Animal behavior; Mating;
    Sexual behavior; Socialization; Social organization;
    Society(ies)]
    aggressive, 544
    appeasing, 544, *545,* 550
    courtship, 544, 545, *546, 547*
    dominant, **545–548, 550–551,** 553
    as ecologically adaptive, 554
    greeting, 544
    grooming, 551
    and hunting of large game, 542
    and imprinting, **543–544**
    of insects, 538, **539–542**
    ritual, *473,* 543, **544,** 547, 548
    of vertebrates, **542–554**
Social bonds, among baboons, **551–552,** 553–554
Social Darwinism, 446
Social dominance
    among baboons, **550–551,** 553
    and sex reversal, *548*
    in vertebrate societies, **545–548**
Socialization
    of early man, 542
    stages of insect, **539**
Social organization
    among baboons, **549–554**
    among vertebrates, 542
Society(ies)
    advantages of, 542
    defined, 538, 554
    insect, 538, **539–542**
    vertebrate, **542–554**
Sodium, 81, *278,* 279, 303, 395, 406, 407
Sodium chloride, 403, 406
Sodium ions, *29, 59,* 354, 355, 360, 406
Sodium-potassium pump, 354
Soil
    composition of, 516
    erosion of, 486, 492, 513, 514, 515–516
Solar energy, *5,* 18, *30,* 84, *222,* 480, **481,** 482, 483
    conversion of, to chemical energy, 79–80
    dependence of man on, *511*
    in food chain, 508
    and life on earth, 77, 79, **82–83**
    in water cycle, *33*
Solar plexus, *351*
Solar system, formation of, 11
Solutions
    acidic, 29, 31
    basic, 29, 31
    defined, 28
    hypertonic, 34, *35*
Solvent, water as, **28–29**
Somatic cells (*see also* Cloning)
    chromosome number of, 121
    DNA in, 169
    effects of radiation on, 142
    of female mammals, 150
Somatic nervous system, **349–350**
Somatotropin [*see* Growth hormone(s), human]
Somites, of human embryo, 338–339
Song thrush, as predator, 455
Sound, frequencies of, 414
Spawning, 468
Special creation, concept of, 432
Specialization
    of arthropod species, 241
    of cells, *21,* 216, 228
    of human brain, 250
Speciation, **463** (*see also* Genetic isolation)
    of Galapagos finches, **468–470**
    genetic variation and, 457